Clinical Neurovirology

NEUROLOGICAL DISEASE AND THERAPY

Advisory Board

1. Handbook of Parkinson's Disease, *edited by William C. Koller*
2. Medical Therapy of Acute Stroke, *edited by Mark Fisher*
3. Familial Alzheimer's Disease: Molecular Genetics and Clinical Perspectives, *edited by Gary D. Miner, Ralph W. Richter, John P. Blass, Jimmie L. Valentine, and Linda A. Winters-Miner*
4. Alzheimer's Disease: Treatment and Long-Term Management, *edited by Jeffrey L. Cummings and Bruce L. Miller*
5. Therapy of Parkinson's Disease, *edited by William C. Koller and George Paulson*
6. Handbook of Sleep Disorders, *edited by Michael J. Thorpy*
7. Epilepsy and Sudden Death, *edited by Claire M. Lathers and Paul L. Schraeder*
8. Handbook of Multiple Sclerosis, *edited by Stuart D. Cook*
9. Memory Disorders: Research and Clinical Practice, *edited by Takehiko Yanagihara and Ronald C. Petersen*
10. The Medical Treatment of Epilepsy, *edited by Stanley R. Resor, Jr., and Henn Kutt*
11. Cognitive Disorders: Pathophysiology and Treatment, *edited by Leon J. Thal, Walter H. Moos, and Elkan R. Gamzu*
12. Handbook of Amyotrophic Lateral Sclerosis, *edited by Richard Alan Smith*
13. Handbook of Parkinson's Disease: Second Edition, Revised and Expanded, *edited by William C. Koller*
14. Handbook of Pediatric Epilepsy, *edited by Jerome V. Murphy and Fereydoun Dehkharghani*
15. Handbook of Tourette's Syndrome and Related Tic and Behavioral Disorders, *edited by Roger Kurlan*
16. Handbook of Cerebellar Diseases, *edited by Richard Lechtenberg*
17. Handbook of Cerebrovascular Diseases, *edited by Harold P. Adams, Jr.*
18. Parkinsonian Syndromes, *edited by Matthew B. Stern and William C. Koller*
19. Handbook of Head and Spine Trauma, *edited by Jonathan Greenberg*

20. Brain Tumors: A Comprehensive Text, *edited by Robert A. Morantz and John W. Walsh*
21. Monoamine Oxidase Inhibitors in Neurological Diseases, *edited by Abraham Lieberman, C. Warren Olanow, Moussa B. H. Youdim, and Keith Tipton*
22. Handbook of Dementing Illnesses, *edited by John C. Morris*
23. Handbook of Myasthenia Gravis and Myasthenic Syndromes, *edited by Robert P. Lisak*
24. Handbook of Neurorehabilitation, *edited by David C. Good and James R. Couch, Jr.*
25. Therapy with Botulinum Toxin, *edited by Joseph Jankovic and Mark Hallett*
26. Principles of Neurotoxicology, *edited by Louis W. Chang*
27. Handbook of Neurovirology, *edited by Robert R. McKendall and William G. Stroop*
28. Handbook of Neuro-Urology, *edited by David N. Rushton*
29. Handbook of Neuroepidemiology, *edited by Philip B. Gorelick and Milton Alter*
30. Handbook of Tremor Disorders, *edited by Leslie J. Findley and William C. Koller*
31. Neuro-Ophthalmological Disorders: Diagnostic Work-Up and Management, *edited by Ronald J. Tusa and Steven A. Newman*
32. Handbook of Olfaction and Gustation, *edited by Richard L. Doty*
33. Handbook of Neurological Speech and Language Disorders, *edited by Howard S. Kirshner*
34. Therapy of Parkinson's Disease: Second Edition, Revised and Expanded, *edited by William C. Koller and George Paulson*
35. Evaluation and Management of Gait Disorders, *edited by Barney S. Spivack*
36. Handbook of Neurotoxicology, *edited by Louis W. Chang and Robert S. Dyer*
37. Neurological Complications of Cancer, *edited by Ronald G. Wiley*
38. Handbook of Autonomic Nervous System Dysfunction, *edited by Amos D. Korczyn*
39. Handbook of Dystonia, *edited by Joseph King Ching Tsui and Donald B. Calne*
40. Etiology of Parkinson's Disease, *edited by Jonas H. Ellenberg, William C. Koller, and J. William Langston*
41. Practical Neurology of the Elderly, *edited by Jacob I. Sage and Margery H. Mark*
42. Handbook of Muscle Disease, *edited by Russell J. M. Lane*
43. Handbook of Multiple Sclerosis: Second Edition, Revised and Expanded, *edited by Stuart D. Cook*
44. Central Nervous System Infectious Diseases and Therapy, *edited by Karen L. Roos*
45. Subarachnoid Hemorrhage: Clinical Management, *edited by Takehiko Yanagihara, David G. Piepgras, and John L. D. Atkinson*
46. Neurology Practice Guidelines, *edited by Richard Lechtenberg and Henry S. Schutta*
47. Spinal Cord Diseases: Diagnosis and Treatment, *edited by Gordon L. Engler, Jonathan Cole, and W. Louis Merton*
48. Management of Acute Stroke, *edited by Ashfaq Shuaib and Larry B. Goldstein*
49. Sleep Disorders and Neurological Disease, *edited by Antonio Culebras*
50. Handbook of Ataxia Disorders, *edited by Thomas Klockgether*
51. The Autonomic Nervous System in Health and Disease, *David S. Goldstein*
52. Axonal Regeneration in the Central Nervous System, *edited by Nicholas A. Ingoglia and Marion Murray*
53. Handbook of Multiple Sclerosis: Third Edition, *edited by Stuart D. Cook*
54. Long-Term Effects of Stroke, *edited by Julien Bogousslavsky*
55. Handbook of the Autonomic Nervous System in Health and Disease, *edited by C. Liana Bolis, Julio Licinio, and Stefano Govoni*
56. Dopamine Receptors and Transporters: Function, Imaging, and Clinical Implication, Second Edition, *edited by Anita Sidhu, Marc Laruelle, and Philippe Vernier*

57. Handbook of Olfaction and Gustation: Second Edition, Revised and Expanded, *edited by Richard L. Doty*
58. Handbook of Stereotactic and Functional Neurosurgery, *edited by Michael Schulder*
59. Handbook of Parkinson's Disease: Third Edition, *edited by Rajesh Pahwa, Kelly E. Lyons, and William C. Koller*
60. Clinical Neurovirology, *edited by Avindra Nath and Joseph R. Berger*

Additional Volumes in Preparation

Neuromuscular Junction Disorders: Diagnosis and Treatment, *by Matthew N. Meriggioli, James F. Howard, Jr., and C. Michel Harper*

Drug-Induced Movement Disorders, *edited by Kapil D. Sethi*

Clinical Neurovirology

edited by

Avindra Nath
The Johns Hopkins University School of Medicine
Baltimore, Maryland, U.S.A.

Joseph R. Berger
University of Kentucky College of Medicine
Lexington, Kentucky, U.S.A.

MARCEL DEKKER

MARCEL DEKKER, INC. NEW YORK • BASEL

Library of Congress Cataloging-in-Publication Data
A catalog record for this book is available from the Library of Congress.

ISBN: 0-8247-4081-5

This book is printed on acid-free paper.

Headquarters
Marcel Dekker, Inc., 270 Madison Avenue, New York, NY 10016, U.S.A.
tel: 212-696-9000; fax: 212-685-4540

Distribution and Customer Service
Marcel Dekker, Inc., Cimarron Road, Monticello, New York 12701, U.S.A.
tel: 800-228-1160; fax: 845-796-1772

Eastern Hemisphere Distribution
Marcel Dekker AG, Hutgasse 4, Postfach 812, CH-4001 Basel, Switzerland
tel: 41-61-260-6300; fax: 41-61-260-6333

World Wide Web
http://www.dekker.com

The publisher offers discounts on this book when ordered in bulk quantities. For more information, write to Special Sales/Professional Marketing at the headquarters address above.

Current printing (last digit):

10 9 8 7 6 5 4 3 2 1

PRINTED IN THE UNITED STATES OF AMERICA

To my many teachers, especially Frank Yatsu, M.D.,
former Chair of the Department of Neurology,
University of Texas Health Sciences Center at Houston—
an outstanding teacher, mentor, and friend

—A.N.

To Peritz Scheinberg, M.D., former Chair of the Department of Neurology,
University of Miami, Florida—a wonderful mentor, role model, and friend

—J.R.B.

Foreword

My friends and colleagues Avi Nath and Joe Berger have made great effort to cover the clinical aspects of neurovirology in this volume. They have included practical problems such as brain biopsy, spinal fluid examination, vaccines, and pharmacotherapeutics. They have recruited a sterling group of contributors. They asked me to write a foreword—not to impart wisdom, but to recall how I stumbled into this fascinating field.

My long journey into what is now called neurovirology began on a winter night in San Francisco in 1957. The Medicine house staff from Stanford University Hospitals (which in those olden days was in San Francisco) were having a party at the Chief Resident's house. In the kitchen, Rod Beard, a young attending physician, lifted his drink and congratulated me for being selected as an assistant resident. At that time I wanted nothing more than to stay at Stanford and practice internal medicine in Pacific Heights. Sadly, I informed him that I had just received my draft notice and would probably be leaving my wife in California and going to Korea. ''You don't want to do that!'' he said. I agreed. He asked why I did not do research in the Army—possibly in virology. I laughed. I had never done any research, and there was no field in which I knew less than virology. ''Then you would be bound to learn something,'' he chided. That off-the-cuff kitchen conversation changed my life.

Several weeks later, a telephone call came from the Walter Reed Army Institute of Research (WRAIR) requesting a meeting in San Francisco the following week—at midnight in the bar of the St. Francis Hotel. The contact turned out to be Geoffry Edsall, head of the Communicable Diseases Division. After brief introductions, he asked if I would be interested in going to Southeast Asia to work on a pediatric service, taking care of children and sending specimens from febrile patients back to WRAIR. He explained that ''if we ever deployed troops in Southeast Asia,'' the military would need to know what infectious diseases might be encountered. Since we had only recently extracted ourselves from Korea, the French had recently lost Vietnam, and our leaders had spoken of the folly of fighting land wars in Asia, the idea seemed absurd. On the other hand, going to Southeast Asia with my wife and in civilian clothes was not an opportunity to pass up. I eagerly volunteered to become a virologist.

I was posted to WRAIR but never assigned to Asia. I arrived at the peak chaos of the 1957 Asian influenza epidemic, and when that settled I concentrated on studies of central nervous system infections. Herpes simplex virus, arthropod-borne virus, and enterovirus infections of the nervous system were my primary concerns; viruses that continue to pose fascinating questions in epidemiology, pathogenesis, diagnosis, and treatment. During those years, I consulted with Webb Haymaker on the neuropathological changes in our human tissues and experimental animals; in addition to infectious disease rounds I often accompanied Des O'Doherty on neurology rounds and Wallie Nauta at weekly neuroanatomy conferences. In those days, the knowledge of virology by pathologists and clinicians was rudimentary, and the knowledge of the nervous system by microbiologists was negligible. That led me to choose the road less traveled; I opted to train in neurology and neuropathology.

After clinical training at the Massachusetts General Hospital, I wanted to return to experimental virology. Encouraged by Elizabeth Hartmann, the wonderful Director of Training Programs for the National Institutes of Neurological Diseases and Blindness, I applied for a special fellowship to work with Frank Fenner at the John Curtin School of Medical Research in Canberra, Australia. My clinical mentor, Raymond Adams, was required to sign my application; he initially balked. No one in neurology was interested in viruses—I would have no colleagues; I would be wasting a potentially successful career. The reaction was not unique. When Donald Harter applied to do virology work the next year at the Rockefeller Institute, he received similar advice from his mentor at the Neurological Institute at Columbia University, Houston Merritt. Don and I both persisted.

In 1961, I prepared my fellowship application, and in the space for "field of study," I typed "Pathogenesis of virus infections of the nervous system," which seriously exceeded the allotted 42 spaces. Betsy Hartmann "whited-out" my area of interest and typed in "neurovirology." I objected to the apparent neologism, and I recall her smile and statement: "We have neuroanatomy, neurophysiology, and neurochemistry—why not neurovirology?"

In the 40 years that have followed, studies of viral infections of the nervous system have opened an exhilarating panorama. Landmarks include the development of effective therapies in many herpesvirus infections, the recovery of viruses from chronic diseases (subacute sclerosing panencephalitis and progressive multifocal leukencephalopathy), the transmission of kuru and Cruetzfeldt-Jakob disease and the emergence of bovine spongiform encephalopathies and the variant human disease, the discovery of novel diseases and modes of pathogeneses of human immunodeficiency virus infections, the association of HTLV-1 virus with tropical spastic paraplegia, and the invasion of North America by West Nile virus from Africa and the potential of other exotic agents to travel great distances with modern transportation.

The term *neurovirology* has not yet been legitimized by inclusion in *Dorland's Illustrated Medical Dictionary*, but I assume it will be included in the next edition. I hope they give credit to Betsy.

Richard T. Johnson, M.D.
Distinguished Service Professor of
Neurology, Microbiology and Neuroscience
The Johns Hopkins University School of Medicine
and Bloomberg School of Public Health
Baltimore, Maryland, U.S.A.

Preface

Despite our increasingly detailed knowledge of the nature of viruses, improvements in hygiene, effective vaccines for many viral illnesses, and the relatively recent emergence of antiviral therapies, viral disorders of the central nervous system remain a very real threat to human populations. In some regions of sub-Saharan Africa, for example, more than one in three people have been infected with HIV, and as many as one-half of those infected will suffer from neurological disorders. In the early 1900s, mankind dreaded influenza, which unexpectedly felled millions, and feared the Parkinson-like illness referred to as von Economo's encephalitis, which arose as a consequence of a still-unknown viral infection. New viral disorders with devastating effects on the nervous system, such as HIV and Nipah virus, have been recognized.

The spread of infections today occurs rapidly across geographical and political boundaries, unlike any ever before experienced by mankind. Diseases that had been believed to be confined to isolated regions of the world are now being reported in unexpected locales, such as West Nile virus, which in a few short years has spread across the continental United States. These emerging illnesses have undoubtedly arisen as a consequence of the greater social intercourse between once-isolated communities, as predicted by Hans Zinsser in his 1934 classic *Rats, Lice and History*. They represent a significant threat to public health. This threat was recognized and articulated by Juvenal before the dawn of the common era: "This plague has come upon us by infection, and it will spread still further, just as in the fields the scab of one sheep or the mange of one pig destroys an entire herd" (*Satires*, II.78). These emerging illnesses are superimposed on the common and uncommon endemic and epidemic viral illnesses of the central nervous system. Similarly, in this era of bioterrorism, the potential for the re-emergence of old scourges such as smallpox has arisen. Concomitant with this potential re-emergence is the risk of neurological complications arising from the re-instituted vaccination programs. Fortunately, we have greatly surpassed the era of descriptive medicine. Today, we understand the mechanisms by which viruses enter the body and the nervous system and how they replicate within cells, and we have, in many instances, developed effective means of harnessing this knowledge to halt the spread of viruses in the community or in the individual patient.

While we have made significant progress in controlling parasitic and bacterial infections, viral infections continue to dazzle mankind. The nervous system is susceptible to a large number of viral infections that manage to effectively penetrate the blood–brain barrier or bypass it by invading the peripheral nerves and then traveling to the central nervous system along the nerve trunks. Viral infections can also remain latent in nervous system tissue for extended periods of time—sometimes spanning several years—and then become activated at an opportune moment, invading distant sites within the nervous system.

The presentations of any viral disorder of the central nervous system can try the diagnostic acumen of even the most skilled and knowledgeable physician. The infectious etiology can be mistaken for any of a number of other processes, including those of autoimmune, neurodegenerative, vascular, or metabolic/toxic nature. *Clinical Neurovirology* is intended to assist the clinician in understanding the basic science of neurovirology and negotiate the diagnostic and therapeutic maze associated with viral CNS disorders. Although particular emphasis is placed on retroviruses, herpesviruses, and arboviruses, this book is designed to deal with a broad range of viral infections of the CNS. Additionally, the consequences of viral CNS infection, such as autoimmune demyelinating disorders and behavioral changes, are addressed. The clinical presentations (both common and uncommon), the measures that are most effective in establishing the diagnosis, and the available therapies (both established and experimental) are addressed in each chapter.

Major developments have recently occurred in the treatment of viral infections of the brain—particularly with the retroviruses, herpesviruses, and arboviruses—and several new antiviral drugs are currently under development; this book provides a detailed discussion of these new and future therapeutic approaches. We hope that this book has a broad appeal to neurologists, infectious disease specialists, internists, family practitioners, pediatricians, physicians-in-training, and all others who confront these patients.

Avindra Nath
Joseph R. Berger

Contents

Contributors

Malcolm Avison, Ph.D. Department of Neurology, University of Kentucky College of Medicine, Lexington, Kentucky, U.S.A.

Alan D. T. Barrett, Ph.D. Department of Pathology, Center for Biodefense and Emerging Infectious Diseases, and Sealy Center for Vaccine Development, The University of Texas Medical Branch at Galveston, Galveston, Texas, U.S.A.

Robert J. Baumann, M.D. Departments of Neurology and Pediatrics, University of Kentucky College of Medicine, Lexington, Kentucky, U.S.A.

Joseph R. Berger, M.D. Department of Neurology, University of Kentucky College of Medicine, Lexington, Kentucky, U.S.A.

John Booss, M.D. Veterans Affairs (VA) Connecticut Healthcare System, West Haven, and Departments of Neurology and Laboratory Medicine, Yale University School of Medicine, New Haven, Connecticut, U.S.A.

Kathryn M. Carbone, M.D. Center for Biologics Evaluation and Research, U.S. Food and Drug Administration, Bethesda, Maryland, U.S.A.

Paola Cinque, M.D., Ph.D. Division of Infectious Diseases, San Raffaele Hospital, Milan, Italy

Bruce A. Cohen, M.D. Department of Neurology, Northwestern University Medical School, Chicago, Illinois, U.S.A.

Kevin M. Coombs, Ph.D. Department of Medical Microbiology and Infectious Diseases, University of Manitoba, Winnipeg, Manitoba, Canada

Sidney Croul, M.D. Center for Neurovirology and Cancer Biology, College of Science and Technology, Temple University, Philadelphia, Pennsylvania, U.S.A.

Alessandro Di Rocco, M.D. Albert Einstein College of Medicine at Beth Israel Medical Center, New York, New York, U.S.A.

Lydia Estanislao, M.D. Mount Sinai Medical Center, New York, New York, U.S.A.

Gerald M. Fenichel, M.D. Department of Neurology, Vanderbilt University Medical Center, Nashville, Tennessee, U.S.A.

David Galey Department of Neurology, The Johns Hopkins University School of Medicine, Baltimore, Maryland, U.S.A.

Anthony Geraci, M.D. Mount Sinai Medical Center, New York, New York, U.S.A.

Donald H. Gilden, M.D. Department of Neurology, University of Colorado Health Sciences Center, Denver, Colorado, U.S.A.

Isabella C. Glitza University of Heidelberg, Heidelberg, Germany

Thiravat Hemachudha, M.D. Neurology Division, Department of Medicine, Chulalongkorn University Hospital, Bangkok, Thailand

Steven Jacobson, Ph.D. Viral Immunology Branch, National Institutes of Neurological Disorders and Stroke, National Institutes of Health, Bethesda, Maryland, U.S.A.

Richard T. Johnson, M.D. Departments of Neurology, Microbiology, and Neuroscience, The Johns Hopkins University School of Medicine and Bloomberg School of Public Health, Baltimore, Maryland, U.S.A.

Burk Jubelt, M.D. Department of Neurology, and Department of Microbiology/Immunology and the Program in Neuroscience, State University of New York (SUNY) Upstate Medical University, Syracuse, New York, U.S.A.

Nick Karabatsos, Ph.D. Division of Vector-Borne Infectious Diseases, National Center for Infectious Diseases, Centers for Disease Control and Prevention, Fort Collins, Colorado, U.S.A.

Kamel Khalili, Ph.D. Center for Neurovirology and Cancer Biology, College of Science and Technology, Temple University, Philadelphia, Pennsylvania, U.S.A.

James J. LaGuardia, M.D. Department of Neurology, Southern Illinois University, Springfield, Illinois, U.S.A.

Robert M. Levy, M.D., Ph.D. Department of Neurosurgery, Northwestern University Medical School, Chicago, Illinois, U.S.A.

Darryl L'Heureux Center for Neurovirology and Cancer Biology, College of Science and Technology, Temple University, Philadelphia, Pennsylvania, U.S.A.

Annika Linde, M.D., Ph.D. Department of Virology, Swedish Institute for Infectious Disease Control, Solna, and Microbiology and Tumor Biology Center, Karolinska Institute, Stockholm, Sweden

Eugene O. Major, Ph.D. National Institute of Neurological Disorders and Stroke, National Institutes of Health, Bethesda, Maryland, U.S.A.

Michael Mayne, Ph.D. Department of Pharmacology and Therapeutics, University of Manitoba, Winnipeg, Manitoba, Canada

Justin C. McArthur, M.B.B.S., M.P.H. Department of Neurology, The Johns Hopkins University School of Medicine, Baltimore, Maryland, U.S.A.

Israel I. Mendez Department of Medical Microbiology and Infectious Diseases, University of Manitoba, Winnipeg, Manitoba, Canada

Erawady Mitrabhakdi, M.D. Neurology Division, Department of Medicine, Chulalongkorn University Hospital, Bangkok, Thailand

Flor M. Munoz, M.D. Department of Pediatrics and Department of Molecular Virology and Microbiology, Baylor College of Medicine, Houston, Texas, U.S.A.

Avindra Nath, M.D. Department of Neurology, The Johns Hopkins University School of Medicine, Baltimore, Maryland, U.S.A.

Mitsuhiro Osame, M.D. Third Department of Internal Medicine, Kagoshima University Faculty of Medicine, Kagoshima, Japan

Claire Pomeroy, M.D.* Division of Infectious Disease, Department of Internal Medicine, University of Kentucky College of Medicine, Lexington, Kentucky, U.S.A.

Julie A. Ribes, M.D. Department of Pathology and Laboratory Medicine, University of Kentucky College of Medicine, Lexington, Kentucky, U.S.A.

Stacie L. Ropka, Ph.D. Department of Neurology, State University of New York (SUNY) Upstate Medical University, Syracuse, New York, U.S.A.

Frank Romanelli, Pharm.D., B.C.P.S. Department of Pharmacy and Physician Assistant Studies, University of Kentucky College of Pharmacy, Lexington, Kentucky, U.S.A.

Steven A. Rubin, M.S. Center for Biologics Evaluation and Research, U.S. Food and Drug Administration, Bethesda, Maryland, U.S.A.

Bruce F. Sabath, B.S. National Institute of Neurological Disorders and Stroke, National Institutes of Health, Bethesda, Maryland, U.S.A.

Jürgen Schneider-Schaulies, Dr. Institute for Virology and Immunobiology, University of Würzburg, Würzburg, Germany

* *Current affiliation*: Office of the Dean, University of California, Davis School of Medicine, Davis, California, U.S.A.

David M. Simpson, M.D. Mount Sinai Medical Center, New York, New York, U.S.A.

Tom Solomon, B.A., B.M., M.R.C.P., D.C.H., D.T.M.H., Ph.D. Department of Medical Microbiology and Department of Neurological Science, University of Liverpool, Liverpool, United Kingdom

Israel Steiner, M.D. Department of Neurology, Hadassah University Hospital, Jerusalem, Israel

Marie Studahl, M.D., Ph.D. Institute of Internal Medicine, Department of Infectious Diseases, Sahlgrenska University Hospital/Östra, Göteborg University, Göteborg, Sweden

Magdalena I. Swanson Department of Medical Microbiology and Infectious Diseases, University of Manitoba, Winnipeg, Manitoba, Canada

Volker ter Meulen, M.D. Institute for Virology and Immunobiology, University of Würzburg, Würzburg, Germany

Chong-Tin Tan, F.R.C.P., M.D. Department of Medicine, University of Malaya, Kuala Lumpur, Malaysia

Alex C. Tselis, M.D., Ph.D. Department of Neurology, Wayne State University/Detroit Medical Center, Detroit, Michigan, U.S.A.

Thomas Weber, M.D. Department of Neurology, Neurologische Klinik, Marienkrankenhaus Hamburg, Hamburg, Germany

Benedikt Weissbrich, Dr. Institute for Virology and Immunobiology, University of Würzburg, Würzburg, Germany

Henry Wilde, M.D. Queen Saovabha Memorial Institute, Thai Red Cross Society, and Department of Medicine, Chulalongkorn University Hospital, Bangkok, Thailand

Kum-Thong Wong, F.R.C.Path. Department of Pathology, University of Malaya, Kuala Lumpur, Malaysia

1

Introduction to Virus Structure, Classification, Replication, and Hosts

Israel I. Mendez, Magdalena I. Swanson, and Kevin M. Coombs
University of Manitoba
Winnipeg, Manitoba, Canada

1 VIRUS STRUCTURE

1.1 General Nature of Viruses

Although the concept of an infectious ''virus'' is only about 100 years old, diseases caused by these agents have been known since ancient times. For example, both rabies and polio, which are discussed in greater detail in later chapters of this volume, appear to have been known in Egypt around 2000 BCE, almost 1000 years before the time of the Pharaoh Tutankhamen. Significant work preceding and during the nineteenth century CE allowed visualization of bacteria and established their disease-causing properties. It became appreciated toward the end of the nineteenth century that some agents capable of causing illness were small enough to pass through filters known to block bacteria. Thus, the term *virus* (Latin for poison) was coined to describe these ''filterable toxins.'' However, it was soon realized that viruses were different from poisons. Toxins can be diluted when serially passaged from one host to another, whereas viruses undergo multiplication. Although humans have been aware of viruses for a relatively short period of time, these agents are probably as old as life itself and have probably coevolved with other forms of life.

Viruses are among the simplest and smallest of currently known living organisms. In fact, because of their simplicity, there is some debate as to whether viruses should be considered living. Most viruses consist of both protein and nucleic acid. Viroids (plant pathogens that consist solely of RNA) and prions (agents that appear to consist solely of protein) are exceptions. Viruses generally exist in two forms. The actively replicating virus inside an infected cell is the form that may be considered ''alive.'' The extracellular form of the virus is known as the *virion*. The virion is analogous to a seed or spore. It generally is a stable ''crystalline'' structure whose primary function is to protect the genetic material until the nucleic acid reaches the interior of a suitable host cell. No virus is capable of growing by itself. All must make use of macromolecular ''building blocks'' (amino acids, nucleotides, and, in some cases, lipids) and employ enzymes found within living cells. Thus, all viruses are obligate intracellular parasites.

1.2 Virus Morphology

Virion Size and Complexity

There is enormous variability in the size of virions. The smallest animal virions are the parvoviruses (e.g., the human parvovirus B19), which belong to the family *Parvoviridae*. As detailed later in Sec. 2, almost all viruses are organized into families, which are designated with an italicized name ending with the suffix *-viridae*. Groups of viruses (families) may be referred to by their italicized family name (e.g., *Parvoviridae*) or may be referred to by a nonitalicized generic name (e.g., parvoviruses). Both conventions are used in this chapter. Parvovirus virions are less than 20 nm in diameter (Figure 1). The largest animal viruses (for example, vaccinia virus and the smallpox agent *Variola major*) are members of the *Poxviridae* family. Poxvirus virions are generally approximately 200 nm $\times$ 300 nm in size (Figure 1) and are barely visible by light microscopy. There also is significant variability in virion complexity. Parvoviruses consist of a small piece of nucleic acid surrounded by 60 copies of a single protein. Other viruses, such as JC virus, a member of the *Polyomaviridae*, and the plant Tobamovirus, tobacco mosaic virus, may be more complex and larger, composed of a larger piece of nucleic acid and more than 60 copies of a single protein. Most virions are even more complicated and larger. Some (e.g., adenoviruses and the bacterial virus T4) contain a single piece of nucleic acid surrounded by multiple different proteins that are not present in the same quantities; some (e.g., rubella virus and various equine encephalitis viruses, which belong to the family *Togaviridae*) contain the same number of various proteins but also contain a lipid membrane (envelope); some [e.g., rabies virus, which belongs to the family *Rhabdoviridae*; measles virus and mumps virus, which belong to the *Paramyxoviridae*; herpes simplex viruses and cytomegalovirus, members of the *Herpesviridae*; and the human immunodeficiency virus (HIV), which belongs to the family *Retroviridae*] contain different numbers of various proteins as well as a membrane; some (e.g., members of the family *Reoviridae*) contain different numbers of various proteins as well as multiple segments of nucleic acid (all of which must be present for the virion to be infectious); and some (e.g., the influenza viruses, members of the *Orthomyxoviridae*) contain different amounts of various proteins, multiple segments of nucleic acid, and an envelope.

Viral Nucleic Acid

With the possible exception of the prion agents, all currently known viruses contain either DNA or RNA as their genetic material. Most viruses use their genetic material both for *replication* and for *transcription*. Replication is the process whereby the viral genetic material is copied into exact replicas that will be packaged into progeny virions (discussed more fully in Sec. 3). Transcription involves the generation of messenger RNA (mRNA), whether from RNA or DNA, for the eventual production of viral proteins. Thus, one convenient way to group viruses, for ease of classification and discussion, that bears directly upon how (and where) the virus replicates and causes pathology is by nucleic

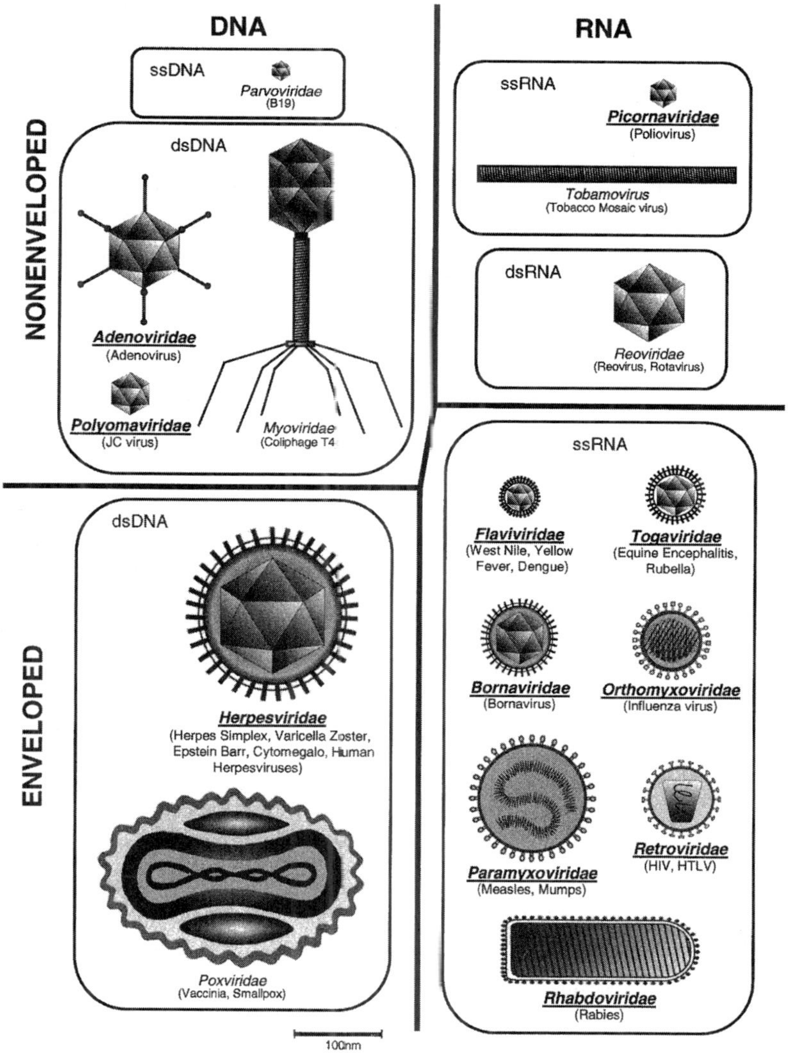

Figure 1 Diagrammatic representations of selected virions. Viruses are divided according to whether their genomic material is DNA (left) or RNA (right) and whether the capsid is nonenveloped (top) or surrounded by an envelope (bottom). Where applicable, each group is further subdivided depending upon whether the nucleic acid is single-stranded (ss) or double-stranded (ds). All viruses are shown at about the same scale to indicate their relative sizes; bar at bottom represents 100 nm (= 0.1 μm). Virus family names (ending in the suffix *-viridae* and italicized) and example members of each family are shown in parentheses. Virus families currently known to contain neurotropic agents that affect humans are highlighted in boldface and underlined. Other non-neurotropic viruses (the animal pathogens parvovirus, poxvirus, and reovirus; the bacterial virus coliphage T4; and the plant virus tobacco mosaic virus) are included for comparative purposes. Enveloped viruses that are fully shaded (e.g., *Bornaviridae, Paramyxoviridae*, and *Retroviridae*) are known to have one or more proteins between the internal capsid and the envelope, whereas partially shaded viruses (e.g., *Flaviviridae* and *Togaviridae*) do not have extra proteins between the capsid and the envelope. (Adapted from Ref. 8.)

acid type (Figure 1). For example, most DNA viruses require enzymes for DNA replication and synthesis. These enzymes, if provided by the host cell, are located within the host's nucleus, so most DNA viruses replicate inside the host nucleus. One group of viruses that are exceptions to this generalization are the *Poxviridae*, large complex viruses that encode all their necessary DNA enzymes and thus can replicate in the cell's cytoplasm. Conversely, RNA viruses do not generally require DNA enzymes (although retroviruses are an exception), so RNA viruses generally replicate in the cell's cytoplasm. In addition to the retroviruses, which use a DNA intermediate, some RNA viruses, such as the influenza viruses, carry out some of their replicative steps in the cell's nucleus because they need to "steal" components from cells before those cellular elements leave the nucleus.

The viral nucleic acid, whether DNA or RNA, may be either single-stranded (ss) or double-stranded (ds). If single-stranded, the genome may be of either positive (+) or negative (−) polarity. By convention, messenger RNA (mRNA), which is "read" by ribosomes to translate the mRNA into protein, is considered (+) polarity. Thus, the template DNA or RNA strand that is transcribed to produce the mRNA is considered (−) polarity. The important ramifications of these differences in viral nucleic acid and polarity with regard to classification and replication are described more fully in later sections of this chapter.

The viral genome may range in size and configuration. The term *genome* refers to all the nucleic acid of a virus, whereas *gene* usually refers to the part of the nucleic acid genome that encodes a specific viral protein. The smallest viruses (e.g., parvoviruses) have genomes of approximately 5000 nucleotides (= 5 kilobases, or 5kb) that contain two genes. The largest viruses (e.g., poxviruses and herpesviruses) can have genomes larger than 200 kilobase pairs (200 kbp; "pairs" because their nucleic acid is double-stranded) and can therefore potentially encode more than 250 proteins. Most viruses have genomes whose sizes fall between these extremes. For most families of viruses, all viral genes are located on a single contiguous linear piece of nucleic acid, with the same gene generally located in the same position on the genome in every virion within that family. A few viruses (e.g., hepatitis B virus, a member of the *Hepadnaviridae*, which, because it is not known to be neurotropic, is not covered in this volume, and the *Polyomaviridae*, which includes the neurotropic JC virus) have a circular rather than linear genome. Some viruses have segmented genomes. For example, the human influenza virus genome consists of eight separate segments of RNA that encode a total of at least 10 different proteins (six segments each encode a single protein, and the two smallest segments each encode two proteins). To be infectious, a virion must contain at least one copy of each of the eight gene segments. Likewise, the *Reoviridae* genome consists of 10–12 segments (depending upon the specific genus of virus) of double-stranded RNA, all of which must be present in a virion for it to be infectious. The segmented nature of these types of viral genomes has dramatic ramifications with regard to their pathogenesis (reviewed in Refs. 1 and 2).

Viral Proteins

Viral proteins can be generally classified as either *structural* or *non structural*. By convention, structural proteins are those that are present within a virion particle. Their identities are usually determined by examining the protein content of highly purified preparations of viral particles. For any given virus there is usually a fixed and characteristic number of proteins located within the virion. For example, poliovirus virions contain a single copy of one protein (called VPg) (VP usually indicates virion protein) and 60 copies each of four other proteins (called VP1, VP2, VP3, and VP4). By contrast, nonstructural proteins are those that are encoded by the virus and are found in infected cells but not within the purified virion. Nonstructural proteins are generally enzymatic. They carry out specific enzymatic functions within the cell but are not included within the viral particle. However, some structural proteins also may be enzymes. Within the virion the viral nucleic acid, whether RNA or DNA, is usually surrounded by a protective protein coat built from structural proteins. Because the genetic code is such that nucleotide base triplets encode each amino acid, it is not possible for any genetic material to encode a single protein sufficiently large to protect the genome that encodes it. Thus, for genetic efficiency the protective protein coat is usually constructed from multiple copies of one or a few proteins.

1.3 Capsid Morphology

Collectively, the nucleic acid and the protein that is closely associated to protect it form a nucleoprotein complex known as the viral *capsid*. There are two general ways in which repeating units of proteins can be organized so that they effectively protect the nucleic acid. One method is to wrap the protein along the nucleic acid, allowing each protein building block to interact with part of the nucleic acid. This results in a helical arrangement. Examples of this type of arrangement include tobacco mosaic virus (Figure 1), an extensively studied plant virus. In addition, most currently known (−) sense animal RNA viruses (e.g., the *Orthomyxoviridae, Paramyxoviridae*, and *Rhabdoviridae*) encase their RNA in a helical configuration. The second general method to protectively surround the nucleic acid is to build a three-dimensional cage. The most efficient way to build a three-dimensional cage that has no holes in it from the smallest number of types of building blocks is to join 20 equilateral triangles to form an icosahedron. The icosahedron is a nearly spherical object, possessing 12 corners (vertices), 20 triangular faces, and 30 edges. Examples of this type of arrangement include poliovirus and JC virus (Figure 1). A few viruses wrap their nucleic acid in a protective coat that is neither helical nor icosahedral. For example, retrovirus capsids are conical and poxvirus capsids are ovoid (Figure 1).

The size of helical capsids is determined by the length of nucleic acid. The size of icosahedral capsids is determined by the size of each protein and by the number of proteins that generate the cage. To increase the number of proteins that participate in building the cage, without introducing holes in the cage, there are only certain sets of protein numbers that can be used. This geometric problem was solved by Caspar and Klug [3], who introduced the concept of triangulation numbers (T values) to describe icosahedral complexity. The simplest icosahedron is built from 60 identical protein subunits (a pentameric aggregate at each of the 12 vertices; $5 \times 12 = 60$) and is described as having a triangulation value of 1 ($T = 1$). Parvoviruses are examples of such an arrangement. To build larger icosahedral capsids, additional proteins have to be added, either on each of the 20 faces, or on each of the 30 edges, or on some combination of faces and edges. Other proteins are usually added as hexameric arrays (although adenoviruses and polyomaviruses are notable exceptions). Thus, the next-most complicated icosahedron would be one in which hexameric arrays were added in the middle of each icosahedral face. The resulting structure would have $(5\times 12)+(6\times 20)$, or 180, proteins and is referred to as a $T = 3$ icosahedron. Strictly speaking, a $T=2$ structure would have holes in it, and so far no $T=2$ structure has been identified. In general, the total number of proteins that form the icosahedral capsid is the triangulation number $\times$ 60. Notable exceptions to this general rule include the adenoviruses, polyomaviruses, and reoviruses. Adenoviruses add trimeric arrays instead of hexameric arrays on icosahedral faces and edges [4], and polyomaviruses are built exclusively from pentameric arrays [5]. Reoviruses are composed of multilayered capsids that omit arrays near their vertices in the outer capsid because internal capsids have proteins that ''block'' the holes present in the outer capsid [6]. In addition, they appear to violate the foregoing rules because their innermost capsids contain 120 copies of major structural proteins. However, rather than using the ''forbidden'' $T=2$ organization, these viruses use a $T=1$ organization of 60 dimers [7].

1.4 Other Virion Components

In some cases the viral capsid structure is surrounded by a lipid membrane (envelope). Thus, presence or absence of an envelope is another convenient way to organize viruses into groups (Figure 1). When an envelope is present, the nucleoprotein structure is then referred to as a *nucleocapsid*. Some viruses, such as the *Orthomyxoviridae* (e.g., influenza virus), the *Paramyxoviridae* (e.g., measles virus), and the *Rhabdoviridae* (e.g., rabies virus) contain helical nucleocapsids surrounded by a membrane. In addition, icosahedral nucleocapsids may be surrounded by an envelope, as seen with the *Flaviviridae* (e.g., dengue virus), *Togaviridae* (e.g., rubella virus), and *Herpesviridae* (e.g., herpes simplex viruses). In contrast to viral proteins and nucleic acid sequence, both of which are specified within the viral genetic code, the lipids within the membrane are provided exclusively by the host cell. For most enveloped viruses, lipid membranes are normally acquired as the nucleocapsid passes through a cellular membrane, whereas a few viruses assemble their envelopes de novo within the cell (discussed in greater detail in Sec. 3).

2 VIRUS CLASSIFICATION

2.1 Classification Schemes

There are currently more than 3600 known virus species organized into approximately 56 families [8]. It is highly probable that this list will increase in the future because there is the potential to discover additional viruses. In order to simplify research on, and discussion of, these viruses, it is essential to classify them in a manner that highlights their similarities and differences. Several classification schemes have been developed to accomplish this task.

The Formal Classification Scheme

Most organisms are classified according to kingdom, phylum, class, order, family, genus, and species. Most viruses, on the other hand, are formally classified into families, subfamilies, genera, and species, based on the similarities and differences between them. For example, members of the family *Retroviridae* all contain a (+)ssRNA genome that is replicated via a dsDNA intermediate. No other family contains viruses with this characteristic. The *Herpesviridae* family also is unique, because it contains viruses with linear dsDNA genomes surrounded by an icosahedral capsid, an envelope with glycoprotein spikes, and a tegument (a space between the capsid and the envelope that contains proteinaceous material). The formal classification of viruses is overseen by the International Committee on Taxonomy of Viruses (ICTV). In addition, based upon immunological or molecular differences, viruses can sometimes be placed in lower levels of classification, which, depending upon the nomenclature for the specific group of viruses, can be known as subspecies, strains, serotypes, clades, or variants. Except for the family names, which are italicized, capitalized, and end in the suffix *-viridae*, there is no formal Latinized binomial nomenclature for viruses. Instead, a virus species is usually known by a nonitalicized, nonunderlined, uncapitalized common name, such as poliovirus.

The Epidemiological and Etiological Classification Schemes

Viral classification schemes can also be based on epidemiology or etiology, with the possibility of a specific virus being placed in several categories. Enteric viruses enter the body by ingestion and usually replicate primarily in the gastrointestinal tract. Examples of such viruses are the *Picornaviridae* (enterovirus genus), the *Reoviridae*, the *Parvoviridae*, and the *Adenoviridae*. Respiratory viruses are acquired by inhalation of aerosols or hand-to-face contact and usually replicate in the respiratory tract. This group includes the *Picornaviridae* (rhinovirus genus), the *Paramyxoviridae*, the *Orthomyxoviridae*, and the *Adenoviridae*. Arboviruses (arthropod-borne viruses), are transmitted by the bites of insects or by inhalation of rodent droppings. Some common arboviruses are found in the *Togaviridae, Flaviviridae, Rhabdoviridae*, and *Reoviridae* families. Sexually transmitted viruses are transmitted through intimate contact that generally involves the exchange of body fluids. They can cause local or general infections. Human immunodeficiency virus (HIV) is a well-known example of a sexually transmitted virus. Oncogenic viruses are often transmitted in much the same manner as sexually transmitted viruses and replicate in specific tissues, often leading to transformation of cells and possibly malignancy. Examples of oncogenic viruses are the *Retroviridae*, the *Hepadnaviridae*, the *Polyomaviridae*, the *Adenoviridae*, and the *Herpesviridae*. Hepatitis viruses target the liver and include the better known hepatitis A, B, C, D, and E viruses as well as other hepatitis viruses that are less well understood. Neuropathogenic viruses that infect, or adversely affect, nervous system cells are the subject of this book.

Classification Based upon Morphology and Composition

Another way to classify viruses is according to their genomic composition and virion morphology (Figure 1). As indicated earlier, the viral genome can be DNA or RNA, single- stranded or double-stranded, negative or positive sense in polarity, linear or circular, segmented or nonsegmented. The presence or absence of a lipid envelope is also important, and the overall shape of the capsid (helical, icosahedral, or otherwise) and virion [spherical, bullet-shaped (e.g., *Rhabdoviridae*), or pleomorphic (lacking a defined shape) (e.g., the myxoviruses *Orthomyxoviridae* and *Paramyxoviridae*)] also may be taken into consideration.

The Baltimore Classification Scheme

The mechanisms by which the viral genome is transcribed to produce mRNA for protein production is the basis for another classification strategy. This classification was devised by Dr. David Baltimore [9] and is known as the Baltimore scheme. In this scheme, viruses are placed in one of six classes (Figure 2). Class I viruses follow the ''central dogma,'' in that dsDNA is used to transcribe mRNA, and the mRNA is then translated into protein. Most of these viruses use host enzymes for the synthesis of mRNA. Examples of such viruses that are discussed in subsequent chapters of this volume are the *Adenoviridae, Herpesviridae*, and *Polyomaviridae*. Class II viruses have an ssDNA genome, which is usually negative sense in currently known animal viruses and positive sense in non-animal viruses. The genome of (−)ssDNA viruses can be directly transcribed into mRNA. However, (+)ssDNA viruses must make a (−)ssDNA copy to act as a template for the synthesis of mRNA (e.g., parvovirus B19). Class III viruses (e.g., *Reoviridae*) have dsRNA genomes, and mRNA is transcribed from the (−) sense strand. Class IV viral genomes are

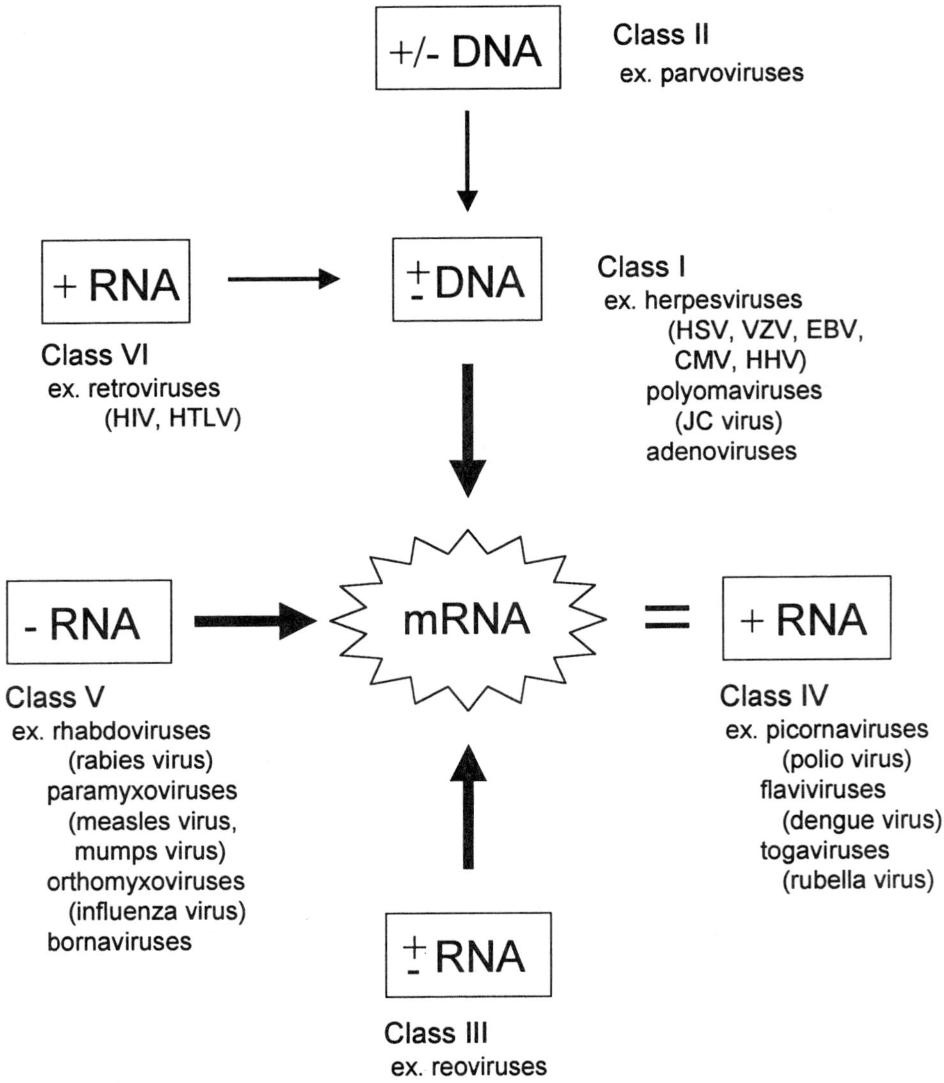

Figure 2 The Baltimore transcription scheme, showing how mRNA is produced from various types of genomic nucleic acid. "+", "−", and "+/−" refer to single-stranded (ss) nucleic acid of positive or negative polarity, respectively. "±" refers to double-stranded nucleic acid. Larger arrows indicate direct synthesis of mRNA from genomic material, and smaller arrows indicate an intermediate step. (Adapted from Ref. 9.)

(+)ssRNA molecules. No transcription is required because the genome can serve directly as mRNA. Class IV viruses relevant to this volume are dengue virus, many enteroviruses, and rubella virus. The class V viral genome is (−)ssRNA, which serves as a template for the synthesis of mRNA. Rabies virus, measles virus, mumps virus, and influenza virus belong to this class. Class VI, the last group, comprises retroviruses that contain (+)ssRNA genomes. The (+)ssRNA, unlike in the viruses of class IV, does not serve as mRNA. The production of mRNA requires the synthesis of a dsDNA intermediate, followed by transcription of the dsDNA.

Although the Baltimore classification scheme is based on the transcription strategies of viruses, usually the various classes also can be distinguished by the manner in which the viral genomes are replicated. Class I viruses use their dsDNA genomes as a template to synthesize more dsDNA during genome replication. As in transcription, this replication is usually carried out by host enzymes. One group of dsDNA viruses (*Hepadnaviridae*) is an exception to this generalization; replication takes place through an RNA intermediate that is longer than the genome. Class II viruses go through a replicative intermediate to copy their genomes. For example, (+)ssDNA viruses must make a (−)ssDNA copy to act as a template for more (+)ssDNA. In the case of class III viruses, the mRNA that was used for protein synthesis is copied by viral enzymes into a (−) sense RNA that remains associated with the mRNA template to form the progeny dsRNA. Class IV viruses have (+)ssRNA genomes. A (−)ssRNA intermediate is produced, which then serves as the template for more (+)ssRNA. Class V viruses use the same replication strategy as class IV viruses, except that they start with a (−)ssRNA genome and use a (+)ssRNA strand as an intermediate. In class VI, (+)ssRNA genomes are replicated by a unique mechanism. The (+)ssRNA is used as a template to synthesize (−)ssDNA, which in turn serves as a template for the synthesis of a (+)ssDNA strand. The resulting dsDNA molecule is transcribed into mRNA or used to synthesize progeny (+)ssRNA genomes.

2.2 Integration of Classification Schemes

DNA Viruses

Any number of classification schemes may be employed by researchers and healthcare professionals, but for our purpose it is simpler to separate viruses according to their genomes (whether the genome is DNA or RNA, ss or ds) and virion morphology (whether the virus has an envelope and the overall shape of the particle). DNA viruses follow the more conventional replication strategy, although the genome shapes, sizes, and organizations as well as the virion structures are quite varied. Double-stranded DNA viruses that belong to class I of the Baltimore classification scheme can be further divided based upon presence or absence of an envelope. Relevant viruses that lack an envelope include JC virus and adenovirus, both of which have icosahedral capsids. Examples of enveloped dsDNA viruses are members of the family *Herpesviridae*, including the herpes simplex viruses, varicella-zoster virus, Epstein-Barr virus, cytomegalovirus, and human herpesviruses. Herpesviruses have an icosahedral nucleocapsid. Single-stranded DNA viruses that belong to class II have only one known member that infects humans. The human parvovirus B19 contains a (−)ssDNA genome encased in a nonenveloped icosahedral capsid. B19 is currently not associated with neuropathology and therefore is not covered in this volume.

RNA Viruses

Of all virus types, RNA viruses are the largest group of infectious agents responsible for causing the most diseases worldwide [10–12]. Like DNA viruses, the RNA viruses vary widely in genome sizes, gene organizations, genome structures, and virion structures. The RNA viruses are more diverse than the DNA viruses, and there is the additional complication of segmented genomes in some human pathogens. Some of the nonenveloped dsRNA viruses are neuropathogenic in animals but are currently not believed to be neuropathogenic in humans. There are no known enveloped dsRNA viruses that infect humans. The ssRNA virus group contains many neuropathogenic members. Relevant nonenveloped examples are the (+)ssRNA icosahedral enteroviruses. Enveloped examples include (+)ssRNA viruses such as the spherical retroviruses HIV and HTLV, rubella virus and dengue virus, and (−)ssRNA viruses such as the bullet-shaped rabies virus and the pleomorphic measles virus, mumps virus, and influenza virus, the last one of which has a segmented genome.

Unclassified Agents

Interestingly, the foregoing schemes are unsuccessful in classifying several infectious agents, namely viroids and prions. As indicated earlier, viroids are composed solely of nucleic acid and infect only plants. Prions are more medically relevant. They appear to consist solely of protein and are believed to be the causative agents of Creutzfeldt-Jakob syndrome, kuru, and various subacute spongiform encephalopathies. Because these agents are so obviously unique in their composition, they remain unclassified.

As described, many classification schemes are used today. Although the various classification strategies have been useful for organizing the large numbers of known viruses, this chapter focuses on neuropathogenic human viruses. This theme is reflected in later chapters of this volume.

3 VIRAL REPLICATION

Viral replication is a complex process that remains poorly understood. Much of the complexity arises from the fact that there is great diversity among the methods employed by different viruses that depend not only on the types of cells they can infect but also on the specific replicative machinery each virus carries. Furthermore, the current inability to culture some viruses in the laboratory contributes to lack of understanding of their replication cycles.

The replication of viruses can occur only within a living cell because, as indicated earlier, they are obligate intracellular parasites that are absolutely dependent on host cell machinery for replication. Despite significant differences in the details of replication [as implied by the Baltimore scheme (see above)], there are several common features. Two approaches can be used to describe the viral life cycle. The first approach employs a growth curve. The second is a more detailed chronological description of viral life cycle events.

3.1 The Viral Growth Curve

A viral growth curve is calculated by measuring the amount of infectious virus present within and/or released from cells over a period of time. Typical growth curves are depicted in Figure 3 for both a nonenveloped (Figure 3A) and an enveloped (Figure 3B) virus. Details of a particular growth curve (such as time between phases and total virus released) will depend on the virus type and cell type. Infectious virus added to cells initially seems to disappear. This period, which corresponds to the virus being uncoated (see below), is termed the *eclipse period* and can range from 3 to 12 h for animal viruses. During the *eclipse period*, most of the infectious virus has entered the host cell and is undergoing replication as indicated by the detection of new viral nucleic acid and proteins prior to detection of progeny infectious virus. Another more general term often used is *latent period*, for the period that begins with the onset of infection and ends with the first newly assembled detectable extracellular infectious virus. This is distinct from the eclipse period, which ends with the first detectable infectious virus that may be found intracellularly. The differences in assembly requirements of enveloped viruses, as contrasted to nonenveloped viruses, also lead to fundamental differences in whether intracellular infectious virions are produced. An enveloped virus that matures by budding through the external membrane does not exist as an infectious virus until it leaves the cell; thus, there is no intracellular infectious virus (Figure 3B). The redetection and subsequent rapid rise of infectious virus, whether enveloped or nonenveloped, and intracellular or extracellular, denotes the end of the eclipse period and the beginning of the *productive* or *rise period*. This period, initiated upon intracellular assembly of virions, is marked by exponential rise in virus numbers and is characteristic of most viruses. The *burst* refers to the time point where cell lysis occurs, which results in the subsequent release and detection of cell-free virus.

Several terms are used to describe the various types of viral infections that can occur in people. The more common type of infection that a virus can cause is an *acute infection*, which is characterized by a rapid onset, visible symptoms, and short duration (e.g., influenza virus). A *chronic* or *persistent infection*, unlike acute infections, which can end in weeks, can last from years to indefinitely. In such cases, the virus often reproduces at a much slower rate and can even result in an inflicted individual being apparently symptom-free (e.g., herpesviruses and hepatitis B virus). A *latent virus infection* refers to viruses that stop reproducing as they enter a state of dormancy, only to become active again at a later time point (e.g., herpesviruses). Like patients with chronic infections, the patient with a latent virus infection may not exhibit any symptoms until the virus becomes activated.

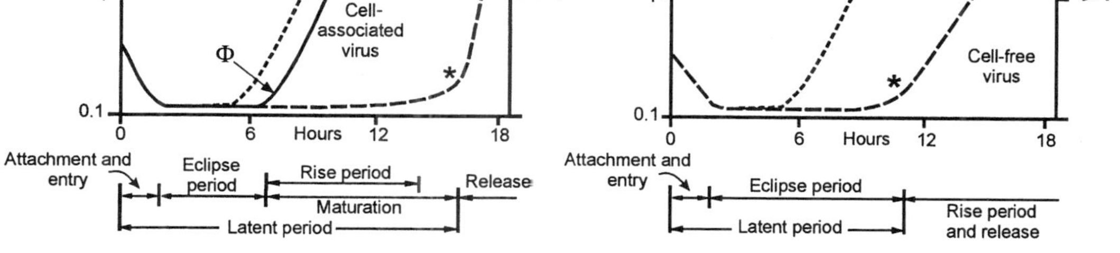

Figure 3 Schematic representations of viral growth curves for (A) a nonenveloped virus or an enveloped virus that acquires its membrane inside the cell (see text) and (B) an enveloped virus that acquires its membrane as it matures through the plasma membrane. (Φ) Intracellular assembly begins at the onset of cell-associated virus detection. (*) The burst, or lysis of the cell, occurs upon detection of cell-free virus. (B) For viruses that acquire an envelope from the plasma membrane, the point of release coincides with the point of maturation, and no infectious cell-associated virus is observed. PFU (plaque-forming unit) is a measure of the amount of infectious units of virus per cell. The coordinates are labeled only as an example, because they will vary with the type of virus.

3.2 Steps in Viral Replication

Overview

There is a general flow of events that occur for virtually all viruses (Figure 4). The early events are (1) attachment, (2) penetration into the cell, and (3) uncoating to release genomic material; the middle events are (4) transcription of genes to produce mRNA, (5) synthesis of proteins, and (6) replication of genomic material; and the late events are (7) assembly to produce mature virions and (8) virion release. These steps may vary greatly with each virus, but for the sake of simplicity will be discussed here in general.

The Early Events

Attachment. To initiate an infection, a virus must first attach to an appropriate host cell. This interaction is mediated by external viral proteins, those found in either the capsid

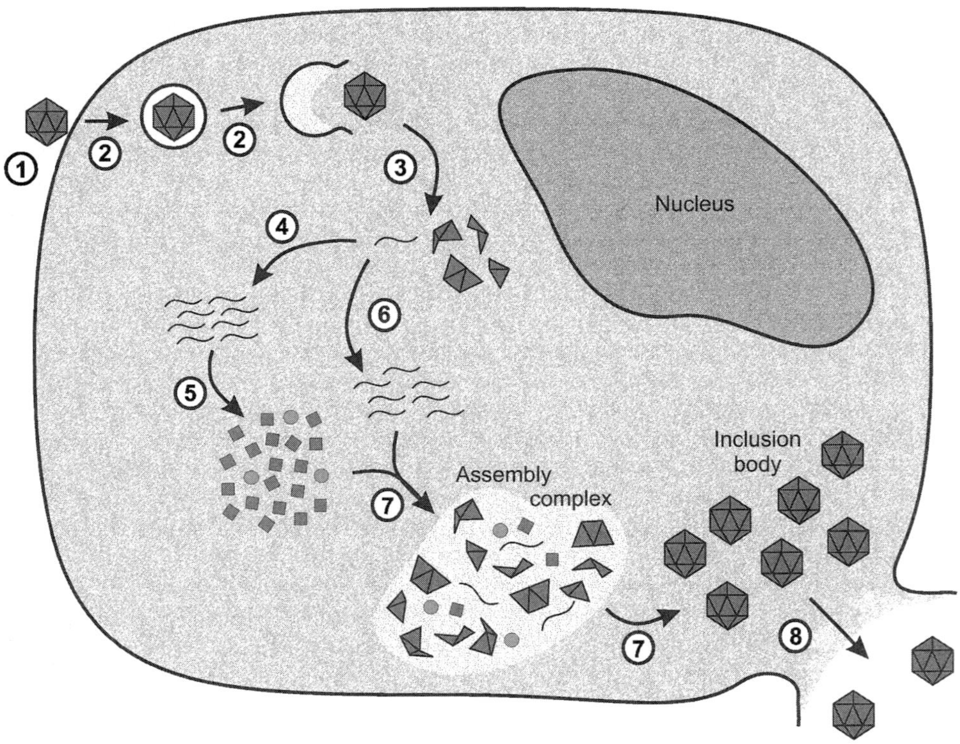

Figure 4 Typical viral replicative cycle for a nonenveloped RNA virus. *1*, Attachment; *2*, penetration into the cell; *3*, uncoating and release of genomic material; *4*, transcription of genes; *5*, translation of mRNA; *6*, replication of genomic material; *7*, assembly and maturation, which can occur within an assembly complex; and *8*, cell lysis and virion release. Note that the locations of the various steps will vary depending on the virus type (e.g., herpesviruses carry out transcription and replication within the nucleus). Envelope acquisition for enveloped viruses can occur at intracellular membranes such as the nucleus, or at the plasma membrane upon release. Short wavy lines represent viral nucleic acid, small squares represent viral structural proteins that assemble into complexes, and small circles represent nonstructural proteins that are present within the cell and assist in viral replication and assembly but are not found within mature virions.

if the virus is nonenveloped or in the envelope if the virus has a membrane. Typically, the viral protein binds noncovalently in a classic lock-and-key fashion to specific cell surface macromolecules (either carbohydrates, proteins, or glycolipids) that serve as *viral receptors*. A virus's host range can vary from narrow to broad and is dependent on the presence of viral receptors that dictate species specificity and can even dictate tissue specificity. Some viruses are known to have more than one type of receptor that depends on the cell type, whereas others (e.g., HIV) have been shown to require coreceptors on a single cell type. Furthermore, some viral receptors (e.g., sialic acid) are common to many viruses.

Penetration (Entry). Once a virus has successfully attached to a cell, it must next penetrate the cell's plasma membrane to permit the subsequent release of the nucleocapsid and genomic material. Evidence supports three modes of entry (Figure 5): (1) Fusion with the plasma membrane (usually seen only with enveloped viruses); (2) receptor-mediated endocytosis, which can be further divided into fusion within the endosome or lysis of the endosome; and (3) direct passage through the plasma membrane (usually seen only with nonenveloped viruses).

Fusion. Fusion is one mechanism of entry employed by enveloped viruses during which the viral lipid envelope fuses with and then becomes part of the plasma membrane, liberating the viral nucleocapsid into the cytoplasm. The fusion between the membranes is often mediated by *fusion proteins* found on the surface of the viral envelope. Such an entry strategy is common to retroviruses (e.g., HIV), herpesviruses, and paramyxoviruses (e.g., measles virus and mumps virus).

Receptor-Mediated Endocytosis. Both enveloped and nonenveloped viruses may be engulfed by the cell in clathrin-coated vesicles to form endosomes. Once the virus is enclosed in this endosome, acid-dependent events are believed in some cases to trigger either the fusion of the membranes, in the case of enveloped viruses, or the lysis of the endosome, in the case of nonenveloped viruses. The endosomal decrease in pH appears to trigger a conformational change in specific viral proteins that mediate the subsequent release of the nucleocapsid. For example, acidification of influenza virus causes a conformational change in the hemagglutinin protein, which may cause internal fusion of the membranes. Sometimes (e.g., adenoviruses), the low pH–induced conformational change in a protein serves to disrupt and in turn lyse the endosomal membrane at key membrane–viral protein contacts.

Direct Transfer. A third method of entry that appears to exist in certain viruses is direct transfer. In such instances, viruses are believed to pass directly through the plasma membrane, bypassing endocytosis. This is a poorly understood mechanism of entry and seems to be limited to nonenveloped viruses.

Uncoating. Uncoating describes the disassembly process during which the virus will shed some or all viral proteins, allowing the transcription and replication of its nucleic acid. Uncoating marks the beginning of the eclipse period. Shedding of viral proteins can occur either at the plasma membrane in conjunction with penetration (e.g., paramyxoviruses), inside the endosome (e.g., adenoviruses and orthomyxoviruses), in the cytoplasm, or at the nuclear membrane (e.g., herpesviruses). There may be complete uncoating to reveal naked nucleic acid (e.g., picornaviruses) or only partial uncoating, which leaves the nucleic acid complexed with specific proteins that are required for subsequent biosynthetic events (e.g., paramyxoviruses such as measles virus and mumps virus).

The Middle Events

The greatest variability among viral life cycles occurs during the middle stage. As indicated earlier, events that take place during this time are transcription (the process whereby the

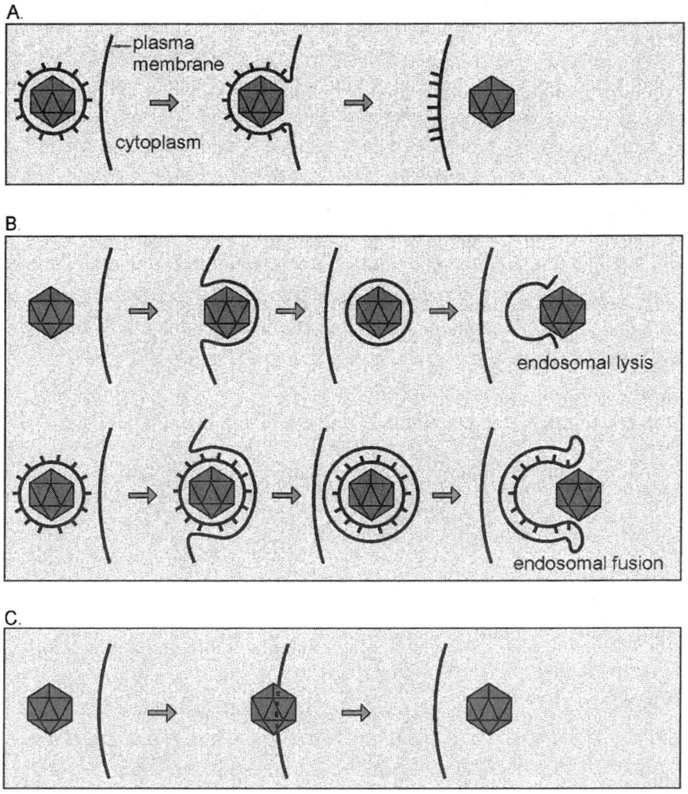

Figure 5 Schematic diagram depicting the varous modes of viral entry. (A) Fusion of an enveloped virus occurs at the plasma membrane interface as the viral membrane becomes part of the plasma membrane. (B) Receptor-mediated endocytosis can occur for nonenveloped or enveloped viruses. Note the subsequent fusion of the membranes to release the nucleocapsid in the case of enveloped viruses. (C) Direct transfer of nonenveloped viruses whereby no endocytosis or membrane damage is thought to occur.

nucleic acid, whether RNA or DNA, is copied to produce complementary positive sense mRNA), translation (''reading'' of the mRNA nucleotide sequence by cellular ribosomes to produce a corresponding order of linked amino acids that form a polypeptide protein), and replication (the copying of parental genomic material that serves as the template to produce an identical copy).

Transcription. After uncoating, most viruses first generate mRNA, either in the cytoplasm or in the nucleus, depending on where the incoming virus was delivered to (a process specific for the virus type). Transcription of the viral genome may occur at a single time (as implied in Figure 4). However, in many cases (particularly for most DNA viruses covered in this volume) transcription can occur at different times and is sometimes divided into two time periods: transcription of early genes and transcription of late genes. In some cases, particularly with the herpesviruses, transcription may be divided into three sequential stages (immediate early, early, and late). As the names suggest, transcription of early genes occurs before transcription of late genes. Transcription of late genes usually occurs after the genome has been replicated and usually cannot occur until specific proteins are generated from the early genes. However, in a few instances (e.g., adenoviruses), the distinction between early genes and late genes may not be absolute, with some genes transcribed at all times but at different efficiencies. In such cases early genes are transcribed most efficiently at early times and poorly at late times, and late genes are transcribed poorly prior to genome replication and most efficiently after genome replication.

DNA viruses generally use the same strategy employed by the host cell, making use of host machinery (cellular enzymes) that can recognize and transcribe any DNA. However, as indicated earlier, some viruses contain genomic material that consists of RNA. This presents a problem for the virus because host cells do not carry specific enzymes that are able to transcribe from RNA. Therefore, RNA viruses must provide the specific machinery required, either in an already synthesized form present in the infecting virion (e.g., rhabdoviruses and myxoviruses) or in a genomic form that can be translated by the host to produce the necessary enzymes (e.g., picornaviruses). One such enzyme is an RNA-dependent RNA polymerase (RdRp) that transcribes the parental RNA to generate mRNA; alternatively, the genome itself may serve as the mRNA. Retroviruses are unique because they make use of an RNA-dependent DNA polymerase (*reverse transcriptase*) to copy their RNA genome into a DNA intermediate. This intermediate is then transcribed to generate mRNA.

Translation. Genomic nucleic acid that serves as mRNA or mRNA that has been produced must become available in the cytoplasm where translation occurs to generate viral proteins. Unlike the mRNA of RNA viruses, which will already be in the cytoplasm, the mRNA of most DNA viruses (except that of poxviruses) must be transported to the cytoplasm from the cell's nucleus where it was initially generated. All viruses make use of the host's translational ribosome apparatus to translate their mRNA, which becomes capped and polyadenylated. Some viruses, such as herpesviruses, shift cellular emphasis to viral translation by inhibiting the translation of host mRNA. Cellular ribosomes will often initiate translation at the initial AUG codon and can sometimes synthesize a large polypeptide that consists of more than one protein [e.g., polycistronic (multiple genes) as opposed to typical mammalian monocistronic (single gene) transcripts]. In such cases, a viral or cellular protein will then serve as a protease to cleave the large polyprotein to generate multiple smaller proteins (e.g., the 3C protein in picornavirus serves as the virus-encoded protease). Such a strategy is a common theme among many viruses. In other instances, ribosomes can initiate at internal start codons (resulting in different proteins), usually with the aid and involvement of viral proteins. One feature prominent in certain viruses such as retroviruses is *ribosomal frameshifting*, where the ribosome may slip one or several nucleotides forward or backward on the same mRNA to generate a different protein. *Splicing* (excising certain areas of the transcript) can also occur (e.g., adenoviruses) to generate different proteins from the same genomic transcript. Like transcription, translation can occur at several time points during the course of the viral life cycle and is sometimes broken up into early and late translational events.

Some newly synthesized proteins may be shuttled (directed by specific signal sequences) through typical eukaryotic host cell pathways, undergoing further modification necessary for proper functioning. Post-translational modification of viral proteins (e.g., glycosylation, phosphorylation, acylation, sulfation, or cleavage) is usually carried out by cellular enzymes, although certain virus-encoded enzymes are sometimes used. The resultant proteins, either post-translationally modified or not, may be structural, in which case they are recruited and grouped for later virion assembly either within the cytoplasm or at the envelope, or they may be nonstructural and function to regulate either a cellular or viral activity such as transcription and replication.

Replication. Replication usually takes place after the synthesis of viral proteins. Both DNA and RNA viruses have the ability to selectively inhibit cellular DNA synthesis while promoting viral DNA or RNA synthesis and will exhibit diverse strategies depending on the makeup of their genome and type of virus. Sometimes a DNA virus will make use of cellular DNA polymerases (e.g., parvoviruses) to carry out the major catalytic process. In other cases virus-encoded enzymes such as helicases (which unwind a double helix), ligases (which join DNA fragments), RNases (which degrade RNA primers), or a virally encoded DNA or RNA polymerase is required. Some DNA viruses (e.g., herpesviruses and poxviruses) are very self-sufficient and encode all of the replication proteins required. With the exception of poxviruses, most currently known animal DNA viruses undergo genomic replication in the nucleus.

In order for RNA viruses to replicate, they must go through an intermediate complementary RNA that will subsequently serve as the template for progeny RNA. For example, if the genome is (−)ssRNA, a (+)ssRNA intermediate must first be made to serve as the template for subsequent generation of (−)ssRNA progeny genomes. In some RNA viruses, the same enzyme that is responsible for transcription (RdRp) also carries out replication. For other viruses, evidence exists that there are two separate enzymes for replication and transcription. As previously noted, retroviruses are unique RNA viruses because they make use of their reverse transcriptase to produce a DNA intermediate first in order to generate progeny RNA.

Late Events

Assembly. Prior to the release of progeny virions from the cell, assembly must first occur in either the nucleus (typical of DNA viruses such as adenoviruses) or the cytoplasm (typical of RNA viruses such as picornaviruses). Only after sufficient amounts of protein and progeny nucleic acid have been generated will assembly take place. Typically in a host cell, viral proteins will be grouped together in *assembly complexes* prior to assembly. Formation of these complexes is usually mediated by one or more viral proteins and occasionally with the aid of host proteins. In some viruses assembly is a purely virus-dependent process and does not require facilitation by the host cell. The specific order of assembly will vary; in some cases, certain proteins complex together before encompassing a genome, whereas in other cases the genome will first complex with certain proteins before being further encased by the outer layers of proteins. In multilayered viruses (e.g., reoviruses), assembly of the inner layer usually precedes the assembly of the outer layer. Clustering of virions within the cell can form an *inclusion body* (sometimes membrane-bound), which eventually leads to the disruption of the cellular membrane.

In the case of enveloped viruses, the inner nucleocapsid is assembled first. In most cases, enveloped viruses acquire their membranes as the newly formed progeny nucleocapsids pass through a cellular membrane. When the virus's nucleocapsid is formed in the cell cytoplasm, the envelope is normally acquired as the nucleocapsid passes from the cell's cytoplasm through the plasma membrane into the extracellular environment (e.g., rhabdoviruses, many togaviruses, and the myxoviruses). When the virus's nucleocapsid is formed in the cell nucleus (e.g., herpesviruses), the viral envelope is acquired as the nucleoprotein complex passes through the nuclear membrane. Some viruses acquire their envelopes as they pass through membranes of the endoplasmic reticulum (e.g., flaviviruses) or Golgi apparatus (e.g., rubella virus). A few viruses do not follow this general rule. Both the poxviruses and the hepadnaviruses assemble their envelopes de novo from lipid components within the cell. Viruses that contain envelopes have virus-specified proteins embedded within the membrane. Most viruses succeed in selectively excluding host proteins from their envelopes, but some (e.g., retroviruses) contain host-specified proteins within their membranes that play important roles in pathogenesis.

Virion Release. There are two processes by which viral particles are released from the cell. For nonenveloped viruses and some enveloped viruses that have acquired their envelope intracellularly, once a critical concentration of virions is achieved inside the cell, release of mature virions typically occurs via rupture or *lysis* of the cell membrane. Cell lysis often results in cell death, although some nonenveloped viruses can exit without killing the cell (e.g., polyomaviruses). The released progeny virions are now free to infect neighboring cells and repeat the replicative cycle.

Many enveloped viruses exit the cell through a process termed *budding*, in which the virus "buds" from the cell, acquiring its envelope at the same time (as described above). Budding may or may not damage the cell, and in some cases the cell subsequently recovers from viral infection. In some cases budding at the outer membrane is restricted to a particular surface of the cell; it can occur at either the apical (top) surface (e.g., orthomyxoviruses and paramyxoviruses) or at the basal (bottom) surface (e.g., rhabdoviruses and some retroviruses).

Lysogeny

Some viruses will bypass the above-described replicative cycle and instead will make use of an alternative pathway termed the *lysogenic cycle*. Among the herpesviruses, this is generally known as *latency*, but *lysogeny* refers to the general phenomenon, originally described for some bacterial viruses. In the lysogenic cycle the viral DNA will undergo some alteration that results in the viral replicative cycle being arrested. In some cases (e.g., the bacterial virus lambda) the viral DNA integrates into the host cell chromosome. This effectively puts the viral replicative cycle "on hold," because no progeny virions will be produced. Once integrated, the viral DNA can persist there indefinitely; each daughter cell will contain one or more copies of the viral DNA, because the incorporated DNA is replicated along with the cell's DNA. As part of the host DNA, the viral DNA can remain silent, can serve to express a low copy number of genes, or can be induced to complete the replicative cycle. Exactly how induction is carried out in the host remains poorly understood, but stress, sunlight, other infections, and certain chemicals can function as inducers. The herpesviruses arrest their replicative cycles and enter latency by a different mechanism. In this case the viral DNA circularizes and persists in the host cell's nucleus, being passed to daughter cells, and may eventually be induced as described above.

The complexity, richness, and variety of virion structure and the means by which these infectious agents replicate themselves are reflected in subsequent chapters in this volume and contribute to the similarities and differences in the mechanisms of neuropathogenesis that various viruses effect.

4 HOSTS

4.1 Cell Tropism

As indicated earlier, there currently are more than 3600 known virus species. Viruses have been detected within every other type of living organism, from bacteria to plants and animals [8]. Some families of viruses are capable of infecting organisms from diverse kingdoms. For example, *Rhabdoviridae* are capable of infecting plants and animals. However, any given species of virus is usually extremely limited in the types of host cells it can infect. This is known as *cell tropism*. Thus, a particular virus capable of infecting some bacteria usually cannot infect all bacteria and also cannot infect any plants or animals. Most plant viruses are capable of infecting some, but not all, species of plants but cannot infect any bacteria or animals. Likewise, the cell tropism of a specific animal virus limits the capacity of the virus to infect only certain species of animals.

There is significant variability in the range of cell tropism for individual species of viruses. Some viruses (e.g., arthropod-borne togaviruses) are capable of infecting vertebrate animals of diverse orders such as humans and horses and of different classes such as birds and humans as well as the phylogenetically unrelated insects that vector the virus from one vertebrate host to the next. Other viruses may be extremely limited in their cell tropism. For example, the human retrovirus HIV is capable of infecting only some human-derived cells.

The basis for cell tropism lies in the ability of a particular virus to enter and replicate within a specific cell. This, as indicated earlier, is primarily mediated by precise lock-and-key interactions between the virus and the host cell that allow the virus to bind to and then enter the appropriate cell. Thus, viruses that are capable of infecting and replicating within a wide range of cells (e.g., arboviruses) usually recognize and bind to a molecule that is common to that wide range. of cells, whereas viruses that are restricted to infecting a limited number of cells (e.g., HIV) generally recognize highly specific cell molecules.

4.2 Tissue Tropism

In addition to cell tropism, which is usually used to describe the type of host organism (e.g., bacterium, plant, bird, human) a particular virus is capable of infecting, some viruses are restricted in the types of cells that can be infected within a susceptible organism. This is known as *tissue tropism*. For example, the herpesviruses generally are capable of replicating in numerous types of tissues (brain, liver, skin, etc.), whereas other viruses (e.g., the hepatitis virus) are restricted to liver tissue. One basis for tissue tropism may be the same as the basis for cell tropism: whether or not the host cell contains molecules on its surface that the virus can recognize and bind to. Another basis for tissue tropism is whether the host cell contains the appropriate enzymes necessary for viral replication once the virus has entered the cell.

The focus of this book, as reflected by its title and in subsequent chapters, is neuropathogenic viruses of human clinical relevance. Thus, whereas this chapter has attempted to provide a general overview of the myriad of diverse viruses with respect to their structure, classification, and various replicative strategies (reviewed in greater detail in Ref. 12), the focus of later chapters is those viruses whose cell tropism includes human cells and those viruses capable of infecting the central nervous system.

ACKNOWLEDGMENTS

Research in our laboratory has been supported by grants from the Manitoba Health Research Council, by the Dr. Paul Thorlakson Foundation, and by grant MT-11630 from the Medical Research Council of Canada. M. I. S. is the recipient of a University of Manitoba Graduate Scholarship, and I. I. M. is the recipient of an MHRC Graduate Scholarship.

REFERENCES

1. Nibert, M. L.; Schiff, L. A.; Fields, B. N. Reoviruses and their replication. In Fields Virology; Fields, B. N., Knipe, D. M., Howley, P. M., Chanock, R. M., Melnick, J. L., Monath, T. P., Roizman, B., Straus, S. E., Eds.; Lippincott-Raven: Philadelphia, 1996, 1557–1596.
2. Murphy, B. R.; Webster, R. G. Orthomyxoviruses. In Fields Virology; Fields, B. N., Knipe, D. M., Howley, P. M., Chanock, R. M , Melnick, J. L., Monath, T. P., Roizman, B., Straus, S. E., Eds.; Lippincott-Raven: Philadelphia, 1996, pp 1397–1445.
3. Caspar, D. L. D.; Klug, A. Physical properties in the construction of regular viruses. Cold Spring Harbor Symp. Quant. Biol. 1962, 27, 1–32.
4. Stewart, P. L.; Burnett, R. M.; Cyrklaff, M.; Fuller, S. D. Image reconstruction reveals the complex molecular organization of adenovirus. Cell. 1991, 67, 145–154.
5. Liddington, R. C.; Yan, Y.; Moulai, J.; Sahli, R.; Benjamin, T. L.; Harrison, S. C. Structure of simian virus 40 at 3.8-A resolution. Nature. 1991, 354, 278–294.
6. Dryden, K. A.; Wang, G.; Yeager, M.; Nibert, M. L.; Coombs, K. M.; Furlong, D. B.; Fields, B. N.; Baker, T. S. Early steps in reovirus infection are associated with dramatic changes in supramolecular structure and protein conformation: analysis of virions and subviral particles by cryoelectron microscopy and image reconstruction. J. Cell. Biol. 1993, 122, 1023–1041.
7. Reinisch, K. M.; Nibert, M. L.; Harrison, S. C. Structure of the reovirus core at 3.6 A resolution. Nature. 2000, 404, 960–967.
8. van Regenmortel, M. H. V.; Fauquet, C. M.; Bishop, D. H. L.; Carstens, E. B.; Estes, M. K.; Lemon, S. M.; Maniloff, J.; Mayo, M. A.; McGeoch, D. J.; Pringle, C. R.; Wickner, R. B. Virus Taxonomy. Seventh Report of the International Committee on Taxonomy of Viruses; Academic Press: San Diego, 2000.
9. Baltimore, D. Expression of animal virus genomes. Bacteriol. Rev. 1971, 35, 235–241.
10. The World Health Report; World Health Organization: Geneva, 1996.
11. Murray, C. J. L.; Lopez, A. D. The Global Burden of Disease. A comprehensive assessment of mortality and disability from diseases, injuries, and risk factors in 1990 and projected to 2020; Harvard School of Public Health: Boston, 1996.
12. Fields, B. N.; Knipe, D. M.; Howley, P. M.; Chanock, R. M.; Melnick, J. L.; Monath, T. P.; Roizman, B.; Straus, S. E. Field's Virology, 3rd ed.; Lippincott-Raven: Philadelphia, 1996.

2

Neuropathogenesis of Viral Infections

Avindra Nath and David Galey
The Johns Hopkins University School of Medicine
Baltimore, Maryland, U.S.A.

Joseph R. Berger
University of Kentucky College of Medicine
Lexington, Kentucky, U.S.A.

1 INTRODUCTION

A large number of viruses are capable of invading the nervous system. The reason for the brain being a preferred site of viral invasion is not entirely clear. Surely, a fulminant encephalitis that leads to the demise of the host is counterproductive to the survival and propagation of the virus. In other less severe forms of encephalitis the brain serves as a sanctuary hidden from the immune system. Because the brain lacks immune-competent cells, it may be an excellent site for the virus to reside in and form a reservoir. Some viruses may mutate and evolve within the brain and then exit the brain through a wide variety of routes.

2 TRANSMISSION OF VIRUSES

Interactions of humans with their environment expose them to a number of pathogens including viruses that invade the nervous system. The nervous system is well encased in the bony skull and the spinal column, shielding it from the environment; however, viruses have evolved to develop specialized mechanisms for invading the nervous system. In some cases an organism may use multiple modes of transmission. Most often viruses are transmitted from humans to humans and have adapted in such a way that they either do not infect other animals or do not produce clinical disease if they do. Other viruses require an intermediate vector for transmission to humans. This vector may be a mosquito or tick or even a mammal, as in the case of rabies. Occasionally viruses that normally reside in animals may infect humans when close contact occurs between the two. For example, Hendra virus [1] and equine morbillivirus [2] are paramyxoviruses that infect horses but may cause an encephalitis in humans.

Viruses gain access to the human body via any surface exposed to the environment, i.e., any mucosal surface (eyes, oral, and gastrointestinal system, genital mucosa, respiratory mucosa). The skin may be breached by insect or animal bites. Iatrogenic spread can occur by blood transfusions and organ transplants (Table 1).

Transmission by air requires that the virus be able to exist at different temperatures and in dry and wet conditions. Thus enveloped viruses are most efficiently spread by this mechanism. Of course, an important prerequisite is that the virus be excreted by oral or respiratory routes. Sneezing and coughing lead to the formation of small droplets. Transmission by oral routes requires that the virus be able to resist the proteolytic enzymes and the wide

Table 1 Modes of Viral Infection or Transmission of Infectious Agents to Humans

Air	Animals
Influenza A	Rabies (dogs and other mammals)
Rubella	vCJD (cows)
Measles	Nipah (pigs)
Mumps	Lymphochoriomeningitis virus (rodents)
Adenoviruses	Hendra (horse)
Varicella-zoster virus (VZV)	Sexual
Oral	Herpes simplex virus type-2
JC virus (JCV)	Human immunodeficiency virus (HIV)
Herpes simplex type 1 (HSV-1)	HTLV-I
Enteroviruses	Blood transfusion
Variant Creutzfeldt- Jakob disease (vCJD)	HIV
Kuru	HTLV-I
Breast milk	Organ transplant
Human T-cell leukemia virus type I (HTLV-I)	Creutzfeldt-Jakob disease
HIV	West Nile encephalitis
Ocular	
Enterovirus 70	
HSV-1	
Vector	
Mosquitoes	
Western equine encephalitis	
Eastern equine encephalitis	
Venezuelan equine encephalitis	
West Nile encephalitis	
California encephalitis	
Japanese B encephalitis	
La Crosse	
Dengue	
Ticks	
Colorado tick fever	
Russian spring-summer encephalitis	

changes in pH in saliva and in gastric and intestinal secretions. Hence some viruses such as JC virus have the ability to gain access to the lymphatic system from the oral cavity by invading the tonsils and thus escaping the gastric and intestinal environment [3].

3 SPREAD TO THE BRAIN

3.1 Localized Infection

All viruses undergo replication within a localized region at the site of entry. It is essential that the virus be able to escape the immune system until it reaches a critical mass that allows it to then travel to the nervous system tissues where it may again escape the immune system. These localized regions include mucosal surfaces and wounds that have poor blood supply.

3.2 Hematogenous Spread

Most often, once the virus has replicated to a critical level it then spreads to the brain hematogenously. It may infect cells that form the immune system, thus evading their onslaught. As they travel through the nervous system, these cells spread the virus to brain cells. For example, JCV infects B cells. HIV infects CD4 T cells and macrophages. HTLV-I infects T lymphocytes. As discussed below, transneuronal spread may also occur. Viruses may enter the brain by crossing the blood capillaries either as free virus or in the leukocytes that traffic the brain parenchyma. The latter form of entry has been termed the ''Trojan horse'' phenomenon. Viruses or virus-infected cells may also enter the brain via the choroid plexus and be disseminated via CSF pathways. This mode of spread most often results in a viral meningitis.

3.3 Transneuronal Spread

Some viruses have a unique mechanism of accessing the nervous system (Table 2). These viruses travel up the axons of nerves and hence escape the immune system. The mechanism of transneuronal spread is not entirely clear. The retrograde transport of viruses involves the transport of intact virions and is a slow transport process. However, anterograde transport of viruses may involve the transport of viral proteins that get assembled into a complete virion at the nerve terminal. Herpesvirus type 1, after causing oral lesions, travels along the trigeminal nerve and lies dormant in the trigeminal ganglia. Following reactivation it may then travel to the temporal or frontal lobe to cause a herpes encephalitis. Varicella-zoster virus, after causing skin lesions, resides in dorsal root ganglia and after reactivation travels along the sensory nerve to the dermatome innervated by the nerve, causing an eruption of zoster or shingles.

Table 2 Viruses and Other Infectious Agents Disseminated via Transneuronal Pathways

Virus or infectious agent	Nerve
HSV	Trigeminal nerve
Rabies	Cutanecus or cranial nerve
VZV	Cutanecus nerve
Enterovirus	Trigeminal nerve
Prions	Splenic nerve

Rabies virus travels retrogradely in the nerve from the site of the bite and then transneuronally to invade the limbic system of the brain. The proximity of the site of rabies virus entry to the brain correlates directly with the latency to disease development; i.e., bites on the head or neck lead to disease development more quickly than bites on the distal lower extremity. Similarly, measles virus may also spread transneuronally within the central nervous system and has been implicated in the pathogenesis of subacute sclerosis panencephalitis [4]. Prion proteins emerging from B cells within the spleen [5] have also been shown to undergo retrograde transport via the splenic nerve to regions of the brainstem and are then disseminated throughout the brain [6]. Enteroviruses such as poliovirus may also be transmitted via retrograde axonal transport [7]. Similarly, enterovirus 71 may spread via the trigeminal nerve following an episode of conjunctivitis.

4 CELL TYPES INFECTED AND REPLICATION IN THE BRAIN

All cell types within the brain are capable of supporting viral replication, although each type of virus selects a certain type of cell as its target. Occasionally a virus may replicate in more than one type of cell (Table 3). For example, CMV is permissive in a wide variety of cells within the brain [8].

Infected leukocytes infect resident brain cells via cell to cell contact, resulting in localized areas of brain infection. Viral replication may be restricted at the level of viral entry such that the virus may infect only those cells that have specific receptors. For example, poliovirus infection is dependent upon expression of its receptor, CD155, in the gut [9] and neurons [10]. Other viruses may easily enter cells, but replication may be determined by the availability of certain host proteins. For example, JCV enters a wide variety of cell types, but it replicates in brain derived cells such as astrocytes that have NF-1D protein [11], whereas, due to the lack of NF-1D, [12] it does not replicate in neurons even if the viral genome is microinjected into the nucleus Similarly, herpesviruses and adenoviruses are capable of entering a large number of cell types, but infection becomes established in only a few cell types. The ability of viruses to infect many types of cells and use multiple receptor and non-receptor mediated mechanisms for invading an organism aids their survival in nature.

Viruses that travel along neurons (Table 3) spread across synapses to other neurons in a predictable neuronal pathway. In fact, because of this property, anatomists have tagged such viruses, in particular the herpesviruses, with markers for mapping some of the neuronal pathways [13].

Table 3 Major Cell Types Infected by Viruses in the Nervous System

Cell	Virus
Neurons in CNS	Rabies, HSV-1, polio, measles, rubella, Borna virus, mumps virus, arboviruses
Dorsal root ganglia neurons	VZV, HSV-1
Oligodendroglia	JCV
Astrocytes	HIV, HTLV-1, JCV, CMV, HSV-1
Microglia	HIV
Choroid plexus and meningeal cells	CMV
Endothelial cells	Human parvovirus B19, CMV

Infection of cells may result in a cytopathic effect (e.g., infection of lymphocytes and macrophages/microglia with HIV) or a persistent infection (e.g., JCV or HIV infection of astrocytes) or may induce a proliferation response leading to tumor formation (e.g., EBV infection of B lymphocytes results in CNS lymphomas) (see Chaps. 7 and 26).

Infection may also spread across cells in the brain without viral assembly taking place at the cell membrane. For example, some paramyxoviruses such as measles in subacute sclerosing encephalitis and a form of late onset encephalitis with Nipah virus, cell-to-cell transmission is likely due to transmission of the viral replicative ribonucleoprotein complex through fusion of the infected cells with the adjacent noninfected cell [14,15]. This has been attributed to mutations in the matrix protein gene that prevent viral morphogenesis.

5 ANATOMICAL REGIONS WITHIN BRAIN INFECTED BY VIRUSES

In large measure dictated by their mode of entry into the CNS, viruses reside in specific anatomical regions, causing focal symptoms referable to that site (Table 4). For example, herpesvirus travels up the trigeminal nerve and resides in the trigeminal ganglia. Once activated it infects the temporal lobe, which is in close proximity to the ganglia.

6 BRAIN DEFENSE MECHANISMS

The brain is not only encased in a bony structure, the skull, to protect it from any mechanical insults, it also has a sophisticated anatomical and physiological barrier at the capillary interface that allows the passage of only selected substances from the blood to the brain parenchyma. The barrier itself is composed of a layer of capillary endothelial cells that have tight junctions between them. These junctions are specific to the CNS [16]. In the rest of the body, the vascular endothelial cells are fenestrated or have low-resistance junctions. The barrier is further reinforced by astrocytes on the abluminal surface, which use their foot processes to add another layer around the capillaries. The effect of this barrier is to limit the availability to the brain parenchyma to circulating cells and other plasma components. Small lipophilic molecules can pass through the barrier directly, whereas other molecules enter by carrier-mediated transport. The specialized endothelial cells also exhibit little endocytosis and transcellular transport. There is also limited extravagation of immune cells through the tight junctions. Owing to its highly selective permeability and the uniqueness of the tight junctions, the blood-brain barrier helps to prevent free

Table 4 Major Anatomical Sites Invaded by Viruses

Virus	Site invaded
JCV	Multifocal white matter in brain
HTLV-I	Thoracic spinal cord
HIV	Basal ganglia
HSV	Trigeminal ganglia, temporal lobes
VZV	Dorsal root ganglia, cutaneous nerves
Rabies	Limbic system
Polio	Anterior horn cells

virus from entering the brain via normal circulation, and also reduces the likelihood that infected cells, such as macrophages and lymphocytes, might bring virus into the brain.

7 VIRAL LATENCY IN THE NERVOUS SYSTEM

Several viruses have been shown to reside in the nervous system for long periods of time. The brain provides an ideal place in which these organisms can hide. The lack of a lymphatic system in the brain means that there is limited immune surveillance of the nervous system. Further, the normal brain does not express major histocompatibility complex (MHC) antigens [17]. Interestingly, neurodegeneration may itself lead to the expression of MHC antigens. Neurons may repress the expression of MHC antigens under normal circumstances, and when neurons are injured, MHC antigen expression is induced in these cells. Neurotropic factors such as NGF, BDNF, and NT3 have been shown to suppress MHC antigen expression (see Ref. 18). However, the expression of MHC antigens may be triggered by cytokines such as gamma-interferon. Hence, once the process of inflammation is initiated in the brain parenchyma, it escalates into a self perpetuating process.

8 HOST DEFENSE

In general the host response to a viral infection is an effort to curtail the infection and consists of a variety of cellular and humoral immune responses. Recovery from the infection is due to the ability of the immune system to successfully clear the virus. Because viruses are intracellular pathogens, clearing them may involve killing the cells infected with them, although experimental systems suggest that under some circumstances viruses may be cleared from cells without killing the cells. Occasionally the host responses may mount a fierce attack, whereby the responses themselves may damage the uninfected cells. Much effort has been devoted in recent years to characterize these detrimental responses. It is possible that there are virus specific host responses, and these specific patterns may serve as signatures of each of the pathogens. Currently, effort is underway to investigate these possibilities using microarrays and proteonomics based technologies that allow for the study of a large number of gene products simultaneously. However, mechanisms of viral clearance may also differ among various cell types. For example, in a Sindibis virus model it was shown that antibody responses were necessary for clearance of virus from cortical neurons whereas gamma interferon was necessary for elimination of virus from brainstem and spinal cord neurons [19].

8.1 Cytokines

Within the brain, microglia are the main contributors to cytokine production, although in the setting of a viral infection the invading mononuclear cells also contribute to the production of cytokines. Astrocytes are also to some extent capable of producing cytokines. These cytokine responses may have antiviral properties [20].

8.2 Interferons

Interferons (IFNs) are a group of proteins that derive their name from their ability to interfere with viral replication in an indirect fashion. There are three major families of interferons: IFN-α IFN-β, and IFN-γ. IFN-α and IFN-β have potent antiviral properties within cells exposed to them, whereas IFN-γ enhances the immune system's ability to clear infected cells, mainly after the induction of the adaptive immune response. Therefore, expression of IFN-α and IFN-β early in the course of infection is crucial to preventing the further spread of the virus. IFN-α and IFN-β may become expressed by the presence of a variety of intracellular inducers, including the presence of foreign nucleic acids. In fact, the presence of double-stranded RNA is a potent inducer of their expression. Recent evidence suggests that IFN-γ is also important in controlling viral infection even in the absence of other cell mediated immune responses. For example, in measles virus infection, IFN-γ can clear the virus from infected neurons without causing neuronal cell loss [21].

Once a cell binds IFN-α or IFN-β, which use a common receptor, a cascade of cellular signaling occurs that results in the transcription of several proteins that aid in conferring a hostile environment to viral infection. Three key antiviral proteins have been identified as a result of this transcriptional activation: 2′ 5′-oligoadenylate synthetase, protein kinase PKR, and Mx protein [22]. The 2′, 5′-oligoadenylate synthetase polymerizes adenosine triphosphate into a series of 2′–5′ linked oligomers, which differ from normal nucleotides that are joined 3′–5′. These oligomers in turn activate RNase L, a constitutive endoribonuclease. This enzyme degrades viral RNA. Protein kinase PKR is activated by the presence of double-stranded RNA. Upon activation, PKR phosphorylates the cellular translation initiation factor eIF-2. The result of this is an inhibition of translation and protein synthesis, which contributes to the inhibition of viral replication. The Mx protein acts in the nucleus of an infected cell to confer resistance to influenza virus by inhibiting the synthesis of the influenza virus mRNA.

8.3 Humoral Immune Responses

The body has a whole repertoire of B cells already in place that cover an extensive array of antigenic determinants. To initiate the humoral immune response, a viral epitope must make contact with the appropriate B cell. This triggers the B cell to undergo proliferation and maturation into antibody-secreting plasma cells. The first antibody type produced is IgM, and later, following further stimulation, IgG antibodies are produced. Antibodies may then bind directly to the viral particle or indirectly by binding to viral antigens expressed on the cell surface. This can lead to several layers of protection. First, the antibody binding to the viral particle can interfere or sterically hinder the virions' ability to enter a host's cell. This is termed *virus neutralization*. This effect is not only on a single virion level but might also form antibody–virus complexes owing to the multivalent nature of certain antibodies, effectively decreasing the viral titer.

In the scenario in which an antibody binds to viral antigens expressed on the host cell surface, the cell is then targeted for destruction to prevent viral replication. This might occur in several different ways. One way is by *complement-mediated cell lysis*, whereby an antibody decorated cell initiates a complement cascade on its cell surface. The cascade culminates in the formation of a membrane attack complex, opening a pore in the cell that disrupts the cell's electrolytic and osmotic balance. Another mechanism whereby an infected cell that has been recognized by an antibody might be eradicated is via *antibody-dependent cell-mediated cytotoxicity*. This cytotoxicity is accomplished by means of natural killer cells, which are specialized lymphoid cells. These cells have receptors on their cell surface that recognize the constant region of antibodies and lead to release of cytotoxic agents. Finally, a cell might be destroyed by *antibody mediated phagocytosis*. This phenomenon is dependent upon the humoral response occurring to target the infected cells [23]. Recent observations suggest that antibodies can catalyze the generation of hydrogen peroxide from singlet molecular oxygen and water, which further leads to the production of ozone [24]. Although this phenomenon has been characterized in bacterial killing, it is possible that similar pathways may be operative in viral defense mechanisms.

8.4 Cellular Immune Responses

Cytotoxic immune responses play an important role in viral clearance from the brain. This may be established through several different mechanisms. Interactions of T-cell receptors with antigenic peptide bound to major histocompatibility complex (MHC) on the surface of antigen presenting cells (APCs) occur in a protected environment called the immunological synapse. The immunological synapse contains at least two functional domains: a central cluster of engaged antigen receptors and a surrounding ring of adhesion molecules [25]. This is critical for the initiation of the cell mediated immune responses. Thus when neurone lack MHC antigen expression, viruses that infect neurons may escape cytotoxic immune responses [26]. However, neurons may trigger the apoptotic pathways in an effort to eliminate the virus [27].

Monocytic infiltation into the brain may have a dual purpose. The ability of activated monocytes to produce cytokines such as TNF-α has been associated with neurotoxic properties [28]. On the other hand, it has also been shown that monocytes can produce neurotrophic factors, thus playing a neuroprotective role. Understanding the regulation of this delicate balance may be critical for therapeutic approaches that aim to target monocytic infiltration within the brain. Monocytes are unique in that they express a large number of chemokine receptors and hence respond to many different chemokines. Hence, not surprisingly, they are important participants in the cellular infiltrates in most viral infections of the brain. Some viruses such as HIV [29], measles virus [30], and human parvovirus B19 [31] can cause fusion of the monocytic cells, resulting in multinucleated giant cells. Some HIV-infected patients with opportunistic CMV or VZV brain infection may have multinucleated giant cells that are coinfected with HIV and VZV or CMV [32]. It is thus possible that the formation of multinucleated giant cells may represent a phenomenon whereby one infected cell is trying to engulf another. The cellular infiltrates may organize themselves into small nodules, called microglial nodules although in addition to microglial cells and infiltrating monocytes these nodules may also have lymphocytes and reactive astrocytes. Microglial nodules are most often seen with HIV infection [29], some of the herpesvirus infections of the brain [33–35], and West Nile encephalitis [36]. They may also be present in Rasmussen's encephalitis, the etiology of which remains obscure [37].

8.5 Chemokines

Because the brain has few immune effector cells, induction of chemoattractant cytokines called chemokines may be an important defense mechanism whereby the brain under attack from viruses may recruit lymphocytic and monocytic cells to itself. However, if the virus is capable of infecting the lymphocytes or monocytes, these cells may carry the pathogen with them. Also, if the uninfected cells become activated and are present in large numbers, they may produce cytotoxic substances that can damage host cells in the brain. All cell types within the brain are capable chemokines. To date nearly 50 different chemokine receptors and an equal number of chemokines have been identified. Multiple terms have been used by different research groups to name the same chemokine, making the field difficult to follow. Further, one chemokine receptor may respond to several different chemokines, and one chemokine may interact with more than one type of chemokine receptor [38]. Nonetheless, some common themes are beginning to emerge. For example, MCP-1 is the major chemokine for monocyte infiltration to the brain. Levels of MCP-1 are elevated in the cerebrospinal fluid (CSF) of patients with HIV dementia [39] and CMV encephalitis [40].

8.6 Host Genetics

Although host susceptibility genes in the context of CNS viral infections have not been well studied, it is likely that they do play a role [41,42]. Animal studies clearly indicate that host genetic factors determine susceptibility to Theiler's murine encephalitis virus and simian immunodeficiency virus, for example.

9 MECHANISMS OF NERVOUS SYSTEM INJURY

9.1 Effect of Viral Infection on Cell Function

The effect on cellular function may be quite variable and largely depends on the degree of replication of the virus within the cell. In a latently infected cell, the virus may have little or no effect on cellular function. At the other extreme, the virus may take over the cellular machinery for its own propagation, resulting in cytopathic changes in the cell. However, within the nervous system, cell–cell interactions are critical in maintaining normal function. Hence disruption of function in a small number of cells due to infection can have far reaching effects on other cells.

9.2 Bystander Effect

Most microorganisms are known to produce toxic substances. Examples of bacterial toxins include cholera toxin, botulinum toxin, and tetanus toxoid. Prion proteins have been studied extensively with regard to their neurotoxic properties. Similarly, viral products may also be toxic. Although virotoxins have been best characterized for HIV gene products (see Chap. 11), it is increasingly clear that several other viruses also produce toxic gene products (Table 5). For example, the rabies virus [43] envelope glycoprotein and the measles virus hemagglutinin glycoprotein [44] have sequence homology to snake venom neurotoxins, and the fusion domain of influenza virus has a striking similarity to the neurotoxic domain of amyloid beta peptide [45]. A common theme emerges among these viruses, in that often it is the envelope and the transactivating viral genes that are toxic. These viral proteins may interact with neurons and glial cells to disrupt their function.

Table 5 Non-HIV Viral Proteins with Neurotoxic Properties

Virus	Gene product	Ref.
Rabies virus	Envelope	43
Influenza virus	Envelope	45
Feline immunodeficiency virus	Envelope	46
Feline leukemia virus	Envelope	47
Measles virus	Hemagglutinin	44
Adenovirus	E4	48
Human foamy virus	Bel	49
HTLV-I	Tax	50
Visna virus	Tat	51

There are several mechanisms by which viral proteins may become available to the extracellular environment.

1. When a cell ruptures, all its contents, including all structural and nonstructural viral proteins, become available to the extracellular environment.
2. There may be a restricted expression of viral genes whereby some proteins are overexpressed but a nonreplicative state of the viral genome is maintained [52]; for example, in HIV-infected astrocytes, regulatory genes *tat, nef*, and *rev* are overexpressed. Furthermore, during the normal course of viral replication, not all structural proteins formed within infected cells get incorporated into the viral structure. These proteins are either degraded by the cells or are available for extracellular release. Viral proteins such as Tat protein of HIV may be actively secreted by the cell [53,54].
3. All viral particles formed by infected cells are not replication competent. Thus, structural proteins of the virus in the form of defective viral particles may have access to and affect uninfected cells [55]. In fact, for animal and plant viruses, most viral particles produced are defective and/or noninfectious.
4. The viral coat protein may be shed upon viral entry [56].
5. Viral proteins may interact with surrounding uninfected cells by cell–cell contact with an infected cell. For example, gp41 is a transmembrane protein of HIV that is expressed on the surface of infected cells, which may induce neuronal injury to cells in close proximity [57].

Prolonged continuous exposure to the viral proteins may not be necessary to disrupt neuro-glial relationships or induce neurotoxicity. Rather, a transient exposure may be sufficient to trigger a cascade of events that eventually results in neuronal damage referred to as the ''hit and run phenomenon'' [58]. Once they are in the extracellular environment, these viral proteins, may cause neurotoxicity either by acting directly on the neuronal cells or by activating glial cells to cause the release of cytokines, chemokines, or neurotoxic substances. These substances initiate several positive feedback loops. For example, release of chemokines such as MCP-1 would lead to the influx of monocytes, which upon activation would lead to further release of cytokines, chemokines, and neurotoxic substances [39]. Thus, viral proteins are able to amplify their neurotoxic potential and cause damage at distant sites. For example, injection of Tat intraventricularly can lead to neuronal cell loss and gliosis in the substantia nigra [59]. These same proteins might also lead to glial dysfunction, which would contribute to a hostile microenvironment for the neurons [60].

9.3 Autoimmune Responses

Occasionally viral infections trigger an immune response against host antigens. Such immune responses have been implicated in the pathogenesis of diseases such as postviral encephalomyelitis, multiple sclerosis, transverse myelitis, and Gullian-Barráe syndrome [61]. Viruses may trigger autoimmune responses by several different mechanisms [62]. Enveloped viruses, as they exit the cell membrane, may carry host antigens from the membrane that become incorporated into the viral envelope. These host antigens may now trigger an immune response. Alternatively, the viral proteins themselves may have sequence homology that is similar to that of host proteins. Hence immune responses directed against the virus may also target the host proteins. This phenomenon has been termed ''molecular mimicry'' [63]. However, the immune responses may be perpetuated by ''epitope spreading'' as demonstrated in a Theiler's murine encephalomyelitis model [64].

Additionally, virus-infected cells can release small immunodominant peptides that can sensitize uninfected cells. These ''bystander sensitized'' cells can then become targets for antigen-specific immune responses by cytotoxic T cells [65].

10 EXIT OF VIRUSES FROM THE BRAIN

Even viruses that infect nervous system tissue exit the human body through various body fluids, including saliva, urine, fecal material, genital secretions, breast milk, and tears. For successful transmission via these fluids, the virus must be able to colonize the organ that produces the bodily fluid and be able to survive the fluid itself. For example, the urine may be acidic, and tears and saliva have proteases. Viruses may exit via hematogenous routes or anterogradely via axons. For example, herpes simplex virus exits via the trigeminal nerve to cause blisters in the mouth and lips; varicella-zoster virus exits via nerves from neurons in the dorsal root ganglia, causing a vesicular eruption on the skin in a dermatomal distribution; human polyoma viruses exit via urine; and rabies virus exits via saliva.

11 EMERGENCE OF NEW VIRUSES

The recent past has seen the emergence of new viral and prion-mediated pathogens that cause encephalitis [66]. This includes HIV infection, West Nile encephalitis, Nipah encephalitis, enterovirus 70 epidemics with poliomyelitis-like disease, the appearance of California virus encephalitis in the midwestern United States, bovine spongiform encephalitis, and variant Creuzfeldt-Jakob disease. Factors that contribute to the emergence of such diseases include evolution of the virus and change in dietary habits with exposure to animal products [67]. The increasing global population provides greater numbers of hosts for mutational evolution and sufficient hosts to ensure maintenance of new agents; the magnitude and modes of modern travel make a larger population of susceptible people accessible and enable the rapid spread of infectious agents [68]. The threat of biological warfare opens the possibility of the use of virulent viral pathogens such as smallpox and the exposure of large populations to such agents, which was previously unimaginable.

REFERENCES

1. McCormack, J. G.; Allworth, A. M. Emerging viral infections in Australia. Med. J. Aust. 2002, 177, 45–49.
2. O'Sullivan, J. D.; Allworth, A. M.; Paterson, D. L., et al. Fatal encephalitis due to novel paramyxovirus transmitted from horses. Lancet. 1997, 349, 93–95.
3. Wei, G.; Liu, C. K.; Atwood, W. J. JC virus binds to primary human glial cells, tonsillar stromal cells, and B-lymphocytes, but not to T lymphocytes. J. Neurovirol. 2000, 6, 127–136.
4. Lawrence, D. M.; Patterson, C. E.; Gales, T. L., et al. Measles virus spread between neurons requires cell contact but not CD46 expression, syncytium formation, or extracellular virus production. J. Virol. 2000, 74 1908–1918.
5. Raeber, A. J.; Montrasio, F.; Hegyi, I., et al. Studies on prion replication in spleen. Dev. Immunol. 2001, 8, 291–304.
6. Blattler, T.; Brandner, S.; Raeber, A. J., et al. PrP-expressing tissue required for transfer of scrapie infectivity from spleen to brain. Nature. 1997, 389, 69–73.

7. Arita, M.; Ohka, S.; Sasaki, Y.; Nomoto, A. Multiple pathways for establishment of poliovirus infection. Virus. Res. 1999, 62, 97–105.
8. Morgello, S.; Cho, E. S.; Nielsen, S., et al. Cytomegalovirus encephalitis in patients with acquired immunodeficiency syndrome: an autopsy study of 30 cases and a review of the literature. Hum. Pathol. 1987, 18, 289–297.
9. Iwasaki, A.; Welker, R.; Mueller, S., et al. Immunofluorescence analysis of poliovirus receptor expression in Peyer's patches of humans, primates, and CD155 transgenic mice: implications for poliovirus infection. J. Infect. Dis. 2002, 186, 585–592.
10. Gromeier, M.; Mueller, S.; Solecki, D., et al. Determinants of poliovirus neurovirulence. J. Neurovirol. 1997, 3(suppl 1), S35–S38.
11. Amemiya, K.; Traub, R.; Durham, L.; Major, E. O. Interaction of a nuclear factor-1-like protein with the regulatory region of the human polyomavirus JC virus. J. Biol. Chem. 1989, 264, 7025–7032.
12. Major, E. O.; Amemiya, K.; Tornatore, C. S., et al. Pathogenesis and molecular biology of progressive multifocal leukoencephalopathy, the JC virus-induced demyelinating disease of the human brain. Clin. Microbiol. Rev. 1992, 5, 49–73.
13. Kuypers, H. G.; Ugolini, G. Viruses as transneuronal tracers. Trends. Neurosci. 1990, 13, 71–75.
14. Cattaneo, R.; Schmid, A.; Rebmann, G., et al. Accumulated measles virus mutations in a case of subacute sclerosing panencephalitis: interrupted matrix protein reading frame and transcription alteration. Virology. 1986, 154, 97–107.
15. Tan, C. T.; Goh, K. J.; Wong, K. T., et al. Relapsed and late-onset Nipah encephalitis. Ann. Neurol. 2002, 51, 703–708.
16. Abbott, N. J. Astrocyte-endothelial interactions and blood-brain barrier permeability. J. Anat. 2002, 200, 629–638.
17. Joly, E.; Mucke, L.; Oldstone, M. B. Viral persistence in neurons explained by lack of major histocompatibility class I expression. Science. 1991, 253, 1283–1285.
18. Neumann, H.; Cavalie, A.; Jenne, D. E.; Wekerle, H. Induction of MHC class I genes in neurons. Science. 1995, 269, 549–552.
19. Binder, G. K.; Griffin, D. E. Interferon-gamma-mediated site-specific clearance of alphavirus from CNS neurons. Science. 2001, 293, 303–306.
20. Cheeran, M. C.; Hu, S.; Gekker, G.; Lokensgard, J. R. Decreased cytomegalovirus expression following proinflammatory cytokine treatment of primary human astrocytes. J. Immunol. 2000, 164, 926–933.
21. Patterson, C. E.; Lawrence, D. M.; Echols, L. A.; Rall, G. F. Immune-mediated protection from measles virus-induced central nervous system disease is noncytolytic and gamma interferon dependent. J. Virol. 2002, 76, 4497–4506.
22. Samuel, C. E. Antiviral actions of interferons. Clin. Microbiol. Rev. 2001, 14, 778–809.
23. Parren, P. W.; Burton, D. R. The antiviral activity of antibodies in vitro and in vivo. Adv. Immunol. 2001, 77, 195–262.
24. Wentworth Jr, P.; McDunn, J. E.; Wentworth, A. D., et al. Evidence for antibody-catalyzed ozone formation in bacterial killing and inflammation. Science. 2002, 298, 2195–2199.
25. Irvine, D. J.; Purbhoo, M. A.; Krogsgaard, M.; Davis, M. M. Direct observation of ligand recognition by T cells. Nature. 2002, 419, 845–849.
26. Griffin, D. E. Arboviruses and the central nervous system. Semin. Immunopathol. 1995, 17, 121–132.
27. Griffin, D. E.; Levine, B.; Tyor, W. R., et al. Age-dependent susceptibility to fatal encephalitis: alphavirus infection of neurons. Arch. Virol. Suppl. 1994, 9, 31–39.
28. Shi, B.; Raina, J.; Lorenzo, A., et al. Neuronal apoptosis induced by HIV-1 Tat protein and TNF-alpha: potentiation of neurotoxicity mediated by oxidative stress and implications for HIV-1 dementia. J. Neurovirol. 1998, 4, 281–290.

29. Nebuloni, M.; Pellegrinelli, A.; Ferri, A., et al. Etiology of microglial nodules in brains of patients with acquired immunodeficiency syndrome. J. Neurovirol. 2000, 6, 46–50.
30. Rahman, S. M.; Eto, H.; Morshed, S. A.; Itakura, H. Giant cell pneumonia: light microscopy, immunohistochemical, and ultrastructural study of an autopsy case. Ultrastruct. Pathol. 1996, 20, 585–591.
31. Isumi, H.; Nunoue, T.; Nishida, A.; Takashima, S. Fetal brain infection with human parvovirus B19. Pediatr. Neurol. 1999, 21, 661–663.
32. Gray, F.; Mohr, M.; Rozenberg, F., et al. Varicella-zoster virus encephalitis in acquired immunodeficiency syndrome: report of four cases. Neuropathol. Appl. Neurobiol. 1992, 18, 502–514.
33. Jay, V.; Hwang, P.; Hoffman, H. J., et al. Intractable seizure disorder associated with chronic herpes infection. HSV1 detection in tissue by the polymerase chain reaction. Childs. Nerv. Syst. 1998, 14, 15–20.
34. Booss, J.; Winkler, S. R.; Griffith, B. P.; Kim, J. H. Viremia and glial nodule encephalitis after experimental systemic cytomegalovirus infection. Lab. Invest. 1989, 61, 644–649.
35. Schmidbauer, M.; Budka, H.; Ambros, P. Herpes simplex virus (HSV) DNA in microglial nodular brainstem encephalitis. J. Neuropathol. Exp. Neurol. 1989, 48, 645–652.
36. Sampson, B. A.; Armbrustmacher, V. West Nile encephalitis: the neuropathology of four fatalities. Ann. NY. Acad. Sci. 2001, 951, 172–178.
37. Bien, C. G.; Urbach, H.; Deckert, M., et al. Diagnosis and staging of Rasmussen's encephalitis by serial MRI and histopathology. Neurology. 2002, 58, 250–257.
38. Karpus, W. J. Chemokines and central nervous system disorders. J. Neurovirol. 2001, 7, 493–500.
39. Conant, K.; Garzino-Demo, A.; Nath, A., et al. Induction of monocyte chemotactic protein-1 in HIV-1 Tat-stimulated astrocytes and elevation in AIDS dementia. Proc. Nat. Acad. Sci. USA. 1998, 95, 3117–3121.
40. Cinque, P.; Vago, L.; Mengozzi, M., et al. Elevated cerebrospinal fluid levels of monocyte chemotactic protein-1 correlate with HIV-1 encephalitis and local viral replication. AIDS. 1998, 12, 1327–1332.
41. Gonzalez, E.; Rovin, B. H.; Sen, L., et al. HIV-1 infection and AIDS dementia are influenced by a mutant MCP-1 allele linked to increased monocyte infiltration of tissues and MCP-1 levels. Proc. Natl. Acad. Sci. USA. 2002, 99, 13795–13800.
42. Corder, E. H.; Robertson, K.; Lannfelt, L., et al. HIV-infected subjects with the E4 allele for APOE have excess dementia and peripheral neuropathy [see comments]. Nat. Med. 1998, 4, 1182–1184.
43. Lentz, T. L.; Wilson, P. T.; Hawrot, E.; Speicher, D. W. Amino acid sequence similarity between rabies virus glycoprotein and snake venom curaremimetic neurotoxins. Science. 1984, 226, 847–848.
44. Yoshikawa, Y.; Yamanouchi, K.; Takasu, T., et al. Structural homology between hemagglutinin (HA) of measles virus and the active site of long neurotoxins. Virus. Genes. 1991, 5, 57–67.
45. Crescenzi, O.; Tomaselli, S.; Guerrini, R., et al. Solution structure of the Alzheimer amyloid beta-peptide (1–42) in an apolar microenvironment. Eur. J. Biochem. 2002, 269, 5642–5648.
46. Bragg, D. C.; Meeker, R. B.; Duff, B. A., et al. Neurotoxicity of FIV and FIV envelope protein in feline cortical cultures. Brain. Res. 1999, 816, 431–437.
47. Mitchell, T. W.; Rojko, J. L.; Hartke, J. R., et al. FeLV envelope protein (gp70) variable region 5 causes alterations in calcium homeostasis and toxicity of neurons. J. Acquir. Immune. Defic. Syndr. Hum. Retrovirol. 1997, 14, 307–320.
48. Lavoie, J. N.; Nguyen, M.; Marcellus, R. C., et al. E4orf4, a novel adenovirus death factor that induces p53-independent apoptosis by a pathway that is not inhibited by zVAD-fmk. J. Cell. Biol. 1998, 140, 637–645.
49. Aguzzi, A.; Bothe, K.; Wagner, E. F., et al Human foamy virus: an underestimated neuropathogen? Brain. Pathol. 1992, 2, 61–69.

50. Arai, M.; Ohashi, T.; Tsukahara, T., et al. Human T-cell leukemia virus type 1 Tax protein induces the expression of lymphocyte chemoattractant SDF-1/PBSF. Virology. 1998, 241, 298–303.
51. Starling, I.; Wright, A.; Arbuthnott, G.; Harkiss, G. Acute in vivo neurotoxicity of peptides from Maedi Visna virus transactivating protein. Tat. Brain. Res. 1999, 830, 285–291.
52. Saito, Y.; Sharer, L. R.; Epstein, L. G., et al. Overexpression of nef as a marker for restricted HIV-1 infection of astrocytes in postmortem pediatric central nervous tissues. Neurology. 1994, 44, 474–481.
53. Ensoli, B.; Buonaguro, L.; Barillari, G., et al. Release, uptake, and effects of extracellular human immunodeficiency virus type-1 Tat protein on cell growth and viral replication. J. Virol. 1993, 67, 277–287.
54. Chang, H. C.; Samaniego, F.; Nair, B. C., et al. HIV-1 tat protein exits from cells via a leaderless secretory pathway and binds to extracellelar matrix-associated heparan sulfate proteoglycan through its basic region. AIDS. 1997, 11, 1421–1431.
55. Daniell, E. Genome structure of incomplete particles of adenovirus. J. Virol. 1976, 19, 685–708.
56. McKeating, J. A.; McKnight, A.; Moore, J. P. Differential loss of envelope glycoprotein gp120 from virions of HIV-1 isolates: effects on infectivity and neutralization. J. Virol. 1991, 65, 852–860.
57. Adamson, D. C.; Wildermann, B.; Sasaki, M., et al. Immunologic NO synthase: elevation in severe AIDS dementia and induction by HIV-1 gp41. Science. 1996, 274, 1917–1920.
58. Nath, A.; Conant, K.; Chen, P., et al. Transient exposure to HIV-1 Tat protein results in cytokine production in macrophages and astrocytes: a hit and run phenomenon. J. Biol. Chem. 1999, 274, 17098–17102.
59. Hayman, M.; Arbuthnott, G.; Harkiss, G., et al. Neurotoxicity of peptide analogues of the transactivating protein tat from Maedi-Visna virus and human immunodeficiency virus. Neuroscience. 1993, 53, 1–6.
60. Mollace, V.; Salvemini, D.; Riley, D. P., et al. The contribution of oxidative stress in apoptosis of human-cultured astroglial cells induced by supernatants of HIV-1-infected macrophages. J. Leukoc. Biol. 2002, 71, 65–72.
61. Johnson, R. T. The virology of demyelinating diseases. Ann. Neurol. 1994, 36(suppl), S54–60.
62. Oldstone, M. B. Viruses and autoimmune diseases. Scand. J. Immunol. 1997, 46, 320–325.
63. Oldstone, M. B. Molecular mimicry and immune-mediated diseases. FASEB J. 1998, 12, 1255–1265.
64. Tompkins, S. M.; Fuller, K. G.; Miller, S. D. Theiler's virus-mediated autoimmunity: local presentation of CNS antigens and epitope spreading. Ann. NY. Acad. Sci. 2002, 958, 26–38.
65. Endo, K.; Tsukamoto, T. Experimental bystander encephalitis induced by immunization with HTLV-I-producing T cells in mice. Acta. Neurol. Scand. 1997, 96, 106–113.
66. Johnson, R. T. The Soriano Award Lecture. Emerging infections of the nervous system. J. Neurol. Sci. 1994, 124, 3–14.
67. Nathanson, N.; McGann, K. A.; Wilesmith, J., et al. The evolution of virus diseases: their emergence, epidemicity, and control. Virus. Res. 1993, 29, 3–20.
68. Johnson, R. T. Emerging viral infections. Arch. Neurol. 1996, 53, 18–22.

3

The Role of Brain Biopsy in the Diagnosis of CNS Viral Infections

Bruce A. Cohen and Robert M. Levy
Northwestern University Medical School
Chicago, Illinois, U.S.A.

1 INTRODUCTION

The need for brain biopsy to define the pathology and specific etiology of central nervous system (CNS) infections has been substantially reduced in recent years as a result of the development of newer, less invasive diagnostic techniques. Current magnetic resonance imaging (MRI) techniques are highly sensitive for demonstrating patterns of anatomic localization of CNS pathology. The application of polymerase chain reaction amplification techniques (PCR) to the analysis of cerebrospinal fluid (CSF) has had an extraordinary impact on the rapid diagnosis of those entities for which such testing is available. Further refinements in these and other noninvasive or minimally invasive diagnostic measures will continue to reduce the need for tissue sampling to establish future diagnoses. Despite these advances, however, there remain clinical circumstances in which diagnostic uncertainty persists and where tissue sampling may be appropriate. Advancements in neurosurgical practice now permit diagnostic brain biopsy to be performed effectively, with less patient discomfort and with reduced morbidity, in centers where neurosurgical expertise in stereotactic biopsy procedures and sophisticated neuropathological resources are available.

The decision to pursue diagnostic brain biopsy is based on two essential criteria. The first is that a specific diagnosis cannot be established by an alternative procedure, and the second is that the information obtained will be of sufficient value to justify the risks of the surgery undertaken. The latter consideration entails not only assessments of the neurological and medical condition of the patient, but also the extent of the differential diagnosis, the likelihood of finding a treatable condition, and/or the impact of a specific diagnosis and its prognostic implications on subsequent medical care and personal affairs.

Improvements in pathological diagnostics and image-directed targeting now allow for a definitive answer in the majority of cases and may lead to successful therapy of an amenable etiology. Other chapters in this volume address in detail the clinical features, imaging, pathology, and treatment of the viral entities selected for mention below. In this chapter, we present several examples to illustrate the potential and changing role of diagnostic biopsy and the decision-making considerations encountered in its application.

2 HERPES SIMPLEX ENCEPHALITIS

Herpes simplex virus type 1 (HSV-1) is the most common cause of sporadic viral encephalitis. The virus causes a necrotizing encephalitis that characteristically involves the temporal and inferior frontal lobes and responds favorably to prompt treatment with acyclovir. The recent history of this disease illustrates both issues in deciding whether to pursue diagnostic brain biopsy as well as the more recent impact of molecular diagnostic techniques. Subjects in a large prospective series of patients with focal encephalitis presumed to have HSV-1 encephalitis, who underwent diagnostic brain biopsy in the era before the availability of current molecular diagnostics, were reviewed to assess the accuracy of clinically based diagnoses and to assess the utility of brain biopsy. In this large series of over 400 biopsies, 45% had the diagnosis of HSV encephalitis confirmed, 22% had an alternative diagnosis established, and 33% had biopsies that failed to reveal a specific pathological diagnosis. The presentation of patients with an alternative or nonspecific diagnosis could not be distinguished clinically from those in whom herpetic encephalitis was confirmed [1]. Although the surgical morbidity was only 1.4% with no mortality, subsequent authors disagreed on the need for biopsy to establish a specific diagnosis on a routine basis. Acyclovir was available and offered an effective therapy with relatively modest risk of complications. Thus authorities differed on whether to use a 10–14 day intravenous therapy on an empirical basis or to pursue a more specific diagnosis with rationales based on the likelihood of finding an alternative treatable entity, risks of surgery, and presence or absence of clinical features suggestive of HSV-1 encephalitis [2,3].

The era of molecular diagnostics rendered many of these issues obsolete when PCR amplification techniques to identify HSV-1 DNA became available, offering high sensitivity (96–98%) and specificity (99%) when applied early in the disease course, thus establishing a new diagnostic standard [4,5]. As a result of the common availability of these effective and minimally invasive diagnostic techniques, consideration of brain biopsy in patients initially suspected to have HSV encephalitis is now limited to those individuals with a localized lesion; normal, atypical, or nondiagnostic CSF, and failure to respond to empirical therapy. This paradigm is increasingly being seen in other forms of meningoencephalitis as the spectrum of molecular diagnostic assays expands.

3 NEUROLOGICAL DISEASE ASSOCIATED WITH HUMAN IMMUNODEFICIENCY VIRUS INFECTION

The emergence of the human immunodeficiency virus (HIV) pandemic with its initially high prevalence of neurological disease, in conjunction with improvements in imaging and computerized localization for stereotactic neurosurgical techniques, led to frequent use of brain biopsy to define specific pathology of cerebral lesions. Several considerations encouraged the selective use of brain biopsy in the setting of AIDS. A number of the opportunistic neurological illnesses known to be associated with HIV are themselves treatable. Specific agents may present with subtle and atypical features in the setting of AIDS, increasing the uncertainty of empirical diagnosis. Concurrent neuropathologies in AIDS are not uncommon, and treatment of an entity may lead to only partial response, subsequently raising the question of whether the physician is encountering an atypical response, a correctly identified but resistant organism, or an alternative pathology.

These diagnostic dilemmas favored the use of less invasive brain biopsy techniques. Small series of AIDS patients undergoing diagnostic stereotactic brain biopsy have generally revealed high diagnostic sensitivities of 87–98% with morbidities of 8–26%, and mortalities of 0–5%, usually due to hemorrhage at the biopsy site [6–12]. The potential value of establishing a specific diagnosis is illustrated by one series in which comparison of the prebiopsy presumptive diagnosis to the diagnosis established by biopsy revealed corroboration in only 52% of the cases [10]. We have treated cases in which successful therapy of one opportunistic process is followed by emergence of an alternative and treatable entity that was previously masked In 1998, the American Academy of Neurology quality standards subcommittee issued an advisory on the evaluation and management of intracranial mass lesions in AIDS. This document recommended that open brain biopsy and decompression be immediately pursued in those patients presenting with mass lesions threatening herniation. Stereotactic biopsy was recommended in individuals with single mass lesions and negative serology for *Toxoplasma gondii*, while empirical therapy for toxoplasmosis was recommended for patients with multiple mass lesions and serological evidence of exposure to *T. gondii*. Emphasis was placed on the need for close follow-up of individuals treated presumptively for toxoplasmosis. It was recommended that persistence or worsening in spite of therapy be considered an indication for stereotactic biopsy to establish a specific diagnosis [13].

The availability of PCR-based assays has enhanced the ability to diagnose opportunistic viral infections from CSF samples; however, the potential for concurrent processes, occasional false negative CSF studies, and the occasional atypical presentation may still require brain biopsy for diagnosis in selected instances. Most solitary mass lesions in AIDS patients are likely to be CNS lymphomas [14]; however, several cases of cytomegalovirus encephalitis presenting as cerebral mass lesions resembling tumors have been reported [15,16]. Biopsy of cerebral mass lesions in AIDS patients may also yield cryptococcus, mycobacterium tuberculosis, seronegative toxoplasmosis, and a variety of other nonviral pathogens.

In additional to cerebral mass lesions, stereotactic brain biopsy may also be of value to identify opportunistic viruses that produce infiltrative lesions predominantly affecting white matter regions of the CNS. A study utilizing brain biopsies and serial imaging demonstrated that varicella zoster virus (VZV) in AIDS patients may initially present as multiple nonenhancing discrete and asymmetrically clustered subcortical lesions, which subsequently coalesce and become enhancing and cavitated [17]. Demyelination and ischemic changes may be seen in the small vessel vasculopathy associated with VZV in AIDS, whereas large and medium vessel pathology presents with infarctions [18]. The availability of PCR assays to detect VZV in CSF will limit the need for tissue samples to make these diagnoses in the future; however, such lesions are still likely to be biopsied when CSF is unrevealing. Leukoencephalopathic patterns in AIDS as well as vasculitis and infarction are nonspecific. Infarctions may also be seen with CMV [19] and *T. gondii* [20]. Similarly, a number of entities may cause abnormalities of white matter regions, including HIV itself, VZV, and JC virus, which causes progressive multifocal leukoencephalopathy (PML).

Progressive multifocal leukoencephalopathy, an infection of oligodendroglia by JC virus, is seen in up to 4% of AIDS patients. PML presents most often as asymmetrical patchy lesions in cerebral or cerebellar white matter. It is discussed in detail elsewhere in this volume. Definitive diagnosis formerly required brain biopsy; however, the availability of PCR analyses with reported sensitivities of 74–92% and specificities of 92–100% [21–23] has allowed for specific diagnosis by less invasive means more recently. When CSF fails to yield a diagnosis, the decision on whether to pursue brain biopsy depends on the second criterion regarding the potential value of the information. As this is written, there is no treatment with proven efficacy for PML other than aggressive anti-retroviral therapy. Therefore, diagnostic biopsy is probably indicated only when an alternative diagnosis is suspected to be likely or to establish the diagnosis of PML in order to permit participation in a clinical treatment trial after CSF sampling with PCR testing has failed to yield a specific diagnosis in a patient with clinical and imaging features suggestive of the disease.

4 CREUTZFELDT-JACOB DISEASE

Creutzfeldt-Jacob disease (CJD) is now attributed to a transmissible pathogen termed a prion, a protease-resistant protein, that in the disease state alters its conformation and accumulates in neurons, causing their death. This results in one of several progressive disease patterns including CJD, which is a dementing illness. The prion diseases are considered in detail in another chapter in this volume. Unlike the viral entities mentioned above, the diagnosis of CJD may be suspected on clinical grounds and may be accompanied by suggestive abnormalities on MRI studies and periodic discharges on electroencephalograms in some cases. There is currently no specific CSF test for CJD. Elevated levels of 14–3–3 protein may be found, but the disease also occurs with normal CSF. When present, elevated levels of 14–3–3 protein are nonspecific and also occur in viral encephalitis. As a result, the definitive diagnosis of CJD may require brain biopsy.

Two considerations pertain to the decision for biopsy diagnosis in a patient suspected to have CJD. First, there is no current therapy for the disease, so the potential benefit of establishing the diagnosis lies in excluding any alternative explanation still considered feasible and the consequent prognostic implications with the opportunity to address end-of-life personal issues. A second consideration pertains to the potential risk to those coming into contact with the biopsied tissue or the instruments used to handle it should they not be properly sterilized. Iatrogenic CJD has been transmitted from contaminated neurosurgical instruments and depth electrodes and from cadaveric dural graft tissue. Guidelines for sterilization require soaking in 1 N sodium hydroxide or undiluted sodium hypochlorite for 1 h and then autoclaving at 134°C for an hour. Tissue samples are soaked in concentrated formic acid for 1 h and then 4% formaldehyde solution for at least 48 h. Archived formalin-fixed tissue embedded in paraffin and stored at room temperature may retain its infectivity for years and must be handled with appropriate caution [24]. Thus biopsy of a patient with suspected CJD should always be undertaken with appropriate preparation and precautions to protect both medical and laboratory personnel and future patients.

5 STEREOTACTIC NEUROSURGICAL PROCEDURE

The evolution of image-guided stereotactic neurosurgery provides an effective means of obtaining a specific diagnosis of a cerebral process, with modest discomfort and low risk of morbidity in experienced hands. The most common neurovirological disease in which such diagnostic procedures are considered today is HIV infection, although the principles of the procedure are essentially the same in other applications. Although frameless stereotactic systems are currently available, stereotactic biopsy procedures in patients with HIV infection are usually performed with frame based systems. These devices, in which an external landmark system is rigidly attached to the patient's head prior to imaging, provide a number of benefits over frameless procedures, including significant improvements in accuracy and precision and rigid fixation of stereotactic instrumentation. The patient is taken to the operating room and, under local anesthesia, has a base ring attached to the head with pins. Two commonly used systems are the Brown-Roberts-Wells stereotactic system (Radionics, Inc., Boston, MA) and the Leksell stereotactic system (AB Elekta Instruments, Decatur, GA). Computerized tomographic (CT) scans can be used for targeting lesions that are adequately defined with these images. Magnetic resonance imaging (MRI) is used for lesions that are not as well defined on CT, such as deep white matter and temporal lobe lesions. Following application of the external fiducial system, the patient undergoes imaging with either double-dose iodine contrast for CT or 0.1 mmol/kg gadopentate dimeglumine (Gd-DPTA, Berlex Laboratories, Cedar Knolls, NJ) for MRI, using 3–5 mm slice thickness and, for MRI, multislice acquisitions with a 30 mm head coil. Both the center and periphery of identified lesions are targeted to ensure that adequate and sufficient tissue is obtained for a specific diagnosis. The center of a lesion may contain predominantly necrotic debris, whereas a section too peripheral to the active process may contain only nonspecific reactive inflammation.

Once obtained, two dimensional CT or MRI coordinates are transformed to three-dimensional stereotactic coordinates using the appropriate computer software. The imaging localizer is then removed and replaced with a stereotactic arc system. Under local anesthesia, a 2 mm skin incision and a twist drill hole are made, the dura is lacerated with a sharp needle, and a stereotactic biopsy needle such as the Nashold side-biting biopsy instrument (Radionics, Inc.) is used to obtain multiple tissue samples.

To maximize the yield of stereotactic biopsy, intraoperative consultation with the neuropathologist and cytopatholgist is pursued to verify that biopsy sections contain sufficient and representative pathological material prior to concluding the operation. In many instances, frozen sections from a single core stained with hematoxyline/eosin (H&E) provide enough information for a preliminary diagnosis. If not, additional cores may need to be taken and frozen to ascertain that identifiable pathological material is present in the biopsied tissue. If PML or lymphoma is suspected, additional frozen sections are cut for specific immunopathological staining. For permanent sections, cores fixed in formalin are used for identifying infectious agents. Specific immunohistopathological stains and molecular diagnostic assays are used as described in other chapters in this volume that discuss the relevant pathogens. When electron microscopy is anticipated, additional cores are fixed in glutaraldehyde. When lymphoma is suspected, cores are fixed in B5.

Ten touch imprint slides are prepared by placing a biopsy fragment on a sterile-gloved fingertip and gently pressing the tissue against alcohol-sterilized glass microscope slides. The remaining tissue is transferred to a test tube containing 2 mL of sterile saline, homogenized, and used to inoculate cultures. All tissue manipulation and culture setup are carried out in a biological safety cabinet. In the setting of HIV infection, smears are Gram stained after air drying, and Giemsa staining for *T. gondii*, periodic acid–Schiff staining for fungi, auramine-rhodamine staining for acid-fast bacteria, and Gomorri silver staining for fungi are performed in addition to the viral diagnostic procedures. The saline homogenate is used to inoculate cultures on chocolate agar and carbon dioxide incubation for aerobic bacteria, laked blood agar and anaerobic incubation for anaerobic bacteria, inhibitory mold agar for fungi, 7H11 agar and ATS medium for mycobacteria, and MRC-5 fibroblast, primary rhesus monkey kidney, A-549, and rabbit kidney tissue cell monolayers for viruses.

Following histopathological verification of abnormal tissue by intraoperative frozen section histopathology or cytological smears, the headframe is removed and a single suture is placed to close the scalp wound. Patients are generally observed overnight, and a CT scan is obtained in the morning. Depending on their medical condition, many can be discharged the day after surgery and return as an outpatient when microbiological and pathological results are available. Morbidity of stereotactic biopsy in HIV-infected patients in our institution has been reported as 2% major morbidity, usually cerebral hemorrhage in patients with CNS lymphomas, and 4% minor morbidity [6].

6 CONCLUSION

The role for diagnostic brain biopsy has been reduced by the availability of new molecular diagnostic techniques applied to cerebrospinal fluid. However, when a specific diagnosis cannot be obtained by less invasive means, image-guided stereotactic biopsy can provide a high diagnostic yield with acceptably low morbidity in the hands of an experienced operator. In pursuing invasive diagnostic measures, the potential for finding a treatable pathology and the value of a specific diagnosis in limiting further diagnostic procedures and establishing a prognosis for the patient must be weighed against the small but not insignificant risks of the procedure. Diagnostic yield of stereotactic brain biopsy is enhanced by multiple target sites within a lesion and intraoperative neuropathological consultation to optimize the sensitivity of acquired tissue prior to conclusion of the operative procedure.

REFERENCES

1. Whitley, R.J.; Cobbs, G.; Alford, C.A., Jr.; Soong, S.-J.; Hirsch, M.S.; Connor, J.D.; Corey, L.; Hanlley, D.F.; Levin, M.; Powell, D.A. and the NIAID Collaborative Antiviral Study Group. Diseases that mimic herpes simplex encephalitis. JAMA. 1989, 262, 234–239.
2. Anderson, N.E.; Willoughby, E.W.; Synek, B.J.L.; Croxson, M.C.; Glasgow, G.L. Brain biopsy in the management of focal encephalitis. J. Neurol Neurosurg Psychiatry. 1991, 54, 1001–1003.
3. Soong, S.-J.; Watson, N.E.; Caddell, G.R.; Alford, C.A., Jr.; Whitley, R.J. and the NIAID Collaborative Antiviral Study Group. Use of brain biopsy for diagnostic evaluation of patients with suspected herpes simplex encephalitis: a statistical model and its clinical implications. J. Infect. Dis. 1991, 163, 17–22.
4. Lakeman, F.D.; Whitley, R.J. and the National Institute of Allergy and Infectious Diseases Collaborative Antiviral Study Group. Diagnosis of herpes simplex encephalitis: application of polymerase chain reaction to cerebrospinal fluid from brain-biopsied patients and correlation with disease. J. Infect. Dis. 1995, 171, 857–863.
5. Tebas, P.; Nease, R.F.; Stjorch, G.A. Use of the polymerase chain reaction in the diagnosis of herpes simplex encephalitis: a decision analysis model. Am. J. Med. 1998, 105, 287–295.
6. Levy, R.M.; Russell, E.; Yungbluth, M.; Hidvegi, D.F.; Brocy, B.A.; Dal Canto, M.C. The efficacy of image-guided stereotactic brain biopsy in neurologically symptomatic acquired immunodeficiency syndrome patients. Neurosurgery. 1992, 30, 186–190.
7. Feiden, W.; Bise, K.; Steude, U.; Pfister, H.-W.; Moller, A.A. The stereotactic biopsy diagnosis of focal intracerebral lesions in AIDS patients. Acta. Neurol. Scand. 1993, 87, 228–233.

8. Nielsen, C.J.; Gjerris, F.; Pedersen, H.; Jensen, F.K.; Wagn, P. Brain biopsy in AIDS: diagnostic value and consequence. Acta. Neurochir. 1994, 127, 99–102.
9. Antinori, A.; Ammassari, A.; De Luca, A.; Cingolani, A.; Murri, R.; Scoppettuolo, G.; Fortini, M.; Tartaglione, T.; Larocca, L.M.; Zannoni, G.; Cattani, P.; Grillo, R.; Roselli, R.; Iacoangeli, M.; Serrati, M.; Ortona, L. Diagnosis of AIDS-related focal brain lesions: a decision-making analysis based on clinical and neuroradiologic characteristics combined with polymerase chain reaction assays in CSF. Neurology. 1997, 48, 687–694.
10. Hornef, M.W.; Iten, A.; Maeder, P.; Villemure, J.-G.; Regli, L. Brain biopsy in patients with acquired immunodeficiency syndrome. Arch. Intern. Med. 1999, 159, 2590–2596.
11. Antinori, A.; Ammassari, A.; Luzzati, R.; Castagna, A.; Maserati, R.; Rizzardini, G.; Ridolfo, A.; Fasan, M.; Vaccher, E.; Landonio, G.; Scerrati, M.; Rocca, A.; Butti, G.; Nicolato, A.; Lazzarin, A.; Tirelli, U. for the Gruppo Italiano Cooperativo AIDS & Tumori. Neurology. 2000, 54, 993–997.
12. Gildenberg, P.L.; Gathe, J.C., Jr.; Kim, J.H. Stereotactic biopsy of cerebral lesions in AIDS. Clin. Infect. Dis. 2000, 30, 491–499.
13. Anonymous, Evaluation and management of intracranial mass lesions in AIDS. Report of the Quality Standards Subcommittee of the American Academy of Neurology. Neurology. 1998, 50, 21–26.
14. Ciricillo, S.F.; Rosenblum, M.L. Use of CT and MR imaging to distinguish intracranial lesions and to define the need for biopsy in AIDS patients. J. Neurosurg. 1980, 73, 720–724.
15. Dyer, J.R.; French, M.A.H.; Mallal, S.A. Cerebral mass lesions due to cytomegalovirus in patients with AIDS: report of two cases. J. Infect. 1995, 30, 147–151.
16. Moulignier, A.; Mikol, J.; Gonzalez-Canali, G.; Polivka, M.; Pialoux, G.; Welker, Y.; Alain, S.; Thiebaut, J.-B.; Dupont, B. AIDS-associated cytomegalovirus infection mimicking central nervous system tumors: a diagnostic challenge. Clin. Infect. Dis. 1996, 22, 626–631.
17. Weaver, S.; Rosenblum, M.K.; deAngelis, L.M. Herpes varicella zoster encephalitis in immunocompromised patients. Neurology. 1999, 52, 193–195.
18. Kleinschmidt-Demasters, B.K.; Amlie-Lefond, C.; Gilden, D.H. The patterns of varicella zoster virus encephalitis. Hum. Pathol. 1996, 27, 927–938.
19. Golden, M.P.; Hammer, S.M.; Wnke, C.A.; Albrecht, M.A. Cytomegalovirus vasculitis. Medicine. 1994, 73, 246–255.
20. Cohen, B.A. Neurologic manifestations of toxoplasmosis in AIDS. Sem. Neurol. 1999, 19, 201–211.
21. Weber, T.; Turner, R.W.; Frye, S.; Ruf, B.; Haas, J.; Schielke, E.; Pohle, H.-D.; Luke, W.; Luer, W.; Felgenhaure, K.; Hunsmann, G. Specific diagnosis of progressive multifocal leukoencephalopathy by polymerase chain reaction. J. Infect. Dis. 1994, 169, 1138–1141.
22. Fong, I.W.; Britton, C.B.; Luinstra, K.E.; Toma, E.; Mahony, J.B. Diagnostic value of detecting JC virus DNA in cerebrospinal fluid of patients with progressive multifocal leukoencephalopathy. J. Clin. Microbiol. 1995, 33, 484–486.
23. McGuire, D.; Barhite, S.; Hollander, H.; Miles, M. JC virus DNA in cerebrospinal fluid of human immunodeficiency virus-infected patients: predictive value for progressive multifocal leukoencephalopathy. Ann. Neurol. 1995, 37, 395–399.
24. Johnson, R.T.; Gibbs, C.J., Jr. Creutzfeldt-Jacob disease and related transmissible spongiform encephalopathies. N. Engl. J. Med. 1998 339, 1994–2004.

4

CSF Analysis in the Diagnosis of Viral Encephalitis and Meningitis

Paola Cinque
San Raffele Hospital
Milan, Italy

Annika Linde
Swedish Institute for Infectious Disease Control
Solna, and
Karolinska Institute
Stockholm, Sweden

1 BACKGROUND

Cerebrospinal fluid (CSF) examination is almost invariably included in the diagnostic workup of patients with suspected central nervous system (CNS) viral infections. Besides providing general information on the nature of diseases, CSF analysis and brain biopsy are the only means to identify a responsible virus and thus lead to an etiological diagnosis.

Until little more than 10 years ago, diagnosis of viral infections of the CNS was often based on the CSF profile and on exclusion of other causes, because current diagnostic techniques were not very sensitive and were time-consuming [1]. Only virus isolation in cell culture was regarded of value for diagnosis of aseptic meningitis, with enteroviruses found by this technique in almost half of the cases. However, virus isolation was insensitive for other viruses, such as herpes simplex virus type 1 (HSV-1) and most arboviruses. Serology had low sensitivity in early disease stages and was often impractical to perform. Virus antigen detection techniques were still in the offing, and molecular methods had not yet been developed.

Over the last decade, nucleic acid (NA) amplification–based techniques, primarily the polymerase chain reaction (PCR), have contributed outstandingly to the diagnosis of many infectious diseases [2]. The molecular analysis of CSF for the identification of microbial genomes found its first successful applications in CNS infections during the early 1990s, when PCR was used for the diagnosis of herpes simplex encephalitis (HSE) and enteroviral and tuberculous meningitis in immunocompetent patients [3–5]. Since these first investigations, NA amplification techniques have been extensively applied to the study of CSF. They have been shown to be reliable for diagnosis in a number of infectious CNS diseases and have become the test of choice in some viral CNS infections, such as HSE [6–9].

In this chapter the diagnostic techniques for CSF analysis are described and the most relevant clinical applications discussed, leaving more details to the virus-specific chapters. Because of their importance, special attention is devoted to molecular techniques.

2 THE CEREBROSPINAL FLUID (CSF)

2.1 The CSF and Its Anatomical Relationships

The brain is structured into compartments: the intracellular space, the extracellular space, blood, and CSF. Barriers between these compartments are the blood-brain barrier, at the brain capillary site, and the blood-CSF barrier, at the choroid plexus. In addition, a third, less tight anatomical barrier is present at the lining of the ventricular and brain surfaces between the CSF and the brain extracellular fluid, where the ependimal or pia mater cells form only a loose interface (Fig. 1) [10,11]. The brain and spinal cord are surrounded by the meninges, consisting, from the outer most to the inner most, of the dura mater, which is tightly adherent to the skull bone; the arachnoid, which covers the brain, spinal cord, and nerves; and the pia mater, which is adherent to the brain surface. CSF is mainly produced by the choroid plexuses, which are projections of vessels and pia mater into the cerebral ventricles. Once formed, the major part of the CSF moves by bulk flow into the subarachnoidal space and around the brain surface, finally exiting into the venous system in the superior sagittal sinus. There it is readsorbed by the arachnoid villi, extensions of the arachnoid membrane. The remaining part flows through the ventricles (the two lateral, third and fourth ventricles). The total volume of CSF is approximately 100–160 mL in adults, and it is replaced four or five-times daily [10]. Its physiological composition is summarized in Table 1 [12].

2.2 Lumbar Puncture

Access to the CSF is generally achieved by lumbar puncture. This procedure, first described toward the end of the 1800s [13], has been of indisputable value in the diagnosis of infectious CNS diseases. The lumbar puncture may provide a great deal of information, from general, e.g., presence of inflammation or the status of the brain barriers, to identification of etiological agents [14,15].

Lumbar puncture is performed with the patient sitting upright with his back toward the operator or lying on his side. Usually, a 20 or 22 gauge or smaller needle is inserted perpendicularly to the patient's back between the third and fourth lumbar vertebrae, corresponding to the cauda equina roots. The needle is passed through the dura mater until the subarachnoid space is reached. Up to 10 mL of CSF is usually obtained; however, a volume of 20 mL or more can safely be drawn from an adult. To guarantee CSF sterility, test tubes are usually filled directly at the bedside.

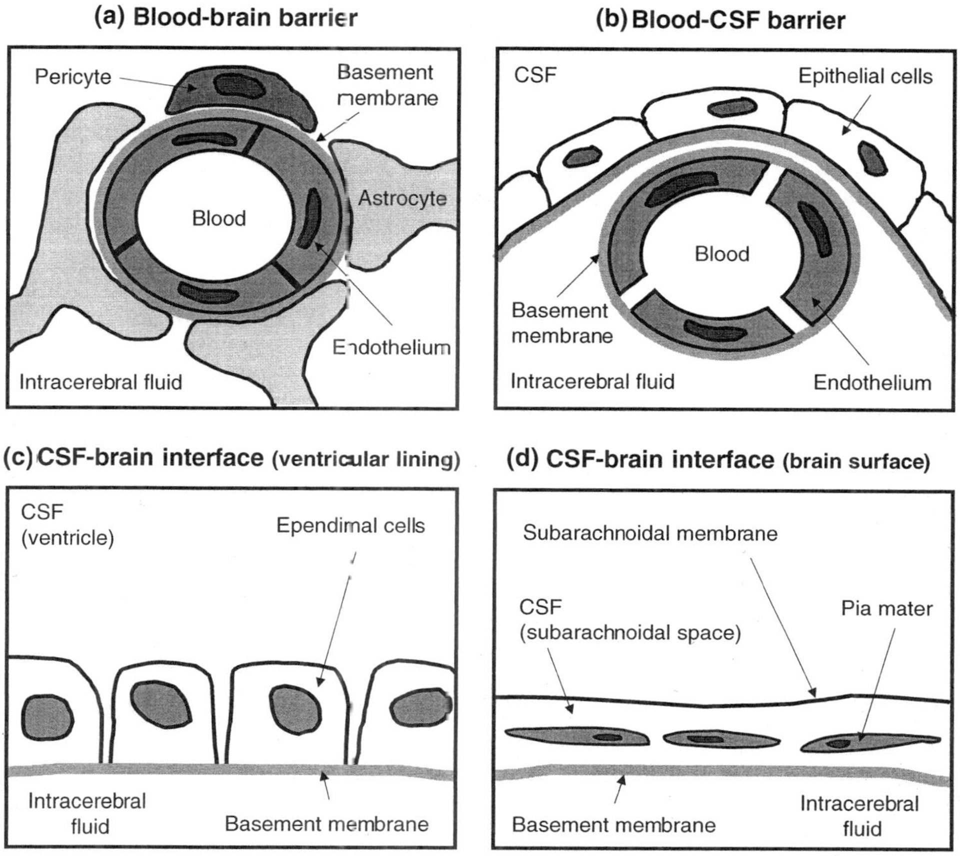

Figure 1 Schematic representation of the brain barriers. (a) Blood-brain barrier (at the brain capillary site) formed by endothelial cells joined by tight junctions, basement membrane, pericytes, and astrocyte processes. (b) Blood-CSF barrier (at the choroid plexa), formed by choroid endothelial cells separated by gap junctions, basement membranes, and choroid epithelial cells, joined by tight junctions. (c) CSF-brain interface (at the ventricular lining), formed by ependymal cells and basement membrane. (d) CSF-brain interface (at the brain surface), formed by pia mater cells separated by gap junctions and basement membrane.

The lumbar puncture carries a low degree of risk, with possible complications ranging from mild, e.g., mild headache to fatal, e.g., brain herniation [14]. Headache is the most common complication, reported in up to 36% of cases [16]. This effect is related to the hole left in the dura after withdrawal of the needle, which allows CSF to leak out of the subarachnoid space. Bed rest has not been proven to be effective [17–19], but certain maneuvers that decrease the size of the hole can reduce the frequency and severity of headache [20–22]. These include orienting the needle bevel parallel to the spinal cord axis [23], using atraumatic ''blunt'' needles that separate rather than cut dural fibers [24–26], or reinserting the stylet after the procedure [27]. The use of finer, e.g., 22 gauge, needles, however, seems to be the most effective procedure [20,28,29]. Up to 26 gauge needles, such as those commonly used in anaesthetic and radiological practice, have also

Table 1 Normal Parameters in Adult CSF and Blood

	CSF	Plasma or serum
Total volume	100–160 mL	
Pressure	7–20 cmH_2O	
pH	7.35–7.40	7.35–7.45
Sodium	136–150 mmol/L	136–145 mmol/L
Potassium[b]	2.5–3.2 mmol/L	3.5–5.1 mmol/L
Calcium, total	1.05–1.35 mmol/L	2.10–2.55 mmol/L
Chloride	118–132 mmol/L	98–107 mmol/L
Glucose	40–70 mg/dL	70–105 mg/dL
Lactate	<25.2 mg/dL	5–12 mg/dL
Lactate dehydrogenase	~10% of serum value	100–190 U/L
Total protein	15–40 mg/dL	6.0–7.8 g/dL
Prealbumin	2–7%	Traces
Albumin	10–30 mg/dL	3.9–5.1 g/dL
	56–76%	
$\alpha 1$	2–7%	0.2–0.4 g/dL
$\alpha 2$	4–12%	0.4–0.8 g/dL
β	8–18%	0.5–1.1 g/dL
γ	3–12%	0.6–1.3 g/dL
IgG	0.8–4.2 mg/dL	650–1600 mg/dL
IgA	0.07±0.03 mg/dL[c]	40–350 mg/dL
IgM	0.016±0.003 mg/dL[c]	50–300 mg/dL

[a] In the horizontal position.
[b] Potassium values in CSF are approximately 70% of those in plasma
[c] Average values ± standard deviation
Source: Ref. 12.

been proposed for the diagnostic lumbar puncture, along with the use of gentle syringe aspiration to speed up CSF collection [30,31].

Less frequent complications of the lumbar puncture are paresthesias (reported in <1–13% of patients), bleeding (<1–2%), spinal infections (<1%), and brain herniation [14]. Bleeding is mainly described in patients with coagulation defects, e.g., thrombocytopenia, or on anticoagulant therapy, in whom it may lead to spinal hematomas [14,32,33]. Such serious bleeding needs to be distinguished from the ''traumatic puncture'' that occurs in up to 20% of patients and is due to injury of the local vessels, i.e., those located along the spinal sac or the cauda equina [14]. Brain herniation is the most serious lumbar puncture complication, especially in patients with elevated intracranial pressure. For this reason, a computed tomographic (CT) scan of the brain to exclude the presence of mass lesions is usually performed before doing a diagnostic lumbar puncture. Although difficult to estimate, the exact risk of this complication should not be higher than 1–2%, even in patients with increased intracranial pressure [14,34].

3 NONVIROLOGICAL CSF ANALYSES

Standard laboratory examination of CSF is almost always performed when a viral CNS infection is suspected. This examination always includes the measurement of glucose and protein content and cell counts. Additional parameters such as CSF pressure or the function of the blood-brain barrier are also often assessed on a routine basis. Furthermore, the array of possible tests is variably extended to exclude the presence of other neurological conditions. Although virological analyses are necessary to establish a definitive diagnosis, standard CSF analysis may provide clues supporting or excluding a viral etiology. In cases of acute meningitis, it has been shown that the CSF glucose levels or the CSF/blood glucose ratio, CSF protein level, and leukocyte or polymorphonuclear leukocyte counts can all be used to rule out a viral etiology [35].

3.1 CSF Glucose

Cerebrospinal fluid glucose is measured by the same enzymatic techniques that are used for its determination in blood, which usually consist of rapid automated procedures. In the absence of pathological CNS conditions, glucose concentrations in the CSF and blood are at equilibrium, resulting in a CSF/blood glucose ratio of approximately 0.60. For a physiological glucose range of 70–110 mg/dL, the corresponding CSF values are between 40 and 70 mg/dL (Tables 1 and 2) [10,12,36].

Glucose levels and CSF/plasma glucose ratios are only occasionally decreased in viral CNS infection, whereas low levels are frequent in meningitis caused by bacteria, mycobacteria, and fungi, presumably as a consequence of increased consumption by microorganisms and inflammatory cells and altered transport through the blood-CSF barrier (Table 2). Hypoglycorrhachia is occasionally observed in meningoencephalitis caused by herpes simplex viruses (HSV), varicella-zoster virus (VZV), mumps, or enterovirus [11,37]. Furthermore, glucose levels below 40 mg/dL are frequent in human immunodeficiency virus (HIV)-infected patients with cytomegalovirus (CMV) ventriculoencephalitis [38]. In viral meningitis, however, it has been observed that either a CSF glucose level less than 34 mg/dL or a CSF/blood glucose ratio less than 0.23 can be useful to exclude a viral etiology [35].

3.2 CSF Protein

Classically, total protein determination in CSF is based on turbidimetry or dye-binding techniques. Modifications of the biuret method, commonly used with serum, are often employed [39]. The normal content of total protein in lumbar CSF of an adult is 15–50 mg/dL, with notable variations in children and ventricular CSF (Tables 1 and 2). More than 80% of CSF proteins originate from the plasma; the remainder are produced intrathecally.

Mild increases of total protein content, i.e., up to 150 mg/dL, can be observed in viral meningitis and encephalitis, whereas bacterial or tubercular meningitis or cerebral abscesses are usually associated with more substantial variations (Table 2). A value of more than 220 mg/dL has been associated with 99% sensitivity for diagnosis of bacterial meningitis as opposed to viral meningitis [35]. An increased total protein content is likely to result from passage of blood proteins into the CSF, following disruption of the brain barriers, and/or from an increased intrathecal release of inflammatory and brain structural proteins.

Total CSF proteins can be separated electrophoretically into the albumin, $\alpha 1$, $\alpha 2$, $\beta 1$, $\beta 2$, and γ fractions (Table 1). Functionally, these include structural brain cell proteins, enzymes, immunoglobulins, cytokines, chemokines, and other inflammatory molecules. In some CNS diseases, the demonstration of abnormal CSF levels of some proteins has been proposed for diagnostic use. Examples are increased lactate levels in bacterial menin-

Table 2 CSF Glucose, Protein, and White Cell Counts: Normal Values and Changes in Patients with Acute Meningitis

		Normal	Viral ("aseptic")[a]	Bacterial ("purulent")[a]
Glucose (mg/dL)	Adult	40–70	Normal	<40
	Infant	60–80		
Glucose (CSF/plasma)	Adult	0.50–0.80	Normal	<0.50
	Infant	0.45–2.45		
Total protein (mg/dL)	Adult	15–40[b]	Normal or <150	100–500
	>1 month	30–100		
	<1 month	40–120		
White blood cells (μL^{-1})	Adult	0–5	<100–1000[b]	100–>10,000
	<6 weeks[c]	3.73 ± 3.40		
	>1 year[c]	1.94 ± 2.72		
Polymorphonuclear cells (μL^{-1})	Adult	Absent	0–>50%[d]	>80%
	<6 weeks[c]	1.87 (50%) ± 2.98		
	>1 year[c]	0.51 (14%) ± 1.41		

[a] Changes in children should be considered in relation to normal values.
[b] Normal values in ventricular fluid: 5–10 mg/dL.
[c] Expressed as means ± standard deviations. Intermediate values are observed between 6 weeks and 12 months (from Ref. 36).
[d] Mainly mononuclear cells, though neutrophils may predominate during early infection.
Source: Refs. 10, 12, and 36.

gitis [40] or the demonstration of the 14–3–3 brain protein in patients with Creutzfeldt-Jakob disease [41]. Examples in neurovirology include γ-interferon, found at high CSF concentrations in the early phases of herpes simplex encephalitis (HSE) but not in later HSE stages, postinfectious viral CNS diseases, or other neurological conditions [42]. In HIV infection, a large number of CSF immune activation molecules have been investigated in order to identify diagnostic markers for AIDS dementia complex (ADC), the most severe consequence of HIV infection of the CNS. Increased levels of β2-microglobulin, neopterin, quinolinic acid, monocyte chemotactic factor-1, and other molecules have been found in patients with ADC or with HIV-induced neuropathology. None of these markers, however, has definitely been proven to be sensitive and specific enough for diagnostic use [43–47].

Immunoglobulins

Examination of CSF immunoglobulins (Ig), particularly IgG, is mainly used to detect an increased permeability of the brain barriers or an intrathecal antibody production. IgG, IgM, and IgA are normally excluded from CSF, with the higher blood/CSF ratio of 500:1 for IgG (Table 1). Therefore their presence in the CSF is revealing of a pathological condition. The IgG content can accurately be measured after protein electrophoresis, by nephelometry, electroimmunodiffusion, or radial immunodiffusion. Pandy's test, which qualitatively detects an increased Ig content in CSF, is still employed in some laboratories because of its simplicity. This reaction is based on the observation of turbidity after one drop of CSF is added to 1.0 mL of saturated aqueous phenol solution. Damage of the blood-brain and/or blood-CSF barrier with consequent passive spread of IgG from the blood occurs in a variety of CNS diseases. An increase in intrathecal IgG production is typically observed in CNS diseases associated with immune dysregulation, such as multiple sclerosis. In viral CNS infections the IgG content may be either normal or increased as a consequence of both intrathecal antibody production and damage of the blood-CSF barrier.

A number of indexes are described that can help discriminate between an intrathecal origin of IgG in CSF and its passive spread from the blood. These were initially designed for use in the diagnosis of multiple sclerosis but can also be applied efficiently to viral CNS infections, provided that virus-specific IgG is measured (see Sec. 8). The simplest index is the CSF/serum albumin ratio. Because albumin is neither synthesized nor metabolized intrathecally, its presence in CSF, i.e., an albumin index of >10, necessarily reflects passive transfer through an impaired barrier. More accurate indices correlate both albumin and IgG concentrations in CSF and plasma compartments. These are based on the principle that in the presence of intrathecal IgG production, the ratios between the Ig CSF/serum quotient and albumin CSF/serum quotient are changed. These include, among others, the Link IgG index [48], Tourtellottes Ig G index [49], and the Reiber hyperbolic discrimination function [50]. Strong evidence supporting intrathecal IgG production is also given by the demonstration of oligoclonal IgG bands in CSF but not in serum by isoelectric focusing (IEF) [51].

3.3 Total and Differential White Blood Cell Count

Both normal and differential CSF cell counts are performed using counting chambers. For an accurate morphological examination, at least 30–40 cells/μL are required. Therefore, depending on the cell number, the CSF needs to be concentrated from relatively large volumes by means of centrifugation, cytocentrifugation, sedimentation, or filtration. It is important that CSF be observed within 30 min of sampling, because of lysis and the tendency of cells to adhere to the tube surface, the latter being only partially reversible with tube agitation. In normal adults, CSF is usually acellular, though it may contain up to four or five white blood cells per microliter, usually lymphocytes [11] (Table 2).

Viral encephalitis and meningitis are characterized by lymphocyte pleocytosis, whereas a selective increase in polymorphonuclear leukocytes (PMNLs) is an indicator of purulent bacterial meningitis (Table 2). A preponderance of neutrophils, however, is not rare in the initial phases of viral meningitis or encephalitis, i.e., during the first 6–48 h [52,53], and the demonstration of a decrease in the percentage of neutrophils on an early repeated lumbar puncture is diagnostically helpful [54]. In patients with acute meningitis, a CSF leukocyte count of more than 2000 cells/μL or more than 1180 PMNL/μL has been shown to be a strong individual predictor of bacterial infection [35]. On the other hand, lymphocytic pleocytosis is also observed in meningitis caused by nonviral pathogens such as *M. tuberculosis, B. burgdorferi, T. pallidum, or C. neoformans* and in neoplastic or drug-induced meningitis.

In enteroviral meningitis, the cell count is usually 100–1000 cells/μL, although up to several thousand cells per microliter can occasionally be observed. A high number of leukocytes in CSF, i.e., greater than 100 cells/μL, is associated with a higher rate of virus isolation [55]. PCR results, in contrast, seem to be less correlated with CSF cell counts [56,57]. In meningitis caused by HSV-2 or mumps virus, the CSF cell count is usually less than 500 cells/μL [58,59]. In HSE, CSF cell counts are variable (0–1000 cells/μL), with a predominant lymphoid reaction that may persist over months or even years following acute infection [60]. However, polymorphonuclear cells may in some cases dominate initially [52]. In HIV infection, mild pleocytosis (5–50 cells/μL) is common through the entire course of infection. Higher cell counts, usually below 200 cells/μL, can accompany acute retroviral infection or ADC [61,62].

4 DIRECT CSF EXAMINATION FOR VIRUSES

4.1 Light Microscopy

Direct examination of CSF by light microscopy is usually nonproductive. The basic methods employed are those used in the differential cell count, and the chances of obtaining diagnostically useful information are limited to the possibility of observing viral inclusions in CSF cells. Following immunocytochemistry or in situ hybridization of CSF cells, a positive reaction for viral antigens or nucleic acids can also be visualized (see Sec. 6). As examples, typical CMV inclusions, CMV antigens, or nucleic acids have all been demonstrated in CSF cells from HIV-infected patients with CMV polyradiculomyelitis and PMNL pleocytosis [63–66].

4.2 Electron Microscopy

Electron microscopy (EM) of CSF is rarely employed for diagnosis of CNS infections. Like light microscopy, its use in neurovirology is almost exclusively restricted to examination of brain tissues. Despite its advantages of being a rapid technique and allowing visualization of multiple possible viruses, its usefulness is limited by the low concentration of viral particles in CSF. In addition to low sensitivity, other problems include the cost and consequent scarcity of EM instrumentation and the need for skilled operators able to recognize and identify viruses. The negative staining technique can be used in CSF, and CSF ultracentrifugation or immune scanning EM can be employed to enhance sensitivity. In immune EM, virus-specific monoclonal antibodies conjugated to microspheres are added to the CSF, and the microspheres are collected on a filter surface and inspected by scanning EM.

Successful examples of EM application to the CSF have been the direct visualization of herpesvirus particles (HSV, VZV) by negative CSF staining [67,68] and of measles virus and HIV by immune scanning microscopy [69,70]. During the 1998 and 1999 outbreak of Nipah virus encephalitis in Malaysia and Singapore [71,72], conventional EM revealed enveloped virus-like structures with characteristics similar to those of paramyxoviruses in the CSF of a patient with this disease [73]. This finding indicates the potential contributions of CSF EM for characterizing new or emerging pathogens for which standardized tests are still lacking [74].

5 VIRUS ISOLATION IN CSF

Virus isolation on cell culture has long been the most widely used approach for identification of viruses in CSF. By this method, 20–40% of aseptic meningitis cases can be assigned a viral etiology. In most of these cases, the virus isolated is an enterovirus, which led to the practical assumption that either enterovirus will grow from CSF culture or it is unlikely that a virus will be isolated [75]. With the exception of enteroviruses and a few other viruses such as the mumps virus, most neurotropic viruses do not grow easily in tissue cultures. Furthermore, days or weeks can be required for their demonstration. Additional potential drawbacks of the routine use of virus isolation include the need to maintain tissue cell systems and the potential hazard of culturing some viruses such as arboviruses [76]. Over recent decades, some these problems have been overcome by developments of tissue culture systems and the use of culture combinations. In addition, rapid staining procedures have reduced the time necessary to obtain a positive result [77,78]. Since more sensitive and rapid molecular techniques became available, cell cultures have lost a large part of their diagnostic importance. Nevertheless, virus isolation maintains the advantage of allowing for further biological analyses, such as assessment of susceptibility to antiviral agents or virus serotyping.

5.1 Methods

In general, three major types of cell systems are used to grow viruses: primary cultures, diploid cell lines, and continuous cell lines. Primary cells are obtained from animal organs and, after mincing and treatment with trypsin, are allowed to attach to plastic or glass to form a monolayer. These can be passaged, that is, used to reproduce a new cell generation, for a few times before they die. An example of primary cells is monkey kidney cells, which are largely used to grow enteroviruses. Diploid or semicontinuous cell lines are also derived from animal organs but can be passaged for up to 50–100 generations. Examples include human lung fibroblasts, the preferred cell type for herpesvirus cultivation. Continuous cell lines are derived from either normal or tumor tissues that are immortalized and can be passaged indefinitely. Examples are HeLa and HEp-2 from human tumor cells and Vero cells from monkey kidney [77,78]. Because no single cell type enables growth of all viruses suspected of being responsible for CNS infections, it is common practice to inoculate the CSF into an array of different tissue cultures. The choice of the cell types employed is based on a number of considerations, including laboratory experience with a given cell system, cell availability, costs, and, ideally, the presence of clues suggesting an etiology. In addition to cell cultures, viruses can also be isolated in animals or embryonated eggs. Although animal inoculation is the only way to isolate viruses like coxsackie A enteroviruses or some arboviruses, this practice is extremely cumbersome and is performed only in research and reference laboratories.

At least 1 mL of CSF is usually required for virus isolation. This is important, because the number of infectious particles in the CSF is crucial for virus growth. To minimize loss of infectivity, CSF samples should be transported as soon as possible to the diagnostic laboratory and inoculated into cell systems. After incubation, cell cultures are examined at regular intervals to detect signs of viral growth, the primary one being a cytopathic effect (CPE) consisting of morphological cell changes such as cell lysis, vacuolization, and the formation of syncitia. A number of viruses can be recognized by their characteristic CPE. For instance, herpesviruses produce foci of enlarged cells, whereas measles virus typically induces formation of multinucleated giant cells. After appearance of CPE, final virus identification requires additional tests such as immunofluorescence (IF) or immunoperoxidase (IP) stainings using virus-specific antibodies. Virus identification is also performed when different serotypes may be implicated, as in the case of enteroviruses. Some viruses may grow in cell cultures without producing a CPE, thus requiring further investigations for their identification. Examples are rubella virus, which is detected by growth inhibition of another challenge virus, generally echovirus type 11, and influenza virus, which can be detected by hemadsorption, i.e., adherence of guinea pig erythrocytes to the surface of infected cells.

There is wide variation in the time required to yield a CPE, depending on the type and amount of virus and what cell type is used. HSV CPE is detected rapidly, often 1–2 days after inoculation, whereas up to 4–8 weeks may be required for cytomegalovirus [78]. New procedures may enhance the speed and sensitivity of virus isolation in tissue cultures. One of the most widely used is the shell vial assay, most frequently employed for CMV. By this method, samples are centrifuged in vials containing a shell-shaped coverslip covered by a human fibroblast cell monolayer. In the case of CMV, immediate early antigens can be demonstrated by monoclonal antibody fluorescence staining after 1 or 2 days of incubation [77,78].

5.2 Clinical Applications

Neurotropic viruses that can be detected relatively easily in cell cultures include enteroviruses, HSV-2 from cases of meningitis, HSV-1 and HSV-2 in neonatal CNS infections, VZV in patients with herpes zoster–related CNS complications, mumps virus, and some arboviruses (Table 3) [1,59,75,79–120].

Before molecular techniques became available, enteroviruses represented up to the 80% of the cases of aseptic meningitis for which an etiology could be determined [75]. However, only 65–80% of confirmed enteroviral meningitis cases can be diagnosed by virus isolation, partly resulting from the inability of many coxsackie virus A serotypes to grow. The chance of yielding an enterovirus also depends on its titer in CSF; it has been estimated that $10–10^3$ tissue culture infectious doses per milliliter are necessary to yield a positive isolation, and the rate of positive isolation decreases rapidly with time after the onset of symptoms [93]. Enteroviruses usually require 3–7 days to show a cytopathic effect, but up to 14 days, or even more, may be needed in the case of ''difficult'' isolates or samples containing mixtures of viruses [121]. Following isolation, one of the 66 enterovirus subtypes can be identified by the use of type-specific hyperimmune antisera and neutralization of infectivity. Enterovirus subtyping is important for epidemiological purposes [94], but it has limited clinical relevance, with the possible exception of suspected cases of poliomyelitis, in which it may be important to distinguish between poliovirus from other enteroviruses or from vaccine virus [122]. It was observed that both length of hospitalization and unnecessary use of antibiotics were decreased as a result of the virus isolation approach in the diagnosis of aseptic meningitis [75,123]. However, it was difficult to establish whether this effect was cost-effective overall, because the high rate of isolation was associated with a high number of CSF specimens unnecessarily sent to the diagnostic laboratory for viral culture [75].

Growth in cell culture is problematic for some viruses that cause CNS infections, and others cannot be isolated at all (Table 3). Examples of the latter are human herpesvirus-6 (HHV-6), parvovirus B19, and JC virus, which has been isolated only from brain tissues. The inability to isolate a virus can be due to a number of factors, including the cell-associated nature of the virus or a limited replication in the CNS, which might occur in immune-mediated diseases. The rapid development of neutralizing antibodies and the presence in the CSF of molecules that inactivate infectious virus have also been hypothesized.

Table 3 Virus Isolation from the CSF for Diagnosis of Viral CNS Infections

Virus	Family[a]	Examples of the most commonly used tissue cultures	Comments	Virus recovery from CSF[b]	Comments	Refs.
Herpes simplex virus type 1 (HSV-1)	*Herpesviridae*	Human diploid fibroblasts, primary human embryonic, primary rabbit kidney, HEp-2, Vero		Rare	≤ 5% sensitivity in HSE, higher in neonates (25–40%) and immunocompromised.	79–81
Herpes simplex virus type 2 (HSV-2)	*Herpesviridae*	Human diploid fibroblasts, primary human embryonic, primary rabbit kidney, HEp-2, Vero		Frequent	In patients with meningitis, neonates, and immunocopromised.	59, 80, 81
Varicella zoster virus (VZV)	*Herpesviridae*	Human diploid fibroblasts, primary human embryonic		Rare	Higher sensitivity in HZ-associated complications and immunocompromised.	82–85
Cytomegalovirus (CMV)	*Herpesviridae*	Human diploid fibroblasts		Variable depending on clinical syndromes	Rare in encephalitis. ≥50% sensitivity in HIV-associated polyradiculopathy.	81, 86–88
Epstein-Barr virus (EBV)	*Herpesviridae*	Human CBL		Occasional		89, 90
Adenovirus	*Adenoviridae*	Human embryonic kidney, HEp-2, HeLa, KB		Occasional		81, 91, 92
Enterovirus (EV)	*Picornaviridae*	Primary monkey kidney, human diploid fibroblasts, RD, Hep-2, HeLa	No single cell system supports the growth of all EV. Coxsackie A viruses require isolation in suckling mice.	Frequent	60–80% sensitivity in EV aseptic meningitis. Type-specific hyperimmune antisera are used to identify the EV by neutralization of infectivity.	1, 75, 93–96
Rubella virus	*Togaviridae*	Primary African green monkey kidney, Vero, RK-13		Rare		97–100
Influenza virus	*Orthomyxoviridae*	Madin-Darby canine kidney, primary monkey kidney	No CPE may be produced; identification possible by hemadsorbance. Can be isolated in embryonated chicken eggs.	Rare		101, 102
Mumps virus	*Paramyxoviridae*	Primary monkey kidney, primary human kidney	No CPE may be produced; identification possible by virus interference or hemadsoption.	Frequent	~40% sensitivity.	103–105

(*continued*)

Table 3 *Continued*

Virus	Family[a]	Examples of the most commonly used tissue cultures	Comments	Virus recovery from CSF[b]	Comments	Refs.
Measles virus	*Paramyxoviridae*	Primary monkey kidney, primary human kidney, Vero		Rare	More frequent in patients with SSPE or immunocompromised patients with subacute encephalitis.	106
Parainfluenza virus	*Paramyxoviridae*	Primary monkey kidney, primary human kidney, Vero		Occasional		107
Nipih virus	*Paramyxoviridae*	Primary monkey kidney, primary human kidney, Vero		Frequent	Positive isolation associated with higher mortality.	108
Lymphocytic choriomeningitis virus (LCMV)	*Arenaviridae*	Vero	Isolation in weanling mice is the standard technique.	Occasional		109
Rabies	*Rhabdoviridae*	Murine neuroblastoma, McCoy cells		Rare		110
Human immnunodeficiency virus type-1 (HIV-1)	*Retroviridae*	Cocultivation with PBL		Yes	Possible at any stage of HIV infection (overall isolation rate: 40–60%), 30% sensitivity and 80% specificity in ADC.	111–114
Human T-lymphotropic virus type I (HTLV-I)	*Retroviridae*	Cocultivation with PBL	Most arboviruses can be isolated in suckling mice.	Occasional		115
Arbovirus[c]	*Togaviridae, Flaviviridae, Bunyaviridae*	Primary hamster kidney, chick or duck embryonic, mosquito cell lines, Vero, BHK-21, LLC-MK2		Variable depending on viruses		116–119

No reports have been found in the literature of CSF isolation for the following viruses: HHV-6 (which grows from other body sites after coculture with PBL); JC virus (which grows slowly from other body sites in selected cell systems); rotaviruses (which hardly grow from any body site); parvovirus B19 (which does not grow in cell systems).

[a] Ref. 120.

[b] "Frequent": virus can be isolated in half of the cases or more; "rare": virus can be isolated in the minority of the cases; "occasional": virus isolation only occasionally reported, frequency difficult to estimate.

[c] Arboviruses (arthropod-borne viruses) do not represent a taxonomic family. Principal arbovirus families causing CNS disease include *Togaviridae* (e.g., eastern, western, and Venezuelan equine encephalitis viruses), *Flaviviridae* (e.g., Japanese encephalitis, yellow fever, dengue, West Nile fever, St. Louis encephalitis, tick-borne encephalitis, Murray Valley encephalitis, Powassan viruses), *Bunyaviridae* (e.g., La Crosse, Jamestown Canyon, Toscana viruses).

Abbreviations: CBL, cord blood lymphocytes; PBL, peripheral blood lymphocytes, CPE, cytopathic effect; HSE, herpes simplex encephalitis; HZ, herpes zoster; SSPE, subacute sclerotizing panencephalopathy; ADC, AIDS dementia complex.

6 ANTIGEN DETECTION IN CSF

Cerebrospinal fluid is not the ideal specimen for viral antigen detection, although this approach has occasionally provided some diagnostic benefit. The main advantages of the use of CSF for antigen detection are its speed and practicality and the fact that it does not require viable virus. On the other hand, antigens must be present in clinical samples in adequate amounts, or sample must be concentrated from large volumes, which may be a problem with CSF. In general, methods for viral antigen detection in CSF have shown satisfactory test specificity but have lacked sensitivity. For this reason, they have not gained a major role in the diagnosis of viral CNS disease in the past and, more recently, they have been replaced almost completely by the more sensitive molecular techniques.

6.1 Methods

Antigen detection is based on the use of antibodies that bind specifically to viral antigens. Various techniques have been developed, including immunofluorescence and immuno peroxidase stainings or solid-phase immunoassays such as agglutination tests, radioimmunoassay (RIA), and enzyme immunoassay (EIA). The former methods generally require infected cells, such as those obtained from the respiratory tract or from tissues, or highly virus-concentrated fluids such as vesicle fluids. In contrast, the more sensitive solid-phase assays, EIA in particular, can be used for antigen detection in serum and other fluids, including CSF (Fig. 2).

6.2 Clinical Applications

In the years preceding the advent of molecular methods, detection of HSV antigen in CSF was regarded as a promising technique for a rapid diagnosis of herpes simplex encephalitis. Early encouraging results were observed by IF or IP staining of CSF lymphocytes [124–126]. However, the overall performance of these procedures was poor when assessed on a larger scale [1]. Immunoassays later developed for the direct detection of HSV glycoproteins in CSF proved to be more reliable, but, despite high specificity, they varied greatly in sensitivity, from 33% to 92% [127–129].

Prior to the availability of methods for HIV-1 RNA quantification, serum HIV-1 p24 core antigen, measured by EIA, was widely used as a marker of HIV replication and disease progression. High HIV-1 p24 antigen levels in CSF were associated with the presence of severe dementia, and, overall, this test was estimated to be 95–98% specific though only 21–47% sensitive for diagnosis of ADC [112,130]. CSF p24 antigen levels were also used to document a local virological response following anti-HIV therapy [131].

Among other viral CNS infections, CMV pp65 antigen has been found in CSF leukocytes from HIV-infected patients with CMV ventriculoencephalitis or polyradiculomyelitis and CSF pleocytosis [66,132]. Japanese encephalitis (JE) virus antigen can be detected in CSF cells of approximately one-half of patients with JE, although the test is far less sensitive than serology for diagnosis of JE [133,134]. Enteroviral antigen detection assays were developed in the past but then abandoned because of the need to perform a separate test for each virus, because no useful group-specific antigen has been identified [135,136]. Occasionally, viral antigens have also been identified in the CSF of patients with VZV or mumps meningitis [126,137], and measles virus antigen in children with subacute sclerotizing panencephalitis (SSPE) [138].

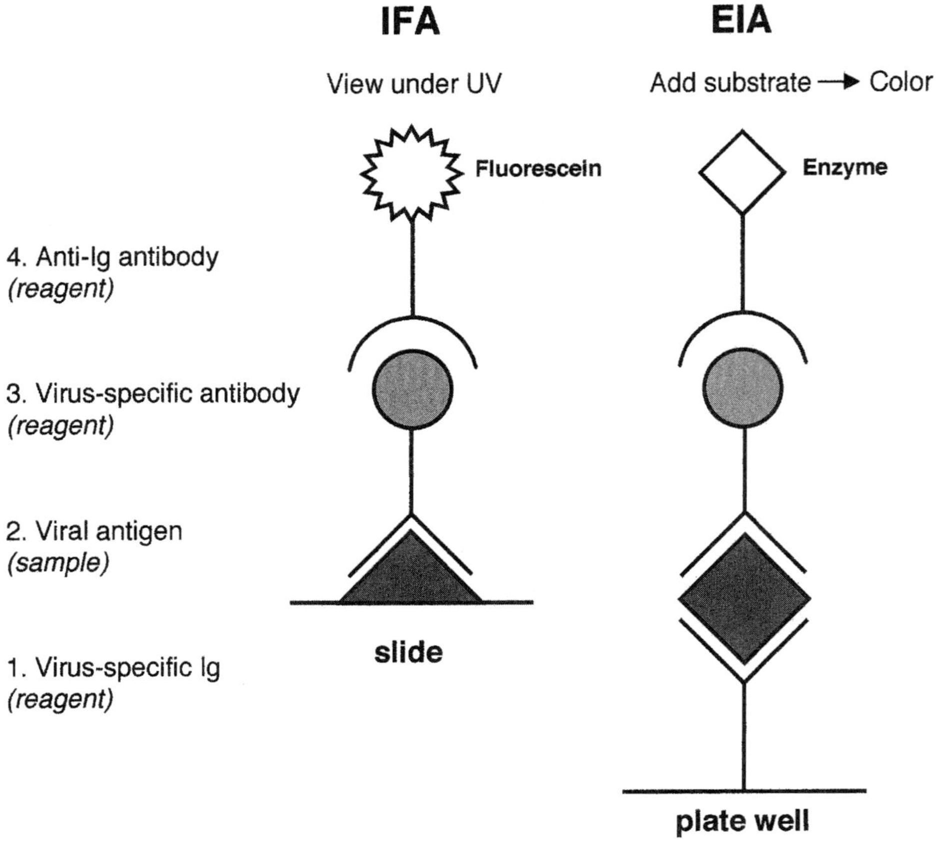

Figure 2 Antigen detection by indirect immunofluorescence assay (IFA) or enzyme immunoassay (EIA). (1,2) Sample containing the antigen is prepared on a slide (IFA) or the antigen binds to a virus-specific antibody (reagent) attached to the microplate well (EIA, or "sandwich" enzyme-linked immunosorbent assay, ELISA). (3) Virus-specific antibody (reagent) binds to the antigen. (4) Anti-Ig antibody (reagent) binds to the virus-specific antibody. In the IFA, the anti-Ig antibody is labeled with a fluorescent molecule, generally fluorescein; the fluorescence is detected by UV illumination. In the EIA, the anti-Ig antibody is labeled with an enzyme; an enzyme substrate is added to develop a colorimetric reaction, which is detected by a spectrophotometer. In the direct versions of IFA or EIA, the virus-specific antibody (step 3) is directly fluorescein- or enzyme-labeled.

7 MOLECULAR TECHNIQUES

Analysis of CSF by molecular methods is essentially based on nucleic acid (NA) amplification techniques. Their application to the study of CSF has revolutionized the diagnosis of CNS infections, especially those caused by viruses. Since the earliest reports, experiences in this field have multiplied and are continuously increasing. A Medline search using "cerebrospinal fluid", "polymerase chain reaction," and "virus" as keywords retrieved almost 800 reports as of December 2002. The most widely studied viruses have been herpes simplex viruses, followed by enteroviruses, CMV, JC virus (JCV), HIV, and Epstein-Barr virus (EBV), reflecting both the relative frequencies of CNS diseases induced by these viruses and the need for rapid and reliable diagnostic tools. Besides the experience largely documented in the literature, molecular analysis of CSF has progressively entered clinical practice and has completely changed the nature of the work in clinical virology laboratories. NA amplification assays are routinely performed in most hospital laboratories, and in several instances these tests have gained an invaluable role in neurovirology diagnostics.

The exquisite sensitivity of NA amplification techniques, primarily PCR, has enabled efficient and rapid detection and identification of viruses in the CSF. Furthermore, CSF PCR has made it possible to establish the viral etiology of neurological syndromes of dubious origin such as Mollaret's meningitis [139] and to recognize unusual or atypical CNS diseases such as mild forms of herpes encephalitis [140–143] or CMV ventriculoencephalitis in HIV-infected patients [144]. Finally, viruses normally causing extracerebral infections, such as rotavirus, parvovirus B19, CMV, or HHV6, have been demonstrated in CSF of patients with neurological symptoms, strongly supporting their etiological role in inducing CNS disease [145–150].

7.2 Methods

Techniques for Nucleic Acid Amplification

The main property accounting for the extraordinary sensitivity of nucleic acid (NA) amplification techniques is their ability to amplify a small quantity of target nucleic acid molecules to considerably larger amounts (over 10^6 DNA copies), which can be visualized by means of common laboratory procedures. A number of techniques have been described, including PCR, the ligase chain reaction (LCR), the strand displacement assay (SDA), transcription-mediated amplification, nucleic acid sequence based amplification (NASBA), branched DNA, and hybrid capture assay [151]. PCR is the most widely used method for CSF analysis and is discussed here in detail. The NASBA and branched DNA techniques, which have also been applied to the CSF, are also described.

Polymerase Chain Reaction. The polymerase chain reaction (PCR) is based on the use of oligonucleotides, or primers, that specifically recognize and anneal to a target DNA, and a thermostable DNA polymerase, which makes new DNA copies starting from single nucleotides. DNA amplification takes place during repeated cycles of heating and cooling, which allow the denaturation of DNA, the annealing of the primer to the denatured DNA strand, and final extension of the DNA itself (Fig. 3) [151,152]. In the case of RNA viruses, it is necessary to first generate a complementary DNA from RNA (cDNA), the suitable target for PCR amplification. This is accomplished by the use of a reverse transcriptase (RT) before the amplification steps (RT-PCR) [151]. The primers are usually designed to recognize highly conserved genome regions, to avoid false negative results due to virus strain variation. A widely employed variant of classical PCR is ''nested'' PCR, which increases the sensitivity and specificity of detection. This procedure consists of two amplification reactions using two primer sets, the primers of the second set being nested between the primers of the first one (Fig. 4).

Other NA Amplification Techniques. Nucleic acid amplification techniques other than PCR have been developed mainly for use in commercial kits that are designed for both NA amplification and the detection of amplified products. Nucleic acid sequence based amplification (NASBA), like PCR, is based on target NA amplification, but it is isothermal and requires three different enzymes. Furthermore, the template consists of RNA (Fig. 5) [153]. NASBA has been applied to CSF to detect the CMV pp67 late gene

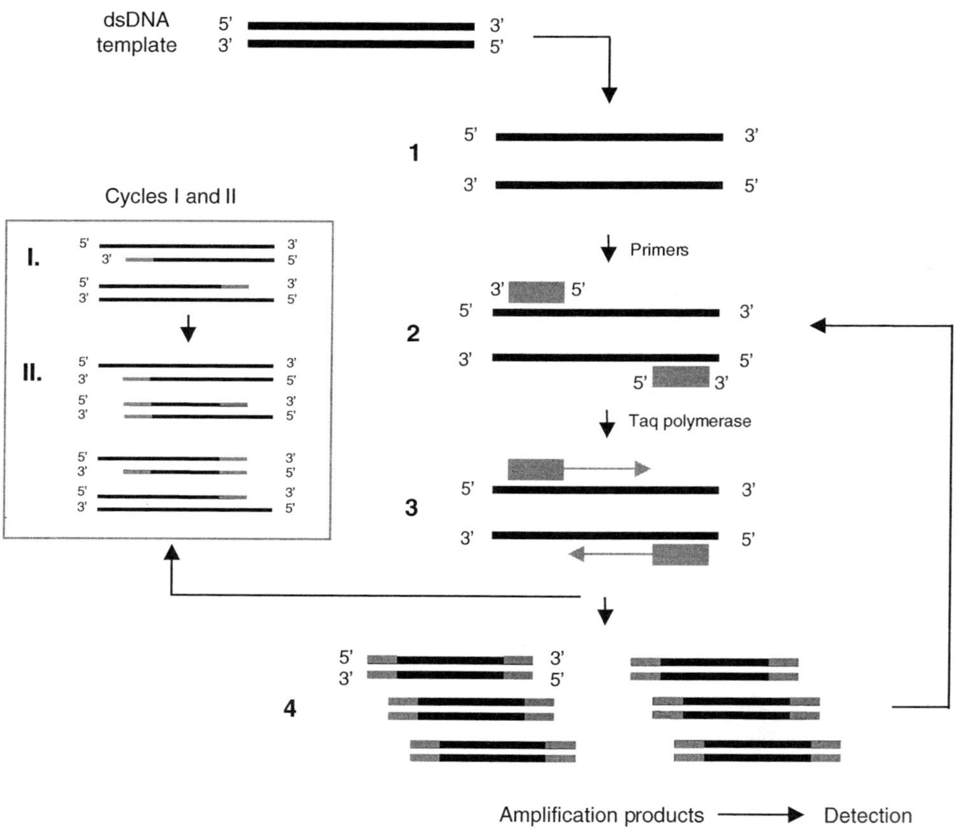

Figure 3 Polymerase chain reaction (PCR). A DNA template is added to a reaction mixture containing a pair of primers complementary to the target, a thermostable enzyme DNA polymerase, and deoxynucleotides in an appropriate buffer. (1) The DNA template is denatured by high temperature (''denaturation,'' usually at 94–95°C). (2) The oligonucleotide primers anneal to target DNA (''annealing,'' usually at 55–70°C). (3) The thermostable enzyme DNA polymerase allows synthesis of new DNA strands (''extension,'' usually at 72°C). (4) Amplification products accumulate exponentially through 20–40 cycles of heating and cooling through steps 1–3. These can be detected after DNA gel electrophoresis or other procedures. The box on the left of the figure shows more in detail the dynamics of amplification during the first two PCR cycles. (I) During the first cycle, one ''long'' PCR product is generated from each DNA strand. (II) During the second cycle, two new ''long'' products are generated, together with one ''short'' product from each of the long fragments previously produced. The long products double at each reaction, whereas the amount of short products will increase exponentially.

transcripts in HIV-infected patients with CMV encephalitis [154,155], or viral DNA in CNS infections caused by RNA viruses, such as HIV-1, enteroviruses, or flaviviruses [156,157b]. In contrast to the previously described techniques, the branched DNA technique is based on signal amplification rather than target amplification (Fig. 6) [158]. Also, this assay has been used to assess the CMV DNA and HIV-1 RNA levels in the CSF of HIV-infected patients [132,159].

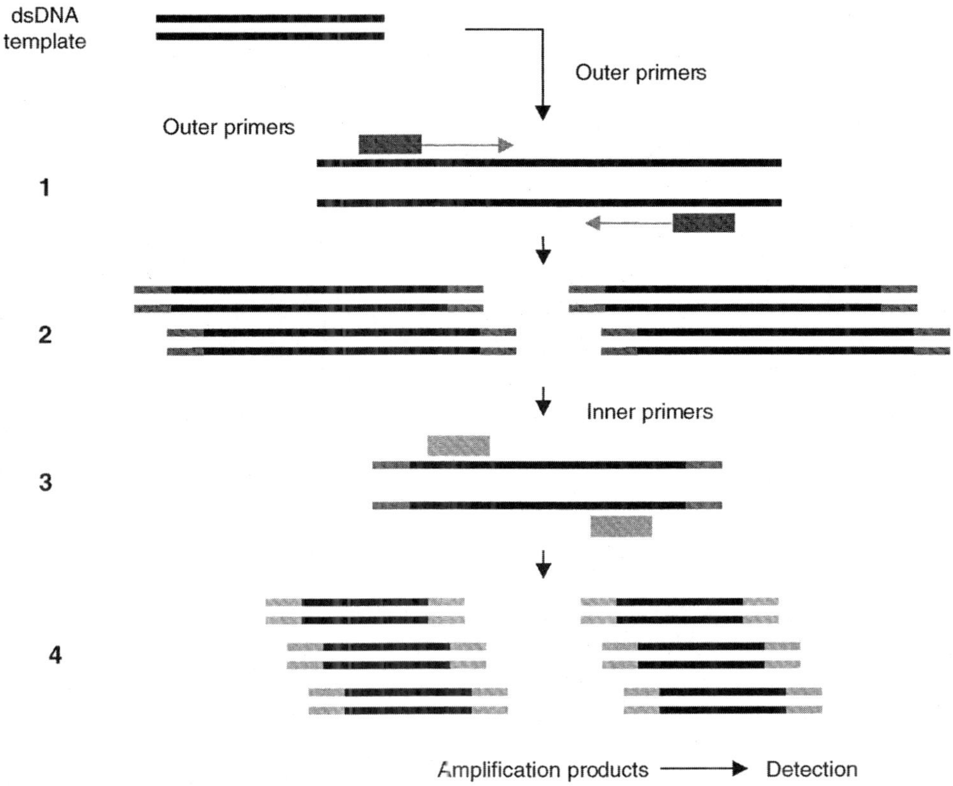

Figure 4 Nested PCR. (1) A double-stranded DNA template is subjected to a first amplification with ''outer primers.'' (2) Amplification products are generated. (3) The amplified products are subjected to a second amplification with ''inner'' primers. (4) Shorter fragments are produced, which can be detected by agarose gel electrophoresis or other procedures.

Detection of Amplified Products

Various procedures can be used to detect PCR-amplified DNA. The simplest consists of visualization of DNA bands of the expected size after electrophoresis of the amplification products in agarose gel stained with ethidium bromide (Fig. 7). Hybridization with DNA probes complementary to the target DNA may follow or be used in place of gel electrophoresis after the transfer of DNA to a filter, tubes, or microplates. The probes are labeled with enzymes or other molecules that, on appropriate stimulation, lead to signal detection (Fig. 8). Colorimetric enzyme-linked immunosorbent assay (ELISA), in which the amplified product is captured by a probe coated on to microplate wells, have proved to be very practical and have largely been adapted to commercial kits.

CSF Preparation

Various protocols are in use for the pre amplification preparation of CSF to release nucleic acids from cells and to remove substances that may degrade nucleic acid or inhibit amplification. The relatively simple CSF composition may obviate, at least for DNA viruses, the need for nucleic acid purification. The simplest approaches include the heating to 95°C

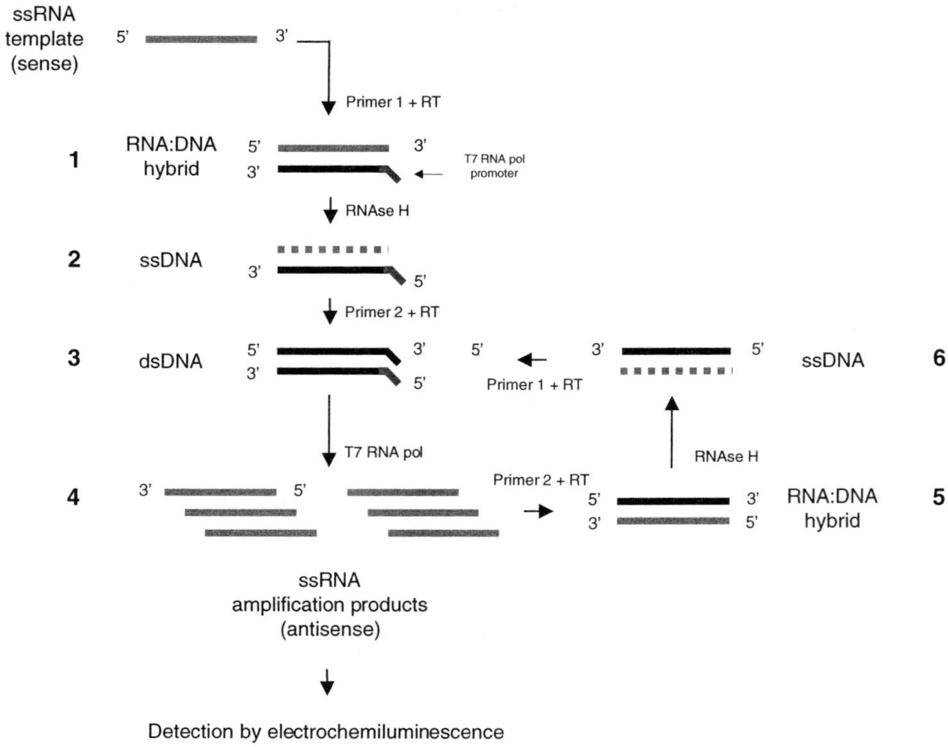

Figure 5 Nucleic acid sequence based amplification (NASBA). An RNA template is amplified through an isothermal reaction at 41°C using three enzymes: avian myeloblastoid virus reverse transcriptase (AMV-RT), RNAse H, and T7 RNA polymerase (T7 RNA pol). (1) The first primer, containing a T7RNA polymerase promoter, anneals to the target and allows RT to form an RNA: DNA hybrid. (2) RNAse degrades the RNA strand. (3) The second primer anneals to the DNA and RT copies a new DNA molecule, forming a double-stranded DNA. (4) T7RNA polymerase synthesizes new RNA molecules. The amplification products accumulate through repeated steps in which the newly formed RNA acts as a template for DNA (5), is digested (6), and then replaced by a new DNA molecule to form a new double DNA strand (3). One or more internal standards (IS) at known copy number are coextracted and amplified with each sample. The amplified RNA products, including the IS, are detected by electrochemoluminescence following hybridization with rutenium-labeled target- or IS-specific probes. To facilitate the detection process, the amplification products are captured onto magnetic beads bound to streptavidin by using a second, biotin-labeled probe. In the quantitative version of the assay, e.g., for quantification of HIV-1 RNA, three IS molecules, or calibrators, are used. The signal produced from both target and calibrators is detected, and a standard curve is obtained for each sample by plotting the known concentrations of the calibrators versus their signal intensity. The amount of target RNA is extrapolated by comparison with the standard curve.

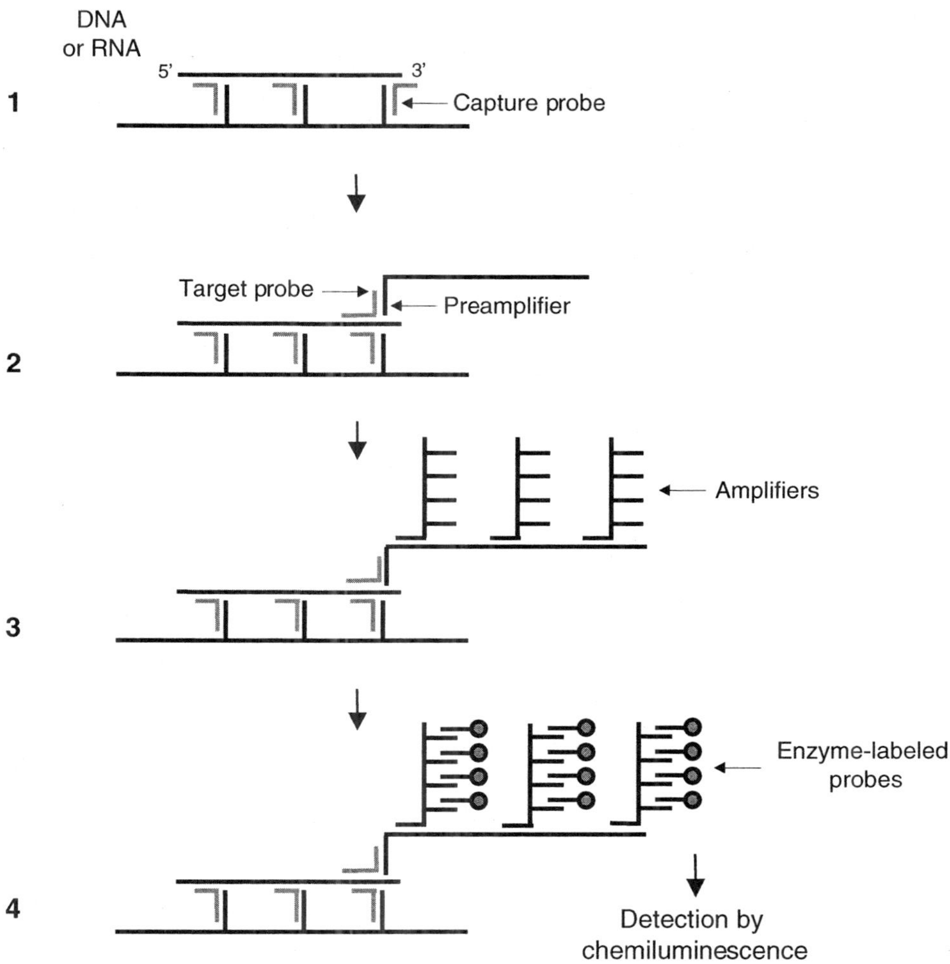

Figure 6 Branched-DNA. (1) A DNA or RNA template is immobilized on microplate wells coated with a capture probe. (2) A ''target'' probe binds to the template, followed by binding of a ''preamplifier'' probe. (3) An ''amplifier'' probe binds to the preamplifier. (4) Enzyme-labeled probes eventually bind the amplifier. A chemiluminescent substrate is added, leading to light emission. Nucleic acid quantification is accomplished by comparison of the signal in the samples with an external calibration curve.

or the repeated freezing and thawing of specimens, procedures that facilitate cell membrane disruption and the release of DNA. These are quicker, require smaller CSF volumes, and reduce the risk of sample contamination during nucleic acid extraction. However, they are inefficient for use with RT-PCR or with certain types of polymerases. Nucleic acids may also be concentrated and/or extracted from CSF by a number of in-house procedures or commercial kits (Table 4). Some extraction methods have performed better than others in comparative studies [160,160a], but none of the known protocols has been shown to be clearly superior for any use. The choice of extraction method is weighted by a number

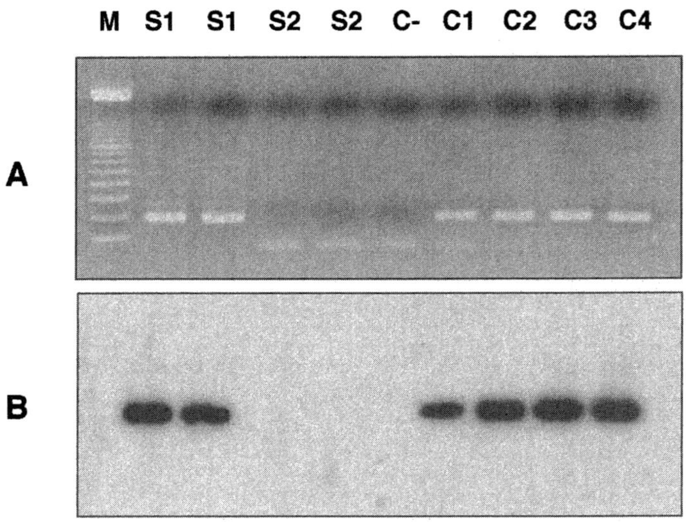

Figure 7 Agarose gel electrophoresis and Southern blot hybridization. A representative example of (A) gel electrophoresis and (B) Southern blot detection of amplified products following PCR. (A) A 173 bp long fragment from JC virus large T antigen has been amplified by PCR. The amplification products are detected after electrophoresis on 2% agarose gel containing 0.5 μg/mL ethidium bromide, and the results are photographed under UV illumination. (B) Following electrophoresis, the DNA is transferred by Southern blot to a nylon filter and hybridized with a JCV-specific internal probe. After hybridization the filter is exposed to X-ray film. M: 100 bp DNA ladder marker; S1: JCV DNA positive CSF sample (in duplicate); S2: JCV DNA negative CSF sample (in duplicate); C-: negative control; C1–C4: positive controls consisting of plasmidic DNA containing 100, 1000, 10,000, and 100,000 JCV genome equivalents per reaction.

of considerations, including practical aspects in the individual laboratory, type of NA target, and amplification protocols employed.

Multiplex PCR and PCR with Consensus Primers

Because CNS infections caused by different viruses may result in similar clinical pictures, PCR assays have been designed to detect more that one virus or infectious agent in the same reaction. Basically, two strategies have been used for this purpose: multiplex PCR and PCR with consensus primers. The most obvious advantage of these approaches is that the number of tests is reduced, with substantial savings in time and cost.

Multiplex PCR enables the identification of more than one DNA sequence in the same PCR reaction by using two or more primer pairs, each specific for a single sequence (Fig. 9). A potential difficulty with this approach is that the primers to be used need to be chosen carefully in order not to compromise amplification efficiency, because each primer pair requires its own conditions of amplification, i.e., reagent mixture composition and thermocycling profile. Duplex PCR protocols for simultaneous detection of HSV-1 and HSV-2 are commonly employed [161–163]. The same strategy is also widely applied to detect a larger number of herpesviruses responsible for CNS infections, including HSV-1, HSV-2, VZV, CMV, EBV, and HHV-6 [164–167]. Multiplex PCR protocols have also been proposed for the simultaneous amplification of herpesvirus and enterovirus sequences

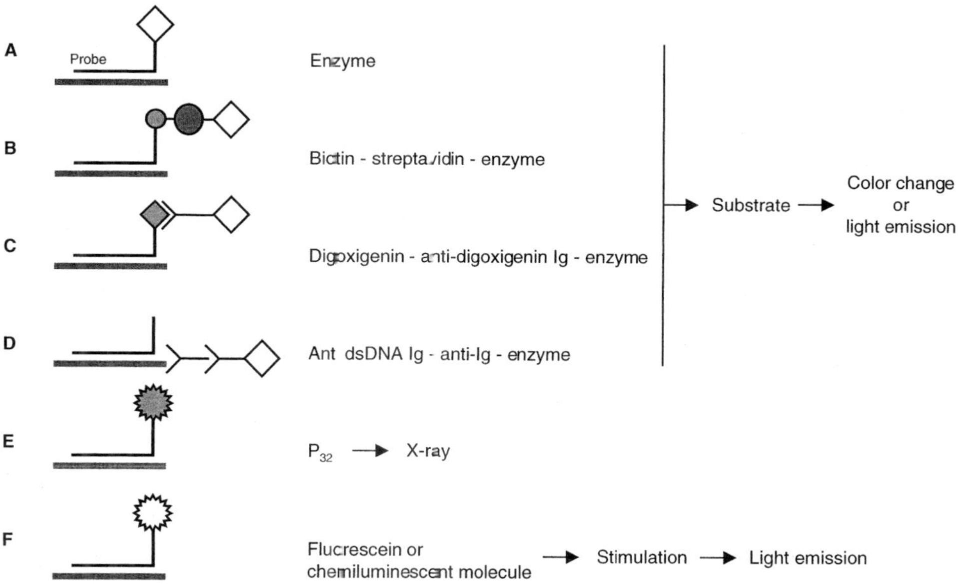

Figure 8 Detection of amplified products. Examples of amplification product detection using probes labeled with different molecules. (A) The probe is directly conjugated with an enzyme (e.g., horseradish peroxidase, alkaline phosphatase). (B) Biotin-labeled probe binds to enzyme-conjugated streptavidin. (C) Digoxigenin-labeled probe binds to enzyme-conjugated antidigoxigenin antibody. (D) The hybrid is detected by an antibody against double-stranded DNA, which is bound by an enzyme-conjugated anti-Ig antibody [DNA enzyme assay (DEIA)]. (E) The probe is labeled with a radioactive molecule (usually ^{32}P). (F) The probe is conjugated with a chemiluminescent molecule (e.g., ruthenium) or a fluorescent dye. Following hybridization, the enzyme reacts with a substrate, leading to color change of the hybridization solution or light emission (A–D). Radioactivity is detected after exposure to X-rays (E). Chemiluminescent or fluorescent molecules emit light under appropriate stimulation (F).

Table 4 Techniques of CSF Preparation for Nucleic Acid Amplification

Principle	Methods[a] (examples)
CSF cell lysis by mechanical procedures	Heating to 95 °C, freezing–thawing
CSF cell lysis–protein digestion	Detergents (SDS), proteases (protease K), chaotropic agents (guanidinium thiocyanate)[b]
Nucleic acid concentration	CSF ultracentrifugation, ethanol precipitation of nucleic acids
Nucleic acid extraction	Phenol–chloroform, spin column, silica adsorption, magnetic separation

[a] Methods for cell lysis, nucleic acid concentration, and extraction can be combined in various ways. Approximate time required varies from 10 min (e.g., by mechanical cell lysis) to ≥1 h (necessary for protease K digestion and/or complex nucleic acid procedures, e.g., phenol–chloroform). Approximate volume required varies from 2 to 5 μL (e.g., by mechanical cell lysis) to ≥1 mL (when CSF concentration procedures are employed).
[b] Commonly used before RNA extraction because of its ability to inactivate ribonucleases.

[168,169] or, in AIDS patients, of EBV and *Toxoplasma gondii* sequences to distinguish CNS lymphoma from toxoplasmosis [170]. Further developments of multiplex PCR might lead to ''universal'' diagnostic PCR protocols, based on the use of several primer pairs with fixed thermocycle programs and reagent compositions [171].

Viruses belonging to the same family can be amplified in the same test tube through the use of ''consensus'' primer pairs that target common regions in the viral genomes (Fig. 10) [172]. Consensus primers that recognized a conserved region of the herpesvirus DNA polymerase gene were initially used for simultaneous amplification of HSV-1, HSV-2, CMV, and EBV [172]. A similar approach, using two PCR assays amplifying separately two groups of herpesviruses (HSV-1, HSV-2, CMV, EBV, and human herpesvirus-8; and VZV, HHV-6 variants A and B, and human herpesvirus-7), has more recently been proposed [68]. Through a PCR variant employing a mixture of primers in which the 5′ end remains constant and the 3′ end is displaced base by base, it has also been possible to amplify six of the human herpesviruses in a single reaction [173,174]. Polyomaviruses, JCV, BK virus, and SV40 are another example of multiple genome amplification by means of a single primer pair [175,176].

Quantitative NA Amplification Techniques

Measuring the amount of nucleic acids in clinical specimens has represented an important goal and a successful achievement in diagnostic molecular biology. Both PCR and other

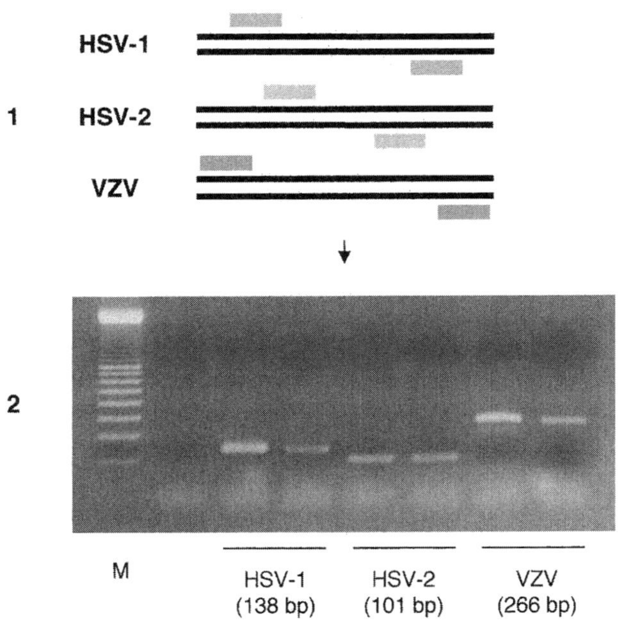

Figure 9 Multiplex PCR. Representative example of a multiplex PCR assay. (1) Three different sequences from HSV-1, HSV-2, and VZV are amplified simultaneously in the same tube by using three different primer pairs, each specific for a single virus. (2) The amplified PCR products have different lengths and produce bands of different sizes on agarose gel (M: 100 bp DNA ladder marker). Alternatively, amplification products can be distinguished by hybridization with specific probes, restriction enzyme analysis, nested PCR with specific internal primers, or DNA sequencing.

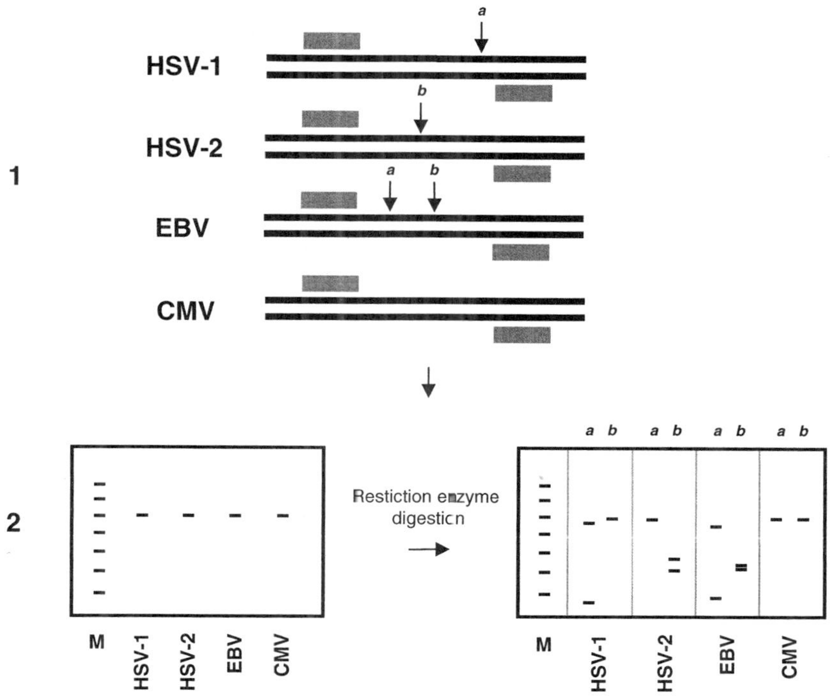

Figure 10 PCR with consensus primers. Representative example of a PCR assay using consensus primers. (1) Conserved DNA sequences from the polymerase genes of HSV-1, HSV-2, EBV, and CMV are amplified simultaneously in the same tube by a consensus primer pair targeting common regions. (2) Following gel electrophoresis, the amplified products have similar lengths (left), but each virus can be distinguished by specific patterns resulting from DNA cleavage with two restriction enzymes (a, *Sma*I and b, *Bam*HI) (right) (M: DNA marker). Alternatively, viruses can be distinguished following hybridization with virus-specific probes, nested PCR using virus-specific internal primers, or DNA sequencing. (From Ref. 172.)

NA amplification techniques have been proven to be reliable for this purpose, and a variety of ''semiquantitative'' and quantitative PCR methods have been described (Table 5) [177–179]. Semiquantitative techniques include limiting dilutions of samples before amplification and methods that compare the extent of amplification between samples and ''external'' standards at known NA concentrations. The main disadvantage of these procedures is that they do not take into account the possible differences in amplification efficiency between different samples and/or standards. Quantitative techniques allow a more accurate estimate of nucleic acid levels, generally achieved through the coamplification in the same tube of target NA and an ''internal'' standard at known NA concentration, which enables control of amplification efficiency (Fig. 11) (see also the following subsection).

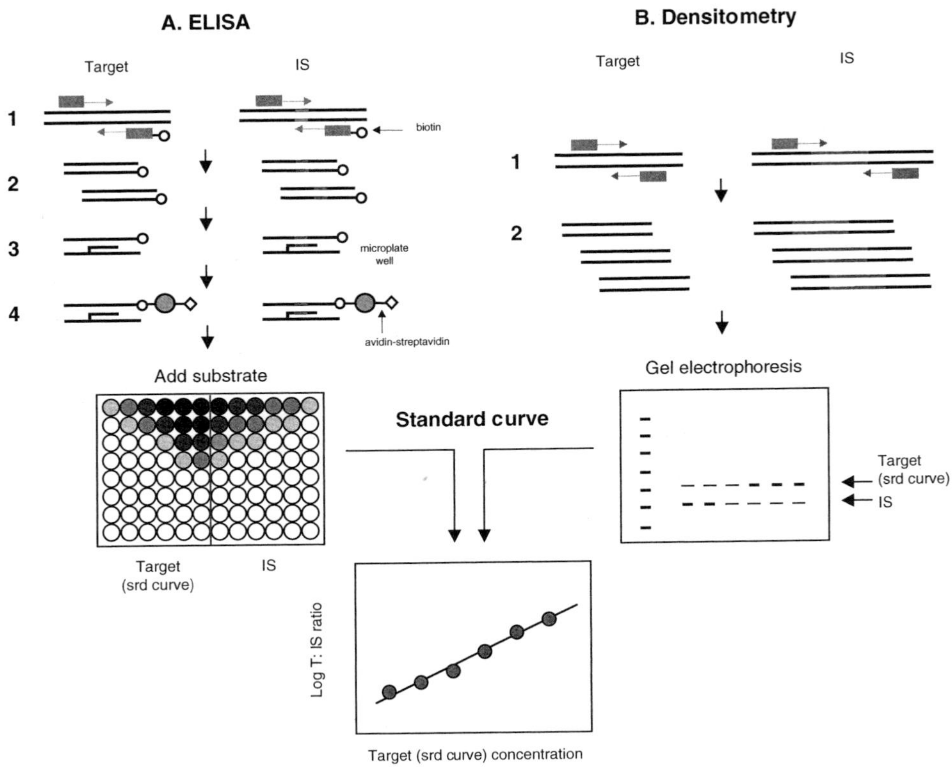

Figure 11 Quantitative PCR. Examples of quantitative PCR assays based on detection of amplified products by (A) enzyme immunoassay (EIA) or (B) densitometry. In both assays an internal DNA standard (IS) is coamplified in the same tube with target DNA and the amounts of target and IS at the end of amplification are calculated. (A) ELISA. (1) The DNA template is coamplified with an IS containing the same primer-binding sites but distinguishable for an internal IS-specific modified sequence. One of the primers is biotinilated at its 5′ end. (2) Both target and IS are amplified. One of the two DNA strands is biotinilated. (3) Following amplification, the DNA is denatured and the biotinylated strand is immobilized onto a microplate well by either a target- or IS-specific probe. (4) Enzyme-conjugated streptavidin and enzyme substrate are added. The colorimetric reaction is detected by a spectrophotometer. (B) Densitometry. (1) The DNA template is coamplified with an IS containing the same primer-binding sites but distinguishable for an IS-specific modified sequence that differs in size from the target sequence. (2) Both target and IS are amplified, resulting in amplification products of different sizes that are distinguished by agarose gel electrophoresis. The intensity of the DNA bands is quantitatively estimated by a densitometer. In both procedures, the ratio between target and IS (OD by EIA or band intensity by densitometry) is calculated. A standard curve is obtained by plotting known amounts of the target DNA versus their target/IS ratios. The DNA amount in each sample is calculated by comparing the sample target/IS ratio with the standard curve ratio.

Table 5 Most Common Methods for Quantification of Nucleic Acids

Target amplification-based techniques
PCR-based techniques
End-point dilutions[a]
Comparison with external standard curve[a]
Coamplification of target with IS and comparison with external standard curve[b]
Coamplification of target with IS and comparison with internal standard curve[b]
Real-time PCR (TaqMan, Light Cycler)
Nucleic acid sequence based amplification (NASBA)
Signal amplification-based techniques
Branched DNA
Hybrid capture

IS, internal standard.
[a] Often referred to as "semiquantitative' techniques.
[b] Often referred to as "competitive" techniques because an internal standard, or competitor, is coamplified with the target.

Recently, new automated procedures based on real-time detection of nucleic acids have been applied in diagnostic virology: TaqMan (Fig. 12) [180–182] and LightCycler [183,184]. The main characteristic of these methods is that they measure the PCR product as it accumulates rather than at the end of amplification when amplification efficiency is reduced. Compared to classical end-point measurement, real-time PCR is therefore more accurate and also expands the dynamic range of quantification. Furthermore, it eliminates post-PCR processing of PCR products, resulting in reducing the risk of contamination, removing potential sources of errors, and increasing the processing speed. These technologies are extremely promising because they also allow simultaneous quantification, in the same tube, of different genomes as well as mutational analysis.

Procedures to Reduce False Positive or False Negative Results

A potential risk of NA amplification techniques is the possibility of producing false positive results by contaminating the samples. Contaminating nucleic acids may originate from clinical specimens or, more commonly, from the products of previous amplification. To minimize this risk, it is necessary to maintain the sterility of CSF before its arrival at the laboratory and to carry out the different laboratory steps—i.e., sample or reagent preparation, the transfer of amplified products in the case of nested PCR, and the detection of amplified products—in separate areas [185]. The use of a number of ''negative controls,'' usually water or known negative samples tested in parallel with the clinical specimens throughout the procedure, and analysis of the samples in duplicate represent a useful way of assessing the occurrence of false positive results. Another way to prevent carryover of amplification products is through the use of the enzyme uracil *N*-glycosylase (UNG), which degrades products from previous amplifications but not native NA templates. This is accomplished by substituting dUTP for dTTP in the amplification mixture and pretreating all subsequent mixtures with UNG prior to amplification [186].

On the other hand, the presence of inhibitors, i.e., substances that may affect correct functioning of enzymes, may cause false negative results. Inhibition of amplification has been reported in 1–5% of CSF specimens [187]. To reveal the presence of inhibition, ''internal'' standards can be added to the amplification mixture to be coamplified with

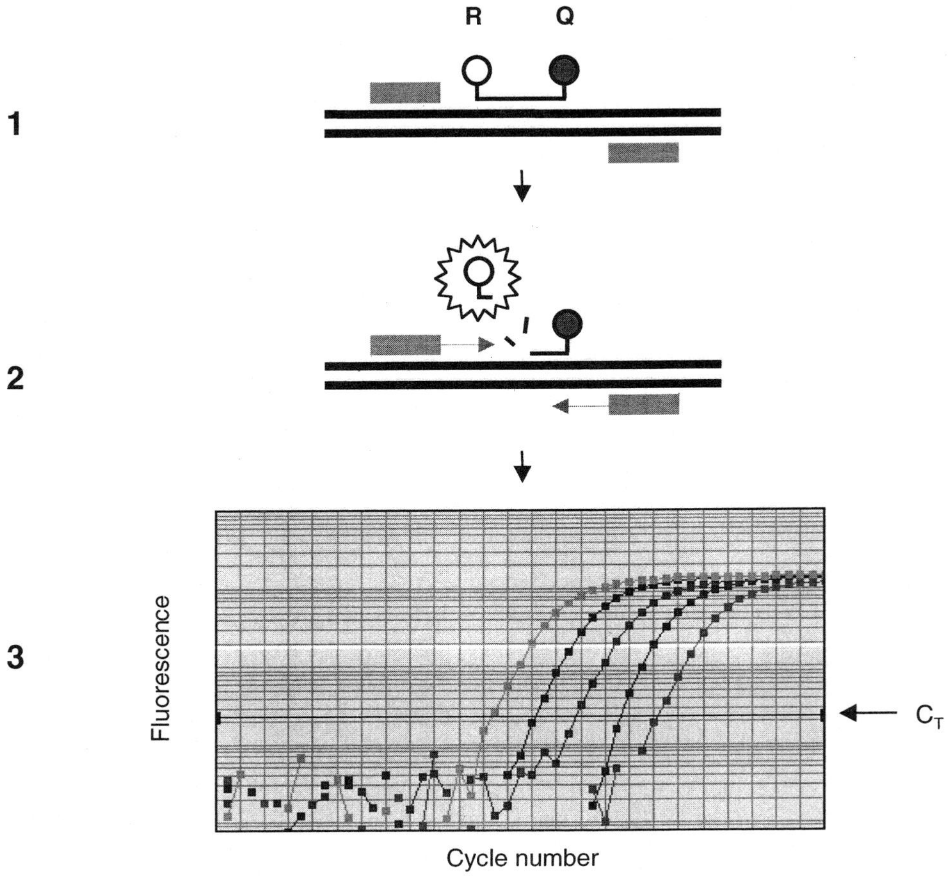

Figure 12 Real-time PCR (TaqMan). (1) A primer pair and a probe labeled with two fluorescent dyes, a quencher (Q) and a reporter (R), anneal to the template DNA. When the primer is intact, the vicinity of the quencher reduces the fluorescence emitted by the reporter. (2) During the DNA extension phase of PCR, the nuclease activity of the Taq DNA polymerase digests the probe and separates the reporter from the quencher, resulting in an increased fluorescence signal. (3) The fluorescence, proportional to the amount of amplified products, is acquired at each PCR cycle by an automated fluorimeter. A threshold cycle (C_T) defines the cycle number at which the fluorescence passes a fixed threshold. Quantification of the amount of target in unknown samples is accomplished by measuring the C_T and using a standard curve in which known initial amounts of target DNA are plotted versus the C_T values.

the template. These molecules are recognized by the same primers and are amplified as efficiently as the target but are somehow distinguishable from the target (Fig. 11). In addition, the use of both weak and strong positive controls in the same run can help monitor amplification efficiency.

CSF Collection and Storage Conditions

For DNA viruses, it is considered safe to send CSF specimens to the laboratory at room temperature, though it is preferable to store them at −20°C if they cannot be delivered within one day [9]. Actually, it has been observed that DNA of HSV is stable in CSF for up to 30 days, not only at −20°C or 2–8°C but also at room temperature [188]. On the other hand, RNA is regarded to be less stable than DNA in plasma, where it seems to be affected by a number of factors, including the type of anticoagulant used for specimen collection and storage temperature [189,190]. However, storage of CSF at 4°C or at room temperature seemed not to affect the recovery of enterovirus RNA after 96 h [191]. Furthermore, measuring HIV-1 RNA load in CSF after up to 96 h of storage at 4°C did not result in any relevant viral load decay [192,192a].

7.2 Clinical Applications

Clinical applications of NA amplification techniques span a large spectrum of CNS infections and involve different patient populations such as immunocompetent children or adults, neonates, and immunocompromised patients. Whereas PCR is by far the most widely used technique, the clinical applications of alternative amplification methods are increasingly being reported. Overviews of their clinical uses in immunocompetent and immunocompromised patients are presented in Table 6 [4,72,96,101,105,120,139,145–150,193–244] and Table 7 [88,163,242,245–267], respectively. The most relevant examples are now briefly discussed.

Herpes Simplex Encephalitis

The detection of HSV-1 DNA in CSF is one of the most powerful examples of the usefulness of molecular CSF analysis. This test is now considered the diagnostic method of choice for HSE, and it has largely replaced the identification of HSV in brain tissue biopsies, which used to be the diagnostic standard [187,195]. A number of retrospective and prospective studies have clearly defined the reliability of PCR, showing that it provides more than 90% sensitivity and virtually 100% specificity [193–195]. The technique is rapid, which allows a diagnosis to be established in time for management decision making. Furthermore, it enables the diagnosis of uncommon forms of HSV CNS infection that may otherwise go unrecognized [140–143]. However, it is important that PCR results be interpreted cautiously in relation to the clinical presentation and the duration of disease and of antiviral therapy. For example, PCR may fail to amplify HSV DNA in CSF samples drawn very early after onset on CNS symptoms; such results are likely to reflect a still limited virus replication [172]. On the other hand, the likelihood of finding a positive CSF PCR result is reduced following a few days of acyclovir treatment and also in untreated patients from whom CSF is obtained late after onset of neurological symptoms [172,193,194].

Enterovirus Meningitis

Diagnosis of enteroviral CNS infections has been greatly improved with the use of molecular techniques. As for viral culture, enteroviruses are the viral agents most frequently detected by PCR in aseptic meningitis cases [220]. Techniques using primers designed to target conserved sequences within the 5′ noncoding region are commonly used. These recognize almost all of the enterovirus serotypes, including enteroviruses that cannot be isolated in cell cultures, with the only exception of echoviruses 22 and 23 which diverge extremely from the other serotypes [268]. One of the advantages of molecular diagnosis is the reduction of the time for diagnosis from 4–10 days for conventional cell cultures to 1 day. Furthermore, NA amplification enables virus identifications in CSF samples obtained some days after onset of symptoms, when virus isolation is infrequent [93]. A commercial PCR assay, based on colorimetric microwell detection [269], as well as a number of protocols developed in-house, have been largely evaluated for diagnostic relia-

Table 6 Diagnostic Use of Nucleic Acid Amplification Techniques in CSF in Viral CNS Infections of Immunocompetent Patients

Virus	Family	Nucleic acid (NA)[a]	Most common clinical syndromes	Significance of NA amplification[b]	Refs.
HSV-1	*Herpesviridae*	dsDNA	Herpes encephalitis (HSE), neonatal infection	Improved diagnosis of HSE: test of choice (>90% sensitivity vs. brain biopsy); diagnostic potential in neonatal infections; identification of atypical HSE forms	194–197 139, 196–199
HSV-2	*Herpesviridae*	dsDNA	Aseptic meninigitis, recurrent meningitis, neonatal infection	Improved diagnosis of aseptic meningitis; diagnostic potential in neonatal infections; identification of recurrent meningitis	
VZV	*Herpesviridae*	dsDNA	Varicella and herpes zoster (HZ) complications	Improved diagnosis; association with uncomplicated HZ	200–202
CMV	*Herpesviridae*	dsDNA	Aseptic meningitis, encephalitis, neonatal infection	Improved diagnosis; diagnostic potential in neonatal infections; identification of CMV neurological syndromes	148, 203, 204
EBV	*Herpesviridae*	dsDNA	Aseptic meningitis, encephalitis	Useful for diagnosis	205, 206, 206a
HHV6	*Herpesviridae*	dsDNA	Febrile seizures, encephalitis	Association with febrile child seizures and encephalitis; potentially useful for diagnosis	145, 146, 150, 207, 208
HHV7	*Herpesviridae*	dsDNA	Febrile seizures	Association with febrile child seizures	209–212
Adenovirus	*Adenoviridae*	dsDNA	Encephalitis	Diagnostic potential	212a
BK virus (BKV)	*Polyomaviridae*	dsDNA	Unknown	Association with CNS disease	213, 213a
Parvovirus B19	*Parvoviridae*	ssDNA	Aseptic meningitis	Identification of parvovirus B19 meningitis; useful for diagnosis	147, 214, 215
Rotavirus	*Reoviridae*	dsRNA	Aspetic meningitis, encephalitis	Identification of rotavirus CNS disease; potentially useful for diagnosis	149, 216–218
Enterovirus	*Picornaviridae*	ss+RNA	Aseptic meningitis	Improved diagnosis: test of choice (>90 sensitivity vs. virus isolation)	4, 96, 219, 220

(continued)

Table 6 *Continued*

Virus	Family	Nucleic acid (NA)[a]	Most common clinical syndromes	Significance of NA amplification[b]	Refs.
Rubella	*Togaviridae*	ss+RNA	Aseptic meningitis, subacute panencephalitis, neonatal infection	Occasional association with encephalitis	221
Influenza	*Orthomyxoviridae*	ss−RNA	Encephalitis	Identification of influenza-associated CNS disease; potentially useful for diagnosis	101, 222–224
Mumps	*Paramyxoviridae*	ss−RNA	Aspetic meningitis	Improved diagnosis (96% sensitivity vs. virus isolation)	105
Measles	*Paramyxoviridae*	ss−RNA	Acute encephalitis, subacute encephalitis, subacute sclerotizing panencephalitis (SSPE)	Diagnostic potential in SSPE and subacute encephalitis	225–227
Nipah	*Paramyxoviridae*	ss−RNA	Encephalitis	Diagnostic potential not known	72
Rabies	*Rhabdoviridae*	ss−RNA	Rabies	Diagnostic potential	228, 229
Lassa	*Arenaviridae*	ss−RNA	Encephalitis	Diagnostic potential	229a
HTLV-I	*Retroviridae*	RNA, DNA (reverse transcription virus)	HTLV-associated myelopathy (HAM)	Diagnostic potential not known	230–233
Jamestown Canyon, La Crosse, Toscana	*Bunyaviruses*	ss−RNA	Encephalitis, meningitis	Diagnostic potential (high sensitivity in Toscana virus aseptic meningitis)	234–237
West Nile, Dengue, Japanese, tick-borne, St. Louis	*Flaviviridae*	ss+RNA	Encephalitis	Diagnostic potential (up to 55% sensitivity in West Nile encephalitis)	157a, 238–240

Nucleic acids from other viruses, e.g., hepatitis C virus (HCV) and coronavirus have also been found in the CSF, but without clear association with CNS disease (Refs. 241–244).

[a] dsDNA, double-stranded DNA virus, ssDNA, single-stranded DNA virus, dsRNA, double-stranded RNA virus, (+)ssRNA positive-stranded RNA virus, (−)ssRNA, negative-stranded RNA virus (Ref. 120).

[b] PCR has been the most commonly employed NA amplification technique.

Table 7 Diagnostic Use of Nucleic Acid Amplification Techniques in CSF in Viral CNS Infections of Immunocompromised Patients

Virus	Family	Nucleic acid[a]	Main clinical syndromes[b]	Significance of NA amplification	Comments	Refs.
HSV-1	*Herpesviridae*	dsDNA	Subacute encephalitis	Improved diagnosis; definition of HSV-associated clinical syndromes in HIV-infected patients.	100% sensitivity, 99% specificity (HIV-infected patients)	163, 245
HSV-2	*Herpesviridae*	dsDNA	Subacute encephalitis	Improved diagnosis; definition of HSV-associated clinical syndromes in HIV-infected patients.	100% sensitivity, 99% specificity (HIV-infected patients)	163, 246
VZV	*Herpesviridae*	dsDNA	Varicella and herpes zoster (HZ) complications	Improved diagnosis; better definition of VZV-associated clinical syndromes in HIV-infected patients.		247–249
CMV	*Herpesviridae*	dsDNA	Subacute encephalitis, polyradiculopathy	Improved diagnosis; better definition of CMV-associated clinical syndromes in HIV-infected patients.	82–100% sensitivity, 89–100% specificity (HIV-infected patients)	88, 250–254
EBV	*Herpesviridae*	dsDNA	Lymphoproliferative disorders (transplanted patients), PCNSL (HIV-infected patients)	Improved diagnosis	88–100% sensitivity, 89–100% specificity (PCNSL in HIV-infected patients)	255–257
HHV6	*Herpesviridae*	dsDNA	Encephalitis (transplant recipients)	Potentially useful for diagnosis in transplant recipients.	No clear association with CNS disease in HIV-infected patients	258–260
JCV	*Polyomaviridae*	dsDNA	Progressive multifocal leukoencephalopathy (PML)	Improved diagnosis; noninvasive method of choice.	72–100% sensitivity, 92–100% specificity (HIV-infected patients)	261–264
BKV	*Polyomaviridae*	dsDNA		Occasional association with meningoencephalitis.		265

HCV RNA has also been found in the CSF of HIV-infected patients but without clear association with CNS disease (Refs. 242, 266, and 267).
[a] See Table 6 footnote a.
[b] PCNSL, primary CNS lymphoma.

bility, showing >90% sensitivity and 48–89% specificity compared to viral isolation, with low specificity just reflecting enterovirus detection in culture-negative CSF samples [93,96,121,219,270,271].

HIV-Related Opportunistic Diseases of the Central Nervous System

Neurological complications have for years afflicted patients with HIV infection. Among these, CNS diseases caused by viruses, including CMV, HSV-1, HSV-2, and VZV encephalitis and progressive multifocal leukoencephalopathy (PML) have played a dominant role. Following widespread use of highly active antiretroviral therapies (HAART), their frequency in the developed world dramatically declined, but they still present a major diagnostic and therapeutic challenge. Molecular detection of CMV-DNA has been shown to be highly sensitive and specific for the diagnosis of CMV encephalitis, a disease reported in as many as one-third of AIDS patients [88,250–254]. The identification of HSV-1, HSV-2, and VZV DNA in CSF has largely contributed to the recognition and clinical characterization of the CNS complications caused by these viruses and has also provided a means for their diagnosis and clinical management [163,245–249]. In PML, the causative agent JCV is demonstrated by PCR in approximately two-thirds of the patients, with higher rates of detection in the advanced stages of disease [261–263,272,273]. CSF PCR for JCV is now used commonly in PML diagnosis, where it has partly replaced the practice of brain biopsy. Recently, however, clearance of JCV DNA from CSF has frequently been observed in patients receiving HAART, in association with stabilization of PML [274,275]. It is thus possible that the rate of JCV DNA detection among PML patients will decrease as a consequence of anti-HIV therapy. Another virus-related CNS disease in HIV-infected patients is primary CNS lymphoma (PCNSL), which is in virtually all cases associated with the presence of EBV in the tumor cells [276]. Studies comparing CSF PCR with histopathological findings at autopsy or on biopsy material reported a striking association between the presence of PCNSL and EBV DNA detection [255–257]. In some patients, EBV DNA could even be detected days or months before the lymphoma manifested itself clinically. Furthermore, EBV DNA in CSF is also associated with CNS localization of systemic non-Hodgkin's lymphomas [255,256,277].

Clinical Applications of Quantitative NA Amplification Techniques

Quantification of viral genomes in the CSF can be important at the time of diagnosis of viral encephalitis or meningitis to obtain prognostic information. In addition, the rapid and continuous development of antiviral compounds has extended the potentiality of molecular techniques to treatment management of patients with viral meningitis or encephalitis. Some of the most significant clinical applications of quantitative molecular techniques are summarized in Table 8 [132,155,156,206a,278–301].

In HSE, the prognostic value of CSF HSV-1 DNA load at the time of diagnosis is still controversial [278–280]. On the other hand, a decrease of DNA levels is commonly observed during acyclovir therapy, indicating that this test could be useful for treatment follow-up of HSE patients [278,280]. A large body of experience has been collected with quantification of HIV-1 RNA in CSF, by the use of PCR, NASBA, and bDNA assays [300]. HIV-1 RNA is detectable at any stage of HIV infection, irrespective of the presence of neurological symptoms, which is the likely consequence of early viral invasion of the CNS. However, CSF viral load is higher in patients with more advanced disease, especially in those with ADC or HIV-related neuropathological abnormalities, presumably resulting from productive HIV infection of brain cells [156,292–294]. CSF viral load is currently used to monitor the local response to anti-HIV therapy [295–297,299,301]. In the majority

Table 8 Examples of Nucleic Acid Quantification in the CSF[a]

Virus	Quantitative technique	Main findings	Refs.
HSV-1	Competitive PCR, real-time PCR	Variable association of DNA levels with HSE outcome; decline of DNA levels following aciclovir therapy in HSE. Higher levels associated with bad prognosis in neonatal encephalitis	278–281b
HSV-2	Real-time PCR	In HSV-2 meningitis: lower DNA levels and narrower range of variation than in HSE	281b
VZV	Real-time PCR	DNA load higher in HZ than in varicella CNS complications	281b
CMV	Semiquantitative PCR, competitive PCR, branched DNA	High DNA levels in VE or PRP and in extensive VE lesions in HIV-infected patients; decline of DNA levels following antiviral therapy in HIV-infected patients.	132, 155, 282–284
EBV	Real-time PCR	High levels in patients with encephalitis or HIV-related SNC lymphoma	206a, 284a
HHV-6	Real-time PCR	Low levels in children with neurological complications	281b
JCV	Semiquantitative PCR, competitive PCR	Association of high DNA levels with poor PML outcome in HIV-infected patients; decline of DNA levels following HAART.	285–289
Enterovirus	Competitive PCR, real-time PCR	Only methodological evaluation, no clinical applications.	290, 291a
HIV-1	Competitive PCR, NASBA, branched DNA	High RNA levels in ADC or HIV-E (59% sensitivity and 93% specificity for HIV-E with a cutoff of 32,000 RNA c/mL); decline of RNA levels following antiretroviral therapy	292–301
HTLV-I	Real-time PCR	Higher proviral DNA load in CSF than in blood in patients with HAM	231

[a] HSE, herpes simplex encephalitis; HZ, herpes zoster; VE, ventriculoencephalitis; PRP, polyradiculopathy; PCNSL, primary CNS lymphoma; NHL, non-Hodgkin's lymphoma; ADC, AIDS dementia complex; HIV-E, HIV encephalitis; PML, progressive multifocal leukoencephalopathy; HAART, highly active antiretroviral therapy; HAM, HTLV-associated myelopathy.

of cases, HAART induces marked decreases of CSF RNA levels. However, a different dynamics of response between CSF and plasma is frequently observed, supporting the hypothesis of compartmentalization of viral replication in the CSF [297,298,302].

7.3 Practical Considerations

It is clear that the study of CSF by molecular techniques has provided an inestimable contribution to diagnosis and clinical management of viral encephalitis and meningitis. Current protocols have in most cases reached satisfactory diagnostic reliability and allowed a diagnosis to be established rapidly, and it is likely that continuous technical development will further improve efficiency and rapidity. On the other hand, important issues concerning interpretation of results and practical aspects are still pending.

Interpretation of NA Amplification Results

An important concern of CSF NA amplification techniques in diagnostics relates to the sporadic finding of nucleic acids without clear association with an underlying CNS disease. An example is the detection of EBV, in both immunocompetent and HIV-infected patients, in concomitance with CNS infections caused by other viruses [272,303–305]. Theoretically, this finding might suggest virus reactivation within the CNS, but sliding of virus through an impaired blood-CSF barrier is possible, especially in the case of a latent virus. Viral genomes have also been found in the CSF of patients with a variety of noninfectious CNS diseases. For instance, JCV, HHV-6, or coronavirus genomes have been demonstrated in patients with multiple sclerosis [244,306,307]. Also in these cases, it is unclear whether these findings are incidental or rather consistent with an etiological role for the virus. Finally, viral nucleic acids can be detected in the presence of small CNS lesions that do not cause overt clinical symptoms [272] This is not infrequent in AIDS patients, in whom more than one CNS disease can be present at the same time. Unlike the above examples, however, these latter observations are consistent with the presence of CNS infection and can be advantageous in allowing an early diagnosis. Taken together, these observations are the likely consequence of the extremely high sensitivity of NA amplification techniques and underline the importance of careful interpretation of NA amplification CSF findings in relation to the individual clinical context. In this regard, it is likely that the use of quantitative methods could become useful in discriminating a fortuitous CSF finding from a clinically significant one.

Although NA amplification techniques have a clearly established diagnostic value in a number of viral CNS infections such as HSE, enterovirus meningitis, or opportunistic diseases in HIV-infected patients, their actual diagnostic potential in less frequent CNS diseases is still unknown. This is the case of CNS disease caused by some arboviruses or of viral encephalitis or meningitis following exanthematic diseases of children, e.g., measles, which are now rarely encountered in the developed world as a result of vaccination. It is hoped that further technical developments and the spread of molecular techniques as well as a systematic collection of rare forms of viral encephalitis and meningitis will help to establish the diagnostic potential of NA amplification techniques also in unusual contexts.

Costs and Savings of NA Amplification Techniques

A potential disadvantage of CSF examination by NA amplification techniques is its cost. If only expenses for technical equipment, reagents, and disposables are taken into account, the cost per sample of a basic PCR may vary between approximately US$20 and US$200, mainly depending on the procedure used. With in-house developed assays, costs can be controlled by avoiding, when possible, expensive procedures for CSF preparation and NA detection and by using assays for the simultaneous examination of multiple viruses. On the other hand, commercially available assays, which have the great advantage of standardization, are quite expensive. However, the costs of NA amplification techniques must be related to the savings of establishing a rapid and correct diagnosis [308]. In HSE, for instance, the savings of molecular techniques compared to the invasive brain biopsy approach are obvious. Furthermore, a PCR-based approach of HSE also seems cost-effective compared to empirical initiation of antiviral therapy. In a recent decision analysis model of HSE treatment, the PCR approach was associated not only with a better outcome but also with significant savings in the use of acyclovir, resulting from a higher rate of correct acyclovir discontinuation in PCR-negative patients [309]. In aseptic meningitis, an early demonstration of an enterovirus as causative agent is associated with a reduction in the number of requests for other diagnostic examinations, in the duration of empirical antibiotic treatments, and in the duration of hospitalization [57,310–312].

Quality Control Assessment

Another important drawback of NA amplification techniques is their lack of standardization. For each virus, different protocols are in use, and this makes it difficult to compare results among laboratories. Quality control assessments, using coded panels of test and control samples distributed to participant laboratories and tested blindly, have been carried out for viruses responsible for CNS infections such as enteroviruses, HSV-1, HSV-2, and JCV [313–317]. Variation in sensitivity of virus detection among laboratories was commonly observed, though there was no or only a weak relationship with technique-related variables. Where commercial techniques were used, their efficiency was comparable to that of in-house methods. Although the majority of the participants seemed to perform satisfactorily, a major problem disclosed by enterovirus and HSV studies was the relatively high rate of false positive results, which was most pronounced in some laboratories [314,317]. Overall, these observations underline the importance of optimization of NA amplification techniques in the individual laboratories and also the need for continuing interlaboratory quality control programs.

7.4 Postamplification Analysis

Besides their use in diagnostics, NA amplification techniques provide the basis for genomic analysis. Following direct amplification from CSF or isolation in cell culture, viruses can be genetically characterized for epidemiological purposes and phylogenetic studies or analyzed for the presence of mutations such as those conferring antiviral drug resistance or neuropathogenic properties. In certain instances, genotypic analysis may be used to recognize CNS diseases caused by unusual viral strains or by new viral pathogens.

Methods

The methods most frequently employed in neurovirological analysis include DNA sequencing, restriction fragment length polymorphism (RFLP), and high stringency hybridization techniques [318]. Other postamplification analysis, e.g., denaturing gradient gel electrophoresis (DGGE), single-strand conformational polymorphism (SSCP), the heteroduplex mobility assay (HMA), and the tracking mobility assay (HMA), can all theoretically be applied to the study of CSF.

Nucleotide sequencing is the most accurate method to acquire information on genome composition. Automated procedures have been developed during recent years that make sequencing relatively easy to carry out (Fig. 13). RFLP is based on nucleic acid cleavage by restriction enzymes to generate DNA fragments of different sizes, which can be visualized by gel electrophoresis [318]. This technique is often employed to distinguish individual viruses or viral strains following PCR with consensus primers flanking common virus regions (Fig. 10) [172,175]. RFLP can also be used for detection of specific point mutations in the genome, provided that the searched for mutation falls within the recognized sequence of the restriction enzyme. There are several examples of hybridization-based methods for the analysis of amplification products, ranging from the classical Southern blot to modern high stringency hybridization techniques. The latter enable recognition of minimal variations in the genome composition such as naturally occurring or drug-induced mutations. An example is the reverse hybridization technique, incorporated into the commercial Line Probe Assay (LIPA), for the study of HIV-1 sequences obtained from the CSF [319]. With this method, probes specific for the codons most frequently involved in drug resistance to RT or protease inhibitors are coated as discrete lines on a nitrocellulose strip. Biotinylated amplification products are captured by the probes, and hybrids are visualized as lines on the strips following their detection by alkaline phosphatase-conjugate streptavidin.

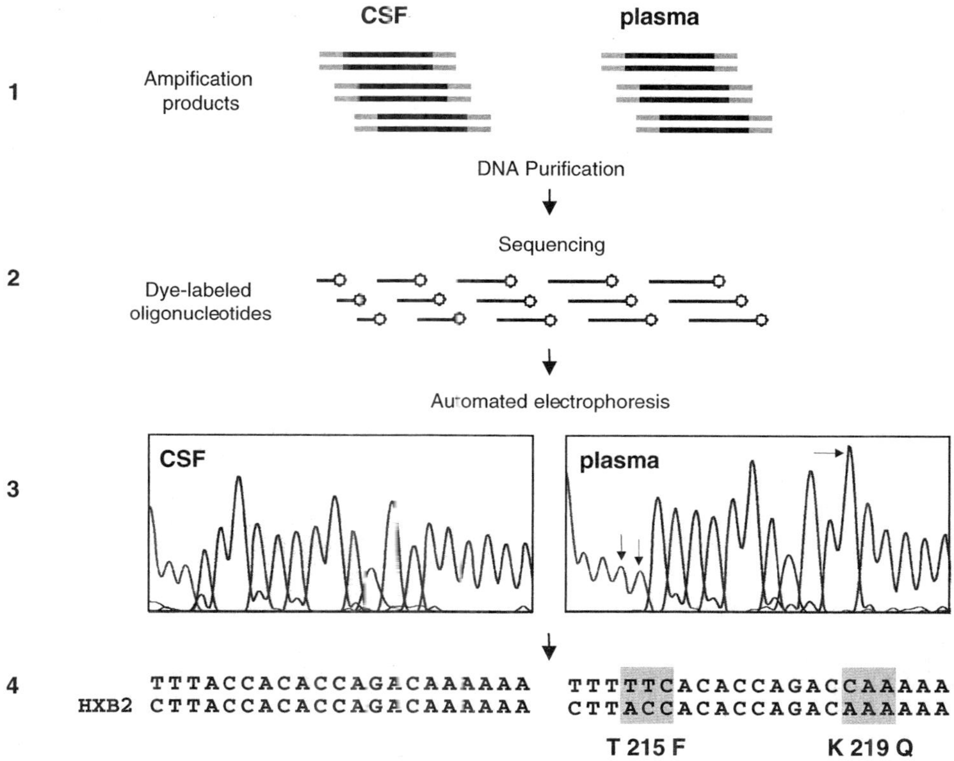

Figure 13 DNA sequencing. A representative example of nucleotide sequencing from CSF using the cycle sequencing procedure. (1) Amplified products are obtained from paired CSF and plasma specimens following nucleic acid extraction, RNA retrotranscription, and PCR amplification of a fragment from the HIV-1 reverse transcriptase (RT) gene. The amplified DNA is purified from unincorporated primers and nucleotides. (2) The purified DNA is added to a reaction mixture containing primers, a DNA polymerase, deoxynucleotides, and dideoxynucleotides (ddNTPs) labeled with four different fluorescent dyes, one for each base, and it is subjected to a new PCR amplification. The ddNTPs act as terminators of growing DNA strands, enabling the production of a large number of dye-labeled oligonucleotides of different lengths. (3) The dye-labeled oligonucleotides are electrophoresed, and each ddNTP or terminator is recognized by a laser scanner. A four-color electropherogram is produced, which is translated into a linear nucleotide sequence by computer software. (4) The final sequence is compared to reference sequences, e.g., HXB2 for HIV-1, subtype B. Three nucleotide mutations, resulting in two amino acid substitutions at codons 215 (threonine → phenylalanine) and 219 (lysine → glutamine) are present in plasma but not in CSF (arrows). Such mutations are known to be associated with zidovudine resistance.

In the near future a valid support for the identification and genomic analysis of viral sequences amplified in the CSF might be represented by DNA microarrays. The principle of the DNA microarrays, or ''DNA chips,'' consists of the placement of a series of probes on a solid surface, such as silicon or glass, in a miniaturized system [320–322]. The use of novel technologies to fix the probes to supports makes it possible to achieve probe densities of 10^4 to 10^6 in a 1 cm^2 chip area, thus allowing rapid analysis of tens to thousands of genes simultaneously (Fig. 14). Despite high costs and the current limited availability of technology and instrumentation, the DNA chip technology is in rapid development in virology, especially in the field of research, e.g., for measuring viral gene expression [323,324]. Sequences from plasma or other clinical samples have initially been tested for epidemiological or diagnostic purposes, such as influenza virus typing [325] or the screen for multiple HIV-1 drug resistance mutations [326].

Clinical Applications

There are a variety of examples of clinical applications of post-PCR analyses in clinical neurovirology, some of which are presented in Table 9 [74,92,122, 159,226,241,302,319,327–352].

Genotypic analyses can be useful for epidemiological studies. With enteroviruses, the development of a molecular typing system based on nucleotide sequencing could

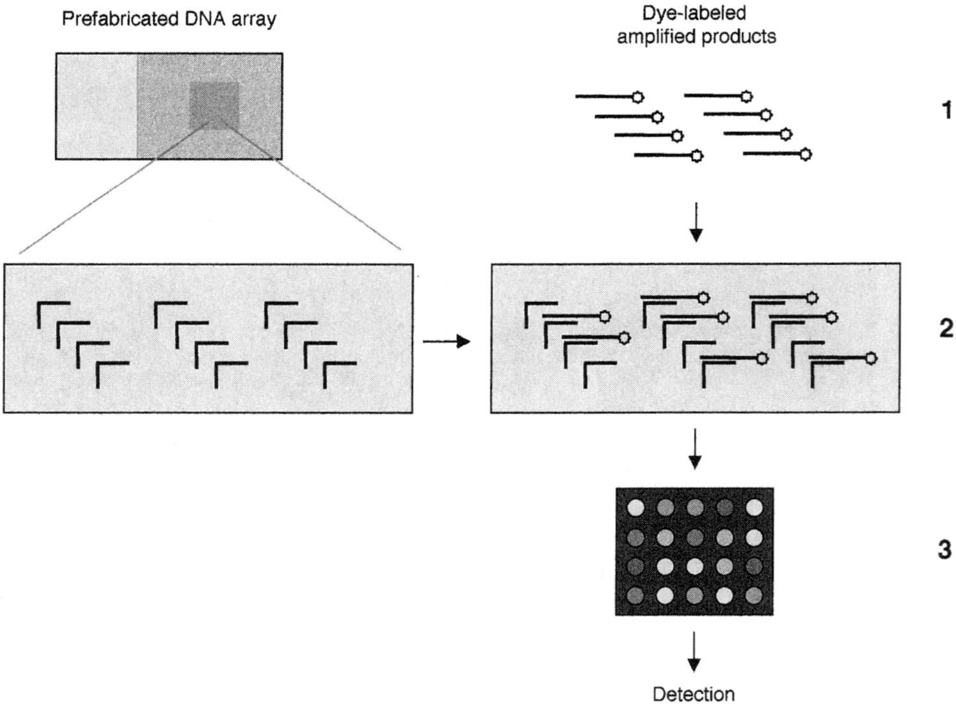

Figure 14 DNA microarrays. (1) Amplified products are labeled with a fluorescent or chemiluminescent molecule. (2) Following denaturation, amplified products are captured by DNA probes fixed on a microchip (e.g., glass, silicon). Microchips are prepared to enable probe densities of up to 10^6 in a 1 cm^2 area. (3) The fluorescence or chemiluminescence is determined by image analysis with an automated instrument or optical microscope.

Table 9 Examples of Postamplification Analysis of CSF

Virus	Genomic region	Method	Use	Main findings	Refs.
HSV-1, HSV-2	gD	DNA sequencing	Identification of possible determinants of neurovirulence	No characteristic signatures found.	327
	Thymidine kinase	DNA sequencing	Identification of possible determinants of neurovirulence	No characteristic signatures found.	328
CMV	UL-97	RFLP; DNA sequencing	Identification of resistance mutations	Detection of resistance mutations in patients with CMV-induced CNS disease on long-term treatment with ganciclovir.	329
Adenovirus	Complete sequence	DNA sequencing	Characterization of a new neurotropic virus	Definition of the complete sequence.	92
JCV	VP-1, large T, intergenic region	DNA sequencing	JCV genotyping (genotypes 1–4)	Geographic distribution of genotypes; association of genotypes 1 and 2 with PML.	330, 331
	Hypervariable noncoding transcriptional control region	DNA sequencing	Distinction of archetypal vs. rearranged virus	Association of rearranged virus with PML; association of rearranged virus with long PML survival.	330, 332–334
Enterovirus	5′ noncoding region, other regions	RFLP; DNA sequencing	Monitoring EV transmission	Detection of common genetic patterns in outbreaks.	335, 336
	5′ noncoding region	RFLP; DNA sequencing	Distinction of poliovirus vs. vaccine virus or non-polio EV	Identification of poliomyelitis, postpolio and post vaccination flaccid paralysis cases.	337–339
	5′ noncoding region, VP-1, other regions	DNA sequencing	Enterovirus genotyping as potential replacement of traditional subtyping	Association between serotypes and genotypes.	340, 341
Mumps	Hemagglutinin neuroaminidase	DNA sequencing	Distinction of vaccine vs. wild-type virus	Identification of CNS disease caused by the vaccine strain Urabe.	342, 343

(continued)

Table 9 *Continued*

Virus	Genomic region	Method	Use	Main findings	Refs.
Measles	Nucleocapsid, hemagglutinin	DNA sequencing	Studies on viral evolution	Genotype switches in viruses circulating during recent decades.	226, 344, 345
Nipah virus	Complete genome	DNA sequencing	Characterization of a new neurotropic virus	Definition of the complete sequence.	74
HIV	pol (RT, protease)	DNA sequencing; LIPA; DNA microarrays	Identification of resistance mutations	Detection of different resistance mutations between CSF and blood in patients on long-term antiretroviral therapy.	159, 302, 319, 346
	env	DNA sequencing	Studies on virus evolution	Detection of different virus evolution between CSF and plasma.	347–349
	env	DNA sequencing	Identification of possible determinants of neurotropism/neurovirulence	Detection of polymorphisms specifically associated with ADC.	350, 351
YFV	Complete genome	DNA sequencing	Distinction of vaccine vs. wild-type virus	Identification of CNS disease caused by vaccine strain 17D.	352

RFLP, restriction fragment length polymorphism; LIPA, line probe assay; ADC, AIDS dementia complex; PML, progressive multifocal leukoencephalopathy; YF, yellow fever.

actually represent an alternative to traditional serotyping, which is time-consuming and labor-intensive and requires isolation of virus. However, serotypes are determined by the presence of critical antigenic epitopes, and there is still incomplete knowledge on their genotypic determinants and therefore of which regions are most suitable to characterize [122]. Nevertheless, a recent analysis of enterovirus VP1 sequences from a large number of clinical isolates showed agreement between antigenic and molecular typing findings [339].

Another field of application of postamplification analyses is pharmacogenomics, the study of genomes for treatment management. In virology, one of the most representative examples is the study of the HIV genome for mutations selected by anti-HIV drugs [353,354]. Drug-resistant viral mutants can be recognized in plasma and in virtually any body site, including the CSF (Fig. 13) [159,302,319,346].

Sequencing of DNA from the CSF has proven useful to recognize CNS diseases caused by attenuated vaccine strains, such as in the meningitis cases caused by the Urabe vaccine in the late 1980s [246,342,343] or in the more recently reported vaccine-induced yellow fever cases [352]. Finally, the genomic sequence of emergent viral pathogens can be defined following their isolation and/or amplification from the CSF. Two recent examples are the identification of a novel paramyxovirus, the Nipah virus, during the 1998 and 1999 encephalitis outbreaks in Malaysia and Singapore [74] and of a novel B adenovirus during the 1997 epidemic of enterovirus 71–associated encephalitis in Malaysia [92].

8 SEROLOGY

Serological techniques can provide indirect evidence of viral CNS infection. Various approaches are used, including the demonstration of an increased IgG titer in plasma specimens collected at distance, the detection of IgM in serum or in CSF, or the detection of an intrathecal IgG synthesis by simultaneous analysis of CSF and serum. The latter procedure is required in CNS infections caused by common or latent viruses such as herpesviruses. With some exceptions, a general disadvantage of serological techniques in the diagnosis of CNS infections is their low sensitivity in the acute stage of disease, due to late appearance of antibodies. Furthermore, serology may lack sensitivity in immunosuppressed patients. On the other hand, these tests may be of unique diagnostic help in CNS diseases with no or low viral replication, including those mainly sustained by immune mechanisms.

Theoretically, all viruses can be investigated by means of serological methods. Practical limitations, however, exist in some instances, such as for enteroviruses, for which the lack of a suitable group-specific antigen would demand a large number of tests.

8.1 Methods

General Serological Procedures

Immune-based assays are currently the most widely used serological techniques. These include IF, RIA, and EIA. Although IF is still used, RIA has largely been replaced by EIA techniques (Fig. 15). The classical procedures, such as neutralization, hemagglutination inhibition (HI), and complement fixation (CF), are less sensitive and more labor-intensive and are therefore less frequently employed. Serological assays can detect any antibody response, irrespective of the Ig class, or be specific for one antibody class, e.g., IgG or IgM.

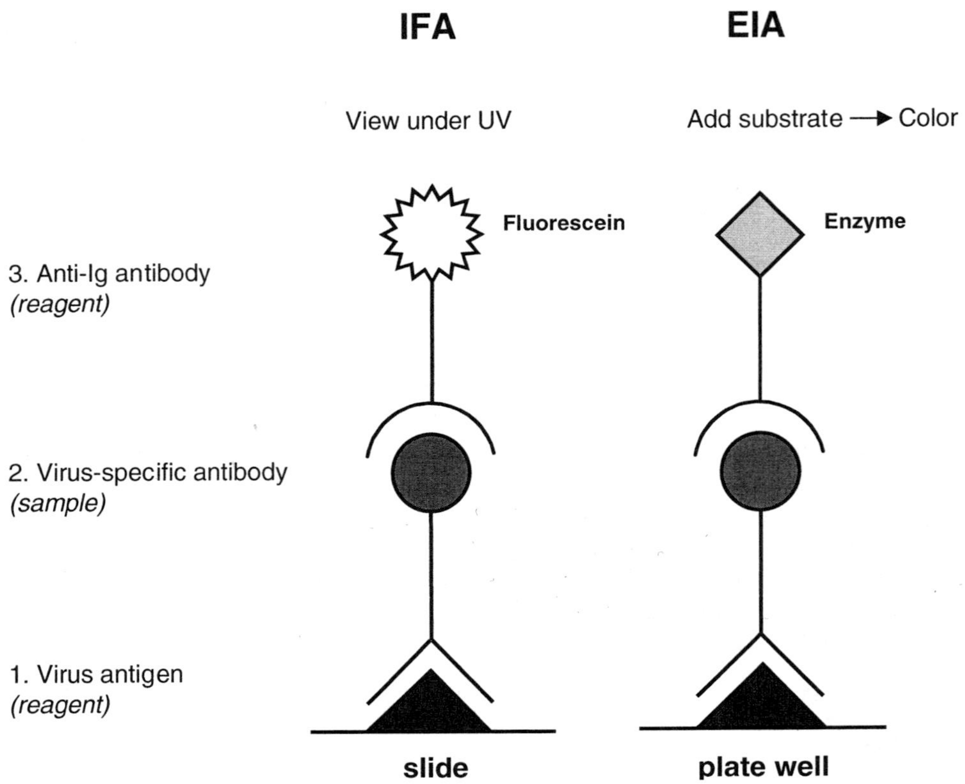

Figure 15 Antibody detection by indirect immunofluorescence (IFA) or enzyme immunoassay (EIA). (1) Virus-specific antigen (reagent) is fixed on a slide (IFA) or attached to a plate microwell (EIA). (2) Virus-specific antibody in the sample binds to the antigen. (3) An anti-Ig antibody (reagent), labeled with either a fluorescence molecule (IFA) or an enzyme (EIA), binds to the virus-specific antibody. In the IFA, fluorescence is detected by UV illumination. In the EIA, an enzyme substrate is added to develop a colorimetric reaction, which is detected by a spectrophotometer. In the direct versions of IFA and EIA, the virus-specific antibody (step 2) is labeled directly with fluorescein or enzyme.

Detection of virus-specific IgM was previously accomplished by measuring the total Ig both before and after procedures that destroy IgM, such as treatment with β-mercaptoethanol. More practical procedures based on the use of an IgM-specific conjugate or the capture assays (Fig. 16) are currently in use. The antibody capture EIA, also referred to as MAC-ELISA, is more sensitive than indirect techniques for IgM detection. Furthermore, this test helps avoid the false positive reactions that occur in the presence of rheumatoid factor, which can form immunocomplexes with virus-specific IgG antibodies and prevent the occurrence of false negative results caused by the presence of high serum titers of virus-specific IgG. Detection of the presence of virus-specific IgM in CSF by this technique is also regarded to reflect intrathecal antibody production.

Intrathecal Antibody Synthesis

For the simultaneous detection and quantification of specific antibody in CSF and serum, sensitive techniques such as EIA are needed. IgG is generally measured, but other antibody

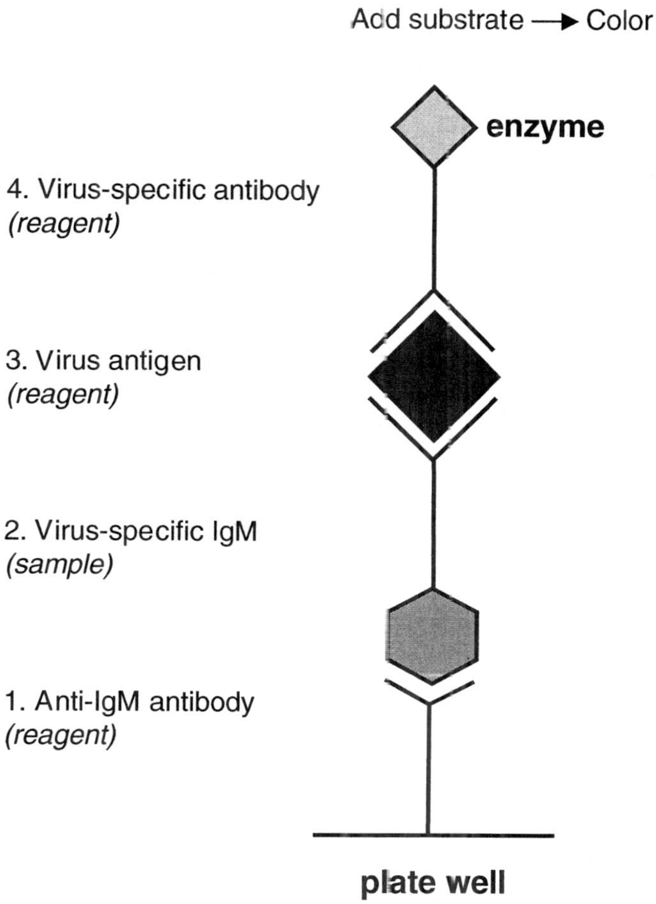

Figure 16 IgM capture enzyme-linked immunosorbent assay (ELISA). (1) IgM-specific antibody (reagent) is bound to the microwell plate. (2) Virus-specific IgM in the sample binds to the anti-IgM antibody. (3) Viral antigen (reagent) binds to the virus-specific antibody. (4) Antigen-specific enzyme-labeled antibody (reagent) binds to the antigen. An enzyme substrate is added to develop a colorimetric reaction, which is detected by a spectrophotometer.

isotypes such as IgM or IgA can be analyzed by the use of specific conjugates. Antibody can be quantified by end-point titration or, using EIA with predetermined dilutions of serum and CSF, by optical density (OD) or by units following interpolation from standard curves. To ascertain whether specific antibodies are produced intrathecally and not passively transferred from serum, it is essential to assess the integrity of the blood-CSF barrier. Antibody titer or OD ratios between CSF and serum are calculated, and these are related to the indicators of blood-CSF barrier damage [355]. An accurate formula has been devised that defines intrathecal antibody production but also allows differentiation of virus-specific from polyclonal locally produced IgG [356]. According to this formula, the ratio between the CSF and serum quotients for specific antibodies (Q_{spec} = CSF Ig $_{spec}$/serum Ig_{spec}) and total IgG (Q_{IgG} = CSF IgG/serum IgG) is calculated and termed the antibody index

(AI = Q_{spec}/Q_{IgG}). To differentiate a local synthesis of polyclonal IgG, Q_{lim}, representing the IgG fraction in CSF originating only from serum, is calculated from the individual albumin quotient ($Q_{lim} = 0.93\ [(Q_{alb} + 6 \times 10^6) - 1.7 \times 10^3]$). In the case of $Q_{IgG} > Q_{lim}$, the AI is corrected likewise (AI = Q_{spec}/Q_{lim}).

A disruption of the brain barriers, or the presence of polyclonal, nonspecific antibody production in the CNS, can also be resolved using IEF, followed by virus-specific antigen-mediated capillary blotting or immunoblotting [357–359]. Immunoblotting and immunoassay methods for detection of intrathecal anti-HSV antibodies have been compared, showing good correlation between the two methods [360].

Alternatively, several viral antigens on one plate or antibody capture techniques are also used. The former method is accomplished by testing the specific CSF/serum antibody ratio for a number of common viruses, e.g. HSV, VZV, CMV, mumps, and measles. In the case of virus-specific intrathecal IgG production, only the virus-specific CSF/serum IgG ratio will be altered. In contrast, polyclonal IgG synthesis or disturbance of the brain barriers will affect more than one or all of the antigen ratios [361]. In the capture assay, the magnitude of the EIA signal is determined by the proportion of specific antibodies in CSF or serum. When this proportion is higher in the CSF, the EIA signal in the CSF will exceed that in the serum, reflecting intrathecal antibody synthesis [362,363]. Because of its simplicity, this test is considered practicable for routine use, and in some instances it has been shown to be superior to indirect EIA indexed against the albumin or IgG ratio [364].

8.2 Clinical Applications

CSF or Serum IgG

In general, the evaluation of a single CSF or serum specimen for IgG titer lacks specificity for diagnosis of CNS diseases. However, testing for IgG can be helpful in the case of unusual viruses. An example is presented by rabies, in which the presence of IgG in CSF or serum is always diagnostic after the first week of illness in patients who have not been vaccinated [365]. In the case of arboviruses that are unusual for a particular geographic area, a single IgG titer is also suggestive of etiology. However, the demonstration of an increased IgG titer in two samples collected at a time distance, i.e., acute and convalescent sera, provides much stronger evidence of recent systemic infection and can thus support a diagnosis of CNS infection. In arboviral encephalitis, a fourfold rise in serum IgG titers by EIA, IFA, CF, HI, or neutralization is regarded as diagnostic of CNS infection [119].

CSF or Serum IgM

Usually the demonstration of IgM in serum provides circumstantial evidence, whereas the presence of virus-specific IgM in CSF is diagnostic of CNS infection. IgM detection of CSF is currently the diagnostic method of choice for most CNS infections caused by arboviruses [119]. Approximately 40% of patients with arbovirus encephalitis or meningitis will show CSF IgM by capture assay within the first 4 days after onset of symptoms, with almost 100% of positive cases by day 10 [76]. On the other hand, arbovirus-specific IgM can be detected in serum for up to 1 year after onset of CNS symptoms [76], and possible cross-reactions between closely related viruses may occur in areas where these are endemic, e.g., dengue virus and Japanese encephalitis (JE) in the Far East. IgM capture ELISA is highly sensitive and specific for diagnosis of a number of arboviral CNS infections, including Japanese encephalitis, La Crosse virus [366,367], West Nile virus [241,368], and tick-borne encephalitis (TBE) [364]. Capture IgM, applied to either CSF or serum, is currently the accepted diagnostic standard in JE. Not only is this test highly reliable, it also reduces the occurrence of cross-reactions between JE and dengue viruses [134,369–371]. IgM detection in CSF has also been shown to be useful in other viral CNS infections, including mumps, enteroviral infection, and rubella [372]. Historical studies revealed the presence of measles-specific IgM antibodies in the CSF of patients with SSPE, with CSF titers higher than those in serum in over one-third of the cases [373]. In the case of suspected CNS involvement during systemic infections such as those caused by CMV or EBV, the demonstration of virus-specific IgM in serum may be diagnostically supportive [195]

Intrathecal Antibody Synthesis

The measurement of virus-specific intrathecal antibody synthesis is, with few exceptions, a powerful method for the diagnosis of viral CNS infections. Although the assays may lack sensitivity at the onset, they may be helpful in later stages, including those cases in which viral replication is no longer detectable by cell culture or PCR. For this reason, this test should be considered as a complement, not an alternative, to the techniques aiming to detect active viral infection.

One of the major applications of the measurement of intrathecal antibody synthesis is in CNS infections caused by herpesviruses. Because herpesviruses are ubiquitous, classical serological procedures lack specificity toward CNS localization. In HSE, virtually all patients develop an intrathecal antibody response to HSV, in most cases detectable within 10–14 days after the onset of neurological symptoms [193,356]. Type-specific EIAs using HSV glycoproteins or synthetic peptides allow the differentiation of HSV-1 and HSV-2 infections. In VZV infections of the CNS, VZV-specific intrathecal antibodies can be detected 5 days or more after the appearance of neurological symptoms [82,374]. In cases of acute neurological disease it is not uncommon to detect the synthesis of intrathecal antibodies against both VZV and HSV, leading one to hypothesize assay cross-reaction or polyclonal B-cell stimulation. However, the possibility of a dual CNS infection is supported by the detection of both HSV and VZV DNA by CSF PCR [375]. Intrathecally produced IgG has also been demonstrated in a variety of neurological conditions such as mumps [376], measles [356,377], rubella [356,378], CMV [356,379,380], PML [381,382], adenoviruses [383], HIV [384], and HTLV-I [115].

10 SUMMARY

An array of virological techniques are nowadays available for CSF analysis to confirm or support an etiological diagnosis of viral encephalitis or meningitis. As a consequence of the molecular revolution in the diagnostic laboratory, the spectrum of conditions that can be recognized has greatly expanded, and diagnostic reliability has significantly improved. Furthermore, molecular techniques have also enabled characterization of neurotropic viruses following recovery of their genomes from the CSF. In addition to NA amplification techniques, serology is maintaining an important diagnostic role, especially in diseases for which the diagnostic potential of the amplification technique is either low or still unknown and in late stages of disease. Viral culture remains a unique option for recovering infectious virus and thus allowing additional biological studies. On the other hand, the diagnostic potential of methods such as antigen detection and CSF cytology is limited to very exceptional instances. Current efforts in diagnostic neurovirology are mainly aimed at further improvement of the diagnostic efficiency of molecular techniques, their speed and standardization, to investigate less common infections and to reduce costs. More ambitiously, CSF diagnostic panels might be available in the near future for rapid investigation of a large number of viral, nonviral, or even noninfectious CNS diseases in the context of various neurological syndromes.

REFERENCES

1. Rubin, S.J. Detection of viruses in spinal fluid. Am. J. Med. 1983, 75, 124–128.
2. Fredricks, D.N.; Relman, D.A. Application of polymerase chain reaction to the diagnosis of infectious diseases. Clin. Infect. Dis. 1999, 29, 475–486; quiz 487–478.
3. Puchhammer-Stockl, E.; Popow-Kraupp, T.; Heinz, F.X.; Mandl, C.W.; Kunz, C. Establishment of PCR for the early diagnosis of herpes simplex encephalitis. J. Med. Virol. 1990, 32, 77–82.
4. Rotbart, H.A. Diagnosis of enteroviral meningitis with the polymerase chain reaction. J. Pediatr. 1990, 117, 85–89.
5. Kaneko, K.; Onodera, O.; Miyatake, T.; Tsuji, S. Rapid diagnosis of tuberculous meningitis by polymerase chain reaction (PCR). Neurology. 1990, 40, 1617–1618.
6. Darnell, R.B. The polymerase chain reaction: application to nervous system disease. Ann. Neurol. 1993, 34, 513–523.
7. Tyler, K.L. Polymerase chain reaction and the diagnosis of viral central nervous system diseases. Ann. Neurol. 1994, 36, 809–811.
8. Weber, T.; Frye, S.; Bodemer, M.; Otto, M.; Luke, W. Clinical implications of nucleic acid amplification methods for the diagnosis of viral infections of the nervous system. J. Neurovirol. 1996, 2, 175–190.
9. Cinque, P.; Cleator, G.M.; Weber, T.; Monteyne, P.; Sindic, C.J.; van Loon, A.M. The role of laboratory investigation in the diagnosis and management of patients with suspected herpes simplex encephalitis: a consensus report. The EU Concerted Action on Virus Meningitis and Encephalitis. J. Neurol. Neurosurg. Psychiatry. 1996, 61, 339–345.
10. Fishman, R.A. Cerebrospinal Fluid in Diseases of the Central Nervous System, 2nd ed.; W.B. Saunders: Philadelphia, 1992.
11. Greenlee, J.E.; Carroll, C. Cerebrospinal fluid in CNS infections. In Infections of the Central Nervous System; Scheld, W.M., Whitley, R.J., Durack, D.T., Eds.; Lippincott-Raven: Philadelphia, 1997, 899–922.
12. Painter, P.C.; Cope, J.Y.; Smith, J.L. Appendix. Reference intervals. Tietz Textbook of Chemical Chemistry; Burtis, C.A., Ashwood, E.R., Eds.; W.B. Saunders: Philadelphia, 1994, 2175–2218.
13. Quincke, H.I. Die Lumbalpunction des Hydrocephalus. Berl. Klin. Wochenschr. 1891, 929–965.
14. Marton, K.I.; Gean, A.D. The spinal tap: a new look at an old test. Ann. Intern. Med. 1986, 104, 840–848.
15. Health and Public Policy Committee, American College of Physicians. The diagnostic spinal tap. Ann. Intern. Med. 1986, 104, 880–886.
16. Kuntz, K.M.; Kokmen, E.; Stevens, J.C.; Miller, P.; Offord, K.P.; Ho, M.M. Post-lumbar puncture headaches: experience in 501 consecutive procedures. Neurology. 1992, 42, 1884–1887.
17. Brocker, R.J. Technique to avoid spinal tap headache. JAMA. 1958, 68, 261–263.
18. Vilming, S.T.; Schrader, H.; Monstad, I. Post-lumbar-puncture headache: the significance of body posture. A controlled study of 300 patients. Cephalalgia. 1988, 78, 75–78.
19. Spriggs, D.A.; Burn, D.J.; French, J.; Cartlidge, N.E.; Bates, D. Is bed rest useful after diagnostic lumbar puncture?. Postgrad. Med. J. 1992, 68, 581–583.
20. Greene, H.M. Lumbar puncture and the prevention of post puncture headache. JAMA. 1926, 86, 391–392.

21. Ready, L.B.; Cuplin, S.; Haschke, R.H.; Nessly, M. Spinal needle determinants of rate of transdural fluid leak. Anesth Analg. 1989, 69, 457–460.
22. Halpern, S.; Preston, R. Postdural puncture headache and spinal needle design. Metaanalyses. Anesthesiology. 1994, 81, 1376–1383.
23. Mihic, D.N. Postspinal headaches, needle surfaces and longitudinal orientation of the dural fibers. Results of a survey. Reg. Anaesth. 1986, 9, 54–56.
24. Braune, H.J.; Huffmann, G.A. A prospective double-blind clinical trial, comparing the sharp Quincke needle (22G) with an ''atraumatic'' needle (22G) in the induction of post-lumbar puncture headache. Acta. Neurol. Scand. 1992, 86, 50–54.
25. Muller, B.; Adelt, K.; Reichmann, H.; Toyka, K. Atraumatic needle reduces the incidence of post-lumbar puncture syndrome. J. Neurol. 1994, 241, 376–380.
26. Thomas, S.R.; Jamieson, D.R.; Muir, K.W. Randomised controlled trial of atraumatic versus standard needles for diagnostic lumbar puncture. Br. Med. J. 2000, 321, 986–990.
27. Strupp, M.; Brandt, T.; Muller, A. Incidence of post-lumbar puncture syndrome reduced by reinserting the stylet: a randomized prospective study of 600 patients. J. Neurol. 1998, 245, 589–592.
28. Tourtellotte, W.W.; Henderson, W.G.; Tucker, R.P.; Gilland, O.; Walker, J.E.; Kokman, E. A randomized, double-blind clinical trial comparing the 22 versus 26 gauge needle in the production of the post-lumbar puncture syndrome in normal individuals. Headache. 1972, 12, 73–78.
29. Carson, D.; Serpell, M. Choosing the best needle for diagnostic lumbar puncture. Neurology. 1996, 47, 33–37.
30. Strachan, A.; Lumbar puncture and headache, J.Train. Aspirating cerebrospinal fluid speeds up procedure. Br. Med. J. 1998, 316, 1018–1019.
31. Serpell, M.G.; Rawal, N. Headaches after diagnostic dural punctures. Br. Med. J. 2000, 321, 973–974.
32. Adler, M.D.; Comi, A.E.; Walker, A.R. Acute hemorrhagic complication of diagnostic lumbar puncture. Pediatr. Emerg. Care. 2001, 17, 184–188.
33. Howard, S.C.; Gajjar, A.; Ribeiro, R.C.; Rivera, G.K.; Rubnitz, J.E.; Sandlund, J.T.; Harrison, P.L.; de Armendi, A.; Dahl, G.V.; Pui, C.H. Safety of lumbar puncture for children with acute lymphoblastic leukemia and thrombocytopenia. JAMA. 2000, 284, 2222–2224.
34. Duffy, G.P. Lumbar puncture in the presence of raised intracranial pressure. Br. Med. J. 1969, 1, 407–409.
35. Spanos, A.; Harrell Jr, F.E.; Differential diagnosis of acute meningitis, D.T.Durack. An analysis of the predictive value of initial observations. JAMA. 1989, 262, 2700–2707.
36. Portnoy, J.M.; Olson, L.C. Normal cerebrospinal fluid values in children: another look. Pediatrics. 1985, 75, 484–487.
37. Adair, C.V.; Gauld, R.L.; Smadel, J.E. Aseptic meningitis, a disease of diverse etiology: clinical and etiological studies on 854 cases. Ann. Intern. Med. 1953, 39, 675–704.
38. Singh, N.; Anderegg, K.A.; Yu, V.L. Significance of hypoglycorrhachia in patients with AIDS and cytomegalovirus meningoencephalitis. Clin. Infect. Dis. 1993, 17, 283–284.
39. Silverman, L.M.; Christenseon, R.H. Amino acids and protein. In Tietz Textbook of Chemical Chemistry; Burtis, C.A., Ashwood, E.R., Eds.; W.B. Saunders: Philadelphia, 1994, 625–725.
40. Bailey, E.M.; Domenico, P.; Cunha, B.A. Bacterial or viral meningitis? Measuring lactate in CSF can help you know quickly. Postgrad. Med. 1990, 88, 217–219, 223.
41. Hsich, G.; Kenney, K.; Gibbs, C.J.; Lee, K.H.; Harrington, M.G. The 14–3–3 brain protein in cerebrospinal fluid as a marker for transmissible spongiform encephalopathies. N. Engl. J. Med. 1996, 335, 924–930.
42. Lebon, P.; Boutin, B.; Dulac, O.; Ponsot, G.; Arthuis, M. Interferon gamma in acute and subacute encephalitis. Br. Med. J. (Clin. Res. Ed.). 1988, 296, 9–11.
43. Brew, B.J.; Bhalla, R.B.; Fleisher, M.; Paul, M.; Khan, A.; Schwartz, M.K.; Price, R.W. Cerebrospinal fluid beta 2 microglobulin in patients infected with human immunodeficiency virus. Neurology. 1989, 39, 830–834.

44. Fuchs, D.; Chiodi, F.; Albert, J.; Asjo, B.; Hagberg, L.; Hausen, A.; Norkrans, G.; Reibnegger, G.; Werner, E.R.; Wachter, H. Neopterin concentrations in cerebrospinal fluid and serum of individuals infected with HIV-1. AIDS. 1989, 3, 285–288.
45. Heyes, M.P.; Rubinow, D.; Lane, C.; Markey, S.P. Cerebrospinal fluid quinolinic acid concentrations are increased in acquired immune deficiency syndrome. Ann. Neurol. 1989, 26, 275–277.
46. Cinque, P.; Vago, L.; Mengozzi, M.; Torri, V.; Ceresa, D.; Vicenzi, E.; Transidico, P.; Vagani, A.; Sozzani, S.; Mantovani, A.; Lazzarin, A.; Poli, G. Elevated cerebrospinal fluid levels of monocyte chemotactic protein-1 correlate with HIV-1 encephalitis and local viral replication. AIDS. 1998, 12, 1327–1332.
47. Conant, K.; Garzino-Demo, A.; Nath, A.; McArthur, J.C.; Halliday, W.; Power, C.; Gallo, R.C.; Major, E.O. Induction of monocyte chemoattractant protein-1 in HIV-1 Tat-stimulated astrocytes and elevation in AIDS dementia. Proc. Natl. Acad. Sci. USA. 1998, 95, 3117–3121.
48. Link, H.; III, G.Tibbling.Principles of albumin and IgG analyses in neurological disorders. Evaluation of IgG synthesis within the central nervous system in multiple sclerosis. Scand. J. Clin. Lab. Invest. 1977, 37, 397–401.
49. Tourtellotte, W.W.; Ma, B.I. Multiple sclerosis: the blood-brain-barrier and the measurement of de novo central nervous system IgG synthesis. Neurology. 1978, 28, 76–83.
50. Reiber, H. The discrimination between different blood-CSF barrier dysfunctions and inflammatory reactions of the CNS by a recent evaluation graph for the protein profile of cerebrospinal fluid. J. Neurol. 1980, 224, 89–99.
51. Andersson, M.; Alvarez-Cermeno, J.; Bernardi, G.; Cogato, I.; Fredman, P.; Frederiksen, J.; Fredrikson, S.; Gallo, P.; Grimaldi, L.M.; Gronning, M. Cerebrospinal fluid in the diagnosis of multiple sclerosis: a consensus report. J. Neurol. Neurosurg. Psychiatry. 1994, 57, 897–902.
52. Koskiniemi, M.; Vaheri, A.; Taskinen, E. Cerebrospinal fluid alterations in herpes simplex virus encephalitis. Rev. Infect. Dis. 1984, 6, 608–618.
53. Mengel, M. The use of the cytocentrifuge in the diagnosis of meningitis. Am. J. Clin. Pathol. 1985, 84, 212–216.
54. Varki, A.P.; Value of second lumbar puncture in confirming a diagnosis of aseptic meningitis, P.Puthuran. A prospective study. Arch. Neurol. 1979, 36, 581–582.
55. Dagan, R.; Jenista, J.A.; Menegus, M.A. Association of clinical presentation, laboratory findings, and virus serotypes with the presence of meningitis in hospitalized infants with enterovirus infection. J. Pediatr. 1988, 113, 975–978.
56. Henquell, C.; Chambon, M.; Bailly, J.L.; Alcaraz, S.; De Champs, C.; Archimbaud, C.; Labbe, A.; Charbonne, F.; Peigue-Lafeuille, H. Prospective analysis of 61 cases of enteroviral meningitis: interest of systematic genome detection in cerebrospinal fluid irrespective of cytologic examination results. J. Clin. Virol. 2001, 21, 29–35.
57. Ramers, C.; Billman, G.; Hartin, M.; Ho, S.; Sawyer, M.H. Impact of a diagnostic cerebrospinal fluid enterovirus polymerase chain reaction test on patient management. JAMA. 2000, 283, 2680–2685.
58. Levitt, L.P.; Rich, T.A.; Kinde, S.W.; Lewis, A.L.; Gates, E.H.; Bond, J.O. Central nervous system mumps. A review of 64 cases. Neurology. 1970, 20, 829–834.
59. Bergstrom, T.; Vahlne, A.; Alestig, K.; Jeansson, S.; Forsgren, M.; Lycke, E. Primary and recurrent herpes simplex virus type 2-induced meningitis. J. Infect. Dis. 1990, 162, 322–330.
60. Aurelius, E.; Forsgren, M.; Skoldenberg, B.; Strannegard, O. Persistent intrathecal immune activation in patients with herpes simplex encephalitis. J. Infect. Dis. 1993, 168, 1248–1252.
61. Appleman, M.E.; Marshall, D.W.; Brey, R.L.; Houk, R.W.; Beatty, D.C.; Winn, R.E.; Melcher, G.P.; Wise, M.G.; Sumaya, C.V.; Boswell, R.N. Cerebrospinal fluid abnormalities in patients without AIDS who are seropositive for the human immunodeficiency virus. J. Infect. Dis. 1988, 158, 193–199.
62. Hollander, H. Cerebrospinal fluid normalities and abnormalities in individuals infected with human immunodeficiency virus. J. Infect. Dis. 1988, 158, 855–858.

63. de Gans, J.; Portegies, P.; Tiessens, G.; Troost, D.; Danner, S.A.; Lange, J.M. Therapy for cytomegalovirus polyradiculomyelitis in patients with AIDS: treatment with ganciclovir. AIDS. 1990, 4, 421–425.
64. Pantoni, L.; Inzitari, D.; Colao, M.G.; De Mayo, E.; Marini, P.; Mazzota, F. Cytomegalovirus encephalitis in a non-immunocompromised patient: CSF diagnosis by in situ hybridization cells. Acta. Neurol. Scand. 1991, 84, 56–58.
65. Musiani, M.; Zerbini, M.; Venturoli, S.; Gentilomi, G.; Borghi, V.; Pietrosemoli, P.; Pecorari, M.; La Placa, M. Rapid diagnosis of cytomegalovirus encephalitis in patients with AIDS using in situ hybridisation. J. Clin. Pathol. 1994, 47, 886–891.
66. Revello, M.G.; Percivalle, E.; Sarasini, A.; Baldanti, F.; Furione, M.; Gerna, G. Diagnosis of human cytomegalovirus infection of the nervous system by pp65 detection in polymorphonuclear leukocytes of cerebrospinal fluid from AIDS patients. J. Infect. Dis. 1994, 170, 1275–1279.
67. Steele, R.W.; Keeney, R.E.; Bracsher, R.W.; Moses, E.B.; Soloff, B.L. Treatment of varicella-zoster meningoencephalitis with acyclovir—demonstration of virus in cerebrospinal fluid by electron microscopy. Am. J. Clin. Pathol. 1983, 80, 57–60.
68. Johnson, G.; Nelson, S.; Petric, M.; Tellier, R. Comprehensive PCR-based assay for detection and species identification of human herpesviruses. J. Clin. Microbiol. 2000, 38, 3274–3279.
69. Andersson, J.; Ehrnst, A.; Larsson, P.H.; Hedlund, K.O.; Norrby, E.; Nybom, R.; Forsgren, M.; Olding-Stenquist, E.; Persson, B. Visualization of defective measles virus particles in cerebrospinal fluid in subacute sclerosing panencephalitis. J. Infect. Dis. 1987, 156, 928–933.
70. Sonnerborg, A.; Nybom, R.; Britton, S.; Ehrnst, A.; Forsgren, M.; Larsson, P.H.; Strannegard, O.; Andersson, J. Detection of cell-free human immunodeficiency virus in cerebrospinal fluid by using immune scanning electron microscopy. J. Infect. Dis. 1989, 159, 1037–1041.
71. Chua, K.B.; Goh, K.J.; Wong, K.T.; Kamarulzaman, A.; Tan, P.S.; Ksiazek, T.G.; Zaki, S.R.; Paul, G.; Lam, S.K.; Tan, C.T. Fatal encephalitis due to Nipah virus among pig-farmers in Malaysia. Lancet. 1999, 354, 1257–1259.
72. Paton, N.I.; Leo, Y.S.; Zaki, S.R.; Auchus, A.P.; Lee, K.E.; Ling, A.E.; Chew, S.K.; Ang, B.; Rollin, P.E.; Umapathi, T.; Sng, I.; Lee, C.C.; Lim, E.; Ksiazek, T.G. Outbreak of Nipah-virus infection among abattoir workers in Singapore. Lancet. 1999, 354, 1253–1256.
73. Chow, V.T.; Tambyah, P.A.; Yeo, W.M.; Phoon, M.C.; Howe, J. Diagnosis of Nipah virus encephalitis by electron microscopy of cerebrospinal fluid. J. Clin. Virol. 2000, 19, 143–147.
74. Chua, K.B.; Bellini, W.J.; Rota, P.A.; Harcourt, B.H.; Tamin, A.; Lam, S.K.; Ksiazek, T.G.; Rollin, P.E.; Zaki, S.R.; Shieh, W.; Goldsmith, C.S.; Gubler, D.J.; Roehrig, J.T.; Eaton, B.; Gould, A.R.; Olson, J.; Field, H.; Daniels, P.; Ling, A.E.; Peters, C.J.; Anderson, L.J.; Mahy, B.W. Nipah virus: a recently emergent deadly paramyxovirus. Science. 2000, 288, 1432–1435.
75. Chonmaitree, T.; Menegus, M.A.; Powell, K.R. The clinical relevance of ''CSF viral culture''. A two-year experience with aseptic meningitis in Rochester, NY. JAMA. 1982, 247, 1843–1847.
76. Calisher, C.H. Medically important arboviruses of the United States and Canada. Clin. Microbiol. Rev. 1994, 7, 89–116
77. McIntosh, K. Diagnostic virology. In Fields Virology; Fields, B.N., Knipe, P.M., Howley, P.M., Eds.; Lippincott-Raven: Philadelphia, 1996, 401–430.
78. Storch, A.G. Methodological overview. In Essentials of Diagnostic Virology; Storch, G.A., Ed.; Churchill Livingstone: New York, 2000, 1–23.
79. Nahmias, A.J.; Whitley, R.J.; Visintine, A.N.; Takei, Y.; Alford Jr, C.A. Herpes simplex virus encephalitis: laboratory evaluations and their diagnostic significance. J. Infect. Dis. 1982, 145, 829–836.
80. Whitley, R.J.; Arvin, A.M Herpes simplex virus infections. In Infectious Diseases of the Fetus and Newborn Infant; Remington, J.S., Klein, J.O., Eds.; W.B. Saunders: Philadelphia, 1995, 354–376.

81. Dix, R.D.; McCarthy, M.; Berger, J.R. Diagnostic value for culture of cerebrospinal fluid from HIV-1-infected individuals for opportunistic viruses: a prospective study. AIDS. 1994, 8, 307–312.
82. Andiman, W.A.; White-Greenwald, M.; Tinghitella, T. Zoster encephalitis. Isolation of virus and measurement of varicella-zoster-specific antibodies in cerebrospinal fluid. Am. J. Med. 1982, 73, 769–772.
83. Peterson, L.R.; Ferguson, R.M. Fatal central nervous system infection with varicella-zoster virus in renal transplant recipients. Transplantation. 1984, 37, 366–368.
84. Snoeck, R.; Gerard, M.; Sadzot-Delvaux, C.; Andrei, G.; Balzarini, J.; Reymen, D.; Ahadi, N.; De Bruyn, J.M.; Piette, J.; Rentier, B. Meningoradiculoneuritis due to acyclovir-resistant varicella zoster virus in an acquired immune deficiency syndrome patient. J. Med. Virol. 1994, 42, 338–347.
85. Echevarria, J.M.; Casas, I.; Martinez-Martin, P. Infections of the nervous system caused by varicella-zoster virus: a review. Intervirology. 1997, 40, 72–84.
86. So, Y.T.; Olney, R.K. Acute lumbosacral polyradiculopathy in acquired immunodeficiency syndrome: experience in 23 patients. Ann. Neurol. 1994, 35, 53–58.
87. Gozlan, J.; el Amrani, M.; Baudrimont, M.; Costagliola, D.; Salord, J.M.; Duvivier, C.; Picard, O.; Meyohas, M.C.; Jacomet, C.; Schneider-Fauveau, V. A prospective evaluation of clinical criteria and polymerase chain reaction assay of cerebrospinal fluid for the diagnosis of cytomegalovirus-related neurological diseases during AIDS. AIDS. 1995, 9, 253–260.
88. Cinque, P.; Cleator, G.M.; Weber, T.; Monteyne, P.; Sindic, C.; Gerna, G.; van Loon, A.M.; Klapper, P.E. Diagnosis and clinical management of neurological disorders caused by cytomegalovirus in AIDS patients. European Union Concerted Action on Virus Meningitis and Encephalitis. J. Neurovirol. 1998, 4, 120–132.
89. Halsted, C.C.; Chang, R.S. Infectious mononucleosis and encephalitis: recovery of EB virus from spinal fluid. Pediatrics. 1979, 64, 257–258.
90. Schiff, J.A.; Schaefer, J.A.; Robinson, J.E. Epstein-Barr virus in cerebrospinal fluid during infectious mononucleosis encephalitis. Yale. J. Biol. Med. 1982, 55, 59–63.
91. Kelsey, D.S. Adenovirus meningoencephalitis. Pediatrics. 1978, 61, 291–293.
92. Cardosa, M.J.; Krishnan, S.; Tio, P.H.; Perera, D.; Wong, S.C. Isolation of subgenus B adenovirus during a fatal outbreak of enterovirus 71-associated hand, foot, and mouth disease in Sibu, Sarawak. Lancet. 1999, 354, 987–991.
93. Yerly, S.; Gervaix, A.; Simonet, V.; Caflisch, M.; Perrin, L.; Wunderli, W. Rapid and sensitive detection of enteroviruses in specimens from patients with aseptic meningitis. J. Clin. Microbiol. 1996, 34, 199–201.
94. Atkinson, P.J.; Sharland, M.; Maguire, H. Predominant enteroviral serotypes causing meningitis. Arch. Dis. Child. 1998, 78, 373–374.
95. Nairn, C.; Clements, G.B. A study of enterovirus isolations in Glasgow from 1977 to 1997. J. Med. Virol. 1999, 58, 304–312.
96. Rotbart, H.A. Enteroviruses. In Manual of Clinical Microbiology; Murray, P.R., Baron, E.J., Pfaller, M.A., Tenover, F.C., Yolken, R.H., Eds.; Am. Soc. Microbiol: Washington, DC, 1999, 990–998.
97. Squadrini, F.; Taparelli, F.; De Rienzo, B.; Giovannini, G.; Pagani, C. Rubella virus isolation from cerebrospinal fluid in postnatal rubella encephalitis. Br. Med. J. 1977, 2, 1329–1330.
98. Abe, T.; Nukada, T.; Hatanaka, H.; Tajima, M.; Hiraiwa, M.; Ushijima, H. Myoclonus in a case of suspected progressive rubella panencephalitis. Arch. Neurol. 1983, 40, 98–100.
99. Dwyer, D.E.; Hueston, L.; Field, P.R.; Cunningham, A.L.; North, K. Acute encephalitis complicating rubella virus infection. Pediatr. Infect. Dis. J. 1992, 11, 238–240.
100. Frey, T.K. Neurological aspects of rubella virus infection. Intervirology. 1997, 40, 167–175.
101. McCullers, J.A.; Facchini, S.; Chesney, P.J.; Webster, R.G. Influenza B virus encephalitis. Clin. Infect. Dis. 1999, 28, 898–900.

102. Hakoda, S.; Nakatani, T. A pregnant woman with influenza A encephalopathy in whom influenza A/Hong Kong virus (H3) was isolated from cerebrospinal fluid. Arch. Intern. Med. 1045, 2000, 160, 1041.
103. Wolontis, S.; Bjorvatn, B. Mumps meningoencephalitis in Stockholm. V. Virus isolations from samples of cerebrospinal fluid and urine—a comparison between some cell systems and typing techniques. Scand. J. Infect. Dis. 1974, 6, 117–123.
104. Donald, P.R.; Burger, P.J.; Becker, W.B. Mumps meningo-encephalitis. S. Afr. Med. J. 1987, 71, 283–285.
105. Poggio, G.P.; Rodriguez, C.; Cisterna, D.; Freire, M.C.; Cello, J. Nested PCR for rapid detection of mumps virus in cerebrospinal fluid from patients with neurological diseases. J. Clin. Microbiol. 2000, 38, 274–278.
106. Wairagkar, N.S.; Gandhi, B.V.; Katrak, S.M.; Shaikh, N.J.; Parikh, P.R.; Wadia, N.H.; Gadkari, D.A. Acute renal failure with neurological involvement in adults associated with measles virus isolation. Lancet. 1999, 354, 992–995.
107. Arisoy, E.S.; Demmler, G.J.; Thakar, S.; Doerr, C. Meningitis due to parainfluenza virus type 3: report of two cases and review. Clin. Infect. Dis. 1993, 17, 995–997.
108. Chua, K.B.; Lam, S.K.; Tan, C.T.; Hooi, P.S.; Goh, K.J.; Chew, N.K.; Tan, K.S.; Kamarulzaman, A.; Wong, K.T. High mortality in Nipah encephalitis is associated with presence of virus in cerebrospinal fluid. Ann. Neurol. 2000, 48, 802–805.
109. Peters, C.J.; Buchmeier, M.; Rollin, P.E.; Ksiazek, T.G. Arenaviruses. In Fields Virology; Fields, B.N., Knipe, P.M., Howley, P.M., Eds.; Lippincott-Raven: Philadelphia, 1996, 1521–1551.
110. Nogueira, Y.L. Morphometric analysis of McCoy cells inoculated with cerebrospinal fluid from patients with rabies. Mem. Inst. Oswaldo Cruz. 1998, 4, 509–514.
111. Spector, S.A.; Hsia, K.; Pratt, D.; Lathey, J.; McCutchan, J.A.; Alcaraz, J.E.; Atkinson, J.H.; Gulevich, S.; Wallace, M. Virologic markers of human immunodeficiency virus type 1 in cerebrospinal fluid, I.Grant. The HIV Neurobehavioral Research Center Group. J. Infect. Dis. 1993, 168, 68–74.
112. Brew, B.J.; Paul, M.O.; Nakajima, G.; Khan, A.; Gallardo, H.; Price, R.W. Cerebrospinal fluid HIV-1 p24 antigen and culture: sensitivity and specificity for AIDS-dementia complex. J. Neurol. Neurosurg. Psychiatry. 1994, 57, 784–789.
113. Pratt, R.D.; Nichols, S.; McKinney, N.; Kwok, S.; Dankner, W.M.; Spector, S.A. Virologic markers of human immunodeficiency virus type 1 in cerebrospinal fluid of infected children. J. Infect. Dis. 1996, 174, 288–293.
114. Andersson, L.M.; Svennerholm, B.; Hagberg, L.; Gisslen, M. Higher HIV-1 RNA cutoff level required in cerebrospinal fluid than in blood to predict positive HIV-1 isolation. J. Med. Virol. 2000, 62, 9–13.
115. McKendall, R.R.; Oas, J.; Lairmore, M.D. HTLV-I-associated myelopathy endemic in Texas-born residents and isolation of virus from CSF cells. Neurology. 1991, 41, 831–836.
116. Sotomayor, E.A.; Josephson, S.L. Isolation of eastern equine encephalitis virus in A549 and MRC-5 cell cultures. Clin. Infect. Dis. 1999, 29, 193–195.
117. Mendoza-Montero, J.; Gamez-Rueca, M.I.; Navarro-Mari, J.M.; de la Rosa-Fraile, M.; Oyonarte-Gomez, S. Infections due to sandfly fever virus serotype Toscana in Spain. Clin. Infect. Dis. 1998, 27, 434–436.
118. Valassina, M.; Meacci, F.; Valensin, P.E.; Cusi, M.G. Detection of neurotropic viruses circulating in Tuscany: the incisive role of Toscana virus. J. Med. Virol. 2000, 60, 86–90.
119. Tsai, T.F. Arboviruses. In Manual of Clinical Microbiology; Murray, P.R., Baron, E.J., Pfaller, M.A., Tenover, F.C., Yolken, R.H., Eds.; Am. Soc. Microbiol: Washington, DC, 1999, 1107–1124.
120. van Regenmortel, M.H.V.; Fauquet, C.M.; Bishop, D.H.L.; Carstens, E.B.; Estes, M.K.; Lemon, S.M.; Maniloff, J.; Mayo, M.A.; McGeoch, D.J.; Pringle, C.R.; Wickner, R.B. Virus

Taxonomy. The Classification and Nomenclature of Viruses. The Seventh Report of the International Committee on Taxonomy of Viruses; Academic Press: San Diego, 2000.
121. Muir, P.; van Loon, A.M. Enterovirus infections of the central nervous system. Intervirology. 1997, 40, 153–166.
122. Muir, P.; Kammerer, U.; Korn, K.; Mulders, M.N.; Poyry, T.; Weissbrich, B.; Kandolf, R.; Cleator, G.M.; Molecular typing of enteroviruses: current status and future requirements, A.M.van Loon. The European Union Concerted Action on Virus Meningitis and Encephalitis. Clin. Microbiol. Rev. 1998, 11, 202–227.
123. Singer, J.I.; Maur, P.R.; Riley, J.P.; Smith, P.B. Management of central nervous system infections during an epidemic of enteroviral aseptic meningitis. J. Pediatr. 1980, 96, 559–563.
124. Lindeman, J.; Muller, W.K.; Versteeg, J.; Bots, G.T.; Peters, A.C. Rapid diagnosis of meningoencephalitis, encephalitis. Immunofluorescent examination of fresh and in vitro cultured cerebrospinal fluid cells. Neurology. 1974, 24, 143–148.
125. Taber, L.H.; Brasier, F.; Couch, R.B.; Greenberg, S.B.; Jones, D.; Knight, V. Diagnosis of herpes simplex virus infection by immunofluorescence. J. Clin. Microbiol. 1976, 3, 309–312.
126. Maltseva, N.; Manovich, Z.; Seletskaya, T.; Kaptsova, T.; Nikulina, V. Rapid diagnosis of viral neuroinfections by immunofluorescent and immunoperoxidase technics. J. Neurol. 1979, 220, 125–130.
127. Coleman, R.M.; Bailey, P.D.; Whitley, R.J.; Keyserling, H.; Nahmias, A.J. ELISA for the detection of herpes simplex virus antigens in the cerebrospinal fluid of patients with encephalitis. J. Virol. Methods. 1983, 7, 117–125.
128. Bos, C.A.; Olding-Stenkvist, E.; Wilterdink, J.B.; Scheffer, A.J. Detection of viral antigens in cerebrospinal fluid of patients with herpes simplex virus encephalitis. J. Med. Virol. 1987, 21, 169–178.
129. Lakeman, F.D.; Koga, J.; Whitley, R.J. Detection of antigen to herpes simplex virus in cerebrospinal fluid from patients with herpes simplex encephalitis. J. Infect. Dis. 1987, 155, 1172–1178.
130. Royal, W.; Selnes, O.A.; Concha, M.; Nance-Sproson, T.E.; McArthur, J.C. Cerebrospinal fluid human immunodeficiency virus type 1 (HIV-1) p24 antigen levels in HIV-1-related dementia. Ann. Neurol. 1994, 36, 32–39.
131. de Gans, J.; Lange, J.M.; Derix, M.M.; de Wolf, F.; Eeftinck Schattenkerk, J.K.; Danner, S.A.; Ongerboer de Visser, B.W.; Cload, P.; Goudsmit, J. Decline of HIV antigen levels in cerebrospinal fluid during treatment with low-dose zidovudine. AIDS. 1988, 2, 37–40.
132. Flood, J.; Drew, W.L.; Miner, R.; Jekic-McMullen, D.; Shen, L.P.; Kolberg, J.; Garvey, J.; Follansbee, S.; Poscher, M. Diagnosis of cytomegalovirus (CMV) polyradiculopathy and documentation of in vivo anti-CMV activity in cerebrospinal fluid by using branched DNA signal amplification and antigen assays. J. Infect. Dis. 1997, 176, 348–352.
133. Mathur, A.; Kumar, R.; Sharma, S.; Kulshreshtha, R.; Kumar, A.; Chaturvedi, U.C. Rapid diagnosis of Japanese encephalitis by immunofluorescent examination of cerebrospinal fluid. Indian J. Med. Res. 1990, 91, 1–4.
134. Gajanana, A.; Samuel, P.P.; Thenmozhi, V.; Rajendran, R. An appraisal of some recent diagnostic assays for Japanese encephalitis. Southeast Asian J. Trop. Med. Public Health. 1996, 27, 673–679.
135. Yolken, R.H.; Torsch, V. Enzyme-linked immunosorbent assay for the detection and identification of coxsackie B antigen in tissue cultures and clinical specimens. J. Med. Virol. 1980, 6, 45–52.
136. Yolken, R.H.; Torsch, V.M. Enzyme-linked immunosorbent assay for detection and identification of coxsackieviruses A. Infect. Immun. 1981, 31, 742–750.
137. Boyd, J.F.; Vince-Ribaric, V. The examination of cerebrospinal fluid cells by fluorescent antibody staining to detect mumps antigen. Scand. J. Infect. Dis. 1973, 5, 7–15.
138. Dayan, A.D.; Stokes, M.I. Immunofluorescent detection of measles-virus antigens in cerebrospinal-fluid cells in subacute sclerosing panencephalitis. Lancet. 1971, 1, 891–892.

139. Tedder, D.G.; Ashley, R.; Tyler, K.L.; Levin, M.J. Herpes simplex virus infection as a cause of benign recurrent lymphocytic meningitis. Ann. Intern. Med. 1994, 121, 334–338.
140. Schlesinger, Y.; Buller, R.S.; Brunstrom, J.E.; Moran, C.J.; Storch, G.A. Expanded spectrum of herpes simplex encephalitis in childhood. J. Pediatr. 1995, 126, 234–241.
141. De Vincenzo, J.P.; Thorne, G. Mild herpes simplex encephalitis diagnosed by polymerase chain reaction: a case report and review. Pediatr. Infect. Dis. J. 1994, 13, 662–664.
142. Domingues, R.B.; Tsanaclis, A.M.; Pannuti, C.S.; Mayo, M.S.; Lakeman, F.D. Evaluation of the range of clinical presentations of herpes simplex encephalitis by using polymerase chain reaction assay of cerebrospinal fluid samples. Clin. Infect. Dis. 1997, 25, 86–91.
143. Fodor, P.A.; Levin, M.J.; Weinberg, A.; Sandberg, E.; Sylman, J.; Tyler, K.L. Atypical herpes simplex virus encephalitis diagnosed by PCR amplification of viral DNA from CSF. Neurology. 1998, 51, 554–559.
144. Arribas, J.R.; Storch, G.A.; Clifford, D.B.; Tselis, A.C. Cytomegalovirus encephalitis. Ann. Intern. Med. 1996, 125, 577–587.
145. Kondo, K.; Nagafuji, H.; Hata, A.; Tomomori, C.; Yamanishi, K. Association of human herpesvirus 6 infection of the central nervous system with recurrence of febrile convulsions. J. Infect. Dis. 1993, 167, 1197–1200.
146. Suga, S.; Yoshikawa, T.; Asano, Y.; Kozawa, T.; Nakashima, T.; Kobayashi, I.; Yazaki, T.; Yamamoto, H.; Kajita, Y.; Ozaki, T. Clinical and virological analyses of 21 infants with exanthem subitum (roseola infantum) and central nervous system complications. Ann. Neurol. 1993, 33, 597–603.
147. Barah, F.; Vallely, P.J.; Chiswick, M.L.; Cleator, G.M.; Kerr, J.R. Association of human parvovirus B19 infection with acute meningoencephalitis. Lancet. 2001, 358, 729–730.
148. Studahl, M.; Bergstrom, T.; Ekeland-Sjoberg, K.; Ricksten, A. Detection of cytomegalovirus DNA in cerebrospinal fluid in immunocompetent patients as a sign of active infection. J. Med. Virol. 1993, 46, 274–280.
149. Ushijima, H.; Xin, K.Q.; Nishimura, S.; Morikawa, S.; Abe, T. Detection and sequencing of rotavirus VP7 gene from human materials (stools, sera, cerebrospinal fluids, and throat swabs) by reverse transcription and PCR. J. Clin. Microbiol. 1993, 32, 2893–2897.
150. McCullers, J.A.; Lakeman, F.D.; Whitley, R.J. Human herpesvirus 6 is associated with focal encephalitis. Clin. Infect. Dis. 1993, 21, 571–576.
151. Tang, Y.W.; Persing, D.H. Molecular detection and identification of microorganisms. In Manual of Clinical Microbiology; Murray, P.R., Baron, E.J., Pfaller, M.A., Tenover, F.C., Yolken, R.H., Eds.; Am. Soc. Microbiol: Washington, DC, 1999, 215–244.
152. Mullis, K.B.; Faloona, F.A. Specific synthesis of DNA in vitro via a polymerase-catalyzed chain reaction. Methods. Enzymol. 1987, 155, 335–350.
153. Kievits, T.; van Gemen, B.; van Strijp, D.; Schukkink, R.; Dircks, M.; Adriaanse, H.; Malek, L.; Sooknanan, R.; Lens, P. NASBA isothermal enzymatic in vitro nucleic acid amplification optimized for the diagnosis of HIV-1 infection. J. Virol. Methods. 1991, 35, 273–286.
154. Zhang, F.; Tetali, S.; Wang, X.P.; Kaplan, M.H.; Cromme, F.V.; Ginocchio, C.C. Detection of human cytomegalovirus pp67 late gene transcripts in cerebrospinal fluid of human immunodeficiency virus type 1-infected patients by nucleic acid sequence-based amplification. J. Clin. Microbiol. 2000, 38, 1920–1925.
155. Bestetti, A.; Pierotti, C.; Terreni, M.; Zappa, A.; Vago, L.; Lazzarin, A.; Cinqu, P. Comparison of three nucleic acid amplification assays of cerebrospinal fluid for diagnosis of cytomegalovirus encephalitis. J. Clin. Microbiol. 2001, 39, 1148–1151.
156. McArthur, J.C.; McClernon, D.R.; Cronin, M.F.; Nance-Sproson, T.E.; Saah, A.J.; St Clair, M.; Lanier, E.R. Relationship between human immunodeficiency virus-associated dementia and viral load in cerebrospinal fluid and brain. Ann. Neurol. 1997, 42, 689–698.
157. Shepard, R.N.; Schock, J.; Robertson, K.; Shugars, D.C.; Dyer, J.; Vernazza, P.; Hall, C.; Cohen, M.S.; Fiscus, S.A. Quantitation of human immunodeficiency virus type 1 RNA in different biological compartments. J. Clin. Microbiol. 2000, 38, 1414–1418.

157a. Lanciotti, R.S.; Kerst, A.J. Nucleic acid sequence-based amplification assays for rapid detection of West Nile and St. Louis encephalitis viruses. J. Clin. Microbiol. 2001, 39, 4506–4513.
157b. Fox, J.D.; Han, S.; Samuelson, A.; Zhang, Y.; Neale, M.L.; Westmoreland, D. Development and evaluation of nucleic acid sequence based amplification (NASBA) for diagnosis of enterovirus infections using the NucliSens Basic Kit. J. Clin. Virol. 2002, 24, 117–130.
158. Urdea, M.S. Branched DNA signal amplification. Biotechnology (NY). 1994, 12, 926–928.
159. Stingele, K.; Haas, J.; Zimmermann, T.; Stingele, R.; Hubsch-Muller, C.; Freitag, M.; Storch-Hagenlocher, B.; Hartmann, M.; Wildemann, B. Independent HIV replication in paired CSF and blood viral isolates during antiretroviral therapy. Neurology. 2001, 56, 355–361.
160. Casas, I.; Powell, L.; Klapper, P.E.; Cleator, G.M. New method for the extraction of viral RNA and DNA from cerebrospinal fluid for use in the polymerase chain reaction assay. J. Virol. Methods. 1995, 53, 25–36.
160a. Fahle, G.A.; Fischer, S.H. Comparison of six commercial DNA extraction kits for recovery of cytomegalovirus DNA from spiked human specimens. J. Clin. Microbiol. 2000, 38, 3860–3863.
161. Kimura, H.; Shibata, M.; Kuzushima, K.; Nishikawa, K.; Nishiyama, Y.; Morishima, T. Detection and direct typing of herpes simplex virus by polymerase chain reaction. Med. Microbiol. Immunol. (Berl). 1990, 179, 177–184.
162. Cassinotti, P.; Mietz, H.; Siegl, G. Suitability and clinical application of a multiplex nested PCR assay for the diagnosis of herpes simplex virus infections. J. Med. Virol. 1996, 50, 75–81.
163. Cinque, P.; Vago, L.; Marenzi, R.; Giudici, B.; Weber, T.; Corradini, R.; Ceresa, D.; Lazzarin, A.; Linde, A. Herpes simplex virus infections of the central nervous system in human immunodeficiency virus-infected patients: clinical management by polymerase chain reaction assay of cerebrospinal fluid. Clin. Infect. Dis. 1998, 27, 303–309.
164. Tenorio, A.; Echevarria, J.E.; Casas, I.; Echevarria, J.M.; Tabares, E. Detection and typing of human herpesviruses by multiplex polymerase chain reaction. J. Virol. Methods. 1993, 44, 261–269.
165. Baron, J.M.; Rubben, A.; Grussendorf-Conen, E.I. Evaluation of a new general primer pair for rapid detection and differentiation of HSV-1, HSV-2, and VZV by polymerase chain reaction. J. Med. Virol. 1996, 49, 279–282.
166. Pozo, F.; Tenorio, A. Detection and typing of lymphotropic herpesviruses by multiplex polymerase chain reaction. J. Virol. Methods. 1999, 79, 9–19.
167. Quereda, C.; Corral, I.; Laguna, F.; Valencia, M.E.; Tenorio, A.; Echeverria, J.E.; Navas, E.; Martin-Davila, P.; Moreno, A.; Moreno, V.; Gonzalez-Lahoz, J.M.; Arribas, J.R.; Guerrero, A. Diagnostic utility of a multiplex herpesvirus PCR assay performed with cerebrospinal fluid from human immunodeficiency virus-infected patients with neurological disorders. J. Clin. Microbiol. 2000, 38, 3061–3067.
168. Read, S.J.; Kurtz, J.B. Laboratory diagnosis of common viral infections of the central nervous system by using a single multiplex PCR screening assay. J. Clin. Microbiol. 1999, 37, 1352–1355.
169. Casas, I.; Pozo, F.; Trallero, G.; Echevarria, J.M.; Tenorio, A. Viral diagnosis of neurological infection by RT multiplex PCR: a search for entero- and herpesviruses in a prospective study. J. Med. Virol. 1999, 57, 145–151.
170. Roberts, T.C.; Storch, G.A. Multiplex PCR for diagnosis of AIDS-related central nervous system lymphoma and toxoplasmosis. J. Clin. Microbiol. 1997, 35, 268–269.
171. Kuno, G. Universal diagnostic RT-PCR protocol for arboviruses. J. Virol. Methods. 1998, 72, 27–41.
172. Rozenberg, F.; Lebon, P. Amplification and characterization of herpesvirus DNA in cerebrospinal fluid from patients with acute encephalitis. J. Clin. Microbiol. 1991, 29, 2412–2417.
173. Minjolle, S.; Michelet, C.; Jusselin, I.; Joannes, M.; Cartier, F.; Colimon, R. Amplification of the six major human herpesviruses from cerebrospinal fluid by a single PCR. J. Clin. Microbiol. 1999, 37, 950–953.

174. Bouquillon, C.; Dewilde, A.; Andreoletti, L.; Lambert, V.; Chieux, V.; Gerard, Y.; Lion, G.; Bocket, L.; Wattre, P. Simultaneous detection of 6 human herpesviruses in cerebrospinal fluid and aqueous fluid by a single PCR using stair primers. J. Med. Virol. 2000, 62, 349–353.
175. Arthur, R.R.; Dagostin, S.; Shah, K.V. Detection of BK virus and JC virus in urine and brain tissue by the polymerase chain reaction. J. Clin. Microbiol. 1989, 27, 1174–1179.
176. Fedele, C.G.; Ciardi, M.; Delia, S.; Echevarria, J.M.; Tenorio, A. Multiplex polymerase chain reaction for the simultaneous detection and typing of polymavirus JC, BK and SV40 DNA in clinical samples. J. Virol. Methods. 1999, 82, 137–144.
177. Clementi, M.; Menzo, S.; Bagnarelli, P.; Valenza, A.; Paolucci, S.; Sampaolesi, R.; Manzin, A.; Varaldo, P.E. Clinical use of quantitative molecular methods in studying human immunodeficiency virus type 1 infection. Clin. Microbiol. Rev. 1996, 9, 135–147.
178. Hodinka, R.L. The clinical utility of viral quantitation using molecular methods. Clin. Diagn. Virol. 1998, 10, 25–47.
179. Preiser, W.; Elzinger, B.; Brink, N.S. Quantitative molecular virology in patient management. J. Clin. Pathol. 2000, 53, 76–83.
180. Holland, P.M.; Abramson, R.D.; Watson, R.; Gelfand, D.H. Detection of specific polymerase chain reaction product by utilizing the 5′→3′ exonuclease activity of Thermus aquaticus DNA polymerase. Proc. Natl. Acad. Sci. USA. 1991, 88, 7276–7280.
181. Heid, C.A.; Stevens, J.; Livak, K.J.; Williams, P.M. Real time quantitative PCR. Genome. Res. 1996, 6, 986–994.
182. Higuchi, R.; Fockler, C.; Dollinger, G.; Watson, R. Kinetic PCR analysis: real-time monitoring of DNA amplification reactions. Biotechnology (NY). 1993, 11, 1026–1030.
183. Wittwer, C.T.; Ririe, K.M.; Andrew, R.V.; David, D.A.; Gundry, R.A.; Balis, U.J. The LightCycler: a microvolume multisample fluorimeter with rapid temperature control. Biotechniques. 1997, 22, 176–181.
184. Wittwer, C.T.; Herrmann, M.G.; Moss, A.A.; Rasmussen, R.P. Continuous fluorescence monitoring of rapid cycle DNA amplification. Biotechniques. 1997, 22, 130–131, 134–138.
185. Persing, D.H. Polymerase chain reaction: trenches to benches. J. Clin. Microbiol. 1991, 29, 1281–1285.
186. Longo, M.C.; Berninger, M.S.; Hartley, J.L. Use of uracil DNA glycosylase to control carry-over contamination in polymerase chain reactions. Gene. 1990, 93, 125–128.
187. Tang, Y.W.; Mitchell, P.S.; Espy, M.J.; Smith, T.F.; Persing, D.H. Molecular diagnosis of herpes simplex virus infections in the central nervous system. J. Clin. Microbiol. 1999, 37, 2127–2136.
188. Wiedbrauk, D.L.; Cunningham, W. Stability of herpes simplex virus DNA in cerebrospinal fluid specimens. Diagn. Mol. Pathol. 1996, 5, 249–252.
189. Moudgil, T.; Daar, E.S. Infectious decay of human immunodeficiency virus type 1 in plasma. J. Infect. Dis. 1993, 167, 210–212.
190. Holodniy, M.; Mole, L.; Yen-Lieberman, B.; Margolis, D.; Starkey, C.; Carroll, R.; Spahlinger, T.; Todd, J.; Jackson J.B. Comparative stabilities of quantitative human immunodeficiency virus RNA in plasma from samples collected in VACUTAINER CPT, VACUTAINER PPT, and standard VACUTAINER tubes. J. Clin. Microbiol. 1995, 33, 1562–1566.
191. Rotbart, H.A.; Levin, M.J.; Villarreal, L.P.; Tracy, S.M.; Semler, B.L.; Wimmer, E. Factors affecting the detection of enteroviruses in cerebrospinal fluid with coxsackievirus B3 and poliovirus 1 cDNA probes. J. Clin. Microbiol. 1985, 22, 220–224.
192. Ahmad, M.; Tashima, K.T.; Caliendo, A.M.; Flanigan, T.P. Cerebrospinal fluid and plasma HIV-1 RNA stability at 4 degrees C. AIDS. 1999, 13, 1281–1282.
192a. Singer, E.J.; Aronow, H.A.; Lee, S.Y.; Hinkin, C.H.; Lazarus, T. Stability of human immunodeficiency virus type 1 RNA in cerebrospinal fluid determined with the AMPLICOR HIV-1 MONITOR test, version 1.5 (ultrasensitive). J. Clin. Microbiol. 2002, 40, 3863–3864.
193. Aurelius, E.; Johnsson, B.; Skoldenberg, B.; Staland, A.; Forsgren, M. Rapid diagnosis of herpes simplex encephalitis by nested polymerase chain reaction assay of cerebrospinal fluid. Lancet. 1991, 337, 189–192.

194. Lakeman, F.D.; Whitley, R.J. Diagnosis of herpes simplex encephalitis: application of polymerase chain reaction to cerebrospinal fluid from brain-biopsied patients and correlation with disease. National Institute of Allergy and Infectious Diseases Collaborative Antiviral Study Group. J. Infect. Dis. 1995, 171, 857–863.
195. Linde, A.; Klapper, P.E.; Monteyne, P.; Echevarria, J.M.; Cinque, P.; Rozenberg, F.; Vestergaard, B.F.; Ciardi, M.; Lebon, P.; Cleator, G.M. Specific diagnostic methods for herpesvirus infections of the central nervous system: a consensus review by the European Union Concerted Action on Virus Meningitis and Encephalitis. Clin. Diagn. Virol. 1997, 8, 83–104.
196. Kimura, H.; Futamura, M.; Kito, H.; Ando, T.; Goto, M.; Kuzushima, K.; Shibata, M.; Morishima, T. Detection of viral DNA in neonatal herpes simplex virus infections: frequent and prolonged presence in serum and cerebrospinal fluid. J. Infect. Dis. 1991, 164, 289–293.
197. Kimberlin, D.W.; Lakeman, F.D.; Arvin, A.M.; Prober, C.G.; Corey, L.; Powell, D.A.; Burchett, S.K.; Jacobs, R.F.; Starr, S.E.; Whitley, R.J. Application of the polymerase chain reaction to the diagnosis and management of neonatal herpes simplex virus disease. National Institute of Allergy and Infectious Diseases Collaborative Antiviral Study Group. J. Infect. Dis. 1996, 174, 1162–1167.
198. Aurelius, E.; Johansson, B.; Skoldenberg, B.; Forsgren, M. Encephalitis in immunocompetent patients due to herpes simplex virus type 1 or 2 as determined by type-specific polymerase chain reaction and antibody assays of cerebrospinal fluid. J. Med. Virol. 1993, 39, 179–186.
199. Schlesinger, Y.; Tebas, P.; Buller, R.S.; Storch, G.A. Herpes simplex virus type 2 meningitis in the absence of genital lesions: improved recognition with use of the polymerase chain reaction. Clin. Infect. Dis. 1995, 20, 842–848.
200. Puchhammer-Stockl, E.; Popow-Kraupp, T.; Heinz, F.X.; Mandl, C.W.; Kunz, C. Detection of varicella-zoster virus DNA by polymerase chain reaction in the cerebrospinal fluid of patients suffering from neurological complications associated with chicken pox or herpes zoster. J. Clin. Microbiol. 1991, 29, 1513–1516.
201. Echevarria, J.M.; Casas, I.; Tenorio, A.; de Ory, F.; Martinez-Martin, P. Detection of varicella-zoster virus-specific DNA sequences in cerebrospinal fluid from patients with acute aseptic meningitis and no cutaneous lesions. J. Med. Virol. 1994, 43, 331–335.
202. Haanpaa, M.; Dastidar, P.; Weinberg, A.; Levin, M.; Miettinen, A.; Lapinlampi, A.; Laippala, P.; Nurmikko, T. CSF and MRI findings in patients with acute herpes zoster. Neurology. 1998, 51, 1405–1411.
203. Darin, N.; Bergstrom, T.; Fast, A.; Kyllerman, M. Clinical, serological and PCR evidence of cytomegalovirus infection in the central nervous system in infancy and childhood. Neuropediatrics. 1994, 25, 316–322.
204. Troendle Atkins, J.; Demmler, G.J.; Williamson, W.D.; McDonald, J.M.; Istas, A.S.; Buffone, G.J. Polymerase chain reaction to detect cytomegalovirus DNA in the cerebrospinal fluid of neonates with congenital infection. J. Infect. Dis. 1994, 169, 1334–1337.
205. Imai, S.; Usui, N.; Sugiura, M.; Osato, T.; Sato, T.; Tsutsumi, H.; Tachi, N.; Nakata, S.; Yamanaka, T.; Chiba, S. Epstein-Barr virus genomic sequences and specific antibodies in cerebrospinal fluid in children with neurologic complications of acute and reactivated EBV infections. J. Med. Virol. 1993, 40, 278–284.
206. Landgren, M.; Kyllerman, M.; Bergstrom, T.; Dotevall, L.; Ljungstrom, L.; Ricksten, A. Diagnosis of Epstein-Barr virus-induced central nervous system infections by DNA amplification from cerebrospinal fluid. Ann. Neurol. 1994, 35, 631–635.
206a. Weinberg, A.; Li, S.; Palmer, M.; Tyler, K.L. Quantitative CSF PCR in Epstein-Barr virus infections of the central nervous system. Ann. Neurol. 2002, 2, 543–548.
207. Caserta, M.T.; Hall, C.B.; Schnabel, K.; McIntyre, K.; Long, C.; Costanzo, M.; Dewhurst, S.; Insel, R.; Epstein, L.G. Neuroinvasion and persistence of human herpesvirus 6 in children. J. Infect. Dis. 1994, 170, 1586–1589.
208. Hall, C.B.; Caserta, M.T.; Schnabel, K.C.; Long, C.; Epstein, L.G.; Insel, R.A.; Dewhurst, S. Persistence of human herpesvirus 6 according to site and variant: possible greater neurotropism of variant A. Clin. Infect. Dis. 1998, 26, 132–137.

209. Torigoe, S.; Koide, W.; Yamada, M.; Miyashiro, E.; Tanaka-Taya, K.; Yamanishi, K. Human herpesvirus 7 infection associated with central nervous system manifestations. J. Pediatr. 1996, 129, 301–305.
210. van den Berg, J.S.; van Zeijl, J.H.; Rotteveel, J.J.; Melchers, W.J.; Gabreels, F.J.; Galama, J.M. Neuroinvasion by human herpesvirus type 7 in a case of exanthem subitum with severe neurologic manifestations. Neurology. 1999, 52, 1077–1079.
211. Yoshikawa, T.; Ihira, M.; Suzuki, K.; Suga, S.; Matsubara, T.; Furukawa, S.; Asano, Y. Invasion by human herpesvirus 6 and human herpesvirus 7 of the central nervous system in patients with neurological signs and symptoms. Arch. Dis. Child. 2000, 83, 170–171.
212. Pohl-Koppe, A.; Blay, M.; Jager, G.; Weiss, M. Human herpes virus type 7 DNA in the cerebrospinal fluid of children with central nervous system diseases. Eur. J. Pediatr. 2001, 160, 351–358.
212a. Steininger, C.; Popow-Kraupp, T.; Laferl, H.; Seiser, A.; Godl, I.; Djamshidian, S.; Puchhammer-Stockl, E. Acute encephalopathy associated with influenza A virus infection. Clin. Infect. Dis. 2003, 36, 567–574.
213. Voltz, R.; Jager, G.; Seelos, K.; Fuhry, L.; Hohlfeld, R. BK virus encephalitis in an immunocompetent patient. Arch. Neurol. 1996, 53, 101–103.
213a. Behzad-Behbahani, A.; Klapper, P.E.; Vallely, P.J.; Cleator, G.M. BK virus DNA in CSF of immunocompetent and immunocompromised patients. Arch. Dis. Child. 2003, 88, 174–175.
214. Druschky, K.; Walloch, J.; Heckmann, J.; Schmidt, B.; Stefan, H.; Neundorfer, B. Chronic parvovirus B-19 meningoencephalitis with additional detection of Epstein-Barr virus DNA in the cerebrospinal fluid of an immunocompetent patient. J. Neurovirol. 2000, 6, 418–422.
215. Okumura, A.; Ichikawa, T. Aseptic meningitis caused by human parvovirus B19. Arch. Dis. Child. 1993, 68, 784–785.
216. Nishimura, S.; Ushijima, H.; Nishimura, S.; Shiraishi, H.; Kanazawa, C.; Abe, T.; Kaneko, K.; Fukuyama, Y. Detection of rotavirus in cerebrospinal fluid and blood of patients with convulsions and gastroenteritis by means of the reverse transcription polymerase chain reaction. Brain. Dev. 1993, 15, 457–459.
217. Keidan, I.; Shif, I.; Keren, G.; Passwell, J.H. Rotavirus encephalopathy: evidence of central nervous system involvement during rotavirus infection. Pediatr. Infect. Dis. J. 1992, 11, 773–775.
218. Abe, T.; Kobayashi, M.; Araki, K.; Kodama, H.; Fujita, Y.; Shinozaki, T.; Ushijima, H. Infantile convulsions with mild gastroenteritis. Brain. Dev. 2000, 22, 301–306.
219. Glimaker, M.; Johansson, B.; Olcen, P.; Ehrnst, A.; Forsgren, M. Detection of enteroviral RNA by polymerase chain reaction in cerebrospinal fluid from patients with aseptic meningitis. Scand. J. Infect. Dis. 1993, 25, 547–557.
220. Jeffery, K.J.; Read, S.J.; Peto, T.E.; Mayon-White, R.T.; Bangham, C.R. Diagnosis of viral infections of the central nervous system: clinical interpretation of PCR results. Lancet. 1997, 349, 313–317.
221. Date, M.; Gondoh, M.; Kato, S.; Fukushima, M.; Nakamoto, N.; Kobayashi, M.; Abe, T. A case of rubella encephalitis: rubella virus genome was detected in the cerebrospinal fluid by polymerase chain reaction. No To. Hattatsu. 1995, 27, 286–290.
222. Fujimoto, S.; Kobayashi, M.; Uemura, O.; Iwasa, M.; Ando, T.; Katoh, T.; Nakamura, C.; Maki, N.; Togari, H.; Wada, Y. PCR on cerebrospinal fluid to show influenza-associated acute encephalopathy or encephalitis. Lancet. 1998, 352, 873–875.
223. Ito, Y.; Ichiyama, T.; Kimura, H.; Shibata, M.; Ishiwada, N.; Kuroki, H.; Furukawa, S.; Morishima, T. Detection of influenza virus RNA by reverse transcription-PCR and proinflammatory cytokines in influenza-virus-associated encephalopathy. J. Med. Virol. 1999, 58, 420–425.
224. Togashi, T.; Matsuzono, Y.; Narita, M. Epidemiology of influenza-associated encephalitis-encephalopathy in Hokkaido, the northernmost island of Japan. Pediatr. Int. 2000, 42, 192–196.

225. Tomoda, A.; Shiraishi, S.; Hosoya, M.; Hamada, A.; Miike, T. Combined treatment with interferon-alpha and ribavirin for subacute sclerosing panencephalitis. Pediatr. Neurol. 2001, 24, 54–59.
226. Nakayama, T.; Mori, T.; Yamaguchi, S.; Sonoda, S.; Asamura, S.; Yamashita, R.; Takeuchi, Y.; Urano, T. Detection of measles virus genome directly from clinical samples by reverse transcriptase-polymerase chain reaction and genetic variability. Virus. Res. 1995, 35, 1–16.
227. Matsuzono, Y.; Narita, M.; Ishiguro, N.; Togashi, T. Detection of measles virus from clinical samples using the polymerase chain reaction. Arch. Pediatr. Adolesc. Med. 1994, 148, 289–293.
228. Wacharapluesadee, S.; Hemachudha, T. Nucleic-acid sequence based amplification in the rapid diagnosis of rabies. Lancet. 2001, 358, 892–893.
229. Crepin, P.; Audry, L.; Rotivel, Y.; Gacoin, A.; Caroff, C.; Bourhy, H. Intravitam diagnosis of human rabies by PCR using saliva and cerebrospinal fluid. J. Clin. Microbiol. 1998, 36, 1117–1121.
229a. Gunther, S.; Weisner, B.; Roth, A.; Grewing, T.; Asper, M.; Drosten, C.; Emmerich, P.; Petersen, J.; Wilczek, M.; Schmitz, H. Lassa fever encephalopathy: Lassa virus in cerebrospinal fluid but not in serum. J. Infect. Dis. 2001, 184, 345–349.
230. Cavrois, M.; Gessain, A.; Gout, O.; Wain-Hobson, S.; Wattel, E. Common human T cell leukemia virus type 1 (HTLV-1) integration sites in cerebrospinal fluid and blood lymphocytes of patients with HTLV-1-associated myelopathy/tropical spastic paraparesis indicate that HTLV-1 crosses the blood-brain barrier via clonal HTLV-1-infected cells. J. Infect. Dis. 2000, 182, 1044–1050.
231. Nagai, M.; Yamano, Y.; Brennan, M.B.; Mora, C.A.; Jacobson, S. Increased HTLV-I proviral load and preferential expansion of HTLV-I Tax-specific CD8 + T cells in cerebrospinal fluid from patients with HAM/TSP. Ann. Neurol. 2001, 50, 807–812.
232. Kompoliti, A.; Gage, B.; Sharma, L.; Daniels, J.C. Human T-cell lymphotropic virus type 1-associated myelopathy, Sjogren syndrome, and lymphocytic pneumonitis. Arch. Neurol. 1996, 53, 940–942.
233. Yang, Y.C.; Hung, T.P.; Wang, C.H.; Lin, M.T.; Hsu, T.Y.; Chen, J.Y.; Chen, Y.C.; Yang, C.S. Establishment and characterization of an HTLV-1 cell line from a Taiwanese patient with HTLV-1-associated myelopathy. J. Neurol. Sci. 1993, 120, 46–53.
234. Valassina, M.; Cusi, M.G.; Valensin, P.E. Rapid identification of Toscana virus by nested PCR during an outbreak in the Siena area of Italy. J. Clin. Microbiol. 1996, 34, 2500–2502.
235. Echevarria, J.M.; de Ory, F.; Guisasola, M.E.; Sanchez-Seco, M.P.; Tenorio, A.; Lozano, A.; Cordoba, J.; Gobernado, M. Acute meningitis due to Toscana virus infection among patients from both the Spanish Mediterranean region and the region of Madrid. J. Clin. Virol. 2003, 26, 79–84.
236. Huang, C.; Campbell, W.; Grady, L.; Kirouac, I.; LaForce, F.M. Diagnosis of Jamestown Canyon encephalitis by polymerase chain reaction. Clin. Infect. Dis. 1999, 28, 1294–1297.
237. Huang, C.; Chatterjee, N.K.; Grady, L.J. Diagnosis of viral infections of the central nervous system. N. Engl. J. Med. 1999, 340, 483–484.
238. Cam, B.V.; Fonsmark, L.; Hue, N.B.; Phuong, N.T.; Poulsen, A.; Heegaard, E.D. Prospective case-control study of encephalopathy in children with dengue hemorrhagic fever. Am. J. Trop. Med. Hyg. 2001, 65, 848–851.
239. Petersen, L.R.; Marfin, A.A. West Nile virus: a primer for the clinician. Ann. Intern. Med. 2002, 6, 173–179.
240. Briese, T.; Glass, W.G.; Lipkin, W.I. Detection of West Nile virus sequences in cerebrospinal fluid. Lancet. 2000, 355, 1614–1615.
241. Laskus, T.; Radkowski, M.; Bednarska, A.; Wilkinson, J.; Adair, D.; Nowicki, M.; Nikolopoulou, G.B.; Vargas, H.; Rakela, J. Detection and analysis of hepatitis C virus sequences in cerebrospinal fluid. J. Virol. 2002, 76, 10064–10068.

242. Maggi, F.; Giorgi, M.; Fornai, C.; Morrica, A.; Vatteroni, M.L.; Pistello, M.; Siciliano, G.; Nuccorini, A.; Bendinelli, M. Detection and quasispecies analysis of hepatitis C virus in the cerebrospinal fluid of infected patients. J. Neurovirol. 1999, 5, 319–323.
243. Dessau, R.B.; Lisby, G.; Frederiksen, J.L. Coronaviruses in spinal fluid of patients with acute monosymptomatic optic neuritis. Acta. Neurol. Scand. 1999, 100, 88–91.
244. Cristallo, A.; Gambaro, F.; Biamonti, G.; Ferrante, P.; Battaglia, M.; Cereda, P.M. Human coronavirus polyadenylated RNA sequences in cerebrospinal fluid from multiple sclerosis patients. New Microbiol. 1997, 20, 105–114.
245. Tan, S.V.; Guiloff, R.J.; Scaravilli, F.; Klapper, P.E.; Cleator, G.M.; Gazzard, B.G. Herpes simplex type 1 encephalitis in acquired immunodeficiency syndrome. Ann. Neurol. 1993, 34, 619–622.
246. Miller, R.F.; Fox, J.D.; Waite, J.C.; Severn, A.; Brink, N.S. Herpes simplex virus type 2 encephalitis and concomitant cytomegalovirus infection in a patient with AIDS: detection of virus-specific DNA in CSF by nested polymerase chain reaction. Genitourin. Med. 1995, 71, 262–264.
247. Burke, D.G.; Kalayjian, R.C.; Vann, V.R.; Madreperla, S.A.; Shick, H.E.; Leonard, D.G. Polymerase chain reaction detection and clinical significance of varicella-zoster virus in cerebrospinal fluid from human immunodeficiency virus-infected patients. J. Infect. Dis. 1997, 176, 1080–1084.
248. Cinque, P.; Bossolasco, S.; Vago, L.; Fornara, C.; Lipari, S.; Racca, S.; Lazzarin, A.; Linde, A. Varicella-zoster virus (VZV) DNA in cerebrospinal fluid of patients infected with human immunodeficiency virus: VZV disease of the central nervous system or subclinical reactivation of VZV infection?. Clin. Infect. Dis. 1997, 25, 634–639.
249. Iten, A.; Chatelard, P.; Vuadens, P.; Miklossy, J.; Meuli, R.; Sahli, R.; Meylan, P.R. Impact of cerebrospinal fluid PCR on the management of HIV-infected patients with varicella-zoster virus infection of the central nervous system. J. Neurovirol. 1999, 5, 172–180.
250. Wolf, D.G.; Spector, S.A. Diagnosis of human cytomegalovirus central nervous system disease in AIDS patients by DNA amplification from cerebrospinal fluid. J. Infect. Dis. 1992, 166, 1412–1415.
251. Cinque, P.; Vago, L.; Brytting, M.; Castagna, A.; Accordini, A.; Sundqvist, V.A.; Zanchetta, N.; Monforte, A.D.; Wahren, B.; Lazzarin, A. Cytomegalovirus infection of the central nervous system in patients with AIDS: diagnosis by DNA amplification from cerebrospinal fluid. J. Infect. Dis. 1992, 166, 1408–1411.
252. Gozlan, J.; Salord, J.M.; Roullet, E.; Baudrimont, M.; Caburet, F.; Picard, O.; Meyohas, M.C.; Duvivier, C.; Jacomet, C.; Petit, J.C. Rapid detection of cytomegalovirus DNA in cerebrospinal fluid of AIDS patients with neurologic disorders. J. Infect. Dis. 1992, 166, 1416–1421.
253. Clifford, D.B.; Buller, R.S.; Mohammed, S.; Robinson, L.; Storch, G.A. Use of polymerase chain reaction to demonstrate cytomegalovirus DNA in CSF of patients with human immunodeficiency virus infection. Neurology. 1993, 43, 75–79.
254. Fox, J.D.; Brink, N.S.; Zuckerman, M.A.; Neild, P.; Gazzard, B.G.; Tedder, R.S.; Miller, R.F. Detection of herpesvirus DNA by nested polymerase chain reaction in cerebrospinal fluid of human immunodeficiency virus-infected persons with neurologic disease: a prospective evaluation. J. Infect. Dis. 1995, 172, 1087–1090.
255. Cinque, P.; Brytting, M.; Vago, L.; Castagna, A.; Parravicini, C.; Zanchetta, N.; D'Arminio Monforte, A.; Wahren, B.; Lazzarin, A.; Linde, A. Epstein-Barr virus DNA in cerebrospinal fluid from patients with AIDS-related primary lymphoma of the central nervous system. Lancet. 1993, 342, 398–401.
256. Arribas, J.R.; Clifford, D.B.; Fichtenbaum, C.J.; Roberts, R.L.; Powderly, W.G.; Storch, G.A. Detection of Epstein-Barr virus DNA in cerebrospinal fluid for diagnosis of AIDS-related central nervous system lymphoma. J. Clin. Microbiol. 1995, 33, 1580–1583.

257. De Luca, A.; Antinori, A.; Cingolani, A.; Larocca, L.M.; Linzalone, A.; Ammassari, A.; Scerrati, M.; Roselli, R.; Tamburrini, E.; Ortona, L. Evaluation of cerebrospinal fluid EBV-DNA and IL-10 as markers for in vivo diagnosis of AIDS-related primary central nervous system lymphoma. Br. J. Haematol. 1995, 90, 844–849.
258. Knox, K.K.; Harrington, D.P.; Carrigan, D.R. Fulminant human herpesvirus six encephalitis in a human immunodeficiency virus-infected infant. J. Med. Virol. 1995, 45, 288–292.
259. Wang, F.Z.; Linde, A.; Hagglund, H.; Testa, M.; Locasciulli, A.; Ljungman, P. Human herpesvirus 6 DNA in cerebrospinal fluid specimens from allogeneic bone marrow transplant patients: does it have clinical significance?. Clin. Infect. Dis. 1999, 28, 562–568.
260. Bossolasco, S.; Marenzi, R.; Dahl, H.; Vago, L.; Terreni, M.R.; Broccolo, F.; Lazzarin, A.; Linde, A.; Cinque, P. Human herpesvirus 6 in cerebrospinal fluid of patients infected with HIV: frequency and clinical significance. J. Neurol Neurosurg Psychiatry. 1999, 67, 789–792.
261. Weber, T.; Turner, R.W.; Frye, S.; Ruf, B.; Haas, J.; Schielke, E.; Pohle, H.D.; Luke, W.; Luer, W.; Felgenhauer, K. Specific diagnosis of progressive multifocal leukoencephalopathy by polymerase chain reaction. J. Infect. Dis. 1994, 169, 1138–1141.
262. Fong, I.W.; Britton, C.B.; Luinstra, K.E.; Toma, E.; Mahony, J.B. Diagnostic value of detecting J. C. virus DNA in cerebrospinal fluid of patients with progressive multifocal leukoencephalopathy. J. Clin. Microbiol. 1995, 33, 484–486.
263. McGuire, D.; Barhite, S.; Hollander, H.; Miles, M. J. C. virus DNA in cerebrospinal fluid of human immunodeficiency virus-infected patients: predictive value for progressive multifocal leukoencephalopathy. Ann. Neurol. 1995, 37, 395–399.
264. Cinque, P.; Scarpellini, P.; Vago, L.; Linde, A.; Lazzarin, A. Diagnosis of central nervous system complications in HIV-infected patients: cerebrospinal fluid analysis by the polymerase chain reaction. AIDS. 1997, 11, 1–17.
265. Bratt, G.; Hammarin, A.L.; Grandien, M.; Hedquist, B.G.; Nennesmo, I.; Sundelin, B.; Seregard, S. B. K. virus as the cause of meningoencephalitis, retinitis and nephritis in a patient with AIDS. AIDS. 1999, 13, 1071–1075.
266. Morsica, G.; Bernardi, M.T.; Novati, R.; Uberti Foppa, C.; Castagna, A.; Lazzarin, A. Detection of hepatitis C virus genomic sequences in the cerebrospinal fluid of HIV-infected patients. J. Med. Virol. 1997, 53, 252–254.
267. Gazzola, P.; Mavilio, D.; Costa, P.; Fogli, M.; Bruzzone, B.; Icardi, G.; Primavera, A.; Cocito, L.; De Maria, A. Possible hepatitis C virus involvement in acute meningoradiculitis/polyradiculitis of HIV-1-co-infected patients. AIDS. 2001, 15, 539–541.
268. Oberste, M.S.; Maher, K.; Pallansch, M.A. Complete sequence of echovirus 23 and its relationship to echovirus 22 and other human enteroviruses. Virus. Res. 1998, 56, 217–223.
269. Rotbart, H.A.; Sawyer, M.H.; Fast, S.; Lewinski, C.; Murphy, N.; Keyser, E.F.; Spadoro, J.; Kao, S.Y.; Loeffelholz, M. Diagnosis of enteroviral meningitis by using PCR with a colorimetric microwell detection assay. J. Clin. Microbiol. 1994, 32, 2590–2592.
270. Lina, B.; Pozzetto, B.; Andreoletti, L.; Beguier, T.; Bourlet, T.; Dussaix, E.; Grangeot-Keros, L.; Gratacap-Cavallier, B.; Henquell, C.; Legrand-Quillien, M.C.; Novillo, A.; Palmer, P.; Petitjean, J.; Sandres, K.; Dubreuil, P. Multicenter evaluating of a commercially available PCR assay for diagnosing enterovirus infection in a panel of cerebrospinal fluid specimens. J. Clin. Microbiol. 1996, 34, 3002–3006.
271. Romero, J.R. Reverse-transcription polymerase chain reaction detection of the enteroviruses. Arch. Pathol. Lab. Med. 1999, 123, 1161–1169.
272. Cinque, P.; Vago, L.; Dahl, H.; Brytting, M.; Terreni, M.R.; Fornara, C.; Racca, S.; Castagna, A.; Monforte, A.D.; Wahren, B.; Lazzarin, A.; Linde, A. Polymerase chain reaction on cerebrospinal fluid for diagnosis of virus-associated opportunistic diseases of the central nervous system in HIV-infected patients. AIDS. 1996, 10, 951–958.
273. de Luca, A.; Cingolani, A.; Linzalone, A.; Ammassari, A.; Murri, R.; Giancola, M.L.; Maiuro, G.; Antinori, A. Improved detection of JC virus DNA in cerebrospinal fluid for diagnosis

of AIDS-related progressive multifocal leukoencephalopathy. J. Clin. Microbiol. 1996, 34, 1343–1346.
274. Miralles, P.; Berenguer, J.; Garcia de Viedma, D.; Padilla, B.; Cosin, J.; Lopez-Bernaldo de Quiros, J.C.; Munoz, L.; Moreno, S.; Bouza, E. Treatment of AIDS-associated progressive multifocal leukoencephalopathy with highly active antiretroviral therapy. AIDS. 1998, 12, 2467–2472.
275. Giudici, B.; Vaz, B.; Bossolasco, S.; Casari, S.; Brambilla, A.M.; Luke, W.; Lazzarin, A.; Weber, T.; Cinque, P. Highly active antiretroviral therapy and progressive multifocal leukoencephalopathy: effects on cerebrospinal fluid markers of JC virus replication and immune response. Clin. Infect. Dis. 2000, 30, 95–99.
276. MacMahon, E.M.; Glass, J.D.; Hayward, S.D.; Mann, R.B.; Becker, P.S.; Charache, P.; McArthur, J.C.; Ambinder, R.F. Epstein-Barr virus in AIDS-related primary central nervous system lymphoma. Lancet. 1991, 338, 969–973.
277. Cingolani, A.; Gastaldi, R.; Fassone, L.; Pierconti, F.; Giancola, M.L.; Martini, M.; De Luca, A.; Ammassari, A.; Mazzone, C.; Pescarmona, E.; Gaidano, G.; Larocca, L.M.; Antinori, A. Epstein-Barr virus infection is predictive of CNS involvement in systemic AIDS-related non-Hodgkin's lymphomas. J. Clin. Oncol. 2000, 18, 3325–3330.
278. Ando, Y.; Kimura, H.; Miwata, H.; Kudo, T.; Shibata, M.; Morishima, T. Quantitative analysis of herpes simplex virus DNA in cerebrospinal fluid of children with herpes simplex encephalitis. J. Med. Virol. 1993, 41, 170–173.
279. Revello, M.G.; Baldanti, F.; Sarasini, A.; Zella, D.; Zavattoni, M.; Gerna, G. Quantitation of herpes simplex virus DNA in cerebrospinal fluid of patients with herpes simplex encephalitis by the polymerase chain reaction. Clin. Diagn. Virol. 1997, 7, 183–191.
280. Domingues, R.B.; Lakeman, F.D.; Mayo, M.S.; Whitley, R.J. Application of competitive PCR to cerebrospinal fluid samples from patients with herpes simplex encephalitis. J. Clin. Microbiol. 1998, 36, 2229–2234.
281. Kessler, H.H.; Muhlbauer, G.; Rinner, B.; Stelzl, E.; Berger, A.; Dorr, H.W.; Santner, B.; Marth, E.; Rabenau, H. Detection of herpes simplex virus DNA by real-time PCR. J. Clin. Microbiol. 2000, 38, 2638–2642.
281a. Kimura, H.; Ito, Y.; Futamura, M.; Ando, Y.; Yabuta, Y.; Hoshino, Y.; Nishiyama, Y.; Morishima, T. Quantitation of viral load in neonatal herpes simplex virus infection and comparison between type 1 and type 2. J. Med. Virol. 2002, 67, 349–353.
281b. Aberle, S.W.; Puchhammer-Stockl, E. Diagnosis of herpesvirus infections of the central nervous system. J. Clin. Virol. 2002, 25(Suppl 1), S79–S85.
282. Arribas, J.R.; Clifford, D.B.; Fichtenbaum, C.J.; Commins, D.L.; Powderly, W.G.; Storch, G.A. Level of cytomegalovirus (CMV) DNA in cerebrospinal fluid of subjects with AIDS and CMV infection of the central nervous system. J. Infect. Dis. 1995, 172, 527–531.
283. Cinque, P.; Baldanti, F.; Vago, L.; Terreni, M.R.; Lillo, F.; Furione, M.; Castagna, A.; Monforte, A.D.; Lazzarin, A.; Linde, A. Ganciclovir therapy for cytomegalovirus (CMV) infection of the central nervous system in AIDS patients: monitoring by CMV DNA detection in cerebrospinal fluid. J. Infect. Dis. 1995, 171, 1603–1606.
284. Shinkai, M.; Spector, S.A. Quantitation of human cytomegalovirus (HCMV) DNA in cerebrospinal fluid by competitive PCR in AIDS patients with different HCMV central nervous system diseases. Scand. J. Infect. Dis. 1995, 27, 559–561.
284a. Bossolasco, O.; Cinque, P.; Ponzoni, M.; Vigano, M.G.; Lazzarin, A.; Linde, A.; Falk, K.I. Epstein-Barr virus DNA load in cerebrospinal fluid and plasma of patients with AIDS-related lymphoma. J. Neurovirol. 2002, 8, 432–438.
285. Taoufik, Y.; Gasnault, J.; Karaterki, A.; Pierre Ferey, M.; Marchadier, E.; Goujard, C.; Lannuzel, A.; Delfraissy, J.F. E. Dussaix, Prognostic value of JC virus load in cerebrospinal fluid of patients with progressive multifocal leukoencephalopathy. J. Infect. Dis. 1998, 178, 1816–1820.

286. Koralnik, I.J.; Boden, D.; Mai, V.X.; Lord, C.I.; Letvin, N.L. J. C. virus DNA load in patients with and without progressive multifocal leukoencephalopathy. Neurology. 1999, 52, 253–260.
287. Yiannoutsos, C.T.; Major, E.O.; Curfman, B.; Jensen, P.N.; Gravell, M.; Hou, J.; Clifford, D.B.; Hall, C.D. Relation of J. C. virus DNA in the cerebrospinal fluid to survival in acquired immunodeficiency syndrome patients with biopsy-proven progressive multifocal leukoencephalopathy. Ann. Neurol. 1999, 45, 816–821.
288. Garcia de Viedma, D.; Alonso, R.; Miralles, P.; Berenguer, J.; Rodriguez-Creixems, M.; Bouza, E. Dual qualitative-quantitative nested PCR for detection of JC virus in cerebrospinal fluid: high potential for evaluation and monitoring of progressive multifocal leukoencephalopathy in AIDS patients receiving highly active antiretroviral therapy. J. Clin. Microbiol. 1999, 37, 724–728.
289. Eggers, C.; Stellbrink, H.J.; Buhk, T.; Dorries, K. Quantification of JC virus DNA in the cerebrospinal fluid of patients with human immunodeficiency virus-associated progressive multifocal leukoencephalopathy—a longitudinal study. J. Infect. Dis. 1999, 180, 1690–1694.
290. Martino, T.A.; Sole, M.J.; Penn, L.Z.; Liew, C.C.; Liu, P. Quantitation of enteroviral RNA by competitive polymerase chain reaction. J. Clin. Microbiol. 1993, 31, 2634–2640.
291. Arola, A.; Santti, J.; Ruuskanen, O.; Halonen, P.; Hyypia, T. Identification of enteroviruses in clinical specimens by competitive PCR followed by genetic typing using sequence analysis. J. Clin. Microbiol. 1996, 34, 313–318.
291a. Verstrepen, W.A.; Kuhn, S.; Kockx, M.M.; Van De Vyvere, M.E.; Mertens, A.H. Rapid detection of enterovirus RNA in cerebrospinal fluid specimens with a novel single-tube real-time reverse transcription-PCR assay. J. Clin. Microbiol. 2001, 39, 4093–4096.
292. Brew, B.J.; Pemberton, L.; Cunningham, P.; Law, M.G. Levels of human immunodeficiency virus type 1 RNA in cerebrospinal fluid correlate with AIDS dementia stage. J. Infect. Dis. 1997, 175, 963–966.
293. Ellis, R.J.; Hsia, K.; Spector, S.A.; Nelson, J.A.; Heaton, R.K.; Wallace, M.R.; Abramson, I.; Atkinson, J.H.; Grant, I.; McCutchan, J.A. Cerebrospinal fluid human immunodeficiency virus type 1 RNA levels are elevated in neurocognitively impaired individuals with acquired immunodeficiency syndrome. HIV Neurobehavioral Research Center Group. Ann. Neurol. 1997, 42, 679–688.
294. Cinque, P.; Vago, L.; Ceresa, D.; Mainini, F.; Terreni, M.R.; Vagani, A.; Torri, W.; Bossolasco, S.; Lazzarin, A. Cerebrospinal fluid HIV-1 RNA levels: correlation with HIV encephalitis. AIDS. 1998, 12, 389–394.
295. Gisslen, M.; Norkrans, G.; Svennerholm, B.; Hagberg, L. The effect on human immunodeficiency virus type 1 RNA levels in cerebrospinal fluid after initiation of zidovudine or didanosine. J. Infect. Dis. 1997, 175, 434–437.
296. Foudraine, N.A.; Hoetelmans, R.M.; Lange, J.M.; de Wolf, F.; van Benthem, B.H.; Maas, J.J.; Keet, I.P.; Portegies, P. Cerebrospinal-fluid HIV-1 RNA and drug concentrations after treatment with lamivudine plus zidovudine or stavudine. Lancet. 1998, 351, 1547–1551.
297. Staprans, S.; Marlowe, N.; Glidden, D.; Novakovic-Agopian, T.; Grant, R.M.; Heyes, M.; Aweeka, F.; Deeks, S.; Price, R.W. Time course of cerebrospinal fluid responses to antiretroviral therapy: evidence for variable compartmentalization of infection. AIDS. 1999, 13, 1051–1061.
298. Ellis, R.J.; Gamst, A.C.; Capparelli, E.; Spector, S.A.; Hsia, K.; Wolfson, T.; Abramson, I.; Grant, I.; McCutchan, J.A. Cerebrospinal fluid HIV RNA originates from both local CNS and systemic sources. Neurology. 2000, 54, 927–936.
299. Gisolf, E.H.; Enting, R.H.; Jurriaans, S.; de Wolf, F.; van der Ende, M.E.; Hoetelmans, R.M.; Portegies, P.; Danner, S.A. Cerebrospinal fluid HIV-1 RNA during treatment with ritonavir/saquinavir or ritonavir/saquinavir/stavudine. AIDS. 2000, 14, 1583–1589.
300. Cinque, P.; Bestetti, A.; Morelli, P.; Presi, S. Molecular analysis of cerebrospinal fluid: potential for the study of HIV-1 infection of the central nervous system. J. Neurovirol. 2000, 6(suppl 1), S95–S102.

301. Price, R.W.; Paxinos, E.E.; Grant, R.M.; Drews, B.; Nilsson, A.; Hoh, R.; Hellmann, N.S.; Petropoulos, C.J.; Deeks, S.G. Cerebrospinal fluid response to structured treatment interruption after virological failure. AIDS. 2001, 15, 1251–1259.
302. Cinque, P.; Presi, S.; Bestetti, A.; Pierotti, C.; Racca, S.; Boeri, E.; Morelli, P.; Carrera, P.; Ferrari, M.; Lazzarin, A. Effect of genotypic resistance on the virological response to highly active antiretroviral therapy in cerebrospinal fluid. AIDS Res. Hum. Retroviruses. 2001, 17, 377–383.
303. Tang, Y.W.; Espy, M.J.; Persing, D.H.; Smith, T.F. Molecular evidence and clinical significance of herpesvirus coinfection in the central nervous system. J. Clin. Microbiol. 1997, 35, 2869–2872.
304. Studahl, M.; Bergstrom, T.; Hagberg, L. Acute viral encephalitis in adults—a prospective study. Scand. J. Infect. Dis. 1998, 30, 215–220.
305. Portolani, M.; Sabbatini, A.M.; Meacci, M.; Pietrosemoli, P.; Cermelli, C.; Lunghi, P.; Golinelli, F.; Stacca, R. Epstein-Barr virus DNA in cerebrospinal fluid from an immunocompetent man with herpes simplex virus encephalitis. J. Neurovirol. 1998, 4, 461–464.
306. Liedtke, W.; Malessa, R.; Faustmann, P.M.; Eis-Hubinger, A.M. Human herpesvirus 6 polymerase chain reaction findings in human immunodeficiency virus associated neurological disease and multiple sclerosis. J. Neurovirol. 1995, 1, 253–258.
307. Ferrante, P.; Omodeo-Zorini, E.; Caldarelli-Stefano, R.; Mediati, M.; Fainardi, E.; Granieri, E.; Caputo, D. Detection of J. C. virus DNA in cerebrospinal fluid from multiple sclerosis patients. Mult. Scler. 1998, 4, 49–54.
308. Ross, J.S. Financial determinants of outcomes in molecular testing. Arch. Pathol. Lab. Med. 1999, 123, 1071–1075.
309. Tebas, P.; Nease, R.F.; Storch, G.A. Use of the polymerase chain reaction in the diagnosis of herpes simplex encephalitis: a decision analysis model. Am. J. Med. 1998, 105, 287–295.
310. Swingler, G.; Delport, S.; Hussey, G. An audit of the use of antibiotics in presumed viral meningitis in children. Pediatr. Infect. Dis. J. 1994, 13, 1107–1110.
311. Rice, S.K.; Heinl, R.E.; Thornton, L.L.; Opal, S.M. Clinical characteristics, management strategies, and cost implications of a statewide outbreak of enterovirus meningitis. Clin. Infect. Dis. 1995, 20, 931–937.
312. Marshall, G.S.; Hauck, M.A.; Buck, G.; Rabalais, G.P. Potential cost savings through rapid diagnosis of enteroviral meningitis. Pediatr. Infect. Dis. J. 1997, 16, 1086–1087.
313. van Vliet, K.E.; Glimaker, M.; Lebon, P.; Klapper, P.E.; Taylor, C.E.; Ciardi, M.; van der Avoort, H.G.; Diepersloot, R.J.; Kurtz, J.; Peeters, M.F.; Cleator, G.M.; van Loon, A.M. Multicenter evaluation of the Amplicor enterovirus PCR test with cerebrospinal fluid from patients with aseptic meningitis. The European Union Concerted Action on Viral Meningitis and Encephalitis. J. Clin. Microbiol. 1998, 36, 2652–2657.
314. Muir, P.; Ras, A.; Klapper, P.E.; Cleator, G.M.; Korn, K.; Aepinus, C.; Fomsgaard, A.; Palmer, P.; Samuelsson, A.; Tenorio, A.; Weissbrich, B.; van Loon, A.M. Multicenter quality assessment of PCR methods for detection of enteroviruses. J. Clin. Microbiol. 1999, 37, 1409–1414.
315. van Loon, A.M.; Cleator, G.C.; Ras, A. External quality assessment of enterovirus detection and typing. European Union Concerted Action on Virus Meningitis and Encephalitis. Bull World Health Org. 1999, 77, 217–223.
316. Weber, T.; Klapper, P.E.; Cleator, G.M.; Bodemer, M.; Luke, W.; Knowles, W.; Cinque, P.; Van Loon, A.M.; Grandien, M.; Hammarin, A.L.; Ciardi, M.; Bogdanovic, G. Polymerase chain reaction for detection of J. C. virus DNA in cerebrospinal fluid: a quality control study. European Union Concerted Action on Viral Meningitis and Encephalitis. J. Virol. Methods. 1997, 69, 231–237.
317. Schloss, L.; van Loon, A.M.; Cinque, P.; Cleator, G.; Echevarria, J.M.; Falk, K.I.; Klapper, P.; Schirm, J.; Vestergaard, B.F.; Niesters, B.; Popow-Kraupp, T.; Quint, W.; Linde, A. European panels for quality control of nucleic acid amplification of herpes simplex virus

(HSV). 5th Annual Meeting of the European Society for Clinical Virology; Lathi, Finland, 2001. An international external quality assessment of nucleic acid amplification of herpes simplex virus. J. Clin. virol. in press.

318. Arens, M. Methods for subtyping and molecular comparison of human viral genomes. Clin. Microbiol Rev. 1999, 12, 612–626.

319. Cunningham, P.H.; Smith, D.G.; Satchell, C.; Cooper, D.A.; Brew, B. Evidence for independent development of resistance to HIV-1 reverse transcriptase inhibitors in the cerebrospinal fluid. AIDS. 2000, 14, 1949–1954.

320. Pease, A.C.; Solas, D.; Sullivan, E.J.; Cronin, M.T.; Holmes, C.P.; Fodor, S.P. Light-generated oligonucleotide arrays for rapid DNA sequence analysis. Proc. Natl. Acad. Sci. USA. 1994, 91, 5022–5026.

321. McGlennen, R.C. Miniaturization technologies for molecular diagnostics. Clin. Chem. 2001, 47, 393–402.

322. Lockhart, D.J.; Winzeler, E.A. Genomics, gene expression and DNA arrays. Nature. 2000, 405, 827–836.

323. Chambers, J.; Angulo, A.; Amaratunga, D.; Guo, H.; Jiang, Y.; Wan, J.S.; Bittner, A.; Frueh, K.; Jackson, M.R.; Peterson, P.A.; Erlander, M.G.; Ghazal, P. DNA microarrays of the complex human cytomegalovirus genome: profiling kinetic class with drug sensitivity of viral gene expression. J. Virol. 1999, 73, 5757–5766.

324. Jenner, R.G.; Alba, M.M.; Boshoff, C.; Kellam, P. Kaposi's sarcoma-associated herpesvirus latent and lytic gene expression as revealed by DNA arrays. J. Virol. 2001, 75, 891–902.

325. Li, J.; Chen, S.; Evans, D.H. Typing and subtyping influenza virus using DNA microarrays and multiplex reverse transcriptase PCR. J. Clin. Microbiol. 2001, 39, 696–704.

326. Wilson, J.W.; Bean, P.; Robins, T.; Graziano, F.; Persing, D.H. Comparative evaluation of three human immunodeficiency virus genotyping systems: the HIV-GenotypR method, the HIV PRT GeneChip assay, and the HIV-1 RT line probe assay. J. Clin. Microbiol. 2000, 38, 3022–3028.

327. Rozenberg, F.; Lebon, P. Analysis of herpes simplex virus type 1 glycoprotein D nucleotide sequence in human herpes simplex encephalitis. J. Neurovirol. 1996, 2, 289–295.

328. Lee, N.Y.; Tang, Y.; Espy, M.J.; Kolbert, C.P.; Rys, P.N.; Mitchell, P.S.; Day, S.P.; Henry, S.L.; Persing, D.H.; Smith, T.F. Role of genotypic analysis of the thymidine kinase gene of herpes simplex virus for determination of neurovirulence and resistance to acyclovir. J. Clin. Microbiol. 1999, 37, 3171–3174.

329. Wolf, D.G.; Lee, D.J.; Spector, S.A. Detection of human cytomegalovirus mutations associated with ganciclovir resistance in cerebrospinal fluid of AIDS patients with central nervous system disease. Antimicrob. Agents. Chemother. 1995, 39, 2552–2554.

330. Agostini, H.T.; Stoner, G.L. Amplification of the complete polyomavirus J. C. genome from brain, cerebrospinal fluid and urine using pre-PCR restriction enzyme digestion. J. Neurovirol. 1995, 1, 316–320.

331. Ferrante, P.; Mediati, M.; Caldarelli-Stefano, R.; Losciale, L.; Mancuso, R.; Cagni, A.E.; Maserati, R. Increased frequency of JCV type 2 and of dual infection with JC virus type 1 and 2 in Italian progressive multifocal leukoencephalopathy patients. J. Neurovirol. 2001, 7, 35–42.

332. Ciappi, S.; Azzi, A.; De Santis, R.; Leoncini, F.; Sterrantino, G.; Mazzotta, F.; Mecocci, L. Archetypal and rearranged sequences of human polyomavirus J. C. transcription control region in peripheral blood leukocytes and in cerebrospinal fluid. J. Gen. Virol. 1999, 80, 1017–1023.

333. Vaz, B.; Cinque, P.; Pickhardt, M.; Weber, T. Analysis of the transcriptional control region in progressive multifocal leukoencephalopathy. J. Neurovirol. 2000, 6, 398–409.

334. Pfister, L.A.; Letvin, N.L.; Koralnik, I.J. JC virus regulatory region tandem repeats in plasma and central nervous system isolates correlate with poor clinical outcome in patients with progressive multifocal leukoencephalopathy. J. Virol. 2001, 75, 5672–5676.

335. Takami, T.; Sonodat, S.; Houjyo, H.; Kawashima, H.; Takei, Y.; Miyajima, T.; Takekuma, K.; Hoshika, A.; Mori, T.; Nakayama, T. Diagnosis of horizontal enterovirus infections in neonates by nested PCR and direct sequence analysis. J. Hosp. Infect. 2000, 45, 283–287.
336. Byington, C.L.; Taggart, E.W.; Carroll, K.C.; Hillyard, D.R. A polymerase chain reaction-based epidemiologic investigation of the incidence of nonpolio enteroviral infections in febrile and afebrile infants 90 days and younger. Pediatrics. 1999, 103, E27.
337. Kammerer, U.; Kunkel, B.; Korn, K. Nested PCR for specific detection and rapid identification of human picornaviruses. J. Clin. Microbiol. 1994, 32, 285–291.
338. Leparc-Goffart, I.; Julien, J.; Fuchs, F.; Janatova, I.; Aymard, M.; Kopecka, H. Evidence of presence of poliovirus genomic sequences in cerebrospinal fluid from patients with postpolio syndrome. J. Clin. Microbiol. 1996, 34, 2023–2026.
339. Furione, M.; Guillot, S.; Otelea, D.; Balanant, J.; Candrea, A.; Crainic, R. Polioviruses with natural recombinant genomes isolated from vaccine-associated paralytic poliomyelitis. Virology. 1993, 196, 199–208.
340. Oberste, M.S.; Maher, K.; Kilpatrick, D.R.; Flemister, M.R.; Brown, B.A.; Pallansch, M.A. Typing of human enteroviruses by partial sequencing of VP1. J. Clin. Microbiol. 1999, 37, 1288–1293.
341. Brown, B.A.; Kilpatrick, D.R.; Oberste, M.S.; Pallansch, M.A. Serotype-specific identification of enterovirus 71 by PCR. J. Clin. Virol. 2000, 2, 107–112.
342. Brown, E.G.; Furesz, J.; Dimock, K.; Yarosh, W.; Contreras, G. Nucleotide sequence analysis of Urabe mumps vaccine strain that caused meningitis in vaccine recipients. Vaccine. 1991, 9, 840–842.
343. Forsey, T.; Mawn, J.A.; Yates, P.J.; Bentley, M.L.; Minor, P.D. Differentiation of vaccine and wild mumps viruses using the polymerase chain reaction and dideoxynucleotide sequencing. J. Gen. Virol. 1990, 71, 987–990.
344. Kreis, S.; Schoub, B.D. Partial amplification of the measles virus nucleocapsid gene from stored sera and cerebrospinal fluids for molecular epidemiological studies. J. Med. Virol. 1998, 56, 174–177.
345. Katayama, Y.; Shibahara, K ; Kohama, T.; Homma, M.; Hotta, H. Molecular epidemiology and changing distribution of genotypes of measles virus field strains in Japan. J. Clin. Microbiol, 35, 2651–2653.
346. Venturi, G.; Catucci, M.; Romano, L.; Corsi, P.; Leoncini, F.; Valensin, P.E.; Zazzi, M. Antiretroviral resistance mutations in human immunodeficiency virus type 1 reverse transcriptase and protease from paired cerebrospinal fluid and plasma samples. J. Infect. Dis. 2000, 181, 740–745.
347. Steuler, H.; Storch-Hagenlocher, B.; Wildemann, B. Distinct populations of human immunodeficiency virus type 1 in blood and cerebrospinal fluid. AIDS Res. Hum. Retroviruses. 1992, 8, 53–59.
348. Kuiken, C.L.; Goudsmit, J.; Weiller, G.F.; Armstrong, J.S.; Hartman, S.; Portegies, P.; Dekker, J.; Cornelissen, M. Differences in human immunodeficiency virus type 1 V3 sequences from patients with and without AIDS dementia complex. J. Gen. Virol. 1995, 76, 175–180.
349. Keys, B.; Karis, J.; Fadeel, B.; Valentin, A.; Norkrans, G.; Hagberg, L.; Chiodi, F. V3 sequences of paired HIV-1 isolates from blood and cerebrospinal fluid cluster according to host and show variation related to the clinical stage of disease. Virology. 1993, 196, 475–483.
350. Power, C.; McArthur, J.C.; Johnson, R.T.; Griffin, D.E.; Glass, J.D.; Perryman, S.; Chesebro, B. Demented and nondemented patients with AIDS differ in brain-derived human immunodeficiency virus type 1 envelope sequences. J. Virol. 1994, 68, 4643–4649.
351. Di Stefano, M.; Gray, F.; Leitner, T.; Chiodi, F. Analysis of ENV V3 sequences from HIV-1-infected brain indicates restrained virus expression throughout the disease. J. Med. Virol. 1996, 49, 41–48.
352. Martin, M.; Tsai, T.F.; Cropp, B.; Chang, G.J.; Holmes, D.A.; Tseng, J.; Shieh, W.; Zaki, S.R.; Al-Sanouri, I.; Cutrona, A.F.; Ray, G.; Weld, L.H.; Cetron, M.S. Fever and multisystem

organ failure associated with 17D-204 yellow fever vaccination: a report of four cases. Lancet. 2001, 358, 98–104.
353. Schinazi, R.F.; Larder, B.A.; Mellors, J.W. Resistance table: mutations in retroviral genes associated with drug resistance. Int. Antiviral. News. 1997, 5, 129–142.
354. Hirsch, M.S.; Brun-Vezinet, F.; D'Aquila, R.T.; Hammer, S.M.; Johnson, V.A.; Kuritzkes, D.R.; Loveday, C.; Mellors, J.W.; Clotet, B.; Conway, B.; Demeter, L.M.; Vella, S.; Jacobsen, D.M.; Richman, D.D. Antiretroviral drug resistance testing in adult HIV-1 infection: recommendations of an International AIDS Society—USA Panel. JAMA. 2000, 283, 2417–2426.
355. Blennow, K.; Fredman, P.; Wallin, A.; Gottfries, C.G.; Frey, H.; Pirttila, T.; Skoog, I.; Wikkelso, C.; Svennerholm, L. Formulas for the quantitation of intrathecal IgG production. Their validity in the presence of blood-brain barrier damage and their utility in multiple sclerosis. J. Neurol. Sci. 1994, 121, 90–96.
356. Reiber, H.; Lange, P. Quantification of virus-specific antibodies in cerebrospinal fluid and serum: sensitive and specific detection of antibody synthesis in brain. Clin. Chem. 1991, 37, 1153–1160.
357. Boucquey, D.; Chalon, M.P.; Sindic, C.J.; Lamy, M.E.; Laterre, C. Herpes simplex virus type 2 meningitis without genital lesions: an immunoblot study. J. Neurol. 1990, 237, 285–289.
358. Pohl-Koppe, A.; Dahm, C.; Elgas, M.; Kuhn, J.E.; Braun, R.W.; ter Meulen, V. The diagnostic significance of the polymerase chain reaction and isoelectric focusing in herpes simplex virus encephalitis. J. Med. Virol. 1992, 36, 147–154.
359. Sindic, C.J.; Monteyne, P.; Laterre, E.C. The intrathecal synthesis of virus-specific oligoclonal IgG in multiple sclerosis. J. Neuroimmunol. 1994, 54, 75–80.
360. Monteyne, P.; Albert, F.; Weissbrich, B.; Zardini, E.; Ciardi, M.; Cleator, G.M.; Sindic, C.J. The detection of intrathecal synthesis of anti-herpes simplex IgG antibodies: comparison between an antigen-mediated immunoblotting technique and antibody index calculations. European Union Concerted Action on Virus Meningitis and Encephalitis. J. Med. Virol. 1997, 53, 324–331.
361. Mathiesen, T.; Fridell, E.; Fredrikson, S.; Linde, A.; Sundqvist, V.A.; Edler, D.; Wahren, B. Combination ELISAs for antiviral antibodies in CSF and serum in patients with neurological symptoms and in healthy controls. J. Virol. Methods. 1988, 19, 169–179.
362. van Loon, A.M.; van der Logt, J.T.; Heessen, F.W.; Postma, B.; Peeters, M.F. Diagnosis of herpes simplex virus encephalitis by detection of virus-specific immunoglobulins A and G in serum and cerebrospinal fluid by using an antibody-capture enzyme-linked immunosorbent assay. J. Clin. Microbiol. 1989, 27, 1983–1987.
363. van Loon, A.M.; van der Logt, J.T.; Heessen, F.W.; Heeren, M.C.; Zoll, J. Antibody-capture enzyme-linked immunosorbent assays that use enzyme-labelled antigen for detection of virus-specific immunoglobulin M, A and G in patients with varicella or herpes zoster. Epidemiol. Infect. 1992, 108, 165–174.
364. Gunther, G.; Haglund, M.; Lindquist, L.; Skoldenberg, B.; Forsgren, M. Intrathecal IgM, IgA and IgG antibody response in tick-borne encephalitis. Long-term follow-up related to clinical course and outcome. Clin. Diagn. Virol. 1997, 8, 17–29.
365. Plotkin, S.A. Rabies. Clin. Infect. Dis. 2000, 30, 4–12.
366. Jamnback, T.L.; Beaty, B.J.; Hildreth, S.W.; Brown, K.L.; Gundersen, C.B. Capture immunoglobulin M system for rapid diagnosis of La Crosse (California encephalitis) virus infections. J. Clin. Microbiol. 1982, 16, 577–580.
367. Dykers, T.I.; Brown, K.L.; Gundersen, C.B.; Beaty, B.J. Rapid diagnosis of LaCrosse encephalitis: detection of specific immunoglobulin M in cerebrospinal fluid. J. Clin. Microbiol. 1985, 22, 740–744.
368. Tardei, G.; Ruta, S.; Chitu, V.; Rossi, C.; Tsai, T.F.; Cernescu, C. Evaluation of immunoglobulin M (IgM) and IgG enzyme immunoassays in serologic diagnosis of West Nile virus infection. J. Clin. Microbiol. 2000, 38, 2232–2239.
369. Innis, B.L. Japanese Encephalitis; Chapman & Hall: London: UK, 1995, 147–174.

370. Burke, D.S.; Nisalak, A.; Hoke, C.H., Jr. Field trial of a Japanese encephalitis diagnostic kit. J. Med. Virol. 1986, 18, 41–49.
371. Cuzzubbo, A.J.; Endy, T.P.; Vaughn, D.W.; Solomon, T.; Nisalak, A.; Kalayanarooj, S.; Dung, N.M.; Warrilow, D.; Aaskov, J.; Devine, P.L. Evaluation of a new commercially available immunoglobulin M capture enzyme-linked immunosorbent assay for diagnosis of Japanese encephalitis infections. J. Clin. Microbiol. 1999, 37, 3738–3741.
372. Xu, Y.; Zhaori, G.; Vene, S.; Shen, K.; Zhou, Y.; Magnius, L.O.; Wahren, B.; Linde, A. Viral etiology of acute childhood encephalitis in Beijing diagnosed by analysis of single samples. Pediatr. Infect. Dis. J. 1996, 15, 1018–1024.
373. Kiessling, W.R.; Hall, W.W.; Yung, L.L.; ter Meulen, V. Measles-virus-specific immunoglobulin-M response in subacute sclerosing panencephalitis. Lancet. 1977, 1, 324–327.
374. Gershon, A.; Steinberg, S.; Greenberg, S.; Taber, L. Varicella-zoster-associated encephalitis: detection of specific antibody in cerebrospinal fluid. J. Clin. Microbiol. 1980, 12, 764–767.
375. Casas, I.; Tenorio, A.; De Ory, F.; Lozano, A.; Echevarria, J.M. Detection of both herpes simplex and varicella-zoster viruses in cerebrospinal fluid from patients with encephalitis. J. Med. Virol. 1996, 50, 82–92.
376. Vandvik, B.; Nilsen, R.E.; Vartdal, F.; Norrby, E. Mumps meningitis: specific and non-specific antibody responses in the central nervous system. Acta. Neurol. Scand. 1982, 65, 468–487.
377. Conrad, A.J.; Chiang, E.Y.; Andeen, L.E.; Avolio, C.; Walker, S.M.; Baumhefner, R.W.; Mirzayan, R.; Tourtellotte, W.W. Quantitation of intrathecal measles virus IgG antibody synthesis rate: subacute sclerosing panencephalitis and multiple sclerosis. J. Neuroimmunol. 1994, 54, 99–108.
378. Felgenhauer, K.; Reiber, H. The diagnostic significance of antibody specificity indices in multiple sclerosis and herpes virus induced diseases of the nervous system. Clin. Invest. 1992, 70, 28–37.
379. Weber, T.; Beck, R.; Stark, E.; Gerhards, J.; Korn, K.; Haas, J.; Luer, W.; Jahn, G. Comparative analysis of intrathecal antibody synthesis and DNA amplification for the diagnosis of cytomegalovirus infection of the central nervous system in AIDS patients. J. Neurol. 1994, 241, 407–414.
380. Studahl, M.; Ricksten, A.; Sandberg, T.; Elowson, S.; Herner, S.; Sall, C.; Bergstrom, T. Cytomegalovirus infection of the CNS in non-compromised patients. Acta. Neurol. Scand. 1994, 89, 451–457.
381. Weber, T.; Trebst, C.; Frye, S.; Cinque, P.; Vago, L.; Sindic, C.J.; Schulz-Schaeffer, W.J.; Kretzschmar, H.A.; Enzensberger, W.; Hunsmann, G.; Luke, W. Analysis of the systemic and intrathecal humoral immune response in progressive multifocal leukoencephalopathy. J. Infect. Dis. 1997, 176, 250–254.
382. Sindic, C.J.; Trebst, C.; Van Antwerpen, M.P.; Frye, S.; Enzensberger, W.; Hunsmann, G.; Luke, W.; Weber, T. Detection of CSF-specific oligocolonal antibodies to recombinant J C virus VP1 in patients with progressive multifocal leukoencephalopathy. J. Neuroimmunol. 1997, 76, 100–104.
383. Antoine, J.C.; Pozetto, B.; Lucht, F.; Michel, D.; Gaudin, O.G.; Rousset, H. Acute adenovirus encephalitis diagnosed by prolonged intrathecal antibody production. Lancet. 1987, 1, 1382.
384. Resnick, L.; diMarzo-Veronese, F.; Schupbach, J.; Tourtellotte, W.W.; Ho, D.D.; Muller, F.; Shapshak, P.; Vogt, M.; Groopman, J.E.; Markham, P.D. Intra-blood-brain-barrier synthesis of HTLV-III-specific IgG in patients with neurologic symptoms associated with AIDS or AIDS-related complex. N. Engl. J. Med. 1985, 313, 1498–1504.

5

Herpes Simplex Viruses

Israel Steiner
Hadassah University Hospital
Jerusalem, Israel

1 INTRODUCTION

The herpesviruses are double-stranded DNA viruses with the unique ability to establish latent infection in their hosts and cause recurrent disease by reactivation. Of the human herpesviruses, herpes simplex virus types 1 and 2 (HSV-1 and 2) and varicella-zoster virus (VZV) are neurotropic, that is, they establish latent infection in the peripheral nervous system (PNS) and the viral genome is maintained in peripheral sensory ganglia (PSG) for the entire life of the host. The PSG are the reservoir from which the neurotropic herpesviruses can reactivate and cause neurological and mucocutaneous disorders.

The three neurotropic herpesviruses vary in the clinical disorders they cause and in their molecular structure. However, they share several features that govern the biology of their infection in the human nervous system: (1) The primary infection involves the mucocutaneous surfaces, which serve as the portal of entry of the viral particles into the PNS; (2) the primary and the infectious recurrent diseases caused by the same virus usually occur within the same cutaneous distribution; (3) under normal (i.e., immunocompetent) conditions, the reactivation infection usually does not spread beyond the anatomic distribution and the vicinity of a single PSG; (4) although primary infection usually takes place during the first two to three decades of life, reactivation may occur at any time in the patient's life, sometimes at a very advanced age. These features can be grouped under a unifying hypothesis that is now the dogma in herpes virology [1,2]: Following primary infection, the virus gets access to axon endings within the mucocutaneous surfaces and is transported to the PSG. The viral genome is maintained within the PSG, which serve as a reservoir for viral nucleic acids. Latent herpetic infection is a lifelong state. Under certain circumstances the virus can reactivate and travel to regions innervated by the respective PSG, causing recurrent disease there.

HSV-1 and 2 are closely related viruses that partially differ in their biochemical composition, cytopathic effect, and genetic information as well as their neurotropic features. Nevertheless, their genetic material is homologous to a considerable extent, and they share cross reactivity between their glycoproteins. Likewise, there is a great deal of similarity between the clinical manifestations of the disorders they cause, although they are usually transmitted via different routes and cause primary and recurrent disease in different parts of the body.

HSV-1 is the causative agent of severe encephalitis and is associated with several peripheral nervous system disorders. In recent years it has also been the focus of intensive research aimed at elucidating the molecular basis of its latent nervous system infection and at harnessing this virus as a potential vector for gene therapy. HSV-2 is responsible for encephalitis in neonates, whereas in adults it causes mainly meningitis, radiculomyelitis, and autonomic failure and is associated with peripheral nervous system disease. The present review aims at covering all these biological, medical, and neurological aspects of herpes simplex viruses.

2 PRIMARY HSV INFECTION

Mucocutaneous surfaces are the site of primary infection. Both HSV-1 and HSV-2 can cause genital and orofacial infections that are clinically indistinguishable. However, the mouth and lips are the common site for HSV-1 primary infection, which usually occurs prior to age 5 and in most of the cases is asymptomatic [3]. When clinically apparent, gingivostomatitis and pharyngitis are the frequent manifestations, and fever and cervical lymphadenopathy are then common. Following primary infection, an immune response is triggered and seroconversion takes place.

3 HERPES SIMPLEX ENCEPHALITIS

3.1 Epidemiology and Pathogenesis

Herpes simplex encephalitis (HSE) is the most common cause of sporadic fatal viral encephalitis in the United States [4]. Its incidence ranges between one and three per 100,000. Some HSE cases may go undiagnosed, and in others the disease may have a relatively mild course with spontaneous resolution. Thus, the current prevalence rates may be underestimated. Case-to-case transmission has not been reported. HSE is associated with 70% mortality in untreated cases and with 30% mortality and a high incidence of severe and permanent neurological sequelae under treatment [5].

The pathogenesis of HSE is still unclear. The mechanism of entry of HSV into the brain could be one of three possibilities. (1) Reactivation of the viral genome in the trigeminal ganglion, a natural reservoir of HSV-1 latent infection [6], with resultant axonal spread via the trigeminal nerve into the frontal and temporal lobes [7]; (2) in situ reactivation of the latent virus from central nervous system tissue, where it can occasionally be identified [8]; and (3) primary infection of the nervous system. The latter is supported by the finding that at least half of the cases of HSE are caused by a different viral strain from the one responsible for cold sores in the same individual [9]. Despite anecdotal reports [10], HSE is generally not a disorder of the immunocompromised host [11], except in the context of bone marrow transplantation [12]. This might be attributed [13] mainly to the type of cell that harbors the latent viral genome. With HSV-1 this is exclusively the neuron, although some accumulative evidence suggests that non-neuronal satellite cells also harbor the latent VZV genome [14]. Practically, all HSE cases beyond the neonatal period are due to HSV-1 (and not HSV-2).

3.2 The Clinical Features

Several studies tried to outline the clinical presentation and characteristic features of HSE [4,15,16]. However, these studies date from the era prior to the routine use of the polymerase chain reaction (PCR) for diagnosis, when, in many studies, the ultimate confirmation of the clinical diagnosis, namely brain histology, was missing. The introduction of PCR technology as a major diagnostic tool did and will enlarge and modify our understanding of the clinical spectrum of HSE (see Sec. 3.4).

The symptoms and signs of HSE are related to nonspecific meningoencephalitis: headache, fever, and neck stiffness associated with signs of brain dysfunction and convulsions. These include alterations in consciousness, personality and behavioral disturbances, focal neurological signs, cognitive disturbances, and all types of seizures. More specific to HSE are prodromal symptoms of upper respiratory tract infection and neurological findings related to dysfunction of the fronto-temporal lobes, sometimes mimicking acute psychiatric conditions. However, besides the typical subacute presentation, HSE may sometimes be explosive at onset, and patients may progress to a state of altered consciousness within a matter of hours.

Several points should be emphasized. Fever is one of the most frequent features at presentation, and its absence should cast doubt upon the diagnosis. Headache, a nonspecific complaint, is present in up to 90% of HSE cases. The disease is of subclinical onset, usually less than a week. Gray matter involvement is a dominant feature: personality changes, confusion, and disorientation are present in about three-fourths of the patients and seizures in half. Focal neurological signs (hemiparesis, dysphasia, etc.) are less frequent and are present in about one-third of all patients.

Occasionally patients are initially referred for psychiatric consultation because of delusions, agitation and personality changes. In the setting of the immunocompromised and AIDS patients, the disease may present in a more benign and indolent form with meningitis and few focal findings.

Introduction of the noninvasive technology of PCR for the diagnosis of viral infections enlarges the clinical spectrum of HSE. Indeed, cumulative data may indicate that there are aborted and relatively benign forms of CNS infections due to HSV-1 [17,18]. A detailed analysis of data combining serological, molecular, and clinical information may eventually enable us to redelineate the clinical spectrum of HSV-1-induced CNS infections.

3.3 Differential Diagnosis

Because HSE is a medical emergency and prognosis is dependent on early initiation of therapy, the need for immediate and accurate diagnosis cannot be overemphasized. However, definite diagnosis of the causative agent of viral encephalitis is still frustrating, and even under optimal circumstances and with molecular and serological measures, the causative pathogen can be identified in only 50–70% of cases.

Although the list of conditions that can mimic HSE is large and anecdotal reports of subacute sclerosing panencephalitis (SSPE), cerebrovascular disease, alcoholic encephalopathy, and other conditions have been documented, the differential diagnosis should usually be limited to infectious, parainfectious, and acute inflammatory conditions. Diagnosis is eventually based on CSF analysis and neuroimaging.

Meticulous history taking and examination are mandatory to enable correct clinical diagnosis. For example, immune status of the patient (HSE incidence is not increased in immunocompromised hosts), time of year, and geographic location (seasonal predilection or an endemic tendency are not features of HSV infections) and the presence of an epidemic or a similar disease in the family (absent with HSE) will suggest an alternative diagnosis. A history of recurrent labial or genital sores is not more common in HSE patients than in the general population.

A thorough systemic clinical examination cannot be overstressed. It should include a check for skin lesions (usually absent in HSE but may be present in other infective conditions such as Lyme disease and Rocky Mountain spotted fever) and other organ involvement (upper respiratory tract infection, typical for herpes infection).

3.4 Auxiliary Studies

In the presence of focal neurological signs, neuroimaging prior to lumbar puncture might sometimes be indicated, provided that it can be obtained immediately. Nevertheless, it is our impression that the introduction of reliable and noninvasive neuroimaging technology, although helping to exclude conditions that can simulate HSE and other CNS infections, may unfortunately sometimes delay diagnosis when immediate therapy is mandatory. Indeed, a recent study addressed this issue with similar conclusions [19]. Therefore, if neuroimaging cannot be obtained within the briefest span of time, lumbar puncture should be postponed only when strict contraindications (such as suspicion of increased intracranial pressure due to a lesion of the posterior fossa space) are present. When such a contraindication exists, antiherpetic therapy should be introduced, though it might eventually be amended according to neuroimaging and CSF findings when those are available.

A CT scan is usually normal within the first 4–6 days of disease [20]. Magnetic resonance imaging (MRI) may demonstrate a lesion of high signal intensity on T2-weighted and flair images at earlier times, typically at the temporal and frontal lobes [21].

The CSF is abnormal in more than 95% of HSE patients. It contains moderate pleocytosis, usually mononuclear, and red blood cells, reflecting the hemorrhagic nature of the infectious process within brain parenchyma. A moderate rise in CSF protein is present in more than 80% of cases, whereas hypoglycorrhachia is the exception (present in less than 5% of biopsy-proven HSE patients [22]). Hypoglycorrhachia should therefore suggest an alternative diagnosis.

Serodiagnosis is based on demonstration of intrathecal production of anti-HSV antibodies. A calculation used to correct for nonspecific passage of such antibodies from the serum [23] is based on the ratio

$$\frac{\text{CSF anti-HSV titer/serum anti-HSV titer}}{\text{CSF albumin/serum albumin}} \tag{5.1}$$

An index greater than 1.9 is considered abnormal [24]. Alternatively, a simpler version may be used:

$$\frac{\text{Serum anti-HSV titer}}{\text{CSF anti-HSV titer}} \tag{5.2}$$

where a ratio of less than 20:1 is abnormal. An increase in serum and CSF anti-HSV-1 titers over time (which requires a repeated lumbar puncture) is also supportive of the diagnosis.

The polymerase chain reaction has become the mainstay of diagnosis, being very sensitive (98%) and specific (94%) compared to the gold standard of histology obtained by brain biopsy [25,26]. False negative results might be present during the first days of disease, and when in doubt a repeat lumbar puncture after 1–2 days is indicated. PCR is an extremely sensitive method, and appropriate controls must be employed to avoid false positive results. The primers used for amplification should be carefully designed to prevent recognition of non-HSV-1 sequences, and the molecular examination has to be performed in a qualified laboratory setting.

We have examined our personal institutional experience with PCR diagnosis of HSE and the correlation of PCR findings with the clinical presentation and course. Our recommendation is that when PCR results and the clinical diagnosis do not agree, repeated lumbar puncture with serological and PCR analysis should be performed 3–5 days following the initial tap and might make it possible to establish the final diagnosis.

Electroencephalography (EEG) is a sensitive but nonspecific diagnostic aid at early stages of the disease [27]. It may enable localization of pathology to fronto-temporal brain regions before any abnormality can be visualized on imaging (mainly CT) studies. The findings are usually those of periodic sharp and slow wave complexes.

Brain biopsy, strongly advocated in the pre-acyclovir, pre-PCR era, is now seldom performed but might be indicated in cases refractory to therapy or in the investigation of a relapsing encephalitic illness. When performed, it should be guided by clinical and radiological localization and aimed at obtaining histology and virus isolation that will establish diagnosis beyond doubt.

3.5 Therapy

Herpes simplex encephalitis is treated with acyclovir 10 mg/kg IV every 8 h for 2 weeks. Acyclovir is an antiviral compound developed for use against herpes simplex infections on the basis of an idea that was one of the reasons that its inventor, Gertrud B. Elion, was granted the Nobel prize [28]. It is phosphorylated to its active form by the virus-specific thymidine kinase enzyme and prevents viral replication by inhibiting the viral (as well as cellular) DNA polymerase in infected cells. Topical, oral, and parenteral preparations are available. Following cases of relapse after a 10 day course of acyclovir therapy [29,30], the current recommended protocol has increased from 10 days of therapy to 2 and even 3 weeks. Although it is unclear for how long HSV DNA can be detected in the CSF after initiation of therapy, a consensus report [31] as well as retrospective assessment [32] suggest that when in doubt, identification of viral DNA by PCR on reexamination of the CSF may indicate the need for an additional 1–2 weeks of acyclovir therapy, or question the initial diagnosis (when negative).

Several new anti-HSV preparations are available. Because acyclovir has only 15–39% oral absorption, valaciclovir, an acyclovir prodrug for oral administration with better bioavailability, was introduced [33,34]. Penciclovir achieves higher intracellular concentrations than acyclovir [35], and famciclovir is a prodrug of penciclovir with better oral bioavailability [36]. So far, however, none of these has replaced acyclovir for the treatment of HSE.

Additional therapeutic considerations include:

Prophylactic therapy with anticonvulsants. Phenytoin is the drug of choice, because it is easily administered and monitored even in the comatose patient.

Respiratory assistance. This may be used to reduce increased intracranial pressure, and therefore early intubation is generally recommended.

ICP monitoring. In cases of increased intracranial pressure, monitoring of ICP via a pressure transducer or other device is advised.

3.6 Prognosis

Survival is dramatically improved by therapy, but the quality of survival is still unsatisfactory. The age of the patient and the level of consciousness at the time of introduction of therapy are prognostic determinants. Cognitive impairment remains the main problem despite early diagnosis and/or treatment and good early outcome [37]. Thus, normal neuropsychiatric function was present in only 1/17 of HSE survivors [4].

4 NEONATAL HSV-2 ENCEPHALITIS

About 90% of neonatal encephalitis cases are due to HSV-2, and most are acquired from a mother who has active genital herpes infection at the time of delivery [38]. Whether primary infection or a recurrent disease, the infection in most mothers is unrecognized either because they do not have a previous history of HSV-2 infection or because the disease at the time of delivery is asymptomatic [39]. Moreover, approximately 0.3% of all women shed HSV around the time of delivery [40].

Neonates present with systemic findings (alterations in body temperature, lethargy, respiratory distress, anorexia, vomiting, cyanosis) and neurological signs (irritability, bulging fontanele, seizures, opisthotonus, and coma) [41]. The infection can take one of three patterns: disseminated infection, an isolated CNS disease, or a focal infection confined to the skin, eye, or mouth. If not treated immediately, the last pattern will develop into a disseminated condition, which carries the worst prognosis. The skin, eye, and/or mouth findings are present in about 80% of all cases and are highly suggestive of the diagnosis.

The condition may resemble bacterial sepsis or meningitis, and therefore laboratory diagnosis is mandatory. This can be done rapidly by isolation of the virus from maternal genital lesions and secretions, by isolation of the virus from the vesicles in the newborn, or by examination of peripheral blood and cerebrospinal specimens from the neonate. Staining of samples for viral antigen and PCR analysis will establish diagnosis within several hours. Serology cannot be relied upon, because IgG might be maternal and IgM is not produced in the infant until 2 weeks after disease onset.

Immediate therapy is mandatory. Although it was shown that vidarabin and acyclovir are equally effective in reducing mortality and minimizing neurological sequelae [42], acyclovir is more easily administered and has fewer side effects and is therefore the current first treatment of choice. It should be given at 30 mg/kg/day intravenously in divided doses every 8 h for 10–14 days.

Even under treatment, prognosis is still very poor. One-third to 54% of all treated babies with disseminated disease die, and about two-thirds of survivors have neurological sequelae.

This prognosis raises two issues:

1. Should a neonate who was discovered postnatally to have been delivered via an HSV-lesioned or culture-positive birth canal be treated prophylactically? Although opinions differ [41,43], it is recommended that even if the infant is untreated, swabs from eyes, nasopharynx, and mouth should be obtained and the infant closely monitored clinically.
2. Is cesarean section recommended in all cases of mothers with a history of HSV genital infection? This procedure does not provide full protection from transmission of infection from mother to baby [44], and some reserve it for cases with active disease at the time of impending delivery.

5 MOLLARET'S MENINGITIS

Mollaret's meningitis is characterized by recurrent self-limiting meningitis in otherwise healthy individuals [45]. Recurrence takes place at intervals of several weeks to months and has been documented after up to 28 years [46]. The attacks last for several days and then resolve. CSF contains from 200 to several thousands of lymphocytes per cubic millimeter, and large endothelial cells termed ''Mollaret cells'' may be present. Protein levels in the CSF are elevated, and glucose may sometimes be low [47]. Complete recovery occurs within several days. Diagnosis is established after other causes of lymphocytic meningitis have been ruled out

Mollaret originally suspected either a hypersensitivity condition or an infectious disease. In some patients the disorder has been associated with Epstein-Barr virus [48], HSV-1 [49], HSV-2 [50], and histoplasmosis [51]. In 1991 Berger reported three patients in whom Mollaret's meningitis followed recrudescence of genital herpes [52]. The attacks responded to acyclovir in two patients, and in one the disorder did not recur following treatment. Prior to the introduction of PCR technology the possibility that the disorder was due to an HSV infection was intriguing because of the recurrent, self-limiting nature of the condition and because of the CSF findings, which were compatible with viral infection. Indeed, PCR made it possible to identify the etiology of the disease in most if not all patients and to attribute it either to HSV-2 e.g., [53–55], or to HSV-1 in selected cases [56,57]. In one study the identification of HSV-2 DNA in the CSF of patients with Mollaret's meningitis was associated with the presence of Mollaret cells in the CSF [54]. Whether to treat patients during the acute attacks remains an open question. Although some reports suggested either shorter episodes [54,55] or resolution of the syndrome [52,53], one could argue that a short course of antiherpetic treatment does not affect the viral reservoir and is not associated with prevention of future mucocutaneous disease [58].

6 MYELORADICULITIS

The reservoir of the latent neurotropic herpesviruses is the PSG. Because the viruses can reactivate and travel antidromically to the periphery and cause recurrent mucocutaneous disease (with HSV-1 and 2) and zoster (in the case of VZV infection), it was postulated that they can also travel to the spinal cord and brainstem [59]. Indeed, zoster is associated with subclinical myelitis in about half of the patients [60].

Herpes simplex virus was shown to be responsible for cases of severe necrotizing myelitis that culminated in death and came to autopsy [61]. PCR technology and magnetic resonance imaging provided the means to prove that both HSV-2 and HSV-1 are associated with a spectrum of myelitic syndromes: mild forms of myelitis, transverse myelitis and ascending myelitis, severe and fatal ascending myelitis, ascending myeloradiculitis, and recurrent myelitis [62–67]. Cases in children were also documented [68]. What determines the severity of the disease is unknown. Thus, it seems reasonable that in every case of acute or subacute myelitis, CSF should be analyzed by PCR and serology for both herpes simplex viruses, and when results are positive, treatment with acyclovir should be introduced.

7 PERIPHERAL NERVOUS SYSTEM DISORDERS

7.1 Primary and Reactivated Infection

Compared with CNS morbidity due to HSV-1 and with VZV-induced peripheral disease, PNS infections due to HSV in the immunocompetent host are of little clinical significance, are rare, and sometimes merit a special report. Thus, radiculomyelopathy following primary HSV-2 infection [69] and acute autonomic neuropathy with favorable prognosis following oral HSV infection [70] were recorded. Sacral radiculitis associated with pain and paresthesias has been reported in genital and rectal primary HSV-1 and HSV-2 infections [69,71]. Acute inflammatory demyelinating polyneuropathy (AIDP), or Guillain-Barre syndrome (GBS), following either primary or reactivated HSV infection seems to be more prevalent [72,73], but there are no exact clinical and epidemiological data to delineate this syndrome.

With both viruses reactivation can be asymptomatic; e.g., viral shedding may not be associated with clinical disease. Reactivation at the orofacial or genital site is caused by the same type of virus that caused the primary infection at the same site. This bears on the frequency of reactivation. In the genital region, HSV-2 has a much higher frequency of reactivation than HSV-1 [74].

In the orofacial region, recurrent disease is almost always due to HSV-1, and it is triggered by exposure to sunlight, severe stress, trauma (such as neurosurgical manipulation), fever, or menses. Herpes labialis (cold sores) is an extremely frequent condition affecting 30–61% of the population [75]. Clinically, lesions are closely clustered vesicles that usually recur at the same site, tend to remain very localized, and do not involve the entire dermatome. They take up to 10–14 days to heal. Some patients can anticipate the appearance of cold sores by mild sensory symptoms such as tingling or burning but only very rarely by neuralgic pain, which is transitory and is not associated with any permanent sensory deficit [76,77].

Genital recurrent infection, mainly due to HSV-2, occurs in up to 60% of patients who have a latent genital herpetic infection [78] and is usually of shorter duration but tends to be more painful than the HSV-1 recurrent infection. HSV-2 has also been isolated from the urethra and urine of women and men without concomitant genital lesions with and without urethritis. Occasionally recurrent herpes genitalis is associated with radiculitis and meningitis [69,71]. Both the primary and recurrent infection may be complicated by radicular pain that can simulate spinal column disease. When sphincter disturbances such as urinary retention or difficulty in initiation of micturition occur, it may be difficult or even impossible to distinguish between polyradiculitis and low segmental myelitis. Most patients have pleocytosis in the CSF and elevated protein.

7.2 Idiopathic Facial (Bell's) Palsy

Idiopathic peripheral facial nerve paralysis (Bell's palsy) is much less common following HSV infection than following VZV, with a prevalence of 0–9% post-HSV compared to 14–25% after VZV [79]. This association, which in recent years has attracted some attention, merits special consideration.

Based on the assumption that HSV-1 resides in the geniculate ganglion in a similar fashion to colonization of other PSG, McCormick hypothesized in 1972 [80] that HSV might be the causative agent of Bell's palsy. Several clinical anecdotes seem to favor the hypothesis: A man developed facial diplegia after having oral sex with a partner who had active genital herpes simplex infection [81]; a patient developed Bell's palsy after having cold sores due to HSV-1 [82]; and HSV DNA was identified by PCR amplification in the geniculate ganglion of a patient who had Bell's palsy 6 weeks prior to his death [83].

More circumstantial evidence followed. HSV nucleic acids were identified in the geniculate ganglion [84,85]. By PCR, HSV DNA was detected in endoneural fluids of the facial nerve or the auricular muscle of nine out of 14 patients with idiopathic facial palsy [86], and the addition of acyclovir to prednisone for the treatment of this condition resulted in a very moderate but statistically significant clinical improvement compared to prednisone therapy alone [87].

However, in humans, HSV-1 reactivation is not associated with motor impairment and HSV-1 causes recurrent reactivation whereas Bell's palsy in the overwhelming majority of cases is a single episode and a rare condition compared to the incidence of cold sores. Moreover, even if HSV-1 has an etiological role in Bell's palsy in a subgroup of patients, this does not imply a straightforward infective pathogenesis. The nerve may be damaged by edema and pressure within a narrow bony canal, by ischemia due to vascular congestion, or by a dysimmune-mediated condition. We therefore suggest, unlike several reviews that addressed the same question [88,89], that the current data are still not sufficient to indicate anti-HSV-1 therapy in Bell's palsy [90].

7.3 Diagnosis and Treatment

The standard laboratory procedure to diagnose peripheral mucocutaneous HSV infection is to grow the virus from the lesions or fluids in tissue culture [91]. Cultures are positive in 60–90% of clinically suspected cases. Additional means may include cytology, histopathology of biopsied lesions, and the use of monoclonal antibodies directed against HSV-1 [92]. Serological diagnosis is not of great clinical value. PCR may become a useful diagnostic tool in the differential diagnosis of cutaneous lesions [93].

Topical, oral, and parenteral preparations of acyclovir have been reported to be effective in controlling recurrent HSV infection in immunocompetent [58] as well as immunocompromised patients [94]. This, however, carries the risk of developing acyclovir-resistant viral strains, a special concern in immunocompromised patients. In immunocompetent individuals, cessation of acyclovir therapy leads to reactivation rates similar to those present prior to therapy, reflecting the fact that the viral reservoir itself is probably not affected by this mode of therapy.

8 LATENT HSV-1 INFECTION

Latent viral infection is defined as the presence of the viral genome in the host tissue without production of infective viral particles. During latency the pathogen maintains the potential to reactivate, resume replication, and cause recurrent disease. In the case of HSV-1, molecular criteria must be added to this definition, because during latency the structure of the viral DNA and the pattern of its gene expression are not the same as during viral replication in cell culture.

The latent HSV-1 infection requires several functions: establishment of the latent state, maintenance of latency in the host cell and organism, protection of the viral reservoir, and reactivation. Because this condition is associated with severe human morbidity, elucidating the molecular and cellular basis of HSV-1 latent infection in the human nervous system has attracted much interest.

HSV-1 is relatively easy to study. It is easily propagated in cells in culture and readily infects experimental animals. It is a lytic virus: During HSV-1 replication in cells in culture the infected cell is destroyed, but replication at the peripheral site of primary infection or in the peripheral sensory ganglia is not mandatory for the establishment of latency [95]. On the contrary, lack of replication carries obvious advantages for the virus because destruction of the host cell would prevent its ability to reside within the cell and establish latent infection [59,96].

The viral particles are transported from the peripheral site of primary infection by retrograde axonal transport to the PSG, where they establish latent infection. During latency the viral DNA changes form and either becomes circular or is present as a concatamer [97] that is maintained in a nonintegrated (episomal) form in the neuronal host cell [98]. The function and metabolism of the infected neuronal cell remain unaltered. Although theoretically the virus can reside latently in non-neuronal cells, all the information suggests that in PSG HSV-1 resides only in neuronal cells.

During latency, viral gene expression is restricted [6,99,100] and consists of two types of RNA species: more abundant RNAs termed latency-associated transcripts (LATs) and less abundant minor LATs (mLATs). The function of the latency-associated genes has been extensively explored. Although the phenotype of aberrant LATs' expression is defective HSV-1 reactivation from latent infection [101–103], cumulative data may suggest that the LATs also participate in the virus's ability to establish latent infection [104,105] and maintain it (our unpublished data). The exact DNA sequences that mediate this function and the potential gene products of the latency-associated genes are unknown, but we have demonstrated that the LATs bind to the neuronal protein synthesis machinery, the polyribosomes [106,107], suggesting that latency-associated proteins might indeed be produced. HSV-1 reactivation, even when multiple and recurrent, is not accompanied by permanent sensory deficit. We have therefore suggested [59] that these events are not associated with neuronal cell destruction. It follows that HSV-1 does not replicate in PSG during reactivation in the same way as during its replication in cultured cells. Whether the latent HSV-1 reservoir in trigeminal ganglia is the source of the virus causing HSE is still doubtful, as discussed earlier.

Because latency is established primarily, if not exclusively, in neurons, it appears that some host and cellular properties of the neuron might favor latency by arresting viral replication. Indeed, neurons, at least in culture, express factors that have an inhibitory effect upon HSV-1 replication at an early stage that is not yet associated with irreversible cell damage [108–110].

Several biological and molecular issues of latency are still not resolved. Do the latency associated genes produce gene products, and how do they function? What functions mandatory for the latent state are mediated by these genes? Is the latent viral reservoir the source of herpes encephalitis? Why is latent HSV-1 infection confined to neurons and the nervous system?

9 THE ASSOCIATION OF HSV-1 WITH ALZHEIMER'S DISEASE

Several neurological conditions have been associated with acute, chronic, or reactivated HSV infections. In some, such as multiple sclerosis [111], there is presently only meager evidence to implicate the virus in the etiology and pathogenesis of the disease. However, in some other conditions such as Alzheimer's disease (AD), intractable focal epilepsy [112], and acute disseminated encephalomyelitis [111,113], the data appear to be quite intriguing and may support the possibility that in a subgroup of patients HSV-1 has a causative etiopathogenetic role. Of special note is the possibility that in some sporadic cases of AD, HSV-1 might be the trigger that initiates or contributes to the pathogenetic cascade.

The characteristic pathological features of Alzheimer's disease (neurofibrillary tangles, senile plaques, granulovacuolar degeneration, Hirano bodies, and neuronal loss) favor the limbic system, the same region that takes the brunt during HSE. This led to the speculation that there might be a cause-and-effect relationship between herpes infection of the CNS and sporadic Alzheimer's disease [114]. Whether following primary infection or due to reactivation, it was based in part on the indications that trigeminal nerve fibers innervate meninges and vessels within the middle and anterior cranial fossae in regions preferentially afflicted during HSE. Indeed, the search for HSV gene products and sequences in brains of AD patients yielded anecdotal reports of evidence for HSV-1 infection in brains of patients with dementia [115,116]. More substantial evidence came from the demonstration that the APOE-4 allele of the APOE gene is a susceptibility factor for development of AD in patients harboring HSV-1 in their brain parenchyma [117,118]. Although this relationship is currently only a basis for speculation [119], one explanation might be that prior damage (such as occurs in HSE, but also during head trauma [120] and other conditions) and a possible defect in repair ability present in APOE-4 carriers [121] may eventually pave the way to the development of AD.

10 THE POTENTIAL USE OF HERPES VECTORS FOR GENE THERAPY TO THE NERVOUS SYSTEM

The "coming of age" of viral gene therapy for the nervous system is the outcome of the increasing knowledge of viral biochemical genetics, the rapidly increasing sophistication of molecular genetic engineering techniques, and the increasing understanding of the molecular basis of both rare and common neurological diseases. The basic idea is to use viruses as targeted vehicles to deliver genetic information across anatomical and cellular barriers and into specific sites and cells. Like other viruses, HSV-1 has the ability to cross cell membranes, a property that is not shared by many gene products such as certain key enzymes.

Advances in the understanding of the molecular biology of HSV-1 latency in the human nervous system has made this virus a particularly promising gene therapy agent [122,123]. Ideally, viral vector therapy should fulfill the following requirements:

1. Because one of the prime objectives of gene delivery is the ability to "package" an additional foreign gene into the vector, the size of the gene is of major

importance. The 152 kb of the HSV-1 genome can harbor at least 30 kb of foreign DNA after the removal of viral sequences not essential for viral replication [124].

2. The vector must be easily delivered to the targeted tissue. HSV spreads naturally within the nervous system, e.g., from the periphery to the PSG and to the CNS via axonal transport and trans-synaptically [125,126], and has been used as a neuronal tracer [126]. Thus, intracranial inoculation with vectors into, for example, the basal ganglia in Parkinson's disease, which clearly involves significant surgical risks, can become unnecessary.
3. The vector must reach its target and continuously express the required gene without any harm to the cells and tissues. This should be possible because HSV-1 establishes lifelong latency in neurons, primarily in PSG (mainly trigeminal ganglia) but also elsewhere in the CNS [8,127]. The latent HSV-1 infection is lifelong and episomal [98]. Neuronal functions including electrophysiological activity remain unaltered, and HSV-induced mutations in neuronal DNA are very unlikely.
4. The virus must not replicate or reactivate in target neurons. Indeed, prior viral DNA replication, and hence damage to the nervous issue, is not required for the establishment of latency [95]. Moreover, the LATs are required for the virus to be able to reactivate, and the deletion of the latency-associated genes will render the virus reactivation incompetent.
5. The vector must reach the appropriate population of neurons. With accumulating knowledge about neuronal function and markers, this difficulty may prove to be largely surmounted by judicious choice of promoter elements within the HSV-1 construct. The regulatory components are required to allow the inserted gene to be transcribed and expressed in particular cells and tissues and therefore determine the cell specificity for the expression of the inserted gene. For example, insertion of the neurone-specific enolase promoter into a vector allows the gene to be expressed only in neurons [128].
6. The foreign gene must be stably expressed in target neurons into which it has been episomally incorporated. Even if the foreign gene is shown to be expressed and its product identified in target tissues, there will be little likelihood of therapeutic benefit if its expression is only transient. The regulatory elements of the LATs have the potential to enable stable expression of any gene whose expression they control.
7. Even if expression of the delivered gene is achieved, the ability to control the level of expression of the gene products will be crucial. Underexpression of the gene will render the procedure ineffective, and overexpression may have severe consequences such as toxicity and the appearance of new cellular malfunctions. Control can be achieved by the use of an inducible promoter [129] using regulatory elements that require additional packaging space. These are available within the large carrying capacity of HSV-1. Alternatively, certain HSV-1 endogenous genes such as thymidine kinase (TK) may, under certain conditions, render the cells that express the viral vector susceptible to the action of antiviral agents [130] and amenable to destruction if the exogenous genes manifest unwarranted effects. These genes are termed suicidal genes.

There are a variety of potential ways in which defective HSV-1 vectors could be used for gene delivery to neurons. The most obvious application is to replace a gene that is known to be missing or defective—the rare single-gene disorders—based on the assumption that the gene delivered will be expressed and the appropriate gene product, e.g., enzyme or other protein, will be produced in situ. HSV-1 vectors may also be used to deliver therapeutic agents such as antimitotic and antiviral agents, antibiotics, and monoclonal antibodies. Another approach is the delivery of an ''antisense'' message to particular cells, e.g., virus-infected cells, which may abolish the activity of selected viral genes that have been incorporated into the host genome or are present episomally or the activity of endogenous deleterious genes.

HSV-1-derived vectors for gene therapy may be used for the treatment of a large spectrum of neurological disorders. These include inherited metabolic brain diseases known to be caused by a specific defect [131]; neurodegenerative disorders, many of which are untreatable at present, that can be approached either by a disease-specific single gene replacement [132] or by the introduction of neurotropic and antiapoptotic factors [133]. Similar nonspecific genes may also help in enhancing brain injury repair in neurological diseases such as head trauma, stroke, or multiple sclerosis [134,135]. Finally, introduction of suicide genes has already been used experimentally to treat brain tumors [136].

ACKNOWLEDGEMENTS

I am greatly indebted to Drs. Bettina Steiner-Birmanns and Itzchak Wirguin for critical review of the manuscript and helpful suggestions. This work was supported in part by grants from the Chief Scientist, Ministry of Health, Israel, the Herman de Stern Fund, and Yad Hanadiv.

REFERENCES

1. Goodpasture, E.W. Herpetic infections with special reference to involvement of the nervous system. Medicine. 1929, 8, 223–243.
2. Hope-Simpson, R.E. The nature of herpes zoster: A long term study and a new hypothesis. Proc. Roy. Soc. Med. 1965, 58, 9–20.
3. Cesario, T.C.; Poland, J.D.; Wulff, H.; Chin, T.D.; Wenner, H.A. Six years experience with herpes simplex virus in a children's home. Am. J. Epidemiol. 1969, 90, 416–422.
4. Skoldenberg, B. Herpes simplex encephalitis. Scan. J. Infect. Dis. 1991, 78(suppl), 40–46.
5. Whitley, R.; Arvin, A.; Prober, C.; Corey, L.; Burchett, S.; Plotkin, S.; Starr, S.; Jacobs, R.; Powell, D.; Nahmias, A. A controlled trial comparing vidarabine with acyclovir in neonatal herpes simplex infection. N. Engl. J. Med. 1991, 324, 444–449.
6. Steiner, I.; Spivack, J.G.; O'Boyle, D.R.; Lavi, E.; Fraser, N.W. Latent herpes virus type 1 transcription in human trigeminal ganglia. J. Virol. 1988, 62, 3493–3496.
7. Davis, L.E.; Johnson, R.T. An explanation for the localization of herpes simplex encephalitis. Ann. Neurol. 1979, 5, 2–5.
8. Fraser, N.W.; Lawrence, N.C.; Wroblewska, Z.; Gilden, D.H.; Koprowski, H. Herpes simplex virus type 1 DNA in human brain tissue. Proc. Natl. Acad. Sci. USA. 1981, 78, 6461–6465.
9. Whitley, R.; Lakeman, A.D.; Nahmias, A.; Roizman, B. DNA restriction-enzyme analysis of herpes simplex virus isolates obtained from patients with encephalitis. N. Engl. J. Med. 1982, 307, 1060–1062.
10. Chretien, F.; Belec, L.; Hilton, D.A.; Flament-Saillour, M.; Guillon, F.; Wingertsmann, L.; Baudrimont, M.; de Truchis, P.; Keohane, C.; Vital, C.; Love, S.; Gray, F. Herpes simplex virus type 1 encephalitis in acquired immunodeficiency syndrome. Neuropathol. Appl. Neurobiol. 1996, 22, 394–404.

11. Whitley, R.J.; Lakeman, F. Herpes simplex virus infections of the central nervous system: therapeutic and diagnostic considerations. Clin. Infect. Dis. 1995, 20, 414–420.
12. Darville, J.M.; Ley, B.E.; Roome, A.P.; Foot, A.B. Acyclovir-resistant herpes simplex virus infections in a bone marrow transplant population. Bone Marrow Transplant. 1998, 22, 587–589.
13. Kennedy, P.G.E.; Steiner, I. A molecular and cellular model to explain the differences in reactivation from latency by herpes simplex and varicella-zoster viruses. Neuropathol. Appl. Neurobiol. 1994, 20, 368–374.
14. Croen, K.D.; Ostrove, J.M.; Dragovic, L.J.; Straus, S.E. Patterns of gene expression and sites of latency in human nerve ganglia are different for varicella-zoster and herpes simplex viruses. Proc. Natl. Acad. Sci. USA. 1988, 85, 9773–9777.
15. Whitley, R.J. Herpes simplex virus infections of the nervous system. Drugs. 1991, 42, 406–427.
16. Kennedy, P.G.E. A retrospective analysis of 46 cases of herpes simplex encephalitis seen in Glasgow between 1962 and 1985. Quart. J. Med. 1988, 168, 533–540.
17. Studahl, M.; Bergstrom, T.; Hagberg, L. Acute viral encephalitis in adults—a prospective study. Scand. J. Infect. Dis. 1998, 30, 215–220.
18. Fodor, P.A.; Levin, M.J.; Weinberg, A.; Sandberg, E.; Sylman, J.; Tyler, K.L. Atypical herpes simplex virus encephalitis diagnosed by PCR amplification of viral DNA from CSF. Neurology. 1998, 51, 554–559.
19. Hasbun, R.; Abrahams, J.; Jekel, J.; Quagliarello, V.J. Computed tomography of the head before lumbar puncture in adults with suspected meningitis. N. Engl. J. Med. 2001, 345, 1727–1733.
20. Hindmarsh, T.; Lindqvist, M.; Olding-Stenkvist, E.; Skoldenberg, B.; Forsgren, M. Accuracy of computed tomography in the diagnosis of herpes simplex encephalitis. Acta. Radiol. 1986, 369(suppl), 192–196.
21. Schroth, G.; Gawehn, J.; Thron, A.; Vallbracht, A.; Voigt, K. Early diagnosis of herpes simplex encephalitis by MRI. Neurology. 1987, 37, 179–183.
22. Sawyer, J.; Eliner, J.; Ransohoff, D.F. To biopsy or not to biopsy in suspected herpes simplex encephalitis: a quantitative analysis. Med. Decis. Making. 1988, 8, 95–101.
23. Klapper, P.E.; Laing, L.; Longson, M. Rapid non-invasive diagnosis of herpes encephalitis. Lancet. 1981, ii, 607–609.
24. Vandvik, B.; Skoldenberg, B.; Forsgren, M.; Stiernstedt, G.; Jeansson, S.; Norrby, E. Long term persistence of intrathecal virus-specific antibody response after herpes simplex virus encephalitis. J. Neurol. 1985, 231, 307–312.
25. Lakeman, F.D.; Whitely, R.J. NIAID Collaborative Antiviral Study Group. Diagnosis of herpes simplex encephalitis: application of polymerase chain reaction to cerebrospinal fluid from brain biopsied patients and correlation with disease. J. Infect. Dis. 1995, 171, 857–863.
26. Hofgartner, W.T.; Huhmer, A.F.; Landers, J.P.; Kant, J.A. Rapid diagnosis of herpes simplex encephalitis using microchip electrophoresis of PCR products. Clin. Chem. 1999, 45, 2120–2128.
27. Dutt, M.K.; Johnston, I.D. Computed tomography and EEG in herpes simplex encephalitis. Their value in diagnosis and prognosis. Arch. Neurol. 1982, 39, 99–102.
28. Elion, G.B. Nobel Lectures in Physiology or Medicine—(1988). The purine path to chemotherapy. In Vitro. Cell. Dev. Biol. 1989, 25, 321–330.
29. VanLandigham, K.E.; Marsteller, H.B.; Roos, G.W. Relapse of herpes simplex encephalitis after conventional acyclovir therapy. JAMA. 1988, 259, 1051–1053.
30. Kimura, H.; Aso, K.; Kuzushima, K.; Hanada, N.; Shibata, M.; Morishima, T. Relapse of herpes simplex encephalitis in children. Pediatrics. 1992, 89, 891–894.
31. Cinque, P.; Cleator, G.M.; Weber, T.; Monteyne, P.; Sindic, C.J.; van Loonm, A.M. The role of laboratory investigation in the diagnosis and management of patients with suspected herpes

simplex encephalitis: a consensus report. J. Neurol. Neurosurg. Psychiatry. 1996, 61, 731–745.
32. Ito, Y.; Kimura, H.; Yabuta, Y.; Ando, Y.; Murakami, T.; Shiomi, M.; Morishima, T. Exacerbation of herpes simplex encephalitis after successful treatment with acyclovir. Clin. Infect. Dis. 2000, 30, 185–187.
33. Patel, R.; Bodsworth, N.J.; Woolley, P.; Peters, B.; Vejlsgaard, G.; Saari, S.; Gibb, A.; Robinson, J. Valaciclovir for the suppression of recurrent genital HSV infection. A placebo controlled study of a once daily therapy. Genitourin. Med. 1997, 73, 105–109.
34. Jacobson, M.A. Valaciclovir (BW256U87): the 1-valyl ester of acyclovir. J. Med. Virol. 1993, 1(suppl), 150–153.
35. Vere Hodge, R.A.; Cheng, Y-C. The mode of action of penciclovir. Antiviral Chem Chemother. 1993, 4(suppl 1), 13–24.
36. Pue, M.A.; Benet, L.Z. Pharmacokinetics of famciclovir in man. Antiviral. Chem. Chemother. 1993, 4(suppl 1), 47–55.
37. Gordon, B.; Selnes, O.A.; Hart Jr, J.; Hanley, D.F.; Whitley, R.J. Long term cognitive sequellae of acyclovir-treated herpes simplex encephalitis. Arch. Neurol. 1990, 47, 646–647.
38. Whitley, R.; Arvin, A.; Prober, C.; Corey, L.; Burchett, S.; Plotkin, S.; Starr, S.; Jacobs, R.; Powell, D.; Nahmias, A. Predictors of morbidity and mortality in neonates with herpes simplex virus infections. The National Institute of Allergy and Infectious Diseases Collaborative Antiviral Study Group. N. Engl J. Med. 1991, 324, 450–454.
39. Brown, Z.A.; Benedetti, J.; Ashley, R.; Burchett, S.; Selke, S.; Berry, S.; Vontver, L.A.; Corey, L. Neonatal herpes simplex virus infection in relation to asymptomatic maternal infection at the time of labor. N. Engl. J. Med. 1991, 324, 1247–1252.
40. Prober, C.G.; Hensleigh, P.A.; Boucher, F.D.; Yasukawa, L.L.; Au, D.S.; Arvin, A.M. Use of routine viral cultures at delivery to identify neonates exposed to herpes simplex virus. N. Engl. J. Med. 1988, 318, 887–891.
41. Overall Jr, J.C. Herpes simplex virus infection of the fetus and newborn. Pediatr. Ann. 1994, 23, 131–136.
42. Whitley, R.; Arvin, A.; Prober, C.; Burchett, S.; Corey, L.; Powell, D.; Plotkin, S.; Starr, S.; Alford, C.; Connor, J. A controlled trial comparing vidarabine with acyclovir in neonatal herpes simplex virus infection. Infectious Diseases Collaborative Antiviral Study Group. N. Engl. J. Med. 1991, 324, 444–449.
43. Overall Jr, J.C.; Whitley, R.J.; Yeager, A.S.; McCracken Jr, G.H.; Nelson, J.D. Prophylactic or anticipatory antiviral therapy for newborns exposed to herpes simplex infection. Pediatr. Infect. Dis. 1984, 3, 193–195.
44. Gibbs, R.S.; Amstey, M.S.; Lezotte, D.C. Role of cesarean delivery in preventing neonatal herpes virus infection. JAMA. 1993, 270, 94–95.
45. Mollaret, P. La meningite endothelio-nucleocytaire multirecurrente benigne: syndrome nouveau ou maladie nouvelle?. Rev. Neurol. 1944, 76, 57–76.
46. Tyler, K.L.; Adler, D. Twenty-eight years of benign recurring Mollaret meningitis. Arch. Neurol. 1983, 40, 42–43.
47. Fredericks, J.A.M.; Bruyn, G.W. Mollaret's meningitis. In Handbook of Clinical Neurology, sr. II, Viral Disease; Vinken, P.J., Bruyn, G.W., Klawans, H.L., Eds.; Elsevier: Amsterdam, 1989; Vol. 12, 627–635.
48. Graman, P.S. Mollaret's meningitis associated with acute Epstein-Barr virus mononucleosis. Arch. Neurol. 1987, 44, 1204–1205.
49. Steel, J.G.; Dix, R.D.; Baringer, J.R. Isolation of herpes simplex virus type I in recurrent (Mollaret) meningitis. Ann. Neurol. 1982, 11, 17–21.
50. Bergstrom, T.; Vahlne, A.; Alestig, K.; Jeansson, S.; Forsgren, M.; Lycke, E. Primary and recurrent herpes simplex virus type 2-induced meningitis. J. Infect. Dis. 1990, 162, 322–330.
51. Haynes, S.F.; Wright, R.; Mcracken, J.P. Mollaret meningitis. A report of three cases. JAMA. 1976, 236, 1967–1969.

52. Berger, J.R. Benign aseptic (Mollaret's) meningitis after genital herpes. Lancet. 1991, 337, 1360–1361.
53. Picard, F.J.; Dekaban, G.A.; Silva, J.; Rice, G.P. Mollaret's meningitis associated with herpes simplex type 2 infection. Neurology. 1993, 43, 1722–1727.
54. Cohen, B.A.; Rowley, A.H.; Long, C.M. Herpes simplex type 2 in a patient with Mollaret's meningitis: demonstration by polymerase chain reaction. Ann. Neurol. 1994, 35, 112–116.
55. Jensenius, M.; Myrvang, B.; Storvold, G.; Bucher, A.; Hellum, K.B.; Bruu, A.L. Herpes simplex virus type 2 DNA detected in cerebrospinal fluid of 9 patients with Mollaret's meningitis. Acta. Neurol. Scand. 1998, 98, 209–212.
56. Yamamoto, L.J.; Tedder, D.G.; Ashley, R.; Levin, M.J. Herpes simplex virus type 1 DNA in cerebrospinal fluid of a patient with Mollaret's meningitis. N. Engl. J. Med. 1991, 325, 1082–1085.
57. Tedder, D.G.; Ashley, R.; Tyler, K.L.; Levin, M.J. Herpes simplex virus infection as a cause of benign recurrent lymphocytic meningitis. Ann. Intern. Med. 1994, 121, 334–338.
58. Straus, S.E.; Takiff, H.E.; Seidlin, M.; Bachrach, S.; Lininger, L.; DiGiovanna, J.J.; Western, K.A.; Smith, H.A.; Lehrman, S.N.; Creagh-Kirk, T. Suppression of frequently recurring genital herpes: a placebo-controlled double blind trial of acyclovir. N. Engl. J. Med. 1984, 310, 1545–1550.
59. Steiner, I.; Kennedy, P.G.E. Herpes simplex virus latency in the nervous system—a new model. Neuropathol. Appl. Neurobiol. 1991, 17, 433–440.
60. Steiner, I.; Steiner-Birmanns, B.; Levin, N.; Hershko, K.; Korn-Lubetzki, I.; Biran, I. Spinal cord involvement in uncomplicated herpes zoster. Clin. Diagn. Lab. Immunol. 2001, 8, 850–851.
61. Wiley, C.A.; VanPatten, P.D.; Carpenter, P.M.; Powell, H.C.; Thal, L.J. Acute ascending necrotizing myelopathy caused by herpes simplex virus type 2. Neurology. 1987, 37, 1791–1794.
62. Folpe, A.; Lapham, L.W.; Smith, H.C. Herpes simplex myelitis as a cause of acute necrotizing myelitis syndrome. Neurology. 1994, 44, 1955–1957.
63. Nakajima, H.; Furutama, D.; Kimura, F.; Shinoda, K.; Ohsawa, N.; Nakagawa, T.; Shimizu, A.; Shoji, H. Herpes simplex virus myelitis: clinical manifestations and diagnosis by the polymerase chain reaction method. Eur. Neurol. 1998, 39, 163–167.
64. Ellie, E.; Rozenberg, F.; Dousset, V.; Beylot-Barry, M. Herpes simplex virus type 2 ascending myeloradiculitis: MRI findings and rapid diagnosis by the polymerase chain method. J. Neurol. Neurosurg. Psychiatry. 1994, 57, 869–870.
65. Gobbi, C.; Tosi, C.; Stadler, C.; Merenda, C.; Bernasconi, E. Recurrent myelitis associated with herpes simplex virus type 2. Eur. Neurol. 2001, 46, 215–218.
66. Kuker, W.; Schaade, L.; Ritter, K.; Nacimiento, W. MRI follow-up of herpes simplex virus (type 1) radiculomyelitis. Neurology. 1999, 52, 1102–1103.
67. Shyu, W.C.; Lin, J.C.; Chang, B.C.; Harn, H.J.; Lee, C.C.; Tsao, W.L. Recurrent ascending myelitis: an unusual presentation of herpes simplex virus type 1 infection. Ann. Neurol. 1993, 34, 625–627.
68. Galanakis, E.; Bikouvarakis, S.; Mamoulakis, D.; Karampekios, S.; Sbyrakis, S. Transverse myelitis associated with herpes simplex virus infection. J. Child. Neurol. 2001, 16, 866–867.
69. Handler, C.E.; Perkin, G.D. Radiculomyelopathy due to genital herpes. Lancet. 1982, ii, 987–988.
70. Neville, B.G.R.; Sladen, G.E. Acute autonomic neuropathy following primary herpes simplex infection. J. Neurol. Neurosurg. Psychiatry. 1984, 47, 648–650.
71. Craig, C.; Nahmias, A.J. Different patterns of neurologic involvement with herpes simplex virus type 1 & 2: isolation of herpes simplex virus type 2 from the buffy coat of two adults with meningitis. J. Infect. Dis. 1973, 127, 365–372.
72. Black, D.; Stewart, J.; Melmed, C. Sacral nerve dysfunction plus generalized polyneuropathy in herpes simplex genitalis. Ann. Neurol. 1983, 14, 692.

73. Olivarius, B.F.; Buhl, M. Herpes simplex virus and Guillain-Barre polyradiculitis. Br. Med. J. 1975, i, 192–193.
74. Reeves, W.C.; Corey, L.; Adams, H G.; Vontver, L.A.; Holmes, K.K. Risk of recurrence after first episodes of genital herpes: relation to HSV type and antibody response. N. Engl. J. Med. 1981, 305, 315–319.
75. Whitley, R.J. Herpes simplex viruses. In Virology, 2nd ed.; Fields, B.N., Knipe, D.M., Eds.; Raven Press: New York, 1990, 1843–1886.
76. Constantine, V.S.; Francis, R.D.; Montes, L.F. Association of recurrent herpes simplex with neuralgia. JAMA. 1988, 205, 181–183.
77. Gominak, S.; Cros, D.; Paydarfar, D Herpes simplex labialis and trigeminal neuropathy. Neurology. 1990, 40, 151–152.
78. Chang, T.W.; Fiumara, N.J.; Weinstein, L. Genital herpes: some clinical and laboratory observations. JAMA. 1974, 229, 544–545.
79. Kukimoto, N.; Ikeda, M.; Yamada, K.; Tanaka, M.; Tsurumachi, H.; Tomita, H. Viral infection in acute peripheral facial paralysis: nationwide analysis centering on CF. Acta. Otolaryngol. 1988, 446(suppl), 17–22.
80. McCormick, D.P. Herpes simplex virus as a cause of Bell's palsy. Lancet. 1972, 1, 937.
81. Santos, D.Q.; Adour, K.K. Bilateral facial paralysis related to sexually transmitted herpes simplex: clinical course and MRI findings. Otolaryngol. Head Neck Surg. 1993, 108, 298–303.
82. Rodriguez, A.S.; Oterino, J.A.M.; Rufz V.A.C. Bell palsy in association with herpes simplex virus infection. Arch. Intern. Med. 1998, 158, 1577–1578.
83. Burgess, R.C.; Michaels, L.; Bale Jr, J.F.; Smith, R.J. Polymerase chain reaction amplification of herpes simplex viral DNA from the geniculate ganglion of a patient with Bell's palsy. Ann. Otol. Rhinol. Laryngol. 1994, 103, 775–779.
84. Takasu, T.; Furuta, Y.; Sato, K.C.; Fukuda, S.; Inuyama, Y.; Nagashima, K. Detection of latent herpes simplex virus DNA and RNA in human geniculate ganglia by the polymerase chain reaction. Acta Otolaryngol. 1992 112, 1004–1011.
85. Schulz, P.; Arbusow, V.; Strupp, M.; Dieterich, M.; Rauch, E.; Brandt, T. Highly variable distribution of HSV-1 specific DNA in human geniculate, vestibular and spiral ganglia. Neurosci. Lett. 1998, 252, 139–142.
86. Murakami, S.; Mizobuchi, M.; Nakashiro, Y.; Doi, T.; Hato, N.; Yanagihara, N. Bell palsy and herpes simplex virus: identification of viral DNA in endoneural fluid and muscle. Ann. Intern. Med. 1996, 127, 27–30.
87. Adour, K.K.; Ruboyianes, J.M.; Von Doersten, P.G.; Byl, F.M.; Trent, C.S.; Quesenberry Jr, C.P.; Hitchcock, T. Bell's palsy treatment with acyclovir and prednisone compared with prednisone alone: a double-blind, randomized, controlled trial. Ann. Otol. Rhinol. Laryngol. 1996, 105, 371–378.
88. Baringer, J.R. Herpes simplex virus and Bell's palsy. Ann. Intern. Med. 1996, 124, 63–65.
89. Knox, W.G. Treatment contraversies in Bell palsy. Arch. Otolaryngol. Head Neck Surg. 1998, 124, 821–823.
90. Steiner, I.; Mattan, Y. Bell's palsy and herpes viruses: to (acyclo)vir or not to (acyclo)vir? J. Neurol. Sci. 1999, 170, 19–23.
91. Corey, L. Laboratory diagnosis of herpes simplex virus infections. Principles guiding the development of rapid diagnostic tests. Diagm. Microbiol. Infect. Dis. 1986, 4, 111S–119S.
92. Goldstein, L.C.; Corey, L.; McDougall, J.K.; Tolentino, E.; Nowinski, R.C. Monoclonal antibodies to herpes simplex viruses: use in antigenetic typing and rapid diagnosis. J. Infect. Dis. 1983, 147, 829–837.
93. Nahass, G.T.; Mandel, M.J.; Cook, S.; Fan, W.; Leonardi, C.L. Detection of herpes simplex and varicella-zoster infection from cutaneous lesions in different clinical stages with the polymerase chain reaction. J. Am. Acad. Dermatol. 1995, 32, 730–733.

94. Straus, S.E.; Seidlin, M.; Takiff, H.; Jacobs, D.; Bowen, D.; Smith, H.A. Oral acyclovir to suppress recurring herpes simplex virus infections in immuno-deficient patients. Ann. Intern. Med. 1984, 100, 522–524.
95. Steiner, I.; Spivack, J.G.; Deshmane, S.L.; Ace, C.I.; Preston, C.M.; Fraser, N.W. A herpes simplex virus type 1 mutant containing a non-transinducing Vmw65 protein establishes latent infection in vivo in the absence of viral replication and reactivates efficiently from explanted trigeminal ganglia. J. Virol. 1990, 64, 1630–1638.
96. Birmanns, B.; Reibstein, I.; Steiner, I. Characterization of an in vivo reactivation model of herpes simplex virus from mice trigeminal ganglia. J. Gen. Virol. 1993, 74, 2487–2491.
97. Rock, D.L.; Fraser, N.W. Detection of HSV-1 genome in the central nervous system of latently infected mice. Nature. 1983, 302, 523–525.
98. Mellerick, D.M.; Fraser, N.W. Physical state of the latent herpes simplex virus genome in mouse model system. Evidence suggesting an episomal state. Virology. 1987, 158, 265–275.
99. Stevens, J.G.; Wagner, E.K.; Devi-Rao, G.B.; Cook, M.L.; Feldman, L.T. RNA complementary to a herpesvirus alpha gene mRNA is prominent in latently infected neurons. Science. 1987, 235, 1056–1059.
100. Spivack, J.G.; Fraser, N.W. Detection of herpes simplex type 1 transcripts during latent infection in mice. J. Virol. 1987, 61, 3841–3847.
101. Javier, R.T.; Stevens, J.G.; Dissette, V.B.; Wagner, E.K. A herpes simplex virus transcript abundant in latently infected neurons is dispensable for establishment of the latent state. Virology. 1988, 166, 254–257.
102. Steiner, I.; Spivack, J.G.; Lirette, R.P.; Brown, S.M.; MacLean, A.R.; Subak-Sharpe, J.; Fraser, N.W. Herpes simplex virus type 1 latency-associated transcripts are evidently not essential for latent infection. EMBO. J. 1989, 8, 505–511.
103. Trousdale, M.D.; Steiner, I.; Spivack, J.G.; Deshmane, S.L.; Brown, S.M.; MacLean, A.R.; Subak-Sharpe, J.H.; Fraser, N.W. In vivo and in vitro reactivation impairment of a herpes simplex virus type 1 latency-associated transcript variant in a rabbit eye model. J. Virol. 1991, 65, 6989–6993.
104. Sawtell, N.M.; Thompson, R.L. Herpes simplex virus type 1 latency-associated transcription unit promotes anatomical site-dependent establishment and reactivation from latency. J. Virol. 1992, 66, 2157–2169.
105. Mador, N.; Goldenberg, D.; Cohen, O.; Panet, A.; Steiner, I. Herpes simplex virus type 1 latency-associated transcripts suppress viral replication and reduce immediate early genes mRNA levels in a neuronal cell line. J. Virol. 1998, 72, 5067–5075.
106. Goldenberg, D.; Mador, N.; Ball, M.J.; Panet, A.; Steiner, I. The abundant latency-associated transcripts of herpes simplex virus type 1 are bound to polyribosomes in cultured neuronal cells and during latent infection in mouse trigeminal ganglia. J. Virol. 1997, 71, 2897–2904.
107. Goldenberg, D.; Mador, N.; Panet, A.; Steiner, I. Tissue specific distribution of the herpes simplex virus type 1 latency-associated transcripts on polyribosomes during latent infection. J. NeuroVirol. 1998, 4, 426–432.
108. Ash, R.J. Butyrate-induced reversal of herpes simplex virus restriction in neuroblastoma cells. Virology. 1986, 155, 584–592.
109. Kemp, L.M.; Dent, C.L.; Latchman, D.S. Octamer motif mediates transcriptional repression of HSV immediate early genes and octamer-containing cellular promoters in neuronal cells. Neuron. 1990, 4, 215–222.
110. Wheatley, S.C.; Dent, C.L.; Wood, J.N.; Latchman, D.S. A cellular factor binding to the TAATGARAT DNA sequence prevents the expression of the HSV immediate-early genes following infection of nonpermissive cell lines derived from dorsal root ganglion neurons. Exp. Cell. Res. 1991, 194, 78–82.
111. Steiner, I.; Nisipiano, P.; Wirguin, I. Infections and the etiology of multiple sclerosis. Curr. Neurol. Neurosci. Rep. 2001, 1, 271–276.

112. Jay, V.; Hwang, P.; Hoffman, H.J.; Becker, L.E.; Zielenska, M. Intractable seizure disorder associated with chronic herpes infection. HSV1 detection in tissue by the polymerase chain reaction. Childs. Nerv. Syst. 1998, 14, 15–20.
113. Kaji, M.; Kusuhara, T.; Ayabe, M.; Hino, H.; Shoji, H.; Nagao, T. Survey of herpes simplex virus infections of the central nervous system, including acute disseminated encephalomyelitis, in the Kyushu and Okinawa regions of Japan. Mult. Scler. 1996, 2, 83–87.
114. Ball, M.J. Limbic predilection in Alzheimer dementia: is reactivated herpesvirus involved?. Can. J. Neurol. Sci. 1982, 9, 303–306.
115. Ball, M.J.; Lewis, E.; Haase, A.T. Detection of herpes virus genome in Alzheimer's disease by in situ hybridization: a preliminary study. J. Neural. Transm. Suppl. 1987, 24, 219–225.
116. Jamieson, G.A.; Maitland, N.J.; Wilcock, G.K.; Yates, C.M.; Itzhaki, R.F. Herpes simplex virus type 1 DNA is present in specific regions of brain from aged people with and without senile dementia of the Alzheimer type. J. Pathol. 1992, 167, 365–368.
117. Itzhaki, R.F.; Lin, W.R.; Shang, D.; Wilcock, G.K.; Faragher, B.; Jamieson, G.A. Herpes simplex virus type 1 in brain and risk of Alzheimer's disease. Lancet. 1997, 349, 241–244.
118. Beffert, U.; Bertrand, P.; Champagne, D.; Gauthier, S.; Poirier, J. HSV-1 in brain and risk of Alzheimer's disease. Lancet. 1998, 351, 1330–1331.
119. Dobson, C.B.; Itzhaki, R.F. Herpes simplex virus type 1 and Alzheimer's disease. Neurobiol. Aging. 1999, 20, 457–465.
120. Plassman, B.L.; Havlik, R.J.; Steffens, D.C.; Helms, M.J.; Newman, T.N.; Drosdick, D.; Phillips, C.; Gau, B.A.; Welsh-Bohmer, K.A.; Burke, J.R.; Guralnik, J.M.; Breitner, J.C. Documented head injury in early adulthood and risk of Alzheimer's disease and other dementias. Neurology. 2000, 55, 1158–1166.
121. Danik, M.; Champagne, D.; Petit-Turcotte, C.; Beffert, U.; Poirier, J. Brain lipoprotein metabolism and its relation to neurodegenerative disease. Crit. Rev. Neurobiol. 1999, 13, 357–407.
122. Costantini, L.C.; Bakowska, J.C.; Breakefield, X.O.; Isacson, O. Gene therapy in the CNS. Gene. Ther. 2000, 7, 93–109.
123. Latchman, D.S. Gene delivery and gene therapy with herpes simplex virus-based vectors. Gene. 2001, 264, 1–9.
124. Roizman, B.; Sears, A.E. Herpes simplex viruses and their replication. In Virology, 2nd ed.; Fields, B.N., Knipe, D.M., Eds.; Raven Press: New York, 1841, 1795.
125. Kristensson, K.; Lycke, E.; Roytta, M. Svennerholm, B.; Vahlne, A. Neuritic transport of herpes simplex in rat sensory neurons in vitro. Effects of substances interacting with microtubular function and axonal flow [nocodazole, taxol, and erythro-9–3–(2-hydroxynonyl)-adenine]. J. Gen. Virol. 1986, 67, 2023–2028.
126. Kuypers, H.G.J.M.; Ugolini, G. Viruses as transneuronal tracers. Trends Neurosci. 1990, 13, 71–75.
127. Steiner, I.; Mador, N.; Reibstein, I.; Spivack, J.G.; Fraser, N.W. Herpes simplex virus type 1 latency in human and mouse central nervous systems. Neuropathol. Appl. Neurobiol. 1994, 20, 253–260.
128. Morelli, A.E.; Larregina, A.T.; Smith-Arica, J.; Dewey, R.A.; Southgate, T.D.; Ambar, B.; Fontana, A.; Castro, M.G.; Lowenstein, P.R. Neuronal and glial cell type-specific promoters within adenovirus recombinants restrict the expression of the apoptosis-inducing molecule Fas ligand to predetermined brain cell types, and abolish peripheral liver toxicity. J. Gen. Virol. 1999, 80, 571–583.
129. Wyman, T.; Rohrer, D.; Kirigiti, P.; Nichols, H.; Pilcher, K.; Nilaver, G.; Machida, C. Promoter-activated expression of nerve growth factor for treatment of neurodegenerative diseases. Gene. Ther. 1999, 6, 1648–1660.
130. Aghi, M.; Hochberg, F.; Breakefield, X.O. Prodrug activation enzymes in cancer gene therapy. J. Gene. Med. 2000, 2, 148–164.
131. Palella, T.D.; Hidaka, Y.; Silverman, L.J.; Levine, M.; Glorioso, J.; Kelley, W.N. Expression of human HPRT mRNA in brains of mice infected with a recombinant herpes simplex virus vector. Gene. 1989, 80, 137–144.

132. During, M.J.; Naegele, J.R.; O'Malley, K.L.; Geller, A.I. Long-term behavioral recovery in Parkinsonian rats by an HSV vector expressing tyrosine hydroxylase. Science. 1994, 266, 1399–1403.
133. Yamada, M.; Oligino, T.; Mata, M.; Goss, J.R.; Glorioso, J.C.; Fink, D.J. Herpes simplex virus vector-mediated expression of Bcl-2 prevents 6-hydroxydopamine-induced degeneration of neurons in the substantia nigra in vivo. Proc. Natl. Acad. Sci. USA. 1999, 96, 4078–4083.
134. Linnik, M.D.; Zahos, P.; Geschwind, M.D.; Federoff, H.J. Expression of bcl-2 from a defective herpes simplex virus-1 vector limits neuronal death in focal cerebral ischemia. Stroke. 1995, 26, 1670–1674.
135. Federoff, H.J.; Geschwind, M.D.; Geller, A.I.; Kessler, J.A. Expression of nerve growth factor in vivo from a defective herpes simplex virus 1 vector prevents effects of axotomy on sympathetic ganglia. Proc. Natl. Acad. Sci. USA. 1992, 89, 1636–1640.
136. Ram, Z.; Culver, K.W.; Oshiro, E.M.; Viola, J.J.; DeVroom, H.L.; Otto, E.; Long, Z.; Chiang, Y.; McGarrity, G.J.; Muul, L.M.; Katz, D.; Blaese, R.M.; Oldfield, E.H. Therapy of malignant brain tumors by intratumoral implantation of retroviral vector-producing cells. Nat. Med. 1997, 3, 1354–1361.

6

Varicella-Zoster Virus Infection

Donald H. Gilden
University of Colorado Health Sciences Center
Denver, Colorado, U.S.A.

James J. LaGuardia
Southern Illinois University
Springfield, Illinois, U.S.A.

1 INTRODUCTION

Varicella-zoster virus (VZV) is an exclusively human herpesvirus that causes varicella (chicken pox), becomes latent in cranial and dorsal root ganglia, and frequently reactivates decades later to produce zoster (shingles) and postherpetic neuralgia. In elderly immunocompetent or immunocompromised individuals, VZV can also produce central nervous system (CNS) disease. New molecular technologies such as polymerase chain reaction (PCR) as well as the presence of antibody to VZV have enabled the detection of VZV in blood vessels and other tissues, widening the recognized clinical spectrum of acute and chronic disorders associated with VZV, including latent infections. Here we review the neurological complications of VZV reactivation, including underemphasized patterns of zoster, pre- and postherpetic neuralgia, and myelitis and encephalitis, including acute, chronic, and recurrent neuropathy, all of which may occur in the absence of zoster rash. Current progress in understanding VZV latency is also summarized.

2 NATURALLY OCCURRING VARICELLA: CLINICAL AND EPIDEMIOLOGICAL FEATURES

Primary varicella-zoster virus (VZV) infection produces varicella (chicken pox), a highly contagious but typically mild disease of childhood. An estimated 4 million cases occurred annually in the United States before 1995, 90% of them in individuals 1–14 years of age [1]. By adult life, nearly everyone in North America is seropositive. In northern regions only 2% of varicella occurs after age 20, but in tropical climates a higher incidence of varicella is seen in adults [2,3]. In temperate climates, the incidence of varicella peaks semiannually, usually in the spring, with another, smaller peak in winter [4,5]. Disease is presumed to be transmitted by direct contact or aerosols containing virus [6–8]. Respiratory infection is probably followed by viral replication in the pharynx and regional lymph nodes [9–11]. The incubation period in healthy children is 9–21 days [5,9,10]. Viremia in immunocompetent varicella patients has been demonstrated 1–11 days prior to rash [12,13], with virus located predominantly in lymphocytes [14–16]. Fever, myalgia, and arthralgia precede or coincide with rash. The exanthem of chickenpox consists of macules and papules that develop into vesicles surrounded by an erythematous halo. Vesicles that reflect degenerative changes of the corium and dermis develop quickly and are characterized by multinucleated giant cells and intranuclear Cowdry type A inclusions [17], a hallmark of herpesvirus family infection. Vesicles contain abundant infectious virus that can be isolated in cell culture [18]. Rash usually begins on the trunk, then spreads to the face, limbs, and often to the buccal and pharyngeal mucosa. Eventually, vesicles burst and their fluid hardens, a process known as ''crusting.'' New vesicles from within the first 4 days after outbreak, whereas crusting begins after 2–3 days; thus, crusting and fresh vesicles may be seen simultaneously. Patients are considered infectious from 2 days before rash until all vesicles have crusted, typically 6 days after the onset of rash. Although most individuals can recount only a single episode of varicella, immunological evidence indicates that subclinical reinfection with VZV is common [19,20]. Immunization of children 12–18 months old with a live attenuated varicella vaccine (Oka strain) may eventually shift the average age of infection to older susceptible individuals [21].

3 VIROLOGY

3.1 Standard

Varicella-zoster virus is an exclusively human pathogen, readily propagated in multiple human and primate cell lines [18,22,23] and maintained in vitro by cocultivation of infected cells with uninfected cells. Virtually every region of the virus genome is transcribed during productive infection [24]. Thus, the low virus yield associated with VZV grown in culture has been attributed to errors in either virion assembly or maturation.

3.2 Molecular

The entire 125,884 base pair (bp) VZV genome has been sequenced and shown to have a high degree of homology to HSV-1, the prototype human alpha-herpesvirus [25–29]. The VZV genome comprises a unique long (UL) segment of 104,836 bp and a unique short (US) segment of 5232 bp. Each unique segment of VZV DNA is bounded by inverted repeats (88 bp inverted repeats around the UL and 7320 bp inverted repeats around the US). There are 71 predicted open reading frames (ORFs) potentially encoding proteins 8–300 kDa in size [25]. Although transcripts mapping to most VZV ORFs have been identified in VZV-infected cells in culture, fewer than 20 VZV genes have been analyzed in detail [24,30–32]. In the VZV genome, the 71 ORFs are separated by an average of 211 bp, indicating that the promoters are closely associated with the genes they regulate. During viral DNA replication, in 50% of viral DNA molecules, the US and its attendant repeats (IRS and TRS) invert with respect to UL [33–36]. The biological significance of VZV genome isomerization has yet to be determined; however, the presence of inverted repeats results in duplication of genes contained within them. Thus VZV genes 62 and 63 are present in two copies per genome.

4 ZOSTER (SHINGLES) EPIDEMIOLOGY

More than 300,000 cases of herpes zoster occur annually in the United States. Zoster generally affects the elderly but is also seen in individuals who are immunocompromised due to HIV infection, malignancy, chemotherapy, or long-term corticosteroid use. The incidence among people over age 50 is double that of people under 50 years [37], translating into an 8–10-fold increased frequency in people over age 60 compared with those under 60. As the aging population increases, the incidence of zoster-associated morbidity and mortality is also expected to increase. Varicella in infancy may predispose to zoster earlier in life [38]. The incidence of recurrent zoster is less than 5% [39]. Although varicella outbreaks occur most often in the spring, zoster may develop at any time of the year. The risk of zoster in vaccinated individuals compared with those who developed naturally occurring chickenpox will not be known for decades. Meanwhile, some investigators predict an increased incidence of zoster with widespread use of the live attenuated varicella vaccine [20,40].

5 PATHOLOGY AND PATHOGENESIS

Despite the ubiquity and frequency of VZV infection, the pathogenesis of zoster remains largely unknown. Our present understanding of virus spread, localization, and replication is based on in vitro studies of virus-infected human or primate cells in tissue culture, correlation of the presence of VZV in human tissues with pathological changes in various clinical situations, and attempts to produce disease experimentally.

Pathological changes in ganglia corresponding to the segmental distribution of rash were first noted by von Barensprung [41] and more extensively detailed by Head and Campbell [42] and Denny-Brown et al. [43]. The older literature accurately reflects the true pathology of zoster, because the lesions described were those of localized zoster in immunocompetent individuals, except perhaps for occasional cases of zoster that developed in syphilis patients treated with arsenic. The cardinal pathological features were inflammation and hemorrhagic necrosis, often associated with neuritis, localized leptomeningitis, unilateral segmental poliomyelitis, and degeneration of related motor and sensory roots. Demyelination was also seen in areas of mononuclear cell infiltration and microglial proliferation.

Later, intranuclear inclusions [44,45] and viral antigen and herpesvirus particles [46,47] were detected in ganglia, and VZV was isolated from ganglia [48]. However, those studies were performed on ganglia from patients with underlying malignancies or other disorders of the immune function who developed disseminated zoster just before death. There is a single report in which VZV antigen was detected and virus isolated from ganglia of a fatal case of bacterial pneumonitis on which acute thoracic zoster was superimposed [49].

Zoster is presumed to reflect reactivation and retrograde transport of virus from ganglia to skin in a host partially immune to VZV. Viremia has also been demonstrated in otherwise immunocompetent zoster patients [50]. Although the significance of viremia in zoster patients remains to be determined, VZV DNA has been detected by in situ hybridization (ISH) in blood mononuclear cells (MNCs) of four uncomplicated zoster patients for 3–7 weeks after rash [51], coinciding with the period of pain in these patients.

In immunocompromised patients with localized and disseminated zoster, VZV can be isolated from blood [52–56], suggesting a role for hematogenous spread in the pathogenesis of zoster in such individuals. VZV in blood is cell-associated [56] and has been detected by electron microscopy in monocytes [57]. A loss of cell-mediated immunity to VZV appears to be responsible for an increased risk of zoster in immunocompromised patients [20].

The detection of VZV in macrophages [58], B cells [59,60], and T cells [50], particularly in activated T lymphocytes [14], provided indirect evidence that blood MNCs are a potential site for VZV persistence. Nucleic acid hybridization studies revealed that VZV DNA did not replicate in human MNCs [50], a finding confirmed by Koropchak et al. [14].

6 ZOSTER (RADICULONEUROPATHY, SHINGLES, GANGLIONITIS)

Herpes zoster is characterized by severe sharp, lancinating, radicular pain and vesicular eruption on an erythematous base in one to three dermatomes. Pain is often associated with itching and dysesthesia. In affected dermatomes, sensation is decreased, yet the skin is exquisitely sensitive to touch (allodynia). All levels of the neuraxis may be involved in zoster. Thoracic zoster is most common, followed by lesions on the face, most often in the ophthalmic division of the trigeminal nerve. The latter is frequently accompanied by zoster keratitis, a potential cause of blindness if unrecognized and not treated promptly. Thus, if visual symptoms are present in patients with ophthalmic distribution zoster, they should have immediate slit-lamp examination by an ophthalmologist, especially if skin lesions extend to the medial side of the nose (Hutchinson's sign). Maxillary and mandibular trigeminal distribution zoster with osteonecrosis and spontaneous tooth exfoliation has also been described in adults [61] and children [62].

The seventh cranial nerve is also commonly involved in zoster. Weakness of all facial muscles of one side develops in conjunction with rash in the ear (zoster oticus) or on the ipsilateral anterior two-thirds of the tongue or hard palate. Vesicles in either site are often overlooked. The combination of zoster oticus and peripheral facial weakness constitutes the Ramsay Hunt syndrome (RHS) (reviewed in Ref. [63]). Compared with Bell's palsy (facial paralysis without rash), patients with RHS often have more severe paralysis at onset and are less likely to recover completely [64].

In the only prospective study of RHS patients, 14% developed vesicles after the onset of facial weakness [65]. Thus, RHS may initially be indistinguishable from Bell's palsy. Further, Bell's palsy is significantly associated with herpes simplex virus (HSV) infection [66]. In light of the known safety and effectiveness of antiviral drugs against VZV and HSV, consideration should be given to early treatment of all RHS or Bell's palsy patients with a 7–10 day course of famciclovir (500 mg, three times daily) or acyclovir (800 mg, five times daily) as well as oral prednisone (60 mg daily for 3–5 days).

Finally, some patients develop peripheral facial paralysis without ear or mouth rash (see zoster sine herpete, Sec. 13), associated with either a four fold rise in antibody to VZV or the presence of VZV DNA in auricular skin, blood mononuclear cells, middle ear fluid, or saliva. This indicates that a proportion of ''Bell's palsy'' patients have RHS zoster sine herpete. Treatment of such patients with acyclovir and prednisone within 7 days of onset has been shown to improve the outcome of recovery from facial palsy [67], although a prospective randomized treatment trial remains to be undertaken.

Zoster may also be accompanied by ophthalmoplegia, most commonly affecting the third cranial nerve [68], optic neuritis [69], or both [70], and less often lower cranial nerve palsies [71,72]. Zoster-associated cranial neuropathy often occurs weeks after acute VZV infection. One explanation for late-onset zoster cranial neuropathy is that virus spreads slowly along trigeminal and other cranial ganglionic afferent fibers to small vessels supplying cranial nerves. This appears to occur in granulomatous arteritis (Sec. 9.2) preceded weeks earlier by trigeminal distribution zoster. Because all cranial nerves receive their blood supply from the carotid circulation via small branches supplying groups of two or three cranial nerves [73], it is likely that VZV spreads transaxonally along trigeminal and other ganglionic afferent fibers from the carotid arteries to the vasa vasorum of small nerves, resulting in small-vessel-mediated infarction [74].

7 POSTHERPETIC NEURALGIA

Most neurological complications of zoster manifest as postherpetic neuralgia (PHN), operationally defined as pain persisting for more than 4–6 weeks after rash. Age is the most important factor that predicts the development of PHN [75,76]. Although PHN is rare before age 50, the incidence is 43–47.5% after age 50, slightly greater in women [77] and after trigeminal distribution zoster [77–79].

The mechanism of PHN is unknown. The detection of VZV-specific proteins in MNCs of patients with PHN [80] suggests that PHN is associated with VZV persistence. Further, the demonstration that VZV DNA persists in blood MNCs of PHN patients [81] compared with zoster patients without PHN provides additional suggestive evidence that VZV infection is more widespread in PHN patients than during latency. It is possible that MNCs trafficking through such ganglia encounter and engulf virus whose DNA can then be amplified by PCR. Although postmortem microscopic analyses of ganglia are limited, analyses of two subjects who suffered from PHN revealed inflammatory infiltrates, often around dying neurons, 1–2 years after acute zoster [82,83]. Further pathological and virological analysis of ganglia obtained at autopsy from individuals with PHN at the time of death is needed. If a greater virus burden could be demonstrated in these ganglia than has been found during latency [84], this would provide a rationale for aggressive treatment of PHN patients with antiviral drugs. Meanwhile, the existence of ganglionitis without rash is supported by the presence of radicular pain up to 100 days preceding zoster [85], so-called preherpetic neuralgia (Sec. 14). Further, a recent report described four patients with acute trigeminal distribution zoster who, after years free from pain, developed severe trigeminal "PHN" [86].

8 ZOSTER PARESIS

Zoster in cervical, thoracic, and lumbosacral dermatome distributions may be associated with muscle weakness, usually developing 1–5 weeks after rash. Cervical distribution zoster has been associated with arm weakness and, rarely, diaphragmatic paralysis [87,88]. The incidence of zoster paresis has been estimated from as low as 0.5% to as high as 31%. The low incidence of thoracic zoster paresis probably reflects the difficulty in diagnosing intercostal muscle weakness at the bedside. Lumbosacral distribution zoster may be associated with leg weakness as well as impairment of bladder and bowel function. Urinary retention, hemorrhagic cystitis, and massive bladder hemorrhage have all been described with sacral distribution zoster [89,90]. Approximately 11% of patients with segmental zoster paresis have malignant disease [68]. Rarely, zoster has developed within a few days to weeks after injury by lightning or injection of foreign material, and a case of zoster that occurred 5 h after spinal anesthesia has been reported [91].

9 VZV ENCEPHALITIS-ARTERITIS

The greatest contribution of modern diagnostic methods to our understanding of VZV pathogenesis has been the demonstration of the virus in large and small blood vessels of the nervous system. Although the ability of VZV to cause vasculitis had long been difficult to document, newer techniques such as PCR and in situ hybridization, as well as immunohistochemistry, have verified the extent to which viral infection of blood vessels causes widely variable clinical syndromes. VZV encephalitis is now recognized as a vasculopathy that affects large or small vessels and often both. Large-vessel artery disease (granulomatous arteritis) occurs predominantly in immunocompetent individuals, and small-vessel artery-mediated encephalitis is found mainly in immunodeficient patients.

9.1 Zoster Small-Vessel Encephalitis

Zoster small-vessel encephalitis is the most common form of CNS involvement. Disease usually develops on a background of cancer, immunosuppression [92], or AIDS. Neurological disease is subacute, and death is common. Zoster encephalitis presents with headache, fever, vomiting, mental changes, seizures, and focal deficit. Brain MRI reveals large and small ischemic or hemorrhagic infarcts, often both, of the cortex and subcortical gray and white matter (Figure 1). Deep-seated white matter lesions often predominate and are ischemic or demyelinative, depending on the size of blood vessels involved and the amount of additional demyelination. The demyelinative lesions are smaller and less coalescent than those seen in progressive multifocal leukoencephalopathy. The CSF shows a mild pleocytosis (predominantly mononuclear), normal or mild elevation of protein, and a normal glucose, findings that do not differ significantly from zoster without encephalitis. Two reports describe hypoglycorrhachia in zoster meningoencephalitis [93,94]. In suspected cases of zoster small-vessel encephalitis, the CSF should be studied for both VZV DNA and antibody to VZV. In the typical clinical setting described above, the presence of either or both in CSF is strong presumptive evidence of VZV small-vessel encephalitis [95]. Zoster small-vessel encephalitis should be treated with acyclovir, 15–30 mg kg^{-1} day^{-1} for 10 days. Longer treatment may be necessary in severely immunocompromised patients. Diagnosis may be particularly difficult in patients without rash unless the clinician is alert to the history of earlier zoster followed by the typical clinical features and multifocal lesions seen on brain MRI [96,97].

9.2 Zoster Large-Vessel Encephalitis (Granulomatous Arteritis)

Zoster large-vessel encephalitis (granulomatous arteritis) is characterized by acute focal deficit (stroke) that develops weeks to months after contralateral trigeminal distribution zoster. A single report describes virologically confirmed VZV large-vessel vasculopathy without previous zoster [98]. Stroke results from bland [99] or, less commonly, hemorrhagic [100] infarction due to arteritis of large cerebral arteries. Disease is uncommon but not rare. Most patients are over age 60, and there is no gender predilection. The mean

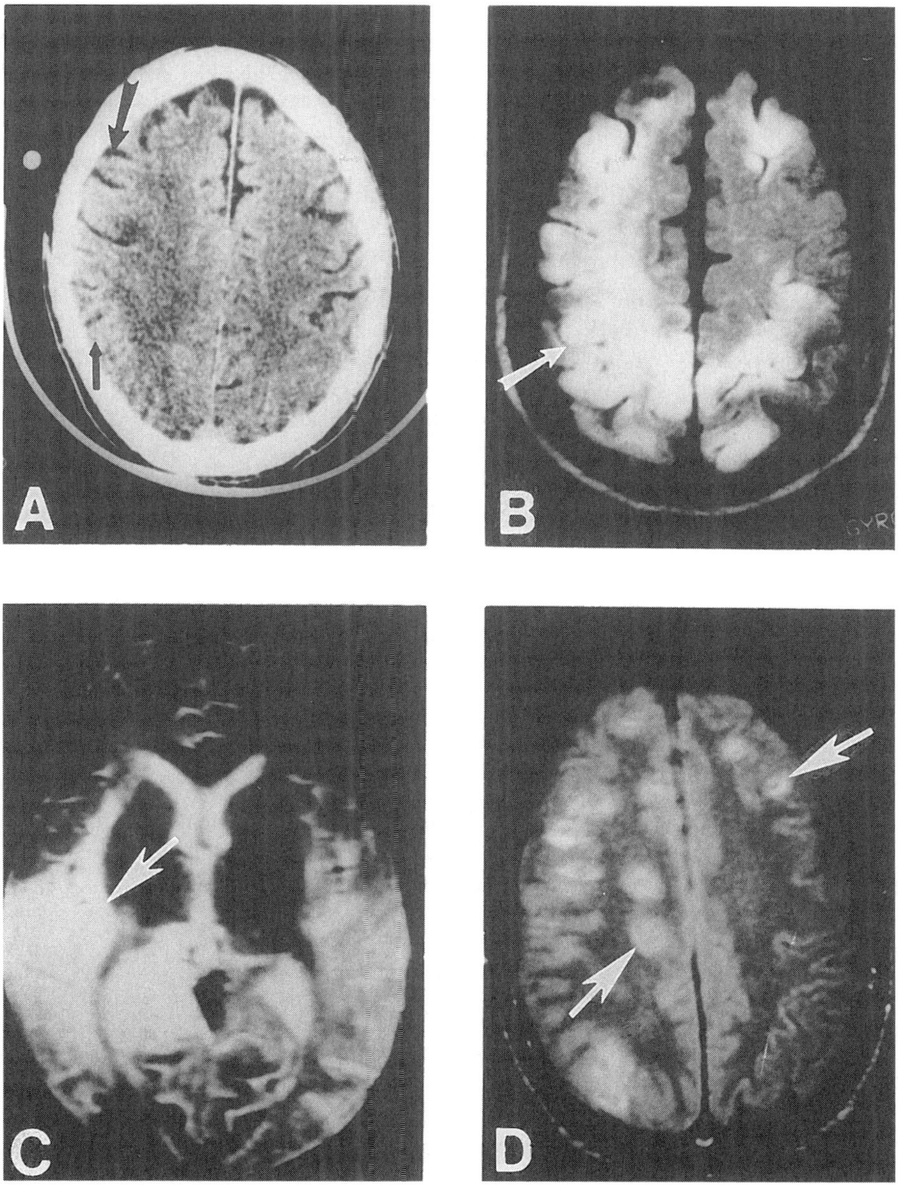

Figure 1 (A) Computed tomographic images showing changes in viral encephalitis. Brain CT scan demonstrates relative effacement of sulci posteriorly in both hemispheres (thin arrow) compared with normal sulcal spaces anteriorly (thick arrow). (B) Changes in viral encephalitis shown by MRI. T2-weighted inversion recovery (fluid-attenuated inversion recovery) MRI brain scan of the same patient demonstrates areas of increased signal in both hemispheres, greater on the right and even more so posteriorly (arrow), reflecting increased water content in mildly swollen brain. (C) Herpes simplex virus encephalitis. T2-weighted MRI brain scan demonstrates bilateral involvement of temporal lobes. The exaggerated signal does not extend beyond the insular cortex (arrow). (D) Varicella-zoster virus encephalitis. Proton-density brain MRI scan shows multiple areas of infarction in both hemispheres (arrows).

onset of neurological disease is 7 weeks after zoster, but intervals of up to 6 months have been recorded. Transient ischemic attacks and mental symptoms are common, and up to 25% of patients die [101]. Most patients have CSF pleocytosis, usually <100 cells (predominantly mononuclear), oligoclonal bands, and an increase in CSF IgG. Angiography reveals focal constriction and segmental narrowing, primarily in the middle and anterior cerebral and internal carotid arteries. Stroke in childhood with the same CSF, MRI, and angiographic changes also occurs after varicella (chickenpox) [102]. Microscopic examination reveals arterial inflammation with multinucleated giant cells, VZV antigen, Cowdry A inclusions, and herpesvirus particles. Recently, PCR detected VZV DNA in affected large cerebral arteries [103].

Other arteries may also be involved in large-vessel granulomatous arteritis. Ipsilateral central retinal artery occlusion may occur after trigeminal distribution zoster [104]. Posterior circulation involvement with brainstem infarction after rash behind the ear [105] or on the neck [106] has been reported, and thalamic infarction occurred after rash on the tongue [107]. There is even a single report of contralateral hemiplegia after thoracic distribution zoster [108]. Afferent trigeminal ganglionic fibers to both intracranial and extracranial arteries [74] provide a pathway for virus spread. The extent to which disease is viral or immunopathological or both is unknown. Owing to the rarity of the condition, large controlled clinical treatment trials have not been possible. Based on pathological studies by our group and others, we recommend that patients receive intravenous acyclovir (10–15 mg/kg three times daily for 7–10 days) to kill persistent virus and a short course of steroids (prednisone 60–80 mg daily for 3–5 days) for their anti-inflammatory effect.

9.3 Predominant VZV Ventriculitis and Meningitis

We have also encountered encephalitis in immunocompromised patients in which VZV predominantly infected the ependyma or meninges [109]. Some patients developed a gait disorder and hydrocephalus with periventricular enhancing lesions. At autopsy, most cases exhibited a necrotizing ventriculitis with preferential VZV infection of ependymal cells and mixed with ovoid ischemic and demyelinative lesions; pure forms of VZV ventriculitis have also been described. Other cases presented as meningoencephalitis in HIV-positive patients, with thousands of cells and grams of protein in the CSF, and enhancing meningeal lesions on MRI, confirmed at autopsy by histopathological evidence of necrotizing vasculitis primarily affecting the meninges. Either the brain or spinal cord may bear the brunt of disease [110,111].

10 VZV MYELITIS

In immunocompetent patients, myelitis may complicate acute varicella or zoster, usually 1–2 weeks after rash. Clinical features are paraparesis with impaired sensory level and sphincter function. The CSF either is normal or shows mild pleocytosis with a normal to mild elevation of protein. MRI reveals T2-weighted hyperintense lesions, sometimes with focal cord swelling. Most patients improve significantly, but some experience persistent lower extremity stiffness and weakness. Because most immunocompetent patients survive, the pathology of this form of VZV-associated transverse myelitis is unknown. Moreover, virological and immunological verification is wanting; VZV cannot be cultured from CSF, although PCR has revealed VZV DNA in CSF, suggesting that the immunocompetent host rapidly clears virus.

In immunocompromised individuals, the development of myelopathy is often more insidious and, progressive and sometimes fatal. Spinal cord MRI scanning shows focal or longitudinal serpiginous enhancing lesions [112]. Autopsy studies have demonstrated spinal cord necrosis and intense inflammation with frank parenchymal invasion by VZV. Long-term low-dose steroids may predispose to VZV myelitis and encephalitis [85,113,114].

VZV myelitis has classically been diagnosed by its close temporal relationship with rash. However, the recent detection of VZV DNA or VZV antibody in CSF has revealed that acute and even recurrent VZV myelopathy can develop without rash [112]. We reported one patient with zoster who developed myelopathy 5 months later, at which time amplifiable VZV DNA was detected in CSF. Another patient developed myelopathy during acute zoster; the myelopathy resolved but recurred 6 months later. Five months after recurrence, the patient's CSF contained both VZV DNA and VZV antibody [112]. Overall, the spectrum of VZV myelopathy is broad, ranging from acute to chronic, and is rarely recurrent. An early search for VZV DNA or VZV antibody in CSF is essential for diagnosis, particularly because aggressive treatment with acyclovir, even in AIDS patients, may produce a favorable response [115].

11 NEUROPATHY

Postinfectious polyneuritis or the Guillain-Barré syndrome (GBS) is an uncommon but well-documented neurological complication of both varicella and zoster. No neurological features distinguish polyneuritis after varicella or zoster from that seen in other clinical settings. The average interval between rash and neurological disease is 12 days [116]. Bilateral facial paralysis is present less often in VZV-associated GBS than in GBS unassociated with varicella [116,117]. GBS after chickenpox is rare, as evidenced by its occurrence in only eight of 2534 cases of chickenpox [118] and by the total absence of associated chickenpox in 50 extensively studied cases of GBS [119].

Guillain-Barré syndrome after zoster is also rare. Only 16 cases have been described since the initial report by Wohlwill [120] and the first compilation of cases by Dayan et al. [121]. Polyneuritis usually occurs many days to a few weeks after rash and in some instances 2 months later. The clinical course is usually acute but occasionally subacute or indolent, especially in cases that developed 1–2 months after rash. Neuropathological examination of teased nerve fibers revealed acute demyelination and remyelination. In addition, fibrinoid necrosis of blood vessels and infarction were seen in severely affected spinal ganglia, suggestive of an Arthus-type immunopathology [121].

Most recently, we encountered three patients with acute, chronic, and recurrent neuropathy associated with VZV infection but without zoster rash [122]. The CSF of all three patients contained VZV IgG antibody but not HSV antibody. Serum/CSF ratios of VZV IgG were reduced compared to normal ratios for total IgG and albumin, and one patient also had VZV IgM in the CSF. All three patients received antiviral therapy and improved. These patients expand the spectrum of neuropathy produced by VZV and emphasize the need to test CSF for VZV antibody, because it may be the only positive finding.

12 REYE'S SYNDROME

Reye's syndrome is a rapidly progressive, often fatal, noninflammatory disorder of children and adolescents. The two most affected organs are the brain (which swells) and the liver (which becomes infiltrated by fat). Disease is associated with infection by the influenza viruses or VZV. Typically, an initial, apparently mild upper respiratory infection or an episode of classic chickenpox is followed by an asymptomatic few days, after which patients develop intractable vomiting followed by seizures, lethargy, coma and often death [123]. Therapeutic amounts of salicylates increase the risk of developing Reye's syndrome [124]. Characteristic laboratory abnormalities include elevated serum transaminase and ammonia levels as well as hypoglycemia in 40% of patients. Brain MRI may reveal sulcal effacement, loss of gray/white matter junctions, and small ventricles, all consistent with brain swelling. The exact metabolic abnormality in Reye's syndrome has not been determined. Patients require intensive care treatment to reverse rapidly developing cerebral edema by hyperventilation and intravenous mannitol. Steroids have not been shown to be effective, probably because brain edema in Reye's syndrome is cytotoxic.

13 ZOSTER SINE HERPETE

The concept of zoster sine herpete (shingles without rash) is nearly 100 years old. The first proposed case was that of a 38-year-old man who developed acute thoracic distribution pain and hyperesthesia, a dilated pupil, a CSF pleocytosis (predominantly mononuclear), and a negative serological test for syphilis [125]. He was presumed to have zoster sine herpete even though serological tests for VZV were not done. Similarly, two patients who experienced segmental pain, hyperesthesia, and focal weakness were presumed to have zoster sine herpete despite the lack of serological confirmation [126]. The notion of zoster sine herpete received further credence when Lewis [127] described numerous zoster patients who, days later, also developed pain without rash in a different dermatome distribution, often on the opposite side.

The first serological evidence of zoster sine herpete appeared in a physician who developed acute trigeminal distribution pain associated with a fourfold rise in complement-fixing antibody to VZV but not to HSV [128]. Virological confirmation of zoster sine herpete did not come until the analysis of two men with thoracic distribution radicular pain that had lasted for months to years revealed PCR-amplifiable VZV DNA but not HSV DNA in their CSF and blood MNCs [129]. After diagnosis, both men were treated successfully with intravenous acyclovir. A third virologically confirmed case of thoracic distribution zoster sine herpete that persisted for years also demonstrated frequent fibrillation potentials restricted to chronically painful thoracic root segments [130]. Unfortunately, the patient did not improve after treatment with intravenous acyclovir and oral famciclovir.

Although the nosological entity of zoster sine herpete as a clinical variant has now been established, its prevalence will not be known until a greater number of patients with prolonged radicular pain have been studied virologically. Analysis should include PCR to amplify VZV DNA in CSF and in blood MNCs as well as a search for antibody to VZV in CSF. The latter, even in the absence of amplifiable VZV DNA, has been useful to support the diagnosis of encephalitis and myelitis produced by VZV without rash [95]. Analysis of serum anti-VZV antibody is of no value in the diagnostic workup of patients with prolonged pain, because such antibodies persist in nearly all adults throughout life, and the presence of serum antibodies to different VZV glycoproteins and nonglycosylated proteins is variable [131].

14 PREHERPETIC NEURALGIA

The existence of ganglionitis without rash is further supported by the presence of radicular pain preceding zoster, so-called preherpetic neuralgia. To our knowledge, ours is the only study of individuals with preherpetic neuralgia reported to date [85]. In the six patients analyzed, pain preceded rash by 7–100 days; was severe, burning, radicular, and was located in dermatomes both outside and inside the area of eventual rash. Two patients ultimately developed disseminated zoster with neurological complications of zoster paresis and fatal zoster encephalitis; both had been taking long-term low-dose steroids. A third case of preherpetic neuralgia developed in a patient with prior metastatic carcinoma, and a fourth was in a patient with an earlier episode of brachial neuritis. Two subjects had no underlying disease. Further documentation of preherpetic neuralgia will determine whether its apparent association with steroid therapy and serious complications is statistically significant.

15 OTHER NON-ZOSTERIFORM VZV INFECTION OF THE NERVOUS SYSTEM WITHOUT RASH

Zoster sine herpete (Sec. 13) is essentially a disorder of the peripheral nervous system (ganglioradiculopathy) produced by VZV without rash. VZV also produces disease of the CNS without rash. Although cases are rare, we at the University of Colorado Health Sciences Center have encountered more cases of VZV infection of the CNS (encephalitis and myelitis) without rash than cases of VZV infection of the peripheral nervous system (ganglioneuropathy) without rash.

In pathologically verified cases of VZV encephalitis without rash, the typical clinical picture has been an immunocompromised individual (usually with AIDS) who develops CNS disease at the time of acute zoster or who may have a history of zoster weeks to months earlier or even recurrent zoster. CNS disease in such patients develops more often in the absence of acute zoster than at the time of its presence [96]. Encephalitis is most often the small-vessel type (Sec. 9.1), and disease is usually protracted. Diagnosis may be difficult unless the clinician is alert to the history of recurrent zoster followed by the typical clinical features and multifocal lesions seen by brain MRI in small-vessel encephalitis. Many patients die of chronic progressive VZV encephalitis without ever having developed rash [96,97]. The most extreme example of VZV infection of the nervous system that we encountered was a 77-year-old man with T-cell lymphoma who developed a fatal meningoradiculitis and died 3 weeks after the onset of neurological disease [132]. He did not develop zoster before or during neurological disease. At autopsy, hemorrhagic inflammatory lesions with Cowdry A inclusions were found in meninges and nerve roots extending from cranial nerve roots to the cauda equina. The same lesions were present in the brain, although to a lesser extent. We detected VZV antigen and nucleic acid, but not HSV or cytomegalovirus antigen or nucleic acid, in infected tissue at all levels of the neuraxis. Thus, VZV should be included in the differential diagnosis of acute encephalomyeloradiculopathy, particularly because antiviral treatment is available.

Acute VZV myelopathy also occurs without rash. Heller et al. [133] described a 31-year-old immunocompetent man who developed transverse myelitis with partial recovery. Disease was attributed to VZV based on the development of antibody in CSF. We encountered two patients with VZV myelopathy in the absence of rash [129] (Sec. 10).

Involvement of VZV of the CNS without rash was verified by the intrathecal synthesis of antibodies to VZV in two patients with aseptic meningitis [134], later in four additional patients with aseptic meningitis [135] and in one patient with acute meningoencephalitis [136]. We also encountered an adult man who was taking low-dose methotrexate and developed acute encephalitis (fever, aphasia, and a profound CSF mononuclear pleocytosis). His CSF contained amplifiable VZV and Epstein-Barr virus DNA. Zosteriform rash never developed, and he recovered completely after treatment with intravenous acyclovir.

There have been two reported cases of polyneuritis cranialis produced by VZV. The first occurred in a 70-year-old man who seroconverted to VZV during acute disease [137]. Another report described a 43-year-old man with acute polyneuritis cranialis who developed antibody in CSF to VZV but not to other human herpesviruses or to multiple ubiquitous paramyxoviruses or togaviruses [138]. Both men were apparently immunocompetent. Finally, cases of acute unilateral facial (Bell's) palsy that developed in the absence of zosteriform rash have been attributed to VZV infection (so-called geniculate zoster sine herpete) based on "a positive serum complement fixation test" [139]. A recent study revealed that VZV was the likely causative agent of acute peripheral facial palsy in 29% of patients clinically diagnosed with Bell's palsy due to a lack of visible vesicles in the external ear or on the palate [140].

Finally, we recently described three patients with acute, chronic, and recurrent neuropathy associated with VZV infection but without zoster rash (Sec. 11).

16 DIAGNOSIS OF VZV INFECTION: AMPLIFIABLE VZV DNA BY PCR AND ANTIBODY TO VZV

Rapid clinical diagnosis of VZV infection and any attendant systemic or neurological complications is essential because antiviral treatment exists. Standard diagnostic methods include histological analysis, attempts to isolate virus from infected tissue, serological assays that measure the humoral or cell-mediated immune response to VZV, and PCR analysis of VZV DNA.

Giemsa-stained smears of varicella vesicle scrapings have been used for histopathological diagnosis of VZV infection to detect multinucleated giant cells and Cowdry A intranuclear inclusions characteristic of any herpesvirus [141]. Inclusions have also been detected in smears of oral mucosa [142–144]. Occasionally, VZV can be isolated by cocultivation of infected pharyngeal tissue [145] or blood MNCs [55] with indicator cells.

Various serological assays (complement fixation, hemagglutination, neutralization) have been used to diagnose VZV infection. The detection of VZV IgG is usually retrospective [146]. However, unlike the detection in serum, demonstration of VZV antibody in CSF, with or without the presence of amplifiable VZV DNA, is useful in the diagnosis of VZV encephalitis, myelitis, and zoster sine herpete [95,112]. VZV IgG has been detected in CSF of four neonates with convulsions, indicating that intrauterine VZV infection can be acquired without skin lesions in the mother [147]. The sensitivity and specificity of different commercially available VZV IgG detection kits have been shown to be comparable [148,149].

In situ hybridization (ISH) has also been used to detect VZV DNA in the brains of patients with VZV encephalitis [150,151], in VZV meningoradiculitis [132], in blood MNCs of patients with varicella and zoster [14,51], and in normal human ganglia [152–157]. Although ISH can identify cells infected with VZV, its sensitivity compared with Southern blot hybridization is unknown.

Polymerase chain reaction technology has provided a means to examine VZV at the molecular level with a specificity and precision not previously attainable. PCR has detected VZV DNA in throat swab samples [158,159], in blood MNCs [158,160], and in vesicles from patients with chickenpox [161]. VZV DNA was detected in all throat swab samples within the first 3 days after the onset of chickenpox [159]. Further, in patients with acute peripheral facial palsy, the detection of VZV DNA by PCR in oropharyngeal swabs was more useful than currently available serological assays for the early diagnosis of zoster sine herpete [162].

Varicella-zoster virus DNA has also been detected by PCR in blood MNCs of elderly patients with postherpetic neuralgia [81], in all of seven zoster patients with various neurological features [163], in CSF from six of 84 (7%) HIV-infected patients presenting with neurological symptoms, and in CSF of three of five (60%) children with post-varicella cerebellitis. The latter finding is particularly important because it suggests that cerebellar ataxia days to weeks after chickenpox, thought to be immune-mediated, is more likely due to frank virus infection.

Recently, PCR was used to detect VZV DNA in vitreous biopsy specimens from patients with viral retinitis [164]. VZV DNA has also been detected in the cornea of seven of 14 patients after herpes zoster ophthalmicus [165]. Also, PCR has shown that VZV is the most likely pathogen of atypical necrotizing herpetic retinopathies [166]. PCR differentiates infection by HSV and VZV [167,168]. Recently, VZV DNA was detected in synovial fluid of one patient with monoarthritis when blood MNCs were PCR-negative [169], suggesting a direct role of VZV in causing the disease.

Overall, the combination of PCR and the detection of antibody to VZV in CSF is extremely useful, not only to evaluate disorders caused by VZV but also to diagnose subclinical reactivation of the virus.

17 TREATMENT

The treatment for acute zoster generally includes the use of analgesics such as extra-strength acetaminophen and/or codeine 30–60 mg every 6 h when necessary [170–174]. Oral acyclovir (800 mg five times daily) or famciclovir (500 mg three times daily) has been reported to decrease new lesion formation and reduce acute pain [175,176]. Although the value of antiviral therapy, especially in immunocompetent individuals under age 50, remains to be definitively proven, we prescribe oral acyclovir or famciclovir for 7 days if any new skin lesions developed within the previous week. All patients with ophthalmic distribution zoster should receive antiviral drugs for at least 7 days.

Two major therapeutic approaches have been used to treat PHN. The first is aimed at preventing PHN by aggressively treating the zoster episode; the second involves treating PHN once it has occurred. Although the efficacy of antiviral prophylaxis to prevent PHN in elderly patients is unknown, acyclovir (800 mg five times daily) or famciclovir (500 mg three times daily) is often given for 7–10 days in addition to medication for zoster pain. Numerous trials have not determined an optimal therapy for preventing PHN. Most studies have focused on antiviral agents or steroids or both. Nonsteroidal therapy studies have included intramuscular injections of leukocyte interferon [177], oral acyclovir (800 mg five times daily for 7 days) [178], a randomized double-blind trial of acyclovir for 7–21 days with or without prednisone [176], a well-controlled study of the dopamine agonist amantadine hydrochloride [179], a controlled study of parenteral adenosine monophosphate [180], and a double-blind study with either oral levodopa and benserazide or placebo [181]. Although these trials demonstrated some efficacy in preventing PHN, they were hampered by factors including potential toxic side effects of interferon, small sample size, and an abnormally high incidence of PHN in control groups. Steroids used to prevent PHN have included oral triamcinolone [182], prednisolone (40 mg daily) [183], and prednisolone with or without acyclovir [184]. No difference between the treatment groups was observed. Again, small study populations and a high incidence of PHN in control groups flaw these trials. Studies with larger numbers of patients are needed to assess the efficacy, if any, of steroids in preventing PHN.

To treat PHN, more than 40 pharmacological; antiseptic, and surgical therapies, including aspirin, hormones, narcotics, vitamins, immunoglobulins, radiotherapy, and various nerve blocks or excision, have been used with limited success [83]. Tricyclic antidepressants, such as amitriptyline or nortriptyline (25–75 mg at night), and the anticonvulsants carbamazepine (400–1200 mg daily) and phenytoin (300–400 mg daily) relieve pain in some PHN patients [172]. Gabapentin (neurontin) 300 mg three times daily and sometimes more may help relieve PHN [185,186]. A short course of steroids, e.g., prednisone (40–60 mg daily for 3–5 days and sometimes longer), may reduce inflammation contributing to pain. Subcutaneous infusion of ketamine reduced PHN [187] but was associated with intolerable side effects. Topical aspirin in chloroform as well as topical lidocaine patches have been shown to relieve zoster pain and PHN [188,189]. A recent trial indicates that intrathecal administration of methylprednisolone is an effective treatment for PHN and may be useful for patients who do not respond to any of the foregoing measures [190].

The more serious CNS and PNS complications of VZV reactivation, including myelitis, encephalitis/arteritis, and radiculitis, are treated with intravenous acyclovir as soon as the diagnosis is suspected. Treatment may be discontinued if CSF proves negative for both anti-VZV antibody and VZV DNA. Otherwise, treatment should be continued for 10–14 days. The rationale for such treatment is that VZV virions, antigen, and DNA are present in arteries of patients with both large-and small-vessel encephalitis [97,106].

18 VZV LATENCY

After primary infection, VZV becomes latent and reactivates with increasing age or immunosuppression; however, the biological mechanisms underlying the transition from latency to active viral replication are still unknown. Many laboratories have devoted major efforts to determining the physical state of virus during latency, because understanding it is essential to predicting or preventing the neurological complications produced by virus reactivation.

18.1 Prevalence, Distribution, Extent, and Configuration of VZV DNA in Human Ganglia

After chickenpox, VZV becomes latent in cranial nerve ganglia, dorsal root ganglia, and autonomic nervous system ganglia [191] along the entire neuraxis of most humans. Unlike HSV, VZV cannot be cultured from human ganglia [192]. Thus, although clinicians had long suspected that ganglia were the site of VZV latency, verification came only after modern molecular techniques were applied. Southern blot and in situ hybridization first detected latent VZV in human trigeminal and thoracic ganglia [153–155]. The detection of multiple regions of the VZV genome indicated that most if not all of the VZV DNA molecule is present during latency [193]. PCR analysis of larger numbers of ganglia revealed VZV DNA in trigeminal ganglia from 13 of 15 subjects and in thoracic ganglia from 9 of 17 subjects [193]. These molecular structural studies validated earlier clinical observations by Hope-Simpson [39] indicating that the thoracic and trigeminal dermatomes were the most common sites of reactivation, i.e., zoster.

The viral burden in latently infected ganglia is low. Competitive PCR revealed 6–31 copies of the viral genome in 10^5 ganglionic cells [84]. Noncompetitive PCR detected 300–5400 copies of VZV DNA in every 10^5 cells [194], an estimate that was closer to the 10^3–10^5 copies of latent HSV-1 DNA in 10^5 cells [195]. The 100-fold difference in VZV copy number reported by Mahalingam et al. [84] and Clarke et al. [194] may reflect the differences in the techniques used. The most recent and accurate analysis used real-time PCR and determined that the number of VZV genomes per subject varied from 37 to 560 copies per 100 ng of DNA [196]. Latent VZV DNA appears to be extrachromosomal (nonintegrated) and possibly in a circular or concatameric (end-to-end) configuration [194] like latent HSV-1 DNA [197]. During latency, no less than four VZV genes are transcribed [198]: two are immediate-early (genes 62 and 63), and two are DNA-binding (genes 21 and 29). A recent study indicated that VZV gene 4 is also expressed in some latently infected human ganglia [199]. Furthermore, at least one protein corresponding to VZV gene 63 has been detected exclusively in the cytoplasm of neurons in latently infected human trigeminal and thoracic ganglia [200].

18.2 Cell Type Harboring Latent VZV

Although the cellular location of latent VZV in ganglia has been controversial for years, most recent studies indicate that neurons are the primary, if not exclusive, site of latent virus. In situ hybridization alone or together with PCR initially detected VZV only in neurons [154,155]. Later, VZV was reported in perineuronal satellite cells [152,201] and then in both neurons and various non-neuronal cells of latently infected human ganglia [202]. Two further studies corroborated the initial finding of VZV latency exclusively in neurons [156,203]. A different strategy using quantitative PCR analysis to study neurons and non-neuronal cells from postmortem ganglion cells sorted by size revealed two to five copies of VZV DNA primarily, if not exclusively, in neurons [204]. Finally, the most extensive in situ hybridization study of latently infected human ganglia to date revealed VZV expression almost entirely in neurons [199].

Despite considerable information accumulated about the physical state of VZV in latently infected ganglia, none is yet directly applicable to the treatment of human disease. A better understanding of virus latency will hopefully lead to testable hypotheses about the prevention of VZV reactivation and its neurological complications. Analogous to vaccination for the prevention of various childhood viral diseases, trials are under way to test the efficacy of vaccination of middle-aged adults in preventing zoster and its complications [205].

19 THE VARICELLA VACCINE: CONCERNS REGARDING REACTIVATION FROM LATENCY

The effect of childhood VZV vaccination on the subsequent development of zoster in elderly adults is an important issue. Because vaccination started in Japan, the answer will be known there first. Nevertheless, some information about VZV reactivation in children is available. In a study to determine whether children immunized with live varicella vaccine were at greater risk of developing zoster than children who had varicella, the incidence of zoster was compared in children with acute lymphocytic leukemia who had had varicella vs. those children who had been vaccinated with live varicella. During a 5 year observation period, 15 of 73 children who had had varicella developed zoster compared to none of 34 vaccinated children [206]. Hardy et al. [207] confirmed a lower incidence of zoster in another group of leukemic children after immunization with live attenuated VZV vaccine than in children who had naturally occurring VZV infection. These well-controlled studies suggest that zoster in the future elderly population might be less frequent in vaccinees than in those who had naturally occurring chickenpox. However, studies by Krause and Klinman [208] revealed that the Oka vaccine strain of VZV frequently reactivates, particularly in individuals with low anti-VZV titers after vaccination in whom the frequency of clinical infection and immunological boosting substantially exceeded the 13% annual rate after exposure to wild-type varicella. Those findings indicate that Oka VZV persists in vivo and reactivates as serum antibody titers decrease after vaccination. Because anti-VZV immunity is weaker in vaccinees than in individuals infected with wild-type VZV, the long- effect of frequent Oka VZV reactivation on the development of clinical zoster in the elderly will not be known for decades. Although it is unclear how Oka VZV reactivation will affect individuals in whom anti-VZV immunity wanes or who become immunosuppressed as adults, results of a controlled study in immunocompromised children indicated that Oka VZV vaccination is safer than wild-type VZV infection [207]. Nevertheless, it will be important to monitor the rates of zoster in childhood vaccine recipients as they become 60 years of age or older. Thus, current trials to boost immunity in middle-aged individuals may be valuable, because vaccination of prior vaccine recipients or of individuals with a history of childhood chickenpox may extend the duration of immunity [208].

REFERENCES

1. Preblud, S. R. Varicella: complications and costs. Pediatrics. 1986, 78, 728–735.
2. Longfield, J. N.; Winn, R. E.; Gibson, R. L.; Juchau, S. V.; Hoffman, P. V. Varicella outbreaks in army recruits from Puerto Rico. Arch. Intern. Med. 1990, 150, 970–973.
3. Nassar, N. T.; Touma, H. C. Susceptibility of Filipino nurses to the varicella-zoster virus. Infect. Control. 1986, 7, 71–72.
4. Preblud, S. R.; D'Angelo, L. J. Chickenpox in the United States. J. Infect. Dis. 1979, 140, 257–260.
5. Preblud, S. R.; Orenstein, W. A.; Bart, K. J. Varicella: clinical manifestations, epidemiology and health impact in children. Pediatr Infect. Dis. 1984, 3, 505–509.
6. Gustafson, T. L.; Lavely, G. B.; Brawner, E. R., Jr.; Hutcheson, R. H.; Wright, P. F.; Schaffner, W. An outbreak of airborne nosocomial varicella. Pediatrics. 1982, 70, 550–556.
7. Josephson, A.; Gombert, M. E. Airborne transmission of nosocomial varicella from localized zoster. J. Infect. Dis. 1988, 158, 238–241.
8. Leclair, J. M.; Zaia, J. A.; Levin, M. J., Jr.; Congdon, R. G.; Goldmann, D. A. Airborne transmission of chickenpox in a hospital. N. Engl. J. Med. 1980, 302, 450–453.
9. Fenner, F. The pathogenesis of the acute exanthems: an interpretation based on experimental investigations with mousepox (infectious ectromelia of mice). Lancet. 1948, 2, 915–920.
10. Grose, C. Variation on a theme by Fenner: the pathogenesis of chickenpox. Pediatrics. 1981, 68, 735–737.
11. Tomlinson, T. H. Giant cell formation in the tonsils in the prodromal stage of chickenpox. Am. J. Pathol. 1939, 15, 523–526.
12. Asano, Y.; Itakura, N.; Hiroishi, Y.; Hirose, S.; Nagai, T.; Ozaki, T.; Yazaki, T.; Yamanishi, K.; Takahashi, M. Viremia is present in incubation period in nonimmunocompromised children with varicella. J. Pediatr. 1985, 106, 69–71.

13. Kallander, C. F.; Gronowitz, J. S.; Olding-Stenkvist, E., Jr. Varicella zoster virus deoxythymidine kinase is present in serum before the onset of varicella. Scand. J. Infect. Dis. 1989, 21, 255–257.
14. Koropchak, C. M.; Solem, S. D.; Diaz, P. S.; Arvin, A. M. Investigation of varicella-zoster virus infection of lymphocytes by in situ hybridization. J. Virol. 1989, 63, 2392–2395.
15. Ozaki, T.; Ichikawa, T.; Matsui, Y.; Kondo, H.; Nagai, T.; Asano, Y.; Yamanishi, K.; Takahashi, M. Lymphocyte-associated viremia in varicella. J. Med. Virol. 1986, 19, 249–253.
16. Vonsover, A.; Leventon-Kriss, S.; Langer, A.; Smetana, Z.; Zaizov, R.; Potaznick, D.; Cohen, I. J.; Gotlieb-Stematsky, T. Detection of varicella-zoster virus in lymphocytes by DNA hybridization. J. Med. Virol. 1987, 21, 57–66.
17. Tyzzer, E. E. The histology of the skin lesions in varicella. Philippine. J. Sci. 1906, 1, 349–372.
18. Weller, T. H.; Witton, H. M.; Bell, E. J. The etiologic agents of varicella and herpes zoster Isolation, propagation, and cultural characteristics in vitro. J. Exp. Med. 1958, 108, 843–868.
19. Baba, K.; Yabuuchi, H.; Takahashi, M. Increased incidence of herpes zoster in normal children infected with varicella zoster virus during infancy: community-based follow-up study. J. Pediatr. 1986, 108, 372–377.
20. Wharton, M. The epidemiology of varicella-zoster virus infections. Infect. Dis. Clin. North. Am. 1996, 10, 571–581.
21. Halloran, M. E. Epidemiologic effects of varicella vaccination. Infect. Dis. Clin. North. Am. 1996, 10, 631–655.
22. Gilden, D. H.; Wroblewska, Z.; Kindt, V.; Warren, K. G.; Wolinsky, J. S. Varicella-zoster virus infection of human brain cells and ganglion cells in tissue culture. Arch. Virol. 1978, 56, 105–117.
23. Gilden, D. H.; Shtram, Y.; Friedmann, A.; Wellish, M.; Devlin, M.; Cohen, A.; Fraser, N.; Becker, Y. Extraction of cell-associated varicella-zoster virus DNA with Triton X-100-Nacl. J. Virol. Meth. 1982, 4, 263–275.
24. Reinhold, W. C.; Straus, S. E.; Ostrove, J. M. Directionality and further mapping of varicella zoster virus transcripts. Virus Res. 1988, 9, 249–261.
25. Davison, A. J.; Scott, J. E. The complete DNA sequence of varicella-zoster virus. J. Gen. Virol. 1986, 67, 1759–1816.
26. McGeoch, D. J.; Dolan, A.; Donald, S.; Rixon, F. J. Sequence determination and genetic content of the short unique region in the genome of herpes simplex virus type 1. J. Mol. Biol. 1985, 181, 1–13.
27. McGeoch, D. J.; Dolan, A.; Donald, S.; Brauer, D. H. K. Complete DNA sequence of the short repeat region in the genome of herpes simplex virus type 1. Nucleic. Acid. Res. 1986, 14, 1727–1745.
28. McGeoch, D. J.; Dalrymple, M. A.; Davison, A. J.; Dolan, A.; Frame, M. C.; McNab, D.; Perry, L. J.; Scott, J. E.; Taylor, P. The complete DNA sequence of the long unique region in the genome of herpes simplex virus type 1. J. Gen. Virol. 1988, 69, 1531–1574.
29. Perry, L. J.; McGeoch, D. J. The DNA sequences of the long repeat region and adjoining parts of the long unique region in the genome of herpes simplex virus type 1. J. Gen. Virol. 1988, 69, 2831–2846.
30. Cohen, J. J.; Straus, S. E. Varicella-zoster virus and its replication. In Fields Virology, 3rd ed.; Fields, B.N., Knipe, D.M., Howley, P.M., Eds.; Lippincott-Raven: Philadelphia, 1996, 2525–2547.
31. Maguire, H. F.; Hyman, R. W. Polyadenylated, cytoplasmic transcripts of varicella-zoster virus. Intervirology. 1986, 26, 181–191
32. Ostrove, J. M.; Reinhold, W.; Fan, C. M.; Zorn, S.; Hay, J.; Straus, S. E. Transcription mapping of the varicella-zoster virus genome. J. Virol. 1985, 56, 600–606.
33. Davison, A. J. Varicella-zoster virus: The Fourteenth Fleming Lecture. J. Gen. Virol. 1991, 72, 475–486.

34. Hayakawa, Y.; Hyman, R. W. Isomerization of the UL region of varicella-zoster virus DNA. Virus Res. 1987, 8, 25–31.
35. Kinchington, P. R.; Reinhold, W. C.; Casey, T. A.; Straus, S. E.; Hay, J.; Ruyechan, W. T. Inversion and circularization of the varicella-zoster virus genome. J. Virol. 1985, 56, 194–200.
36. Straus, S. E.; Owens, J.; Ruyechan, W. T.; Takiff, H. E.; Casey, T. A.; Vande Woude, G. F.; Hay, J. Molecular cloning and physical mapping of varicella-zoster virus DNA. Proc. Natl. Acad. Sci. USA. 1982, 79, 993–997.
37. Harnisch, J. P. Zoster in the elderly: clinical, immunologic and therapeutic considerations. J. Am. Geriatr. Soc. 1984, 32, 789–793.
38. Guess, H. A.; Broughton, D. D.; Melton, L. J.; Kurland, L. T. Epidemiology of herpes zoster in children and adolescents: a population-based study. Pediatrics. 1985, 76, 512–517.
39. Hope-Simpson, R. E. The nature of herpes zoster: a long-term study and a new hypothesis. Proc. Roy. Soc. Med. 1965, 58, 9–20.
40. Garnett, G. P.; Grenfell, B. T. The epidemiology of varicella-zoster virus infections: the influence of varicella on the prevalence of herpes zoster. Epidemiol. Infect. 1992, 108, 513–528.
41. von Barensprung, F. G. F. Beitrage zur Kenntnis des zoster. Ann. Chir. Krankenh. 1863, 11, 96–104.
42. Head, H.; Campbell, A. W. The pathology of herpes zoster and its bearing on sensory localization. Brain. 1900, 23, 353–523.
43. Denny-Brown, D.; Adams, R. D.; Fitzgerald, P. J. Pathologic features of herpes zoster: a note on ''geniculate herpes.''. Arch. Neurol. Psychiatry. 1944, 51, 216–231.
44. Cheatham, W. J.; Weller, T. H.; Dolan, T. F.; Dower, J. C. Varicella: report on two fatal cases with necropsy, virus isolation, and serologic studies. Am. J. Pathol. 1956, 32, 1015–1035.
45. Ghatak, N. R.; Zimmerman, H. M. Spinal ganglion in herpes zoster. Arch. Pathol. 1973, 95, 411–415.
46. Esiri, M. M.; Tomlinson, A. H. Herpes zoster: demonstration of virus in trigeminal nerve and ganglion by immunofluorescence and electron microscopy. J. Neurol. Sci. 1972, 15, 35–48.
47. Nagashima, K.; Nakazawa, M.; Endo, H. Pathology of the human spinal ganglia in varicella-zoster virus infection. Acta. Neuropathol. 1975, 33, 105–117.
48. Bastian, F. O.; Rabson, A. S.; Yee, C. L.; Tralka, T. S. Herpesvirus varicellae: isolated from human dorsal root ganglia. Arch. Pathol. 1974, 97, 331–332.
49. Shibuta, H.; Ishikawa, T.; Hondo, R.; Aoyama, Y.; Kurata, K.; Matumoto, M. Varicella virus isolation from spinal ganglion. Arch. Virusforsch. 1974, 45, 382–385.
50. Gilden, D. H.; Hayward, A. R.; Krupp, J.; Hunter-Laszlo, M.; Huff, J. C.; Vafai, A. Varicella-zoster virus infection of human mononuclear cells. Virus Res. 1987, 7, 117–129.
51. Gilden, D. H.; Devlin, M. E.; Wellish, M.; Mahalingam, R.; Huff, C.; Hayward, A.; Vafai, A. Persistence of varicella-zoster virus DNA in blood mononuclear cells of patients with varicella or zoster. Virus Genes. 1988, 2, 299–305.
52. Feldman, S.; Chaudary, S.; Ossi, M.; Epp, E. A viremic phase for herpes zoster in children with cancer. J. Pediatr. 1977, 91, 597–600.
53. Feldman, S.; Epp, E. Isolation of varicella-zoster virus from blood. J. Pediatr. 1976, 88, 265–267.
54. Gershon, A. A.; Steinberg, S.; Silber, R. Varicella-zoster viremia. J. Pediatr. 1978, 92, 1033–1034.
55. Gold, E. Serologic and virus-isolation studies of patients with varicella or herpes-zoster infection. N. Engl. J. Med. 1996, 274, 181–185.
56. Myers, M. G. Viremia caused by varicella-zoster virus: association with malignant progressive varicella. J. Infect. Dis. 1979, 140, 229–233.
57. Twomey, J. J.; Gyorkey, F.; Norris, S. M. The monocyte disorder with herpes zoster. J. Lab. Clin. Med. 1974, 83, 768–777.

58. Arbeit, R. D.; Zaia, J. A.; Valerio, M. A.; Levin, M. J. Infection of human peripheral blood mononuclear cells by varicella-zoster virus. Intervirology. 1982, 18, 56–65.
59. Cauda, R.; Chatterjee, S.; Tiden, A. B.; Grossi, C. E.; Whitley, R. J. Replication of varicella zoster virus in Raji cells. Virus Res. 1986, 4, 337–342.
60. Leventon-Kriss, S.; Gotlieb-Stematsky, T.; Vonsover, A.; Smetana, Z. Infection and persistence of varicella-zoster virus in lymphoblastoid Raji cell line. Med. Microbiol. Immunol. 1979, 167, 275–283.
61. Manz, H. J.; Canter, H. G.; Melton, J. Trigeminal herpes zoster causing mandibular osteonecrosis and spontaneous tooth exfoliation. South Med. J. 1986, 79, 1026–1028.
62. Garty, B.-Z.; Dinari, G.; Sarnat, H.; Cohen, S.; Nitzan, M. Tooth exfoliation and osteonecrosis of the maxilla after trigeminal herpes zoster. J. Pediatr. 1985, 106, 71–73.
63. Sweeney, C. J.; Gilden, D. H. Ramsay Hunt syndrome. J. Neurol. Neurosurg Psychiatry. 2001, 71, 149–154.
64. Robillard, R. B.; Hilsinger, R. L., Jr; Adour, K. K. Ramsay Hunt facial paralysis: clinical analyses of 185 patients. Otolaryngol Head Neck Surg. 1986, 95, 292–297.
65. Murakami, S.; Honda, N.; Mizobuchi, M.; Nakashiro, Y.; Hato, N.; Gyo, K. Rapid diagnosis of varicella zoster virus infection in acute facial palsy. Neurology. 1998, 51, 1202–1205.
66. Murakami, S.; Mizobuchi, M.; Nakashiro, Y.; Doi, T.; Hato, N.; Yanagihara, N. Bell palsy and herpes simplex virus: identification of viral DNA in endoneural fluid and muscle. Ann. Intern. Med. 1996, 124, 27–30.
67. Furuta, Y.; Ohtani, F.; Mesuda, Y.; Fukuda, S.; Inuyama, Y. Early diagnosis of zoster sine herpete and antiviral therapy for the treatment of facial palsy. Neurology. 2000, 55, 708–710.
68. Thomas, E. J.; Howard, F. M. Segmental zoster paresis—a disease profile. Neurology. 1972, 22, 459–466.
69. Miller, D. H.; Kay, R.; Schon, F.; McDonald, W. I.; Haas, L. F.; Hughes, R. A. Optic neuritis following chickenpox in adults. J. Neurol. 1986, 233, 182–184.
70. Carroll, W. M.; Mastaglia, F. L. Optic neuropathy and ophthalmoplegia in herpes zoster oticus. Neurology. 1979, 29, 726–729.
71. Crabtree, J. A. Herpes zoster oticus. Laryngoscope. 1968, 78, 1853–1879.
72. Steffen, R.; Selby, G. ''Atypical'' Ramsay Hunt syndrome. Med. J. Aust. 1972, 1, 227–230.
73. Lapresle, J.; Lasjuanias, P. Cranial nerve ischemic arterial syndromes. Brain. 1986, 109, 207–215.
74. Mayberg, M. R.; Zervas, N. T.; Moskowitz, M. A. Trigeminal projections to supratentorial pial and dural blood vessels in cats demonstrated by horseradish peroxidase histochemistry. J. Comp. Neurol. 1984, 223, 46–56.
75. Brown, G. R. Herpes zoster: correlation of age, sex, distribution, neuralgia and associated disorders. South Med. J. 1976, 69, 576–578.
76. Ragozzino, M. W.; Melton, III, L. J.; Kurland, L. T.; Chu, C. P.; Perry, H. O. Population-based study of herpes zoster and its sequelae. Medicine. 1982, 61, 310–316.
77. Hope-Simpson, R. E. Postherpetic neuralgia. J. Roy. College Gen. Pract. 1975, 25, 571–575.
78. DeMoragas, J. M.; Kierland, R. R. The outcome of patients with herpes zoster. Arch. Dermatol. 1957, 75, 193–196.
79. Rogers, R. S.; Tindall, J. P. Geriatric herpes zoster. J. Am. Geriatr. Soc. 1971, 19, 495–503.
80. Vafai, A.; Murray, R. S.; Wellish, M.; Devlin, M.; Gilden, D. H. Expression of varicella-zoster virus and herpes simplex virus in normal human trigeminal ganglia. Proc. Natl. Acad. Sci. USA. 1988, 85, 2362–2366.
81. Mahalingam, R.; Wellish, M.; Brucklier, J.; Gilden, D. H. Persistence of varicella-zoster virus DNA in elderly patients with postherpetic neuralgia. J. NeuroVirol. 1995, 1, 130–133.
82. Smith, F. P. Pathological studies of spinal nerve ganglia in relation to intractable intercostal pain. Surg. Neurol. 1978, 10, 50–53.
83. Watson, C. P. N.; Deck, J. H.; Morshead, C.; Van der Kooy, D.; Evans, R. J. Post-herpetic neuralgia: further postmortem studies of cases with and without pain. Pain. 1991, 44, 105–117.

84. Mahalingam, R.; Wellish, M.; Lederer, D.; Forghani, B.; Cohrs, R.; Gilden, D. H. Quantitation of latent varicella-zoster virus DNA in human trigeminal ganglia by polymerase chain reaction. J. Virol. 1993, 67, 2381–2384.
85. Gilden, D. H.; Dueland, A. N.; Cohrs, R.; Martin, J. R.; Kleinschmidt-DeMasters, B. K.; Mahalingam, R. Preherpetic neuralgia. Neurology. 1991, 41, 1215–1218.
86. Schott, G. D. Triggering of delayed-onset postherpetic neuralgia. Lancet. 1998, 351, 419–420.
87. Brostoff, J. Diaphragmatic paralysis after herpes zoster. Br. Med. J. 1966, 2, 1571–1572.
88. Stowasser, M.; Cameron, J.; Oliver, W. A. Diaphragmatic paralysis following cervical herpes zoster. Med. J. Aust. 1990, 153, 555–556.
89. Izumi, A. I.; Edwards, J. Herpes zoster and neurogenic bladder dysfunction. JAMA. 1973, 224, 1748–1749.
90. Jellinek, E. H.; Tulloch, W. S. Herpes zoster with dysfunction of bladder and anus. Lancet. 1976, 2, 1219–1222.
91. Arnold, D. G. Herpes zoster as a sequel of spinal anesthesia. J. Int. Coll. Surgeons. 1941, 4, 66–67.
92. Horton, B.; Price, R. W.; Jimenez, D. Multifocal varicella-zoster virus leukoencephalitis temporally remote from herpes zoster. Ann. Neurol. 1981, 9, 251–266.
93. Reimer, L. G.; Reller, L. B. CSF in herpes zoster meningoencephalitis. Arch. Neurol. 1981, 38, 668.
94. Wolf, S. M. Decreased cerebrospinal fluid glucose level in herpes zoster meningitis. Arch. Neurol. 1974, 30, 109.
95. Gilden, D. H.; Bennett, J. L.; Kleinschmidt-DeMasters, B. K.; Song, D. D.; Yee, A. S.; Steiner, I. The value of cerebrospinal fluid antiviral antibody in the diagnosis of neurologic disease produced by varicella zoster virus. J. Neurol. Sci. 1998, 159, 140–144.
96. Amlie-Lefond, C.; Kleinschmidt-DeMasters, B. K.; Mahalingam, R.; Davis, L. E.; Gilden, D. H. The vasculopathy of varicella zoster virus encephalitis. Ann. Neurol. 1995, 37, 784–790.
97. Gilden, D. H.; Kleinschmidt-DeMasters, B. K.; Wellish, M.; Hedley-Whyte, E. T.; Rentier, B.; Mahalingam, R. Varicella zoster virus, a cause of waxing and waning vasculitis NEJM case 5–1995 revisited. Neurology. 1996, 47, 1441–1446.
98. Nau, R.; Lantsch, M.; Stiefel, M.; Polak, T.; Reiber, H. Varicella zoster virus-associated focal vasculitis without herpes zoster: recovery after treatment with acyclovir. Neurology. 1998, 51, 914–915.
99. Kuroiwa, Y.; Furukawa, T. Hemispheric infarction after herpes zoster ophthalmicus: computed tomography and angiography. Neurology. 1981, 31, 1030–1032.
100. Eible, R. J. Intracerebral hemorrhage with herpes zoster ophthalmicus. Ann. Neurol. 1983, 14, 591–592.
101. Hilt, D. C.; Buchholz, D.; Krumholz, A.; Weiss, H.; Wolinsky, J. S. Herpes zoster ophthalmicus and delayed contralateral hemiparesis caused by cerebral angiitis: diagnosis and management approaches. Ann. Neurol. 1983, 14, 543–553.
102. Singhal, A. B.; Singhal, B. S.; Ursekar, M. A.; Koroshetz, W. J. Serial MR angiography and contrast-enhanced MRI in chickenpox-associated stroke. Neurology. 2001, 56, 815–817.
103. Melanson, M.; Chalk, C.; Georgevich, L.; Fett, K.; Lapierre, Y.; Duong, H.; Richardson, J.; Marineau, C.; Rouleau, G. A. Varicella-zoster virus DNA in CSF and arteries in delayed contralateral hemiplegia: evidence for viral invasion of cerebral arteries. Neurology. 1996, 47, 569–570.
104. Hall, S.; Carlin, L.; Roach, S. E.; McLean, W. T. Herpes zoster and central retinal artery occlusion. Ann. Neurol. 1983, 13, 217–218.
105. Ross, M. H.; Abend, W. K.; Schwartz, R. B.; Samuels, M. A. A case of C2 herpes zoster with delayed bilateral pontine infarction. Neurology. 1991, 41, 1685–1686.
106. Fukumoto, S.; Kinjo, M.; Hokamura, K.; Tanaka, K. Subarachnoid hemorrhage and granulomatous angiitis of the basilar artery: demonstration of the varicella-zoster virus in the basilar artery lesions. Stroke. 1986, 17, 1024–1028.

107. Geny, C.; Yulis, J.; Azoulay, A.; Brugieres, P.; Saint-Val, C.; Degos, J. D. Thalamic infarction following lingual herpes zoster. Neurology. 1991, 41, 1846.
108. Rawlinson, W. D.; Cunningham, A. L. Contralateral hemiplegia following thoracic herpes zoster. Med. J. Aust. 1991, 155, 344–346.
109. Kleinschmidt-DeMasters, B. K.; Amlie-Lefond, C.; Gilden, D. H. The patterns of varicella zoster virus encephalitis. Hum. Pathol. 1996, 27, 927–938.
110. Devinsky, O.; Cho, E.-S.; Petito, C. K.; Price, R. W. Herpes zoster myelitis. Brain. 1991, 114, 1181–1196.
111. Kleinschmidt-DeMasters, B. K.; Mahalingam, R.; Shimek, C.; Marcoux, H. L.; Wellish, M.; Tyler, K. L. Profound cerebrospinal fluid pleocytosis and Froin's syndrome secondary to widespread necrotizing vasculitis in an HIV-positive patient with varicella zoster virus encephalomyelitis. J. Neurol. Sci. 1998, 159, 213–218.
112. Gilden, D. H.; Beinlich, B. R.; Rubinstein, E. M.; Stommel, E.; Swenson, R.; Rubinstein, D.; Mahalingam, R. VZV myelitis: an expanding spectrum. Neurology. 1994, 44, 1818–1823.
113. Hogan, E. L.; Krigman, M. R. Herpes zoster myelitis: evidence for viral invasion of spinal cord. Arch. Neurol. 1973, 29, 309–313.
114. Tako, J.; Rado, J. P. Zoster meningoencephalitis in a steroid-treated patient. Arch. Neurol. 1965, 12, 610–612.
115. de Silva, S. M.; Mark, A. S.; Gilden, D. H.; Mahalingam, R.; Balish, M.; Sandbrink, F.; Houff, S. Zoster myelitis: improvement with antiviral therapy in two cases. Neurology. 1996, 47, 929–931.
116. Underwood, E. A. The neurological complications of varicella: a clinical and epidemiological study. Br. J. Child. Dis. 1935, 32, 83–107.
117. Miller, H. G.; Stanton, J. B.; Gibbons, J. Para-infectious encephalomyelitis and related syndromes: a critical review of the neurological complications of certain specific fevers. Quant. J. Med. 1956, 25, 427–505.
118. Bullowa, J. G. M.; Wishik, S. M. Complications of varicella. I. Their occurrence among 2,534 patients. Am. J. Dis. Child. 1935, 49, 923–926.
119. Haymaker, W.; Kernohan, J. W. The Landry-Guillain-Barre syndrome. Medicine. 1949, 28, 59–141.
120. Wohlwill, F. Zur pathologischen anatomie des Nervensystems beim herpes zoster. Zentralbl. Gesamte. Neurol. Psychiatr. 1924, 89, 171–212.
121. Dayan, A. D.; Ogul, E.; Graveson, G. S. Polyneuritis and herpes zoster. J. Neurol. Neurosurg. Psychiatry. 1972, 35, 170–175.
122. Fox, R. J.; Galetta, S. L.; Mahalingam, R.; Wellish, M.; Forghani, B.; Gilden, D. H. Acute, chronic and recurrent varicella zoster virus (VZV) neuropathy without zoster rash. Neurology. 2001, 57, 351–354.
123. Hurwitz, E. S.; Nelson, D. B.; Davis, C.; Morens, D.; Schonberger, L. B. National surveillance for Reye syndrome: a five-year review. Pediatrics. 1982, 70, 895–900.
124. Remington, P. L.; Rowley, D.; McGee, H.; Hall, W. N.; Monto, A. S. Decreasing trends in Reye syndrome and aspirin use in Michigan, 1979 to 1984. Pediatrics. 1986, 77, 93–98.
125. Widal, A. S. J. Med. Chiropractic. Pract. 1907, 78, 12.
126. Weber, F. P. Herpes zoster: its occasional association with a generalized eruption and its occasional connection with muscular paralysis—also an analysis of the literature of the subject. Int. Clin. 1916, 3, 185–202.
127. Lewis, G. W. Zoster sine herpete. Br. Med. J. 1958, 2, 418–421.
128. Easton, H. G. Zoster sine herpete causing acute trigeminal neuralgia. Lancet. 1970, 2, 1065–1066.
129. Gilden, D. H.; Wright, R. R.; Schneck, S. A.; Gwaltney, J. M., Jr; Mahalingam, R. Zoster sine herpete, a clinical variant. Ann. Neurol. 1994, 35, 530–533.
130. Amlie-Lefond, C.; Mackin, G. A.; Ferguson, M.; Wright, R. R.; Mahalingam, R.; Gilden, D. H. Another case of virologically confirmed zoster sine herpete, with electrophysiologic correlation. J. NeuroVirol. 1996, 2, 136–138.

131. Vafai, A.; Mahalingam, R.; Zerbe, G.; Wellish, M.; Gilden, D. H. Detection of antibodies to varicella-zoster virus proteins in sera from the elderly. Gerontology. 1988, 34, 242–249.
132. Dueland, A. N.; Devlin, M.; Martin, J. R.; Mahalingam, R.; Cohrs, R.; Manz, H.; Trombley, I.; Gilden, D. Fatal varicella zoster virus meningoradiculitis without skin involvement. Ann. Neurol. 1991, 29, 569–572.
133. Heller, H. M.; Carnevale, N. T.; Steigbigel, R. T. Varicella zoster virus transverse myelitis without cutaneous rash. Am. J. Med. 1990, 88, 550–551.
134. Martinez-Martin, P.; Garcia-Saiz, A.; Rapun, J. L.; Echevarria, J. M. Intrathecal synthesis of IgG antibodies to varicella-zoster virus in two cases of acute aseptic meningitis syndrome with no cutaneous lesions. J. Med. Virol. 1985, 16, 201–209.
135. Echevarria, J. M.; Martinez-Martin, P.; Tellez, A.; de Ory, F.; Rapun, J. L.; Bernal, A.; Estevez, E.; Najera, R. Aseptic meningitis due to varicella-zoster virus: antibody levels and local synthesis of specific IgG, IgM and IgA. J. Infect. Dis. 1987, 155, 959–967.
136. Vartdal, F.; Vandvik, B.; Norby, E. Intrathecal synthesis of virus-specific oligoclonal IgG, IgA and IgM antibodies in a case of varicella-zoster meningoencephalitis. J. Neurol. Sci. 1982, 57, 121–132.
137. Mayo, D. R.; Booss, J. Varicella zoster-associated neurologic disease without skin lesions. Arch. Neurol. 1989, 46, 313–315.
138. Osaki, Y.; Matsubayashi, K.; Okumiya, K.; Wada, T.; Doi, Y. Polyneuritis cranialis due to varicella-zoster virus in the absence of rash. Neurology. 1995, 45, 2293.
139. Aitken, R. S.; Brain, R. T. Facial palsy and infection with zoster virus. Lancet. 1933, 1, 19–22.
140. Furuta, Y.; Ohtani, F.; Kawabata, H.; Fukuda, S.; Bergstrom, T. High prevalence of varicella-zoster virus reactivation in herpes simplex virus-seronegative patients with acute peripheral facial palsy. Clin. Infect. Dis. 2000, 30, 529–533.
141. Taylor-Robinson, D.; Caunt, A. E. Varicella virus. In Virology Monographs; Gard, S., Hallauer, C., Meyer, K.F., Eds.; Springer-Verlag: New York, 1972; Vol. 12, 13–17.
142. Cooke, B. E. D. Epithelial smears in diagnosis of herpes simplex and herpes zoster affecting the oral mucosa. Br. Dent. J. 1960, 109, 83–96.
143. Cooke, B. E. D. Exfoliative cytology in evaluating oral lesions. J. Dent. Res. 1963, 42, 343–347.
144. Williams, B.; Capers, T. H. The demonstration of intranuclear inclusion bodies in sputum from a patient with varicella pneumonia. Am. J. Med. 1959, 27, 836–839.
145. Ozaki, T.; Matsui, Y.; Asano, Y.; Okuno, T.; Yamanishi, K.; Takahashi, M. Study of virus isolation from pharyngeal swabs in children with varicella. Am. J. Dis. Child. 1989, 143, 1448–1450.
146. Schmidt, N. Y.; Arvin, A. M. Sensitivity of different assay systems for immunoglobulin in responses to varicella zoster virus in reactivated infections (zoster). J. Clin. Microbiol. 1986, 19, 310–316.
147. Mustonen, K.; Mustakangas, P.; Smeds, M.; Mannonen, L.; Uotila, L.; Vaheri, A.; Koskiniemi, M. Antibodies to varicella zoster virus in the cerebrospinal fluid of neonates with seizures. Arch. Dis. Child. Fetal. Neonatal. Ed. 1998, 78, F57–F61.
148. Doern, G. V.; Robbie, L.; St. Armand, R. Comparison of the Vidas and Bio-Whittaker enzyme immunoassays for detecting IgG reactive with varicella-zoster virus and mumps virus. Diagn. Microbiol. Infect. Dis. 1997, 28, 31–34.
149. Gleaves, C. A.; Schwarz, K. A.; Campbell, M. B. Determination of varicella-zoster virus (VZV) immune status with the VIDAS immunoglobulin G automated immunoassay and the VZV Scan latex agglutination assay. Clin. Diagn. Lab. Immunol. 1996, 3, 365–367.
150. Gilden, D. H.; Murray, R. S.; Wellish, M.; Kleinschmidt-DeMasters, B. K.; Vafai, A. Chronic progressive varicella-zoster virus encephalitis in an AIDS patient. Neurology. 1988, 38, 1150–1153.

151. Ryder, J. W.; Croen, K.; Kleinschmidt-DeMasters, B. K.; Ostrove, J. M.; Straus, S. E.; Cohn, D. L. Progressive encephalitis three months after resolution of cutaneous zoster in a patient with AIDS. Ann. Neurol. 1986, 19, 182–188.
152. Croen, K. D.; Ostrove, J. M.; Dragovic, L. J.; Straus, S. E. Patterns of gene expression and sites of latency in human nerve ganglia are different for varicella-zoster and herpes simplex viruses. Proc. Natl. Acad. Sci. USA. 1988, 85, 9773–9777.
153. Gilden, D. H.; Vafai, A.; Shtram, Y.; Becker, Y.; Devlin, M.; Wellish, M. Varicella-zoster virus DNA in human sensory ganglia. Nature. 1983, 306, 478–480.
154. Gilden, D. H.; Rozemann, Y.; Murray, R.; Devlin, M.; Vafai, A. Detection of varicella-zoster virus nucleic acid in neurons of normal human thoracic ganglia. Ann. Neurol. 1987, 22, 377–380.
155. Hyman, R. W.; Ecker, J. R.; Tenser, R. B. Varicella-zoster virus RNA in human trigeminal ganglia. Lancet. 1983, 2, 814–816.
156. Kennedy, P. G.; Grinfeld, E.; Gow, J. W. Latent varicella-zoster virus is located predominantly in neurons in human trigeminal ganglia. Proc. Natl. Acad. Sci. USA. 1998, 95, 4658–4662.
157. Meier, J. L.; Holman, R. P.; Croen, K. D.; Smialek, J. D.; Straus, S. E. Varicella-zoster virus transcription in human trigeminal ganglia. Virology. 1993, 193, 193–200.
158. Koropchak, C. M.; Graham, G.; Palmer, J.; Winsberg, M.; Ting, S. F.; Wallace, M.; Prober, C. G.; Arvin, A. M. Investigation of varicella-zoster virus infection by polymerase chain reaction in the immunocompetent host with acute varicella. J. Infect. Dis. 1991, 163, 1016–1022.
159. Ozaki, T.; Miwata, H.; Matsui, Y.; Kido, S.; Yamanishi, K. Varicella zoster virus DNA in throat swabs. Arch. Dis. Child. 1991, 66, 333–334.
160. Devlin, M. E.; Gilden, D. H.; Mahalingam, R.; Dueland, A. N.; Cohrs, R. Peripheral blood mononuclear cells of the elderly contain varicella-zoster virus DNA. J. Infect. Dis. 1992, 165, 619–622.
161. Kido, S.; Ozaki, T.; Asada, H.; Higashi, K.; Kondo, K.; Hayakawa, Y.; Morishima, T.; Takahashi, M.; Yamanishi, K. Detection of varicella-zoster virus (VZV) DNA in clinical samples from patients with VZV by the polymerase chain reaction. J. Clin. Microbiol. 1991, 29, 76–79.
162. Furuta, Y.; Fukuda, S.; Suzuki, S.; Takasu, T.; Inuyama, Y.; Nagashima, K. Detection of varicella-zoster virus DNA in patients with acute peripheral facial palsy by the polymerase chain reaction, and its use for early diagnosis of zoster sine herpete. J. Med. Virol. 1997, 52, 316–319.
163. Puchhammer-Stockl, E.; Popow-Kraupp, T.; Heinz, F. X.; Mandl, C. W.; Kunz, C. Detection of varicella-zoster virus DNA by polymerase chain reaction in the cerebrospinal fluid of patients suffering from neurological complications associated with chickenpox or herpes zoster. J. Clin. Microbiol. 1991, 29, 1513–1516.
164. Knox, C. M.; Chandler, D.; Short, G. A.; Margolis, T. P. Polymerase chain reaction-based assays of vitreous samples for the diagnosis of viral retinitis Use in diagnostic dilemmas. Ophthalmology. 1998, 105, 37–44.
165. Mietz, H.; Eis-Hubinger, A. M.; Sundmacher, R.; Font, R. L. Detection of varicella-zoster virus DNA in keratectomy specimens by use of the polymerase chain reaction. Arch. Ophthalmol. 1997, 115, 590–594.
166. Garweg, J.; Bohnke, M. Varicella-zoster virus is strongly associated with atypical necrotizing herpetic retinopathies. Clin. Infect. Dis. 1997, 24, 603–608.
167. Beards, G.; Graham, C.; Pillay, D. Investigation of vesicular rashes for HSV and VZV by PCR. J. Med. Virol. 1998, 54, 155–157.
168. Rubben, A.; Baron, J. M.; Grussendorf-Conen, E. I. Routine detection of herpes simplex virus and varicella-zoster virus by polymerase chain reaction reveals that initial herpes zoster is frequently misdiagnosed as herpes simplex. Br. J. Dermatol. 1997, 137, 259–261.

169. Stebbings, S.; Highton, J.; Croxson, M. C.; Powell, K.; McKay, J.; Rietveld, J. Chickenpox monoarthritis: demonstration of varicella-zoster in joint fluid by polymerase chain reaction. Br. J. Rheumatol. 1998, 37, 311–313.
170. Bowsher, D. Post-herpetic neuralgia in older patients: incidence and optimal treatment. Drugs Aging. 1994, 5, 411–418.
171. Gilden, D. H.; Kleinschmidt-DeMasters, B. K.; LaGuardia, J. J.; Mahalingam, R.; Cohrs, R. Neurologic complications of the reactivation of varicella-zoster virus. N. Engl. J. Med. 2000, 342, 635–645.
172. Kost, R. G.; Straus, S. E. Postherpetic neuralgia B pathogenesis, treatment, and prevention. N. Engl. J. Med. 1996, 335, 32–42.
173. Portenoy, R. K.; Duma, C.; Foley, K. M. Acute herpetic and postherpetic neuralgia: clinical review and current management. Ann. Neurol. 1986, 20, 651–664.
174. Watson, C. P. N. Pain Research and Clinical Management, Vol 8, Herpes Zoster and Postherpetic Neuralgia; Elsevier: Amsterdam, 1993.
175. Tyring, S.; Barbarash, R. A.; Nahlik, J. E.; Cunningham, A.; Marley, J.; Heng, M.; Jones, T.; Rea, T.; Boon, R.; Saltzman, R. and the collaborative Famciclovir Herpes Zoster Study Group. Famciclovir for the treatment of acute herpes zoster: effects on acute disease and postherpetic neuralgia. A randomized, double-blind, placebo-controlled trial. Ann. Intern. Med. 1995, 123, 89–96.
176. Wood, M. J.; Johnson, R. W.; McKendrick, M. W.; Taylor, J.; Mandal, B. K.; Crooks, J. A randomized trial of acyclovir for 7 days or 21 days with and without prednisolone for treatment of acute herpes zoster. N. Engl. J. Med. 1994, 330, 896–900.
177. Merigan, T. C.; Rand, K. H.; Pollard, R. B.; Abdallah, P. S.; Jordan, G. W.; Fried, R. P. Human leukocyte interferon for the treatment of herpes zoster in patients with cancer. N. Engl. J. Med. 1978, 298, 981–987.
178. McKendrick, M. W.; McGill, J. I.; Wood, M. J. Lack of effect of acyclovir on postherpetic neuralgia. Br. Med. J. 1989, 298, 431.
179. Galbraith, A. W. Prevention of post-herpetic neuralgia by amantadine hydrocholoride (Symmetrel). Br. J. Clin. Pract. 1983, 37, 304–306.
180. Sklar, S. H.; Blue, W. T.; Alexander, E. J.; Bodian, C. A. Herpes zoster: the treatment and prevention of neuralgia with adenosine monophosphate. JAMA. 1985, 253, 1427–1430.
181. Kernbaum, S.; Hauchecorne, J. Administration of levodopa for relief of herpes zoster pain. JAMA. 1981, 246, 132–134.
182. Eaglstein, W. H.; Katz, R.; Brown, J. A. The effects of early corticosteroid therapy on the skin eruption and pain of herpes zoster. JAMA. 1970, 211, 1681–1683.
183. Keczkes, K.; Basheer, A. M. Do corticosteroids prevent post-herpetic neuralgia?. Br. J. Dermatol. 1980, 102, 551–555.
184. Esmann, V.; Geil, J. P.; Kroon, S.; Fogh, H.; Peterslund, N. A.; Petersen, C. S.; Ronne-Rasmussen, J. O.; Danielsen, L. Prednisolone does not prevent post-herpetic neuralgia. Lancet. 1987, 2, 126–129.
185. Segal, A. Z.; Rordorf, G. Gabapentin as a novel treatment for postherpetic neuralgia. Neurology. 1996, 46, 1175–1176.
186. Rowbotham, M.; Harden, N.; Stacey, B.; Bernstein, P.; Magnus-Miller, L. Gabapentin for the treatment of postherpetic neuralgia: a randomized controlled trial. JAMA. 1998, 280, 1837–1842.
187. Eide, K.; Stubhaug, A.; Oye, I.; Breivik, H. Continuous subcutaneous administration of the N-methyl-D-aspartic acid (NMDA) receptor antagonist ketamine in the treatment of post-herpetic neuralgia. Pain. 1995, 61, 221–228.
188. Galer, B. S.; Rowbotham, M. C.; Perander, J.; Friedman, E. Topical lidocaine patch relieves postherpetic neuralgia more effectively than a vehicle topical patch: results of an enriched enrollment study. Pain. 1999, 80, 533–538.
189. King, R. B. Topical aspirin in chloroform and the relief of pain due to herpes zoster and postherpetic neuralgia. Arch. Neurol. 1993, 50, 1046–1053.

190. Kotani, N.; Kushikata, T.; Hashimoto, H.; Kimura, F.; Muraoka, M.; Yodono, M.; Asai, M.; Matsuki, A. Intrathecal methylprednisolone for intractable postherpetic neuralgia. N. Engl. J. Med. 2000, 343, 1514–1519.
191. Gilden, D. H.; Gesser, R.; Smith, J.; Wellish, M.; LaGuardia, J. J.; Cohrs, R. J.; Mahalingam, R. Presence of VZV and HSV-1 DNA in nodose and celiac ganglia. Virus Genes; 2001; Vol. 23, 145–147.
192. Plotkin, S. A.; Stein, S.; Synder, M.; Immesoete, P. Attempts to recover varicella zoster from ganglia. Ann. Neurol. 1977, 2, 249.
193. Mahalingam, R.; Wellish, M.; Wolf, W.; Dueland, A. N.; Cohrs, R.; Vafai, A.; Gilden, D. H. Latent varicella-zoster viral DNA in human trigeminal and thoracic ganglia. N. Engl. J. Med. 1990, 323, 627–631.
194. Clarke, P.; Beer, T.; Cohrs, R.; Gilden, D. H. Configuration of latent varicella-zoster virus DNA. J. Virol. 1995, 69, 8151–8154.
195. Efstathiou, S.; Minson, A. C.; Field, H. J.; Anderson, J. R.; Wildy, P. Detection of herpes simplex virus-specific sequences in latently infected mice and in humans. J. Virol. 1986, 57, 446–455.
196. Cohrs, R. J.; Randall, J.; Smith, J.; Gilden, D. H.; Dabrowski, C.; van der Keyl, H.; Tal-Singer, R. Analysis of individual human trigeminal ganglia for latent herpes simplex virus type 1 and varicella-zoster virus nucleic acids using real-time PCR. J. Virol. 2000, 74, 11464–11471.
197. Mellerick, D. M.; Fraser, N. W. Physical state of the latent herpes simplex virus genome in a mouse model system: evidence suggesting an episomal state. Virology. 1987, 158, 265–275.
198. Cohrs, R.; Barbour, M.; Gilden, D. H. VZV transcription during latency in human ganglia: detection of transcripts mapping to genes 21, 29, 62, and 63 in a cDNA library enriched for VZV RNA. J. Virol. 1996, 70, 2789–2796.
199. Kennedy, P. G. E.; Grinfeld, E.; Bell, J. E. Varicella-zoster virus gene expression in latently infected and explanted human ganglia. J. Virol. 2000, 74, 11893–11898.
200. Mahalingam, R.; Wellish, M.; Cohrs, R.; Debrus, S.; Piette, J.; Rentier, B.; Gilden, D. H. Expression of protein encoded by varicella-zoster virus open reading frame 63 in latently infected human ganglionic neurons. Proc. Natl. Acad. Sci. USA. 1996, 93, 2122–2124.
201. Meier, J. L.; Straus, S. E. Varicella-zoster virus DNA polymerase and major DNA-binding protein genes have overlapping divergent promoters. J. Virol. 1993, 67, 7573–7581.
202. Lungu, O.; Annunziato, P. W.; Gershon, A.; Staugaitis, S. M.; Josefson, D.; LaRussa, P.; Silverstein, S. J. Reactivated and latent varicella-zoster virus in human dorsal root ganglia. Proc. Natl. Acad. Sci. USA. 1995, 92, 10980–10984.
203. Dueland, A. N.; Ranneberg-Nilsen, T.; Degre, M. Detection of latent varicella zoster virus DNA and human gene sequences in human trigeminal ganglia by in situ amplification combined with in situ hybridization. Virology. 1995, 140, 2055–2066.
204. LaGuardia, J. J.; Cohrs, R. J.; Gilden, D. H. Prevalence of varicella-zoster virus DNA in dissociated human trigeminal ganglia neurons and non-neuronal cells. J. Virol. 1999, 73, 8571–8577.
205. Levin, M. J.; Barber, D.; Goldblatt, E.; Jones, M.; LaFleur, B.; Chang, C.; Stinson, D.; Zerbe, G. O.; Hayward, A. R. Use of a live attenuated varicella vaccine to boost varicella-specific immune responses in seropositive people 55 years of age and older: duration of booster effect. J. Infect. Dis. 1998, 178(suppl 1), S109–S112.
206. Brunell, P. A.; Taylor-Wiedeman, J.; Geiser, C. F.; Frierson, L.; Lydick, E. Risk of herpes zoster in children with leukemia: varicella vaccine compared with history of chickenpox. Pediatrics. 1986, 77, 53–56.
207. Hardy, I.; Gershon, A. A.; Steinberg, S. P.; LaRussa, P. The Varicella Vaccine Collaborative Study Group. The incidence of zoster after immunization with live attenuated varicella vaccine. N. Engl. J. Med. 1991, 325, 1545–1550.
208. Krause, P. R.; Klinman, D. M. Varicella vaccination: evidence for frequent reactivation of the vaccine strain in healthy children. Nature Med. 2000, 6, 451–454.

7

Epstein-Barr Virus and the Nervous System

Alex C. Tselis
Wayne State University/Detroit Medical Center
Detroit, Michigan, U.S.A.

1 INTRODUCTION

The Epstein-Barr virus (EBV) is a member of the herpesvirus family and infects more than 90% of the world's population. The results of infection with this virus depend on age and degree of immunocompetence and can lead to an astonishing array of systemic and neurological manifestations. In this chapter, we discuss these after some introductory material.

2 HISTORY

The syndrome of fatigue, malaise, fever, sore throat, and cervical lymphadenopathy with splenomegaly was first described in the late 1800s. Although the syndrome was etiologically heterogeneous, a more distinct clinical entity with characteristic clinical and laboratory findings gradually emerged as experience accumulated. The first formal descriptions of infectious mononucleosis were by Filatov in 1885 and by Pfeiffer in 1889 [1]. In 1920, Sprunt and Evans introduced the term "infectious mononucleosis" and described the characteristic hematological finding of "atypical lymphocytes" [2] The observation that infectious mononucleosis gave rise to antibodies that coincidentally agglutinated sheep red blood cells (heterophile antibodies) was reported by Paul and Bunnell in 1932 [3]. This gave rise to the modern monospot test for EBV-associated infectious mononucleosis.

Initial attempts to isolate the etiological agent were hampered by the primitive state of virology at the beginning of the twentieth century. Transmission of the disease by inoculation of serum from infectious mononucleosis patients to normal volunteers was unsuccessful, probably because the volunteers had been previously exposed to EBV and were therefore immune to the acute disease. The tale of the discovery of EBV took an unusual turn, after much frustrating work.

In 1958, Denis Burkitt published a report of an unusual tumor he saw in African children, a lymphoma that was confined to the jaw and face of the patients [4]. The incidence of Burkitt's lymphoma was confined to certain geographical regions, and it was noted that these coincided with areas in which malaria was endemic, suggesting that perhaps the disease was vector-borne. Burkitt sent samples of tumor tissue to M. A. Epstein in London, who was able to establish cell lines from the tumor. In 1964, electron microscopy showed herpes-like viral particles in the cells. Samples were then sent to the laboratory of W. Henle and G. Henle in Philadelphia for more precise identification of the virus. They found that antibodies to the virus were present in sera from Burkitt's lymphoma patients but also from many others, including normal laboratory staff. The first hint connecting the virus to a disease came when one of the Henles' laboratory technicians, whose serum was used as a negative control for antibodies to EBV, developed infectious mononucleosis. Her serum then became strongly positive for EBV antibodies [5]. Further epidemiological studies stimulated by this observation established the etiological link between Epstein-Barr virus and infectious mononucleosis [6].

3 THE VIRUS

Epstein-Barr virus is a member of the herpesvirus family. It is a double-stranded DNA virus of length 172 kilobase pairs (kbp) that codes for about 100 proteins. The viral genome is contained in an icosahedral capsid, which is surrounded by an amorphous tegument. This in turn is bounded by a viral envelope. The structure of the genome is similar to that of other herpesviruses, with a unique short and a unique long segment separated by a segment of multiple (6–12) tandem repeats of a stretch of 3071 bp. The unique long segment is further broken up into four smaller segments by tandem internal repeats [7]. The number of repeats is conserved in each strain of EBV and can be used for molecular epidemiological tracing [8]. The viral genes form two broad groups, those expressed during latency and those expressed during the lytic cycle (Figure 1).

The Epstein-Barr virus has tropism for B lymphocytes, with the C3b complement receptor serving as the viral receptor on the cell. The virus causes a mostly latent, but occasionally also lytic, infection of B lymphocytes. Approximately 90 viral proteins are expressed in lytic infection, but only about nine (EBNA1–6, LMP-1, 2a, and 2b) in latent infection, in which two RNA molecules, EBER1 and EBER2, are also expressed. Latent infection of B cells results in their immortalization, proliferation, activation, and infiltration

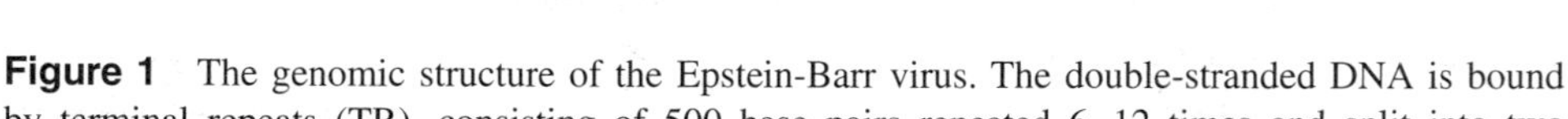

Figure 1 The genomic structure of the Epstein-Barr virus. The double-stranded DNA is bound by terminal repeats (TR), consisting of 500 base pairs repeated 6–12 times and split into two segments, a unique small (US or U1) and a unique large (UL or U2–U5) by a large internal repeat (IR1), consisting of 6–12 repeats of a 3071 bp sequence. The unique long segment is further broken up by internal repeats (IR2, IR3, IR4) into several unique segments (U2, U3, U4, U5). The latent and lytic cycle genes are scattered throughout the genome, and not explicitly shown in the diagram.

into systemic tissues. The infected B cells express activation antigens and synthesize immunoglobulin, which gives rise to the polyclonal hypergammaglobulinemia noted during the acute illness.

Occasionally, T cells, epithelial cells, and smooth muscle cells are infected, although such infections do not appear to play a major part in the viral life cycle. Such infections can result in disease, however. Thus, oral hairy leukoplakia is an infection of oral epithelial cells, and nasopharyneal carcinoma results from infection of epithelial cells in the nasopharynx. In rare cases, lymphoproliferation of EBV-infected T lymphocytes can occur and cause clinical illness (see Sec. 4). Leiomyosarcomas occasionally occur in immunosuppressed patients, especially those with advanced HIV disease.

4 EBV-RELATED DISEASES AND BASIC PATHOGENESIS

Epstein-Barr virus is the cause of infectious mononucleosis, an acute febrile illness characterized by fever, fatigue, sore throat, cervical lymphadenopathy, hepatosplenomegaly, and occasionally rash. The illness usually lasts no more than a few weeks and is followed by complete recovery. It is of variable severity, and fatalities are rare. Many of the fatalities in the normal host are caused by the neurological manifestations of the infection. Occasionally, there is a disproportionate involvement of certain organs in the disease, such as the liver, leading to a hepatitis, or the brain, leading to encephalitis. Splenomegaly may occur, and rupture of the spleen can lead to fatal hemorrhage.

Typical laboratory findings in infectious mononucleosis include leukocytosis (occasionally to very high levels, causing a ''leukemoid'' reaction), the presence of atypical lymphocytes in the peripheral blood smear, hypergammaglobulinemia, increased liver enzyme levels with occasional hyperbilirubinemia, and hemolytic anemia. The disease gets its name from the prominent leukocytosis present, with atypical lymphocytes seen in peripheral blood smears.

As mentioned above, B lymphocytes are the main cells infected by EBV, and this results in the proliferation of immortalized, activated B cells, which infiltrate lymphoid and systemic tissues [9]. A small number of the B cells are lytically infected and produce more virus, which in turn infects other B cells. This further amplifies the burden of latently EBV-infected B cells. A T-cell-mediated immune reaction against EBV antigens is induced, and cytotoxic T cells also proliferate to eliminate the EBV-infected cells. These reactive T cells form the ''atypical lymphocytes'' that are seen during acute infection, as discussed above [10]. Eventually, a dynamic equilibrium between latently infected B cells and EBV-specific cytotoxic T cells results. In situations during which this equilibrium is upset (i.e., with the suppression of EBV-specific T cells by immunosuppression secondary to transplantation or cancer chemotherapy), the EBV-infected B cells begin to proliferate, resulting in a reactivated chronic active EBV infection such as post-transplant lymphoproliferative disease. The lymphoproliferation is initially polyclonal but then evolves into an oligoclonal and finally, if unopposed, a monoclonal form, resulting in lymphoma. Lymphoblastoid cell lines consisting of EBV-immortalized B cells can be isolated from the blood of EBV-seropositive patients if T-cell activity in the leukocyte fraction is suppressed (e.g., by cyclosporin). B cells from EBV-seronegative patients do not survive very long in vitro.

The precise pathogenesis of infectious mononucleosis and its immediate complications (including neurological) are unknown but undoubtedly involve the interplay between the activated, immortalized B cells, the T-cell proliferative response, and the location where the ''battle'' is most intense, which suggests a ''bystander'' pathogenesis. Indeed,

EBV does not infect neurons or other specifically neural cells. A particularly heavy infiltration of infected B cells and reactive T cells in the brain might therefore give rise to an encephalitis associated with EBV mononucleosis. Similarly, an infiltration of the meninges may cause an aseptic meningitis (which is seen in EBV mononucleosis) or cranial nerve palsies coinciding with or following the acute febrile illness. It is tempting to speculate that the timing of the insult to neural tissues from the local immune activation resulting from B cell–T cell interaction may be variable, and so the same basic immunopathological mechanisms may underlie an acute EVB neurological manifestation as well as a postinfection neurological disease. Thus, ''infectious'' and ''postinfectious'' may not be completely distinct from each other, and ''parainfectious'' may be the best description.

As implied by the fact that EBV induces immortalization of B cells and its association with Burkitt's lymphoma, the virus is oncogenic. EBV is the cause (or a major contributing factor) of a number of other neoplastic or lymphoproliferative diseases, including post-transplant lymphoproliferative disease (PTLD); the rare X-linked lymphoproliferative disease (XLPD), which has been identified in a few families; a proportion of cases of Hodgkin's disease; and primary central nervous system lymphomas in patients with the acquired immunodeficiency syndrome. EBV-associated neoplasms in cells other than B lymphocytes have been reported. These include epithelial cells in nasopharyngeal carcinoma in southeast Asia, in myocytes in leiomyomas, and in T cells of certain T-cell lymphomas. Many of these occur in patients with immunosuppression due to HIV disease (primary CNS lymphoma, leiomyosarcoma), cancer chemotherapy, or organ transplantation (PTLD, leiomyosarcoma). Others occur in those with an incompletely understood, specific congenital inability to clear EBV (XLPD). Many of these tumors, such as Hodgkin's lymphoma, occur without any clear defect in the immune system.

5 DIAGNOSIS OF ACUTE EBV INFECTION

Given the protean clinical manifestations of acute EBV infection, laboratory confirmation is essential for appropriate diagnosis. Even the classical presentation of acute infectious mononucleosis, with fever, sore throat, malaise, cervical lymphadenopathy, and splenomegaly, can be mimicked by other diseases such as heterophile negative mononucleosis (usually caused by cytomegalovirus), toxoplasmosis, acute CMV infection, acute HIV infection, and lymphoma. Serological methods are most commonly used to confirm the diagnosis of acute EBV infection. In most cases, a neurological manifestation of EBV infection is diagnosed by the coincidence of EBV seroconversion with the neurological syndrome, although more recently detection of EBV genome in cerebrospinal fluid has been used to diagnose EBV encephalitis.

Acute EBV infection is usually confirmed by the detection of heterophile antibodies. During the acute disease, agglutinating antibodies reactive against sheep erythrocytes are detectable. The precise nature of the antigen on sheep cells is unknown. Normal serum can contain small amounts of nonspecific sheep cell agglutinins, and these nonspecific antibodies (Forssman antibodies) must be absorbed out of the serum, leaving the EBV-specific heterophile antibodies behind. Accordingly, serum to be tested for heterophile antibodies is first incubated with guinea pig kidney, which contains the Forssman antigen (which has been identified as lipopolysaccharide–protein complexes present on cell surfaces of many different tissues, particularly in guinea pig kidney cells and horse red cells). If the resultant serum can still agglutinate sheep cells, then the serum contains EBV-specific heterophile antibodies, and acute EBV infection is confirmed. Heterophile antibodies are present only in the acute infection and fall to undetectable levels over a few weeks. The phenomenon of heterophile antibody production in infectious mononucleosis is the basis of the monospot test, which is performed on commercially available prepared slides.

More specific testing for EBV infection uses the detection of antibodies to specific classes of EBV antigens, namely viral capsid antigen (VCA), early antigen (EA), and Epstein-Barr nuclear antigen (EBNA) [11]. The first antibody produced in infectious mononucleosis is IgM antibody against VCA (VCA IgM). VCA IgM is present only transiently and may disappear by the time the first symptoms occur. VCA IgM is then replaced by VCA IgG, which persists for life. The next antibody to appear is the one reacting with EA. Anti-EA antibodies are seen in over 70% of patients with infectious mononucleosis and are detectable for 3–6 months. Finally, late in the course of the disease, antibodies against EBNA appear, and these last for life. Thus, in an acute EBV infection, the serum is positive for VCA IgM or VCA IgG and anti-EA but negative for anti-EBNA antibodies. Patients who have had infectious mononucleosis in the remote past have serum positive for VCA IgG and anti-EBNA antibodies. The interpretation of EBV panel results at my institution are shown in Table 1.

Direct detection of virus in CSF is good evidence of its involvement in neurological disease. Detection by culture is not easy to do, because the virus does not usually cause lytic infection, so there is no cytopathogenic effect that can be used to identify its presence. The virus can be detected, however, through its effects in inducing immortalization [12], which requires specialized expertise not easily available in most laboratories. Alternatively, it can be detected by direct amplification through the polymerase chain reaction (Sec. 6.2) [13].

6 EBV-RELATED NEUROLOGICAL DISEASES

The neurological manifestations of EBV disease were first noted by Epstein and Dameshek [14], who reported a case of encephalitis, and by Johansen [15], who reported aseptic meningitis, in patients with infectious mononucleosis. Since then, a number of neurological manifestations of EBV infection have been documented.

Epstein-Barr virus causes a spectrum of neurological diseases that affect both the central and peripheral nervous systems. These diseases may be divided into several broad categories, which are discussed below. Neurological complications of infectious mononucleosis are not rare. In a series of 109 cases of patients with infectious mononucleosis admitted to a London hospital between May 1957 and May 1964, neurological manifestations were seen in eight (7.3%) [16]. Of these, five patients had encephalitis, one had

Table 1 Serology in EBV Infection[a]

EBV status	VCA IgM	VCA IgG	EA	EBNA
Seronegative	−	−	−	−
Recent primary	+	+	+/−	−
Seropositive (remote infection)	−	+	+/−	+
Infectious mononucleosis	+	+	+	−
Reactivated infection	+/−	+++	+++	+

[a] (−) No antibody; (+/−) either positive or negative; (+) detectable antibody; (+++) high titer antibody.

meningitis, and one each had polyneuropathy and mononeuropathy. In another series of 144 hospitalized infectious mononucleosis patients, 5.5% had neurological problems as the prominent or major presentation [17]. In a Mayo Clinic series of 1285 cases of infectious mononucleosis, 12 had confirmed neurological problems directly attributed to the disease [18].

6.1 Aseptic Meningitis

Aseptic meningitis is a common complication of acute EBV infection and is probably underestimated. Headaches are not uncommon in the acute illness, and it is likely that some of these are due to a mild aseptic meningitis. One of the first mentions of a neurological complication of EBV infection was the report of Johansen [13] of a case of aseptic meningitis. In a review of the neurological complications of infectious mononucleosis, 14 out of 34 cases (41%) reported in the literature as of 1950 had aseptic meningitis [19]. The presentation is that of any other meningitis, with headache, fever, and stiff neck, usually with the systemic illness present (though not always). The meningitis is self-limiting.

6.2 EBV Encephalitis

Characteristic symptoms of EBV-associated encephalitis are fever, headache, confusion, seizures, and paresis, as in any other form of viral encephalitis. The encephalitis often occurs in the context of a clinical infectious mononucleosis, with fever, sore throat, malaise, and lymphadenopathy, but it can occur without systemic signs [13,17,18,20,21]. Focal features are often seen, and occasionally EBV encephalitis resembles herpes encephalitis [22]. Three cases of brainstem encephalitis have been reported, with one patient recovering completely, one left with mild residual gait ataxia and nystagmus, and one expiring. All three cases were diagnosed by serology [23–25]. Occasionally, the onset of EBV encephalitis is slow and insidious and can consist of behavioral and focal neurological deficits [26]. A few rare cases of relapsing-remitting disease, satisfying the criteria of multiple sclerosis, following acute EBV infection with neurological manifestations (such as the ones discussed later in this chapter) have been described [27]. The relation ship between the acute EBV disease and the subsequent MS-like illness is not clear, but recent serological studies have suggested a contributory role of EBV in MS [28].

Pathological findings are scarce, because death from EBV encephalitis is rare. Reports have described variable findings on pathological examination of the brain, which points to several possible pathogenetic processes, including typical viral encephalitis and postinfectious acute disseminated encephalomyelitis. Thus, perivascular infiltrates of lymphocytes, as well as diffuse parenchymal infiltrates consisting of both lymphocytes and microglia, have been found in the cortex, as is typical in viral encephalitis [29]. In one patient, both meningeal and diffuse parenchymal white matter perivascular infiltrates of lymphocytes and lymphoblastoid cells (some of which showed mitotic figures, reminiscent of neoplasm) were found. Most of these cells were EBV-infected B cells, but a few T cells and microglia/macrophages were found [21]. Some patients have had typical histopathological findings of acute disseminated encephalomyelitis, with perivenular infiltrates of lymphocytes in the white matter, with lipid-laden macrophages and demyelination [30,31]. Finally, a peculiar case of a fatal EBV-associated acute encephalopathy in an adult was described, with pathological findings of scattered neuronal pyknosis, diffuse cortical edema, and visual cortical perivascular edema but no perivascular infiltrates or microglial nodules [32]. These findings are reminiscent of what has been called a ‘‘toxic encephalopathy,'' which is a poorly understood parainfectious process seen most often in children [33].

Radiographic and Neurophysiological Findings

The imaging findings in EBV encephalitis are nonspecific. In one case the brain MRI showed normal parenchyma, but there was leptomeningeal enhancement, particularly in the basal cisterns [13]. An abnormal signal in the basal ganglia has also been described [26].

Electroencephalography of EBV-associated encephalitis usually shows nonspecific abnormalities such as focal and diffuse slowing [18]. Periodic EEG complexes, reminiscent of those seen in herpes encephalitis, have been described [34,35].

CSF Findings

The CSF in EBV encephalitis shows variable pleocytosis and normal to mildly increased protein. Occasionally, the atypical lymphocytes characteristic of infectious mononucleosis are seen in the CSF [13,36]. Oligoclonal bands in the CSF have been reported, and in one case these appeared about 3 weeks after the onset of infectious mononucleosis [13]. Specific antibodies against EBV viral capsid protein have been detected in the CSF in a case of EBV encephalitis [37]. Similar CSF abnormalities can be seen in other EBV-associated neurological disease. The CSF glucose is normal.

Polymerase chain reaction methods have been used to detect EBV DNA in the CSF of patients with EBV encephalitis, although this has not been validated as a diagnostic test, because it is possible that EBV may be nonspecifically reactivated in the CSF. Furthermore, EBV PCR is positive in primary CNS lymphoma in AIDS patients (see below). However, several cases in which EBV was detected by PCR in the CSF of patients with CNS disease and concurrent acute EBV serology have been reported. In one patient, the CSF PCR for EBV DNA was found to be positive, coincident with an encephalopathic illness and EBV serology consistent with acute EBV mononucleosis [13]. In two patients, one with encephalitis and one with myelitis coinciding with acute EBV mononucleosis, EBV was detected in the CSF by PCR [38]. These considerations suggest that the best way to diagnose EBV encephalitis is by obtaining an EBV serum panel that shows acute EBV infection, with EBV PCR positivity in the CSF providing further support for the diagnosis. It would also be reasonable to do a simultaneous PCR for other viruses (such as HSV) to help exclude the possibility of nonspecific reactivation.

Epstein-Barr virus cannot be detected by conventional viral culture methods, because the virus causes little cytopathic effect in lymphocytes, but a lymphocyte transformation assay has been used to detect EBV in the CSF of patients with EBV encephalitis [2,39]. This form of EBV detection is cumbersome and requires specialized techniques not readily available in most clinical laboratories.

6.3 Guillain-Barré Syndrome and Other Forms of Peripheral Neuropathy

Epstein-Barr virus infection can be associated with the Guillain-Barré syndrome (GBS), as first described by Zohman and Silverman [40]. Grose and Feorino [41] compared EBV antibody titers of five patients with GBS to those of age-matched controls and found that the GBS patients had considerably higher levels, usually seen in acute infectious mononucleosis. Two of the patients had positive heterophile antibodies, indicating an acute EBV infection. Both of those patients had generalized adenopathy, and one of them had a pleocytosis of eight cells [41]. Although EBV-associated GBS is well documented, it is not a common complication of infectious mononucleosis. In a series of 109 hospitalized patients with infectious mononucleosis, one had Guillain-Barré syndrome, coincident with fever, headache, lymphadenopathy, and appropriate serology [16]. This complication can be fatal. In one fatal case, the patient's illness was characterized by cranial nerve palsies progressing to areflexia and complete flaccid paralysis necessitating intubation after 3 days. Autopsy showed inflammatory demyelination of both dorsal and ventral roots as well as the cranial nerves and cauda equina [42].

Other forms of peripheral nerve involvement have been reported in conjunction with EBV infection. Lumbosacral radiculoplexopathy with pain and lower extremity weakness has been reported in five patients. In all cases, pain (in the gluteal area and the thigh) was an early complaint, followed by leg weakness, which was severe enough for the patients to require ambulatory assistance, two being wheelchair-bound. All patients recovered completely or nearly completely. Electromyography showed acute denervation and mild slowing of motor nerve conduction. Serology showed acute EBV infection in all patients. Cerebrospinal fluid was examined in the five patients and showed mild elevation in protein in three and a very mild pleocytosis in two. Two patients received oral prednisone and seemed to improve on it. All patients were independently ambulatory several months after onset [43]. Infectious mononucleosis has also preceded brachial plexopathy. In one case, a 19-year-old man developed acute pain in the shoulders about 2 weeks after developing infectious mononucleosis, diagnosed by a positive heterophile test. Several days later he was unable to lift his arms above his head, and he developed atrophy of the shoulder girdle muscles. Electromyography showed bilateral brachial plexopathy. Complete recovery occurred over the next 4 months [44]. Another patient developed a bilateral brachial plexopathy with pain and weakness of the arms, along with a unilateral Bell's palsy, about a week after a febrile pharyngitis. Infectious mononucleosis was diagnosed by a positive heterophile test [45].

Acute autonomic neuropathy, with blurred vision, orthostatic hypotension, constipation, and burning dysesthesias has also been reported in infectious mononucleosis [46].

A very rare form of T-cell lymphoproliferative disease has been reported to infiltrate peripheral nerves, causing paresthesias and motor weakness of the arms and legs, as well as progressive dilated cardiomyopathy in a young man without any previous history of immune disease. He was treated with acyclovir, methylprednisolone, cyclophosphamide, and polyglobulin N, with improvement in the neuropathy and ejection fraction. At autopsy, multiple organs were found to be infiltrated by monoclonal and polyclonal atypical T-cell populations that contained EBV [47].

6.4 Cranial Nerve Palsy

The classical cranial nerve palsy associated with EBV infection is Bell's palsy. In three cases of young adults with infectious mononucleosis, diagnosed serologically, unilateral peripheral facial palsy was noted [48]. Bell's palsy in very young children has also been reported in association with infectious mononucleosis [49]. In several of these patients, the Bell's palsy was the presenting and sole symptom of infectious mononucleosis, and the findings of lymphadenopathy and splenomegaly led to blood studies that confirmed the diagnosis. The Bell's palsy can be bilateral. In one case, a facial diplegia occurred 2 weeks after clinical infectious mononucleosis, with fever, malaise, and cervical lymphadenopathy [50]. Bell's palsy can occur with involvement of other cranial nerves. In one patient, clinical infectious mononucleosis was followed by left-side deafness and then by left-side Bell's palsy. Examination revealed left facial numbness as well, so this patient had involvement of the left Vth, VIIth, and VIIIthnerves. Nine months later, the patient had recovered completely [51]. Hypoglossal nerve palsy was reported in a patient 6 days after the onset of a febrile pharyngitis and malaise and was diagnosed by a heterophile antibody test [52]. Other cranial palsies have been reported to occur with infectious mononucleosis.

Optic neuritis has also been reported to occur with EBV infection, with several cases of bilateral optic nerve involvement. Several of these cases occurred before infectious mononucleosis was diagnosed [53–55].

6.5 Transverse Myelitis

Transverse myelitis is a rare complication of EBV infection. The myelitis may be very rapid in onset. In one case, a young woman had a 2 week history of fever, sore throat, and malaise, followed by dysesthesias in the legs, which became weak. She was unable to walk 3 days later. Examination showed a spinal sensory level and upgoing toes. Her CSF had a protein of 100 mg/dL and a cell count of 2 μL^{-1}. CSF viral culture was negative. The heterophile screen was positive, and a blood film showed atypical lymphocytes. She was treated with ACTH and prednisone and had a slow, almost complete recovery over 6 months [56]. In another case, a young woman noted difficulty in voiding, which was followed by paresthesias and weakness in the legs within 24 h, leading to flaccid paraplegia 2 days later, with a thoracic sensory level. There was no systemic illness. The CSF had an increased protein (106 mg/dL) and pleocytosis (249 cells/μL). A recent EBV infection was diagnosed by very high anti-EBV antibody titers in the blood. A slow recovery over several months ensued [57]. Recently, EBV DNA was detected in the CSF of someone with EBV-associated myelitis [58]. In another case, a young man was diagnosed with infectious mononucleosis 10 days before developing a transient tetraparesis. Examination showed a spinal sensory level, a bilateral Babinski sign, and a normal gait. Serology was consistent with an acute EBV infection. The CSF showed minor pleocytosis (27 cells/μL) and normal protein. EBV genome was detected in the CSF by PCR, in a higher concentration than in blood or saliva. A month later, there were only mild residua.

6.6 Cerebellar Ataxia

Cerebellar ataxia has been reported to occur with EBV infection, although the most common cause of acute cerebellar ataxia in children is varicella-zoster virus. Although the syndrome has typically been thought to affect children [59,60] it has also been seen in both young and older adults [61–63]. In most cases, the patients had a systemic illness, often mild, before developing gait ataxia and dysarthria. All were found to have atypical lymphocytes in the blood and positive heterophile screens. Pleocytosis was absent or mild (up to 15 cells/μL), and CSF protein was at most modestly elevated. Recovery was complete within a few weeks. One of the patients was treated with ACTH and improved. Usually remission is permanent, but relapses have been reported. One patient developed scanning speech, ataxic gait, and dysmetria, coincident with a positive EBV VCA IgM, which resolved after a course of oral prednisone. A year later, these symptoms recurred and resolved spontaneously after 2 months [64]. Given the uniformly good prognosis of the neurological complications of EBV infection, it is likely that he would have improved in any case. No pathological findings are available to explain the pathogenesis of EBV-associated cerebellar ataxia, and it is unknown whether this is a manifestation of a direct viral cerebellitis or a postinfectious demyelination, or whether there is even a clear distinction between the two.

6.7 Psychiatric Manifestations and the Alice-in-Wonderland Syndrome

Occasionally, EBV infection is complicated by prominent psychiatric symptoms that occur in the course of the illness. One such patient, a 25-year-old married college student, developed aggressive, impulsive, unpredictable, and sexually inappropriate behavior, delusional thinking, and auditory and visual hallucinations during the course of clinical infectious mononucleosis, which was diagnosed by a positive heterophile test. Atypical lymphocytes were seen in the blood. He was fully oriented, and CSF examination was normal. His clinical picture was felt to resemble an acute schizophrenic episode [65]. Two other patients, both teenagers, developed severe depression during a bout of infectious mononucleosis. Neither patient had any premorbid psychiatric history, and both were well-adjusted and doing well in school. The depression persisted after the clinical illness resolved. The neurological exam was remarkable for normal cognition in both patients and soft signs in one. The depression led to suicidal ideation in both. The EEG showed diffuse slowing during the depression in both patients. The depression resolved in a few months in one case but persisted for several years in the other [66]. Two other patients with acute depression coincident with infectious mononucleosis, requiring electroconvulsive therapy, have been reported [67]. The pathogenesis of this depression is unknown.

A very interesting (and characteristic) syndrome has been reported to occur with infectious mononucleosis, the so-called Alice-in-Wonderland syndrome, in which metamorphopsia (bizarre distortions of spatial sense) occurs, similar to that in migraine. This was first reported in three patients, two teenagers and one 9-year-old boy. The syndrome consisted of several anxiety-provoking episodes a day, each lasting up to a half hour, of distortions in the sizes, shapes, and orientations of objects in the environment. These episodes were coincident with or shortly followed infectious mononucleosis. One of the patients reported bumping into objects while she walked. EEG was normal in one case and had only minor abnormalities in another. Neurological examination was normal or showed only soft signs. One patient was given a single dose of corticosteroid, which caused improvement in the infectious mononucleosis but did not affect the metamorphopsia. Another patient was put on phenytoin, without effect. The metamorphopsia resolved after several weeks in all cases [68]. In another case, a 6-year-old boy had similar intermittent episodes of metamorphopsia beginning several days after the onset of a febrile sore throat. He was noted to have fever, a reddened throat, lymphadenopathy, and hepatosplenomegaly. Neurological examination was normal. Liver function studies were mildly abnormal, and atypical lymphocytes were found in the blood. EBV serology showed acute EBV infection. The metamorphopsia gradually resolved over the next 3 weeks [69]. The resemblance of the syndrome to hemiplegic migraine is noteworthy. A study of the visual evoked responses in five children with Alice-in-Wonderland syndrome showed a high amplitude of the P100-N145 wave complex, compared to normal controls [70]. Another study of children with Alice-in-Wonderland syndrome, in some of whom it was associated with EBV infection, using HMPAO single-photon emission computed tomography (SPECT) (which measures cerebral perfusion), showed decreased perfusion near the visual tract and visual cortex [71].

6.8 Acute Hemiplegia

Acute hemiplegia, resembling an acute vascular event, has been reported in EBV infection. Hemiplegia of childhood, a recognized clinical entity, often does not have a clear etiology. A few such cases have been associated with acute EBV infection. In one case, a 14-year-old girl had a left-sided hemiplegia develop over several hours, accompanied by right-sided headache, photophobia, and emesis. Examination showed left hemiplegia, left-side numbness, and left hyperreflexia. These resolved over several hours but recurred later on the same day and resolved again. Two days later, she had two seizures, fever, and cervical adenopathy. CSF examination showed a moderate pleocytosis of 103 cells/μL. Several days later, the patient became confused and ataxic, with diffuse slowing on the EEG. EBV serology was consistent with acute primary EBV infection. She recovered completely after 3 months [72]. In another case, a 9-year-old girl with fever and sore throat developed a right-sided headache, fever, and vomiting and left-side hemiparesis, with left hyperreflexia and left homonymous hemianopsia. CSF examination showed 63 cells/μL, and brain CT was normal. EEG showed diffuse slowing. EBV serology was consistent with acute primary EBV infection. The hemiplegia resolved completely over the next few days [73]. A similar case has been reported in an adult. A 32-year-old man had fever, sore throat, and headache and developed left-side weakness several days later. Examination showed mild left hemiparesis with hyperreflexia, as well as fever, adenopathy, and splenomegaly. A slide test for heterophile antibodies was positive. A CT scan of the brain was normal. He was given oral dexamethasone, with resolution of the hemiparesis over the next 24 h [74].

6.9 Primary CNS Lymphoma

Primary CNS lymphoma (PCNSL) is a neoplasm of the brain usually seen in the elderly and in the immunosuppressed. With the advent of the HIV-AIDS epidemic, this tumor has become much more common, especially in patients with advanced HIV disease. In a series of 20 cases of PCNSL reported in 1986, all but one had one or more opportunistic infections or neoplasms (such as Kaposi's sarcoma) [75]. The known oncogenic effects of EBV and its association with systemic lymphomas suggests that the virus may play an important role in this tumor, and in fact the virus is found in all AIDS-associated PCNSL [76,77] but only in about 50% of systemic lymphomas in HIV patients. In situ hybridization studies have shown that the neoplastic cells all express the latency molecules EBER and LMP, which are associated with immortalization of infected lymphocytes. Control tissues from brains of both HIV-positive and HIV-negative patients with other diagnoses showed no such expression [76]. In another study, primary CNS lymphoma samples from 26 AIDS and 22 HIV-negative patients were tested for EBV by in situ hybridization for EBER and immunostaining for LMP-1. All AIDS-associated PCNSLs were positive for EBV infection, whereas none of the HIV-negative cases were [77]. Rare instances of EBV-positive PCNSL in HIV-negative patients have been reported [78].

Clinically, PCNSL presents with subacute progressive mental status changes (such as apathy and confusion) with a variable combination of focal weakness, seizures, and headaches. The MRI typically shows a deeply situated ring-enhancing lesion (Figure 2), with a thick rim of enhancement (Figure 3) and a nodularity of the rim (Figure 4). Often the lesion is periventricular, and occasionally there is periventricular spread, with a lumpy, bumpy appearance [79]. The lesions can strongly resemble those seen in toxoplasmosis. Points of possible differentiation include the presence of multiple lesions in toxoplasmosis, whereas a single lesion is more suggestive of lymphoma. Seronegativity for toxoplasma

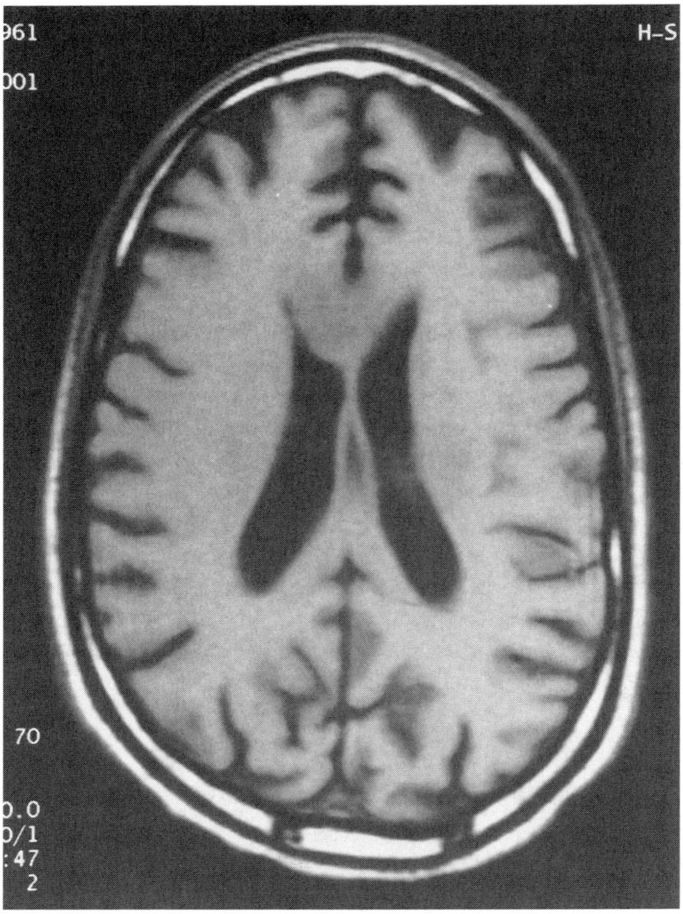

Figure 2 Magnetic resonance image of brain of AIDS patient with primary CNS lymphoma, T1-weighted, noncontrast. Note area of decreased intensity in the anterior corpus callosum, medial to the right frontal horn, with perilesional edema. The CSF was positive for EBV DNA by PCR. (Courtesy of Dr. I. Zak, Division of Neuroradiology, Department of Radiology, Harper University Hospital, Detroit, Michigan.)

antibodies makes toxoplasmosis very unlikely. A lack of response to antitoxoplasma therapy strongly suggests another diagnosis, with lymphoma becoming more likely. The detection of EBV DNA in the CSF by PCR is very strongly suggestive of PCNSL in HIV patients [80]. Recently, thallium-201-SPECT scans to differentiate between PCNSL and toxoplasmosis (or other ring-enhancing mass lesions) has been used. Thallium is a potassium analog that is taken up by tumor but not by inflammatory cells. Typically PCNSL ''lights up'' on thallium scans, but toxoplasmosis or other types of brain abscesses do not (see Figure 5) [79]. Biopsy of the lesions typically shows an angiocentric distribution of neoplastic cells (Figure 6) that stain positively for B-cell markers (Figure 7) and for latency-associated proteins such as LMP-1 (Figure 8).

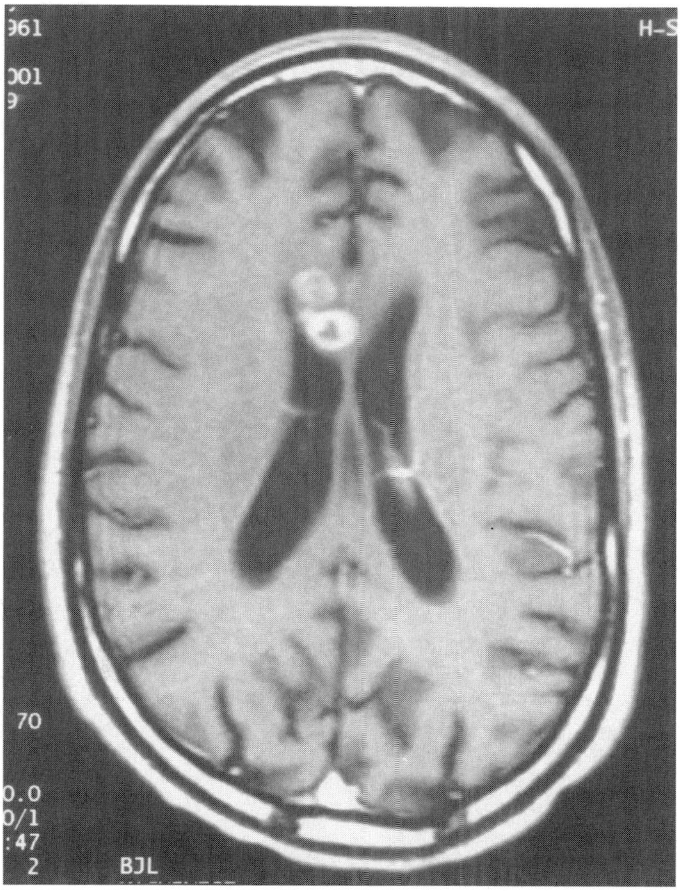

Figure 3 T1-weighted MRI showing contrast enhancement of the lesions shown in Figure 2. Note that two distinct lesions, one anterolateral and the other posteromedial, are resolved on this image. (Courtesy of Dr. I. Zak, Division of Neuroradiology, Department of Radiology, Harper University Hospital, Detroit, Michigan.)

The prognosis of primary CNS lymphoma is poor. In the days before highly active antiretroviral therapy (HAART), the average survival of 20 patients with AIDS PCNSL was less than 2 months [75]. Whole brain radiation therapy may increase survival. Ten AIDS patients given whole brain radiation had a median survival of 5.5 months. Deaths were due to progression of disease and to development of other AIDS-associated complications [81]. Cases of significant improvement after hydroxyurea [82] and HAART [83] have also been reported. One small study demonstrated that decrease in the burden of EBV in the CSF, as measured by quantitative EBV PCR, correlated with clinical improvement [84].

6.10 Lymphoproliferative Disease

Post-transplant lymphoproliferative disease is an EBV-driven polyclonal B-cell proliferation seen in patients who were immunosuppressed after solid organ transplants. The early

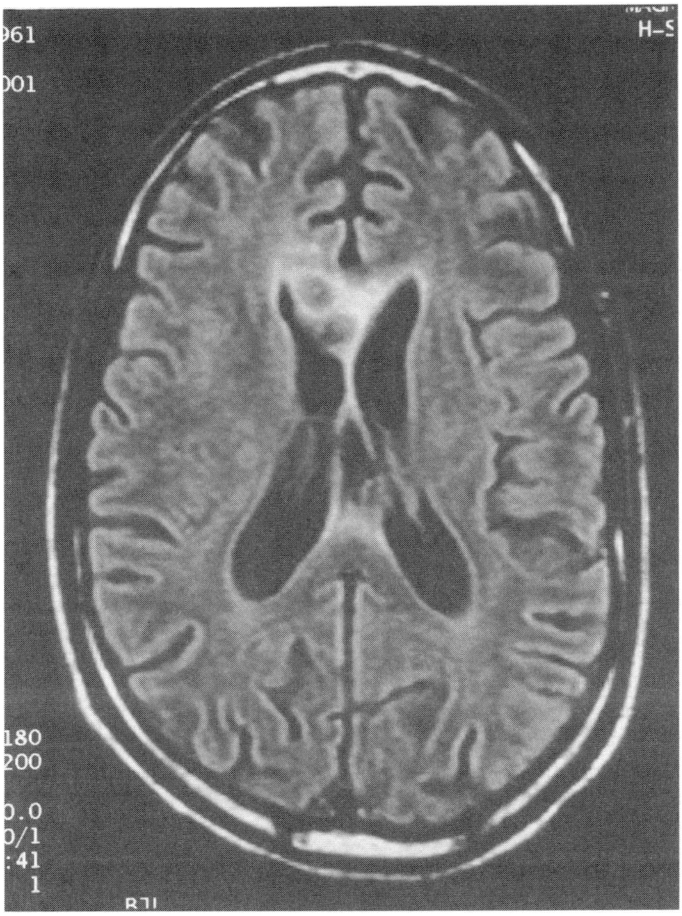

Figure 4 Fluid attenuated inversion recovery (FLAIR) image of the lesions shown in Figure 2. (Courtesy of Dr. I. Zak, Division of Neuroradiology, Department of Radiology, Harper University Hospital, Detroit, Michigan.)

form of the disease, beginning between 6 and 12 months post-transplant, presents as a rapidly progressive severe form of infectious mononucleosis, which can evolve into a sepsis-like syndrome. Central nervous system involvement is rarely seen and is mostly present in patients with very advanced disease [85].

7 TREATMENT

There are no studies of the use of any antiviral drugs in the neurological complications of EBV. Clearly, supportive care is very important, because death due to neurological complications of EBV is uncommon, although residual deficits are not rare. The role of antiviral drugs in the treatment of infectious mononucleosis is unclear, because the pathogenesis of the disease is not completely clear.

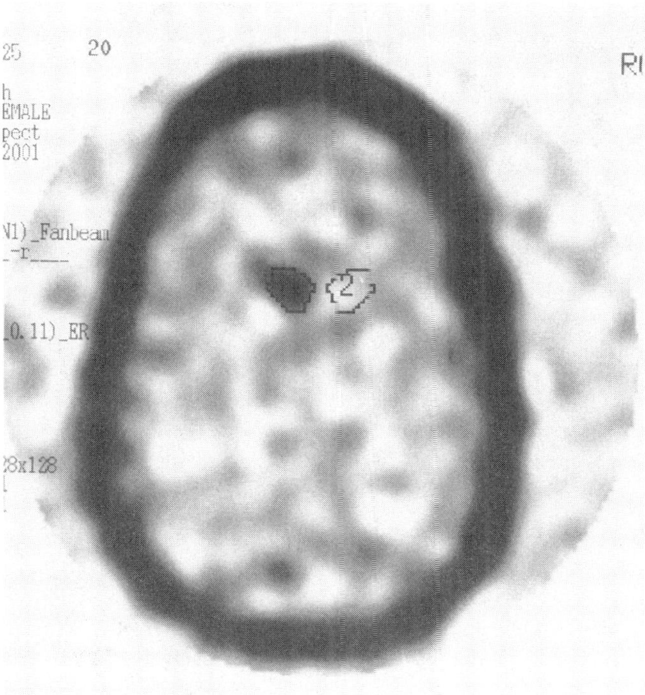

Figure 5 Thallium-201 single-photon emission computed tomography (SPECT) image of the lesions seen in Figure 2. There is an increased uptake of the tracer in the locations coinciding with the lesions seen in Figures 2–4. The lesion-to-background ratio of tracer uptake in the posteromedial lesion is 4.9, and that in the anterolateral lesion is 2.1. Any ratio greater than 2.0 suggests malignancy. (Courtesy of Dr. Larry Davis, Division of Nuclear Medicine, Department of Radiology, Harper University Hospital, Detroit, Michigan.)

Corticosteroids and acyclovir have been used to treat infectious mononucleosis. Studies of acyclovir alone in infectious mononucleosis have shown little benefit, apart from reduction of viral shedding in the treatment groups, although the studies tend to have small numbers of patients [86]. Acyclovir inhibits EBV DNA polymerase in vitro, although viral production returns to control levels after discontinuing the drug, even after 11 months [87,88]. Furthermore, although it inhibited viral replication in productive infection, acyclovir did not affect latent viral burden, implying that latent virus replicates by host-dependent enzymes [87]. Because the effects of latent virus are probably pathogenetically relevant, it is no surprise that acyclovir has little overall effect on acute EBV infection. Indeed, a meta-analysis of five randomized controlled clinical trials involving 339 patients found only a nonstatistically significant trend toward clinical improvement and a significant reduction in viral shedding in the oropharynx [89]. The role of ganciclovir, another nucleoside analog, is even less well established. Two cases of EBV encephalitis in transplant patients, diagnosed in both cases (a renal transplant patient and a bone marrow transplant patient) by a positive EBV PCR in the CSF, were treated with ganciclovir and made good clinical recoveries. One patient was given ganciclovir 100 mg intravenously for 4 weeks, followed by 4 weeks of oral ganciclovir. The other patient was given ganciclovir 5

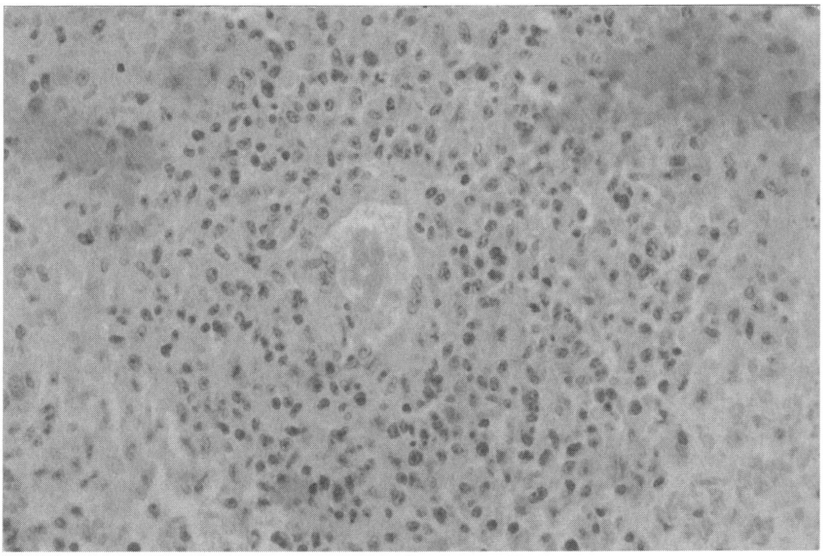

Figure 6 Primary CNS lymphoma in an AIDS patient. Biopsy specimen. Note the angiocentric distribution of the neoplastic cells. Hematoxylin and eosin stain. 100×. (Courtesy of Dr. William Kupsky, Division of Neuropathology, Harper University Hospital, Detroit, Michigan.)

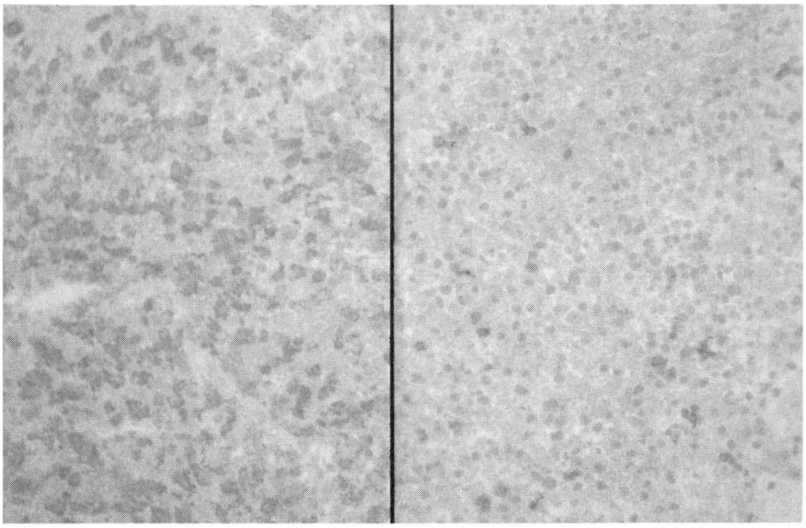

A **B**

Figure 7 Primary CNS lymphoma in an AIDS patient. Same patient as in Figure 6. (A) Stained for L26, a B-lymphocyte marker. (B) Stained for CD3, a T-lymphocyte marker. This shows that the neoplastic cells are B lymphocytes, in accord with the known tropism of EBV. (Courtesy of Dr. William Kupsky, Division of Neuropathology, Harper University Hospital, Detroit, Michigan.)

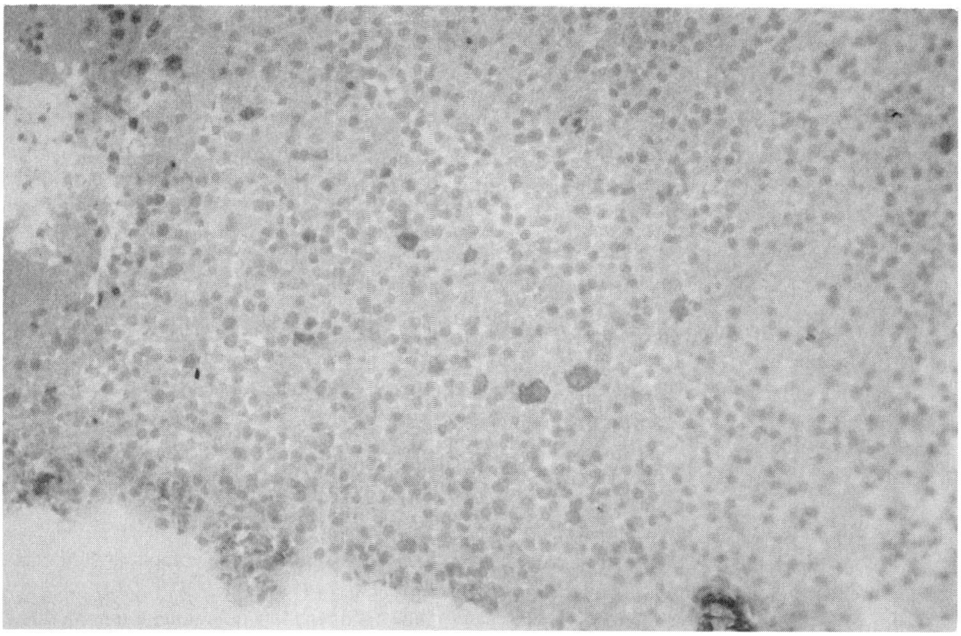

Figure 8 Primary CNS lymphoma in an AIDS patient. Same patient as in Figure 6. Immunostained for the EBV antigen LMP-1, which is a marker of latent infection. Note expression in the cytoplasm of the neoplastic cells. (Courtesy of Dr. William Kupsky, Division of Neuropathology, Harper University Hospital, Detroit, Michigan.)

mg/kg intravenously for 2 weeks. Both patients had resolution of CSF and MRI abnormalities [90,91]. Although these reports are encouraging, the role of the drug in a disease the natural history of which is not completely known is unclear.

The combination of corticosteroids and acyclovir was examined in one study in which 11 patients with fulminant infectious mononucleosis requiring hospitalization were treated with a combination of acyclovir (10 mg/kg every 8 h intravenously for 3–7 days, followed by 800 mg by mouth five times a day for a total course of 10 days) and prednisolone (0.7 mg/kg per day for 4 days, then reduced by 50% every second day for a total course of 6 days) in fulminant disease. There was a reduction in viral shedding and oropharyngeal symptoms compared to historical acyclovir and placebo-treated controls [92]. In another study it was demonstrated that acyclovir reduces oral viral shedding but that immortalized B cells can still be readily isolated from the blood and there was little apparent effect on clinical symptoms [93]. These results suggest that the pathogenesis of infectious mononucleosis is not due simply to direct infection of B cells and that immunopathology plays a role in the pathogenesis of the disease. It is probably reasonable to treat significant EBV infections with acyclovir with the possible addition of prednisone, after a careful consideration of the risks and benefits of the use of these drugs. It must be emphasized, however, that there has been no randomized controlled study that showed significant clinical benefit from the use of any drug in the neurological complications of EBV infection.

For the treatment of lymphoproliferative disease, which can rarely affect the nervous system, the best regimen is unknown. Reduction of immunosuppression can result in regression of disease [94]. More recently, immune-based approaches were explored in preliminary clinical trials, in particular the use of EBV-specific cytotoxic T cells infused intravenously [95]. In vitro proliferation of B cells infected with EBV was shown to be inhibited by mycophenolate, although there are no reports of its clinical use [96]. Other modalities, such as monoclonal antibodies, etoposide, and cyclosporin, have been tried in individual cases of systemic lymphoproliferative disease with apparent benefit [97].

ACKNOWLEDGMENTS

I thank John Booss, MD, for discussions through the years and Carol E. Jackson for editorial help, also through the years.

REFERENCES

1. Evans, A.S. The history of infectious mononucleosis. Am. J. Med. Sci. 1974, 267, 189–194.
2. Sprunt, T.P.; Evans, F.A. Mononucleosis leukocytosis in reaction to acute infections (infectious mononucleosis). Johns Hopkins Hosp. Bull. 1920, 31, 409–417.
3. Paul, J.R.; Bunnell, W. The presence of heterophile antibodies in infectious mononucleosis. Am. J. Med. Sci. 1932, 183, 191–194.
4. Burkitt, D. A sarcoma involving the jaws in African children. Br. J. Surg. 1958, 46, 218–223.
5. Henle, G.; Henle, W. Epstein-Barr virus: past, present and future. In Epstein-Barr Virus and Associated Diseases; Levine, P.H., Ablashi, D.V., Pearson, G.R., Kottaridis, S.D., Eds.; Martinus Nijhoff: Dordrecht, 1985, chap. 63.
6. Niederman, J.C.; Evans, A.E. Epstein-Barr virus. In Viral Infections of Humans; Evans, A.E., Kaslow, R.A., Eds.; Plenum: New York, 1997, Chap. 10.
7. Kieff, E.; Dambaugh, T.; Heller, M.; King, W.; Cheung, A.; van Santen, V. The biology and chemistry of Epstein-Barr virus. J. Infect. Dis. 1982, 146, 506–517.
8. Gratama, J.W.; Oosterveer, M.A.P.; Klein, G.; Ernberg, I. EBNA size polymorphism can be used to trace Epstein-Barr virus spread within families. J. Virol. 1990, 64, 4703–4708.
9. Pattengale, P.K.; Smith, R.W.; Gerber, P. B-cell characteristics of human peripheral and cord blood lymphocytes transformed by Epstein-Barr virus. J. Natl. Cancer Inst. 1974, 52, 1081–1086.
10. Papamichail, M.; Sheldon, P.J.; Holborow, E.J. T and B cell subpopulations in infectious mononucleosis. Clin. Exp. Immunol. 1974, 18, 1–11.
11. Henle, W.; Henle, G.E.; Horwitz, C.A. Epstein-Barr virus specific diagnostic tests in infectious mononucleosis. Hum. Pathol. 1974, 5, 551–565.
12. Halsted, C.C.; Chang, R.S. Infectious mononucleosis and encephalitis: recovery of EB virus from spinal fluid. Pediatrics. 1979, 64, 257–258.
13. Tselis, A.; Duman, R.; Storch, G.A.; Lisak, R.P. Epstein-Barr virus encephalomyelitis diagnosed by polymerase chain reaction: detection of the genome in the CSF. Neurology. 1997, 48, 1351–1355.
14. Epstein, S.H.; Dameshek, W. Involvement of the central nervous system in a case of glandular fever. N. Engl. J. Med. 1931, 205, 128–1241.
15. Johansen, A.J. Serous meningitis and infectious mononucleosis. Acta. Med. Scand. 1931, 76, 269.
16. Gautier-Smith, P.C. Neurological complications of glandular fever (infectious mononucleosis). Brain. 1965, 88, 323–334.
17. Silverstein, A.; Steinberg, G.; Nathanson, M. Nervous system involvement in infectious mononucleosis. The heralding and/or major manifestation. Arch. Neurol. 1972, 26, 353–358.

18. Schnell, R.G.; Dyck, P.J.; Bowie, E.J.W.; Klass, D.W.; Taswell, H.F. Infectious mononucleosis: neurologic and EEG findings. Medicine. 1966, 45, 51–63.
19. Bernstein, T.C.; Wolff, H.G. Involvement of the nervous system in infectious mononucleosis. Ann. Intern. Med. 1950, 33, 1120–1138.
20. Walsh, F.; Poser, C.M.; Carter, S. Infectious mononucleosis encephalitis. Pediatrics. 1954, 13, 536–543.
21. Schellinger, P.D.; Sommer, C.; Leithauser, F.; Schwab, S. Epstein-Barr virus meningoencephalitis with a lymphoma-like response in an immunocompetent host. Ann. Neurol. 1999, 45, 659–662.
22. Thomson, D.J. Focal encephalitis in infectious mononucleosis simulating herpes simplex encephalitis: case report. Mil. Med. 1975. 140, 188–189.
23. Shian, W.J.; Chi, C.S. Fatal brainstem encephalitis caused by Epstein-Barr virus. Pediatr. Radiol. 1994, 24, 596–597.
24. North, K.; de Silva, L.; Procopis, P. Brainstem encephalitis caused by Epstein-Barr virus. J. Child Neurol. 1993, 8, 40–42.
25. Angelini, L.; Bugiani, M.; Zibordi, F.; Cinque, P.; Bizzi, A. Brainstem encephalitis resulting from Epstein-Barr virus mimicking an infiltrating tumor in a child. Pediatr. Neurol. 2000, 22, 130–132.
26. Caruso, J.M.; Tung, G.A.; Gascon, G.G.; Rogg, J.; Davis, L.; Brown, W.D. Persistent preceding focal neurologic deficits in children with chronic Epstein-Barr virus encephalitis. J. Child Neurol. 2000, 15, 791–796.
27. Bray, P.F.; Culp, K.W.; McFarlin, D.E.; Panitch, H.S. Demyelinating disease after neurologically complicated primary Epstein-Barr virus infection. Neurology. 1992, 42, 278–282.
28. Ascherio, A.; Munger, K.L.; Lennette, E.T.; Spiegelman, D.; Hernan, M.A. Epstein-Barr virus antibodies and risk of multiple sclerosis. A prospective study. JAMA. 2001, 286, 3083–3088.
29. Sworn, M.J.; Urich, H. Acute encephalitis in infectious mononucleosis. J. Pathol. 1970, 100, 201–205.
30. Ambler, M.; Stoll, J.; Tzamaloukas, A.; Albala, M.M. Focal encephalomyelitis in infectious mononucleosis. A report with pathological description. Ann. Intern. Med. 1971, 75, 579–583.
31. Paskavitz, J.F.; Anderson, C.A.; Filley C.M.; Kleinschmidt-DeMasters, B.K.; Tyler, K.L. Acute arcuate fiber demyelinating encephalopathy following Epstein-Barr virus infection. Ann. Neurol. 1995, 38, 127–131.
32. Bergin, J.D. Fatal encephalopathy in glandular fever. J. Neurol. Neurosurg. Psych. 1960, 23, 69–73.
33. Tselis, A.; Lisak, R.P. Acute disseminated encephalomyelitis. In Clinical Neuroimmunology; Antel, J.P., Birnbaum, G., Hartung, H.-P., Eds.; Blackwell: London, 1997, chap 9.
34. Greenberg, D.A.; Weinkle, D.J.; Aminoff, M.J. Periodic EEG complexes in infectious mononucleosis encephalitis. J. Neurol. Neurosurg. Psych. 1982, 45, 648–651.
35. Russell, J.; Fisher, M.; Zivin, J.A.; Sullivan, J.; Drachman, D.A. Status epilepticus and Epstein-Barr virus encephalopathy. Diagnosis by modern serologic techniques. Arch. Neurol. 1985, 42, 789–792.
36. Hollister, L.E.; Houck, G.H.; Dunlap, W.A. Infectious mononucleosis of the central nervous system. Demonstration of atypical lymphocytes in the cerebrospinal fluid. Am. J. Med. 1956, 20, 643–646.
37. Joncas, J.H.; Chicoine, L.; Thivierge, R; Bertrand, M. Epstein-Barr virus antibodies in the CSF. A case of infectious mononucleosis with encephalitis. Am. J. Dis. Child. 1974, 127, 282–285.
38. Landgren, M.; Kyllerman, M.; Bergstrom, T.; Dorevall, L. Diagnosis of Epstein-Barr virus-induced central nervous system infections by DNA amplification from cerebrospinal fluid. Ann. Neurol. 1994, 35, 631–635.
39. Schiff, J.A.; Schaefer, J.A.; Robinson, J.E. Epstein-Barr virus in cerebrospinal fluid during infectious mononucleosis encephalitis. Yale J. Biol. Med. 1982, 55, 59–63.

40. Zohman, B.L.; Silverman, E.G. Infectious mononucleosis and encephalomyelitis. Ann. Intern. Med. 1942, 16, 1233–1239.
41. Grose, C.; Feorino, P.M. Epstein-Barr virus and Guillain-Barre syndrome. Lancet. 1972, 2, 1285–1287.
42. Davie, J.C.; Ceballos, R.; Little, S.C. Infectious mononucleosis with fatal neuronitis. Arch. Neurol. 1963, 9, 265–272.
43. Sharma, K.R.; Sriram, S.; Fries, T.; Bevan, H.J.; Bradley, W.G. Lumbosacral radiculoplexopathy as a manifestation of Epstein-Barr virus infection. Neurology. 1993, 43, 2550–2554.
44. Watson, P.; Ashby, P. Brachial plexus neuropathy associated with infectious mononucleosis. Can. Med. Assoc. J. 1976, 114, 758–759.
45. Mohanaruban, K.; Fisher, D.J.H. A combination of cranial and peripheral nerve palsies in infectious mononucleosis. Postgrad. Med. J. 1986, 62, 1129–1130.
46. Bennett, J.L.; Mahalingam, R.; Wellish, M.C.; Gilden, D.H. Epstein-Barr virus-associated acute autonomic neuropathy. Ann. Neurol. 1996, 40, 453–455.
47. Hauptmann, S.; Meru, N.; Schewe, C.; Jung, A.; Hiepe, F. Fatal atypical T-cell proliferation associated with Epstein-Barr virus infection. Br. J. Haematol. 2001, 112, 377–380.
48. Grose, C.; Feorino, P.M.; Dye, L.A.; Rand, J. Bell's palsy and infectious mononucleosis. Lancet. 1973, 2, 231–232.
49. Snyder, R.D. Bell's palsy and infectious mononucleosis. Lancet. 1973, 2, 917–918.
50. Egan, R.W. Facial diplegia in infectious mononucleosis in the absence of Landry-Guillain-Barre syndrome. N. Engl. J. Med. 1960, 262, 1178–1179.
51. Taylor, L.; Parsons-Smith, G. Infectious mononucleosis, deafness and facial palsy. J. Laryngol. Otol. 1969, 83, 613–616.
52. DeSimone, P.A.; Snyder, D. Hypoglossal nerve paralysis in infectious mononucleosis. Neurology. 1978, 28, 844–847.
53. Ashworth, J.; Motto, S.A. Infectious mononucleosis complicated by bilateral papilloretinal edema. N. Engl. J. Med. 1947, 237, 544–545.
54. Blaustein, A.; Caccavo, A. Infectious mononucleosis complicated by bilateral papilloretinal edema. Arch. Ophthalmol. 1950, 43, 853–856.
55. Bonynge, T.W.; Van Hagen, K.O. Severe optic neuritis in infectious mononucleosis. JAMA. 1952, 145, 933–934.
56. Cotton, P.B.; Webb-Peploe, M.M. Acute transverse myelitis as a complication of glandular fever. Lancet. 1966, 1, 654–655.
57. Grose, C.; Feorino, P.M. Epstein-Barr virus and transverse myelitis. Lancet. 1973, 1, 892.
58. Clevenbergh, R.; Brohee, P.; Velu, T.; Jacobs, E.; Liesnard, C.; Deneft, E.; Thys, J.P. Infectious mononucleosis complicated by transverse myelitis: detection of the viral genome by polymerase chain reaction in the cerebrospinal fluid. J. Neurol. 1997, 244, 592–594.
59. Bergen, D.; Grossman, H. Acute cerebellar ataxia of childhood associated with infectious mononucleosis. J. Pediatr. 1975, 87, 832–833.
60. Cleary, T.G.; Henle, W.; Pickering, L.K. Acute cerebellar ataxia associated with Epstein-Barr virus infection. JAMA. 1980, 243, 148–149.
61. Bennett, D.R.; Peters, H.A. Acute cerebellar syndrome secondary to infectious mononucleosis in a fifty-two year old man. Ann. Intern. Med. 1961, 55, 147–149.
62. Gilbert, J.A.; Culebras, A. Cerebellitis in infectious mononucleosis. JAMA. 1972, 220, 727.
63. Lascelles, R.G.; Longson, M.; Johnson, P.J.; Chiang, A. Infectious mononucleosis presenting as acute cerebellar syndrome. Lancet. 1973, 2, 707–709.
64. Shoji, H.; Goto, Y.; Yanase, Y.; Sato, Y.; Nakashima, K.; Natori, H.; Kaji, M. Recurrent cerebellitis. A case report of a possible relationship with Epstein-Barr virus infection. Kurume Med. J. 1983, 30, 23–26.
65. Raymond, R.W.; Williams, R.I. Infectious mononucleosis with psychosis. Report of a case. N. Engl. J. Med. 1948, 239, 542–544.

66. Hendler, N.; Leahy, W. Psychiatric and neurologic sequelae of infectious mononucleosis. Am. J. Psychiatry. 1978, 135, 842–844.
67. White, P.D.; Lewis, S.W. Delusional depression after infectious mononucleosis. Br. Med. J. 1982, 295, 97–98.
68. Copperman, S.M. ''Alíce in Wonderland'' syndrome as a presenting symptom of infectious mononucleosis in children. Clin. Pediatr. 1977, 16, 143–146.
69. Eshel, G.M.; Eyov, A.; Lahat, E.; Brauman, A. Alice in Wonderland syndrome, a manifestation of acute Epstein-Barr virus infection. Pediatr. Infect. Dis. 1987, 6, 68–69.
70. Lahat, E.; Berkovitch, M.; Barr, J.; Peret, G.; Barzilai, A. Abnormal visual evoked potentials in children with ''Alice in Wonderland'' syndrome due to infectious mononucleosis. J. Child. Neurol. 1999, 14, 732–735.
71. Kuo, Y.T.; Chiu, N.C.; Shen, C.Y. Cerebral perfusion in children with Alice in Wonderland syndrome. J. Child. Neurol. 1998. 19, 105–108.
72. Leavell, R.; Ray, C.G.; Ferry, P.; Minnich, L. Unusual acute neurological presentations with Epstein-Barr virus infection. Arch. Neurol. 1986, 43, 186–188.
73. Baker, F.J.; Kotchmar, G.S.; Foshee, W.S.; Sumaya, C.V. Acute hemiplegia of childhood associated with Epstein-Barr virus infection. Pediatr. Infect. Dis. 1983, 2, 136–138.
74. Adamson, D.J.A.; Gordon, P.M. Hemiplegia—a rare complication of acute Epstein-Barr virus infection. Scan. J. Infect. Dis. 1992, 24, 379–380.
75. So, Y.T.; Beckstead, J.H.; Davis, R.L. Primary central nervous system lymphoma in acquired immune deficiency syndrome: a clinical and pathological study. Ann. Neurol. 1986, 20, 566–572.
76. MacMahon, E.M.E.; Glass, J.D.; Hayward, S.D. Epstein-Barr virus in AIDS-related primary CNS lymphoma. Lancet. 1991, 338, 969–973.
77. Larocca, L.M.; Capello, D.; Rinelli, A.; Nori, S.; Antinori, A. The molecular and phenotypic profile of primary central nervous system lymphoma identifies distinct categories of the disease and is consistent with histogenetic derivation from germinal center-related B-cells. Blood. 1998, 92, 1011–1019.
78. Hochberg, F.H.; Miller, G.; Schooley, R.T.; Hirsch, M.S.; Feorino, P.; Henle, W. Central nervous system lymphoma related to Epstein-Barr virus. N. Engl. J. Med. 1983, 309, 745–748.
79. Ruiz, A.; Post, M.J.D.; Bundschu, C.; Ganz, W.I.; Georgiou, M. Primary central nervous system lymphoma in patient with AIDS. In Neuroimaging of AIDS 1. Neuroimaging Clinics of North America; Post, M.J.D., Ed.; 1997; Vol. 7, 281–296.
80. Cinque, P.; Brytting, M.; Vago, L.; Castagna, A.; Parravicini, C. Epstein-Barr virus DNA in cerebrospinal fluid from patients with AIDS-related primary lymphoma of the central nervous system. Lancet. 1993, 342, 398–401.
81. Formenti, S.C.; Gill, P.S.; Lean E.; Rarick, M.; Meyer, P.R. Primary central nervous system lymphoma in AIDS. Results of radiation therapy. Cancer. 1989, 63, 1101–1107.
82. Slobod, K.S.; Taylor, G.H.; Sandlund, J.T.; Furth, P.; Helton, K.J.; Sixbey, J.W. Epstein-Barr virus-targeted therapy for AIDS-related primary lymphoma of the central nervous system. Lancet. 2000, 356, 1493–1494.
83. McGowan, J.P.; Shah, S. Long-term remission of AIDS-related primary central nervous system lymphoma associated with highly active antiretroviral therapy. AIDS. 1998, 12, 952–954.
84. Antinori, A.; Cingolani, A.; De Luca, A.; Gaidano, G.; Ammassari, A. Epstein-Barr virus in monitoring the response to therapy of acquired immunodeficiency syndrome-related primary central nervous system lymphoma. Ann. Neurol. 1999, 5, 259–261.
85. Swinnen, L.J. Posttransplant lymphoproliferative disorder. In Infectious Causes of Cancer; Goedert, J.J., Ed.. Targets for Intervention; Humana Press: Totowa: NJ, 2000, Chap. 4.
86. Strauss, S.E.; Cohen, J.I.; Tosato. G.; Meier, J. Epstein-Barr virus infection: biology, pathogenesis and management. Ann. Intern. Med. 1993, 18, 45–58.
87. Colby, B.M.; Shaw, J.E.; Elion, G.B.; Pagano, J.S. Effect of acyclovir on Epstein-Barr virus DNA replication. J. Virol. 1980, 34, 560–568.

88. Colby, B.M.; Shaw, J.E.; Datta, A.K.; Pagano, J.S. Replication of Epstein-Barr virus DNA in lymphoblastoid cells treated for extended periods with acyclovir. Am. J. Med. 1982, 73(suppl), 77–81.
89. Torre, D.; Tambini, R. Acyclovir for treatment of infectious mononucleosis: a meta-analysis. Scan. J. Infect. Dis. 1999, 31, 543–547.
90. Dellemijn, P.L.I.; Brandenbrug, A.; Niesters, H.G.M. Successful treatment with ganciclovir of presumed Epstein-Barr meningoencephalitis following bone marrow transplant. Bone Marrow Transplant. 1995, 16, 311–312.
91. Garamendi, I.; Montejo, M.; Cancelo, L.; Lopez, L. Encephalitis caused by Epstein-Barr virus in a renal transplant recipient. Clin. Infect. Dis. 2002, 34, 287–288.
92. Andersson, J.; Ernberg, I. Management of Epstein-Barr virus infections. Am. J. Med. 1988, 85(suppl 2A), 107–115.
93. Pagano, J.S.; Sixbey, J.W.; Lee, J.-C. Acyclovir and Epstein-Barr virus infection. J. Antimicrob Chemother. 1983, 12(suppl B), 113–121.
94. Nalesnik, M.A.; Makowka, L.; Starzl, T.E. The diagnosis and treatment of posttransplant lymphoproliferative disorders. Curr. Prob. Surg. 1988, 25, 367–472.
95. Papadopoulos, E.B.; Ladanyi, C.; Emanuel, D.; Mackinnon, S.; Boulad, F.; Carabasi, M.H. Infusions of donor leukocytes to treat Epstein-Barr virus-associated lymphoproliferative disorders after allogeneic bone marrow transplantation. N. Engl. J. Med. 1994, 330, 1185–1191.
96. Alfieri, C.; Allsion, A.C.; Kieff, E. Effect of mycophenolic acid on Epstein-Barr virus infection of human B-lymphocytes. Antimicrob Agents Chemother. 1994, 38, 126–129.
97. Okano, M. Epstein-Barr virus in patients with immunodeficiency disorders. Biomed Pharmacother. 2001, 55, 353–361.

8

Cytomegalovirus

Claire Pomeroy* **and Julie A. Ribes**
University of Kentucky College of Medicine
Lexington, Kentucky, U.S.A.

1 DESCRIPTION OF CYTOMEGALOVIRUS

Human cytomegalovirus (CMV) is a member of the *Herpesviridae* family in the β herpesvirus subfamily (Fig. 1) [1]. The large double-stranded DNA genome encodes for 230 proteins, including DNA polymerase and a viral kinase that functions to phosphorylate the therapeutic agent ganciclovir to an active form. The genome is contained in a capsid, which in turn is surrounded by matrix proteins such as pp65 that are detected by many of the current diagnostic antigen detection tests. The virion has a proteolipid envelope derived from the host cell's nucleus and containing glycoproteins that play a role in viral entry into the cell [2].

2 ILLNESS

2.1 History of the Illness

Although human CMV was first isolated from salivary glands in the mid-1950s, it was not until 1965 that a CMV disease syndrome in healthy hosts was reported [1]. CMV was first recognized as a cause of a mononucleosis-like syndrome in patients treated with blood transfusions, and it soon became clear that a wide variety of disease manifestations are possible. In the 1980s, effective therapies first became available to treat this disease. The discovery of ganciclovir was a major breakthrough in disease management and was followed by other treatment options, including innovative modes of delivery of ganciclovir

* *Current affiliation:* University of California, Davis School of Medicine, Davis, California, U.S.A.

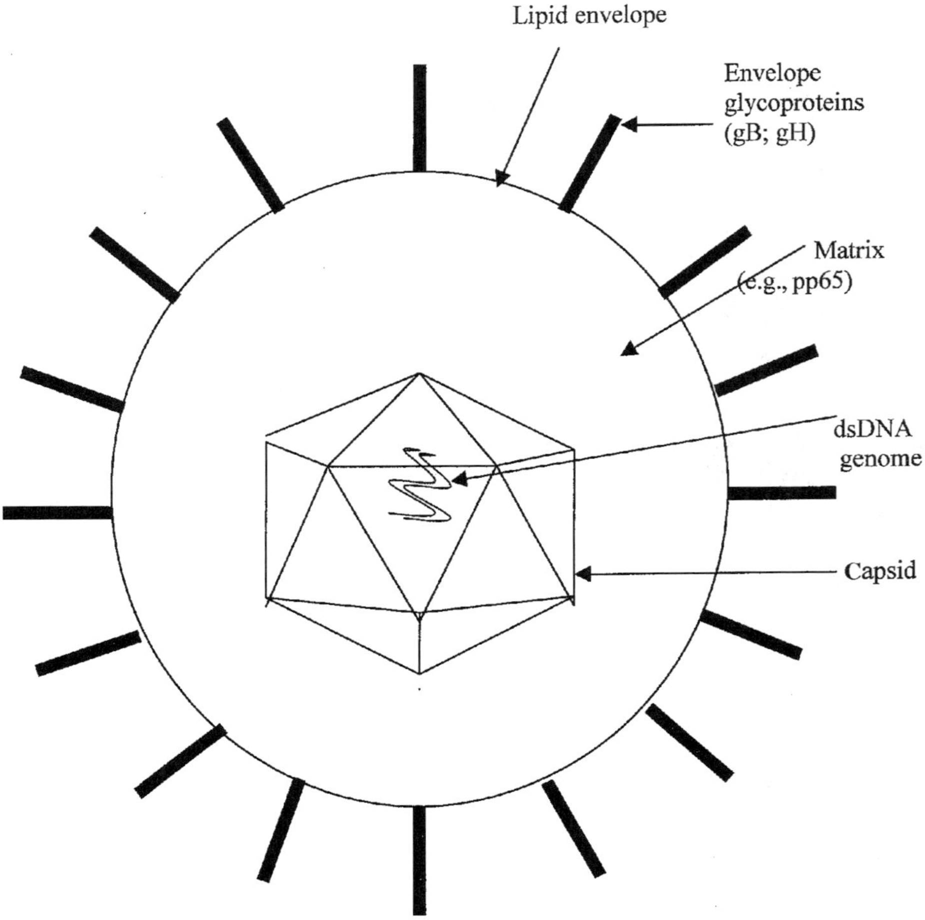

Figure 1 Schematic diagram of cytomegalovirus (CMV). CMV is a dsDNA virus in which the nucleocapsid is surrounded by matrix and a lipid envelope.

and other drugs such as foscarnet and cidofovir [3]. Unfortunately, despite extensive research efforts, no effective vaccine has been developed to prevent CMV infection.

2.2 Epidemiology

Cytomegalovirus is a ubiquitous virus with a worldwide distribution [4]. Whereas seroprevalence rates among children exceed 90% in many parts of the world, the rates in the United States among children are lower, reaching a seroprevalence of 40–70% for adults. Transmission occurs via transfer of infected body substances, resulting in higher seroprevalence rates in areas with overcrowding and poor socioeconomic conditions. Mothers can transmit the virus to their babies during pregnancy or delivery and later by breastfeeding. Intimate contact with oral secretions or other body fluids is a common mode of spread between young children, especially in day care settings. CMV is also transmitted by sexual contact, and rates of CMV infection are very high in sexually active populations, including gay men [5]. Finally, CMV can be transmitted via white blood cell–containing blood transfusions, by transplantation of latently infected organs, and via donor semen used for artificial insemination. CMV is not spread to others via aerosol. CMV is transmitted by individuals in whom virus is actively replicating, as a reflection of either primary or reactivated infection. Most of the people from whom virus is transmitted are asymptomatic. As a result, infection control efforts that rely on isolating patients with diagnosed CMV disease are unlikely to be effective [6].

Cytomegalovirus causes primary disease in both immunocompetent and immunocompromised hosts. Primary CMV infection can occur at any age but is most common in children. Although primary infection is most often asymptomatic, a wide variety of manifestations may occur, even in immunocompetent hosts [1]. Infection acquired in utero can be devastating, with manifestations ranging from asymptomatic infection to severe congenital cytomegalic inclusion disease [4]. Symptomatic congenital disease nearly always occurs in babies born to mothers who suffered primary infection during pregnancy. However, it has recently been recognized that women who are seropositive for CMV can be reinfected with a different strain of CMV, with consequent symptomatic congenital infection in their babies [7]. Infants born to mothers with reactivated CMV disease during pregnancy are generally asymptomatic at birth [4]. Primary infection during childhood is common and only rarely causes symptomatic illness. With improved living conditions, disease acquisition in the developed world is increasingly likely to occur later in life. Indeed, significant concern has been raised about the phenomenon of a mother acquiring CMV from her young child who is in day care, resulting in congenital infection of the next child if the mother is pregnant when she develops primary disease [8].

Secondary or reactivated infection occurs when the host develops abnormal cell-mediated immunity, and disease manifestations often vary depending upon the specific immune defects. CMV infection is especially prevalent in the HIV/AIDS population [9]. Up to half of these patients have evidence of subclinical CMV infection at the time of autopsy [10]. Indeed, CMV retinitis was the most common AIDS-defining opportunistic infection in the era prior to highly active antiretroviral therapy (HAART), and CMV disease remains a frequent complication in HIV-infected individuals. Interestingly, CMV typically causes different manifestations in the various populations of immunocompromised individuals [11]. For example, CMV disease in AIDS patients most frequently manifests as retinitis and less frequently as CMV colitis. In contrast, patients undergoing bone marrow transplantation more often develop CMV pneumonia, occasionally suffer from CMV colitis, but are rarely diagnosed with CMV retinitis.

Neurological manifestations of CMV can occur in immunocompetent individuals and include meningoencephalitis or polyradiculopathy, usually in association with a mononucleosis-like syndrome. Neurological manifestations are most common in patients with HIV infection or AIDS and can range from meningoencephalitis to polyradiculopathy to mononeuritis multiplex to peripheral neuropathy. Less frequently, comparable neurological syndromes develop in the patient immunocompromised because of transplantation, malignancies, or other diseases characterized by defective cell-mediated immunity.

2.3 Pathophysiology

Cells infected with CMV have altered morphology, becoming swollen and round and containing intranuclear and intracytoplasmic inclusions (Fig. 2). Peripheralization of nuclear chromatin imparts a characteristic ''owl's eyes'' appearance of infected cells, due to the eccentric location of the nucleus in the enlarged cell. Cytopathic changes may also be seen in the cytoplasm of infected cells. The presence of eosinophilic granules on hematoxylin and eosin–stained tissue sections help to differentiate infections with other viruses from those seen in cytomegalovirus infection.

Like other herpesviruses, a hallmark characteristic of CMV is the ability to establish latency after acute infection. Latent virus exists in a nonreplicating form that can reactivate and cause disease during periods of immunosuppression. Thus, CMV disease can reflect either primary infection or reactivated infection in immunocompromised individuals, such as transplant recipients, oncology patients, or human immunodeficiency virus (HIV)-seropositive individuals. The cellular site of latent CMV infection has not been definitively established, but current evidence suggests that mononuclear blood leukocytes are a likely location [12].

During primary infection or as a result of reactivation of previously latent virus, CMV disseminates to all organs of the body, predominantly via polymorphonuclear leukocytes. CMV causes disease either via direct destruction of tissues by replicating virus or by inducing immune responses that mediate organ damage. CMV infection elicits CMV-

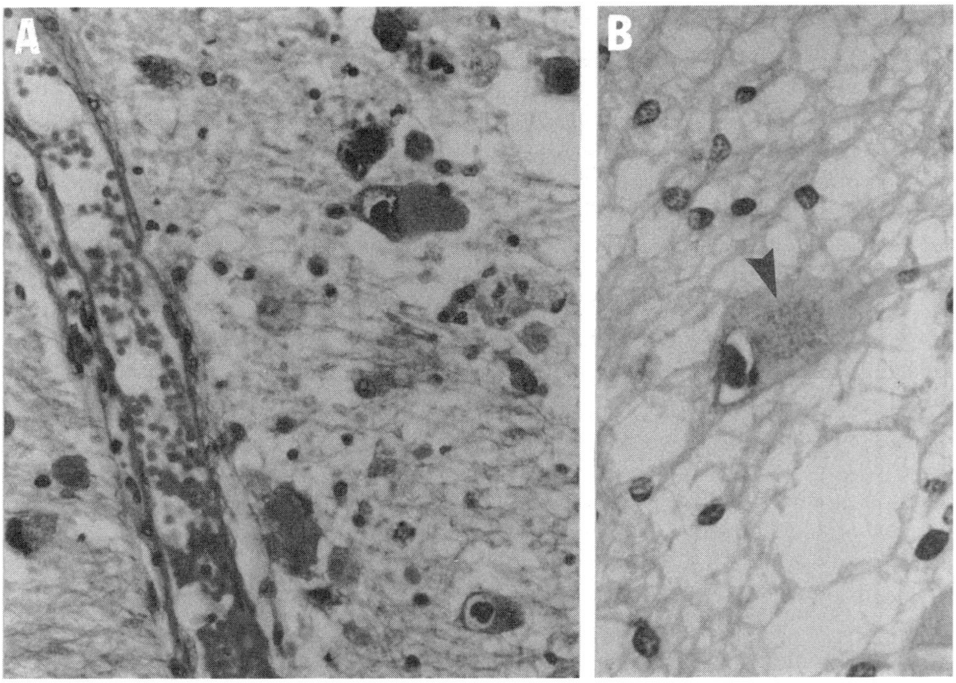

Figure 2 Histopathological specimen of brain tissue demonstrating CMV encephalitis. Neurons infected with CMV demonstrate the classical cytopathic effect on H&E staining. (A) The blotlike intranuclear inclusion surrounded by a halo is clearly seen at low power (40×). (B) At higher magnification (60×) the cytoplasmic inclusions are also evident (arrowhead). The surrounding brain tissue demonstrates extensive vacuolization and disintegration in the most heavily infected areas of the tissue sections.

specific cytotoxic T-lymphocyte responses and cytokine production that appear to be important in recovery. CMV-specific antibodies are also produced and are often used to assess status of infection, although the importance of the role of antibodies in host defense remains unclear. Furthermore, CMV infection can actually cause suppression of some host defenses, included T-lymphocyte responses and production of some cytokines [13].

2.4 Clinical Manifestations

Cytomegalovirus causes a wide variety of neurological diseases (Table 1). Congenital cytomegalic inclusion disease can be characterized by a heartbreaking array of abnormalities of the eye and brain, and even subclinical infection acquired in utero can be associated with neurological sequelae. Rarely, primary infection in immunocompetent hosts can result in neurological complications. More frequently, CMV disease of the neurological system occurs in immunocompromised hosts with defective cell-mediated immunity. Disease can manifest in the eye, brain, spinal cord, or peripheral nerves.

Congenital Cytomegalic Inclusion Disease

Cytomegalovirus is the most frequent infectious cause of birth defects, with up to 40,000 infected infants born annually in the United States. Thus, about 1% of babies born in the United States have CMV infection acquired in utero, as evidenced by shedding of virus in the urine or saliva [14]. The vast majority of these babies are asymptomatic at birth, but the remaining 5–10% have potentially devastating symptomatic cytomegalic inclusion disease [4,15]. Pathological examination of the brain in cases of congenital infection will reveal ventriculoencephalitis. Neuropathological findings may include cerebral necrosis or periventricular leukomalacia

Symptomatic congenital cytomegalic inclusion disease is characterized by disease of the central nervous system, including intracranial calcifications, microcephaly, and ventriculomegaly, with or without ophthalmic and auditory damage [4,16]. Ill babies may have an array of clinical findings, including jaundice, hepatitis, hepatosplenomegaly, pneumonitis, petechiae, thrombocytopenia, hemolytic anemia, chorioretinitis, dental defects,

Table 1 Major Neurological Diseases Caused by CMV

Congenital cytomegalic inclusion diseases
Cerebral dysgenesis
Mental retardation
Motor deficits
Sensorineural hearing loss
Optic atrophy
Seizures, including infantile spasms
Other neurological and neurobehavioral diseases
Postnatal acquisition
Retinitis
Meningoencephalitis
Transverse myelitis
Polyradiculopathy
Mononeuritis multiplex
Demyelinating polyneuropathy (Guillain-Barré)
Distal symmetrical polyneuropathy

and fetal growth retardation. Neurological findings include mental retardation, microcephaly and other brain defects, abnormal eye development including ocular atrophy and chorioretinitis, motor deficits including paresis and paralysis, and seizures including the infantile spasm syndrome [14]. Mortality may be as high as 30%, and most survivors will suffer permanent neurodevelopmental sequelae, including retardation, behavioral problems, visual defects, and hearing loss.

Of the asymptomatic babies, most will never suffer any clinically apparent consequences, but 10–15% of them will develop sensorineural hearing loss or other subtle neurological abnormalities later in life [17]. Infants with congenital CMV infections should be monitored with serial audiograms to detect hearing loss that develops later during childhood.

The differential diagnosis of congenital CMV disease includes other congenital TORCH infections such as rubella, toxoplasmosis, and neonatal herpes simplex virus infections as well as syphilis and varicella virus infections.

Retinitis

Although CMV retinitis can occur in other immunocompromised patients, individuals with HIV/AIDS are most often affected by this disease [18,19]. Prior to the era of HAART, CMV was the most common viral opportunistic infection in AIDS patients, and retinitis is the most common manifestation of the virus in this population. Most patients do not develop CMV retinitis until the $CD4^+$ cell count falls below 50–100 cells/mm^3. Detection of viremia by PCR has been shown to predict risk for retinitis in AIDS patients [20–23].

History will reveal complaints of ''floaters'' and peripheral blind spots with progressive loss of vision. Severe visual disturbance is usually caused either by extension of the infection centrally to the fovea or by retinal detachment. On physical examination, fundoscopic evaluation will demonstrate characteristic white, fluffy retinal infiltrates with surrounding areas of hemorrhage. CMV retinitis results in retinal destruction and blindness within several months if left untreated.

Since the advent of HAART, the incidence of CMV retinitis has decreased dramatically [24–27]. The introduction of HAART has also had a major impact on the natural history of CMV retinitis with improved survival time and decreased risk of progression following diagnosis [25,28,29]. However, patients who have low $CD4^+$ counts when HAART is initiated and those who fail HAART therapy remain at high risk of developing disease [30–32]. In addition, immune reconstitution may be associated with inflammatory complications that can result in significant visual loss [33], even in the absence of active CMV disease [34].

Differential diagnosis of CMV retinitis includes cotton wool spots; these are small, white lesions that are not associated with exudates or hemorrhage and are usually clinically benign. Other infections must also be considered in the differential diagnosis, including toxoplasmosis, syphilis, herpes simplex virus, varicella-zoster virus, tuberculosis, fungal infections and even malignancies.

Meningoencephalitis

In rare cases, meningoencephalitis can complicate the course of CMV mononucleosis in the immunocompetent patient [14,35–39]. In these patients, symptoms may include fever, headache, photophobia, lethargy, seizures, and pyramidal tract abnormalities. CMV encephalitis occurs most frequently in immunocompromised patients, especially those with HIV/AIDS and occasionally in transplant recipients [9]. Symptoms are generally indistinguishable from those caused by many other etiologies of encephalitis and include mental slowness, confusion, difficulty concentrating, personality changes, forgetfulness, somnolence, and coma [40]. Fever and focal neurological abnormalities such as cranial nerve palsies and seizures may occur. Hearing loss has been reported [41]. CMV encephalitis is a sign of poor prognosis in the patient with AIDS, and most patients survive only for weeks to months, even with therapy.

Two forms of encephalitis are now recognized—microglial nodule encephalitis and ventriculoencephalitis [42,43]. These reflect the two routes by which CMV can disseminate to the brain. In the first, dissemination occurs via the cerebrospinal fluid (CSF) through the ependymal cells and results in ventriculoencephalitis. This is the form most commonly observed in congenital CMV disease. In the second, dissemination occurs via the blood through endothelial cells and results in microglial nodular encephalitis. This is the form most frequently seen in transplant recipients. Both forms occur in patients with HIV/AIDS.

The two forms of encephalitis have been linked to distinct clinical patterns in AIDS patients [42,44,45]. Microglial nodular encephalitis appears to manifest earlier, the onset is usually more insidious, and the course is characterized by confusion and delirium. A high incidence of fluid and electrolyte abnormalities has been associated with microglial nodular encephalitis. Death usually occurs within months. In contrast, ventriculoencephalitis occurs later in the course of HIV disease, onset is more acute, and symptoms include cranial nerve deficits, nystagmus, and ventricular enlargement. The CSF chemistries tend to be more abnormal and the pleocytosis greater in ventriculoencephalitis. Death often occurs within weeks of the diagnosis of CMV ventriculoencephalitis [44,45].

Microglial nodules are the most common lesions in patients with CMV encephalitis. They are dense cellular aggregates of macrophages and rod cells that are observed more commonly in gray matter than in white matter [46]. Although few nodules contain cytomegalic cells, CMV has been detected by in situ hybridization in one-third and by PCR in two-thirds of microglial nodules. Focal parenchymal necrosis can also occur in patients with CMV encephalitis but is less common than the microglial nodules. Ventriculoencephalitis is manifested by focal or diffuse destruction of the ependymal lining and necrosis of periventricular parenchymal tissue associated with dense accumulation of cytomegalic cells in the ependymal and periependymal areas. Cytomegalic cells in brain tissue can occur without other associated pathological findings, usually in patients without symptoms.

It may be quite difficult to distinguish CMV encephalitis from other forms of encephalitis. This is especially true for AIDS patients, in whom HIV may cause a similar clinical presentation. Furthermore, HIV can cause microglial nodules similar to those seen with CMV, making it difficult to distinguish the two etiologies even when histopathological samples are available. In these patients, it is unclear if CMV is merely a "bystander" or if it plays a true pathogenic role.

Transverse Myelitis

Transverse myelitis has been associated with CMV infection in both immunocompetent and immunocompromised hosts [47–49]. Symptoms are comparable to those of transverse myelitis caused by other viral etiologies and include sensory abnormalities and motor deficits at a specific spinal level. The disease process is limited to the spinal cord and usually does not involve other parts of the nervous system. Additional symptoms may be caused by disseminated CMV infection, such as a mononucleosis-like syndrome or end-organ dysfunction due to viral infection.

Other potential etiologies of transverse myelitis should be excluded. Although many viruses can cause transverse myelitis, the exact etiology is never established in many of these patients. The differential diagnosis also includes transverse myelitis secondary to multiple sclerosis or spinal cord ischemia.

Polyradiculopathy

Cytomegalovirus has recently been recognized as the cause of a new syndrome called polyradiculopathy, which to date appears to be confined to patients with advanced immunosuppression in AIDS [50–55]. In contrast to most viral infections characterized by lymphocytic responses, CMV polyradiculopathy is associated with a marked neutrophil pleocytosis in the CSF [56–58]. The syndrome is characterized by rapidly ascending motor weakness associated with pain and paresthesias in the cauda equina distribution and loss of bowel and bladder function, and ultimately the development of flaccid paralysis of the lower extremities [44,45]. There may be sensory abnormalities, and rarely, especially in severe disease, the upper extremities may become involved. Clinical findings include leg pain and weakness, spasticity, hyporeflexia, urinary retention, fecal incontinence, paresthesias, and sensory loss. Many patients with CMV-associated polyradiculopathy have evidence of CMV disease elsewhere, most commonly CMV retinitis [44,45]. About one-half of patients will have an associated myelitis. Survival is only a few weeks in untreated cases and may be extended to an average of a few months in treated patients.

The etiological role of CMV is supported by the finding of CMV inclusions in the nuclei of endothelial, ependymal, and Schwann cells as well as the ability to detect viral genome in the CSF by PCR [59]. Polymorphonuclear and mononuclear infiltrates, axon destruction, and CMV inclusions can be observed in the cauda equina and lumbosacral nerve roots [60]. In some cases, small vessels show focal inflammatory necrosis of the vessels walls, with thrombosis and segmental infarction.

The most common etiology of polyradiculopathy in AIDS patients is CMV, but a similar clinical presentation can occur with neurosyphilis or lymphomatous meningitis. MRI or CT scans should be obtained to exclude the possibility of a structural lesion in the cauda equina.

Mononeuritis Multiplex

Cytomegalovirus infection of the nervous system can manifest as mononeuritis multiplex [61]. This entity most often occurs in patients with advanced HIV infection or other forms of severe immunosuppression [44,45,62]. When caused by CMV, the syndrome usually presents with involvement of multiple nerves in the extremities and/or cranial nerve abnormalities. Symptoms include multifocal sensory and motor abnormalities in the distribution of more than one peripheral or cranial nerve. In patients with severe immunosuppression, the disease may be rapidly progressive.

Specimens from patients with CMV-associated mononeuritis multiplex demonstrate mixed demyelination and axonal degeneration and virions in mononuclear cells and endothelial cells. Nerve biopsy may demonstrate areas of endoneurial necrosis with polymorphonuclear and mononuclear inflammatory infiltrates associated with CMV inclusion in Schwann cells and endothelial cells. Damage appears to be secondary to direct viral invasion and destruction of the involved nerves.

The syndrome should be distinguished from a form of mononeuritis multiplex that occurs in early HIV infection and is not caused by CMV but rather appears to be mediated by immune mechanisms. The latter disease is characterized by self-limited motor and sensory deficits in the distribution of individual peripheral nerves and is treated with immunomodulatory drugs alone.

Demyelinating Polyneuropathy (Guillain-Barré)

Cytomegalovirus has long been implicated as one of the causes of Guillain-Barré disease [14,63]. The disease manifests as a polyneuritis with sensory and motor weakness of the extremities and may include cranial nerve abnormalities and respiratory failure. Well-documented cases in which Guillain-Barré occurred in immunocompetent patients suffering from a classic CMV mononucleosis syndrome with atypical lymphocytosis, and characteristic serologies support this association. These cases most likely are cases of Guillain-Barré syndrome caused by primary CMV infection.

Cytomegalovirus-associated Guillain-Barré syndrome also occurs in immunocompromised individuals, especially patients with HIV infection and rarely in transplant patients. The ascending weakness and in severe cases paralysis may be inexorable or relapsing. Cases in which CMV inclusions have been found in Schwann cells, CMV DNA is detected by PCR, and virions are detected on nerve biopsy support the conclusion that at least some cases of demyelinating polyneuropathy are attributable to the virus. However, it is unclear if the neurological damage is due to direct viral destruction or if the virus triggers an immune reaction, possibly against components of peripheral nerve myelin [64].

Cases in immunocompromised patients must be distinguished from the more recently recognized CMV-induced polyradiculopathy that occurs in AIDS patients. It also should be distinguished from the demyelinating polyneuropathy that occurs in conjunction with acute HIV infection around the time of seroconversion. The latter is characterized by a more rapid and progressive ascending weakness and paralysis, with a CSF lymphocytic pleocytosis. Because seroconversion demyelinating polyneuropathy appears to be caused by immune mechanisms rather than CMV, it is treated with immunomodulatory therapy.

Distal Symmetrical Polyneuropathy

Cytomegalovirus is one of the many causes of distal symmetricial polyneuropathy and occurs predominantly in individuals immunocompromised by HIV infection or other causes of defective cell-mediated immunity [61,62]. These patients usually present with symptoms that are clinically indistinguishable from those present in other polyneuropathies. Patients will note numbness, paresthesias, and dysesthesias in the distal extremities, usually the feet and sometimes the hands. Physical examination may reveal hypersensitivity or decreased sensitivity on sensory testing and depressed reflexes. Pain and temperature sensation may be impaired, but proprioception and motor function are generally relatively preserved.

2.5 Radiographic Findings

In congenital CMV disease, CT or other imaging studies will reveal intracranial calcifications in up to 50% of symptomatic infants and in a small percentage of children with asymptomatic infection [14]. Symptomatic babies may also have cerebral dysgenesis, evidenced by pachygyria, polymicrogyria, porencephaly, lissencephaly, schizencephaly, cerebellar hypoplasia, or hydrocephalus [15].

Computed tomography and MRI studies are usually included in the evaluation of the patient with symptoms of encephalitis. However, in cases caused by CMV, findings are generally normal or the tests reveal only nonspecific abnormalities [40,65]. Periventricular enhancement has been considered suggestive of CMV encephalitis [66,67] but may also be observed in HIV encephalitis, limiting its usefulness in distinguishing the two entities. MRI may show increased signal intensity on T2-weighted images [68], and periventricular enhancement correlates with CMV ventriculoencephalitis but has limited sensitivity. On

MRI testing, ependymal enhancement with gadolinium is significantly more common on scans of patients whose CSF samples are positive for CMV genome by PCR [69]. Rarely, CMV infection of the CNS can manifest as ring-enhanced, space-occupying lesions that mimic tumors [70]. The use of these radiographic modalities is important for excluding other CNS diseases in the differential diagnosis of CMV encephalitis.

In CMV associated polyradiculopathy, MRI may reveal enlargement of the conus medullaris and clumping of the lumbosacral rootlets. Gadolinium enhancement of the meninges may be noted on MRI in about one-third of cases, and abnormalities may be observed by myelogram in about one-sixth of patients [71].

2.6 Diagnostic Strategies

Laboratory diagnosis of CMV infection requires the growth of the virus in cell culture, detection of characteristic histopathological findings on tissue specimens, or demonstration of viral antigens or DNA by one of several newer diagnostic assays. It is critical to remember that many individuals shed virus during periods of immunosuppression induced by other diseases. Furthermore, low-level viremia may be detected in the white blood cells of healthy blood donors. Therefore, although detection of virus is evidence of infection, it does not prove that the patient's current illness is caused by CMV.

The diagnosis of CMV disease in the nervous system must be based on a combination of clinical presentation and supportive imaging and laboratory findings [72]. Evidence of CMV by culture, serological testing, and CSF chemistries and cell counts can provide important clues to the correct diagnosis. Antigen detection assays and PCR to demonstrate viral genome in peripheral blood samples, CSF, and tissues now facilitate the diagnosis (Table 2).

Cultures of CMV are traditionally performed in human fibroblast cultures, but the availability of new "shell vial" cultures has markedly decreased the amount of time necessary to detect a "positive" culture, from several weeks to a few days [73,74]. The shell vial technique employs a slow centrifugation of the specimen onto a cell monolayer to enhance the rate of cellular infection. The cell monolayer is typically stained with

Table 2 Diagnostic Laboratory Tools to Detect CMV

Viral culture
Routine tissue culture
Shell vial with monoclonal antibody staining for detection of early antigens
Histopathology
Routine light and electron microscopy
Immunofluorescence, immunoperoxidase
Hybridization techniques
Serology
IgM, IgG, paired acute and convalescent sera to demonstrate seroconversion
Antigen detection assays (i.e. pp65 antigenemia)
Detection of viral DNA
Gene amplification by PCR
Branched DNA assay
Hybrid capture CMV DNA assay
Detection of late viral mRNA transcripts

fluorescence-labeled monoclonal antibodies to immediate-early antigen after 1–5 days of incubation to detect evidence of viral replication (Fig. 3). Tube cultures remain the gold standard for the isolation of cytomegalovirus, however, because low levels of the virus may not be detected with the shell vial technique. The presence of virus is indicated by the development of a characteristic cytopathic effect (Fig. 4). Tube culture diagnosis, although more sensitive, requires significantly more time to detect virus; cultures must be held for up to 21 days for maximum recovery. Most laboratories routinely perform both types of cultures for optimal sensitivity. CMV cultures of the urine or sputum will be positive in virtually all cases of congenital CMV disease. In contrast, CMV cultures of CSF will be positive in only about one-half of patients with evidence of CNS disease caused by the virus [44,45,72,75].

When available, histopathological examination of samples of brain, spinal cord, or peripheral nerve can also help establish CMV as the etiology of neurological diseases (Fig. 2) [46,76]. Infection can be confirmed by the detection of classic histopathological findings on tissue specimens. The finding of intranuclear inclusions with an ''owl's eye'' appearance is characteristic of CMV, and the viral infection can be confirmed with specific stains for viral antigens. Detection of viral antigen or genome in the biopsied tissues, as

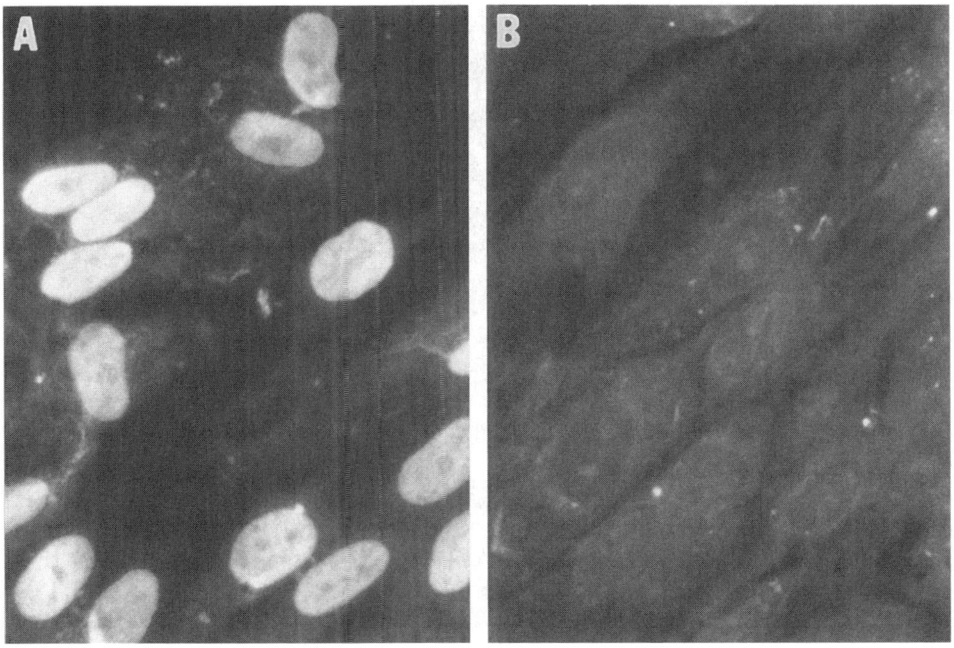

Figure 3 CMV-infected cells in shell vial assay with immunostaining. The rate of viral infection is enhanced by slowly centrifuging the specimen onto the MRC-5 cell monolayer. Early detection of CMV-infected cells is accomplished by staining the monolayer with fluorescently labeled monoclonal antibodies against the immediate-early antigen of CMV. Infected cells (A) demonstrate the typical nuclear staining pattern. Use of this stain permits identification of infected cells well before a cytopathic effect is produced, usually within 2–5 days. Note that the cobblestone pattern of the monolayer is intact, similar to the uninfected cells (B). Uninfected cells (B) demonstrate no nuclear staining. (Fluorescently stained with antibodies against the immediate-early antigen, 100×.)

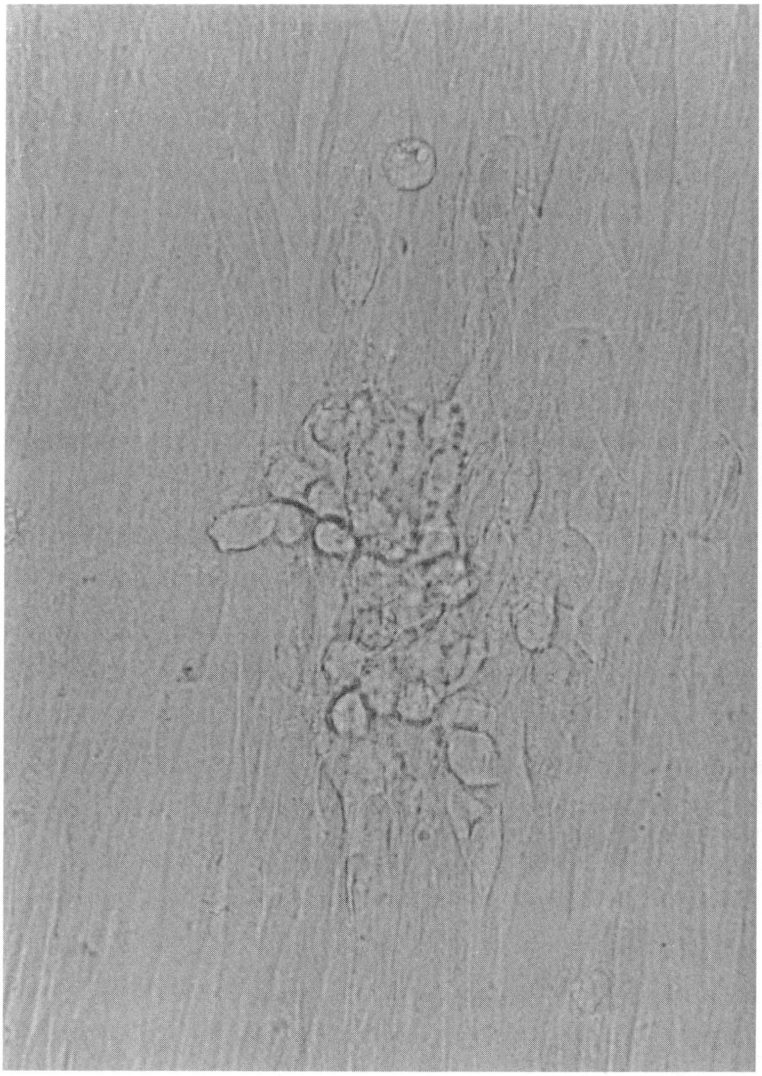

Figure 4 CMV cytopathic effect (CPE) in MRC–5 cell culture. Growth of CMV in tube culture with MRC-5 cells is typically slow, developing after one or more weeks of incubation. The CPE is characterized by collections of rounded, refractile, and granular cells forming well-defined plaques of 5–12 cells. (Unstained cells, 40× magnification.)

evidenced by immunofluorescence with monoclonal antibodies and hybridization techniques, establishes infection and further supports the identification of CMV as the likely etiology of the patient's clinical syndrome. Biopsy of peripheral nerves in mononeuritis multiplex or peripheral neuropathy is not routinely performed but can reveal evidence of virions or characteristic pathological abnormalities in the involved nerves. Electron microscopy (EM) can also demonstrate the presence of viral particles in infected cells (Fig. 5).

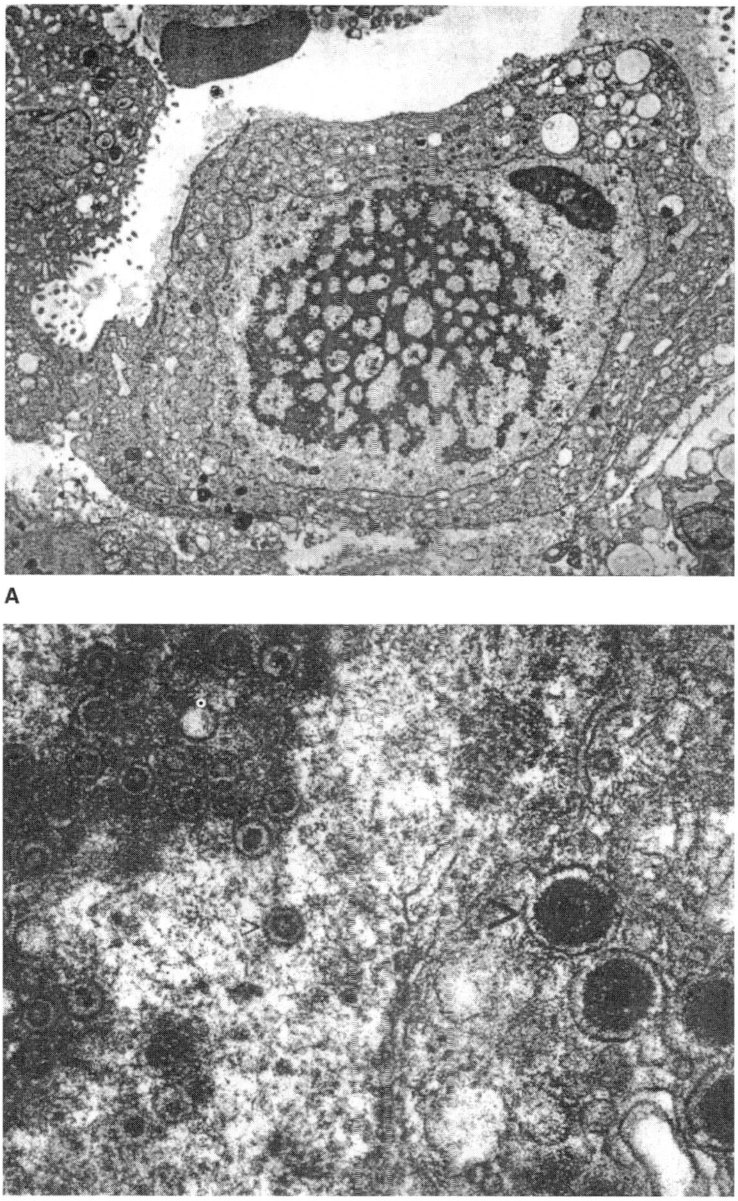

A

B

Figure 5 CMV by electron microscopy. (A) Low power magnification (10,000×) demonstrates the composition of the intranuclear inclusion seen in CMV infection. What appears to be a homogeneous inclusion by light microscopy is actually quite inhomogeneous when viewed by electron microscopy. This inclusion is shown to be composed of both electron-dense areas where the viral capsids are focally packed quite tightly and electron-lucent areas where virus is relatively sparse. (B) At higher magnification (165,000×), the icosohedral symmetry of the DNA-containing capsids is seen (small arrowhead). Once assembled in the nucleus, the virus must pass through the cell's nuclear membrane, where it acquires its membrane envelope. This viral particle has been captured in transit through the nuclear membrane (large arrowhead) on its way to becoming an infective virion.

Serology can detect antibody formation and may indicate current or past infection. However, serological tests can be difficult to interpret, especially in some immunocompromised hosts. Classically, CMV-specific IgM suggests recent infection and IgG suggests past infection. Demonstration of seroconversion with development of IgG antibodies in paired acute and convalescent sera supports the diagnosis of primary cytomegalovirus infection. However, some immunocompromised patients may fail to mount a normal antibody response, transfusion recipients can acquire exogenous antibody that is detected on serological testing, and some patients can periodically develop IgM antibody responses during bouts of limited reactivation of the virus. Finally, IgM antibodies can persist for weeks or even months, in some cases long after the patient has recovered from the primary infection.

Cerebrospinal Fluid chemistries and cell counts can be very helpful in the diagnosis of CMV-associated diseases of the CNS. In CMV encephalitis, CSF findings may be entirely normal, but most often a mild mononuclear cellular response, mildly elevated protein, and normal glucose can be expected. The hallmark characteristic of CMV-associated polyradiculopathy is the finding of a marked CSF neutrophil pleocytosis, accompanied by increased protein or decreased glucose in the CSF. Patients with mononeuritis multiplex often have a mild to moderate mononuclear pleocytosis in the CSF, with mildly elevated protein and a normal glucose.

Viral infection can be diagnosed by antigenemia testing in which monoclonal antibody testing is used to detect the presence of pp65 matrix protein in samples of neutrophils from peripheral blood or CSF [77,78]. Most often this test is performed on buffy coat preparations of peripheral blood to ensure that an adequate number of neutrophils are examined for signs of infection. For patients with CMV infection of the nervous system, the antigenemia assay appears to be most useful in those with polyradiculopathy, likely due to the increased numbers of white cells in the CSF samples of these patients [78,79]. The sensitivity and specificity of these tests appears to be quite good, and semiquantitative results may be obtained to suggest the degree of viremia. The antigenemia test can be performed in clinical laboratories, including those without PCR facilities [72]. However, the assay is of limited usefulness for detecting evidence of CMV in the CSF of patients without a CSF pleocytosis.

The use of gene amplification by PCR techniques allows the detection of viral genome in many body fluids (Fig. 6), including CSF [72,77,80–82]. Detection of CMV DNA by PCR is currently the method of choice for the laboratory diagnosis of CMV infection of the CNS in immunocompromised patients. Detection of CMV in CSF by PCR appears to be a sensitive and specific method for diagnosing CMV encephalitis and polyradiculopathy [44,45]. PCR with a sensitivity of 20 genomes per 10 μL of CSF has been designated as the current reference method for diagnosis of CNS infection in AIDS patients by the European Union Concerted Action on Virus Meningitis and Encephalitis [83]. Findings on CMV PCR correlate with histological abnormalities of CNS viral infection and with response to therapy [50,84–86]. State-of-the-art gene amplification techniques make PCR quite sensitive but not specific for CMV disease because the virus can be present without causing disease in immunocompromised patients [72,87]. PCR assays that can simultaneously amplify genome for several viruses likely to cause disease of the CNS hold future promise [88–90].

Recent studies indicate that quantitative PCR or branched DNA (bDNA) assays can be used to identify patients more likely to have clinical disease; i.e., patients with higher viral loads are more likely to have symptoms attributable to the virus [72,85,91–93].

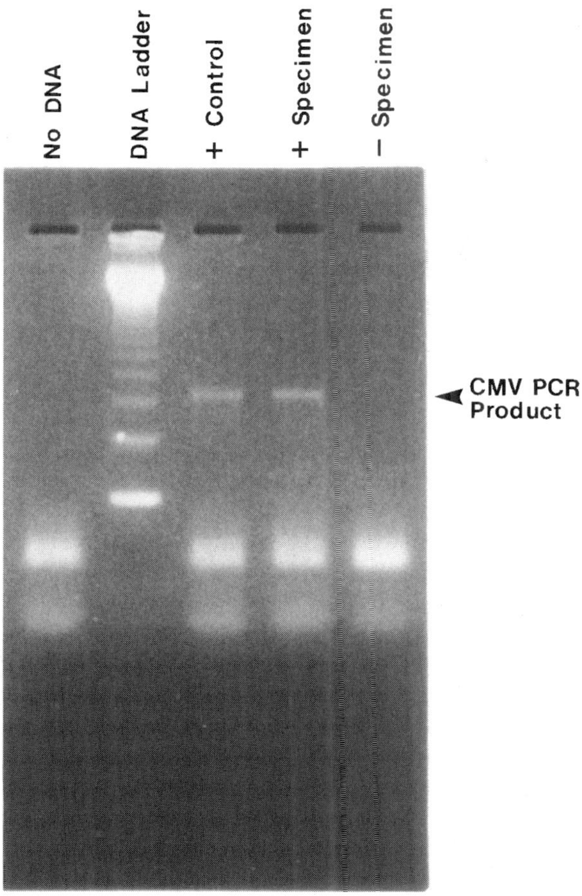

Figure 6 Polymerase chain reaction (PCR) detection of CMV in fluid specimens. Although there are no FDA-approved assays on the market at this time, PCR assays are available experimentally through several commercial laboratories. Fluid samples lacking significant red blood cell contamination such as CSF, BAL, serum, or urine may be extracted for PCR amplification. Assays developed "in house," such as the one used for this figure, use CMV-specific primers to produce PCR products of known size that are visualized by ethidium bromide or labeled probes following gel electrophoresis.

Furthermore, viral loads fall with effective therapy, suggesting that quantitative methods can be useful for monitoring response to therapy [94,95]. Levels of viral genome in CSF appear to be higher in patients with polyradiculopathy than in those with encephalitis [95]. However, these techniques are still predominantly available only in research settings (often developed "in house") and have not been widely adapted to the routine clinical setting. Several PCR based commercial assays (Amplicor PCR, the SHARP Signal Detection System, and PrimeCapture) offer more options to the clinician [72,75,96]; additional studies are needed to determine the optimal approach to interpreting results obtained with these systems. Procedures to detect late gene transcripts are being developed [72,97].

Although the use of CSF PCR has enhanced the ability to detect CMV infection in the central nervous system, it is critical to be mindful of the caveat that the presence of viral genome establishes infection but does not prove that symptoms are due to the virus. In addition, patients without clinical CMV disease may also have a positive PCR. However, studies provide conflicting data about the true efficacy of these methodologies, and the results of these assays must be interpreted with caution and within the clinical context.

2.7 Treatment

Overview

Three drugs have been approved for the treatment of CMV infection: ganciclovir, foscarnet, and cidofovir (Table 3). All three drugs exert their antiviral effect by inhibiting the viral DNA polymerase enzyme. In addition, an antisense inhibitor of CMV, fomivirsen, was recently approved for treatment of CMV retinitis and is administered by direct intravitreal injection.

Ganciclovir

Ganciclovir is a nucleoside analog drug with antiviral activity against sensitive strains of CMV [98]. For activity, the drug must be phosphorylated by a CMV phosphotransferase enzyme encoded by the UL97 gene. The phosphotransferase converts ganciclovir to ganciclovir monophosphate, which in turn is phosphorylated to ganciclovir triphosphate by cellular enzymes. The triphosphate form exerts antiviral activity via inhibition of the DNA polymerase enzyme, by interfering with elongation of viral DNA. Because of the critical role of the phosphotransferase, mutations in the UL97 gene result in ganciclovir resistance [99]. These resistant CMV isolates have been reported in a variety of patient populations, especially in individuals who have received prolonged antiviral therapy. Ganciclovir resistance can also occur in strains with mutations in the CMV DNA polymerase gene.

Ganciclovir was the first drug to have demonstrated efficacy against CMV disease and is the agent with which practitioners have the most experience. Efficacy in the treatment of CMV retinitis has been well established, and numerous reports support use of the drug in other forms of CMV disease, including neurological syndromes.

Ganciclovir can be delivered in several forms—intravenous, direct intravitreal instillation, ganciclovir ocular implants, and as an oral drug. Intravenous ganciclovir is indicated for the treatment of severe CMV disease, including retinitis and other neurological syndromes. The drug must be administered at full "induction" doses for the first 2 weeks of therapy. When used for CMV retinitis in AIDS patients, extensive studies have documented that relapse will usually occur quickly unless the drug is continued in lower "maintenance" doses. Prior to the HAART era, treatment was always lifelong. However, if antiretroviral therapy can result in effective immune reconstitution with significant sustained increases in $CD4^+$ cell counts to greater than 100 mm^{-3}, some studies have indicated that patients may be able to safely discontinue therapy. The role of maintenance therapy for other forms of CMV neurological disease is less well studied.

Other forms of ganciclovir can be used to treat CMV retinitis. Ganciclovir can be directly instilled into the eye by intravitreal injection. Although effective in treating retinitis, this modality does not prevent the development of disease in the other eye and will not prevent the development of systemic CMV disease. The availability of an intraocular implant that continuously releases ganciclovir into the eye affords an improved approach and in fact has become the treatment of choice for many patients [100,101]. However, concerns about the possibility of spreading the infection to the contralateral eye and the

Table 3 Antiviral Drugs for Treatment of Neurological Disease Caused by CMV

Drug	Brand name	Route	Induction dose	Maintenance dose	Major toxicity/concerns
Ganciclovir	Cytovene	IV	5 mg/kg every 12 h for 14 days[a]	5 mg/kg/day	Bone marrow suppression (anemia, neutropenia, thrombocytopenia); central venous catheter infection/thrombosis.
		Oral	Not efficacious	1–2 g TID[a,b]	Bone marrow suppression (usually less severe than with IV ganciclovir).
		Intraocular implant[c,d] (Vitrasert)	Surgical placement	Replacement every 6–8 months	Retinal detachment, endophthalmitis; risk of retinitis in other eye/systemic CMV disease.
		Intraocular injection[c,d]	400 μg twice a week	400 μg/week	Retinal detachment, endophthalmitis; subconjunctival/intravitreal hemorrhage; risk of retinitis in other eye/systemic CMV disease.
Valganciclovir	Valcyte	Oral	900 mg every 12 h for 21 days	900 mg/day	Similar to IV ganciclovir.
Fomivirsen	Vitravene	Intravitreal injection	330 μg/week for 2 weeks	330 μg every 4 weeks	Inflammation of anterior and posterior chambers (uveitis); increased intraocular pressure; peripheral retinopathy; cataracts.
Combination therapy—ganciclovir plus foscarnet	Cytovene plus Foscavir	IV	Same as for drugs given individually		See individual drugs; increased chance of not tolerating therapy with combination treatment; see individual drugs.
Oral ganciclovir plus intraocular ganciclovir implant[c,d]	Cytovene plus Vitrasert	Oral plus surgical placement	Same as for drugs given individually		See individual drugs.

(*continued*)

Table 3 *Continued*

Drug	Brand name	Route	Induction dose	Maintenance dose	Major toxicity/concerns
Foscarnet	Foscavir	IV	90 mg/kg every 12 h for 14 days[a]	90–120 mg/kg/day[a]	Fluid and electrolyte abnormalities (calcium, phosphorus, magnesium, potassium); Nephrotoxicity; seizures; genital ulcers; central venous catheter infection/thrombosis.
		Intraocular injection[c,d]	2400 μg twice weekly	2400 μg/week	Retinal detachment; endophthalmitis; subconjunctival/intravitreal hemorrhage; risk of retinitis in other eye/CMV disease.
Cidofovir	Vistide	IV	5 mg/kg/week for 2 weeks[f,g]	5 mg/kg/week[f,g]	Nephrotoxicity (Fanconi's syndrome); neutropenia; fluid overload; uveitis/intraocular hypotony; neuropathy; adverse reactions to probenecid (fever, rash, nausea).

[a] Reduced dose needed in patients with abnormal renal function.
[b] Efficacy controversial; most authorities recommend using only in conjunction with ganciclovir intraocular implant for retinitis; not studied in other neurological diseases.
[c] Best results achieved if intraocular therapy is combined with concomitant systemic antiviral therapy, e.g., oral ganciclovir.
[d] Not indicated for CMV neurological diseases other than retinitis.
[e] Requires infusion pump and 2 h infusion time, as well as careful monitoring of blood chemistries with replacement as necessary.
[f] Given with probenecid 2 g orally 3 h prior, 2 h after, and 8 h after cidofovir, and infusion of 1 L normal saline 1 h prior and again with cidofovir.
[g] Contraindicated if $\geq$ 2+ protein on urinalysis, serum creatinine >1.5, or creatinine clearance $\leq$ 55 mL/min.

development of other systemic CMV disease have led authorities to recommend combining oral ganciclovir therapy with use of the implant [102]. Oral ganciclovir alone is not currently advised for the treatment of retinitis because extensive studies suggested a higher failure rate than with intravenous delivery of the drug, most likely reflecting the poor oral bioavailability of the drug [103]. Oral ganciclovir therapy has not been studied in other neurological syndromes. Ongoing studies are in progress to determine if oral ganciclovir might have a role in preventing CMV disease in high-risk patient populations [104,105].

The use of ganciclovir may be limited by adverse effects, most commonly bone marrow suppression. Patients should be monitored for evidence of anemia, leukopenia, or thrombocytopenia, and a few patients may require concomitant therapy with growth factors. Other complications reflect the use of central venous access and include a high risk of infection and sepsis as well as thrombosis or mechanical problems with the catheter. The drug dosage must be adjusted in patients with renal insufficiency.

Recently, an oral prodrug of ganciclovir, valganciclovir, was approved for treatment of CMV retinitis in AIDS patients. In a randomized, controlled study, valganciclovir was as effective as intravenous ganciclovir in induction therapy for CMV retinitis [106]. Pharmacokinetic studies demonstrate that the bioavailability of this prodrug is tenfold that of oral ganciclovir. Adverse events observed in patients taking valganciclovir are comparable to those seen with IV ganciclovir therapy. This drug appears to offer an exciting new option for the treatment of CMV retinitis that eliminates the need for intravenous line placement.

Foscarnet

Foscarnet is a pyrophosphate analog with antiviral activity against sensitive strains of CMV [107]. The drug inhibits DNA polymerase by directly binding to the enzyme and acting as a reversible competitive inhibitor. Because this mechanism of action does not require the presence of the protein kinase gene to phosphorylate the antiviral drug (as is necessary with ganciclovir), foscarnet is active against many ganciclovir-resistant strains of CMV. However, other mutations, especially of the viral DNA polymerase, can result in foscarnet resistance.

The efficacy of foscarnet has been well established for CMV retinitis. Like ganciclovir, it must initially be given in ''induction'' doses and then later continued in lower ''maintenance'' doses. Studies have indicated that ganciclovir and foscarnet have comparable efficacy in the treatment of CMV retinitis. Although one study indicated a possible survival advantage to foscarnet, this has not been observed in other studies. Furthermore, foscarnet therapy requires more aggressive medical monitoring and for this reason is often a second choice for therapy. Foscarnet has not been extensively studied for other forms of CMV neurological disease, but clinical experience suggests that its efficacy is comparable to that of ganciclovir. Adverse effects associated with foscarnet include nephrotoxicity and metabolic abnormalities such as hypophosphatemia, hypomagnesemia, and hypocalcemia. Patients require careful monitoring of their fluid and electrolyte status during therapy.

Several studies have suggested that the combination of ganciclovir and foscarnet may provide additional benefit, especially in patients who have failed treatment using either drug alone. Because the side effects of ganciclovir and foscarnet are distinct, it is possible to administer the combination therapy safely in many patients.

Cidofovir

Cidofovir is a nucleotide analog with activity against CMV via direct inhibition of the viral DNA polymerase after the drug is phosphorylated to the triphosphate form by cellular enzymes [108]. Like foscarnet, it is active against strains of CMV that are resistant to ganciclovir because of mutations in the UL97 gene. However, strains with mutations in both the UL97 gene and the DNA polymerase gene may be cross-resistant to both ganciclovir and cidofovir.

Cidofovir has been approved only for treatment of CMV retinitis in patients with AIDS [109] but is increasingly being used by practitioners for treatment of other forms of CMV disease. A major advantage of this drug is that it has an extremely long half-life and therefore needs to be administered only once per week during ''induction'' therapy and every 2 weeks for maintenance therapy. Probenecid (2 g po) should be given before each dose of cidofovir. Use of the drug has been limited, however, by concerns about nephrotoxicity [109]. The drug can cause a potentially irreversible destruction of proximal convoluted renal tubular cells, especially if it is not given with probenecid.

Fomivirsen

Fomivirsen is a phosphorothioate oligonucleotide that inhibits CMV replication through an antisense mechanism [110]. The drug has activity against strains of CMV that are resistant to ganciclovir and foscarnet. It is administered by intravitreal injection for treatment of CMV retinitis. Like other anti-CMV therapy, it must be given in induction doses followed by maintenance therapy. Fomivirsen is approved for the local treatment of CMV retinitis in patients who are unable to take other anti-CMV drugs or who have failed treatment. Side effects of fomivirsen include ocular inflammation (which may require therapy with topical steroids) and increased intraocular pressure.

Approaches to Specific Clinical Diseases

Unfortunately, there is no known effective therapy for congenital CMV disease. A role for antiviral therapy has not yet been established. Phase II studies of ganciclovir therapy in babies with symptomatic congenital cytomegalovirus disease demonstrated that drug therapy can decrease the amount of CMV excreted in the urine but that viruria returned to pretreatment levels after cessation of the drug [111]. Encouragingly, further hearing loss was prevented in 16% of babies, suggesting possible efficacy; studies of more prolonged therapy are ongoing.

Treatment of CMV retinitis requires aggressive antiviral therapy with ganciclovir, foscarnet, or cidofovir [112–115]. Initial therapy with ganciclovir can be delivered via the intravenous route, by direct intraocular instillation, or via a ganciclovir implant. Recent consensus statements favor the combination of ganciclovir implant plus oral ganciclovir (the latter for preemptive therapy of possible systemic disease) or, if the patient prefers, intravenous administration of the drug [18,27,116]. Intravenous foscarnet has been shown to have comparable efficacy for treatment of CMV retinitis and in one major study actually conferred a survival advantage [117,118]. However, many practitioners and patients relegate it to a second line option because of the difficulties in administration and potential adverse effects. Cidofovir has also been shown to have efficacy in CMV retinitis, but the potential for nephrotoxicity precludes its designation as a first line agent [109]. For AIDS patients with relapse of CMV retinitis, combination therapy with ganciclovir and foscarnet appears to be the most effective therapy [119]. Oral ganciclovir alone is not adequate for the treatment of CMV retinitis [115], presumably because of the failure to achieve adequate levels in the retina. Valganciclovir appears to offer an excellent option for treatment of CMV retinitis in AIDS patients.

Regardless of the drug chosen for initial therapy, it is critical that maintenance therapy be prescribed to prevent relapse. It was traditionally thought that maintenance therapy was required for life. However, new studies indicate that maintenance antiviral therapy may be safely discontinued for some, but not all, patients in whom the initiation of HAART results in effective immune reconstitution as reflected in rises in the $CD4^+$ cell count above 100 cells/mm^3 [29,120–122].

A trial of antiviral therapy should be considered for patients in whom a diagnosis of CMV encephalitis is suspected. It is important to maintain a high index of suspicion and initiate therapy with anti-CMV drugs as early as possible [40,115]. Unfortunately, response to therapy is generally suboptimal and of limited duration. There is no evidence that treatment of CMV encephalitis with foscarnet or ganciclovir extends survival [115]. Further studies to determine whether treatment with the combination of ganciclovir plus foscarnet improves prognosis are indicated [123]. Because of the difficulty in ascertaining the true role of the virus in many immunocompromised patients with encephalitis, it has been difficult to design trials of therapy for encephalitis.

For CMV-associated polyradiculopathy, treatment with ganciclovir or other antiviral drugs effective against CMV has been associated with improvement in some patients in whom therapy was initiated early in disease. Rapid recognition of the correct diagnosis and early initiation of anti-CMV drugs are necessary to prevent permanent neurological deficits secondary to nerve root necrosis [124,125]. Combination therapy with ganciclovir and foscarnet may be necessary [126–129], and cases of the syndrome caused by ganciclovir-resistant strains of CMV have been reported [130]. Although not well studied, it has been recommended that anti-CMV drugs be continued as maintenance after induction treatment is completed [131].

In patients with advanced immunosuppression, especially AIDS patients, a trial of anti-CMV therapy should be given to patients presenting with mononeuritis multiplex. Because the viral etiology may be difficult to establish, an empirical trial of ganciclovir or foscarnet should be considered in at risk populations with characteristic clinical findings, even if CMV has not been definitively shown to be the causative agent.

Treatment of Guillain Barré syndrome caused by CMV must focus on both supportive care and definitive antiviral therapy. For cases of Guillain Barré in which CMV is suspected as an etiology, a trial of antiviral therapy, usually with ganciclovir or foscarnet, should be given.

For patients with CMV-associated peripheral neuropathy, treatment with anti-CMV drugs may be efficacious if the neuropathy is indeed related to the virus. However, in many cases there is an incomplete response and treatment revolves around pain control and other supportive therapies. Pain relief with acetaminophen, nonsteroidal agents, and narcotics is often necessary. The use of tricyclic antidepressant drugs such as amitriptyline provides additive relief in some patients. Anticonvulsant drugs such as gabapentin or carbamazepine have been demonstrated to be efficacious in other forms of neuropathic pain and are often used in these patients. More trials are needed to determine the efficacy of topical agents such as topical capsaicin or topical lidocaine, and recombinant human nerve growth factor is being studied.

2.8 Prevention

Given the widespread prevalence of CMV, it is probably not reasonable to attempt to prevent primary disease in healthy hosts. However, handwashing and other routine approaches to minimizing exchange of secretions are wise general principles. The importance of these practices for preventing spread to CMV seronegative pregnant women and immunocompromised patients has been emphasized. In the hospital and other healthcare settings, the use of standard precautions will serve to protect healthcare workers and patients. However, there is no indication for instituting special infection control procedures for patients with known CMV infection, because the preponderance of individuals excreting CMV and capable of transmitting the virus to others do not have a diagnosis of CMV infection.

The risk of CMV infection in recipients of transplants can be minimized by the use of seronegative blood transfusions and seronegative transplanted organs. Because seronegative blood products may not be routinely available, the use of seronegative blood for transfusions has focused on the highest risk groups, specifically premature babies and CMV-positive recipients of bone marrow transplants from CMV-negative donors. The use of leukocyte-poor blood has also been shown to decrease the risk of CMV in immunocompromised patients [132].

Cytomegalovirus disease can also be reduced in transplant recipients by the use of prophylactic or preemptive ganciclovir or other antiviral therapy. Prophylactic regimens prescribe the use of the antiviral drug in all at-risk patients to reduce the risk of CMV disease or at least delay it until some immune recovery has occurred. With preemptive regimens, patients are monitored for early evidence of CMV infection, often with antigenemia or PCR tests, and therapy is instituted before clinical disease develops. Both of these approaches can substantially reduce the risk of symptomatic, life-threatening CMV disease.

Several trials have focused on preventive therapies of CMV disease in HIV patients. Unfortunately, treatment with oral ganciclovir did not show a definite survival advantage, and current guidelines do not recommend CMV prophylaxis for patients with HIV/AIDS.

Given the limitations of therapeutic and preventive interventions, the desirability of a vaccine against CMV is obvious. Despite extensive research, an effective vaccine has remained elusive. However, new advances in vaccine development hold hope for the future.

REFERENCES

1. Crumpacker, C.S. Cytomegalovirus. In Principles and Practices of Infectious Diseases, 5th ed.; Mandell, G.L., Bennett, J.E., Dolin, R., Eds.; Churchill Livingston: Philadelphia, 2000, 1586–1598.
2. Griffiths, P.D.; Emery, V.C. Cytomegalovirus. In Clinical Virology; Richman, D.D., Whitley, R.J., Hayden, F.G., Eds.; Churchill Livingstone: New York, 1997, 445–470.
3. Drew, W.L.; Lalezari, J.P. Cytomegalovirus: disease syndromes and treatment. Curr. Clin. Top. Infect. Dis. 1999, 19, 16–29.
4. Stagno, S. Cytomegalovirus. In Infectious Diseases of the Fetus and Newborn Infant, 4th ed; Remington, J.S., Klein, J.O., Eds.; WB Saunders: Philadelphia, 1995, 312–353.
5. Lange, M.; Klein, E.B.; Kornfield, H.; Cooper, L.Z.; Grieco, M.H. Cytomegalovirus isolation from healthy homosexual men. JAMA. 1984, 252, 1908–1910.
6. Pomeroy, C.; Englund, J.A. Cytomegalovirus: epidemiology and infection control. Am. J. Infect. Control. 1987, 15, 107–119.
7. Boppana, S.B.; Rivera, L.B.; Fowler, K.B.; Mach, M.; Britt, W.J. Intrauterine transmission of cytomegalovirus to infants of women with preconceptional immunity. N. Engl. J. Med. 2001, 344, 1366–1371.
8. Pickering, L.K. Infections in day care. Pediatric. Infect. Dis. J. 1987, 6, 614–617.

9. Jellinger, K.A.; Setinek, U.; Drlicek, M.; Bohm, G.; Steurer, A.; Lintner, F. Neuropathology and general autopsy findings in AIDS during the last 15 years. Acta. Neuropathol. 2000, 100, 213–220.
10. Wiley, C.A.; Schrier, R.D.; Denaro, F.J.; Nelson, J.A.; Lampert, P.W.; Oldstone, M.B. Localization of cytomegalovirus proteins and genome during fulminant central nervous system infection in an AIDS patients. J. Neuropathol. Exp. Neurol. 1986, 45, 127–139.
11. Emory, V.C. Investigation of CMV disease in immunocompromised patients. J. Clin. Pathol. 2001, 54, 84–88.
12. Soderberg-Naucler, C.; Fish, K.N.; Nelson, J.A. Reactivation of latent human cytomegalovirus by allogeneic stimulation of blood cells from healthy donors. Cell. 1997, 91, 119–126.
13. Naniche, D.; Oldstone, M.B. Generalized immunosuppression: how viruses undermine the immune response. Cell. Mol. Life Sci. 2000, 57, 1399–1407.
14. Bale, J.F. Human herpesviruses and neurological disorders of childhood. Semin. Pediatr. Neurol. 1999, 6, 247–287.
15. Bale, J.F. Human cytomegalovirus infection and disorders of the nervous system. Arch. Neurol. 1984, 41, 310–320.
16. Brown, H.L.; Abernathy, M.P. Cytomegalovirus. Semin. Perinatal. 1998, 22, 260–266.
17. Ahlfors, K.; Ivarsson, S.A.; Harris, S. Report on a long-term study of maternal and congenital CMV infection in Sweden. Scand. J. Infect. Dis. 1999, 31, 443–457.
18. Jacobson, M.A. Treatment of cytomegalovirus retinitis in patients with the acquired immunodeficiency syndrome. N. Engl. J. Med. 1997, 337, 105–114.
19. Baumal, C.R.; Levin, A.V.; Read, S.E. Cytomegalovirus retinitis in immunosuppressed children. Am. J. Opthalmol. 1999, 127, 550–558.
20. Bowen, E.F. Cytomegalovirus reactivation in patients infected with HIV: the use of polymerase chain reaction in prediction and management. Drugs. 1999, 57, 735–741.
21. Bowen, E.F.; Emery, V.C.; Wilson, P. Johnson, M.A.; Davey, C.C.; Sabin, C.A.; Farmer, D.; Griffiths, P.D. Cytomegalovirus polymerase chain reaction viraemia in patients receiving ganciclovir maintenance therapy for retinitis. AIDS. 1998, 12, 605–611.
22. Casado, J.L.; Arrizabalaga, J.; Montes, M.; Marti-Belda, P.; Tural, C.; Pinilla, J.; Gutierrez, C.; Portu, J.; Schuurman, R.; Aguirrebengoa, K. Incidence and risk factors for developing cytomegalovirus retinitis in HIV-infected patients receiving protease inhibitor therapy.Spanish CMV-AIDS Study Group AIDS. 1999, 13, 1497–1502.
23. Jabs, D.A.; Forman, M.; Enger, C.; Jackson, J.B. Comparison of cytomegalovirus loads in plasma and leukocytes of patients with cytomegalovirus retinitis. The Cytomegalovirus Retinitis and Viral Resistance Study Group. J. Clin. Microbiol. 1999, 37, 1431–1435.
24. Doan, S.; Cochereau, I.; Guvenisik, N.; Diraison, M.C.; Mousalatti, H.; Thanh, H.X. Cytomegalovirus retinitis in HIV-infected patients with and without highly active antiretroviral therapy. Am. J. Ophthalmol. 1999, 128, 250–251.
25. Casado, J.L.; Perez-Elias, M.J ; Marti-Belda, P.; Antela, A.; Suarez, M.; Ciancas, E.; Frutus, B.; Perez, M.D.; Guerrero, A. Improved outcome of cytomegalovirus retinitis in AIDS patients after introduction of protease inhibitors. J. Acquir. Immune. Defic. Syndr. Hum. Retrovirol. 1998, 19, 130–134.
26. Reed, J.B.; Schwab, I.R.; Gordon, J.; Morse, L.S. Regression of cytomegalovirus retinitis associated with protease inhibitor treatment in patients with AIDS. Am. J. Ophthalmol. 1997, 124, 199–205.
27. Martin, D.; Dunn, J.P.; Davis, J.L.; Duker, J.S.; Engstrom, R.E., Jr; Friedberg, G.J.; Jaffe, D.N.; Kuppermann, B.D.; Polis, M.A.; Whitley, R.J.; Wolitz, R.A.; Benson, C.A. Use of ganciclovir implant for the treatment of cytomegalovirus retinitis in the era of potent antiretroviral therapy: recommendations of an International AIDS Society-USA panel. Am. J. Ophthalmol. 1999, 127, 329–339.
28. Deayton, J.R.; Wilson, P.; Sabin, C.A.; Davey, C.C.; Johnson, M.A.; Emery, V.C.; Griffiths, P.D. Changes in the natural history of cytomegalovirus retinitis following the introduction of highly active antiretroviral therapy. AIDS. 2000, 14, 1163–1170.

29. Whitcup, S.M.; Fortin, E.; Lindblad, A.S.; Griffiths, P.; Metcalf, J.A.; Robinson, M.R.; Masnischewitz, M.R.; Baird, B.; Perry, C.; Kidd, I.M.; Vrabec, T.; Davey, R.T.; Falloon, J.; Walker, R.E.; Kovacs, J.A.; Lane, H.C.; Nussenblatt, R.B.; Smith, J.; Masur, H.; Polis, M.A. Cytomegalovirus retinitis after initiation of highly active antiretroviral therapy. Lancet. 1997, 349, 1443.
30. Jacobson, M.A.; Zegans, M.; Pavan, P.R.; O'Donnell, J.J.; Sattler, F.; Rao, N.; Owens, S.; Pollard, R. Cytomegalovirus retinitis after initiation of highly active antiretroviral therapy. Lancet. 1997, 349, 1443.
31. Jacobson, M.A.; Stanley, H.; Holtzer, C.; Margolis, T.P.; Cunningham, E.T. Natural history and outcome of new AIDS-related cytomegalovirus retinitis diagnosed in the era of highly active antiretroviral therapy. Clin. Infect. Dis. 2000, 30, 231–233.
32. Torriani, F.J.; Freeman, W.R.; Macdonald, J.C.; Karavellas, M.P.; Durand, D.M.; Jeffrey, D.D.; Meylan, P.R.A.; Schrier, R.D. CMV retinitis recurs after stopping treatment in virological and immunological failures of potent antiretroviral therapy. AIDS. 2000, 14, 173–180.
33. Zegans, M.E.; Walton, R.C.; Holland, G.N.; O'Donnell, J.J.; Jacobson, M.A.; Margolis, T.P. Transient vitreous inflammatory reactions associated with combination antiretroviral therapy in patients with AIDS and cytomegalovirus retinitis. Am. J. Ophthalmol. 1998, 125, 292–300.
34. Karavellas, M.P.; Lowder, C.Y.; Macdonald, C.; Torriani, F.J.; Shufelt, C.L.; Azen, S.P.; Freeman, W.R. Immune recovery vitritis associated with inactive cytomegalovirus retinitis: a new syndrome. Arch. Ophthalmol. 1998, 116, 169–175.
35. Eddleston, M.; Peacock, S.; Juniper, M.; Warrel, D.A. Severe cytomegalovirus infection in immunocompetent patients. Clin. Infect. Dis. 1997, 24, 52–56.
36. Phillips, A.C.; Fanning, W.L.; Diter, W.G.; Phillips, C.V. Cytomegalovirus encephalitis in immunologically normal adults. JAMA. 1977, 238, 2299–2300.
37. Prosch, S.; Schielke, E.; Reip, A.; Meisel, H.; Volk, H-D.; Einhaupl, K.M.; Kruger, D.H. Human cytomegalovirus (HCMV) encephalitis in an immunocompetent young person and diagnostic reliability of HCMV DNA PCR using cerebrospinal fluid of nonimmunosuppressed patients. J. Clin. Microbiol. 1998, 36, 3636--3640.
38. Studahl, M.; Ricksten, A.; Sandberg, T.; Elowson, S.; Herner, S.; Sall, C.; Berstrom, T. Cytomegalovirus infection of the CNS in non-compromised patients. Acta. Neurol. Scand. 1994, 89, 451–457.
39. Arribas, J.R.; Storch, G.A.; Clifford, D.B.; Tselis, A.C. Cytomegalovirus encephalitis. Ann. Intern. Med. 1996, 125, 577–587.
40. Roullet, E. Opportunistic infections of the central nervous system during HIV-1 infection (emphasis on cytomegalovirus disease). J. Neurol. 1999, 246, 237–243.
41. Milazzo, L.; Di Marco, A.; Gervasoni, C.; Rusconi, S.; Negri, C. Hearing loss as a presentation of CMV encephalitis in an HIV-1-infected patient. Infect. Med. 2000, 17, 495–499.
42. Grassi, M.P.; Clerici, F.; Perin, C.; Monforte, A.D.; Vago, L.; Borella, M.; Boldorini, R.; Mangoni, A. Microglial nodular encephalitis and ventriculoencephalitis due to cytomegalovirus infection in patients with AIDS: two distinct clinical patterns. Clin. Infect. Dis. 1998, 27, 504–508.
43. Kalayjian, R.C.; Cohen, M.L.; Bonomo, R.A.; Flanigan, T.P. Cytomegalovirus ventriculoencephalitis in AIDS: a syndrome with distinct clinical and pathologic features. Medicine. 1993, 72, 67–77.
44. McCutchan, J.A. Clinical impact of cytomegalovirus infections of the nervous system in patients with AIDS. Clin. Infect. Dis. 1995, 21((suppl 2)), S196–S201.
45. McCutchan, J.A. Cytomegalovirus infections of the nervous system in patients with AIDS. Clin. Infect. Dis. 1995, 20, 747–754.
46. Morgella, S.; Cho, E-S.; Nielsen, S.; Devinsky, O.; Petito, C.K. Cytomegalovirus encephalitis in patients with acquired immunodeficiency syndrome: an autopsy study of 30 cases and a review of the literature. Hum. Pathol. 1987, 18, 289–297.

47. Miles, C.; Hoffmann, W.; Lai, C-W. Cytomegalovirus-associated transverse myelitis. Neurology. 1993, 43, 2143–2145.
48. Giobbia, M.; Carniato, A. Scotton, P.G.; Marchiori, G.C.; Vaglia, A. Cytomegalovirus-associated transverse myelitis in a non-immunocompromised patient. Infection. 1999, 27, 228–230.
49. Shahid, M.B.; Ata Kan, M. Cytomegalovirus-associated transverse myelitis in a non-immunocompromised patient. J. Neurol. Sci. 1995, 134, 210–211.
50. Cohen, B.A.; McArthur, J.C.; Grohman, S.; Patterson, B.; Glass, J.D. Neurologic prognosis of cytomegalovirus polyradiculomyelopathy in AIDS. AIDS. 1993, 43, 493–499.
51. Said, G.; Lacroix, C.; Chemouilli, P.; Goulon-Goeau, C.; Roullet, E.; Penaud, ; Brouker, T.; Meduri, G.; Vincent, D.; Torchet, M. Cytomegalovirus neuropathy in acquired immunodeficiency syndrome: a clinical and pathological study. Ann. Neurol. 1991, 29, 139–145.
52. Roullet, E.; Assuerus, V.; Gozlan, J.; Ropert, A.; Said, G.; Baudrimont, M.; Jacomet, C.; Duvivier, C.; Gonzales-Canali, G. Cytomegalovirus multifocal neuropathy in AIDS: analysis of 15 consecutive cases. Neurology. 1994, 44, 2174–2182.
53. So, Y.T.; Olney, R.K. Acute lumbosacral polyradiculopathy in acquired immunodeficiency syndrome: experience in 23 patients. Ann. Neurol. 1994, 35, 53–58.
54. Behar, R.; Wiley, C.; McCutchan, A. Cytomegalovirus polyradiculoneuropathy in acquired immune deficiency syndrome. Neurology. 1987, 37, 557–561.
55. Olney, R.K. Acute lumbosacral polyradiculopathy in AIDS: experience in 23 patients. Ann. Neurol. 1994, 35, 53–58.
56. deGans, J.; Portegies, P.; Tiessens, G.; Troost, D.; Danner, S.A.; Lange, J.M. Therapy for cytomegalovirus polyradiculomyelitis in patients with AIDS: treatment with ganciclovir. AIDS. 1990, 4, 421–425.
57. Granter, S.R.; Doolittle, M.H.; Renshaw, A.A. Predominance of neutrophils in the cerebrospinal fluid of AIDS patients with cytomegalovirus radiculopathy. Am. J. Clin. Pathol. 1996, 105, 364–366.
58. Miller, R.F.; Fox, J.D.; Thomas, P.; Waite, J.C.; Sharvell, Y.; Gazzard, B.G.; Harrison, M.J.; Brink, N.S. Acute lumbosacral polyradiculopathy due to cytomegalovirus in advanced HIV disease: cerebrospinal fluid findings in 17 patients. J. Neurol. Neurosurg. Psychiatr. 1996, 61, 456–460.
59. Clifford, D.B.; Buller, R.S.; Mohammed, S.; Robison, L.; Storch, G.A. Use of polymerase chain reaction to demonstrate cytomegalovirus DNA in CSF of patients with human immunodeficiency virus infection. Neurology. 1993, 43, 75–79.
60. Mahieux, F.; Gray, F.; Fenelon, G.; Gherardi, R.; Admas, D.; Guillard, A. Acute myeloradiculitis due to cytomegalovirus as the initial manifestation of AIDS. J. Neurol. Neurosurg. Psychiatry. 1989, 52, 270–274.
61. Fuller, G.N. Cytomegalovirus and the peripheral nervous system in AIDS. J. Acquired. Immune. Defic. Syndr. 1992, 5(suppl 1), S33–S36.
62. Wolf, D.G.; Spector, S.A. Diagnosis of human cytomegalovirus central nervous system disease in AIDS patients by DNA amplification from cerebrospinal fluid. J. Infect. Dis. 1992, 166, 1412–1415.
63. Schmitz, H.; Enders, G. Cytomegalovirus as a frequent cause of Guillain-Barre syndrome. J. Med. Virol. 1977, 1, 21–27
64. deMaar, E.F.; Kas-Deelen, D.M.; de Jager, A.E.J.; The, H.; Tegzess, A.M.; van Son, W.J. Inflammatory demyelinating polyneuropathy in a kidney transplant patient with cytomegalovirus infection. Nephrol. Dial. Transplant. 1999, 14, 2228–2230.
65. Clifford, D.B.; Arribas, J.R.; Storch, G.A.; Tourtellote, W.; Wippold, F.J. Magnetic resonance brain imaging lacks sensitivity for AIDS associated cytomegalovirus encephalitis. J. Neurovirol. 1996, 2, 397–403.
66. Holland, N.R.; Power, C.; Mathews, V.P.; Glass, J.D.; Forman, M.; McArthur, J.C. Cytomegalovirus encephalitis in acquired immunodeficiency syndrome (AIDS). Neurology. 1994, 44, 507–514.

67. Price, T.A.; Digioia, R.A.; Simon, G.L. Ganciclovir treatment of cytomegalovirus ventriculitis in a patient infected with human immunodeficiency virus. Clin. Infect. Dis. 1992, 15, 606–608.
68. Mussini, C.; Mongiardo, N.; Manicardi, G.; Trenti, F.; Alessandri, A.; Paolillo, F.; Catania, A.; Portolani, M.; Pecorari, M.; Borghi, V.; Ficarra, G.; Cossarizza, A.; Rienzo, B. Relevance of clinical and laboratory findings in the diagnosis of cytomegalovirus encephalitis in patients with AIDS. Eur. Clin. Microbiol. Infect. Dis. 1997, 16, 437–444.
69. Burke, D.G.; Leonard, D.G.B.; Imperiale, T.F.; Valdez, H.; Karaman, B.; Shick, E.; Kalayjian, R.C. The utility of clinical and radiographic features in the diagnosis of cytomegalovirus central nervous system disease in AIDS patients. Mol. Diagn. 1999, 4, 37–43.
70. Moulignier, A.; Mikol, J.; Gonzalez-Canali, G.; Polivka, M.; Pialoux, G.; Welker, Y.; Alain, S.; Thiebaut, J.B.; Dupont, B. AIDS-associated cytomegalovirus infection mimicking central nervous system tumors: a diagnostic challenge. Clin. Infect. Dis. 1996, 22, 626–631.
71. Anders, H.J.; Goebel, F.D. Cytomegalovirus polyradiculopathy in patients with AIDS. Clin. Infect. Dis. 1998, 2, 345–352.
72. Ginocchio, C.C. Laboratory diagnosis of human cytomegalovirus (HCMV) central nervous system disease in AIDS patients. Int. J. Antimicrob Agents. 2000, 16, 447–453.
73. Gleaves, C.A.; Smith, T.F.; Shuster, A.; Pearson, G.R. Rapid detection of cytomegalovirus in MRC-5 cells inoculated with urine specimens by using low speed centrifugation and monoclonal antibody to early antigen. J. Clin. Microbiol. 1984, 19, 917–919.
74. Rabella, N.; Drew, W.L. Comparison of conventional and shell vial cultures for detecting cytomegalovirus infection. J. Clin. Microbiol. 1990, 28, 806–807.
75. Weinberg, A.; Spiers, D.; Cai, G.-Y.; Long, C.M.; Sun, R.; Tevere, V. Evaluation of a commercial PCR kit for diagnosis of cytomegalovirus infection of the central nervous system. J. Clin. Microbiol. 1998, 36, 338–384.
76. Vinters, H.V.; Kwok, M.K.; Ho, H.W.; Anders, K.H.; Tomiyasu, U.; Wolfson, W.L.; Robert, F. Cytomegalovirus in the nervous system of patients with the acquired immune deficiency syndrome. Brain. 1989, 112, 245–268.
77. Boeckh, M.; Boivin, G. Quantitation of cytomegalovirus: methodologic aspects and clinical applications. Clin. Microbiol. Rev. 1998, 11, 533–554.
78. Revello, M.G.; Percivalle, E.; Sarasini, A.; Baldanti, F.; Furione, M.; Gerne, G. Diagnosis of human cytomegalovirus infection of the nervous system by pp65 detection in polymorphonuclear leukocytes of cerebrospinal fluid from AIDS patients. J. Infect. Dis. 1994, 170, 1275–1279.
79. Flood, J.; Drew, W.L.; Miner, R.; Jekic-McMullen, D.; Shen, L.P.; Kolberg, J.; Garvey, J.; Follansbee, S.; Poscher, M. Diagnosis of cytomegalovirus polyradiculopathy and documentation of in vivo anti-CMV activity in cerebrospinal fluid by using branched DNA signal amplification and antigen assays. J. Infect. Dis. 1997, 176, 348–352.
80. Cinque, P.; Vago, L.; Brytting, M.; Castagna, A.; Accordini, A.; Sundqvist, V.-A.; Zanchetta, N.; Monforte, A.D.; Wahren, B.; Lazzarin, A.; Linde, A. Cytomegalovirus infection of the central nervous system in patients with AIDS: diagnosis by DNA amplification from cerebrospinal fluid. J. Infect. Dis. 1992, 166, 1408–1411.
81. Gozlan, J.; Amrani, M.E.; Baudrimont, M.; Costagliola, D.; Salord, J.M.; Duvivar, C.; Picard, O.; Meyohas, M.-C.; Jacomet, C.; Schneider-Fauveau, V.; Petit, J.-C.; Roullet, E. A prospective evaluation of clinical criteria and polymerase chain reaction assay of cerebrospinal fluid for the diagnosis of cytomegalovirus-related neurological diseases during AIDS. AIDS. 1995, 9, 253–260.
82. Gozlan, J.; Salord, J.-M.; Roullet, E.; Baudrimont, M.; Caburet, F.; Picard, O.; Meyohas, M.-C.; Duvivier, C.; Jacomet, C.; Petit, J.-C. Rapid detection of cytomegalovirus DNA in cerebrospinal fluid of AIDS patients with neurologic disorders. J. Infect. Dis. 1992, 166, 1416–1421.

83. Linde, A.; Klapper, P.E.; Monteyne, P.; Echevarri, J.M.; Cinque, P.; Rozenberg, F.; Vestergaard, B.F.; Ciardi, M.; Lebon, P.; Cleator, G.N. and the EU Concerted Action on Virus Meningitis and Encephalitis. Specific diagnostic methods for herpesvirus infections of the central nervous system: a consensus review by the European Union Concerted Action on Virus Meningitis and Encephalitis. Clin. Diagn. Virol. 1997, 8, 83–104.
84. Cohen, B.A. Prognosis and response to therapy of cytomegalovirus encephalitis and meningomyelitis in AIDS. Neurology. 1996, 46, 444–450.
85. Kuhn, J.E.; Wendland, T.; Eggers, H.J.; Lorentzen, E.; Wieland, U.; Eing, B.; Kiessling, M.; Gass, P. Quantitation of human cytomegalovirus genomes in the brain of AIDS patients. J. Med. Virol. 1995, 47, 70–82.
86. Wolf, D.G.; Spector, S.A. Diagnosis of human cytomegalovirus central nervous system disease in AIDS patients by DNA amplification from cerebrospinal fluid. J. infect. Dis. 1992, 166, 1412–1415.
87. Achim, C.L.; Nagra, R.M.; Wang, R.; Nelson, J.A.; Wiley, C.A. Detection of cytomegalovirus in cerebrospinal fluid autopsy specimens from AIDS patients. J. Infect. Dis. 1994, 169, 623–627.
88. Minjolle, S.; Michelet, C.; Jusselin, I.; Joannes, M.; Cartier, F.; Colimon, R. Amplification of the six major human herpesviruses from cerebrospinal fluid by a single PCR. J. Clin. Microbiol. 1999, 37, 950–953.
89. Quereda, C.; Corral, I.; Laguna, F.; Valencia, M.E.; Tenorio, A.; Echeverria, J.E.; Navas, E.; Martin-Davila, P.; Moreno, A.; Moreno, V.; Gonzalez-Lahoz, J.M.; Arribas, J.R.; Guerrero, A. Diagnostic utility of a multiplex herpesvirus PCR assay performed with cerebrospinal fluid from human immunodeficiency virus-infected patients with neurological disorders. J. Clin. Microbiol. 2000, 38, 3061–3067.
90. Yamamoto, T.; Nakamura, Y. A single tube PCR assay for simultaneous amplification of HSV-1/-2, CMV, HHV-6A/-6B, and EBV DNAs in cerebrospinal fluid from patients with virus related neurological diseases. J. Neurovirol. 2000, 6, 410–417.
91. Arribas, J.R.; Clifford, D.B.; Fichtenbaum, C.J.; Commins, D.L.; Powderly, W.G.; Storch, G.A. Level of cytomegalovirus (CMV) DNA in cerebrospinal fluid of subjects with AIDS and CMV infection of the central nervous system. J. Infect. Dis. 1995, 172, 527–531.
92. Fillet, A.M.; Katlama, C.; Camilleri, S.; Rogequx, O.; Huraux, J.M. Human CMV infection of the CNS: concordance between PCR detection in CSF and pathologic examination. AIDS. 1993, 7, 1016–1018.
93. Wildemann, B.; Hass, J.; Lynen, N.; Stingele, K.; Storch-Hagenlocher, B. Diagnosis of cytomegalovirus encephalitis in patients with AIDS by quantitation of cytomegalovirus genomes in cells of cerebrospinal fluid Neurology. 1998, 50, 693–697.
94. Cinque, P.; Baldanti, F.; Vago, L.; Terreni, M.R.; Lillo, F.; Furione, M.; Castagna, A.; Monforte, A.D.; Lazzarin, A.; Linde, A. Ganciclovir therapy for cytomegalovirus infection of the central nervous system in AIDS patients: monitoring by CMV DNA detection in cerebrospinal fluid. J. Infect. Dis. 1995, 171, 1603–1606.
95. Shinkai, M.; Spector, S.A. Quantitation of human cytomegalovirus DNA in cerebrospinal fluid by competitive PCR in AIDS patients with different HCMV central nervous system diseases. Scand. J. Infect. Dis. 1995, 27, 559–561.
96. Wattanamano, P.; Clayton, J.L.; Kopicko, J.J.; Kissinger, P.; Elliot, S.; Jarrott, C.; Rangan, S.; Beilke, M.A. Comparison of three assays for cytomegalovirus detection in AIDS patients at risk for retinitis. J. Clin. Microbiol. 2000, 38, 727–732.
97. Zhang, F.; Tetali, S.; Wang, X.P.; Kaplan, M.H.; Cromme, F.V.; Ginocchio, C.C. Detection of human cytomegalovirus pp67 late gene transcripts in cerebrospinal fluid of human immunodeficiency virus type 1-infected patients by nucleic acid sequence-based amplification. J. Clin. Microbiol. 2000, 38, 1920–1925.
98. Crumpacker, C.S. Ganciclovir. N. Engl. J. Med. 1996, 335, 721–729.

99. Erice, A. Resistance of human cytomegalovirus to antiviral drugs. Clin. Microbiol. Rev. 1999, 12, 286–297.
100. Martin, D.F.; Parks, D.J.; Mellow, S.D.; Ferris, F.L.; Walton, R.C.; Remaley, N.A.; Chew, E.Y.; Ashton, P.; Davis, M.D.; Nussenblatt, R.B. Treatment of cytomegalovirus retinitis with intraocular sustained–release ganciclovir implant. A randomized controlled clinical trial. Arch. Ophthalmol. 1994, 112, 1531–1539.
101. Musch, D.C.; Martin, D.F.; Gordon, J.F.; Davis, M.D.; Kuppermann, B.D. and the Ganciclovir Implant Study Group. Treatment of cytomegalovirus retinitis with a sustained release ganciclovir implant. N. Engl. J. Med. 1997, 337, 83–90.
102. Martin, D.F.; Kuppermann, B.D.; Wolitz, R.A.; Palestine, A.G.; Li, H.; Robinson, C.A. and the Roche Ganciclovir Study Group. Oral ganciclovir for patients with CMV retinitis treated with a ganciclovir implant. N. Engl. J. Med. 1999, 340, 1063–1070.
103. Drew, W.L.; Ives, D.; Lalezari, J.P.; Crumpacker, C.; Follansbee, S.E.; Spector, S.A.; Benson, C.A.; Friedberg, D.N.; Hubbard, L.; Stempien, et al., M.J. Oral ganciclovir as maintenance treatment for cytomegalovirus retinitis in patients with AIDS. N. Engl. J. Med. 1995, 333, 615–620.
104. Brosgart, C.L.; Louis, T.A.; Hillman, D.W.; Craig, C.P.; Alston, B.; Fisher, E.; Abrams, D.I.; Luskin-Hawk, R.L.; Sampson, J.H.; Ward, D.J.; Thompson, M.A.; Torres, R.A. A. randomized placebo-controlled trial of the safety and efficacy of oral ganciclovir for prophylaxis of cytomegalovirus disease in HIV–infected individuals. AIDS. 1998, 12, 269–277.
105. Spector, S.A.; McKinley, G.F.; Lalezari, J.P.; Samo, T.; Andruczk, R.; Follansbee, S.; Sparti, P.D.; Havlir, D.V.; Simpson, G.; Buhles, W.; Wong, R.; Stempien, M. Oral ganciclovir for the prevention of cytomegalovirus disease in persons with AIDS. N. Engl. J. Med. 1996, 334, 1491–1497.
106. Jabs, D. CMV and herpesvirus infections. Presented at the 8th Conference on Retroviruses and Opportunistic Infections: Chicago; Feb 4–8, 2001, Session 66.
107. Jayaweera, D.T. Minimising the dosage-limiting toxicities of foscarnet induction therapy. Drug. Saf. 1997, 16, 258–266.
108. Plosker, G.L.; Noble, S. Cidofovir: a review of its use in cytomegalovirus retinitis in patients with AIDS. Drugs. 1998, 58, 325–345.
109. Studies of the Ocular Complications of AIDS Research Group, AIDS Clinical Trials Group. Long term follow–up of patients with AIDS treated with parenteral cidofovir for cytomegalovirus retinitis: the HPMPC Peripheral Cytomegalovirus Retinitis Trial. AIDS. 2000, 14, 1571–1581.
110. Hoffman, V.F.; Skiest, D.J. Therapeutic developments in cytomegalovirus retinitis. Expert. Opin. Investig. Drugs. 2000, 9, 207–220.
111. Whitley, R.J.; Cloud, G.; Gruber, W.; Storch, G.A.; Demmler, G.J.; Jacobs, R.F.; Dankner, W.; Spector, S.A.; Starr, S.; Pass, R.F.; Stagno, S.; Britt, W.J.; Alford, C.; Soong, S.-J.; Zhou, X.-J.; Sherrill, L.; FitzGerald, J.M.; Sommadossi, J.-P. and the National Institute of Allergy and Infectious Diseases Collaborative Antiviral Study Group. Ganciclovir treatment of symptomatic congenital cytomegalovirus infection: results of a phase II study. J. Infect. Dis. 1997, 175, 1080–1086.
112. Akerele, T.; Lightman, S. Current and novel agents for the treatment of cytomegalovirus retinitis. Drugs. 1999, 2, 289–297.
113. Holbrook, J.T.; Davis, M.D.; Hubbard, L.D.; Martin, B.K.; Holland, G.N.; Jabs, D.A.; Gilpin, A.K.; Meinert, C.; Reshef, D.S. Risk factors for advancement of cytomegalovirus retinitis in patients with acquired immunodeficiency syndrome. Studies of Ocular Complications of AIDS Research Group. Arch. Ophthalmol. 2000, 118, 1196–1204.
114. Ives, D.V. Cytomegalovirus disease in AIDS. AIDS. 1997, 11, 1791–1797.
115. Whitley, R.J.; Jacobson, M.A.; Friedberg, D.N.; Holland, G.N.; Jabs, D.A.; Dieterich, D.T.; Hardy, W.D.; Polis, M.A.; Deutsch, T.A.; Feinberg, J.; Spector, S.A.; Walmsley, S.; Drew, W.L.; Powderly, W.G.; Griffiths, P.D.; Benson, C.A.; Kessler, H.A. Guidelines for the treat-

ment of cytomegalovirus diseases in patients with AIDS in the era of potent antiretroviral therapy: recommendations of an international panel. International AIDS Society—USA. Arch. Intern. Med. 1998, 158, 957–969.
116. Phillips, P. Management of cytomegalovirus retinitis in AIDS. Eur. J. Clin. Microbiol. Infect. Dis. 2000, 19, 571–584.
117. Studies of the Ocular Complications of AIDS Research Group, AIDS Clinical Trials Group. Mortality in patients with the acquired immunodeficiency syndrome treated with either foscarnet or ganciclovir for cytomegalovirus retinitis. N. Engl. J. Med. 1992, 326, 213–220.
118. Studies of the Ocular Complications of AIDS Research Group, AIDS Clinical Trials Group. Foscarnet-Ganciclovir Cytomegalovirus Retinitis Trial. 4. Visual outcomes. Ophthalmology. 1994, 101, 1250–1261.
119. Studies of the Ocular Complications of AIDS Research Group, AIDS Clinical Trials Group. Combination foscarnet and ganciclovir therapy vs monotherapy for treatment of relapsed cytomegalovirus retinitis in patients with AIDS: the Cytomegalovirus Retreatment Trial. Arch. Ophthalmol. 1996, 114, 23–33.
120. Whitcup, S.M. Cytomegalovirus retinitis in the era of highly active antiretroviral therapy. JAMA. 2000, 283, 653–657
121. Johnson, S.C.; Benson, C.A.; Johnson, D.W.; Weinberg, A. Recurrences of cytomegalovirus retinitis in a HIV-infected patient, despite antiretroviral therapy and apparent immune reconstitution. Clin. Infect. Dis. 2001, 32, 815–819.
122. Komanduri, K.V.; Feinberg, J.; Hutchins, R.K.; Frame, R.D.; Schmidt, D.K.; Viswanathan, M.N.; Kalezari, J.P.; McCune, J.M. Loss of cytomegalovirus-specific CD4 + T cell responses in human immunodeficiency type 1-infected patients with high CD4 + T cell counts and recurrent retinitis. J. Infect. Dis. 2001, 183, 1285–1289.
123. Anduze-Faris, B.M.; Fillet, A.-M.; Gozlan, J.; Lancar, R.; Boukli, N.; Gasnault, J.; Caumes, E.; Livartowsky, J.; Matheron, S.; Leport, C.; Salmon, D.; Costagliola, D.; Katlama, C. Induction and maintenance therapy of cytomegalovirus central nervous system infection in HIV-infected patients. AIDS. 2000, 14, 517–524.
124. deGans, J.; Portegies, P.; Tiessens, G.; Troost, D.; Danner, S.A.; Lange, J.M. Therapy for cytomegalovirus polyradiculomyelitis in patients with AIDS: treatment with ganciclovir. AIDS. 1990, 4, 421–425.
125. Miller, R.G.; Storey, J.R.; Greco, C.M. Ganciclovir in the treatment of progressive AIDS-related polyradiculopathy. Neurology. 1990, 40, 569–574.
126. Anders, H.J.; Weiss, N.; Bogner J.R.; Goebel, F.D. Ganciclovir and foscarnet efficacy in AIDS-related CMV polyradiculopathy. J. Infect. 1998, 36, 29–33.
127. Decker, D.F.; Tarver, J.H.; Murray, D.F.; Martin, G.J. Prolonged concurrent use of ganciclovir and foscarnet in the treatment of polyradiculopathy due to cytomegalovirus in a patient with AIDS. Clin. Infect. Dis. 1994, 19, 548–549.
128. Jacobson, M.A.; Mills, J.; Rush, J. O'Donnell, J.J.; Miller, R.G.; Greco, C. Failure of antiviral therapy for acquired immunodeficiency syndrome-related cytomegalovirus myelitis. Arch. Neurol. 1988, 45, 1090–1092.
129. Karmochkine, M.; Molina, J.M.; Scieux, C. Combined therapy with ganciclovir and foscarnet for cytomegalovirus polyradiculopathy in patients with AIDS. Am. J. Med. 1994, 97, 196–197.
130. Tokumoto, J.I.N.; Hollander, H. Cytomegalovirus polyradiculopathy caused by a ganciclovir resistant strain. Clin. Infect. Dis. 1993, 17, 854–856.
131. Kim, Y.S.; Hollander, H. Polyradiculopathy due to cytomegalovirus: report of two cases in which improvement occurred after prolonged therapy and review of the literature. Clin. Infect. Dis. 1993, 17, 32–37.
132. Kader, J.G. Leukocyte-reduced blood products. Curr. Opin. Hematol. 1998, 5, 381–385.

9

Role of Human Herpesvirus Type 6 in Neurological Disease

Michael Mayne
University of Manitoba
Winnipeg, Manitoba, Canada

Steven Jacobson
National Institutes of Health
Bethesda, Maryland, U.S.A.

1 INTRODUCTION

The human herpesvirus type 6 (HHV-6) is a lymphotropic beta-herpesvirus first isolated from the peripheral blood of immunocompromised patients with lymphoproliferative disorders [1]. Infection with HHV-6 typically occurs under the age of 3, and primary infection can account for 10–40% of hospitalizations of children in this age group. HHV-6 is a clinically relevant virus and is the causative agent of exanthem subitum, a pediatric fever, and skin rash (*roseola*) [2] that can have serious and fatal complications [3]. Seroprevalence in the general population is greater than 90% [4,5]; however, reactivation usually occurs only in immunocompromised adults. Reactivation can be associated with serious consequences and like the closely related cytomegalovirus (CMV) has been associated with organ, bone marrow, and peripheral blood cell (PBC) transplantation failure and engraftment inhibition (for reviews see Refs. 4 and 6). HHV-6 infection or reactivation has also been implicated as a pathogenic agent during HIV replication and has been suggested to play a role in HIV/AIDS progression [7,8].

In addition to its role in pediatric febrile illness and transplant rejection, HHV-6 infection has also been associated with central nervous system (CNS) complications including neuroinflammation, febrile seizures, and encephalitis/encephalopathy [6]. In immunocompetent adults, HHV-6 is considered a commensal virus of the CNS [9,10]. However, HHV-6 has been linked with the pathogenesis of two chronic progressive demyelinating diseases of the CNS, multiple sclerosis (MS) and progressive multifocal leukoencephalopathy (PML) [9]. The findings in MS are based on immunological, molecular, and histological studies [9,11–19]. Despite the association of HHV-6 with these clinical disorders, the pathological mechanisms regulated by HHV-6 have yet to be defined. Virushost interactions that are currently being explored include analysis of viral variants, genetic susceptibility loci, and virus-specific immune responses including virus-specific activation of proinflammatory events.

2 HHV-6A AND 6B TAXONOMY, GENETIC ANALYSIS, AND TROPISM

2.1 Taxonomy

During the search for novel viruses associated with hematalogical malignancies or acquired immune deficiency syndrome (AIDS), Salahuddin and colleagues isolated a novel human herpesvirus that they called human B-cell lymphotropic virus (HBLV) [1]. Since this discovery, HBLV has been renamed human herpesvirus type 6 (HHV-6) and determined to have a wide tropic range predominantly because HHV-6 uses the ubiquitously expressed CD46 as its receptor [20]. HHV-6 is a member of the *Herpesviridae* and most likely descended from a common ancestral virus that is widespread in vertebrates including fish, snakes, birds, and mammals. Almost exclusively, herpesviruses are specific for one host; however, one strain of monkey herpesvirus, herpesvirus B, can cross the species barrier and infect humans, causing acute encephalitis and death [21,22].

2.2 Genetic Analysis

Genomic analysis shows that HHV-6 is a member of the *Herpesviridae* family and is related to cytomegalovirus (CMV) and human herpesvirus type 7 (HHV-7). The two HHV-6 variants share approximately 97% sequence homology [23], and specific regions within the genome can vary at the nucleotide level by as much as 25% [24], suggesting that these viruses may encode proteins with varying functions. HHV-6 is an enveloped extracellular virus with an approximate diameter of 160–200 nm [25] (Fig. 1). The HHV-6 genome for both variants has been sequenced recently [23,26]. The HHV-6 genome is approximately 160 kilobases (kb) in length and encodes at least 115 open reading frames (ORFs) [26]. However, recent evidence indicates that HHV-6 uses nonconserved splice sites to generate unique mRNA constructs [27], suggesting that herpesvirus may encode more than the estimated 115 genes. Further extensive reviews of the HHV-6 genomes can be found elsewhere [26,28].

2.3 Tropism

Two major viral subgroups of HHV-6 have been defined and are designated variants A and B. Although there is significant DNA sequence homology between the two variants, each has distinctive genomic, antigenic, and biological properties [29–31]. The prototypical HHV-6B variant is Z29, which was isolated from a Zairian AIDS patient [32]. Variant B can be further subcategorized into groups 1 and 2. HHV-6B is found primarily in the peripheral blood, saliva, and lymph nodes of healthy individuals and has been detected in serum of children with roseola [2]. HHV-6A is detected less frequently than HHV-6B in healthy adults. A highly studied, lab-adapted strain of HHV-6A is U1102, a strain that was isolated from a Ugandan AIDS patient [1]. Variant A is found primarily in the skin,

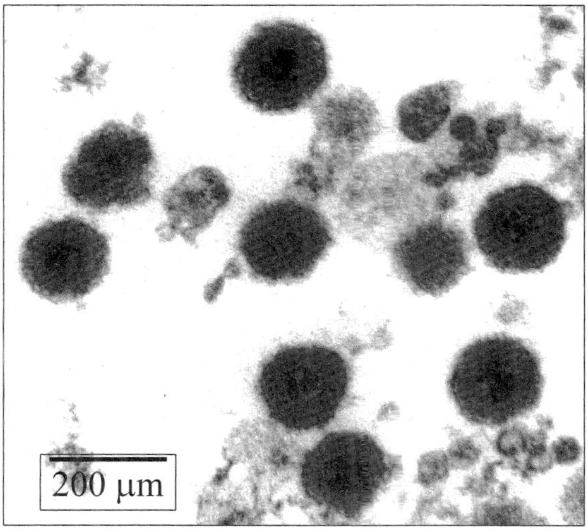

Figure 1 HHV-6 virions of approximately 200 nm are detected by electron microscopy in human lymphocyte cell line Sup T1 infected with HHV-6 B (Z-29). Samples were prepared according to Frenkel et al. (Ref. 79).

brain, and cerebrospinal fluid (CSF). Little is known about the epidemiological and geographical distribution of HHV-6A [7]. A greater neurotropism of the HHV-6A variant has been suggested because of the detection of HHV-6A in the CSF of children and adults [33] and because it was detected recently in the CNS of AIDS patients with areas of demyelination [9]. Increased HHV-6A-specific immune responses as well as the detection of HHV-6A-specific DNA sequences in the serum, urine, and PBL of MS patients support involvement of the HHV-6A variant in this disorder [16,34,35].

3 ROLE IN CLINICAL DISEASE

3.1 Epidemiology

Most adults in North America and Europe are seropositive for HHV-6, and the virus appears to be common in most populations throughout the world [36]. HHV-6 is often shed in the saliva of asymptomatic seropositive children and adults, and this is the most likely mode of transmission [37]. The peak age of acquisition in North America is 6–12 months (median 9 months), and the mother and baby usually have the same strain of virus. Perinatal or congenital infection has not been detected [38]. HHV-6 infection accounts for the majority of roseola infections in the United States (97%) [39]. The closely related beta-herpesvirus human herpesvirus type 7 (HHV-7) has a median age of acquisition of 26 months, and acquisition of this virus is associated with different risk factors than HHV-6 [40].

3.2 Roseola

The only disease for which HHV-6 has been clearly shown to be the causative agent is exanthem subitum in children [2]. This is a common childhood disease first described in

1910 by Zahorsky as roseola infantum and later described in 1921 by Veeder and colleagues as a specific pathological entity called exanthem subitum [41]. Roseola is characterized in children with a constant or intermittent fever of at least 40°C (104°F) for 3–5 days in a child who appears to be relatively well. There may be mild otitis media and pharyngeal mucosa. Roseola is so named because of the appearance of a rose pink macular rash on the neck and trunk of the infant. This rash appears following the return to normal temperature and routinely fades within 24–48 h.

Over 50 years ago, a possible viral etiology for roseola was proposed; however, it was not until the late 1980s that Yamanishi et al. [2] showed that HHV-6 was the viral agent that caused roseola. Variant B is responsible for almost all roseola-associated HHV-6 infection [39,42], and at least 50% of primary episodes of infant fever are due to HHV-6 infection [43]. Primary HHV-6 infections are associated with a rash in greater than 60% of infants in Japan, whereas in the United States only one in four infants appears to develop a rash [41]. Importantly, in a study conducted in 1994 by Hall and colleagues, it was found that 20% of all visits to emergency rooms for febrile illnesses among infants 6–12 months of age were due to HHV-6 infections [44], and moreover 1 in 10 of these patients required hospitalization.

3.3 HHV-6 Reactivation in Immunocompromised Adults

Although the majority of the western population is infected with HHV-6 as children, infection in healthy adults is rare. However, reactivation or primary infection in adults, when it occurs, can be associated with significant consequences. In immunocompromised individuals, specifically organ or stem cell transplant recipients or in cancer patients, HHV-6 often reactivates or new infection occurs as indicated by increased antibody titers or increased frequency of detection of HHV-6 DNA using the polymerase chain reaction (PCR). In organ transplant recipients, HHV-6 reactivation usually occurs 1 month following transplantation and can be associated frequently with CMV [45] and HHV-6 reactivation has been associated with organ rejection. In severe cases of HHV-6 reactivation in transplant recipients, encephalitis can occur, sometimes with fatal consequences (see detailed information in Sec. 4). However, in prospective studies, HHV-6 reactivation in transplant recipients is often limited and may cause a mild fever or rash [46,47]. HHV-6 reactivation is clearly opportunistic in immunosuppressed patients and can infect or reactivate in AIDS patients and have serious implications in recipients of major organ or stem cell transplants, where HHV-6 viremia can lead to organ rejection or failure and can in some cases be fatal [6,48,49].

4 HHV-6 ASSOCIATED WITH NEUROLOGICAL DISEASE

4.1 HHV-6-Associated Encephalitis

In children, central nervous system manifestations were recognized long before HHV-6 was clearly implicated as the etiological agent of roseola. Infants may show symptoms including bulging fontanels, irritability, febrile seizures, meningoencephalitis, and residual encephalopathy (for a review see Ref. 41). However, the prognosis for HHV-6 infection in young children is excellent and only rarely does significant neurological disease occur. In a prospective examination of 2716 children with primary HHV-6 infection, the HHV-6A variant was identified more frequently in CSF samples from children with acute febrile illness than in peripheral blood mononuclear cell (PBMC) isolates from children with primary infection [33], suggesting a greater neurotropism of HHV-6A than HHV-6B.

Several recent reports have implicated HHV-6 as a pathogenic agent following solid organ or bone marrow transplantation [46,47,50–52], and in rare cases patients have developed acute encephalitis or other neurological disease [52–55]. Clearly, reactivation and infection from donor organs can occur, and although HHV-6A and HHV-6B have been detected post-transplantation, HHV-6B is the predominant variant that is detected, suggesting that reactivation of HHV-6 may occur in the host. Prospective studies show that there is little clinical impact on the success of organ transplants [50,56]. However, subgroup analysis has linked HHV-6 reactivation with delayed engraftment, graft-versus-host disease, and rash [7]. We and others have postulated that HHV-6 involvement in organ rejection and potentially death is via induction of proinflammation (see current research, Sec. 4.2). In support of this hypothesis, it was reported recently that liver transplantation induced adhesion molecule expression [57], and we reported recently that HHV-6 infection (variant A or B) activates several proinflammatory mechanisms [58].

4.2 HHV-6 and HIV

The role of HHV-6 as a cofactor in HIV disease and AIDS has received considerable attention. It has been suggested that HHV-6 directly enhances HIV replication by breaking latency [8] and also acts as a pathogenic factor in HIV replication through several mechanisms [7,8,37,59,60]. There have been multiple reports of direct interactions between HIV and HHV-6. HHV-6 infection increases CD4 expression [60] (see Sec. 5), possibly rendering cells vulnerable to further HIV infection. Upon autopsy, HIV victims and often found to have higher HHV-6 viral load throughout the body than non-HIV-infected controls [7]. However, HHV-6 is lower in the peripheral blood of HIV patients with lower CD4 cell counts, suggesting that HHV-6 levels in blood may not play an important role in HIV replication [61,62]. HHV-6 on its own cannot induce immuno deficiency, whereas HIV can; thus, it is likely that HHV-6 may act only as an opportunistic agent in immunocompromised HIV patients, much as it does in organ transplant patients.

4.3 HHV-6 Association with Multiple Sclerosis

There is accumulating evidence that links HHV-6 with MS pathogenesis. Etiologically, genetic factors including race, sex, ethnicity, family history, and HLA haplotype and nongenetic factors such as age, weather, diet, and socioeconomic status are all linked with development of MS [63]. However, there is clear evidence that one or more infectious agents may be associated with MS. Perhaps the strongest epidemiological evidence that the rate of concordance is eight times greater in monozygotic than in dizygotic twins [64]; however, the concordance rate among monozygotic twins remains only 25%. Experimentally, oligoclonal bands specific for virus or bacterial epitopes can be found in high frequencies in cerebrospinal fluid (CSF) of MS patients, suggesting the presence of an infectious agent. Elevations in CSF immunoglobulins, which demonstrate an oligoclonal pattern when separated by electrophoretic methods, are found in >90% MS patients [65]. When observed in CNS infections, oligoclonal bands are specific for the infectious agent, and oligoclonal bands have therefore long been considered something of an MS ''holy grail'' that may aid in the identification of agents that cause MS. The two most recent infectious agents linked with MS are *C. pneumoniae* and HHV-6, and both have been postulated to be involved in disease development in specific subsets of MS patients.

One of the first reports of HHV-6 as a possible causative agent in MS came from the detection of viral DNA in MS plaques [13] by means of nonbiased research using representational difference analysis (RDA) that allowed the selection and amplification of previously unknown DNA sequences present by means of successive cycles of subtractive hybridization and subsequent PCR amplification. When applied to material from MS brains, RDA amplified a DNA fragment that was determined to be similar to a specific gene from the Z29 isolate of HHV-6B. Consistent with data demonstrating CNS as a site of HHV-6 latency [66], the percentage of HHV-6 DNA–positive MS brains was not significantly different from that of control brains (78% vs, 74%, respectively). However, monoclonal antibodies against the HHV-6 virion proteins 101K and p41 were able to detect HHV-6 antigen expression in MS plaques and not in control brains. This study had a significant impact in the MS research community and opened the way for new avenues of research investigating the role of HHV-6 as an etiological agent in MS.

Recently we extended the work of Challoner and colleagues by using a nested PCR approach to identify HHV-6 DNA from laser-dissected plaque regions identified within MS autopsy specimens (Fig. 2). Because the nested PCR approach can be fraught with contamination problems, we used a blinded protocol to determine the frequency of HHV-6 DNA in these samples. We reasoned that if the PCR specimens were to become contaminated we would not see a significant difference between our test groups. Following sample isolation and nested PCR, samples were decoded, and below we outline and review our data. Using this approach, we found that HHV-6 DNA was detected in 57% of MS lesions, a significantly greater number than in normal-appearing white matter from MS material or brain specimens from non-MS control groups with other inflammatory diseases and normal brains. Other neurological controls included stroke, Alzheimer's cerebral vasculitis, schizophrenia, brain astrocytoma and lymphoma, fatal gunshot to the cerebrum, septicemia-induced encephalitis, septicemia-induced cerebral infarction, and subacute sclerosing panencephalitis. Because it was reported recently that HHV-6-specific proteins are immunolocalized in MS lesions [67], we sought to determine if HHV-6 mRNA may be present in these plaques. We reasoned that the presence of U83 mRNA, a late expressing chemokine-like gene of HHV-6 [68], in MS plaques indicates that HHV-6 is expressing late proteins and may be active. Further, the presence of U83 mRNA is a first indication of whether U83 protein is present in MS lesions and is thus involved in HHV-6-associated inflammation. U83 has sequence similarities with human CC type chemokines including MIP and RANTES, is released from COS cells transfected with a construct encoding the U83 open reading frame, and can induce calcium mobilization and chemotactic activation of THP-1 (mononuclear) cells [68]. We tested four individual autopsy samples of freshly frozen MS brain from the NIH Brain Banks (Bethesda, MD and Rocky Mountain, MT) for the presence of U83 mRNA. A neuropathologist identified several MS plaques and normal-appearing white matter regions from each brain specimen. To detect U83 mRNA, nucleic acid sequence base amplification (NASBA) was performed [69]. This technique amplifies target mRNA by generating antisense RNA via a T_7 promoter that is attached to a U83-specific primer sequence. NASBA amplification showed that 8/9 plaques were positive for U83 mRNA whereas only 1/6 normal white matter was positive (Fig. 3). We could not, however, detect U83 mRNA using NASBA in PBMCs from MS patients and controls (not shown), supporting the hypothesis that detection of HHV-6 RNA in MS plaques was not due to contaminating inflammatory cells in these lesions. Together, these results suggest that HHV-6 was present during the development and progression of the

MS lesion and thus may be involved in the pathogenesis of MS. Current research is focused on determining the extent to which HHV-6 is active in MS lesions.

A recent study by Blumberg et al. [9] demonstrated the presence of HHV-6 DNA in chronic MS white matter plaques using a two-step in situ PCR (ISPCR) technique. This technique was able to colocalize the virus to specific CNS cell types including oligodendrocytes. HHV-6 DNA was also unexpectedly detected in white matter lesions from PML brains at higher amounts than JC virus, the etiological agent of PML. These findings added to the debate on whether HHV-6 can be considered a commensal agent in the brain or a potentially active virus in MS. Another investigation performed on brain tissues from patients having either secondary progressive or relapsing/remitting MS demonstrated the presence of HHV-6 DNA in 17 of 19 diseased tissue sections and in three of 23 uninvolved regions [67]. The presence of viral DNA was statistically significant in tissues from MS patients (8 of 11) with respect to control CNS tissues (2 of 28). Furthermore, 54% of total blood samples from MS patients were positive for active HHV-6 infection, as demonstrated by a rapid culture assay [67]. Interestingly, the incidence of active HHV-6 viremia decreased in patients with longer duration of the disease, possibly reflecting a shift in pathogenic or host mechanisms.

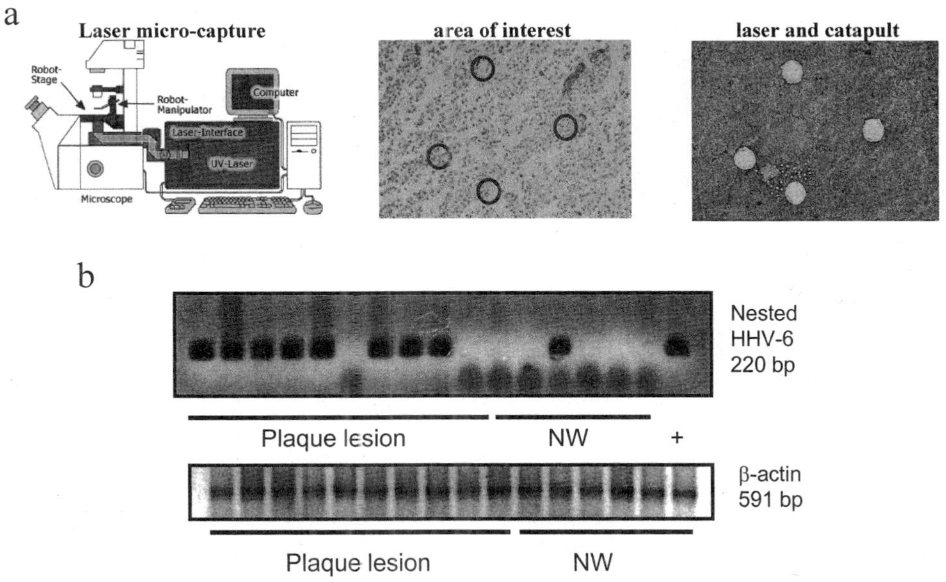

Figure 2 PCR amplification of DNA extracted from LAMC dissected formalin-fixed, paraffin-embedded brain tissue. (a) Cartoon demonstrating LAMC apparatus used to collect laser-dissected sections. Areas of interest were identified in the embedded tissue and dissected using microscopy-assisted laser dissection (a; right panel). (b) Only DNA of sufficient quality, as determined by amplification of β-actin sequences, was used in further analysis for HHV-6 DNA. Representative gel showing nested PCR products (220 bp) in laser-dissected plaque regions (8/10) compared to normal white matter (NWM; 1/6). +, DNA from Sup T1 cells infected with Z-29 HHV-6B. PCR products for β-actin (591 bp) show that approximately equal amounts of starting DNA were used in each reaction

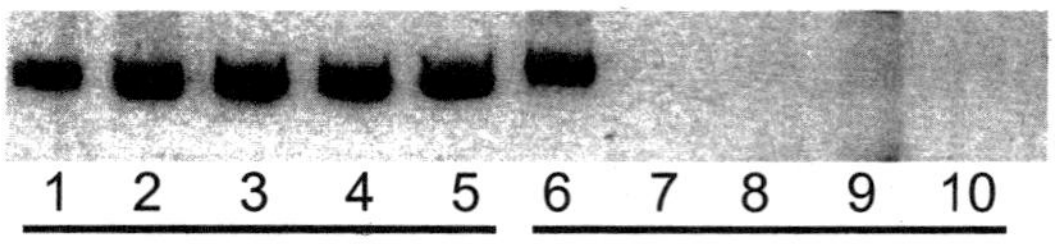

Figure 3 Nucleic acid sequence base amplification (NASBA) detection of U83 mRNA in MS plaques. Total RNA (200 ng) was used as a starting template for NASBA amplification according to the manufacture's protocol (Qiagen). Shown are amplified RNA products from MS plaques from two patients (lanes 1–5) compared to RNA amplification products from normal white matter from the same patients (lanes 6–10). Lanes 1–3 and 6–8 are from the same patient, and lanes 4, 5, 9, and 10 are from the second patient. Figure is representative of two independent experiments. Results demonstrate that U83 is present with higher frequency in plaque lesions than in normal white matter within the same patient

The detection of HHV-6 in MS lesions is an important step in an attempt to make an association between this agent and the pathogenesis of MS. However, additional studies will be required to support this observation. Investigations have also focused on the humoral and cellular immune response to HHV-6 in MS patients and controls. There is a significant increase in the HHV-6 anti-p41/38 early antigen IgM response in MS patients compared to healthy subjects and other neurological disease controls [18]. Two additional studies confirmed the presence of increased IgM responses to HHV-6 in patients with MS [11,19], but no correlation was demonstrated in another report [70].

A recent study examined the T-cell lymphoproliferative responses of healthy controls and patients with MS to HHV-6A, HHV-6B, and HHV-7 [16]. There was no difference in either the frequency or magnitude of proliferative responses between MS patients and healthy controls to either the HHV-6B variant or HHV-7. However, a significantly higher percentage of patients with MS (66%) than of healthy controls (33%) had proliferative responses to the HHV-6A variant. This supports the hypothesis that HHV-6, particularly the A variant, plays a role in MS. It is unknown whether the increased frequency of lymphoproliferative response to the HHV-6A lysate in patients with MS is the result of a higher seroprevalence of the A variant in MS patients or in an altered host immune response [16]. Moreover, a majority of the studies that have focused on the immune response to HHV-6 appear to support an association of this virus in MS.

In contrast to the analysis of humoral and cellular immune responses to HHV-6 in MS, there is less consensus in the literature when PCR-based studies were used to discriminate MS patients from controls (Table 1). HHV-6 DNA was demonstrated in the serum of a subset of MS patients (15 of 50) and not in serum from normal subjects and patients with other inflammatory and other neurological diseases (0 of 47) [18]. Detection of HHV-6 DNA in serum had been shown previously to correlate with active HHV-6 infection [71]. This observation was extended to a larger cohort of MS patients ($n = 167$) in which HHV-6 DNA sequences continue to be detected in a subset of MS patients. This may reflect the disease status of patients, because there appears to be a correlation with MS exacerbations and the presence of HHV-6 DNA in serum. However, attempts to support these results have given equivocal results [72–74] and may be due to several factors, including the ability to distinguish between latent and active infection, differences in the populations studied, different sensitivities in the technologies used, and DNA template sequences for PCR amplification. As outlined in Table 1, the range in the literature for the frequency of HHV-6 in PBMC from normal donors is 5–98%. Given such a disparity in results using PCR-based systems, it is not surprising that consensus among different groups has been difficult to obtain.

The presence of different HHV-6 variants may also account for contrasting results. HHV-6B variant has been detected in cell-associated compartments such as saliva and PBMC at comparable frequencies from normal donors and MS patients. In contrast, the HHV-6A variant has been demonstrated in serum and urine of MS patients but not controls [34]. Ablashi et al. [72] demonstrated that PBMCs from MS patients were mostly variant B (87%), similar to isolates from healthy donors (67%) These data are in agreement with the higher frequency of HHB-6B variant infection in the normal population. Because HHV-6A has been suggested to be more neurotropic [33], it is reasonable to hypothesize that the association of HHV-6 and MS might be variant-specific [16,34,35]

Table 1 Frequency of Detection of HHV-6 in Human Tissues

Tissue	Researchers (year)/Ref.	MS cases (%)	Controls (%)
Brain	Challoner *et al.* (1995)/13	25/32 (78)	40/54 (74)
		Oligos (IHC)	Oligos (—)
	Sanders *et al.* (1996)/15	21/47 (57)	6/37 (43)
		Active plaques 7/22 (32)	
		Inactive plaques 5/29 (17)	
		Normal matter 9/37 (24)	
	Blumberg *et al.* (2000)/9	White matter plaques 11/13	6/8
		high (ISPCR)	low (ISPCR)
CSF	Wilborn *et al.* (1994)	3/21 (14)	0/26 (0)
	Liedetke *et al.* (1995)	4/36 (11)	0/24 (0)
	Martin *et al.* (1997)	0/20 (0)	0/6 (0)
	Fillet *et al.* (1998)	2/32 (6)	0/21 (0)
	Ablashi *et al.* (1998)/11	2/12 (17)	0/4 (0)
	Goldberg *et al.* (1999)	0/6 (0)	0/14 (0)
	Locatelli *et al.* (2000)	10/29 (34)	0/7 (0)
	Taus *et al.* (2000)	0/7 (0)	0/9 (0)
PBMC	Sola *et al.* (1993)	1/31 (3)	1/24 (4)
	Mayne *et al.* (1998)/73	10/40 (25)	5/24 (21)
	Kim *et al.* (2000)/35	7/34 (21)	0/20 (0)
	Jarrett *et al.* (1990)	NT	18/20 (90)
	Sandhoff *et al.* (1991)	NT	4/44 (9)
	Cone *et al.* (1993)	NT	18/20 (90)
	Aubin *et al.* (1994)	NT	15/300 (5)
	Alberle *et al.* (1996)	NT	43/44 (98)
	Hall *et al.* (1998)/33	NT	40/170 (24)
Serum	Wilborn *et al.* (1994)	0/21 (0)	0/21 (0)
	Soldan *et al.* (1997)/18	15/50 (30)	0/47 (0)
	Martin *et al.* (1997)	0/20 (0)	0/20 (0)
	Merelli *et al.* (1997)	3/56 (5)	0/20 (0)
	Fillet *et al.* (1998)	2/32 (6)	1/34 (3)
	Goldberg *et al.* (1998)	1/24 (4)	0/30 (0)
	Locatelli *et al.* (2000)	5/28 (18)	0/22 (0)

5 CURRENT RESEARCH

5.1 HHV-6-Induced Changes in Gene Expression

To characterize the cellular response to HHV-6 infection, we evaluated the molecular gene expression profile of HHV-6-infected T cells using a novel immunomicroarray system. This customized human immunomicroarray was composed almost entirely of known gene sequences that were selected from the human cDNA library from Research Genetics. The immunomicroarray consisted of interleukin ligands and receptors, chemokines, and cellular signaling molecules. An entire listing can be found at www.grc.nia.nih.gov/branches/rrb/dna/dna.htm. We summarize and review our findings here that, independent of the HHV-6 variant studied, HHV-6 induced the gene expression of multiple proinflammatory molecules, in particular IL-18 and CD4, at the mRNA and protein levels. In contrast, anti-inflammatory cytokines, specific chemokine receptors, and members of the presenilin and amyloid beta-processing pathway were downregulated [58]. HHV-6 variant–specific gene expression profiles were also identified.

HHV-6 Infection of SupT1 Cells Induces a TH1-Type Immune Response

To determine the changes in gene expression by HHV-6 infection, we conducted in vitro experiments in which the T-cell lymphoblast line SupT1 was infected with the HHV-6A (GS) and 6B (Z29) variants. SupT1 cells were infected with equal amounts of HHV-6 as determined by cytopathic effect (CPE), percentage of cells expressing HHV-6 antigens as defined by immunofluorescence assay (IFA), and quantitative TaqMan analysis of HHV-6 viral DNA [58]. Immunomicroarray gene expression profiles were determined from HHV-6A- and 6B-infected T cells and compared to that of uninfected cells. Changes in gene expression were reported as Z-ratios that reflected the representative change in gene expression compared to uninfected control cells. As shown in Table 2, a number of genes were elevated by at least 2 Z-ratios in HHV-6-infected cells. Several proinflammatory genes had increased mRNA expression including IL-18, MAPK and JAK family members, TNF-α receptors, HLA class members, and BCL-6. Upregulation of IL-18, also known as IFN-γ-inducing protein, is of interest because MS patients have increased IFN-γ levels [75] and increased activation in caspase-1, the converting enzyme that regulates IL-18 [76,77]. Genes that were downregulated by at least 2 Z-ratios during HHV-6 infection included the anti-inflammatory cytokines IL-10 and IL-14, CXCR5, members of the presenilin and amyloid precursor protein family, and the recently identified measles receptor SLAM (CDw150) [78] (Table 3). Importantly, we found that IL-18 protein levels, as demonstrated by ELISA, were elevated significantly in supernatants from HHV-6A- and HHV-6B-infected T-cell cultures and that IL-10 protein levels (anti-inflammatory cytokine) was decreased significantly in supernatants from T cells infected with HHV-6A or 6B [58]. These findings support the use of microarray technology as an accurate indicator of cell physiology and also suggest that HHV-6 infection of SupT1 cells induces a TH1 type of immune response.

6 CONCLUSION

Although there has been considerable basic and clinical study in the field of MS research, the pathogenesis of MS has yet to be well defined. Etiologically, MS has been considered a multifactorial disease that develops in a susceptible host and may lead to aberrant immune responses triggered by environmental factors. Several lines of evidence suggest that a virus may comprise this environmental component. Although many viruses have been

Table 2 HHV-6-Induced Increase in Proinflammatory Gene Expression

Z-ratio			
6A/CON	6B/CON	Gene	Description
2.47	2.66	PFKP	Phosphofructokinase, platelet
1.55	2.62	PRKM1	Protein kinase, mitogen-activated 1 (MAP kinase 1)
1.86	2.60	TRAF3	TNF receptor-associated factor 3; CD40 binding protein
2.47	2.55	CD4	CD4
2.78	2.46	WSX-1	Class I cytokine receptor
1.60	2.44	SSI-1	JAK binding protein
1.96	2.38	TNFRSF11A	Tumor necrosis factor receptor superfamily, member 11a, NFKB activator
2.25	2.30	MAPK11	Mitogen-activated protein kinase 11
3.12	2.20	SCYB11	Small inducible cytokine subfamily B (Cys-X-Cys), I-TAC
2.03	2.02	IL18	Interleukin 18 (interferon-gamma-inducing factor)
1.73	1.91	BCL6	B-cell CLL/lymphoma 6
1.58	1.73	HLA-DQB1	Major histocompatibility complex, class II, DQ beta 1
2.12	1.63	HLA-A	Major histocompatibility complex, class I, A
1.94	1.53	HLA-DMB	Major histocompatibility complex, class II, DM beta 1

Z-ratios shown represent averages from two independent experiments performed in duplicate. Table is sorted based on high Z-ratios comparing SupT1 cells infected with HHV-6B (Z29) or variant 6A (GS) compared to uninfected cells (CON).
Source: Ref. 58.

proposed as etiological agents in MS, it is of interest that no one virus has ever been definitively associated with disease pathogenesis. Because MS may not be one disease but rather a syndrome with multiple etiologies, so too might there be multiple triggers. How such agents trigger disease is not known. It is possible that individual viral agents may induce a virus-specific and/or a cross-reactive autoimmune process resulting in clinical disease in a subset of genetically susceptible individuals. The involvement of multiple infectious agents in MS was suggested originally over 100 years ago and may explain the difficulty in identifying a single viral agent responsible for this highly variable and chronic disease. In addition to an overview of HHV-6, we have attempted to present the evidence for the role of HHV-6 in the pathogenesis of MS, not to suggest that this may be the ‘‘cause’’ of the disease but rather that this virus may be one of the environmental factors that may be associated with MS disease pathogenesis. Mechanisms by which virus host interactions may lead to demyelination are currently being investigated, particularly in murine models of MS. Moreover, caution is warranted when making associations between a virus and a chronic, progressive neurological disorder such as MS, because it is difficult to distinguish cause from effect, particularly when ubiquitous viral agents are involved. Uniformity in assay design, viral isolation techniques, and research reagents must be employed by different research groups on a large number of MS cohorts to confirm these virus associations. Perhaps, only through well-controlled clinical antiviral therapeutic trials with defined clinical, virological, and radiographic outcome measures will we be able to determine the role, if any, that viruses play in the pathogenesis of MS.

Table 3 HHV-6 Regulated Decreases in Proinflammatory-Associated Gene Expression

Z-ratio			
6A/CON	6B/CON	Gene	Description
−3.02	−3.41	RYK	RYK receptor-like tyrosine kinase
−1.83	−3.40	APPBP1	Amyloid beta precursor protein-binding protein 1, 59 kDa
−2.76	−3.34	ATF2	Activating transcription factor 4; CREB2
−3.56	−3.23	MDM2	Mouse double minute 2, human homolog of p53-binding protein
−2.98	−3.23	CXCR5	Burkitt lymphoma receptor 1, GTP-binding protein; CXCR5
−2.95	−3.19	IL14	Interleukin-14 mRNA
−2.94	−2.92	APLP1	Amyloid beta (A4) precursor-like protein 1
−2.38	−2.85	SCYA25	Small inducible cytokine subfamily A (Cys-Cys), MPIF1
−2.89	−2.76	PSEN2	Presenilin 2
−2.27	−2.74	PSEN1	Presenilin 1
−2.70	−2.42	HSPA1A	Heat shock 70 kDa protein 1
−2.22	−2.35	CDKN1A	Cyclin-dependent kinase inhibitor 1A
−2.11	−2.27	TEFS2	Transcription elongation factor S-II, hS-II-T1, complete cds
−1.66	−2.25	SLAM	Signaling lymphocytic activation molecule
−1.51	−2.18	IL10	Interleukin 10
−2.13	−2.06	APLP2	Amyloid beta (A4) precursor-like protein 2
−2.28	−2.03	APBA1	Amyloid beta (A4) precursor protein-binding, family A, member 1; X11

Z-ratios shown represent averages from two independent experiments performed in duplicate. Table is sorted based on low Z-ratios comparing SupT1 cells infected with HHV-6 (Z29 or GS) compared to uninfected cells (CON).
Source: Ref. 58.

REFERENCES

1. Salahuddin, S.Z.; Ablashi, D.V.; Markham, P.D. Isolation of a new virus, HBLV, in patients with lymphoproliferative disorders. Science. 1986, 234, 596–601.
2. Yamanishi, K.; Okuno, T.; Shiraki, K. Identification of human herpesvirus-6 as a causal agent for exanthem subitum [see comments]. Lancet. 1988, 1, 1065–1067.
3. Asano, Y.; Yoshikawa, T.; Kajita, Y. Fatal encephalitis/encephalopathy in primary human herpesvirus-6 infection. Arch. Dis. Child. 1992, 67, 1484–1485.
4. Campadelli-Fiume, G.; Mirandola, P.; Menotti, L. Human herpesvirus 6: an emerging pathogen. Emerg. Infect. Dis. 1999, 5, 353–366.
5. Di Luca, D.; Mirandola, P.; Ravaioli, T.; Bigoni, B.; Cassai, E. Distribution of HHV-6 variants in human tissues. Infect. Agents Dis. 1996, 5, 203–214.
6. Yoshikawa, T.; Asano, Y. Central nervous system complications in human herpesvirus-6 infection. Brain. Dev. 2000, 22, 307–314.
7. Clark, D.A. Human herpesvirus 6. Rev. Med. Virol. 2000, 10, 155–173.
8. Knox, K.K.; Carrigan, D.R. Active HHV-6 infection in the lymph nodes of HIV-infected patients: in vitro evidence that HHV-6 can break HIV latency. J. Acquir. Immune. Defic. Syndr. Hum. Retrovirol. 1996, 11, 370–378.

9. Blumberg, B.M.; Mock, D.J.; Powers, J.M. The HHV6 paradox: ubiquitous commensal or insidious pathogen? A two-step in situ PCR approach. J. Clin. Virol. 2000, 16, 159–178.
10. Cuomo, L.; Trivedi, P.; Cardillo, M.R. Human herpesvirus 6 infection in neoplastic and normal brain tissue. J. Med. Virol. 2001, 63, 45–51.
11. Ablashi, D.V.; Lapps, W.; Kaplan, M.; Whitman, J.E.; Richert, J.R.; Pearson, G.R. Human herpesvirus-6 (HHV-6) infection in multiple sclerosis: a preliminary report. Mult. Scler. 1998, 4, 490–496.
12. Akhyani, N.; Berti, R.; Brennan, M.B. Tissue distribution and variant characterization of human herpesvirus (HHV)-6: increased prevalence of HHV-6A in patients with multiple sclerosis. J. Infect. Dis. 2000, 182, 1321–1325.
13. Challoner, P.B.; Smith, K.T.; Parker, J.D. Plaque-associated expression of human herpesvirus 6 in multiple sclerosis. Proc. Natl. Acad. Sci. USA. 1995, 92, 7440–7444.
14. Ongradi, J.; Rajda, C.; Marodi, C.L.; Csiszar, A.; Vecsei, L. A pilot study on the antibodies to HHV-6 variants and HHV-7 in CSF of MS patients. J. Neurovirol. 1999, 5, 529–532.
15. Sanders, V.J.; Felisan, S.; Waddell, A.; Tourtellotte, W.W. Detection of herpesviridae in postmortem multiple sclerosis brain tissue and controls by polymerase chain reaction. J. Neurovirol. 1996, 2, 249–258.
16. Soldan, S.S.; Leist, T.P.; Juhng, K.N.; McFarland, H.F.; Jacobson, S. Increased lymphoproliferative response to human herpesvirus type 6A variant in multiple sclerosis patients. Ann. Neurol. 2000, 47, 306–313.
17. Soldan, S.S.; Jacobson, S. Role of viruses in etiology and pathogenesis of multiple sclerosis. In: Advances in Virus Research; Buchmeier, M., Campbell, I., Eds.; Academic Press: San Diego, 2001, 511–552.
18. Soldan, S.S.; Berti, R.; Salem, N. Association of human herpes virus 6 (HHV-6) with multiple sclerosis: increased IgM response to HHV-6 early antigen and detection of serum HHV-6 DNA [see comments]. Nat. Med. 1997, 3, 1394–1397.
19. Friedman, J.E.; Lyons, M.J.; Cu, G. The association of the human herpesvirus-6 and MS. Mult. Scler. 1999, 5, 355–362.
20. Santoro, F.; Kennedy, P.E.; Locatelli, G.; Malnati, M.S.; Berger, E.A.; Lusso, P. CD46 is a cellular receptor for human herpesvirus 6. Cell. 1999, 99, 817–827.
21. Fierer, J.; Bazely, P.; Braude, A.I. Herpes B virus encephalomyelitis presenting as ophthalmic zoster. A possible latent infection reactivated. Ann. Intern. Med. 1973, 79, 225–228.
22. Kaplan, J.E. Herpesvirus simiae (B virus) infection in monkey handlers. J. Infect. Dis. 1988, 157, 1090.
23. Gompels, U.A.; Nicholas, J.; Lawrence, G. The DNA sequence of human herpesvirus-6: structure, coding content, and genome evolution. Virology. 1995, 209, 29–51.
24. Chou, S.; Marousek, G.I. Analysis of interstrain variation in a putative immediate-early region of human herpesvirus 6 DNA and definition of variant-specific sequences. Virology. 1994, 198, 370–376.
25. Biberfeld, P.; Kramarsky, B.; Salahuddin, S.Z.; Gallo, R.C. Ultrastructural characterization of a new human B lymphotropic DNA virus (human herpesvirus 6) isolated from patients with lymphoproliferative disease. J. Natl. Cancer. Inst. 1987, 79, 933–941.
26. Isegawa, Y.; Mukai, T.; Nakano, K. Comparison of the complete DNA sequences of human herpesvirus 6 variants A and B. J. Virol. 1999, 73, 8053–8063.
27. French, C.; Menegazzi, P.; Nicholson, L.; Macaulay, H.; DiLuca, D.; Gompels, U.A. Novel, nonconsensus cellular splicing regulates expression of a gene encoding a chemokine-like protein that shows high variation and is specific for human herpesvirus 6. Virology. 1999, 262, 139–151.
28. Inoue, N.; Dambaugh, T.R.; Pellett, P.E. Molecular biology of human herpesviruses 6A and 6B. Infect Agents Dis. 1993, 2, 343–360.
29. Ablashi, D.V.; Balachandran, N.; Josephs, S.F. Genomic polymorphism, growth properties, and immunologic variations in human herpesvirus-6 isolates. Virology. 1991, 184, 545–552.

30. Aubin, J.T.; Collandre, H.; Candotti, D. Several groups among human herpesvirus 6 strains can be distinguished by Southern blotting and polymerase chain reaction [published erratum appears in J. Clin. Microbiol. 30(9):2524 1992.]. J. Clin. Microbiol. 1991, 29, 367–372.
31. Pellett, P.E.; Sanchez-Martinez, D.; Dominguez, G. A strongly immunoreactive virion protein of human herpesvirus 6 variant B strain Z29: identification and characterization of the gene and mapping of a variant-specific monoclonal antibody reactive epitope. Virology. 1993, 195, 521–531.
32. Black, J.; Sanderlin, K.; Goldsmith, C.; Gary, H.; Lopez, C.; Pellett, P. Growth properties of human herpesvirus-6 strain Z29. J. Virol. Methods. 1989, 26, 133–145.
33. Hall, C.B.; Caserta, M.T.; Schnabel, K.C. Persistence of human herpesvirus 6 according to site and variant: possible greater neurotropism of variant A. Clin. Infect. Dis. 1998, 26, 132–137.
34. Akhyani, N.; Berti, R.; Brennan, M.B. Tissue distribution and variant characterization of human herpesvirus (HHV)-6: increased prevalence of HHV-6A in patients with multiple sclerosis. J. Infect. Dis. 2000, 182, 1321–1325.
35. Kim, J.S.; Lee, K.S.; Park, J.H.; Kim, M.Y.; Shin, W.S. Detection of human herpesvirus 6 variant A in peripheral blood mononuclear cells from multiple sclerosis patients. Eur. Neurol. 2000, 43, 170–173.
36. Ranger, S.; Patillaud, S.; Denis, F. Seroepidemiology of human herpesvirus-6 in pregnant women from different parts of the world. J. Med. Virol. 1991, 34, 194–198.
37. Levy, J.A.; Ferro, F.; Greenspan, D.; Lennette, E.T. Frequent isolation of HHV-6 from saliva and high seroprevalence of the virus in the population. Lancet. 1990, 335, 1047–1050.
38. Stoeckle, M.Y. The spectrum of human herpesvirus 6 infection: from roseola infantum to adult disease. Annu. Rev. Med. 2000, 51, 423–430.
39. Dewhurst, S.; McIntyre, K.; Schnabel, K.; Hall, C.B. Human herpesvirus 6 (HHV-6) variant B accounts for the majority of symptomatic primary HHV-6 infections in a population of U. S. infants. J. Clin. Microbiol. 1993, 31, 416–418.
40. Caserta, M.T.; Hall, C.B.; Schnabel, K.; Long, C.E.; D'Heron, N. Primary human herpesvirus 7 infection: a comparison of human herpesvirus 7 and human herpesvirus 6 infections in children. J. Pediatr. 1998, 133, 386–389.
41. Kimberlin, D.W.; Whitley, R.J. Human herpesvirus-6: neurologic implications of a newly-described viral pathogen. J. Neurovirol. 1998, 4, 474–485.
42. Dewhurst, S.; Chandran, B.; McIntyre, K.; Schnabel, K.; Hall, C.B. Phenotypic and genetic polymorphisms among human herpesvirus-6 isolates from North American infants. Virology. 1992, 190, 490–493.
43. Asano, Y.; Yoshikawa, T.; Suga, S. Clinical features of infants with primary human herpesvirus 6 infection (exanthem subitum, roseola infantum). Pediatrics. 1994, 93, 104–108.
44. Hall, C.B.; Long, C.E.; Schnabel, K.C. Human herpesvirus-6 infection in children. A prospective study of complications and reactivation. N. Engl. J. Med. 1994, 331, 432–438.
45. DesJardin, J.A.; Gibbons, L.; Cho, E. Human herpesvirus 6 reactivation is associated with cytomegalovirus infection and syndromes in kidney transplant recipients at risk for primary cytomegalovirus infection. J. Infect. Dis. 1998, 178, 1783–1786.
46. Cone, R.W.; Huang, M.L.; Corey, L.; Zeh, J.; Ashley, R.; Bowden, R. Human herpesvirus 6 infections after bone marrow transplantation: clinical and virologic manifestations. J. Infect. Dis. 1999, 179, 311–318.
47. Lau, Y.L.; Peiris, M.; Chan, G.C.; Chan, A.C.; Chiu, D.; Ha, S.Y. Primary human herpes virus 6 infection transmitted from donor to recipient through bone marrow infusion. Bone Marrow Transplant. 1998, 21, 1063–1066.
48. Buchbinder, S.; Elmaagacli, A.H.; Schaefer, U.W.; Roggendorf, M. Human herpesvirus 6 is an important pathogen in infectious lung disease after allogeneic bone marrow transplantation. Bone Marrow Transplant. 2000, 26, 639–644.
49. Desachy, A.; Ranger-Rogez, S.; Francois, B. Reactivation of human herpesvirus type 6 in multiple organ failure syndrome. Clin. Infect. Dis. 2001, 32, 197–203.

50. Imbert-Marcille, B.M.; Tang, X.W.; Lepelletier, D. Human herpesvirus 6 infection after autologous or allogeneic stem cell transplantation: a single-center prospective longitudinal study of 92 patients. Clin. Infect. Dis. 2000, 31, 881–886.
51. Kadakia, M.P. Human herpesvirus 6 infection and associated pathogenesis following bone marrow transplantation. Leuk Lymphoma. 1998, 31, 251–266.
52. Rogers, J.; Rohal, S.; Carrigan, D.R. Human herpesvirus-6 in liver transplant recipients: role in pathogenesis of fungal infections, neurologic complications, and outcome. Transplantation. 2000, 69, 2566–2573.
53. Bosi, A.; Zazzi, M.; Amantini, A. Fatal herpesvirus 6 encephalitis after unrelated bone marrow transplant. Bone Marrow Transplant. 1998, 22, 285–288.
54. Paterson, D.L.; Singh, N.; Gayowski, T.; Carrigan, D.R.; Marino, I.R. Encephalopathy associated with human herpesvirus 6 in a liver transplant recipient. Liver Transplant Surg. 1999, 5, 454–455.
55. Singh, N.; Paterson, D.L. Encephalitis caused by human herpesvirus-6 in transplant recipients: relevance of a novel neurotropic virus. Transplantation. 2000, 69, 2474–2479.
56. Schmidt, C.A.; Wilbron, F.; Weiss, K. A prospective study of human herpesvirus type 6 detected by polymerase chain reaction after liver transplantation. Transplantation. 1996, 61, 662–664.
57. Lautenschlager, I.; Hockerstedt, K.; Linnavuori, K.; Taskinen, E. Human herpesvirus 6 infection increases adhesion molecule expression in liver allografts. Transplant. Proc. 1999, 31, 479–480.
58. Mayne, M.; Cheadle, C.; Soldan, S.S. Gene expression profile of herpesvirus-infected T cells obtained using immunomicroarrays: induction of proinflammatory mechanisms. J. Virol. 2001, 75, 11641–11650.
59. Lusso, P.; Ensoli, B.; Markham, P.D. Productive dual infection of human CD4 + T lymphocytes by HIV-1 and HHV-6. Nature. 1989, 337, 370–373.
60. Lusso, P.; De Maria, A.; Malnati, M. Induction of CD4 and susceptibility to HIV-1 infection in human CD8 + T lymphocytes by human herpesvirus 6. Nature. 1991, 349, 533–535.
61. Dolcetti, R.; Di Luca, D.; Carbone, A. Human herpesvirus 6 in human immunodeficiency virus-infected individuals: association with early histologic phases of lymphadenopathy syndrome but not with malignant lymphoproliferative disorders. J. Med. Virol. 1996, 48, 344–353.
62. Fabio, G.; Knight, S.N.; Kidd, I.M Prospective study of human herpesvirus 6, human herpesvirus 7, and cytomegalovirus infections in human immunodeficiency virus-positive patients. J. Clin. Microbiol. 1997, 35, 2657–2659.
63. Ebers, G.C.; Kukay, K.; Bulman, D.E. A full genome search in multiple sclerosis [see comments]. Nat. Genet. 1996, 13, 472–476.
64. Ebers, G.C.; Sadovnick, A.D. The role of genetic factors in multiple sclerosis susceptibility. J. Neuroimmunol. 1994, 54, 1–17.
65. Paty, D.W.; Noseworthy, J.H.; Ebers, G.C. Diagnosis of multiple sclerosis. In: Multiple Sclerosis; Paty, D.W., Ebers, G.C.,. Eds.; F. A. Davis: Philadelphia, 1998, 93–97.
66. Caserta, M.T.; Hall, C.B.; Schnabel, K. Neuroinvasion and persistence of human herpesvirus 6 in children. J. Infect. Dis. 1994, 170, 1586–1589.
67. Knox, K.K.; Brewer, J.H.; Henry, J.M.; Harrington, D.J.; Carrigan, D.R. Human herpesvirus 6 and multiple sclerosis: systemic active infections in patients with early disease. Clin. Infect. Dis. 2000, 31, 894–903.
68. Zou, P.; Isegawa, Y.; Nakano, K ; Haque, M.; Horiguchi, Y.; Yamanishi, K. Human herpesvirus 6 open reading frame U83 encodes a functional chemokine. J. Virol. 1999, 73, 5926–5933.
69. Romano, J.W.; Gemen, van B.; Kievits, T. NASBA: a novel, isothermal detection technology for qualitative and quantitative HIV-1 RNA measurements. Clin. Lab. Med. 1996, 16, 89–103.
70. Enbom, M.; Wang, F.Z.; Fredrikson, S.; Martin, C.; Dahl, H.; Linde, A. Similar humoral and cellular immunological reactivities to human herpesvirus 6 in patients with multiple sclerosis and controls. Clin. Diagn. Lab. Immunol. 1999, 6, 545–549.

71. Secchiero, P.; Zella, D.; Crowley, R.W.; Gallo, R.C.; Lusso, P. Quantitative PCR for human herpesviruses 6 and 7. J. Clin. Microbiol. 1995, 33, 2124–2130.
72. Ablashi, D.V.; Eastman, H.B.; Owen, C.B. Frequent HHV-6 reactivation in multiple sclerosis (MS) and chronic fatigue syndrome (CFS) patients. J. Clin. Virol. 2000, 16, 179–191.
73. Mayne, M.; Krishnan, J.; Metz, L. Infrequent detection of human herpesvirus 6 DNA in peripheral blood mononuclear cells from multiple sclerosis patients. Ann. Neurol. 1998, 44, 391–394.
74. Rotola, A.; Cassai, E.; Tola, M.R.; Granieri, E.; Di Luca, D. Human herpesvirus 6 is latent in peripheral blood of patients with relapsing-remitting multiple sclerosis. J. Neurol. Neurosurg. Psychiatry. 1999, 67, 529–531.
75. Pouly, S.; Antel, J.P. Multiple sclerosis and central nervous system demyelination. J. Autoimmun. 1999, 13, 297–306.
76. Furlan, R.; Martino, G.; Galbiati, F. Caspase-1 regulates the inflammatory process leading to autoimmune demyelination. J. Immunol. 1999, 163, 2403–2409.
77. Furlan, R.; Filippi, M.; Bergami, A. Peripheral levels of caspase-1 mRNA correlate with disease activity in patients with multiple sclerosis; a preliminary study. J. Neurol. Neurosurg. Psychiatry. 1999, 67, 785–788.
78. Tatsuo, H.; Okuma, K.; Tanaka, K. Virus entry is a major determinant of cell tropism of Edmonston and wild-type strains of measles virus as revealed by vesicular stomatitis virus pseudotypes bearing their envelope proteins. J. Virol. 2000, 74, 4139–4145.
79. Frenkel, N.; Schirmer, E.C.; Wyatt, L.S. Isolation of a new herpesvirus from human CD4+ T cells. Proc. Natl. Acad. Sci. USA. 1990, 87, 748–752.

10

JC Virus: Progressive Multifocal Leukoencephalopathy

Joseph R. Berger
University of Kentucky College of Medicine
Lexington, Kentucky, U.S.A.

Eugene O. Major and Bruce F. Sabath
National Institutes of Health
Bethesda, Maryland, U.S.A.

1 INTRODUCTION

Progressive multifocal leukoencephalopathy (PML) was first recognized to be a distinct disorder by Aström and coworkers in 1958 [1]. The syndrome was identified primarily on the basis of its unique pathological features of demyelination, abnormal oligodendroglial nuclei, and giant astrocytes. The three described cases also shared similar clinical manifestations, including dementia, visual impairment, and hemiparesis. Upon their review of the literature, Aström and coworkers discovered several independent descriptions of this entity dating to 1930 [2–4]. By 1984, a comprehensive review of PML found only 230 published and unpublished cases [5], of which 69 cases were pathologically confirmed and 40 cases both virologically and pathologically confirmed. In this series, only two cases were associated with acquired immunodeficiency syndrome (AIDS) [6,7]. This number represented 3.0% of all cases in which an underlying disease was identified [5]. PML occurring in association with AIDS was reported within a year [6] of the initial recognition of AIDS in 1981 [8–10]. Since then, this formerly rare disease has become remarkably common. Berger et al. [11] describe 154 cases of AIDS-associated PML observed between 1981 and 1994 in a tertiary referral center in southern Florida; during that same period, only two cases of PML unassociated with AIDS were seen.

2 ETIOLOGY AND PATHOGENESIS

Progressive multifocal leukoencephalopathy is a demyelinating disease of the central nervous system that results from infection of oligodendrocytes with JC virus (JCV), a polyomavirus. Approximately a decade after its initial clinical and pathological description [1,12,13], evidence of viral etiology was provided by electron microscopic studies demonstrating virus-like particles in the nuclei of abnormal oligodendrocytes [14,15]. This observation was subsequently confirmed by viral isolation. In 1971, Padgett et al. [16] isolated a human *Polyomavirus* (double-stranded DNA-containing virus with an icosahedral symmetry) from long-term cultures composed chiefly of glial cells. This virus was capable of hemagglutination of human type O erythrocytes, allowing for detection of antibody in patients. In the overwhelming majority of cases of PML in which virological confirmation has been obtained, JCV is the cause. In only three cases [17,18] has another polyomavirus, SV40, been implicated. However, these cases have not been well characterized, and in some instances reexamination of these brain tissues by in situ DNA hybridization revealed JCV, not SV40 [19]. A third polyomavirus, BK virus, has not been proven to be neuropathogenic. Our understanding of the illness has improved considerably in the past two decades for two main reasons. First, because PML occurs predominantly in immunocompromised individuals, the AIDS pandemic has led to an increased incidence of PML and consequently more opportunity to study the disease. Second, highly sensitive molecular techniques have been developed that allow detection of very few copies of viral genome, including advances in in situ hybridization and DNA amplification using polymerase chain reaction (PCR) [20,21]. Application of these techniques to tissues from PML patients has directed recent investigations toward determining mechanisms of viral multiplication [22–24], cellular regulation of viral gene expression [25,26], and dissemination of virus to the central nervous system [21,27]. JCV is a widespread infectious agent that infects a sizeable proportion of the population by antibody measures, yet PML is rare. The ability of the virus to target a highly specialized cell in the nervous system, the myelinating oligodendrocyte, accounts for its pathogenesis. A brief description of the biology and molecular regulation of JCV follows.

2.1 Biology of JC Virus

JC virus belongs to the *Polyomavirus* family of primate polyomaviruses. JCV has a simple double-stranded, supercoiled genome of 5.1 kilobases (kb) encapsidated in an icosahedral protein structure measuring 40 nm in diameter. The DNA codes for one nonstructural but multifunctional protein (T), three capsid proteins (VP1, VP2, VP3), and a regulatory protein termed agnoprotein (Fig. 1). The mRNAs that code for these proteins are transcribed from opposite strands of the DNA genome (T protein versus capsid proteins and agnoproteins) and in opposite directions starting near the viral origin of replication. Cellular splicing accounts for nonstructural proteins, t, T'_{135}, T'_{136}, and T'_{165}, originating from the same DNA strand as the T protein. The smaller t protein does not seem to play a role in the multiplication of the virus and consequently is not considered important for pathogenicity. The T′ proteins contribute to JCV DNA replication and are known to interact with several cellular proteins [28].

The T protein is a DNA-binding protein and is responsible for initiation of viral DNA replication and transcription of the capsid proteins. In certain rodent and nonhuman primate cells, JCV T-protein expression is consistent with a malignant transformation or tumor induction, particularly of astroglial cells into astrocytomas [29–31]. In these cells,

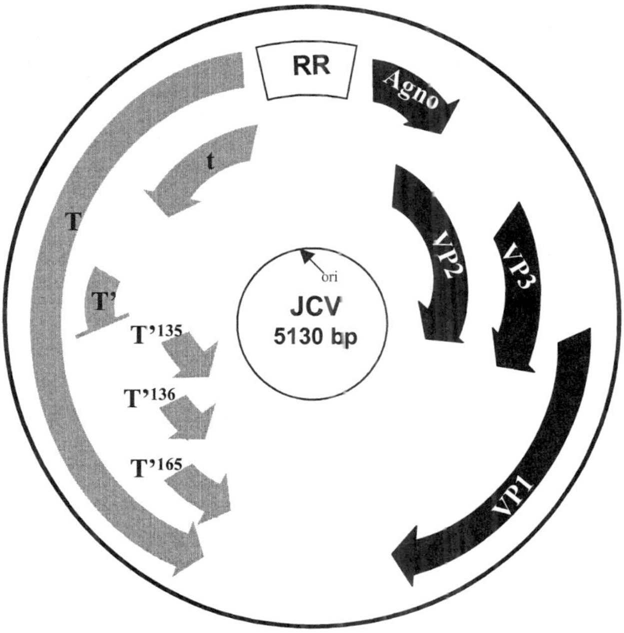

Figure 1 Organization of the 5.1 kilobase JC virus (JCV) genome. The regulatory region (RR) directs transcription of T RNA followed by transcription of agnoprotein and capsid protein (VP1, VP2, VP3) RNA. T RNA and its splice variants yielding t and T′ transcripts are produced from the opposite strand and in the opposite direction of agnoprotein and capsid protein RNA.

only the T protein (named for its tumor-promoting function) is expressed. There has not been any evidence in human gliomas for the presence of JC virions or free or integrated JCV DNA [32,33].

Capsid proteins VP1, VP2, and VP3 compose the virion particle, with the largest, VP1, being responsible for the icosahedral structure [34]. VP1 also enables viral entry into the cell and hemagglutination. Interestingly, this protein can also self-assemble into an icosahedral particle without either of the two minor capsid proteins [35]. The agnoprotein can interact directly with T protein and downregulate T-protein-mediated JCV DNA replication and transcription [36]. Agnoprotein can also suppress basal levels of transcription when T protein is absent. This protein is also involved in viral capsid assembly [34].

The *regulatory region* of the JCV genome comprises approximately 200 nucleotide base pairs of noncoding sequence located between the two coding sequence areas. This region of the genome contains the signals for DNA replication as well as for promotion and enhancement of transcription [37–38]. Moreover, it is considered the primary area of the genome responsible for the cellular tropism of JCV [39–40]. This is also the region of the viral genome found in the brain tissue of many PML patients that demonstrates the most sequence variability resulting from deletions and rearrangements (perhaps acquired during propagation in brain or in extraneural host tissues [41–42]. In fact, the differences in the structure of the regulatory region has been used to distinguish viral variants. From these observations, a nomenclature system has recently been developed that can be useful for describing JCV variants. [43].

The idea that JCV was a neurotropic virus exclusive to glial cells was derived largely from the initial descriptions of JCV host range. From its propagation and first isolation in cultures derived from human fetal brain [14,16], early studies of JCV host range emphasized the almost exclusive nature of growth in glial cells. JCV does not infect neurons in PML brain tissue or in cultures from human adult or fetal brain but does infect both oligodendrocytes and astrocytes [44,45]. Experiments using infectious clones of viral DNA reinforced a glial-specific host range for transcription of JCV [37,46]. Other studies used recombinant DNA constructs of the viral regulatory sequences linked to a reporter gene, chloramphenicol acetyl transferase (CAT), to achieve a quantitative measurement of transcriptional activity. Activity of the CAT gene was greatest in human glial cells [47].

However, in recent years, several studies have demonstrated JCV infection in a variety of other cell types. Human $CD34^+$ hematopoietic progenitor cells and related cell lines, KG-1 and KG-1a, are susceptible to JCV infection, as are tonsillar stromal cells [48]. Primary B lymphocytes and B-cell lines can also support JCV, opening the possibility that these cells may serve as carriers of JCV to the brain [48,49]. JC virions can also infect kidney cell lines and primary cultures of amnion cells and vascular endothelium [34]. Finally, JCV DNA has been detected in human colonic tissue as well [50].

Considering that JCV could induce a malignant transformation in rodent and monkey glial cells but could not multiply in glial cells of these animals, it was known that viral DNA replication was controlled at the species level. Unlike many other human viral pathogens, susceptibility to JCV infection has not been clearly associated with viral attachment to specific cell receptors and penetration. One study demonstrated JCV binding to a variety of cell lines regardless of their susceptibility to JCV infection, but more specific binding with primary cells [51]. Thus, there may be specificity of viral binding in vivo that does not occur in vitro in tumor cell lines.

What is clearer is that susceptibility to JCV infection is controlled by cell type–specific factors for transcription and species-specific factors for replication [52]. These studies have since directed experiments to delineate the molecular factors responsible for JCV growth and have established an early understanding of its pathogenesis at the molecular level.

2.2 Molecular Control of JCV Gene Expression

To explain the neurotropism of JCV for glial cells, experiments concentrated on identifying nuclear DNA-binding proteins that selectively interacted with the regulatory region of the genome. Such proteins, it is assumed, would bind specific cis-acting nucleotide base pairs (np) for control of JCV DNA replication and transcription. Using techniques of gel retention, protein DNA cross-linking, and DNase footprint analysis of cis-acting DNA, sequences were identified in the regulatory region at nt 33–58 on the viral genome map [53]. Because the regulatory region comprises direct tandem repeats of 98 nucleotide pairs each, these binding sites exist twice, as expected, in both repeats. Two other series of nucleotide pairs were also identified as binding sites for nuclear proteins, located on either side of nt 33–58. One of these areas, directly next to the sequences necessary for DNA replication, in the direction of the T-protein coding region, is rich in repeated AT sequences that are known to function as RNA start sites [25,38,55]. The protein that binds these sequences is the TFIID transcription factor represented in almost all eukaryotic cells [56]. The other protein binding site covers an area that includes the transcriptional enhancer for JCV inasmuch as its sequence is similar, but not identical, to sequences previously described for this function [57]. However, nuclear extracts from nonpermissive cells demonstrated some binding to these regions. The functional consequence of this binding appeared to downregulate JCV activity in these cells [26]. Therefore, it seems that there are proteins that positively regulate JCV expression and those that block expression in a cell-type-specific manner.

The region of the regulatory sequences most efficiently bound by protein factors from permissive glial cells was the area of nt 33–58. This area covers the consensus sequence for the binding of nuclear factor 1 (NF-1), a protein that functions in both transcriptional control and replication of DNA. The NF-1 family is composed of four classes of proteins (NF-1 A, B, C, and D, or X) with conserved DNA-binding domains but distinct transcriptional activation domains. Differential splicing results in mRNA transcripts encoding several subclasses of NF-1 proteins. Several reports from independent laboratories have identified an NF-1-like protein found in glial cells that binds to this region [25,53,58,59]. Another report identified cDNA for a glial-specific factor, GF1, that also binds to this region and may be related to NF-1 [60]. Further implications of NF-1 in JCV regulation come from experiments using purified NF-1 proteins, not simply nuclear extracts, that would compete with extracts for binding to the target sequences in the JCV promoter/enhancer region [25,61].

The most apparent evidence for NF-1 regulation of JCV centers on NF-1 class D. A positive correlation has been demonstrated in different cell types between the levels of expression of this protein and susceptibility to JCV infection [62]. Moreover, when susceptible KG-1 cells differentiate to nonsusceptible cells of a macrophage lineage upon phorbol ester stimulation, NF-1 D expression levels are downregulated. Susceptibility to JCV is restored when NF-1 D expression levels are restored by transfection of the NF-1 gene. This correlation has also been observed in susceptible human glial cells and nonsusceptible HeLa cells [63].

Another protein has also been implicated along with NF-1. Additional DNase footprinting data resolved a protein binding immediately adjacent to the NF-1 binding site. This site has been identified as the binding domain for the c-jun protein of the activator protein (AP-1) family, another ubiquitous transcription factor. The combination of binding sites for NF-1 and c-jun separated by only a few nucleotides appears to be a common feature for expressions of many brain-specific genes such as glial fibrillary acidic protein, neurofilament, human and mouse MBP, S100b, proteolipid protein, and proenkephalin [25]. Whether there is direct interaction between these proteins is not yet known. Because both NF-1 and c-jun proteins are found in many other cells besides glial cells, these data suggest a family of such JCV regulatory proteins. Glial cell susceptibility to JCV infection may be determined by the presence of specific members of this family of transcription proteins that are present only in permissive cells. Cells that are not permissive to JCV infection probably do not have these same protein factors and/or have other proteins that bind the JCV regulatory sequences and block transcription [25,26].

2.3 Molecular Interactions Between JCV and HIV-1

The clinical descriptions of the course of PML and the identification of JCV in oligodendrocytes have been well documented from the first recognition of the disease [13,28,64]. The explanation of how JCV enters the brain and initiates infection, however, came much later. Current evidence implicates viral latency in lymphocytes in bone marrow or other lymphoid tissues that can be activated during immune suppression and enter the peripheral blood [64]. Circulating infected lymphocytes may be able to cross the blood-brain barrier and pass infection to astrocytes at the border of vessels, which in turn augments infection through multiplication to eventually infect oligodendrocytes. Using in situ DNA hybridization, JCV-infected cells are frequently found near blood vessels in the brain, in B lymphocytes in bone marrow [27], and in brain tissue [24]. In a report of 19 patients with biopsy-proven PML, PCR technology showed that over 90% had JCV DNA in peripheral blood lymphocytes [65]. The number of cases of PML with viral DNA in their peripheral blood lymphocyte circulation is now greater, but the percentage remains at approximately 95%. These studies were done using a series of paired primers representative of three regions of the viral genome to eliminate possible nonspecific amplification of closely related DNA sequences. Data derived from other groups of individuals without PML revealed that 60% of HIV-1-seropositive individuals, 30% of renal transplant recipients, and approximately 5% of normal, healthy volunteers also had JCV DNA in their peripheral circulation. Individuals whose immune systems may be compromised through immunosuppressive therapies or other diseases would be considered at risk for the development of PML.

There is an incidence of approximately 5–10% of the population of JCV excretion in urine, detected by either PCR or virus isolation. This includes pregnant women, older individuals, and some organ transplant patients [20,66–68]. These observations led to the suggestion that the kidney is the site of viral latency. The DNA sequence of the JCV regulatory region from kidney or urine in these individuals is markedly different from the sequence found in the brain of PML patients [69]. Because the regulatory DNA sequence chiefly governs infectivity of JCV, a number of JCV isolates or clones of DNA have been examined. The most prominent DNA sequence arrangement found related to kidney has been described as an archetype sequence [70]. The archetype sequence contains 187 nucleotide pairs with no tandem repeats but also contains the origin of viral replication and the TATA sequence as an RNA start site. It also contains a 23 nucleotide pair insert next to the TATA site that is found in many regulatory region sequences from PML brain [42,71]. These sequences probably serve as a functional binding site for the Sp1 DNA transcription factor [71,72]. To convert the archetype sequence to the one most often found in PML brain tissue, however, would require deletions, substitutions, and duplications [69,73]. Currently there is no evidence for any biological activity for the archetype sequence or viral isolates that contain these sequences. Several regulatory region sequences have been identified from JCV DNA in peripheral blood of PML patients that are not related to the archetype but are closely related to sequences found in PML brain [65]. Further examination of the distribution and importance of the archetype sequence is needed to understand its role in the pathogenesis of PML. The existence of many variations of the DNA sequence of the regulatory region of JCV highlights the genomic diversity of JCV.

These data have identified several regions within the regulatory sequences that are always represented, however, and by inference are thought to be critical for viral multiplication and perhaps pathogenicity. The origin of replication, the TATA sequence, the 23-nucleotide insert, and sequences found just next to this insert that extend to nucleotides 58–68 are heavily represented in DNA sequences in PML brain [42,65,70,71]. These sequences are responsible for the duplication of the viral genome (the origin), the initiation of RNA synthesis (TATA sequences), and transcriptional control (the putative NF-1 binding and possibly the c-jun binding sites).

As mentioned above, present evidence suggests that lymphocytes, particularly B cells, may harbor JCV in a latent state and, upon activation, carry virus to the brain. If this is correct, then B cells must also have nuclear DNA-binding proteins that recognize these important sites on the JCV regulatory region. Using both gel retention assays and DNase footprinting experiments, several human B cell lines were described as permissive for JCV multiplication and possessed DNA-binding proteins that recognized the same sequences on the JCV genome as the highly permissive human glial cells [49]. Although these proteins may not be identical, they may represent similar members of an entire family of transcription factors such as NF-1, Sp1, c-jun (and perhaps others not yet identified) that regulate cellular susceptibility to JCV.

3 ASSOCIATION WITH AIDS

As suggested by Stoner et al. [74] and confirmed in population studies, the incidence of PML in immunosuppressive conditions other than AIDS does not appear to approach the 4–5% incidence observed with HIV-1 infection. This apparent increase in the incidence of PML with HIV-1 infection is likely to be the result of a combination of cellular immunodeficiency and central nervous system (CNS) inflammation (typically subclinical) resulting from HIV-1 infection. The cellular immunosuppression leads to the expression of JCV in B lymphocytes, and the HIV-1-associated CNS inflammation results in facilitation of the entry of these cells into the CNS. PML is the result of subsequent infection of glial cells by JCV.

Soon after HIV-1 infection, and often long before the development of significant immunosuppression, there is evidence of HIV-1 infection of the CNS, including recovery of HIV-1 from the cerebrospinal fluid (CSF) [75], intrathecal synthesis of antibody to HIV-1 [85], and abnormalities of cerebrospinal fluid [76–78]. HIV-1 has been demonstrated by PCR in the brain of a patient dying within 15 days of infection [79]. Viral cultures [75,80], immunostaining [81], and in situ hybridization [82] confirm the presence of HIV-1 infection in the brain. The infection of the CNS with HIV-1 has been postulated to result from a trafficking into the brain of HIV-1-infected monocytes/macrophages that subsequently establish residence within the brain [83]. There is some evidence of an activated state of cellular and humoral immunity in the HIV-1-infected CNS despite a coexistent systemic immunodeficiency. Cells expressing MHC class I and class II antigens, the prerequisite for antigen presentation to the immunocompetent T cells, are relatively increased in CNS tissue derived from AIDS patients compared with normal CNS tissue [84,85]. B cell activation in the CNS compartment, perhaps in part as a response to increased levels of interleukin-6 (IL-6, B-cell stimulatory factor-2) [86], is evidenced by the intra-blood brain barrier production of antibodies to HIV-1 antigens [87]. Although the predominant cell types in cellular infiltrates observed in the tissue of HIV-1 encephalitis are macrophages and microglia [55,85,88], some T and B cells have been demonstrated by immunohistochemical technique [84,85]. In addition, several cytokines have been demonstrated in CNS tissue from AIDS patients using immunohistochemical techniques: tumor necrosis factor alpha (TNF-α) [85,89], IL-1-B, IL-6, and interferon gamma [85].

The selective homing of B lymphocytes to the various lymphoid organs (peripheral lymph nodes, Peyer's patches) is a dynamic process. The migration of B cells into non-lymphoid tissues has been less well characterized. Entry of B cells into such organs as the brain may result from both selective homing and less specific mechanisms. Cytokines released from macrophages and T cells may be chemotactic for B cells [90], increase the adherence of endothelial cells for B cells [91], or alter the blood brain barrier to allow passage of a variety of inflammatory cells into the brain parenchyma [92]. TNF-α, IL-1,

IL-4, and gamma-interferon, cytokines produced by macrophages, and other inflammatory cells (e.g., T cells [93]), have been demonstrated to increase the adhesion of vascular endothelium for B lymphocytes [91,94,95]. The overall environment of ''immune activation'' that has been described in the CNS [85] will possibly result in chemoattraction, enhanced adhesion of B cells to brain endothelia, and activation of B cells latently infected by JCV. The increased B-cell adhesion to brain endothelia is likely to result from elaboration of IL-1 and TNF-α described in the brains of HIV-1-infected individuals. The stimulation of B cells [96,97] is possibly mediated by IL-6, IL-4 (B cell stimulatory factor-1), and other substances, such as transforming growth factor beta, a cytokine with potent chemotactic activity that has been immunohistochemically demonstrated in CNS tissue from patients with AIDS [98]. The adhesion of circulating lymphocytes to target tissue vasculature precedes their subsequent migration into the organ. Under physiological conditions, lymphocyte traffic into the brain is limited. Two explanations have been proposed to explain why lymphocyte traffic into CNS is so low: Adhesion of lymphocytes to cerebral endothelium may be low, or the tight junctions may restrict transendothelial migration of any adherent cells. These proposals are not mutually exclusive. It is likely that the entry of JC virus–infected B lymphocytes into the brain is facilitated by an upregulation of adhesive molecules on brain endothelial cells in response to HIV-1 infection of the CNS. It is possible that JCV antigen–specific lymphocytes appear in the CNS early, rarely migrating far from the vasculature. Because monocytes and astrocytes can present antigen in situ, it is at this site that the antigen-specific B cells are further activated, most likely in an antigen-specific manner. The cytokines secreted locally by microglial cells and monocytes activate the endothelium and surrounding perivascular cells, leading to the expression of adhesion molecules. For example, due to the expression of an avidly binding form of LFA-1, activated B cells are preferentially recruited to tissue expressing ICAM-1 or other ligands and present in PML. ICAM is expressed in culture—and possibly in vivo—by astrocytes, particularly human fetal astrocytes. Lesions of PML, like those of multiple sclerosis and experimental allergic encephalitis, are centered on blood vessels and are formed by specific subsets of inflammatory cells.

4 EPIDEMIOLOGY OF JCV INFECTION

The ability of JCV to cause hemagglutination of type O erythrocytes has enabled the use of antibody studies to determine evidence of prior infection. To date, no disease has been convincingly associated with acute infection, although Blake et al. [99] reported chronic meningoencephalitis occurring with acute JCV infection in a 13-year-old girl. The acute infection in this patient was identified by a rise of immunoglobulin M (IgM) titers to JCV and not by viral isolation [99]. Viral spread is speculated to be by respiratory means. Between the ages of 1 and 5 years, approximately 10% of children demonstrate antibody to JCV, and by age 10, 40–60% of the population does so [100–102]. By adulthood, this figure rises almost sevenfold [101]. Rates of seroconversion to JCV have exceeded 90% in some urban areas [101].

The high prevalence of antibodies in the adult population and the rarity of PML in children support the contention that PML is the consequence of JCV reactivation in individuals who become immunosuppressed. Additionally, high titers of IgM antibody specific for JCV would be anticipated in patients with PML if it were the result of acute infection. However, one study revealed that the sera of only 1 of 21 patients with PML had IgM specific for JCV, whereas 20 of 21 had IgG antibody specific for JCV [103]. Some investigators have argued that the latter study does not exclude the possibility of PML resulting from acute JCV infection, because many of these patients were studied late in the course of their disease [104].

4.1 Host Factors and Underlying Diseases

In virtually all patients with PML, an underlying immunosuppressive condition has been recognized. Typically, the abnormality is one of cell-mediated immunity. The first three patients described by Astrom et al. [1] had either chronic lymphocytic leukemia or lymphoma as the underlying illness. In a review by Brooks and Walker published in 1984 [5], lymphoproliferative diseases were the most common underlying disorders, accounting for 62.2% of the cases. Other predisposing illnesses included myeloproliferative diseases in 6.5%, carcinoma in 2.2%, granulomatous and inflammatory diseases such as tuberculosis and sarcoidosis in 7.4%, and other immunodeficiency states in 16.1%. Although AIDS was included in the latter category, there were only two reported cases of PML complicating AIDS at that time [5].

Indeed, until the AIDS epidemic, PML remained a rare disease. To most practicing neurologists, it remained a medical curiosity about which one learned from the textbooks. However, the AIDS pandemic changed the incidence of this formerly rare illness quite remarkably. In the quarter century between the clear definition of the disorder by Astrom et al. [1] to the extensive review of PML by Brooks and Walker in 1984, only 230 previously published cases were identified. Following the advance of the AIDS pandemic, the spectrum of underlying illness changed dramatically. From the University of Miami Medical Center and the Broward County Medical Examiner's Office in Florida [11], for example, 156 cases of PML were identified over the 14 year interval 1980–1994, with all but two of these cases related to HIV infection. Comparing the four-year intervals 1980–1984 and 1990–1994 in this series, there was a twentyfold increase in the number of cases of PML [11]. Overall, AIDS has been estimated to be the underlying cause of immunosuppression in from 55% to more than 85% of all current cases of PML [64]. However, that may reflect an underestimate of the true incidence of HIV/AIDS as the underlying immunosuppressive condition predisposing to PML.

4.2 Clinical Epidemiology in the Era of AIDS

The first description of PML complicating AIDS followed the description of AIDS in 1981 by one year [6]. By the late 1980s, AIDS was reported to be the most common underlying disorder predisposing to the development of PML at institutions in New York [105] and Miami [106]. As noted above, the subsequent evolution of the AIDS pandemic significantly changed the epidemiology of PML. Gillespie et al. [107], studying the prevalence of AIDS-related illnesses in the San Francisco Bay area, estimated a prevalence for PML of 0.3%. They acknowledged that this may have been a significant underestimate [107]. Based on death certificate reporting of AIDS to the Centers for Disease Control and Prevention (CDC) between 1981 and June 1990, 971 (0.72%) of 135,644 individuals with AIDS were reported to have PML [108]. Due to the notorious inaccuracies in death certificate reporting [109] and the requirement of pathological confirmation for inclusion in this study, this is also likely a significant underestimate of the true prevalence. Other studies have suggested that the prevalence of PML in AIDS cases is substantially higher than that reported by the CDC, with most estimates ranging between 1% and 5% in clinical studies and as high as 10% in pathological series [105,106,110–112]. In 1987, a large retrospective, hospital-based clinical study [106] found PML in approximately 4% of patients hospitalized with AIDS.

Four percent of all patients dying of AIDS had PML in a combined series of seven separate neuropathological studies comprising a total of 926 patients with AIDS [111]. Two other large neuropathological series found PML in 7% [112] and 9.8% [110] of autopsied AIDS patients. The authors of the latter study acknowledged that an unusually high estimate may have resulted from numerous referral cases from outside the study center and the inherent bias imposed [110]. However, a study of 548 consecutive, unselected autopsies between 1983 and 1991 performed on HIV-seropositive individuals by the Broward County (Florida) Medical Examiner revealed that 29 (5.3%) had PML confirmed at autopsy [113]. Similarly, the Multicenter AIDS Cohort Study also identified a dramatic rise in the incidence of PML over a similar time period. Specifically, the Multicenter AIDS Cohort Study identified 22 cases of PML among the cohort of AIDS cases studied from 1985 to 1992; the average annual incidence of PML was 0.15 per 100 person-years, with a yearly rate of increase of 24% between 1985 and 1992 [114]. Although these estimates may be susceptible to selection and other biases, there is an indisputable markedly increased incidence and prevalence of PML since the inception of the AIDS pandemic. Indeed, it appears that the incidence of PML complicating HIV/AIDS is higher than that of any other immunosuppressive disorder relative to their frequency. Possible explanations include differences in the degree and duration of the cellular immunosuppression in HIV infection, facilitation of the entry into the brain of JC virus–infected B lymphocytes [27] by alterations in the blood-brain barrier due to HIV [115], or the upregulation of adhesion molecules on the brain vascular endothelium due to HIV infection [116,117] and the potential for the HIV Tat protein to transactivate JC virus [118].

Concomitant with the increase in PML in association with AIDS has been the not unexpected alteration in the demographics of the affected population. Prior to the AIDS epidemic, males and females were affected in a ratio of 3:2 [5]. The incidence of PML increased steadily from middle age [5]. Children, regardless of the cause of underlying immunosuppression, rarely develop PML. As exposure to JC virus occurs sometime during childhood, a minority of young children are at risk for the disease. However, it has been observed in HIV-infected children [119–121]. As cited, lymphoproliferative diseases were the most common underlying etiology [5]. Currently, HIV infection is the most common underlying cause of immunosuppression, and the disorder chiefly affects homosexual/bisexual men between the ages of 25 and 50 years, with a correspondingly high male/female ratio of 7.6: 1.0 [119]. One should bear in mind that as the demography of HIV infection changes, one would anticipate a parallel change in the demography of AIDS-related PML. Curiously, there seems to be a higher degree of prevalence of PML in white males than in African American males [78]. Additionally, there may be some geographical differences in the prevalence of PML. For example, PML is considered rare in Africa, and a neuropathological study from southern India suggests an incidence of 1% [122]. The population differences observed may be the consequence of the nature of medical care rendered. PML is typically observed in advanced HIV infection, and therefore it is not unlikely that patients succumbing to other AIDS-related disorders early in the course of their infection may not live long enough to develop PML.

In 1996, there was an expansion of available antiretroviral therapies with the development of highly active antiretroviral therapies (HAART). Opportunistic infections, e.g., cytomegalovirus infection [123,124] and toxoplasmosis [125], and primary central nervous system lymphomas [126] have been reported to have declined significantly following their introduction. The effect of this therapy on the incidence of PML remains uncertain. D'Arminio Monforte et al. [127] detected a 95% risk reduction in all CNS AIDS–related conditions following the adoption of HAART in their cohort. Twenty cases of PML were identified in this study, but a specific analysis for PML was not undertaken. Others [125] have similarly noted a decline in HIV-related CNS disorders but have had too few cases to comment specifically about PML.

In summary, the epidemiology of PML suggests that dysfunction of cellular immune response is the most important determining factor that predisposes to the development of PML, although gender, genetic factors, and viral strains may also play a role. Therapies, in particular HAART, with a restoration of immune function may result in a decline in the frequency of this disorder.

5 CLINICAL FEATURES

In as many as 1% of all HIV-infected persons, PML is the initial AIDS-defining illlness [11]. The occasional patient seen with PML that antedates knowledge of HIV infection may lead to considerable diagnostic confusion with respect to the otherwise healthy individual. Typically, however, the disorder occurs in the setting of advanced immunosuppression and is preceded by other AIDS related illnesses.

Focal weakness is the most common symptom reported by patients with PML or their caregivers [11]. Other common symptoms include cognitive abnormalities, speech and language disorders, headaches, gait disorders, visual impairment, and sensory loss. In a large series with more than 150 AIDS-related PML patients [11], each of these symptoms was seen in more than 15% of patients. In general, these symptoms are similar to those identified in a series of non-HIV-associated PML cases. Compared to the series of Brooks and Walker [5], headaches were significantly more common in the HIV-infected population and visual disturbances were more common in the non-HIV-infected. Seizures are seen in up to 10% of patients and are usually focal in nature although secondary generalization may occur. Seizures in AIDS-associated PML may reflect involvement of the cortical astrocytes by the JC virus [128] or may be secondary to some other process or HIV infection of the brain itself [129].

Limb weakness, seen in over 50% of patients, is the most common sign observed with HIV-associated PML [11]. Cognitive disturbances and gait disorders are seen in approximately one-fourth to one-third of patients [11]. Diplopia, noted by 9% of patients, is usually the consequence of involvement of the third, fourth, or sixth cranial nerve and is typically observed in association with other brainstem findings [11]. Visual field loss due to involvement of the retrochiasmal visual pathways is significantly more common than diplopia or other visual disturbances [5,11,130]. Optic nerve disease does not occur with PML, and although the lesions of PML have been detected in the spinal cord of HIV-infected patients [131], clinical myelopathy secondary to PML must be vanishingly rare. PML does not involve the peripheral nervous system. While the virus resides in the lymphoid tissue and bone marrow, it remains asymptomatic in these reservoirs. Recent evidence of latent infection of tonsillar tissue [132] supports the hypothesis that JC virus infection is initially acquired as an oropharyngeal or respiratory infection; however, despite evidence of seropositivity for JC virus of up to 80% of the population, no clinical illness has been convincingly established with primary infection to date.

6 NEUROIMAGING

In the appropriate clinical context, radiographic imaging may strongly support the diagnosis of PML. Computed tomography (CT) of the brain in PML reveals hypodense lesions of the affected white matter (Fig. 2). On CT scan, typically, the lesions of PML exhibit no mass effect and only rarely contrast enhancement. A ''scalloped'' appearance beneath the cortex is noted when there is involvement of the subcortical arcuate fibers [113]. Cranial magnetic resonance imaging (MRI) is far more sensitive to the presence of the white matter lesions of PML than CT scans [113]. MRI shows a hyperintense lesion on T2-weighted images and fluid attenuated inversion recovery sequences (FLAIR) in the affected regions (Fig. 3). Contrast enhancement may be observed in 5–10% of cases with both MRI and CT and is usually faint and peripheral in location [113]. In rare instances, mass effect may be observed and may be the consequence of immunoreconstitution with highly active antiretroviral therapy [133]. The lesions of PML may occur virtually anywhere in the brain. The frontal lobes and parieto-occipital regions are commonly affected, presumably as a consequence of their volume. However, isolated or associated involvement of the basal ganglia, external capsule, and posterior fossa structures (cerebellum and brainstem) may be seen as well [113]. Other diseases may affect the white matter in a similar manner in association with HIV infection. Particularly notable in this regard are AIDS dementia and cytomegalovirus infection. With respect to AIDS dementia, radiographic distinctions include a greater propensity of PML lesions to involve the subcortical white matter, its hypointensity on T1W1 images, and its rare enhancement [113]. Cytomegalovirus lesions are typically located in the periventricular white matter and centrum semiovale. Subependymal enhancement is often observed as a consequence of CMV infection. Other potentially HIV-associated disorders that may result in hyperintense signal abnormalities of the white matter resembling PML include varicella-zoster leukoencephalitis [134], a multiple sclerosis–like illness [135], acute disseminated encephalomyelitis [136,137], CNS vasculitis [138], and white matter edema associated with primary or metastatic brain tumors. Almost always, the clinical features, laboratory findings, and associated radiographic features enable the correct diagnosis.

Magnetization transfer MRI studies have been suggested as an effective means of monitoring the degree of demyelination in PML [139,140]. Magnetic resonance spectroscopy reveals a decrease in *n*-acetyl aspartate and creatine and increased choline products, myoinositol, and lactate in the lesions of PML [35]. These changes were interpreted as reflective of the neuronal loss and cell membrane and myelin breakdown observed in PML [35]. Cerebral angiography is not routinely performed but exhibited arteriovenous shunting and a parenchymal blush in the absence of contrast enhancement on MRI in four of six patients in one study [141]. Pathological studies suggested that small-vessel proliferation and perivascular inflammation were the explanation for these unexpected angiographic features [141]. Thallium-201 single photon emission computed tomography (^{201}Tl SPECT) generally reveals no uptake in the lesions of PML [142]; however, a single case report of a contrast-enhancing lesion with a positive ^{201}Tl SPECT has been reported [143].

7 LABORATORY STUDIES

In the overwhelming majority of HIV-infected patients with PML, severe cellular immunosuppression, as defined by CD4 lymphocyte counts below 200 cells/mm^3, is observed. In

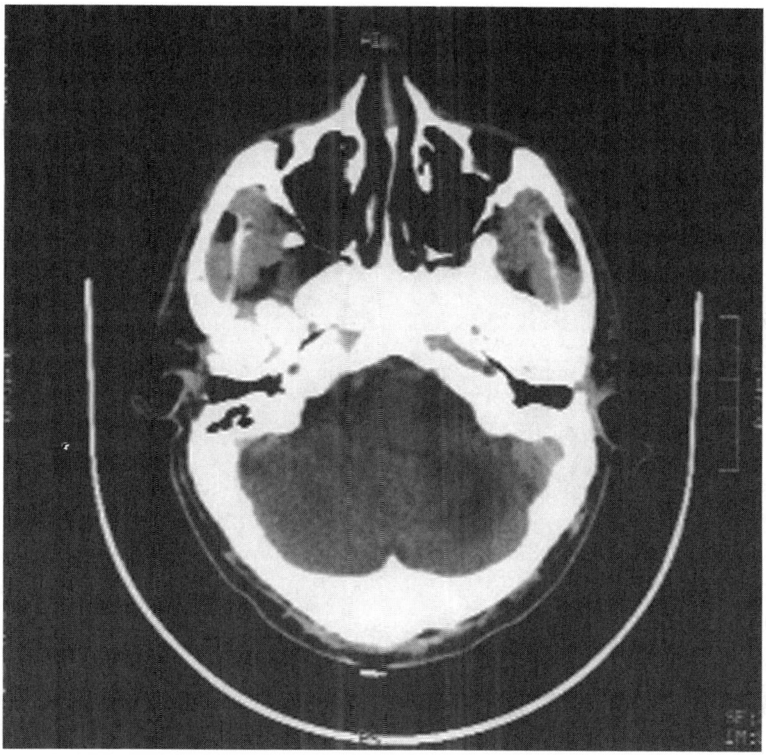

Figure 2 CT scan of the brain. CT scan of the brain reveals a large hypodense lesion of the left cerebellar hemisphere in an HIV-infected patient presenting with hemiataxia due to PML.

three separate series of AIDS-related PML [11,144,145], the mean CD4 count ranged from 84 to 104 cells/mm^3. However, in the largest series of AIDS-related PML [11], 10% or more of patients had CD4 lymphocyte counts in excess of 200 cells/mm^3.

Cerebrospinal fluid examination is very helpful in excluding other diagnoses. Cell counts are usually less than 20 cells/mm^3 [11]. In one large study, the median cell count was 2 cells/mm^3 and the mean was 7.7 cells/mm^3 [11]. In that same study, 55% had an abnormally elevated CSF protein [11], with the highest recorded value being 208 mg/dL (2.08 g/L). Hypoglycorrhachia was observed in less than 15%. These abnormalities are not inconsistent with that previously reported to occur with HIV infection alone [76,77,146]. Several studies [144,147,148] demonstrated a high sensitivity and specificity of CSF PCR for JC virus in PML. Many authorities have regarded the demonstration of JC viral DNA coupled with the appropriate clinical and radiological features sufficiently suggestive of PML to be diagnostic, thus obviating the need for brain biopsy. Quantitative PCR techniques for JC virus in biological fluids continues to be refined [149]. Antibody titers in serum or CSF are not useful because most individuals become seropositive for JC virus before adulthood. Additionally, because PML occurs in the context of immunosuppression, the individual may not be able to mount an antibody response.

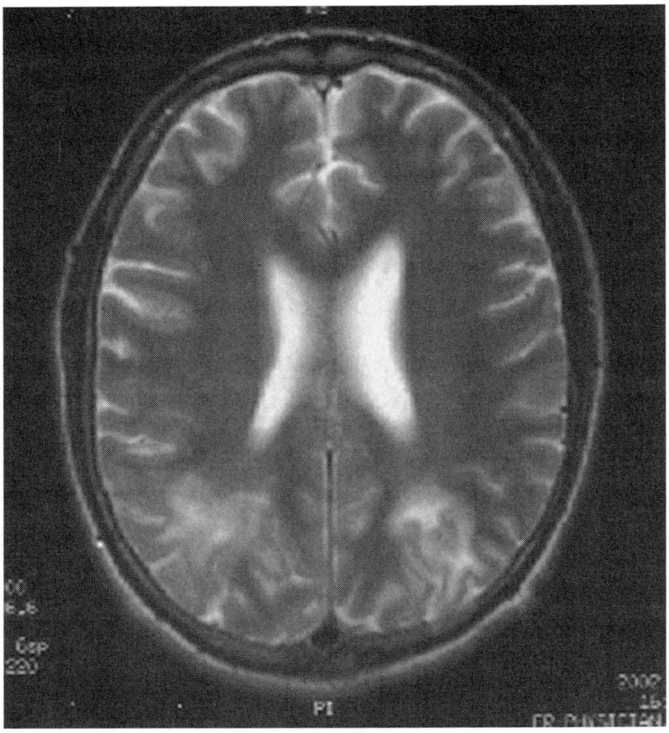

Figure 3 MRI of the brain. Bilateral hyperintense signals in the white matter of the posterior hemispheres are observed on T2 weighted MR image in a patient presenting with visual disturbance due to PML.

8 PATHOLOGY

Macroscopically, the cardinal feature of PML is demyelination. Demyelination may on rare occasions be monofocal, but it typically occurs as a multifocal process, suggesting hematological spread of the virus. These lesions may occur in any location in the white matter and range in size from 1 mm to several centimeters [1,150]; larger lesions are not infrequently the result of coalescence of multiple smaller lesions. The histopathological hallmarks of PML are a triad [1,150] of multifocal demyelination, hyperchromatic, enlarged oligodendroglial nuclei (Fig. 4); and enlarged bizarre astrocytes with lobulated hyperchromatic nuclei (Fig. 5). The latter may be seen to undergo mitosis and appear quite malignant. Electron microscopic examination will reveal the JC virus in the oligodendroglial cells. These virions measure 28–45 nm in diameter and appear singly or in dense crystalline arrays [1,150]. Less frequently, the virions are detected in reactive astrocytes, and they are uncommonly observed in macrophages that are engaged in removing the affected oligodendrocytes [151,152]. The virions are generally not seen in the large, bizarre astrocytes [152]. In situ hybridization and in situ PCR for JCV antigen allow for detection of the virion in the infected cells even in formalin-fixed archival tissue [153].

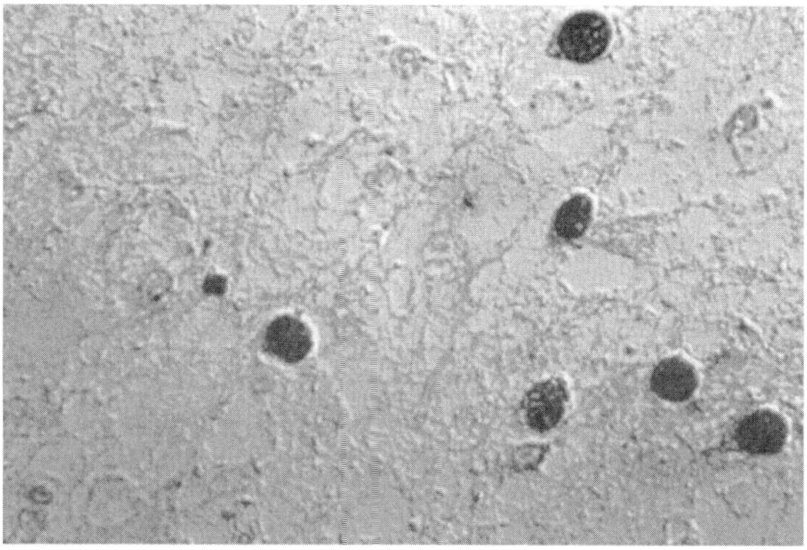

Figure 4 Enlarged oligodendroglial nuclei in biopsy tissue from a PML patient. In situ DNA hybridization using a JCV-specific oligonucleotide probe detects JCV DNA in the nuclei of affected cells. The chromophore diaminobenzidene (DAB) reveals viral DNA as a brown precipitate.

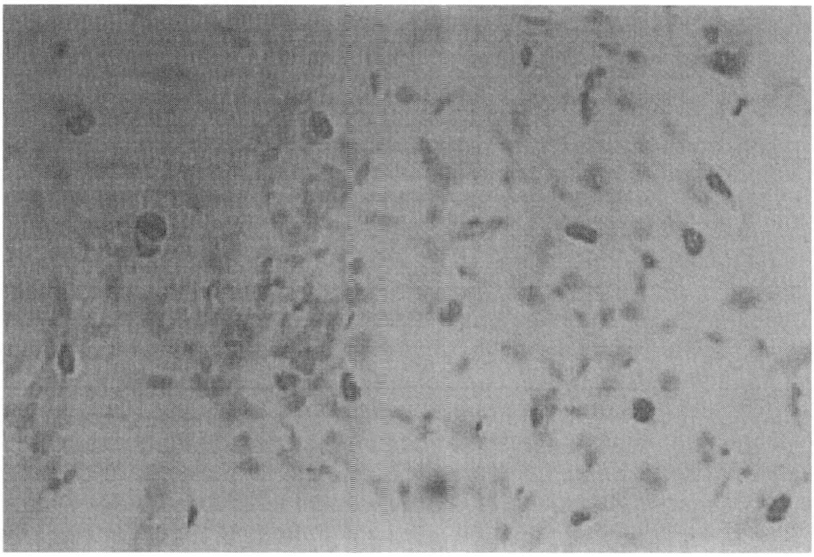

Figure 5 Enlarged astrocytes due to infection with JCV. Viral DNA is detected by in situ DNA hybridization with a JCV-specific probe. DAB was used as the chromophore. Hematoxylin stain identifies nuclei in uninfected cells.

9 PROGNOSIS

The median survival of PML-complicated AIDS is 6 months, and the mode is 1 month [11]. The survival of patients with PML is not significantly different with AIDS than with other immunosuppressive disorders, and survival has not changed measurably in the AIDS era from that in the pre-AIDS era. Recovery of neurological function, improvement of PML lesions in radiographic imaging, and survival exceeding 12 months have been observed in as many as 10% of patients with AIDS-associated PML [11]. Factors that appear to be associated with prolonged survival include PML as the presenting manifestation of AIDS, higher CD4 lymphocyte counts (>300 cells/mm^3), and contrast enhancement on radiographic imaging [154,155], although another study failed to link any radiological features with prognosis [156]. Additionally, a correlation between low titers of JC viral DNA in the CSF and prolonged survival has also been demonstrated [157–159]. Cell-mediated immunity to JCV has been increasingly demonstrated to be important to survival [160–162]. Survival exceeding 90 months from onset of illness in the setting of AIDS [11] and "burned out" cases have been reported with PML occurring in other immunosuppressive conditions. Presumably, the lack of recurrence of PML in some of the patients exhibiting long-term survival and recovery reflects clearance of the JC virus [157].

10 TREATMENT

The treatment of PML remains frustrating. To date there are no unequivocally successful therapeutic modalities whether in patients with AIDS or those with other underlying conditions. Restoration of immune function is likely critical to recovery. Stabilization and prolonged survival have been demonstrated in some patients in whom immunosuppressive medications, prescribed for other reasons, could be discontinued. Unfortunately, most of the extant literature consists of anecdotal reports. In the setting of HIV infection, antiretroviral therapy may be associated with prolonged survival and recovery. This has been most evident since the introduction of highly active antiretroviral therapy (HAART). Small, retrospective series have strongly suggested the value of HAART in HIV-infected patients with PML [163–168]. The benefit of HAART in AIDS-associated PML has not been universally observed, however [169,170]. An inflammatory response may be occasioned by the institution of HAART [171].

Nucleoside analogs have been employed because they impede the synthesis of DNA [172]. In vitro studies [173] have clearly demonstrated the ability of cytosine arabinoside (Cytarabine, ARA-C), a cytosine analog, to inhibit JC virus replication, and anecdotal reports of intravenous and intrathecal administration suggest the value of this therapy in PML [174–180]. However, a carefully conducted clinical trial of AIDS-related PML failed to show any value of either intravenous or intrathecal administration of ARA-C compared to placebo [181]. Theoretically, neither method of administration permitted adequate concentrations of the drug to reach the disease sites, and trials with intraparenchymal convection–enhanced delivery are under way. The delivery of cytosine arabinoside in this fashion appears to be well tolerated [182]. Despite anecdotal reports of the value of other nucleoside analogs such as adenine arabinoside (Vidarabine, ARA-A) [179,183,184] in PML, none has been convincingly demonstrated to ameliorate the disease course.

Interferons have also had occasional positive results both subcutaneously [185] and intrathecally [179] when used in conjunction with ARA-C. The antiretroviral activity of the interferons may be the consequence of their ability to stimulate natural killer (NK) cells. In a pilot study of 17 patients with AIDS and PML treated with alpha 2a interferon and zidovudine, two had long-term clinical stabilization, though none improved [186]. A retrospective study that compared patients with AIDS-associated PML receiving a minimum treatment of 3 weeks of 3 million units of α-IFN daily to untreated historical controls suggested that α-IFN treatment delayed the progression of the disease, palliated symptoms, and significantly prolonged survival [187]. However, reanalysis of this data accounting for HAART revealed no significant difference in those receiving α-IFN [188].

The antineoplastic drug camptothecin, a DNA topoisomerase I inhibitor, was demonstrated to block JC virus replication in vitro when administered in pulsed doses in amounts non toxic to cells [189]. Its therapeutic usefulness in PML has been entirely anecdotal [190,191]. Another antineoplastic drug, topotecan, may also inhibit JC virus replication [189]. However, both these drugs display significant systemic toxicity, and their value in the treatment of PML remains open to question.

Cidofovir [HPMPC; (*S*)-1-(3-hydroxy-2-phosphonylmethoxypropyl)cytosine] and its cyclic counterpart have demonstrated selective anti-polyomavirus activity [192]. The 50% inhibitory concentrations for HPMPC were in the range of 4–7 μg/mL, and its selectivity index varied from 11 to 20 for mouse polyomavirus and from 23 to 33 for SV40 strains in confluent cell monolayers [192]. It has been proposed as an agent for the treatment of PML [193]. Anecdotal cases suggest cidofovir's value in treating PML [194–198]. The drug causes ocular hypotony, bone marrow depression, and renal disorders. Currently, a well-designed AIDS Clinical Trials Group study is addressing the value of cidofovir in the same fashion as it studied ARA-C earlier. Preliminary results regarding the efficacy of cidofovir from this study have been disappointing [198a].

Increased understanding of the molecular biology of JC virus and new technologies will likely result in novel strategies. One possibility is the use of antisense oligonucleotides. An antisense oligonucleotide that is properly designed with a specific complementary base sequence that binds selectively to a targeted region of mRNA can prevent the translation of the mRNA into protein. Antisense oligonucleotide directed to JC virus T antigen may reduce viral expression by 80% [199]. Targeting transcription sites is difficult, however, during to the changes that occur in the viral cycle. Antisense oligonucleotides may also be designed to inhibit genes via triplex formation between the synthetic oligonucleotides and the double helical DNA [200]. Genetic manipulation or certain proteins that bind to a purine-rich domain may also result in inhibition of transcription and downregulate viral expression [201].

11 CONCLUSION

The increased understanding of the pathophysiology of JCV and PML portends well for the development of future curative strategies. The advent of AIDS with its large population of persons developing PML has allowed the organization of carefully designed therapeutic trials to address this issue and provided an additional incentive for the study of this disorder.

REFERENCES

1. Astrom, K.; Mancall, E.; Richardson Jr., E. Profressive multifocal leukoencephalopathy: a hitherto unrecognized complication of chronic lymphocytic leukemia and lymphoma. Brain. 1958, 81, 93–127.

2. Hallervorden, J. Eigennartige und nicht rubriziebare Prozesse. In Handbuch der Geiteskranheiten; Brumke, O., Ed.; Springer: Berlin, 1930, 1063–1107.
3. Winkelman, N.; Moore, M. Lymphogranulomatosis (Hodgkin's disease) of the nervous system. Arch Neurol Psychiat. 1941, 45, 304.
4. Bateman Jr., D.; Squires, G.; Thannhauser, S. Hodgkin's disease associated with Schindler's disease. Ann Intern Med. 1945, 22, 426–431.
5. Brooks, B.R.; Walker, D.L. Progressive multifocal leukoencephalopathy. Neurol Clin. 1984, 2(2), 299–313.
6. Miller, J.R.; Barrett, R.E.; Britton, C.B. Progressive multifocal leukoencephalopathy in a male homosexual with T-cell immune deficiency. N Engl J Med. 1982, 307(23), 1436–8.
7. Bedri, J.; Weinstein, W.; DeGregorio, P.; Verity, M.A. Progressive multifocal leukoencephalopathy in acquired immunodeficiency syndrome. N Engl J Med. 1983, 309(8), 492–493.
8. Gottlieb, M.S.; Schroff, R.; Schanker, H.M. Pneumocystis carinii pneumonia and mucosal candidiasis in previously healthy homosexual men: evidence of a new acquired cellular immunodeficiency. N Engl J Med. 1981, 305(24), 1425–1431.
9. Masur, H.; Michelis, M.A.; Greene, J.B. An outbreak of community-acquired Pneumocystis carinii pneumonia: initial manifestation of cellular immune dysfunction. N Engl J Med. 1981, 305(24), 1431–1438.
10. Siegal, F.P.; Lopez, C.; Hammer, G.S. Severe acquired immunodeficiency in male homosexuals, manifested by chronic perianal ulcerative herpes simplex lesions. N Engl J Med. 1981, 305(24), 1439–1444.
11. Berger, J.R.; Pall, L.; Lanska, D.; Whiteman, M. Progressive multifocal leukoencephalopathy in patients with HIV infection. J Neurovirol. 1998, 4(1), 59–68.
12. Cavanaugh, J.; Greenbaum, D.; Marshall, A.; Rubinstein, L. Cerebral demyelination associated with disorders of the reticuloendothelial system. Lancet. 1959, 2, 524–529.
13. Richardson Jr., E. Progressive multifocal leukoencephalopathy. N Engl J Med. 1961, 265, 815–823.
14. ZuRhein, G. Particles resembling papovavirions in human cerebral demyelinating disease. Science. 1965, 148, 1477–1479.
15. Silverman, L.; Rubinstein, L.J. Electron microscopic observations on a case of progressive multifocal leukoencephalopathy. Acta Neuropathol (Berl). 1965, 5(2), 215–224.
16. Padgett, B.L.; Walker, D.L.; ZuRhein, G.M.; Eckroade, R.J.; Dessel, B.H. Cultivation of papova-like virus from human brain with progressive multifocal leucoencephalopathy. Lancet. 1971, 1(7712), 1257–1260.
17. Narayan, O.; Penney Jr., J.B.; Johnson, R.T.; Herndon, R.M.; Weiner, L.P. Etiology of progressive multifocal leukoencephalopathy. Identification of papovavirus. N Engl J Med. 1973, 289(24), 1278–1282.
18. Walker, D. Progressive multifocal leukoencephalopathy: an opportunistic viral infection of the central nervous system. In Handbook of Clinical Neurology; Vinken, P., Bruyn, G., Eds.; Elsevier North-Holland: Amsterdam, 1978, 307–329.
19. Stoner, G.; Ryschkewitsch, C. Evidence of JC virus in two progressive multifocal leukoencephalopathy (PML) brains previously reported to be infected with SC40. J Neuropathol Exp Neurol. 1991, 50, 342.
20. Arthur, R.R.; Dagostin, S.; Shah, K.V. Detection of BK virus and JC virus in urine and brain tissue by the polymerase chain reaction. J Clin Microbiol. 1989, 27(6), 1174–1179.
21. Weber, T.; Turner, R.; Ruf, B. JC virus detected by polymerase chain reaction in cerebrospinal fluid of AIDS patients with progressive multifocal leukoencephalopathy. In Neurological and Neuropsychological Complications of HIV Infection; Berger, J., Levy, R., Eds.; Proceedings from the Satellite Meeting of the International Conference on AIDS. Kenness Canada, Inc.: Montreal, 1990, 100.
22. Lynch, K.J.; Frisque, R.J. Factors contributing to the restricted DNA replicating activity of JC virus. Virology. 1991, 180(1), 306–317.

23. Lynch, K.J.; Frisque, R.J. Identification of critical elements within the JC virus DNA replication origin. J Virol. 1990, 64(12), 5812–5822.
24. Major, E.O.; Amemiya, K. Elder, G.; Houff, S.A. Glial cells of the human developing brain and B cells of the immune system share a common DNA binding factor for recognition of the regulatory sequences of the human polyomavirus, JCV. J Neurosci Res. 1990, 27(4), 461–471.
25. Amemiya, K.; Traub, R.; Durham, L.; Major, E.O. Adjacent nuclear factor-1 and activator protein binding sites in the enhancer of the neurotropic JC virus. A common characteristic of many brain-specific genes. J Biol Chem. 1992, 267(20), 14204–14211.
26. Tada, H.; Lashgari, M.; Rappaport, J.; Khalili, K. Cell type-specific expression of JC virus early promoter is determined by positive and negative regulation. J Virol. 1989, 63(1), 463–466.
27. Houff, S.A.; Major, E.O.; Katz, D.A. Involvement of JC virus-infected mononuclear cells from the bone marrow and spleen in the pathogenesis of progressive multifocal leukoencephalopathy. N Engl J Med. 1988, 318(5), 301–305.
28. Frisque, R.J. Structure and function of JC virus T′ proteins. J Neurovirol. 2001, 7, 293–297.
29. ZuRhein, G. Polyoma-like virions in a human demyelinating disease. Acta NeuroPathol (Berl). 1967, 8, 57–68.
30. London, W.; Houff, S.; Houff, D.L. Brain tumors in owl monkeys inoculated with a human polyomavirus (JC virus). Science. 1978, 105, 227–237.
31. Walker, D.L.; Padgett, B.L.; ZuRhein, G.M.; Albert, A.E.; Marsh, R.F. Human papovavirus (JC): induction of brain tumors in hamsters. Science. 1973, 181(100), 674–676.
32. Dorries, K.; Loeber, G.; Meixensberger, J. Association of polyomaviruses JC, SV40, and BK with human brain tumors. Virology. 1987, 160(1), 268–270.
33. Major, E.; Vacante, D.; Houff, S. Human papovaviruses: JC virus, progressive multifocal leukoencephalopathy, and model systems for tumors of the central nervous system. In Neuropathogenic Viruses and Immunity; Spector, S., Bendinelli, M., Friedman, H., Eds.; Plenum Press: New York, 1992, 207–229.
34. Major, E.O. Human polyomaviruses. In Fields Virology; Knipe, D.M.H., Ed.; Lippincott Williams & Wilkins: Philadelphia, 2001, 2175–2196.
35. Chang, L.; Ernst, T.; Tornatore, C. Metabolite abnormalities in progressive multifocal leukoencephalopathy by proton magnetic resonance spectroscopy. Neurology. 1997, 48(4), 836–845.
36. Safak, M.; Khalili, K. Physical and functional interaction between viral and cellular proteins modulate JCV gene transcription. J Neurovirol. 2001, 7(4), 288–292.
37. Frisque, R.J.; Martin, J.D.; Padgett, B.L.; Walker, D.L. Infectivity of the DNA from four isolates of JC virus. J Virol. 1979, 32(2), 476–482.
38. Frisque, R.J.; Bream, G.L.; Cannella, M.T. Human polyomavirus JC virus genome. J Virol. 1984, 51(2), 458–469.
39. Khalili, K.; Rappaport, J.; Khoury, G. Nuclear factors in human brain cells bind specifically to the JCV regulatory region. EMBO J. 1988, 7(4), 1205–1210.
40. Vacante, D.; Traub, R.; Major, E. Extension of JC virus host range to monkey cells by insertion of a simian virus 40 enhancer into the JC virus regulatory region. Virology. 1988, 7, 1205–1210.
41. Dorries, K. Progressive multifocal leucoencephalopathy: analysis of JC virus DNA from brain and kidney tissue. Virus Res. 1984, 1(1), 25–38.
42. Martin, J.D.; King, D.M.; Slauch, J.M.; Frisque, R.J. Differences in regulatory sequences of naturally occurring JC virus variants. J Virol. 1985, 53(1), 306–311.
43. Jensen, P.N.; Major, E.O. A classification scheme for human polyomavirus JCV variants based on the nucleotide sequence of the noncoding regulatory region. J Neurovirol. 2001, 7(4), 280–287.

44. Aksamit, A.; Proper, J. JC virus replicates in primary adult astrocytes in culture. Ann Neurol. 1988, 24, 471.
45. Wroblewska, Z.; Wellish, M.; Bilden, D. Growth of JC virus in adult human brain cell cultures. Arch Virol. 1980, 65, 141–148.
46. Martin, J.D.; Padgett, B.L.; Walker, D.L. Characterization of tissue culture-induced heterogeneity in DNAs of independent isolates of JC virus. J Gen Virol. 1983, 64(10), 2271–2280.
47. Kenney, S.; Natarajan, V.; Strike, D.; Khoury, G.; Salzman, N.P. JC virus enhancer-promoter active in human brain cells. Science. 1984, 226(4680), 1337–1339.
48. Monaco, M.C.; Atwood, W.J.; Gravell, M.; Tornatore, C.S.; Major, E.O. JC virus infection of hematopoietic progenitor cells, primary B lymphocytes, and tonsillar stromal cells: implications for viral latency. J Virol. 1996, 70(10), 7004–7012.
49. Atwood, W.; Amemiya, K.; Traub, R.; Harms, J.; Major, E. Interactions of the human polyomavirus, JCV, with human B lymphocytes. Virology. 1992, 190, 716–723.
50. Laghi, L.; Randolph, A.E.; Chauhan, D.P. JC virus DNA is present in the mucosa of the human colon and in colorectal cancers. Proc Natl Acad Sci USA. 1999, 96(13), 7484–7489.
51. Wei, G.; Liu, C.K.; Atwood, W.J. JC virus binds to primary human glial cells, tonsillar stromal cells, and B-lymphocytes, but not to T lymphocytes. J Neurovirol. 2000, 6(2), 127–136.
52. Feigenbaum, L.; Khalili, K.; Major, E.; Khoury, G. Regulation of the host range of human papovavirus JCV. Proc Natl Acad Sci USA. 1987, 84(11), 3695–3698.
53. Amemiya, K.; Traub, R.; Durham, L.; Major, E.O. Interaction of a nuclear factor-1-like protein with the regulatory region of the human polyomavirus JC virus. J Biol Chem. 1989, 264(12), 7025–7032.
54. Miyamura, T.; Jikuya, H.; Soeda, E.; Yoshiike, K. Genomic structure of human polyoma virus JC: nucleotide sequence of the region containing replication origin and small-T-antigen gene. J Virol. 1983, 45(1), 73–79.
55. Nandi, A.; Das, G.; Salzman, N.P. Characterization of a surrogate TATA box promoter that regulates in vitro transcription of the simian virus 40 major late gene. Mol Cell Biol. 1985, 5(3), 591–594.
56. Dynan, W.S.; Tjian, R. Control of eukaryotic messenger RNA synthesis by sequence-specific DNA-binding proteins. Nature. 1985, 316(6031), 774–778.
57. Gruss, P.; Khoury, G. The SV 40 tandem repeats as an element of the early promotor. Proc Natl Acad Sci USA. 1981, 78, 943–947.
58. Ahmed, S.; Rappaport, J.; Tada, H.; Kerr, D.; Khalili, K. A nuclear protein derived from brain cells stimulates transcription of the human neurotropic virus promoter, JCVE, in vitro. J Biol Chem. 1990, 265(23), 13899–13905.
59. Tamura, T.; Inoue, T.; Nagata, K.; Mikoshiba, K. Enhancer of human polyoma JC virus contains nuclear factor I-binding sequences; analysis using mouse brain nuclear extracts. Biochem Biophys Res Commun. 1988, 157(2), 419–425.
60. Kerr, D.; Khalili, K. A recombinant cDNA derived from human brain encodes a DNA binding protein that stimulates transcription of the human neurotropic virus JCV. J Biol Chem. 1991, 266(24), 15876–15881.
61. Sock, E.; Wegner, M.; Grummt, F. DNA replication of human polyomavirus JC is stimulated by NF-I in vivo. Virology. 1991, 182(1), 298–308.
62. Monaco, M.C.; Sabath, B.F.; Durham, L.C.; Major, E.O. JC virus multiplication in human hematopoietic progenitor cells requires the NF-1 class D transcription factor. J Virol. 2001, 75(20), 9687–9695.
63. Sumner, C.; Shinohara, T.; Durham, L.; Traub, R.; Major, E.O.; Amemiya, K. Expression of multiple classes of the nuclear factor-1 family in the developing human brain: differential expression of two classes of NF-1 genes. J Neurovirol. 1996, 2(2), 87–100.
64. Major, E.O.; Amemiya, K.; Tornatore, C.S.; Houff, S.A.; Berger, J.R. Pathogenesis and molecular biology of progressive multifocal leukoencephalopathy, the JC virus-induced demyelinating disease of the human brain. Clin Microbiol Rev. 1992, 5(1), 49–73.

65. Tornatore, C.; Berger, J.R.; Houff, S.A. Detection of JC virus DNA in peripheral lymphocytes from patients with and without progressive multifocal leukoencephalopathy. Ann Neurol. 1992, 31(4), 454–462.
66. Coleman, D.V.; Wolfendale, M.R.; Daniel, R.A. A prospective study of human polyomavirus infection in pregnancy. J Infect Dis. 1980, 142(1), 1–8.
67. Flaegstad, T.; Sundsfjord, A.; Arthur, R.R.; Pedersen, M.; Traavik, T.; Subramani, S. Amplification and sequencing of the control regions of BK and JC virus from human urine by polymerase chain reaction. Virology. 1991, 180(2), 553–560.
68. Myers, C.; Frisque, R.J.; Arthur, R.R. Direct isolation and characterization of JC virus from urine samples of renal and bone marrow transplant patients. J Virol. 1989, 63(10), 4445–4449.
69. Yogo, Y.; Kitamura, T.; Sugimoto, C. Sequence rearrangement in JC virus DNAs molecularly cloned from immunosuppressed renal transplant patients. J Virol. 1991, 65(5), 2422–2428.
70. Yogo, T.; Kitamura, T.; Sugimoto, C. Isolation of a possible archetypal JC virus DNA sequence from nonimmunocompromised individuals. J Virol. 1990, 64, 3139–3143.
71. Henson, J.; Saffer, J.; Furneaux, H. The transcription factor Sp1 binds to the JC virus promoter and is selectively expressed in glial cells in human brain. Ann Neurol. 1992, 32(1), 72–77.
72. Briggs, M.R.; Kadonaga, J.T.; Bell, S.P.; Tjian, R. Purification and biochemical characterization of the promoter-specific transcription factor, Sp1. Science. 1986, 234(4772), 47–52.
73. Tominaga, T.; Yogo, Y.; Kitamura, T.; Aso, Y. Persistence of archetypal JC virus DNA in normal renal tissue derived from tumor-bearing patients. Virology. 1992, 186(2), 736–741.
74. Stoner, G.L.; Ryschkewitsch, C.F.; Walker, D.L.; Webster, H.D. JC papovavirus large tumor (T)-antigen expression in brain tissue of acquired immune deficiency syndrome (AIDS) and non-AIDS patients with progressive multifocal leukoencephalopathy. Proc Natl Acad Sci USA. 1986, 83(7), 2271–2275.
75. Ho, D.; Rota, T.; Schooley, R. Isolation of HTLV-III from cerebrospinal fluid and neural tissues of patients with neurologic syndromes related to the acquired immunodeficiency syndrome. N Engl J Med. 1985, 313, 1493–1497.
76. Marshall, D.W.; Brey, R.L.; Cahill, W.T.; Houk, R.W.; Zajac, R.A.; Boswell, R.N. Spectrum of cerebrospinal fluid findings in various stages of human immunodeficiency virus infection. Arch Neurol. 1988, 45(9), 954–958.
77. Elovaara, I.; Iivanainen, M.; Valle, S.L.; Suni, J.; Tervo, T.; Lahdevirta, J. CSF protein and cellular profiles in various stages of HIV infection related to neurological manifestations. J Neurol Sci. 1987, 78(3), 331–342.
78. Hollander, H. Cerebrospinal fluid normalities and abnormalities in individuals infected with human immunodeficiency virus. J Infect Dis. 1988, 158(4), 855–858.
79. Davis, L.; Hjelle, B.; Miller, V. Early viral invasion of brain in iatrogenic human immunodeficiency virus infection. Ann Neurol. 1991, 30, 314.
80. Levy, J.A.; Shimabukuro, J.; Hollander, H.; Mills, J.; Kaminsky, L. Isolation of AIDS-associated retroviruses from cerebrospinal fluid and brain of patients with neurological symptoms. Lancet. 1985, 2(8455), 586–588.
81. Ward, J.M.; O'Leary, T.J.; Baskin, G.B. Immunohistochemical localization of human and simian immunodeficiency viral antigens in fixed tissue sections. Am J Pathol. 1987, 127(2), 199–205.
82. Shaw, G.M.; Harper, M.E.; Hahn, B.H. HTLV-III infection in brains of children and adults with AIDS encephalopathy. Science. 1985, 227(4683), 177–182.
83. Koenig, S.; Gendelman, H.E.; Orenstein, J.M. Detection of AIDS virus in macrophages in brain tissue from AIDS patients with encephalopathy. Science. 1986, 233(4768), 1089–1093.
84. Vazeux, R.; Brousse, N.; Jarry, A. AIDS subacute encephalitis. Identification of HIV-infected cells. Am J Pathol. 1987, 126(3), 403–410.
85. Tyor, W.R.; Glass, J.D.; Griffin, J.W. Cytokine expression in the brain during the acquired immunodeficiency syndrome. Ann Neurol. 1992, 31(4), 349–360.

86. Gallo, P.; Frei, K.; Rordorf, C.; Lazdins, J.; Tavolato, B.; Fontana, A. Human immunodeficiency virus type 1 (HIV-1) infection of the central nervous system: an evaluation of cytokines in cerebrospinal fluid. J Neuroimmunol. 1989, 23(2), 109–116.
87. Resnick, L.; Berger, J.R.; Shapshak, P.; Tourtellotte, W.W. Early penetration of the blood-brain barrier by HIV. Neurology. 1988, 38(1), 9–14.
88. Kure, K.; Lyman, W.D.; Weidenheim, K.M.; Dickson, D.W. Cellular localization of an HIV-1 antigen in subacute AIDS encephalitis using an improved double-labeling immunohistochemical method. Am J Pathol. 1990, 136(5), 1085–1092.
89. Yoshioka, M.; Bradley, W.G.; Shapshak, P.; Nagano, I.; Stewart, R.V.; Xin, K.Q.; Srivastava, A.K.; Nakamura, S. Role of immune activation and cytokine expression in HIV-1-associated neurologic diseases. Adv Neuroimmunol. 1995, 5(3), 335–358.
90. Oppenheim, J. Lymphokines. In Cellular Functions in Immunity and Inflammation; Oppenheim, J., Rosenstreich, D., Potter, M., Eds.; Elsevier: New York, 1981, 259–282.
91. Cavender, D.E.; Haskard, D.O.; Joseph, B.; Ziff, M. Interleukin 1 increases the binding of human B and T lymphocytes to endothelial cell monolayers. J Immunol. 1986, 136(1), 203–207.
92. Brightman, M.W.; Reese, T.S. Junctions between intimately apposed cell membranes in the vertebrate brain. J Cell Biol. 1969, 40(3), 648–677.
93. Durum, S.; Oppenheim, J. Macrophage-derived mediators: interleukin 1, tumor necrosis factor, interleukin 6, interferon, and related cytokines. In Fundamental Immunology; Paul, W.E., Ed.; Raven Press: New York, 1989, 639–661.
94. Bevilacqua, M.P.; Pober, J.S.; Wheeler, M.E.; Cotran, R.S.; Gimbrone Jr., M.A. Interleukin 1 acts on cultured human vascular endothelium to increase the adhesion of polymorphonuclear leukocytes, monocytes, and related leukocyte cell lines. J Clin Invest. 1985, 76(5), 2003–2011.
95. Pryce, G.; Male, D.K.; Sarkar, C. Control of lymphocyte migration into brain: selective interactions of lymphocyte subpopulations with brain endothelium. Immunology. 1991, 72(3), 393–398.
96. Kishimoto, T. The biology of interleukin-6. Blood. 1989, 74, 1–10.
97. Snapper, C.M.; Finkelman, F.D.; Paul, W.E. Regulation of IgG1 and IgE production by interleukin 4. Immunol Rev. 1988, 102, 51–75.
98. Wahl, S.M.; Allen, J.B. N McCartney-Francis. Macrophage- and astrocyte-derived transforming growth factor beta as a mediator of central nervous system dysfunction in acquired immune deficiency syndrome. J Exp Med. 1991, 173(4), 981–991.
99. Blake, K.; Pillay, D.; Knowles, W.; Brown, D.W.; Griffiths, P.D.; Taylor, B. JC virus associated meningoencephalitis in an immunocompetent girl. Arch Dis Child. 1992, 67(7), 956–957.
100. Taguchi, F.; Kajioka, J.; Miyamura, T. Prevalence rate and age of acquisition of antibodies against JC virus and BK virus in human sera. Microbiol Immunol. 1982, 26(11), 1057–1064.
101. Walker, D.; Padgett, B. The epidemiology of human polyomaviruses. In Polyomaviruses and Human Neurological Disease; Sever, J. eds., Madden, D., Eds.; Alan R Liss: New York, 1983, 99–106.
102. Walker, D.; Padgett, B. Progressive multifocal leukoencephalopathy. In Comprehensive Virology; Fraenkel-Conrat, H., Wagner, R., Eds.; Plenum: New York, 1983, 161–193.
103. Padgett, B.; Walker, D. Virologic and serologic studies of progressive multifocal leukoencephalopathy. In Polyomaviruses and Human Neurological Disease; Sever, J., Madden, D., Eds.; Alan R Liss: New York, 1983, 107–118.
104. Gibson, P.E.; Field, A.M.; Gardner, S.D.; Coleman, D.V. Occurrence of IgM antibodies against BK and JC polyomaviruses during pregnancy. J Clin Pathol. 1981, 34(6), 674–679.
105. Krupp, L.B.; Lipton, R.B.; Swerdlow, M.L.; Leeds, N.E.; Llena, J. Progressive multifocal leukoencephalopathy: clinical and radiographic features. Ann Neurol. 1985, 17(4), 344–349.
106. Berger, J.R.; Kaszovitz, B.; Post, M.J.; Dickinson, G. Progressive multifocal leukoencephalopathy associated with human immunodeficiency virus infection. A review of the literature with a report of sixteen cases. Ann Intern Med. 1987, 107(1), 78–87.

107. Gillespie, S.M.; Chang, Y.; Lemp, G. Progressive multifocal leukoencephalopathy in persons infected with human immunodeficiency virus, San Francisco, 1981–1989. Ann Neurol. 1991, 30(4), 597–604.
108. Holman, R.C.; Janssen, R.S.; Buehler, J.W.; Zelasky, M.T.; Hooper, W.C. Epidemiology of progressive multifocal leukoencephalopathy in the United States: analysis of national mortality and AIDS surveillance data. Neurology. 1991, 41(11), 1733–1736.
109. Messite, J.; Stellman, S.D. Accuracy of death certificate completion: the need for formalized physician training. JAMA. 1996, 275(10), 794–796.
110. Kuchelmeister, K.; Gullotta, F.; Bergmann, M.; Angeli, G.; Masini, T. Progressive multifocal leukoencephalopathy (PML) in the acquired immunodeficiency syndrome (AIDS). A neuropathological autopsy study of 21 cases. Pathol Res Pract. 1993, 189(2), 163–173.
111. Kure, K.; Llena, J.F.; Lyman, W.D. Human immunodeficiency virus-1 infection of the nervous system: an autopsy study of 268 adult, pediatric, and fetal brains. Hum Pathol. 1991, 22(7), 700–710.
112. Lang, W.; Miklossy, J.; Deruaz, J.P. Neuropathology of the acquired immune deficiency syndrome (AIDS): a report of 135 consecutive autopsy cases from Switzerland. Acta Neuropathol. 1989, 77(4), 379–390.
113. Whiteman, M.L.; Post, M.J ; Berger, J.R.; Tate, L.G.; Bell, M.D.; Limonte, L.P. Progressive multifocal leukoencephalopathy in 47 HIV-seropositive patients: neuroimaging with clinical and pathologic correlation. Radiology. 1993, 187(1), 233–240.
114. Bacellar, H.; Munoz, A.; Miller, E.N. Temporal trends in the incidence of HIV-1-related neurologic diseases: Multicenter AIDS Cohort Study, 1985–1992. Neurology. 1994, 44(10), 1892–1900.
115. Power, C.; Kong, P.A.; Crawford, T.O. Cerebral white matter changes in acquired immunodeficiency syndrome dementia: alterations of the blood-brain barrier. Ann Neurol. 1993, 34(3), 339–350.
116. Hofman, F.M.; Dohadwala, M.M.; Wright, A.D.; Hinton, D.R.; Walker, S.M. Exogenous tat protein activates central nervous system-derived endothelial cells. J Neuroimmunol. 1994, 54(1–2), 19–28.
117. Sasseville, V.G.; Newman, W.A.; Lackner, A.A. Elevated vascular cell adhesion molecule-1 in AIDS encephalitis induced by simian immunodeficiency virus. Am J Pathol. 1992, 141(5), 1021–1030.
118. Tada, H.; Rappaport, J.; Lashgari, M.; Amini, S.; Wong-Staal, F.; Khalili, K. Transactivation of the JC virus late promoter by the tat protein of type 1 human immunodeficiency virus in glial cells. Proc Natl Acad Sci USA. 1990, 87(9), 3479–3483.
119. Berger, J.R.; Scott, G.; Albrecht, J.; Belman, A.L.; Tornatore, C.; Major, E.O. Progressive multifocal leukoencephalopathy in HIV-1-infected children. AIDS. 1992, 6(8), 837–841.
120. Morriss, M.C.; Rutstein, R.M.; Rudy, B.; Desrochers, C.; Hunter, J.V.; Zimmerman, R.A. Progressive multifocal leukoencephalopathy in an HIV-infected child. Neuroradiology. 1997, 39(2), 142–144.
121. Singer, C.; Berger, J.R.; Bowen, B.C.; Bruce, J.H.; Weiner, W.J. Akinetic-rigid syndrome in a 13-year-old girl with HIV-related progressive multifocal leukoencephalopathy. Mov Disord. 1993, 8(1), 113–116.
122. Satishchandra, P.; Nalini, A.; Gourie-Devi, M. Profile of neurologic disorders associated with HIV/AIDS from Bangalore, south India (1989–96). Indian J Med Res. 2000, 111, 14–23.
123. Baril, L.; Jouan, M.; Agher, R. Impact of highly active antiretroviral therapy on onset of Mycobacterium avium complex infection and cytomegalovirus disease in patients with AIDS. AIDS. 2000, 14(16), 2593–2596.
124. Verbraak, F.D.; Boom, R.; Wertheim-van Dillen, P.M.; van den Horn, G.J.; Kijlstra, A.; de Smet, M.D. Influence of highly active antiretroviral therapy on the development of CMV disease in HIV positive patients at high risk for CMV disease. Br J Ophthalmol. 1999, 83(10), 1186–1189.

125. Maschke, M.; Kastrup, O.; Esser, S.; Ross, B.; Hengge, U.; Hufnagel, A. Incidence and prevalence of neurological disorders associated with HIV since the introduction of highly active antiretroviral therapy (HAART). J Neurol Neurosurg Psychiatry. 2000, 69(3), 376–380.
126. Sparano, J.A.; Anand, K.; Desai, J.; Mitnick, R.J.; Kalkut, G.E.; Hanau, L.H. Effect of highly active antiretroviral therapy on the incidence of HIV- associated malignancies at an urban medical center. J Acquir Immune Defic Syndr. 1999, 21(suppl 1), S18–S22.
127. d'Arminio Monforte, A.; Duca, P.G.; Vago, L.; Grassi, M.P.; Moroni, M. Decreasing incidence of CNS AIDS-defining events associated with antiretroviral therapy. Neurology. 2000, 54(9), 1856–1859.
128. Sweeney, B.J.; Manji, H.; Miller, R.F.; Harrison, M.J.; Gray, F.; Scaravilli, F. Cortical and subcortical JC virus infection: two unusual cases of AIDS associated progressive multifocal leukoencephalopathy. J Neurol Neurosurg Psychiatry. 1994, 57(8), 994–997.
129. Wong, M.C.; Suite, N.D.; Labar, D.R. Seizures in human immunodeficiency virus infection. Arch Neurol. 1990, 47(6), 640–642.
130. Omerud, L.; Rhodes, R.; Gross, S.; Crane, L.; Houchkin, K. Ophthalmologic manifestations of acquired immune deficiency syndrome-associated progressive multifocal leukoencephalopathy. Ophthalmology. 1996, 103, 899–906.
131. Henin, D.; Smith, T.W.; De Girolami, U.; Sughayer, M.; Hauw, J.J. Neuropathology of the spinal cord in the acquired immunodeficiency syndrome. Hum Pathol. 1992, 23(10), 1106–1114.
132. Monaco, M.C.; Jensen, P.N.; Hou, J.; Durham, L.C.; Major, E.O. Detection of JC virus DNA in human tonsil tissue: evidence for site of initial viral infection. J Virol. 1998, 72(12), 9918–9923.
133. Thurnher, M.; Post, M. Initial and follow-up MR imaging findings in AIDS-related progressive multifocal leukoencephalopathy treated with highly active antiretroviral therapy. AJNR Am J Neuroradiol. 2001, 22(5), 977–984.
134. Gray, F.; Belec, L.; Lescs, M.C. Varicella-zoster virus infection of the central nervous system in the acquired immune deficiency syndrome. Brain. 1994, 117(5), 987–999.
135. Berger, J.R.; Tornatore, C.; Major, E.O. Relapsing and remitting human immunodeficiency virus-associated leukoencephalomyelopathy. Ann Neurol. 1992, 31(1), 34–38.
136. Bhigjee, A.I.; Patel, V.B.; Bhagwan, B.; Moodley, A.A.; Bill, P.L. HIV and acute disseminated encephalomyelitis. S Afr Med J. 1999, 89(3), 283–284.
137. Chetty, K.G.; Kim, R.C.; Mahutte, C.K. Acute hemorrhagic leukoencephalitis during treatment for disseminated tuberculosis in a patient with AIDS. Int J Tuberc Lung Dis. 1997, 1(6), 579–581.
138. Scaravilli, F.; Daniel, S.E.; Harcourt-Webster, N.; Guiloff, R.J. Chronic basal meningitis and vasculitis in acquired immunodeficiency syndrome. A possible role for human immunodeficiency virus. Arch Pathol Lab Med. 1989, 113(2), 192–195.
139. Brochet, B.; Dousset, V. Pathological correlates of magnetization transfer imaging abnormalities in animal models and humans with multiple sclerosis. Neurology. 1999, 53(5), S12–S17.
140. Dousset, V.; Armand, J.P.; Lacoste, D. Magnetization transfer study of HIV encephalitis and progressive multifocal leukoencephalopathy. Groupe d'Epidemiologie Clinique du SIDA en Aquitaine. AJNR Am J Neuroradiol. 1997, 18(5), 895–901.
141. Nelson, P.K.; Masters, L.T.; Zagzag, D.; Kelly, P.J. Angiographic abnormalities in progressive multifocal leukoencephalopathy: an explanation based on neuropathologic findings. AJNR Am J Neuroradiol. 1999, 20(3), 487–494.
142. Iranzo, A.; Marti-Fabregas, J.; Domingo, P. Absence of thallium-201 brain uptake in progressive multifocal leukoencephalopathy in AIDS patients. Acta Neurol Scand. 1999, 100(2), 102–105.
143. Port, J.D.; Miseljic, S.; Lee, R.R. Progressive multifocal leukoencephalopathy demonstrating contrast enhancement on MRI and uptake of thallium-201: a case report. Neuroradiology. 1999, 41(12), 895–898.

144. Fong, I.W.; Britton, C.B.; Luinstra, K.E.; Toma, E.; Mahony, J.B. Diagnostic value of detecting JC virus DNA in cerebrospinal fluid of patients with progressive multifocal leukoencephalopathy. J Clin Microbiol. 1995, 33(2), 484–486.
145. von Einsiedel, R.W.; Fife, T.D.; Aksamit, A.J. Progressive multifocal leukoencephalopathy in AIDS: a clinicopathologic study and review of the literature. J Neurol. 1993, 240(7), 391–406.
146. Katz, R.L.; Alappattu, C.; Glass, J.P ; Bruner, J.M. Cerebrospinal fluid manifestations of the neurologic complications of human immunodeficiency virus infection. Acta Cytol. 1989, 33(2), 233–244.
147. McGuire, D.; Barhite, S.; Hollander, H.; Miles, M. JC virus DNA in cerebrospinal fluid of human immunodeficiency virus-infected patients: predictive value for progressive multifocal leukoencephalopathy. Ann Neurol. 1995, 37(3), 395–399.
148. Weber, T.; Turner, R.W.; Frye, S. Progressive multifocal leukoencephalopathy diagnosed by amplification of JC virus-specific DNA from cerebrospinal fluid. AIDS. 1994, 8(1), 49–57.
149. Drews, K.; Bashir, T.; Dorries, K. Quantification of human polyomavirus JC in brain tissue and cerebrospinal fluid of patients with progressive multifocal leukoencephalopathy by competitive PCR. J Virol Methods. 2000, 84(1), 23–36.
150. Richardson, E. Progressive multifocal leukoencephalopathy. In Handbook of Clinical Neurology; Vinken, P., Bruyn, G., Eds.; Elsevier: New York, 1970, 485–499.
151. Mazlo, M.; Herndon, R. Progressive multifocal leukoencephalopathy: ultrastructural findings in two brain biopsies. Neuropathol Appl Neurobiol. 1977, 3, 323–339.
152. Mazlo, M.; Tariska, I. Are astrocytes infected in progressive multifocal leukoencephalopathy (PML)?. Acta Neuropathol. 1982, 56(1), 45–51.
153. Samorei, I.W.; Schmid, M.; Pawlita, M. High sensitivity detection of JC-virus DNA in postmortem brain tissue by in situ PCR. J Neurovirol. 2000, 6(1), 61–74.
154. Arbusow, V.; Strupp, M.; Pfister, H W.; Seelos, K.C.; Bruckmann, H.; Brandt, T. Contrast enhancement in progressive multifocal leukoencephalopathy: a predictive factor for long-term survival?. J Neurol. 2000, 247(4), 306–308.
155. Berger, J.R.; Levy, R.M.; Flomenhoft, D.; Dobbs, M. Predictive factors for prolonged survival in acquired immunodeficiency syndrome-associated progressive multifocal leukoencephalopathy. Ann Neurol. 1998, 44(3), 341–349.
156. Post, M.J.; Yiannoutsos, C.; Simpson, D. Progressive multifocal leukoencephalopathy in AIDS: are there any MR findings useful to patient management and predictive of patient survival? AIDS Clinical Trials Group, 243 Team. AJNR Am J Neuroradiol. 1999, 20(10), 1896–1906.
157. De Luca, A.; Giancola, M.; Cingolani, A. Clinical and virological monitoring during treatment with intrathecal cytarabine in patients with AIDS-associated progressive multifocal leukoencephalopathy. Clin Infect Dis. 1999, 28(3), 624–628.
158. Taoufik, Y.; Gasnault, J.; Karaterki, A. Prognostic value of JC virus load in cerebrospinal fluid of patients with progressive multifocal leukoencephalopathy. J Infect Dis. 1998, 178(6), 1816–1820.
159. Yiannoutsos, C.T.; Major, E.O.; Curfman, B. Relation of JC virus DNA in the cerebrospinal fluid to survival in acquired immunodeficiency syndrome patients with biopsy-proven progressive multifocal leukoencephalopathy. Ann Neurol. 1999, 45(6), 816–821.
160. Koralnik, I.J.; Du Pasquier, R.A.; Kuroda, M.J. Association of prolonged survival in HLA-A2+ progressive multifocal leukoencephalopathy patients with a CTL response specific for a commonly recognized JC virus epitope. J Immunol. 2002, 168(1), 499–504.
161. Du Pasquier, R.A.; Clark, K.W.; Smith, P.S. JCV-specific cellular immune response correlates with a favorable clinical outcome in HIV-infected individuals with progressive multifocal leukoencephalopathy. J Neurovirol. 2001, 7(4), 318–322.
162. Koralnik, I.J.; Pasquier RA, D.u.; Letvin, N.L. JC virus-specific cytotoxic T lymphocytes in individuals with progressive multifocal leukoencephalopathy. J Virol. 2001, 75(7), 3483–3487.

163. Clifford, D.B.; Yiannoutsos, C.; Glicksman, M. HAART improves prognosis in HIV-associated progressive multifocal leukoencephalopathy. Neurology. 1999, 52(3), 623–625.
164. Albrecht, H.; Hoffmann, C.; Degen, O. Highly active antiretroviral therapy significantly improves the prognosis of patients with HIV-associated progressive multifocal leukoencephalopathy. AIDS. 1998, 12(10), 1149–1154.
165. Inui, K.; Miyagawa, H.; Sashihara, J. Remission of progressive multifocal leukoencephalopathy following highly active antiretroviral therapy in a patient with HIV infection. Brain Dev. 1999, 21(6), 416–419.
166. Miralles, P.; Berenguer, J. D Garcia de Viedma. Treatment of AIDS-associated progressive multifocal leukoencephalopathy with highly active antiretroviral therapy. AIDS. 1998, 12(18), 2467–2472.
167. Tantisiriwat, W.; Tebas, P.; Clifford, D.B.; Powderly, W.G.; Fichtenbaum, C.J. Progressive multifocal leukoencephalopathy in patients with AIDS receiving highly active antiretroviral therapy. Clin Infect Dis. 1999, 28(5), 1152–1154.
168. Giudici, B.; Vaz, B.; Bossolasco, S. Highly active antiretroviral therapy and progressive multifocal leukoencephalopathy: effects on cerebrospinal fluid markers of JC virus replication and immune response. Clin Infect Dis. 2000, 30(1), 95–99.
169. De Luca, A.; Ammassari, A.; Cingolani, A.; Giancola, M.; Antinori, A. Disease progression and poor survival of AIDS-associated progressive multifocal leukoencephalopathy despite highly active antiretroviral therapy [letter]. AIDS. 1998, 12, 1937–1938.
170. Cinque, P.; Pierotti, C.; Vigano, M.G. The good and evil of HAART in HIV-related progressive multifocal leukoencephalopathy. J Neurovirol. 2001, 7(4), 358–363.
171. DeSimone, J.A.; Pomerantz, R.J.; Babinchak, T.J. Inflammatory reactions in HIV-1-infected persons after initiation of highly active antiretroviral therapy. Ann Intern Med. 2000, 133(6), 447–454.
172. Goodman, A.; Goodman, L.; Rall, T.; Murad, F. The Pharmacological Basis of Therapeutics; Macmillan: New York, 1985.
173. Hou, J.; Major, E.O. The efficacy of nucleoside analogs against JC virus multiplication in a persistently infected human fetal brain cell line. J Neurovirol. 1998, 4(4), 451–456.
174. Bauer, W.R.; Turel Jr., A.P.; Johnson, K.P. Progressive multifocal leukoencephalopathy and cytarabine. Remission with treatment. JAMA. 1973, 226(2), 174–176.
175. Buckman, R.; Wiltshaw, E. Letter: Progressive multifocal leucoencephalopathy successfully treated with cytosine arabinoside. Br J Haematol. 1976, 34(1), 153–158.
176. Conomy, J.P.; Beard, N.S.; Matsumoto, H.; Roessmann, U. Cytarabine treatment of progressive multifocal leukoencephalopathy. Clinical course and detection of virus-like particles after antiviral chemotherapy. JAMA. 1974, 229(10), 1313–1316.
177. Lidman, C.; Lindqvist, L.; Mathiesen, T.; Grane, P. Progressive multifocal leukoencephalopathy in AIDS. AIDS. 1991, 5(8), 1039–1041.
178. O'Riordan, T.; Daly, P.A.; Hutchinson, M.; Shattock, A.G.; Gardner, S.D. Progressive multifocal leukoencephalopathy-remission with cytarabine. J Infect. 1990, 20(1), 51–54.
179. Tashiro, K.; Doi, S.; Moriwaka, F.; Maruo, Y.; Nomura, M. Progressive multifocal leucoencephalopathy with magnetic resonance imaging verification and therapeutic trials with interferon. J Neurol. 1987, 234(6), 427–429.
180. Van Horn, G.; Bastian, F.; Moake, J. Progressive multifocal leukoencephalopathy: failure of response to transfer factor and cytarabine. Neurology. 1978, 28, 794–797.
181. Hall, C.; Dafni, U.; Simpson, D. Failure of cytarabine in progressive multifocal leukoencephalopathy associated with human immunodeficiency virus infection. AIDS Clinical Trials Group 243 Team [see comments]. N Engl J Med. 1998, 338, 1345–1351.
182. Levy, R. Personal communication, 2002.
183. Rand, K.H.; Johnson, K.P.; Rubinstein, L.J. Adenine arabinoside in the treatment of progressive multifocal leukoencephalopathy: use of virus-containing cells in the urine to assess response to therapy. Ann Neurol. 1977, 1(5), 458–462.

184. Wolinsky, J.S.; Johnson, K P.; Rand, K.; Merigan, T.C. Progressive multifocal leukoencephalopathy: clinical pathological correlates and failure of a drug trial in two patients. Trans Am Neurol Assoc. 1976, 101, 81–82.
185. Steiger, M.J.; Tarnesby, G.; Gabe, S.; McLaughlin, J.; Schapira, A.H. Successful outcome of progressive multifocal leukoencephalopathy with cytarabine and interferon. Ann Neurol. 1993, 33(4), 407–411.
186. Berger, J.; Pall, L.; McArthur, J. A pilot study of recombinant alpha 2a interferon in the treatment of AIDS-related progressive multifocal leukoencephalopathy (abstract). Neurology. 1992, 42(suppl 3), 257.
187. Huang, S.S.; Skolasky, R.L ; Dal Pan, G.J.; Royal 3rd, W.; McArthur, J.C. Survival prolongation in HIV-associated progressive multifocal leukoencephalopathy treated with alpha-interferon: an observational study. J Neurovirol. 1998, 4(3), 324–332.
188. Geschwind, M.D.; Skolasky, R.I.; Royal, W.S.; McArthur, J.C. The relative contributions of HAART and alpha-interferon for therapy of progressive multifocal leukoencephalopathy in AIDS. J Neurovirol. 2001, 7(4), 353–357.
189. Kerr, D.A.; Chang, C.F.; Gordon, J.; Bjornsti, M.A.; Khalili, K. Inhibition of human neurotropic virus (JCV) DNA replication in glial cells by camptothecin. Virology. 1993, 196(2), 612–618.
190. Vollmer-Haase, J.; Young, P.; Ringelstein, E.B. Efficacy of camptothecin in progressive multifocal leucoencephalopathy. Lancet. 1997, 349(9062), 1366.
191. O'Reilly, S. Efficacy of camptothecin in progressive multifocal leucoencephalopathy. Lancet. 1997, 350(9073), 291.
192. Andrei, G.; Snoeck, R.; Vandeputte, M. E De Clercq. Activities of various compounds against murine and primate polyomaviruses. Antimicrob Agents Chemother. 1997, 41(3), 587–593.
193. Sadler, M.; Chinn, R.; Healy, J.; Fisher, M.; Nelson, M.R.; Gazzard, B.G. New treatments for progressive multifocal leukoencephalopathy in HIV-1- infected patients. AIDS. 1998, 12(5), 533–535.
194. De Luca, A.; Fantoni, M.; Tartaglione, T.; Antinori, A. Response to cidofovir after failure of antiretroviral therapy alone in AIDS-associated progressive multifocal leukoencephalopathy. Neurology. 1999, 52, 891–892.
195. Brambilla, A.M.; Castagna, A.; Novati, R. Remission of AIDS-associated progressive multifocal leukoencephalopathy after cidofovir therapy. J Neurol. 1999, 246(8), 723–725.
196. Dodge, R.T. A case study: the use of cidofovir for the management of progressive multifocal leukoencephalopathy. J Assoc Nurses AIDS Care. 1999, 10(4), 70–74.
197. Blick, G.; Whiteside, M.; Griegor, P.; Hopkins, U.; Garton, T.; LaGravinese, L. Successful resolution of progressive multifocal leukoencephalopathy after combination therapy with cidofovir and cytosine arabinoside. Clin Infect Dis. 1998, 26(1), 191–192.
198. Meylan, P.R.; Vuadens, P.; Maeder, P.; Sahli, R.; Tagan, M.C. Monitoring the response of AIDS-related progressive multifocal leukoencephalopathy to HAART and cidofovir by PCR for JC virus DNA in the CSF. Eur Neurol. 1999, 41(3), 172–174.
198a. Marra, C.M.; Rajicic, N.; Barker, D.E.; Cohen, B.A.; Clifford, D.; Donovan Post, M.J.; Ruiz, A.; Bowen, B.C.; Huang, M.L.; Queen-Baker, J.; Andersen, J.; Kelly, S.; Shriver, S. Adult AIDS Clinical Trials Group 363 Team: A pilot study of cidofovir for progressive multifocal leukoencephalopathy in AIDS. AIDS. 2002 Sep, 6(16), 1791–1797.
199. Khalili, K. Personal communication, 1994.
200. Hadler, N. Antisense oligonucleotide therapies: are they ''magic bullets''? Ann Intern Med. 1994, 120, 161–164.
201. Khalili, K. Personal communication, 1995.

11

HIV Meningitis and Dementia

Malcolm Avison and Joseph R. Berger
University of Kentucky College of Medicine
Lexington, Kentucky, U.S.A.

Justin C. McArthur and Avindra Nath
The Johns Hopkins University School of Medicine
Baltimore, Maryland, U.S.A.

1 INTRODUCTION

Neurological disease is observed in 40–70% of human immunodeficiency virus (HIV)-infected patients. These neurological deficits generally present against a background of profound immunosuppression but may be the heralding event acquired immunodeficiency syndrome (AIDS) in 10–20% of infected individuals. Subtle central nervous system (CNS) abnormalities probably occur in an even higher percentage of HIV-infected individuals, because careful neurological and neuropsychological examination in the absence of specific complaints by the patient often show evidence of CNS dysfunction. Furthermore, some studies have identified neuropathological abnormalities in as many as 90% of AIDS patients at autopsy.

The range of CNS complications associated with HIV infection is extremely broad and can be divided into primary neurological illnesses that arise as a direct result of HIV infection of the CNS and secondary illnesses of distinct non-HIV etiology whose incidence may nonetheless be increased or whose course may be modified by the presence of HIV. These secondary illnesses are believed to generally be a consequence of immunosuppression, although it has been suggested that in some cases (e.g., progressive multifocal leukoencephalopathy) other factors related to HIV itself may play a role.

2 HIV: STRUCTURE AND REPLICATION

Human immunodeficiency virus-1 (HIV-1) belongs to the retrovirus family—''retro'' because these viruses have a unique enzyme called reverse transcriptase that converts viral

RNA to DNA upon viral entry into the cell. Viral replication occurs after proviral DNA is integrated into host cell chromosomal DNA. Broadly, the viral genome encodes for two classes of proteins: structural and regulatory (Fig. 1).

The structural proteins form the envelope, the core, and the matrix of the virus. Three regions within the HIV genome, namely, *env, pol*, and *gag*, encode all the structural proteins. The *env* gene codes for gp160, which is cleaved to form the two major envelope glycoproteins, gp120 and gp41. gp120 forms the surface spikes on the virion, and gp41 is a transmembrane glycoprotein. The *pol* gene codes for reverse transcriptase, a protease that cleaves the polyproteins coded by the *pol* and *gag* genes into their active forms, and an endonuclease that is responsible for viral integration into the host genome. The *gag* gene codes for all the core proteins. Regulatory proteins encoded by the viral genome control viral genome expression at the level of either the proviral DNA or the viral mRNA. At least six genes (*tat*, *rev*, *nef*, *vif*, *vpu*, and *vpr*) code for proteins that are involved in the regulation of viral replication (Table 1). These regulatory proteins do not get incorporated into the viral particle but regulate viral replication and release at multiple levels. For example, Tat, Rev, and Nef are targeted to the nucleus of the cell. However, Nef can also be trapped within the cytoplasm of the cell (e.g., in astrocytes), and Tat may be actively released into the extracellular environment. Some of the structural and regulatory proteins have been shown to cause neuronal dysfunction and/or death and thus may be referred to as virotoxins.

2.1 Pathogenesis

The human immunodeficiency virus enters the CNS within days to weeks of primary infection [1,2]. Although semiquantitative immunohistochemical and quantitative mRNA assays reveal a broad distribution of HIV throughout the CNS, the virus exhibits a particular predilection for the basal ganglia [3–5], particularly the putamen and thalamus, deep white matter, and hippocampus [5]. This subcortical distribution is consistent with the clinical picture observed in patients with HIV dementia (HIVD), the most common primary CNS disease [6].

Principal cellular targets and/or locations of HIV in the CNS are perivascular macrophages [7]. Other potential cellular reservoirs for HIV include astrocytes [8–10] (which

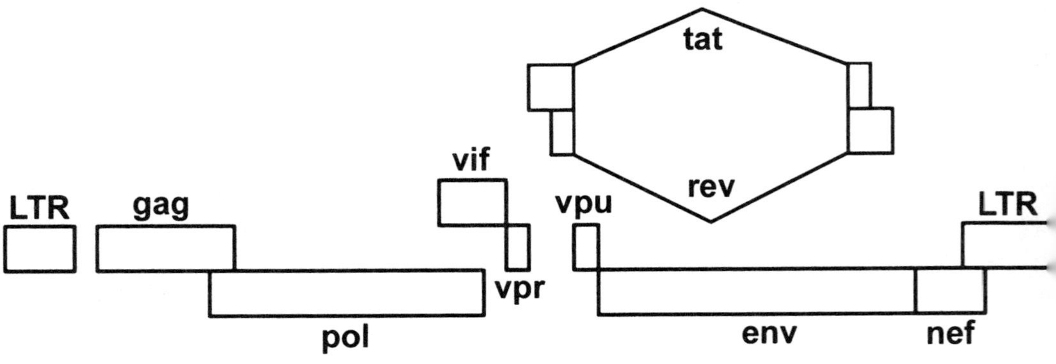

Figure 1 Pictorial representation of the HIV genome showing the encoding of the structural and regulatory proteins. The messages for the regulatory proteins are encoded as spliced messages.

Table 1 HIV Gene Products and Their Neurotoxic Potential

Viral gene	Protein products	Neurotoxicity	Mechanism
Structural			
env	Envelope proteins (gp 120, gp41)	Yes	Acts on microglia and astrocytes, blocks glutamate uptake, releases cytokines, and induces iNOS.
gag	Core protein (p24)	No	
pol	Reverse transcriptase, protease, endonuclease	Unknown	
Regulatory			
tat	Transactivator of transcription (Tat)	Yes	Direct excitation of neurons. Activation of host genes following cellular uptake by glial cells, release of cytokines and chemokines.
rev	Regulator of viral RNA splicing and transport (Rev)	Yes	Basic region interacts with cell membrane causing its disruption.
vpu	Viral protein U (Vpu)	Unknown	
vpr	Viral protein R (Vpr)	Yes	Inserts in cell membrane to form ion channels.
nef	Negative factor (Nef)	Yes	Has sequence similarity to scorpion neurotoxins. Inhibits K channels.
vif	Viral infectivity protein (Vif)	Unknown	

contain only a restricted infection), choroid plexus cells, and microvascular endothelial cells, although evidence for productive infection in humans is scant [10]. Evidence for direct neuronal infection is sparse and contradictory [9,10], suggesting that the level of neuronal infection by HIV, if it occurs at all, is very low.

2.2 Mechanism

Initial entry of HIV into the CNS has been postulated to proceed via a Trojan horse mechanism as a result of trafficking of HIV-infected macrophages across the blood-brain barrier (BBB) [11,12]. Upon entering the CNS, HIV may entrain multiple pathways that have been demonstrated to directly or indirectly lead to neuronal stress and death in vitro. These can be broadly classified as noninflammatory or inflammatory in nature, suggesting two largely independent axes of HIV dementia (HIVD) pathogenesis in vivo.

Noninflammatory pathways include interrupted production of trophic factors essential for neuronal health resulting from HIV infection of microglia [13,14] and astrocytes [8,15,16] and release of viral proteins such as Tat and gp120, which are both directly neurotoxic and can block astrocytic glutamate uptake, activating the arachidonic acid pathway and initiating a neurotoxic cascade (reviewed in Ref. 17).

There is clearly a relationship between the proportions of circulating activated monocytes (bearing CD16 or CD69) and the risk of developing dementia [18,19]. Inflammatory pathways have their genesis in the entry of infected or activated macrophages and in infection and immune activation of perivascular macrophages, astrocytes, endothelial cells, and microglia, which may result in release of neurotoxic inflammatory factors such as nitric oxide (NO), tumor necrosis factor-alpha (TNF-α), interleukin-1β (IL-1beta), and interleukin-6 (IL-6) [20–23], and chemoattractants, particularly monocyte chemoattractant protein-1 (MCP-1) [24–27]. These events, termed the ''hit and run'' phenomenon [20], initiate several positive feedback loops, resulting in a series of self-perpetuating cascades that ultimately disrupt neuroglial relationships and cause neuronal injury. Although the inflammatory and noninflammatory pathways may be initiated by a few infected cells and by viral proteins released from these cells, it is likely that host factors help to determine which of the two major cascades, i.e., inflammatory or noninflammatory, predominate.

Interestingly, host factors may play a role in regulating the susceptibility to neurotoxicity and in regulating the content of the inflammatory response. Polymorphisms in TNF-α (codon 308) were detected four times more frequently in HIVD subjects than in nondemented subjects [28]. This is the same polymorphism that has been associated with a higher risk of death from cerebral malaria [29]. ApoE4 has also been proposed as a genetic risk factor [30], as well as CCR5 [31] and most recently MCP-1. In a large and well-controlled study, specific polymorphisms in MCP-1 were found to increase the risk for HIVD almost five fold [31].

3 HIV MENINGITIS

3.1 Clinical Findings

The acute symptoms of HIV infection are indistinguishable from those of many other viral infections: fever, generalized lymphadenopathy, pharyngeal injection, splenomegaly and splenic tenderness, maculopapular rash, and urticaria. In a small number of individuals, an acute meningitis [32] or meningoencephalitis [33] may supervene, with headache, meningismus, photophobia, generalized seizures, and altered mental state. This acute viral meningitis is observed 3–6 weeks following primary HIV infection [32,34] (and thus prior to seroconversion, which generally occurs 8–12 weeks following exposure). There have been occasional reports of HIV meningoencephalitis developing in chronically infected individuals, even after prolonged antiretroviral therapy. The usual scenario is ''escape'' of CNS virological control because of poorly penetrant antiretroviral regimens. The occurrence of clinically evident HIV meningitis is lower than the high incidence of otherwise unexplained cerebrospinal fluid (CSF) abnormalities [35–37] and the relatively common finding of meningeal inflammation at the time of autopsy might suggest.

3.2 Diagnostic Studies

The diagnosis of HIV meningitis requires CSF examination, including complete and thorough microbiological studies to rule out other infectious agents and cytology to eliminate lymphomatous meningitis. CSF examination reveals an increased protein concentration (greater than 100 mg/dL), mononuclear pleocytosis (more than 200 cells/mm^3), increased IgG, oligoclonal bands, and normal or mildly depressed [38] glucose levels [39]. The presence of myelin basic protein is unexpected. These CSF abnormalities do not appear to be predictive of the subsequent development of neurological disease [37,38], and their interpretation requires caution because CSF abnormalities are found in up to 60% of clinically asymptomatic seropositive subjects [38]. HIV may be isolated from the blood and CSF, but viral cultures are notoriously insensitive. The presence of HIV can be more reliably demonstrated by finding p24 antigen in the CSF or by HIV reverse transcriptase polymerase chain reaction (RT-PCR).

The subarachnoid space often shows pronounced diffuse or focal contrast enhancement on computed tomography (CT) [40] and magnetic resonance imaging (MRI). However, absence of contrast enhancement does not preclude a diagnosis of HIV meningitis nor does its presence exclude other meningitides (see below).

3.3 Neuropathological Findings

The neuropathology and natural history of early CNS changes following HIV infection remain poorly described, owing to the paucity of neuropathological series addressing acute HIV meningitis. The limited studies available have found evidence for early vasculitis and leptomeningitis [41]. Furthermore, many asymptomatic subjects show low-grade leptomeningeal T-cell infiltrates in the absence of HIV encephalitis, which have been interpreted as reflecting the early stages of CNS involvement [42]. Although the precise timing of these events remains unclear in humans, data from simian and feline immunodeficiency virus studies suggest that these changes are indeed a hallmark of the primary infection [41].

3.4 Treatment

Human immunodeficiency virus meningitis is typically a self-limited disease. However, when identified, early aggressive antiretroviral therapy with CNS penetrant antiretroviral agents may be warranted, because a window may exist for the substantial reduction of the virus from this compartment. At the very least, early treatment may help limit trafficking of virus into the CNS. To date, case reports have rarely addressed this issue in the clinical setting and have provided little evidence for this approach [1].

4 HIV DEMENTIA

4.1 Epidemiology

Also known as subacute encephalitis, AIDS dementia complex, HIV encephalopathy, and HIV-associated major cognitive/motor disorder [43], HIV dementia is a unique, progressive, dementing illness that was recognized shortly after the initial description of AIDS [44,45]. Prior to the widespread introduction of effective antiretroviral therapy (ART); 7.3% of AIDS patients reported to the Center for Disease Control and Prevention (CDC) had HIV encephalopathy [46]. The Multicenter AIDS Cohort Study (MACS) of 492 homosexual men found that in the first 2 years after AIDS, HIV dementia developed at an annual rate of 7%, and overall 15% of the cohort followed through death developed dementia [47]. The highest proportion of AIDS patients with HIV encephalopathy is found at the extremes of age: in a pre-ART era study, 13% of AIDS patients less than 15 years old and 19% of patients aged 75 years or older, compared with only 6% of patients aged 15–34 years old [46]. In this cohort, HIV encephalopathy was the initial AIDS-defining illness in 2.8% of adults and 5.3% of children. Increasing age at AIDS diagnosis [48,49] and injection drug use independently increase the risk of HIV dementia [49].

Introduction of zidovudine (azidothymidine, AZT) monotherapy had a marked impact on the incidence of HIV dementia. In an early placebo controlled trial, HIV infected patients showed significant neuropsychological improvement following 16 weeks of AZT therapy [50]. Similarly, in AIDS patients studied between 1982 and 1988 in an academic medical center AIDS clinic, 36% of patients not taking AZT developed dementia compared with only 2% of patients receiving antiretroviral therapy [51]. Perhaps more striking, in the latter study the incidence of dementia fell from 53% in the 6 months prior to introduction of AZT to 3% in the year following its introduction.

The effect of highly active antiretroviral therapy (HAART) relative to monotherapy on the incidence of HIV dementia is less clear. Trials of combination therapies suggest that HAART is effective and may be superior to monotherapy [52]. Consistent with this, data from the MACS over the period 1990–1998 suggest a decreased incidence of HIV dementia [53] while revealing an increasing proportion of new cases occurring in less immunocompromised patients (CD4 cell count 201–350/mm^3). However, comparison of a cohort of patients studied between 1994 and 1995 (Dana Consortium) and a cohort studied between 1998 and 1999 (Northeastern AIDS Dementia Consortium) found no difference in the incidence of HIV dementia or in neuropsychological status in the pre-HAART and HAART eras [54]. Whether the incidence of HIV dementia is declining in the era of HAART or not, it nonetheless remains a serious problem in end-stage AIDS. In a study of the clinical profile of HIV patients (predominantly African American men 35 years and older) who attended an HIV outpatient clinic and died between 1996 and 2001, 92% of patients were newly diagnosed with HIV dementia within 12 months prior to death [55].

4.2 Clinical Findings

Most commonly, HIVD presents in patients with advanced immunosuppression and coexistent systemic disease [56–62] who exhibit hallmarks of advanced AIDS—namely, wasting, alopecia, seborrheic dermatitis, and generalized lymphadenopathy. However, HIV dementia may be the presenting or even sole manifestation of HIV infection before the infected individual exhibits any other illnesses characteristic of impaired immunity [63–66].

Onset of HIVD is often insidious, with early symptoms and signs frequently too subtle to establish a definitive clinical diagnosis. The Memorial Sloan Kettering staging system for HIV dementia classifies patients from normal (grade 0) to end-stage vegetative state (grade 4). Subclinical dementia or minor cognitive motor disorder (grade 0.5) presents a diagnostic challenge but may still have functional consequences on work or driving performance or medication adherence. Individuals so classified typically present with equivocal cognitive complaints accompanied by a relatively normal neurological examination. When HIVD is suspected, neuropsychological tests are useful in demonstrating early cognitive dysfunction and also provide quantitative markers of disease progression.

When clinically evident, symptoms fall into three main categories: cognitive, motor, and behavioral. The primary cognitive symptom is forgetfulness associated with slowed mental abilities. Impaired concentration is common, and patients often complain of difficulty reading [63,67,68]. Lower extremity weakness and impaired balance are among early motor signs. Other early features of the illness that may be observed include abnormal saccadic and pursuit eye movements [69,70] (though these are may be found in otherwise neurologically normal HIV-infected patients [71,72], tremors of the upper extremities, impaired coordination [73], slow and imprecise fine motor movements, slow and clumsy gait, and increased motor tone. These patients are particularly susceptible to neuroleptics, suggesting a disturbance in the dopaminergic system. In many respects, this disorder shares many features characteristic of Parkinson's disease, and some investigators maintain that it is chiefly a disorder affecting the basal ganglia and forebrain [6,73,74]. The most commonly observed behavioral symptoms are apathy and social withdrawal. Sleep disturbances are not uncommon [75], and fatigue, malaise, headaches, and loss of sexual drive are also noted. These symptoms, as well as those of facial masking and hypophonia are often mistakenly diagnosed as signs of depression, which is typically absent [57]. Occasionally organic psychoses such as acute mania may be a primary manifestation of HIVD [76–78]. Such psychoses may be responsive to antiretroviral drugs [76], although anti-retroviral therapy has been reported to precipitate acute mania [79,80]. Both focal and generalized seizures have also been described [81–83].

Studies of CSF show a mononuclear pleocytosis in one-fifth of individuals, with counts of usually less than 50 cells/mm^3 [63]. Increased protein concentration, but usually less than 200 mg/Dl, is observed in two-thirds. Intrathecal synthesis of HIV-specific antibody [84–86] and oligoclonal bands [87] are frequently present but are not predictive of the development of CNS disease [39]. Potential surrogate markers in the CSF for HIV encephalopathy include an elevated HIV p24 antigen [86,88] and elevated levels of beta2-microglobulin [85,89,90], matrix metalloproteinases [91], neopterin [92,93], and quinolinic acid [92,94,95]. The isolation of HIV from CSF is not a useful marker for HIV-related neurological disease [96]. However, quantitative HIV RNA PCR does appear to correlate with disease severity, at least in patients who have not been exposed to antiretroviral therapy [97,98]. The latter is not regarded as a diagnostic marker but may be helpful in explaining whether advancing dementia in the face of systemic viral suppression by antiretroviral agents is the consequence of "CNS viral escape."

The rate of HIVD progression is quite variable and does not appear to be correlated with age, ethnicity, gender, adherence, or predicted CNS penetrance of antiretroviral agents [99,100]. Rapid progressors seem to have more pronounced mental slowing at diagnosis than do slow or nonprogressors [99]. Rapid progression has also been associated with drug abuse [99–101] and lower CD4 count at diagnosis [99], as well as increased levels of CNS markers of macrophage activation [99]. Whether individuals with equivocal features of HIVD invariably progress to a more severe form of dementia remains uncertain.

4.3 Radiological Findings

The most commonly reported abnormality on CT scan of the brain of an HIVD patient is cerebral atrophy (Fig. 2) [102–104]. Up to 50% of patients with mild neurological deficits exhibit atrophy and/or white matter abnormalities on MRI (Fig. 3) [105,106]. Early atrophy is principally subcortical [107]. The white matter changes may be extensive and confluent, patchy, or even punctate [108]. The larger confluent abnormalities appear to be correlated neuropathologically with perivascular macrophages and extravasation of serum protein [109], suggestive of blood-brain barrier compromise secondary to cerebral inflammation. Any BBB breakdown, if present, must be subtle, however, because these white matter abnormalities are generally considered to be nonenhancing (but see below).

Atrophy and white matter abnormalities are generally not considered diagnostic, because they are also seen in 10–20% of neurologically normal seropositive patients [105,106]. Furthermore, with the possible exception of the degree of caudate atrophy, these changes are poorly correlated with the severity of dementia [110], although there is some evidence that white matter changes in HIVD patients may be reversed by HAART

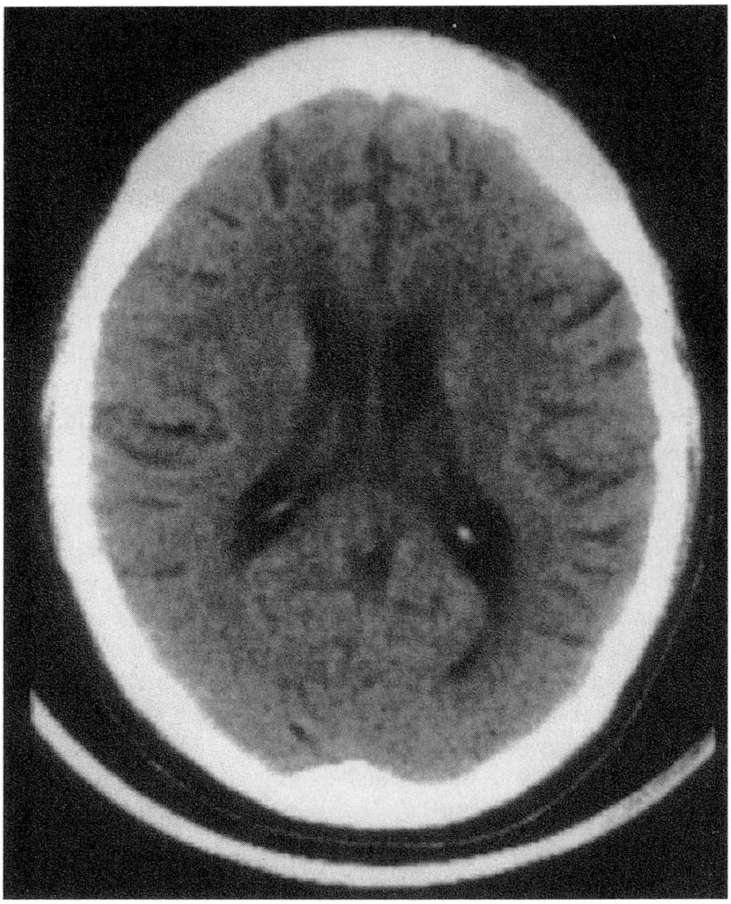

Figure 2 Computed tomographic scan of the brain in HIV dementia showing generalized atrophy greater in the central areas.

and that these changes may be associated with neurological improvement [111]. Profound cortical as well as subcortical atrophy together with extensive confluent white matter abnormalities are common features of advanced dementia. In summary, routine radiological studies serve principally to exclude other neurological disorders in seropositive patients but have limited utility in the unequivocal diagnosis of HIV dementia.

Quantitative MRI at 1.5 T reveals no significant white matter enhancement and a small but significant enhancement in subcortical gray matter consistent with BBB breakdown, which is correlated with dementia severity [112]. Similar studies at 3 T do find increased contrast enhancement in frontal white matter, suggesting that the earlier lower field study may have been limited by scanner sensitivity (L. Chang, personal communication). Among other advanced MRI techniques, diffusion tensor imaging (DTI), which probes microstructural changes in white matter, has demonstrated loss of fiber anisotropy in normal appearing white matter in neurologically normal seropositive subjects [113],

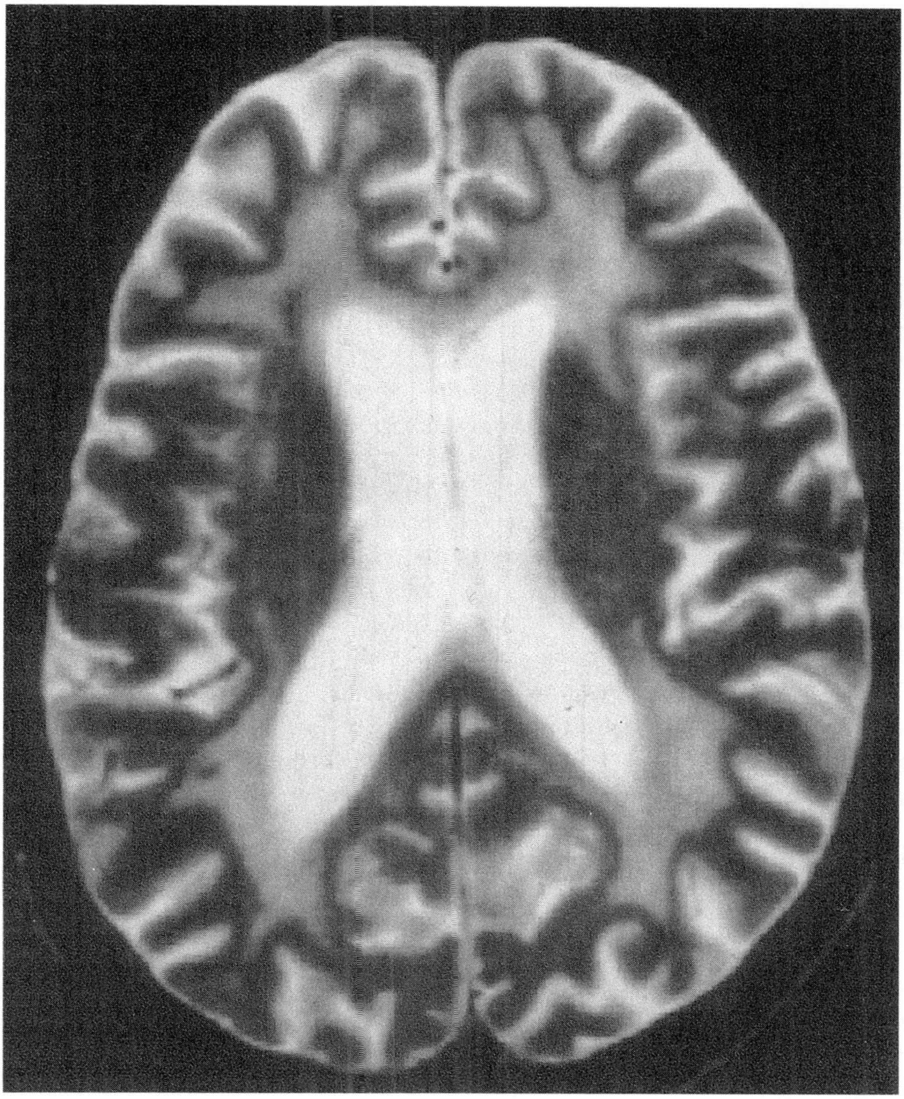

Figure 3 Magnetic resonance image of the brain in HIV dementia showing both generalized atrophy and significant white matter hyperintensities on T2-weighted image.

and magnetization transfer contrast (MTC) imaging reveals a significant reduction in the MT ratio in white matter lesions, possibly reflecting gliosis [114].

Metabolic imaging studies suggest that early changes in subcortical gray and frontal white matter may precede the onset of cognitive motor decline. Thus positron emission tomography (PET) reveals striatal hypermetabolism in patients prior to and early in the development of mild cognitive/motor impairment, with subsequent basal ganglia hypometabolism in the setting of more advanced disease [115–117]. Magnetic resonance spectroscopy (MRS) reveals increased frontal white matter myoinositol, a marker of microglial activation, and decreased basal ganglia *N*-acetyl aspartate, a neuronal marker, in patients with mild cognitive motor decline as well as those with more advanced disease [118]. These metabolic changes suggest that neuronal stress and/or drop out in the basal ganglia, either in parallel with or as a result of activation of CNS inflammatory pathways, underlies the development of dementia [119].

4.4 Other Diagnostic Tests

Despite assertions to the contrary, neither electroencephalography nor other electrophysiological studies are particularly useful diagnostic tools in HIV-infected individuals with HIV encephalopathy. In contrast, detailed neuropsychological test batteries are often invaluable. Whereas earlier studies were somewhat equivocal as to whether a correlation exists between neurocognitive impairment and neuropathologically confirmed HIV encephalitis [120–122], recent work suggests that careful testing has 67% sensitivity, 92% specificity, and 95% positive predictive value for detection of HIVE [123]. Neuropsychological testing is also useful in determining whether there is an associated depression and in gauging the extent of the impairment and the response to therapy.

4.5 Neuropathological Findings

Brain atrophy characterized by sulcal widening and ventricular dilatation is commonly observed at autopsy in patients with HIV encephalopathy; however, it is also found in many nondemented patients, suggesting that it may be a feature of HIV infection rather than a specific marker for HIVD [124]. Meningeal fibrosis may also be present. Histologically, the most common and most distinctive feature of this illness is pallor of the white matter, chiefly seen in a paravascular distribution and often accompanied by an astrocytic reaction [109,125]. This pallor is not associated with significant demyelination but rather seems to be a consequence of subtle changes in BBB [109]. The favored locations for this white matter pallor are the periventricular and central white matter. Multinucleate giant cells, the pathological hallmark of the disease, probably result from direct virus-induced cell fusion [57]. Other microscopic features include microglial nodules, diffuse astrocytosis, and perivascular mononuclear inflammation [125,126] (Fig. 4). These features are most prominent in the basal ganglia [127] and the hippocampus [128], areas for which the virus appears to have a particular predilection (Fig. 4). Quantitative PCR reveals a wide range of HIV RNA levels in brains of nondemented and demented HIV patients, although in general demented subjects have higher levels, particularly in subcortical areas [123], but DNA assays have not demonstrated differences between nondemented and demented individuals [122], suggesting that viral strain or host differences, rather than burden, may be a key determinant [129]. Thinning of the neocortex [130] and neuronal loss on quantitative assessment in specific brain regions [131] have also been noted.

4.6 Treatment

Antiretroviral Therapy

CNS Penetration. There have been several important advances in the past few years that have led to concrete improvements in the care and prognosis of HIV-infected individuals. The first is an understanding of the direct relationship between viral replication and immunological disease progression, which reinforces the need to suppress viral replica-

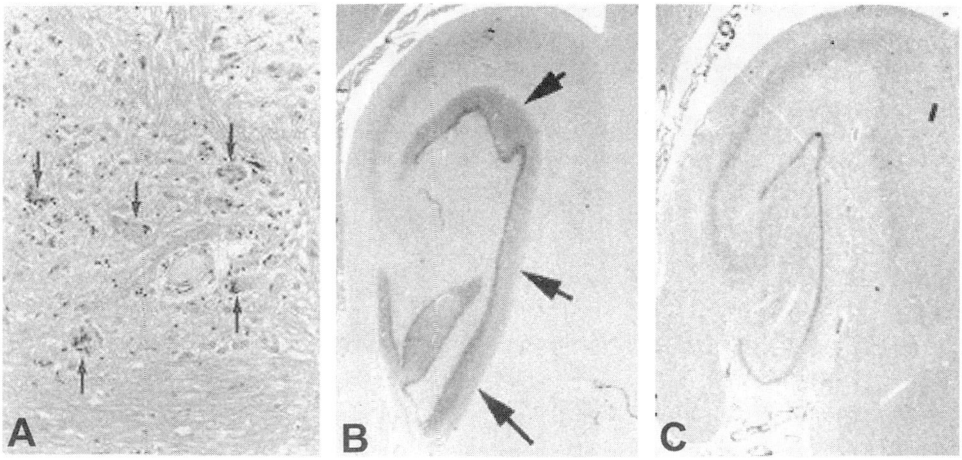

Figure 4 Histopathological features of HIV encephalitis. (A) Several multinucleated giant cells are noted in the perivascular region (arrows). (B) Normal immunostaining pattern of neurites in a dentate gyrus of the hippocampus of a non-HIV infected patient is compared to (C) the loss of neurites in a patient with HIV encephalitis.

tion at the earliest point to control the infection. This has led to the "hit early, hit hard" philosophy arising out of the 1996 reports showing that HAART could suppress viral replication in patients who began therapy early in the course of their HIV disease. The second is the wider availability of multiple potent antiretroviral regimes that can be combined in various ways to provide effective suppression of HIV. The third major change is the ability to monitor the response to therapy through regular measurement of plasma HIV RNA levels, which, with CD4 counts, has become a routine part of clinical care. In addition, resistance to antiretroviral drugs can now be relatively easily measured with genotypic or phenotypic assays. In patients who fail to achieve HIV suppression with antiretroviral therapy, it is often uncertain whether this reflects development of resistance, incomplete adherence, or inadequate delivery of antiretroviral agents to the target site. This is potentially of even greater importance for the treatment of CNS infection given the relatively limited penetrance of most of the available antiretroviral agents.

Today, a typical antiretroviral regimen consists of at least three agents: one or two protease inhibitors or a nonnucleoside reverse transcriptase inhibitor combined with two nucleoside analogs. The goal of therapy is to reduce the measurable plasma viral burden to "below the level of detection." Viral load testing has made it possible to individualize therapy and to more accurately determine the best time to initiate or change therapy, long before declining $CD4^+$ cell counts would have given evidence of active viral replication. The published pharmacokinetic data for CSF penetration for available agents is shown in Table 2.

Effect on HIV Levels in the CSF. There is still relatively little information about the effects of antiretroviral therapy on CSF HIV levels. In patients with HIVD, measurement of response to therapy has traditionally relied upon changes in neuropsychological tests. As new therapies are considered and tested, alternative methods to measure neurological response are also being developed. CSF HIV RNA levels have not yet been validated as

Table 2 Antiretroviral Drugs, Generic and Trade Names, Characteristics

Generic name	Drug class	Abbrev	Trade name	Usual dosage	Common side effects (Comments)	CSF plasma ratio
Zidovudine	Nucleoside RT inhibitor	AZT, ZDV	Retrovir	300 mg BID	Bone marrow suppression, GI upset, headache, myopathy	0.3–1.3
Didanosine	Nucleoside RT inhibitor	ddI	Videx	200 mg BID (125 mg BID if <60 kg) or 300–400 mg qd	Peripheral neuropathy, pancreatitis, diarrhea (take on empty stomach)	0.2
Zalcitabine	Nucleoside RT inhibitor	ddC	HIVID	0.75 mg TID	Peripheral neuropathy, pancreatitis, oral ulcers	0.1–0.4
Stavudine	Nucleoside RT inhibitor	d4T	Zerit	40 mg BID (30 mg BID if <60 kg)	Peripheral neuropathy	0.2
Lamivudine	Nucleoside RT inhibitor	3TC	Epivir	150 mg BID	Anemia, GI upset	0.1
Abacavir	Nucleoside RT inhibitor	ABC	Ziagen	300 mg bid	GI upset, hypersensitivity reaction	0.3
Adefovir	Nucleoside RT inhibitor	ADV	Preveon	60–120 mg QD	GI upset, elevated transaminases, nephrotoxicity (must take with L-carnitine 500 mg/day)	
Nevirapine	Non-nucleoside RT inhibitor	NVP	Viramune	200 mg qd × 14 days, then 200 mg BID	Rash	0.5
Delavirdine	Non-nucleoside RT inhibitor	DLV	Rescriptor	400 mg TID	Rash	<0.05
Efavirenz	Non-nucleoside RT inhibitor	EFV	Sustiva	600 mg QD	Dizziness, nightmares, "disconnectedness" rash	<0.05
Saquinavir	Protease inhibitor	SQV	Invirase Fortovase	600 mg TID 1200 mg TID	(Take with a fatty meal or up to 2 h after meal)	<0.05
Indinavir	Protease inhibitor	IDV	Crixivan	800 mg q 8 h	Kidney stones, hyperbilirubinemia (take on an empty stomach)	0.14
Ritonavir	Protease inhibitor	RTV	Norvir	600 mg BID	GI upset, circumoral paresthesias, diarrhea, fatigue	<0.05
Nelfinavir	Protease inhibitor	NFV	Viracept	750 mg TID or 1250 mg BID	Diarrhea (take with food)	Undete
Amprenavir	Protease inhibitor	141W94	Agenerase	1200 mg BID	Rash, headache, GI upset	<0.05

a measure of treatment effect, but changes appear to bear a relationship to neurological function [132]. For example, Ellis et al. [133] noted that CSF HIV RNA levels were predictive of HIVD. LeTendre and McCutchan assessed CSF HIV levels in 15 patients with HIV-associated minor cognitive motor disorder (i.e., with definite neuropsychological deficits but not frank dementia) in whom HAART was initiated. Improvements in neuropsychological performance correlated in declines in CSF HIV RNA levels. Those who did not show CSF HIV RNA decreases did not have neuropsychological improvement [134].

Table 3 Drugs that Should Be Avoided with Protease Inhibitors[a]

Drug category	Indinavir	Ritonavir[b]	Saquinavir	Nelfinavir	Alternatives
Analgesics	(None)	Meperidine, piroxicam, propoxyphene	(None)	(None)	Acetylsalicylic acid, oxycodon, acetaminophen
Antimycobacterial	Rifampin	Rifabutin[c]	Rifampin, rifabutin	Rifampin	For rifabutin (as alternative for *Mycobacterium avium* intercellular treatment): clarithromycin, ethambutol (treatment not prophylaxis), or azithromycin
Antihistamine	Astemizol, terfenadine	Astemizol, terfenadine	Astemizol, terfenadine	Astemizol, terfenadine	Loratadine
GI	Cisapride	Cisapride	Cisapride	Cisapride	Limited experience
Antidepressant	(None)	Bupropion	(None)	(None)	Fluoxetine, desipramine
Neuroleptic	(None)	Clozapine, pimozide	(None)	(None)	Limited experience
Psychotropic	Midazolam, triazolam	Clorazepate, diazepam, estazolam, flurazepam, midazolam, triazolam, zolpidem	(None)	Midazolam, triazolam	Temazepam, lorazepam
Ergot alkaloid	Dihydroergotamine, Ergotamine (various forms)	dihydroergotamine, ergotamine (various forms)	Dihydroergotamine, ergotamine (various forms)	Dihydroergotamine, ergotamine (various forms)	Limited experience
Miscellaneous	Grapefruit juice reduces indinavir levels by 26%	Desipramine increased 145%: reduce dose Theophylline levels decreased: dose increase	Grapefruit juice increases saquinavir levels[d]		

[a] The contraindicated drugs listed are based on theoretical considerations. Thus, drugs with low therapeutic indices yet with suspected major metabolic contribution from cytochrome P450 3A, CYP2D6, or unknown pathways are included in this table. Actual interactions may or may not occur in patients.
[b] Reduce rifabutin dose to one-fourth of the standard dose.
[c] This is likely a class effect.
[d] Reduce rifabutin dose to one-fourth of the standard dose (150 mg qd).

Drug Interactions. Drug interactions involving antiretroviral agents have become increasingly important with the introduction of combination therapy involving protease inhibitors and nonnucleoside reverse transcriptase inhibitors (NNRTIs) (Table 3); (http://aidsinfo.nih.gov/drugs). In fact, one of the most widely used protease inhibitor combinations uses this interaction to boost levels of lopinavir in combination with low dose ritonavir. All protease inhibitors are substrates and inhibitors of the hepatic cytochrome p450 enzyme system. Ritonavir is the most powerful inhibitor, saquinavir the weakest, and indinavir and nelfinavir are intermediate. Examples of drug interactions arising from inhibition of cytochrome p450 include the increases in rifampin and rifabutin levels with ritonavir and to a lesser degree with the other protease inhibitors. Some protease inhibitors are also inducers of cytochrome p450. Examples of this type of interaction include the lowering of ethinyl estradiol, triptans, Viagra and zidovudine levels by nelfinavir and ritonavir.

Dual protease inhibitor regimens make use of drug interactions to increase drug levels and/or prolong half-lives. Ritonavir increases saquinavir levels by more than tenfold, allowing saquinavir to be given at a reduced dose twice daily. Nelfinavir also increases saquinavir levels, but the effect is less dramatic and does not allow dose reduction. Ritonavir also increases drug levels of nelfinavir and indinavir.

The NNRTIs are also metabolized through the CYP3A pathway, leading to significant drug interactions with protease inhibitors. Nevirapine induces cytochrome p450 enzymes, leading to reductions in protease inhibitor levels. In contrast, delavirdine inhibits cytochrome p450 and increases protease inhibitor levels. With both drugs, the effect is greatest with saquinavir, intermediate with indinavir, and negligible with ritonavir. There are conflicting data on drug interactions between nevirapine and nelfinavir. Efavirenz is both a modest inhibitor and a modest inducer of the cytochrome p450 system. Although it decreases indinavir levels and reduces saquinavir levels by 61%, it increases the AUC of nelfinavir by 20%. Of importance to neurologists are the antconvulsants, because of their ability to induce p450 enzymes. Drugs such as phenytoin, cabamazepine and phenobarbital should be used with caution, and if they are used drug levels and viral loads should be closely monitored. Alternative anticonvulsants such as topiramate and gabapentin may be considered. Protease inhibitors may also induce withdrawal symptoms in patients on methadone [135].

CNS Penetration of ART. Groothuis and Levy [136] summarized some of the issues in relating plasma CSF and brain concentrations of antiretroviral agents. They stress that drug concentrations in the different compartments may be quite different and that using CSF concentrations to estimate brain extracellular fluid levels may overestimate the latter. The estimation of brain tissue levels of drug needs to take into account not only the plasma concentration but also the degree of protein binding and lipophilicity. Additional factors that will tend to lower tissue levels include diffusion in brain tissue and a lack of correction for drug that is within the intravascular space [137]. The CSF/plasma ratios for available antiretroviral agents are very variable (Table 2), reflecting individual drug differences in lipid solubility, molecular size, and state of ionization. In fact, the relevance of this ratio to actual brain concentrations is uncertain, and relatively few data are available. As an example, the CSF/ plasma ratios for several antiretroviral drugs are given in Table 2; however, many of these data are based on only a few patient samples and usually do not include patients with HIVD, whose blood-brain barrier may be more permeable [138]. Other important factors include the active efflux of antiretroviral drugs through transporters including p-glycoprotein [139].

The Effects of Monotherapy on Dementia

From 1987 (when zidovudine, the first licensed antiretroviral agent, was introduced) to around 1992 (when studies of dual therapy were published), antiretroviral therapy usually consisted of one agent—monotherapy. The more widespread and earlier use of AZT appeared to reduce the frequency and/or incidence of HIVD dramatically from 53% to 10% [140]. Early open label studies with zidovudine showed promising improvements in clinical functioning and neuropsychological performance [141] and metabolism on positron emission tomography scans [142]. However, neurological improvement occurred in most patients with mild neurological abnormalities, but there appears to be no relationship between treatment response and CSF AZT concentration, cumulative AZT dose, or HIV isolation from CSF [143]. ddI was shown to improve IQ scores in children. Plasma concentrations correlated with both IQ improvement and HIV levels determined by p24 antigen [144]. In an observational study, when patients with dementia were switched to stavudine, many showed dramatic improvements in psychomotor speed function. Stavudine also does not show any toxic effects on peripheral nerves [145].

Surprisingly, only two placebo-controlled monotherapy antiretroviral trials for HIVD were completed, but these provided important information on changes in neurological function with ART. First, evidence from the multicenter licensing trial of AZT in patients with HIV infection suggested that this drug improved neuropsychological function [50]. This study was not specifically a trial of HIVD (in fact, several demented individuals were excluded), and it was several more years before a controlled trial of zidovudine in HIVD was completed. This study suggested that there was a dose effect, with more improvement with very high doses of AZT (2000 mg daily) [146].

In 1998, a reverse transcriptase inhibitor, abacavir, was tested in a placebo-controlled, double-blinded study in 99 patients with HIVD. Abacavir had been shown to have good CSF penetration and was active in macrophages, HIV's principal target cells within the brain [147]. The patients were heavily pretreated with ART, and in fact only 10% had wild-type virus at entry. Perhaps surprisingly, very few subjects showed neurological deterioration during the 12 weeks of the study: only two in the placebo group and none on abacavir. Overall, both groups showed improvements in neuropsychological performance on standardized tests, with a trend favoring abacavir. The more severely impaired group on abacavir showed greater improvement than placebo recipients. The CSF virological response favored abacavir with a 0.64 log drop during the study, whereas the placebo group showed a rise of 0.25 log. Improvement in neuropsychological performance was also seen in both treatment groups. Abacavir did reduce CSF viral load to a greater extent than background ART. There are multiple important implications from this study, including (1) Single changes in ART are not likely to be very effective; (2) the progression of dementia may be different in an era when combination therapies are used; and (3) other types of outcome measures in addition to neuropsychological testing may be needed to detect changes.

The Effects of Combination ART on Dementia

There have been very few systematic studies of the effects of combination ART on neurocognitive deficits in HIVD [148]. In part, this reflects the difficulty and expense of performing controlled clinical trials in this disorder. It also reflected the dynamic nature of drug development in AIDS research, where new agents may be introduced before dementia trials can be completed or even planned. Graham et al. [149], using data from the MACS, suggested that the use of combination ART (not including protease inhibitors) was associated with a reduced risk of developing HIVD. The effects of adding a nonnucleoside reverse transcriptase inhibitor, nevirapine, to background ART were studied in a blinded clinical trial (ACTG193a) in very advanced HIV infection. Neuropsychological performance was improved in the nevirapine group [148]. Since the introduction of the potent protease inhibitors in 1996, several groups have completed open-label studies of combination regimes including PIs. In one study, 23 patients with HIV-associated cognitive impairment received neuropsychological testing before and after the initiation of combination regimes including protease inhibitors, and improvement was noted in 78% of the patients [52]. Furthermore, the effect of combination antiretroviral therapy including protease inhibitors on neuropsychological testing performance was compared to that of combination antiretroviral therapy without protease inhibitors, monotherapy, and no treatment in subjects in the Multicenter AIDS Cohort Study [52]. Subjects using combination ART either with or without protease inhibitors had better neuropsychological testing performance than subjects using monotherapy or no treatment. In another cross-sectional study it was found that combination therapy with protease inhibitors was associated with a lower prevalence of neuropyschological impairment in HIV-infected patients [150].

Suggestions for Antiretroviral Therapy in HIVD

At this point it is impossible to make definitive recommendations about the optimum antiretroviral therapy for HIV dementia. Stavudine and abacavir appear to be useful alternatives to include in a combination regime for patients with dementia, based upon their pharmacokinetic properties, tolerability, and twice-a-day dosing. Stavudine may have a role in treating neurological disease, as indicated by favorable pharmacokinetic studies in CSF. Nucleoside reverse transcriptase inhibitor nevirapine [151] and protease inhibitor indinavir achieve good CSF levels and may also be useful to include in ART regimes for patients with HIVD, according to accumulating clinical experience [152,153].

There are two main difficulties with antiretroviral therapy of HIVD:

1. Drug resistance. The role of resistance testing may become important in selection of ART combinations for demented patients, most of whom are heavily pretreated and are likely to have multiple resistance mutations (as in the abacavir trial, where 90% of subjects had resistance mutations at baseline). There is no direct utility to examining resistance patterns in CSF, because there is generally concordance between the CSF and plasma with respect to major genotypic resistance mutations.
2. Drug adherence. Adherence is important in maintaining virological suppression, particularly in patients with cognitive impairment. New techniques for improving adherence including directly observed therapy, pill counts, intensive education, and electronic monitors are being applied to this problem.

Neuroprotective Therapies for HIVD

The delineation of the pathophysiological steps contributing to HIV dementia has suggested that antiretroviral therapies alone may not be sufficient to prevent the self-sustaining macrophage activation and the subsequent release of neurotoxic factors. This concept has led to the development of several adjunctive therapies aimed at interrupting or blocking these aberrant pathways (Table 4). These include inflammatory antagonists such as lexipafant, an antagonist of platelet activating factor [154], and neuroprotective agents such as memantine, an open-channel NMDA antagonist [155]. Other compounds have been tested in small phase I/II trials in patients receiving stable ART and have shown promising results. In two studies of the licensed agent deprenyl (Selegiline), significant improvements in memory were seen [156,157]. Larger scale studies within the AIDS Clinical Trials Group are under way, and results are expected in 2004. Investigation of neuroprotective agents has been impeded owing to the need for a large number of patients which, requires complex multicenter studies. The development of MRS as a sensitive surrogate marker of HIV dementia offers hope for rapid screening of potential neuroprotective agents.

Symptomatic Treatment

Patients with HIV dementia are extremely susceptible to the adverse effects of psychoactive drugs, so hypnotic and anxiolytic drugs should be avoided [158,159].

Table 4 Placebo Controlled Trials of Neuroprotective Therapies for HIVD

Therapy	Status
Calcium channel blockers	
Nimodipine	Trend for improvement in cognition at highest dose (60 mg 5 times/day)
Antioxidants	
Selegiline	Neuroprotective (small study)
Thioctic acid	No effect
OPC14117	No effect
Glutamate antagonists	
Memantine	Improvement in cognition in follow-up after completion of double-blind phase. Dosage: 40 mg/day
Platelet-activating factor antagonist	
Lexipafant	Trend for improvement in cognition (small study). Dosage: 500 mg/day
TNF antagonist	
CPI-1189	No effect on neurocognition. Significant improvement in pegboard test ($P<0.01$) at highest dose (100 mg/day).
Miscellaneous	
Peptide T	No effect

Neuroleptics. Due to its selective action on D3 and D4 receptors, clozapine is the prefered neuroleptic drug. Small doses of neuroleptic agents such as haloperidol (Haldol) 0.5 mg may be needed in the agitated or combative patient.

Antidepressants. If marked inertia is present, a tricyclic antidepressant or fluoxetine (Prozac) can be tried, in doses of 25–50% of the usual dose, or methylphenidate (Ritalin). Full doses of tricyclics may precipitate delirium, and serum levels should be monitored frequently.

Anticonvulsants. As discussed above, gabapentin or topiramate are the preferred anticonvulsants because of the lack of drug–drug interactions. Valproate should be particularly avoided because in vitro studies suggest that it can induce viral replication [160].

Headaches. Intractable vascular headaches may develop in some patients. Initial treatment should include migraine prophylaxis [161]. Nonresponders may require treatment with opiates.

Parkinsonism. Dopamine agonists may be used for patients that manifest parkinsonism. However, the response is usually poor.

In patients with progressive dementia, medico-legal issues such as establishing a power of attorney, completion of a living will, and arrangement for the dispersal of assets should be discussed at any early stage before the dementia becomes too severe.

REFERENCES

1. Davis, L.E.; Hjelle, B.L.; Miller, V.E.; Palmer, D.L.; Llewellyn, A.L.; Merlin, T.L. Early viral brain invasion in iatrogenic human immunodeficiency virus infection. Neurology. 1992, 42(9), 1736–1739.

2. Resnick, L.; Berger, J.R.; Shapshak, P.; Tourtellotte, W.W. Early penetration of the blood-brain barrier by HIV. Neurology. 1988, 38(1), 9–14.
3. Neuen-Jacob, E.; Arendt, G.; Wendtland, B.; Jacob, B.; Schneeweis, M.; Wechsler, W. Frequency and topographical distribution of CD68-positive macrophages and HIV-1 core proteins in HIV-associated brain lesions. Clin Neuropathol. 1993, 12(6), 315–324.
4. Pumarola Sune, T.; Navia, B.A.; Cordon Cardo, C.; Cho, E.S.; Price, R.W. HIV antigen in the brains of patients with the AIDS dementia complex. Ann Neurol. 1987, 21(5), 490–496.
5. Wiley, C.A.; Soontornniyomkij, V.; Radhakrishnan, L.; Masliah, E.; Mellors, J.; Hermann, S.A. Distribution of brain HIV load in AIDS. Brain Pathol. 1998, 8(2), 277–284.
6. Berger, J.R.; Nath, A. HIV dementia and the basal ganglia. Intervirology. 1997, 40(2–3), 122–131.
7. Budka, H. Human immunodeficiency virus (HIV) envelope and core proteins in CNS tissues of patients with the acquired immune deficiency syndrome (AIDS). Acta Neuropathol. 1990, 79(6), 611–619.
8. Tornatore, C.; Chandra, R.; Berger, J.R.; Major, E.O. HIV-1 infection of subcortical astrocytes in the pediatric central nervous system. Neurology. 1994, 44(3 Pt 1), 481–487.
9. Takahashi, K.; Wesselingh, S.L.; Griffin, D.E.; McArthur, J.C.; Johnson, R.T.; Glass, J.D. Localization of HIV-1 in human brain using polymerase chain reaction/in situ hybridization and immunocytochemistry. Ann Neurol. 1996, 39(6), 705–711.
10. Bagasra, O.; Lavi, E.; Bobroski, L.; Khalili, K.; Pestaner, J.P.; Tawadros, R. Cellular reservoirs of HIV-1 in the central nervous system of infected individuals:identification by the combination of in situ polymerase chain reaction and immunohistochemistry. AIDS. 1996, 10(6), 573–585.
11. Meltzer, M.S.; Skillman, D.R.; Gomatos, P.J.; Kalter, D.C.; Gendelman, H.E. Role of mononuclear phagocytes in the pathogenesis of human immunodeficiency virus infection. Annu Rev Immunol. 1990, 8, 169–194.
12. Nath, A. Pathobiology of human immunodeficiency virus dementia. Semin Neurol. 1999, 19(2), 113–127.
13. Koenig, S.; Gendelman, H.E.; Orenstein, J.M.; Dal Canto, M.C.; Pezeshkpour, G.H.; Yungbluth, M. Detection of AIDS virus in macrophages in brain tissue from AIDS patients with encephalopathy. Science. 1986, 233(4768), 1089–1093.
14. Brew, B.J.; Rosenblum, M.; Cronin, K.; Price, R.W. AIDS dementia complex and HIV-1 brain infection:clinical-virological correlations [See comments]. Ann Neurol. 1995, 38(4), 563–570.
15. Saito, Y.; Sharer, L.R.; Epstein, L.G.; Michaels, J.; Mintz, M.; Louder, M. Overexpression of nef as a marker for restricted HIV-1 infection of astrocytes in postmortem pediatric central nervous tissues. Neurology. 1994, 44(3 Pt 1), 474–481.
16. Schweighardt, B.; Atwood, W.J. HIV type 1 infection of human astrocytes is restricted by inefficient viral entry. AIDS Res Hum Retroviruses. 2001, 17(12), 1133–1142.
17. Nath, A.; Geiger, J. Neurobiological aspects of human immunodeficiency virus infection: neurotoxic mechanisms. Prog Neurobiol. 1998, 54(1), 19–33.
18. Pulliam, L.; Gascon, R.; Stubblebine, M.; McGuire, D.; McGrath, M.S. Unique monocyte subset in patients with AIDS dementia. Lancet. 1997, 349(9053), 692–695.
19. Gartner, S. HIV infection and dementia. Science. 2000, 287(5453), 602–604.
20. Nath, A.; Conant, K.; Chen, P.; Scott, C.; Major, E.O. Transient exposure to HIV-1 Tat protein results in cytokine production in macrophages and astrocytes. A hit and run phenomenon. J Biol Chem. 1999, 274(24), 17098–17102.
21. New, D.R.; Maggirwar, S.B.; Epstein, L.G.; Dewhurst, S.; Gelbard, H.A. HIV-1 Tat induces neuronal death via tumor necrosis factor-alpha and activation of non-N-methyl-D-aspartate receptors by a NFkappaB-independent mechanism. J Biol Chem. 1998, 273(28), 17852–17858.

22. Shi, B.; Raina, J.; Lorenzo, A.; Busciglio, J.; Gabuzda, D. Neuronal apoptosis induced by HIV-1 Tat protein and TNF-alpha: potentiation of neurotoxicity mediated by oxidative stress and implications for HIV-1 dementia. J Neurovirol. 1998, 4(3), 281–290.
23. Chen, P.; Mayne, M.; Power, C.; Nath, A. The Tat protein of HIV-1 induces tumor necrosis factor-alpha production. Implications for HIV-1-associated neurological diseases. J Biol Chem. 1997, 272(36), 22385–22388.
24. Kelder, W.; McArthur, J.C.; Nance-Sproson, T.; McClernon, D.; Griffin, D.E. Beta-chemokines MCP-1 and RANTES are selectively increased in cerebrospinal fluid of patients with human immunodeficiency virus-associated dementia. Ann Neurol. 1998, 44(5), 831–835.
25. Conant, K.; Garzino Demo, A.; Nath, A.; McArthur, J.C.; Halliday, W.; Power, C. Induction of monocyte chemoattractant protein-1 in HIV-1 Tat-stimulated astrocytes and elevation in AIDS dementia. Proc Natl Acad Sci USA. 1998, 95(6), 3117–3121.
26. Kutsch, O.; Oh, J.; Nath, A.; Benveniste, E.N. Induction of the chemokines interleukin-8 and IP-10 by human immunodeficiency virus type 1 tat in astrocytes. J Virol. 2000, 74(19), 9214–9221.
27. Weiss, J.M.; Nath, A.; Major, E.O.; Berman, J.W. HIV-1 Tat induces monocyte chemoattractant protein-1-mediated monocyte transmigration across a model of the human blood-brain barrier and up-regulates CCR5 expression on human monocytes. J Immunol. 1999, 163(5), 2953–2959.
28. Quasney, M.W.; Zhang, Q.; Sargent, S.; Mynatt, M.; Glass, J.; McArthur, J. Increased frequency of the tumor necrosis factor-alpha-308 A allele in adults with human immunodeficiency virus dementia. Ann Neurol. 2001, 50(2), 157–162.
29. McGuire, W.; Hill, A.V.; Allsopp, C.E.; Greenwood, B.M.; Kwiatkowski, D. Variation in the TNF-alpha promoter region associated with susceptibility to cerebral malaria. Nature. 1994, 371(6497), 508–510.
30. Corder, E.H.; Robertson, K.; Lannfelt, L.; Bogdanovic, N.; Eggertsen, G.; Wilkins, J. HIV-infected subjects with the E4 allele for APOE have excess dementia and peripheral neuropathy [See comments]. Nat Med. 1998, 4(10), 1182–1184.
31. Gonzalez, E.; Rovin, B.H.; Sen, L.; Cooke, G.; Dhanda, R.; Mummidi, S. HIV-1 infection and AIDS dementia are influenced by a mutant MCP-1 allele linked to increased monocyte infiltration of tissues and MCP-1 levels. Proc Natl Acad Sci USA. 2002, 99(21), 13795–13800.
32. Ho, D.D.; Sarngadharan, M.G.; Resnick, L.; Dimarzoveronese, F.; Rota, T.R.; Hirsch, M.S. Primary human T-lymphotropic virus type III infection. Ann Intern Med. 1985, 103(6 Pt 1), 880–883.
33. Newton, P.J.; Newsholme, W.; Brink, N.S.; Manji, H.; Williams, I.G.; Miller, R.F. Acute meningoencephalitis and meningitis due to primary HIV infection. Br Med J. 2002, 325(7374), 1225–1227.
34. Cooper, D.A.; Gold, J.; Maclean, P.; Donovan, B.; Finlayson, R.; Barnes, T.G. Acute AIDS retrovirus infection. Definition of a clinical illness associated with seroconversion. Lancet. 1985, 1(8428), 537–540.
35. Elovaara, I.; Iivanainen, M.; Valle, S.L.; Suni, J.; Tervo, T.; Lahdevirta, J. CSF protein and cellular profiles in various stages of HIV infection related to neurological manifestations. J Neurol Sci. 1987, 78(3), 331–342.
36. Appleman, M.E.; Marshall, D.W.; Brey, R.L.; Houk, R.W.; Beatty, D.C.; Winn, R.E. Cerebrospinal fluid abnormalities in patients without AIDS who are seropositive for the human immunodeficiency virus. J Infect Dis. 1988, 158(1), 193–199.
37. Marshall, D.W.; Brey, R.L.; Butzin, C.A.; Lucey, D.R.; Abbadessa, S.M.; Boswell, R.N. CSF changes in a longitudinal study of 124 neurologically normal HIV-1-infected U.S. Air Force personnel. J Acquir Immune Defic Syndr. 1991, 4(8), 777–781.
38. Marshall, D.W.; Brey, R.L.; Cahill, W.T.; Houk, R.W.; Zajac, R.A.; Boswell, R.N. Spectrum of cerebrospinal fluid findings in various stages of human immunodeficiency virus infection. Arch Neurol. 1988, 45(9), 954–958.

39. Hollander, H.; McGuire, D.; Burack, J.H. Diagnostic lumbar puncture in HIV-infected patients: analysis of 138 cases. Am J Med. 1994, 96(3), 223–228.
40. Post, M.J.; Tate, L.G.; Quencer, R.M.; Hensley, G.T.; Berger, J.R.; Sheremata, W.A. CT, MR, and pathology in HIV encephalitis and meningitis. Am J Roentgenol. 1988, 151(2), 373–380.
41. Gray, F.; Scaravilli, F.; Everall, I.; Chretien, F.; An, S.; Boche, D. Neuropathology of early HIV-1 infection. Brain Pathol. 1996, 6(1), 1–15.
42. Bell, J.E. The neuropathology of adult HIV infection. Rev Neurol (Paris). 1998, 154(12), 816–829.
43. Nomenclature and research case definitions for neurologic manifestations of human immunodeficiency virus-type 1 (HIV-1) infection. Report of a Working Group of the American Academy of Neurology AIDS Task Force. Neurology. 1991, 41(6), 778–785.
44. Snider, W.D.; Simpson, D.M.; Nielsen, S.; Gold, J.W.; Metroka, C.E.; Posner, J.B. Neurological complications of acquired immune deficiency syndrome: analysis of 50 patients. Ann Neurol. 1983, 14(4), 403–418.
45. Britton, C.B.; Miller, J.R. Neurologic complications in acquired immunodeficiency syndrome (AIDS). Neurol Clin. 1984, 2(2), 315–339.
46. Janssen, R.S.; Nwanyanwu, O.C.; Selik, R.M.; Stehr Green, J.K. Epidemiology of human immunodeficiency virus encephalopathy in the United States. Neurology. 1992, 42(8), 1472–1476.
47. McArthur, J.C.; Hoover, D.R.; Bacellar, H.; Miller, E.N.; Cohen, B.A.; Becker, J.T. Dementia in AIDS patients: incidence and risk factors. Multicenter AIDS Cohort Study. Neurology. 1993, 43(11), 2245–2252.
48. Inungu, J.N.; Mokotoff, E.D.; Kent, J.B. Characteristics of HIV infection in patients fifty years or older in Michigan. AIDS Patient Care STDS. 2001, 15(11), 567–573.
49. Wang, F.; So, Y.; Vittinghoff, E.; Malani, H.; Reingold, A.; Lewis, E. Incidence proportion of and risk factors for AIDS patients diagnosed with HIV dementia, central nervous system toxoplasmosis, and cryptococcal meningitis. J Acquir Immune Defic Syndr Hum Retrovirol. 1995, 8(1), 75–82.
50. Schmitt, F.A.; Bigley, J.W.; McKinnis, R.; Logue, P.E.; Evans, R.W.; Drucker, J.L. Neuropsychological outcome of zidovudine (AZT) treatment of patients with AIDS and AIDS-related complex. N Engl J Med. 1988, 319(24), 1573–1578.
51. Portegies, P.; de Gans, J.; Lange, J.M.; Derix, M.M.; Speelman, H.; Bakker, M. Declining incidence of AIDS dementia complex after introduction of zidovudine treatment [published erratum appears in Br Med J, 299(6708):1141, 1989] [See comments]. Br Med J. 1989, 299(6703), 819–821.
52. Sacktor, N.C.; Lyles, R.H.; Skolasky, R.L.; Anderson, D.E.; McArthur, J.C.; McFarlane, G. Combination antiretroviral therapy improves psychomotor speed performance in HIV-seropositive homosexual men. Multicenter AIDS Cohort Study (MACS). Neurology. 1999, 52(8), 1640–1647.
53. Sacktor, N.; Lyles, R.H.; Skolasky, R.; Kleeberger, C.; Selnes, O.A.; Miller, E.N. HIV-associated neurologic disease incidence changes:Multicenter AIDS Cohort Study, 1990–1998. Neurology. 2001, 56(2), 257–260.
54. Sacktor, N.; McDermott, M.P.; Marder, K.; Schifitto, G.; Selnes, O.A.; McArthur, J.C. HIV-associated cognitive impairment before and after the advent of combination therapy. J Neurovirol. 2002, 8(2), 136–142.
55. Welch, K.; Morse, A. The clinical profile of end-stage AIDS in the era of highly active antiretroviral therapy. AIDS Patient Care STDS. 2002, 16(2), 75–81.
56. Price, R.W.; Brew, B.; Sidtis, J.; Rosenblum, M.; Scheck, A.C.; Cleary, P. The brain in AIDS: central nervous system HIV-1 infection and AIDS dementia complex. Science. 1988, 239(4840), 586–592.
57. Price, R.W.; Brew, B.J. The AIDS dementia complex. J Infect Dis. 1988, 158(5), 1079–1083.

58. Brew, B.J.; Rosenblum, M.; Price, R.W. AIDS dementia complex and primary HIV brain infection. J Neuroimmunol. 1988, 20(2–3), 133–140.
59. Van Gorp, W.G.; Miller, E.N.; Satz, P.; Visscher, B. Neuropsychological performance in HIV-1 immunocompromised patients: a preliminary report. J Clin Exp Neuropsychol. 1989, 11(5), 763–773.
60. Selnes, O.A.; Miller, E.; McArthur, J.; Gordon, B.; Munoz, A.; Sheridan, K. HIV-1 infection: no evidence of cognitive decline during the asymptomatic stages. The Multicenter AIDS Cohort Study [See comments]. Neurology. 1990, 40(2), 204–208.
61. Sinforiani, E.; Mauri, M.; Bono, G.; Muratori, S.; Alessi, E.; Minoli, L. Cognitive abnormalities and disease progression in a selected population of asymptomatic HIV-positive subjects. AIDS. 1991, 5(9), 1117–1120.
62. Stern, Y.; Marder, K.; Bell, K.; Chen, J.; Dooneief, G.; Goldstein, S. Multidisciplinary baseline assessment of homosexual men with and without human immunodeficiency virus infection. III. Neurologic and neuropsychological findings. Arch Gen Psychiatry. 1991, 48(2), 131–138.
63. Navia, B.A.; Jordan, B.D.; Price, R.W. The AIDS dementia complex: I. Clinical features. Ann Neurol. 1986, 19(6), 517–524.
64. Navia, B.A.; Price, R.W. The acquired immunodeficiency syndrome dementia complex as the presenting or sole manifestation of human immunodeficiency virus infection. Arch Neurol. 1987, 44(1), 65–69.
65. Chermann, J.C. HIV-associated diseases: acute and regressive encephalopathy in a seropositive man. Res Virol. 1990, 141(2), 137–141.
66. Beckett, A.; Summergrad, P.; Manschreck, T.; Vitagliano, H.; Henderson, M.; Buttolph, M.L. Symptomatic HIV infection of the CNS in a patient without clinical evidence of immune deficiency. Am J Psychiatry. 1987, 144(10), 1342–1344.
67. Brew, B.J. AIDS dementia complex. Neurol Clin. 1999, 17(4), 861–881.
68. Power, C.; Johnson, R.T. HIV-1 associated dementia: clinical features and pathogenesis. Can J Neurol Sci. 1995, 22(2), 92–100.
69. Currie, J.; Ramsden, B.; McArthur, C.; Lunch, J.; Maruff, P.; Benson, E. High-resolution eye movement recording in the assessment of neurologic complications in HIV-1 infection. (In German.). EEG EMG Z Elektroenzephalogr Elektromyogr Verwandte Geb. 1989, 20(4), 273–279.
70. Johnston, J.L.; Miller, J.D.; Nath, A. Ocular motor dysfunction in HIV-1-infected subjects: a quantitative oculographic analysis. Neurology. 1996, 46(2), 451–457.
71. Sweeney, J.A.; Brew, B.J.; Keilp, J.G.; Sidtis, J.J.; Price, R.W. Pursuit eye movement dysfunction in HIV-1 seropositive individuals. J Psychiatry Neurosci. 1991, 16(5), 247–252.
72. Castello, E.; Baroni, N.; Pallestrini, E. Neurotological auditory brain stem response findings in human immunodeficiency virus-positive patients without neurologic manifestations. Ann Otol Rhinol Laryngol. 1998, 107(12), 1054–1060.
73. Arendt, G.; Hefter, H.; Elsing, C.; Strohmeyer, G.; Freund, H.J. Motor dysfunction in HIV-infected patients without clinically detectable central-nervous deficit. J Neurol. 1990, 237(6), 362–368.
74. Koutsilieri, E.; Sopper, S.; Scheller, C.; ter Meulen, V.; Riederer, P. Parkinsonism in HIV dementia. J Neural Transm. 2002, 109(5–6), 767–775.
75. Norman, S.E.; Chediak, A.D.; Kiel, M.; Cohn, M.A. Sleep disturbances in HIV-infected homosexual men. AIDS. 1990, 4(8), 775–781.
76. Mijch, A.M.; Judd, F.K.; Lyketsos, C.G.; Ellen, S.; Cockram, A. Secondary mania in patients with HIV infection: are antiretrovirals protective?. J Neuropsychiatry Clin Neurosci. 1999, 11(4), 475–480.
77. Ellen, S.R.; Judd, F.K.; Mijch, A.M.; Cockram, A. Secondary mania in patients with HIV infection. Aust NZJ Psychiatry. 1999, 33(3), 353–360.
78. Lyketsos, C.G.; Schwartz, J.; Fishman, M.; Treisman, G. AIDS mania. J Neuropsychiatry Clin Neurosci. 1997, 9(2), 277–279.

79. Brouillette, M.J.; Chouinard, G.; Lalonde, R. Didanosine-induced mania in HIV infection. Am J Psychiatry. 1994, 151(12), 1839–1840.
80. Wright, J.M.; Sachdev, P.S.; Perkins, R.J.; Rodriguez, P. Zidovudine-related mania. Med J Aust. 1989, 150(6), 339–341.
81. Wong, M.C.; Suite, N.D.; Labar, D.R. Seizures in human immunodeficiency virus infection. Arch Neurol. 1990, 47(6), 640–642.
82. Parisi, A.; Strosselli, M.; Pan, A.; Maserati, R.; Minoli, L. HIV-related encephalitis presenting as convulsant disease. Clin Electroencephalogr. 1991, 22(1), 1–4.
83. Holtzman, D.M.; Kaku, D.A.; So, Y.T. New-onset seizures associated with human immunodeficiency virus infection: causation and clinical features in 100 cases. Am J Med. 1989, 87(2), 173–177.
84. Resnick, L.; diMarzo Veronese, F.; Schupbach, J.; Tourtellotte, W.W.; Ho, D.D.; Muller, F. Intra-blood-brain-barrier synthesis of HTLV-III-specific IgG in patients with neurologic symptoms associated with AIDS or AIDS-related complex. N Engl J Med. 1985, 313(24), 1498–1504.
85. Royal, W.; Selnes, O.A.; Concha, M.; Nance Sproson, T.E.; McArthur, J.C. Cerebrospinal fluid human immunodeficiency virus type 1 (HIV-1) p24 antigen levels in HIV-1-related dementia. Ann Neurol. 1994, 36(1), 32–39.
86. Goudsmit, J.; Epstein, L.G.; Paul, D.A.; van der Helm, H.J.; Dawson, G.J.; Asher, D.M. Intra-blood-brain barrier synthesis of human immunodeficiency virus antigen and antibody in humans and chimpanzees. Proc Natl Acad Sci USA. 1997, 84(11), 3876–3880.
87. Fainardi, E.; Contini, C.; Benassi, N.; Bedetti, A.; Castellazzi, M.; Vaghi, L. Assessment of HIV-intrathecal humoral immune response in AIDS-related neurological disorders. J Neuroimmunol. 2001, 119(2), 278–286.
88. Epstein, L.G.; Goudsmit, J.; Paul, D.A.; Morrison, S.H.; Connor, E.M.; Oleske, J.M. Expression of human immunodeficiency virus in cerebrospinal fluid of children with progressive encephalopathy. Ann Neurol. 1987, 21(4), 397–401.
89. Brew, B.J.; Bhalla, R.B.; Paul, M.; Sidtis, J.J.; Keilp, J.J.; Sadler, A.E. Cerebrospinal fluid beta 2-microglobulin in patients with AIDS dementia complex: an expanded series including response to zidovudine treatment. AIDS. 1992, 6(5), 461–465.
90. McArthur, J.C.; Nance Sproson, T.E.; Griffin, D.E.; Hoover, D.; Selnes, O.A.; Miller, E.N. The diagnostic utility of elevation in cerebrospinal fluid beta 2-microglobulin in HIV-1 dementia. Multicenter AIDS Cohort Study. Neurology. 1992, 42(9), 1707–1712.
91. Conant, K.; McArthur, J.C.; Griffin, D.E.; Sjulson, L.; Wahl, L.M.; Irani, D.N. Cerebrospinal fluid levels of MMP-2, 7, and 9 are elevated in association with human immunodeficiency virus dementia. Ann Neurol. 1999, 46(3), 391–398.
92. Heyes, M.P.; Brew, B.J.; Saito, K.; Quearry, B.J.; Price, R.W.; Lee, K. Inter-relationships between quinolinic acid, neuroactive kynurenines, neopterin and beta 2-microglobulin in cerebrospinal fluid and serum of HIV-1-infected patients. J Neuroimmunol. 1992, 40(1), 71–80.
93. Brew, B.J.; Dunbar, N.; Pemberton, L.; Kaldor, J. Predictive markers of AIDS dementia complex: CD4 cell count and cerebrospinal fluid concentrations of beta 2-microglobulin and neopterin. J Infect Dis. 1996, 174(2), 294–298.
94. Gendelman, H.E.; Zheng, J.; Coulter, C.L.; Ghorpade, A.; Che, M.; Thylin, M. Suppression of inflammatory neurotoxins by highly active antiretroviral therapy in human immunodeficiency virus-associated dementia. J Infect Dis. 1998, 178(4), 1000–1007.
95. Heyes, M.P.; Saito, K.; Lackner, A.; Wiley, C.A.; Achim, C.L.; Markey, S.P. Sources of the neurotoxin quinolinic acid in the brain of HIV-1-infected patients and retrovirus-infected macaques. FASEB J. 1998, 12(10), 881–896.
96. Buffet, R.; Agut, H.; Chieze, F.; Katlama, C.; Bolgert, F.; Devillechabrolle, A. Virological markers in the cerebrospinal fluid from HIV-1-infected individuals. AIDS. 1991, 5(12), 1419–1424.

97. Ellis, R.J.; Hsia, K.; Spector, S.A.; Nelson, J.A.; Heaton, R.K.; Wallace, M.R. Cerebrospinal fluid human immunodeficiency virus type 1 RNA levels are elevated in neurocognitively impaired individuals with acquired immunodeficiency syndrome. HIV Neurobehavioral Research Center Group. Ann Neurol. 1997, 42(5), 679–688.
98. McArthur, J.C.; McClernon, D.R.; Cronin, M.F.; Nance Sproson, T.E.; Saah, A.J.; St Clair, M. Relationship between human immunodeficiency virus-associated dementia and viral load in cerebrospinal fluid and brain [See comments]. Ann Neurol. 1997, 42(5), 689–698.
99. Bouwman, F.H.; Skolasky, R.L.; Hes, D.; Selnes, O.A.; Glass, J.D.; Nance Sproson, T.E. Variable progression of HIV-associated dementia. Neurology. 1998, 50(6), 1814–1820.
100. Dougherty, R.H.; Skolasky, R.L., Jr; McArthur, J.C. Progression of HIV-associated dementia treated with HAART. AIDS Read. 2002, 12(2), 69–74.
101. Nath, A.; Maragos, W.F.; Avison, M.J.; Schmitt, F.A.; Berger, J.R. Acceleration of HIV dementia with methamphetamine and cocaine. J Neurovirol. 2001, 7(1), 66–71.
102. Levy, R.M.; Bredesen, D.E.; Rosenblum, M.L. Neurological manifestations of the acquired immunodeficiency syndrome (AIDS): experience at UCSF and review of the literature. J Neurosurg. 1985, 62(4), 475–495.
103. Levy, R.M.; Rosenbloom, S.; Perrett, L.V. Neuroradiologic findings in AIDS: a review of 200 cases. Am J Roentgenol. 1986, 147(5), 977–983.
104. Bursztyn, E.M.; Lee, B.C.; Bauman, J. CT of acquired immunodeficiency syndrome. Am J Neuroradiol. 1984, 5(6), 711–714.
105. Post, M.J.; Berger, J.R.; Quencer, R.M. Asymptomatic and neurologically symptomatic HIV-seropositive individuals: prospective evaluation with cranial MR imaging [See comments]. Radiology. 1991, 178(1), 131–139.
106. Post, M.J.; Berger, J.R.; Duncan, R.; Quencer, R.M.; Pall, L.; Winfield, D. Asymptomatic and neurologically symptomatic HIV-seropositive subjects: results of long-term MR imaging and clinical follow-up. Radiology. 1993, 188(3), 727–733.
107. Arendt, G.; Hefter, H.; Neuen Jacob, E.; Wist, S.; Kuhlmann, H.; Strohmeyer, G. Electrophysiological motor testing, MRI findings and clinical course in AIDS patients with dementia. J Neurol. 1993, 240(7), 439–445.
108. Olsen, W.L.; Longo, F.M.; Mills, C.M.; Norman, D. White matter disease in AIDS: findings at MR imaging. Radiology. 1988, 169(2), 445–448.
109. Power, C.; Kong, P.A.; Crawford, T.O.; Wesselingh, S.; Glass, J.D.; McArthur, J.C. Cerebral white matter changes in acquired immunodeficiency syndrome dementia: alterations of the blood-brain barrier. Ann Neurol. 1993, 34(3), 339–350.
110. Paul, R.; Cohen, R.; Navia, B.; Tashima, K. Relationships between cognition and structural neuroimaging findings in adults with human immunodeficiency virus type-1. Neurosci Biobehav Rev. 2002, 26(3), 353–359.
111. Thurnher, M.M.; Schindler, E.G.; Thurnher, S.A.; Pernerstorfer-Schon, H.; Kleibl-Popov, C.; Rieger, A. Highly active antiretroviral therapy for patients with AIDS dementia complex: effect on MR imaging findings and clinical course. Am J Neuroradiol. 2000, 21(4), 670–678.
112. Berger, J.R.; Nath, A.; Greenberg, R.N.; Andersen, A.H.; Greene, R.A.; Bognar, A. Cerebrovascular changes in the basal ganglia with HIV dementia. Neurology. 2000, 54(4), 921–926.
113. Pomara, N.; Crandall, D.T.; Choi, S.J.; Johnson, G.; Lim, K.O. White matter abnormalities in HIV-1 infection: a diffusion tensor imaging study. Psychiatry Res. 2001, 106(1), 15–24.
114. Ernst, T.; Chang, L.; Witt, M.; Walot, I.; Aronow, H.; Leonido Yee, M. Progressive multifocal leukoencephalopathy and human immunodeficiency virus-associated white matter lesions in AIDS: magnetization transfer MR imaging. Radiology. 1999, 210(2), 539–543.
115. von Giesen, H.J.; Antke, C.; Hefter, H.; Wenserski, F.; Seitz, R.J.; Arendt, G. Potential time course of human immunodeficiency virus type 1-associated minor motor deficits: electrophysiologic and positron emission tomography findings. Arch Neurol. 2000, 57(11), 1601–1607.
116. Rottenberg, D.A.; Sidtis, J.J.; Strother, S.C.; Schaper, K.A.; Anderson, J.R.; Nelson, M.J. Abnormal cerebral glucose metabolism in HIV-1 seropositive subjects with and without dementia. J Nucl Med. 1996, 37(7), 1133–1141.

117. Hinkin, C.H.; van Gorp, W.G.; Mandelkern, M.A.; Gee, M.; Satz, P.; Holston, S. Cerebral metabolic change in patients with AIDS: report of a six-month follow-up using positron-emission tomography. J Neuropsychiatry Clin Neurosci. 1995, 7(2), 180–187.
118. Laubenberger, J.; Haussinger, D.; Bayer, S.; Thielemann, S.; Schneider, B.; Mundinger, A. HIV-related metabolic abnormalities in the brain: depiction with proton MR spectroscopy with short echo times. Radiology. 1996, 199(3), 805–810.
119. Avison, M.; Nath, A.; Berger, J. Understanding pathogenesis and treatment of HIV dementia: a role for magnetic resonance?. Trends Neurosci. 2002, 25(9), 468.
120. Glass, J.D.; Fedor, H.; Wesselingh, S.L.; McArthur, J.C. Immunocytochemical quantitation of human immunodeficiency virus in the brain: correlations with dementia. Ann Neurol. 1995, 38(5), 755–762.
121. Glass, J.D.; Wesselingh, S.L.; Selnes, O.A.; McArthur, J.C. Clinical-neuropathologic correlation in HIV-associated dementia [See comments]. Neurology. 1993, 43(11), 2230–2237.
122. Johnson, R.T.; Glass, J.D.; McArthur, J.C.; Chesebro, B.W. Quantitation of human immunodeficiency virus in brains of demented and nondemented patients with acquired immunodeficiency syndrome. Ann Neurol. 1996, 39(3), 392–395.
123. Cherner, M.; Masliah, E.; Ellis, R.J.; Marcotte, T.D.; Moore, D.J.; Grant, I. Neurocognitive dysfunction predicts postmortem findings of HIV encephalitis. Neurology. 2002, 59(10), 1563–1567.
124. Subbiah, P.; Mouton, P.; Fedor, H.; McArthur, J.C.; Glass, J.D. Stereological analysis of cerebral atrophy in human immunodeficiency virus-associated dementia. J Neuropathol Exp Neurol. 1996, 55(10), 1032–1037.
125. Navia, B.A.; Cho, E.S.; Petito, C.K.; Price, R.W. The AIDS dementia complex: II. Neuropathology. Ann Neurol. 1986, 19(6), 525–535.
126. Everall, I.; Luthert, P.; Lantos, P. A review of neuronal damage in human immunodeficiency virus infection: its assessment, possible mechanism and relationship to dementia. J Neuropathol Exp Neurol. 1993, 52(6), 561–566.
127. Everall, I.; Barnes, H.; Spargo, E.; Lantos, P. Assessment of neuronal density in the putamen in human immunodeficiency virus (HIV) infection. Application of stereology and spatial analysis of quadrats. J Neurovirol. 1995, 1(1), 126–129.
128. Masliah, E.; Ge, N.; Achim, C.L.; Hansen, L.A.; Wiley, C.A. Selective neuronal vulnerability in HIV encephalitis. J Neuropathol Exp Neurol. 1992, 51(6), 585–593.
129. Cunningham, A.L.; Naif, H.; Saksena, N.; Lynch, G.; Chang, J.; Li, S. HIV infection of macrophages and pathogenesis of AIDS dementia complex: interaction of the host cell and viral genotype. J Leukoc Biol. 1997, 62(1), 117–125.
130. Wiley, C.A.; Masliah, E.; Morey, M.; Lemere, C.; DeTeresa, R.; Grafe, M. Neocortical damage during HIV infection. Ann Neurol. 1991, 29(6), 651–657.
131. Everall, I.P.; Heaton, R.K.; Marcotte, T.D.; Ellis, R.J.; McCutchan, J.A.; Atkinson, J.H. Cortical synaptic density is reduced in mild to moderate human immunodeficiency virus neurocognitive disorder. HNRC Group. HIV Neurobehavioral Research Center. Brain Pathol. 1999, 9(2), 209–217.
132. Collier, A.C.; Marra, C.; Coombs, R.W.; Claypoole, K.; Cohen, W.; Longstreth, W.T., Jr Central nervous system manifestations in human immunodeficiency virus infection without AIDS. J Acquir Immune Defic Syndr. 1992, 5(3), 229–241.
133. Ellis, R.J.; Moore, D.J.; Childers, M.E.; Letendre, S.; McCutchan, J.A.; Wolfson, T. Progression to neuropsychological impairment in human immunodeficiency virus infection predicted by elevated cerebrospinal fluid levels of human immunodeficiency virus RNA. Arch Neurol. 2002, 59(6), 923–928.
134. McArthur, J.C.; McClernon, D.R.; Cronin, M.F.; Nance-Sproson, E.E.; Saah, A.J.; St. Clair, M. Relationship between human immunodeficiency virus associated dementia and viral load in cerebrospinal fluid and brain. Ann Neurol. 1997, 42, 689–698.

135. Altice, F.L.; Friedland, G.H.; Cooney, E.L. Nevirapine induced opiate withdrawal among injection drug users with HIV infection receiving methadone. AIDS. 1999, 13(8), 957–962.
136. Groothuis, D.R.; Levy, R.M. The entry of antiviral and antiretroviral drugs into the central nervous system [See comments]. J Neurovirol. 1997, 3, 387–400.
137. Blasberg, R.G.; Groothuis, D.R. Chemotherapy of brain tumors: physiological and pharmacokinetic considerations. Semin Oncol. 1986, 13(1), 70–82.
138. Hurwitz, A.A.; Berman, J.W.; Lyman, W.D. The role of the blood-brain barrier in HIV infection of the central nervous system. Adv Neuroimmunol. 1994, 4(3), 249–256.
139. Kim, R.B.; Fromm, M.F.; Wandel, C.; Leake, B.; Wood, A.J.; Roden, D.M. The drug transporter P-glycoprotein limits oral absorption and brain entry of HIV-1 protease inhibitors. J Clin Invest. 1998, 101(2), 289–294.
140. Portegies, P. Review of antiretroviral therapy in the prevention of HIV-related AIDS dementia complex (ADC). Drugs. 1995, 49(suppl 1), 25–31; discussion 38–40.
141. Yarchoan, R.; Broder, S. Development of antiretroviral therapy for the acquired immunodeficiency syndrome and related disorders. A progress report. N Engl J Med. 1987, 316(9), 557–564.
142. Rottenberg, D.A.; Moeller, J.R.; Strother, S.C.; Sidtis, J.J.; Navia, B.A.; Dhawan, V. The metabolic pathology of the AIDS dementia complex. Ann Neurol. 1987, 22(6), 700–706.
143. Tartaglione, T.A.; Collier, A.C.; Coombs, R.W.; Opheim, K.E.; Cummings, D.K.; Mackay, S.R. Acquired immunodeficiency syndrome. Cerebrospinal fluid findings in patients before and during long-term oral zidovudine therapy. Arch Neurol. 1991, 48(7), 695–699.
144. Butler, K.M.; Husson, R.N.; Balis, F.M.; Brouwers, P.; Eddy, J.; el-Amin, D. Dideoxyinosine in children with symptomatic human immunodeficiency virus infection. N Engl J Med. 1991, 324(3), 137–144.
145. Arendt, G.; von Giesen, H.J.; Hefter, H.; Theisen, A. Therapeutic effects of nucleoside analogues on psychomotor slowing in HIV infection. AIDS. 2001, 15(4), 493–500.
146. Sidtis, J.J.; Gatsonis, C.; Price, R.W.; Singer, E.J.; Collier, A.C.; Richman, D.D. Zidovudine treatment of the AIDS dementia complex: results of a placebo-controlled trial. AIDS Clinical Trials Group. Ann Neurol. 1993, 33(4), 343–349.
147. Lanier, E.R.; Sturge, G.; McClernon, D.; Brown, S.; Halman, M.; Sacktor, N. HIV-1 reverse transcriptase sequence in plasma and cerebrospinal fluid of patients with AIDS dementia complex treated with abacavir. AIDS. 2001, 15(6), 747–751.
148. Price, R.W.; Yiannoutsos, C.T.; Clifford, D.B.; Zaborski, L.; Tselis, A.; Sidtis, J.J. Neurological outcomes in late HIV infection: adverse impact of neurological impairment on survival and protective effect of antiviral therapy. AIDS Clinical Trial Group and Neurological AIDS Research Consortium study team. AIDS. 1999, 13(13), 1677–1685.
149. Graham, N.M.; Hoover, D.R.; Park, L.P.; Stein, D.S.; Phair, J.P.; Mellors, J.W. Survival in HIV-infected patients who have received zidovudine: comparison of combination therapy with sequential monotherapy and continued zidovudine monotherapy. Multicenter AIDS Cohort Study Group. Ann Intern Med. 1996, 124(12), 1031–1038.
150. Ferrando, S.; van Gorp, W.; McElhiney, M.; Goggin, K.; Sewell, M.; Rabkin, J. Highly active antiretroviral treatment in HIV infection: benefits for neuropsychological function. AIDS. 1998, 12(8), F65–F70.
151. von Giesen, H.J.; Koller, H.; Theisen, A.; Arendt, G. Therapeutic effects of nonnucleoside reverse transcriptase inhibitors on the central nervous system in HIV-1-infected patients. J Acquir Immune Defic Syndr. 2002, 29(4), 363–367.
152. Martin, C.; Sonnerborg, A.; Svensson, J.O.; Stahle, L. Indinavir-based treatment of HIV-1 infected patients: efficacy in the central nervous system. AIDS. 1999, 13(10), 1227–1232.
153. Haas, D.W.; Arathoon, E.; Thompson, M.A.; de Jesus Pedro, R.; Gallant, J.E.; Uip, D.E. Comparative studies of two-times-daily versus three-times-daily indinavir in combination with zidovudine and lamivudine. AIDS. 2000, 14(13), 1973–1978.

154. Schifitto, G.; Sacktor, N.; Marder, K.; McDermott, M.P.; McArthur, J.C.; Kieburtz, K. Randomized trial of the platelet-activating factor antagonist lexipafant in HIV-associated cognitive impairment. Neurological AIDS Research Consortium. Neurology. 1999, 53(2), 391–396.
155. Lipton, S.A. Memantine prevents HIV coat protein-induced neuronal injury in vitro. Neurology. 1992, 42, 1403–1405.
156. Consortium, D. A randomized, double-blind, placebo-controlled trial of deprenyl and thioctic acid in human immunodeficiency virus-associated cognitive impairment. Dana Consortium on the Therapy of HIV Dementia and Related Cognitive Disorders. Neurology. 1998, 50, 645–651.
157. Sacktor, N.; Schifitto, G.; McDermott, M.P.; Marder, K.; McArthur, J.C.; Kieburtz, K. Transdermal selegiline in HIV-associated cognitive impairment: pilot, placebo-controlled study. Neurology. 2000, 54(1), 233–235.
158. Hriso, E.; Kuhn, T.; Masdeu, J.C.; Grundman, M. Extrapyramidal symptoms due to dopamine-blocking agents in patients with AIDS encephalopathy. Am J Psychiatry. 1991, 148(11), 1558–1561.
159. Mirsattari, S.M.; Power, C.; Nath, A. Parkinsonism with HIV infection. Mov Disord. 1998, 13(4), 684–689.
160. Romanelli, F.; Jennings, H.R.; Nath, A.; Ryan, M.; Berger, J. Therapeutic dilemma: the use of anticonvulsants in HIV-positive individuals. Neurology. 2000, 54(7), 1404–1407.
161. Mirsattari, S.M.; Power, C.; Nath, A. Primary headaches with HIV infection. Headache. 1999, 39, 3–10.

12

HIV Myelopathy, Peripheral Neuropathy, and Myopathy

Lydia Estanislao, Anthony Geraci, and David M. Simpson
Mount Sinai Medical Center
New York, New York, U.S.A.

Alessandro Di Rocco
Albert Einstein College of Medicine at Beth Israel Medical Center
New York, New York, U.S.A.

1 HIV MYELOPATHY

First described at the beginning of the AIDS epidemic [1], the spinal cord disease defined as AIDS-associated vacuolar myelopathy (VM) is one of the most disabling neurological complications of HIV infection. To date, VM is one of the least understood and least studied of the neurological manifestations of AIDS. VM is considered the most common cause of spinal cord disease in AIDS, with a high prevalence of pathological involvement of the spinal cord reported at autopsy. Most AIDS-related postmortem pathological series report a prevalence of 15–55% [2–4], although a prevalence as low as 1% has been reported [5].

1.1 Clinical Features

Vacuolar myelopathy frequently presents in early or late stages of HIV disease. VM develops insidiously over months. Urinary urgency and frequency are often the first symptoms. Erectile dysfunction is a common early symptom. Although there are many other reasons for impotence in AIDS, myelopathy should be a consideration, especially in the setting of a normal testosterone level. Lower extremity weakness may follow. Because VM typically involves thoracic segments initially, arms are usually spared at presentation. In other cases, lower extremity stiffness is more prominent than weakness and results in gait difficulty. Patients may complain of occasional fleeting paresthesias in the lower extremities, but they are rarely severe.

Examination reveals paraparesis, lower extremity spasticity or both. Increased thresholds of vibration and proprioception with positive Romberg's sign result from posterior column impairment. Pain and temperature are usually preserved, unless there is superimposed peripheral neuropathy. Deep tendon reflexes, particularly the patellar, are hyperactive, with bilateral ankle clonus and extensor plantar responses. Gait ataxia with impaired tandem may also be seen. In advanced cases, patients may be unable to ambulate and may eventually be wheelchair-bound.

1.2 Laboratory Studies

Magnetic resonance imaging (MRI) of the spine may be normal or may reveal nospecific abnormalities. Mild atrophy of the spinal cord and areas of increased signal on T2-weighted images are common (see Fig. 1) [6]. CSF studies often reveal mild pleiocytosis (5–10 cells/mm^3) and mild protein elevation, but these are nonspecific findings in HIV infection. Cell counts of >30 cells/mm^3 should raise the suspicion of other etiologies [7,8], including neurosyphilis, TB, and CMV myelitis.

1.3 Electrophysiological Data

Electrophysiological studies, particularly somatosensory evoked potentials (SSEPs), provide an objective measure of spinal cord dysfunction. Prolongation of the tibial central

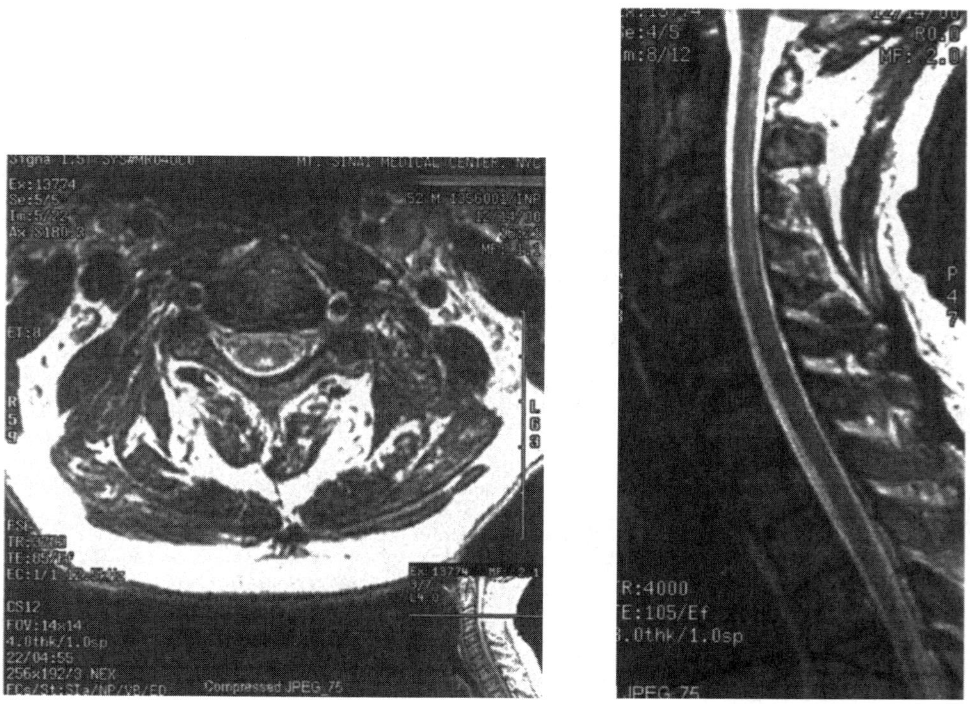

Figure 1 Axial and saggital T2-weighted magnetic resonance images of the spinal cord showing increased signal in the posterior columns.

conduction time correlates with the clinical diagnosis of myelopathy [9]. SSEPs can be used to diagnose subclinical or asymptomatic myelopathy [10] and to follow disease progression [11].

1.4 Pathology

Pathology is most often seen in mid to low thoracic segments and consists of intramyelinic and periaxonal patchy vacuolization in the lateral and posterior columns. The presence of lipid-laden macrophages within the vacuoles is required to avoid confusion with postmortem artifact [2]. Cervical segments may also be involved, but lumbar segments are rarely affected.

Clinico-pathological correlations of spinal cord lesions with neurological symptoms and signs of myelopathy have been difficult because of the retrospective nature of the major series in the literature [2,12]. Moreover, the neurological evaluation may be difficult in patients with advanced AIDS who have severe systemic disease and coexisting peripheral neuropathy. Clear myelopathic signs are infrequent in patients with mild (grade I) VM [2], where only a few vacuoles are present. Patients with mild VM rarely have motor signs and may present with fatigue, mild weakness, increased reflexes, and sphincter abnormalities [2,4,13]. Symptomatic myelopathy clearly correlates in several studies with moderate and severe VM, where numerous (grade II) and confluent (grade III) vacuoles involve the posterior and lateral columns of the spinal cord [2,4,13]. Approximately 20–60% of patients with grade II (moderate) VM and almost all patients with grade III (severe) VM show symptoms and signs of myelopathy with spastic weakness, ataxia, and incontinence [2,4,13]. Since the first pathological descriptions of VM [2,14], a striking similarity with the myelopathy of vitamin B12 deficiency (subacute combined degeneration) has been noted. These changes are usually observed in adults, whereas in children the myelopathy is usually associated with diffuse loss of myelin, axonal loss, and the presence of multinucleated giant cells, and prominent inflammatory infiltrates are commonly observed [13].

1.5 Pathogenesis

Vacuolar myelopathy is not due to direct viral infection of the spinal cord but is most likely the result of a metabolic disorder induced by viral or immunological factors. With a few exceptions [15], most studies have not found evidence of a direct effect of HIV viral infection on the spinal cord in subjects with VM. HIV is found within macrophages but not in neuronal cells or microglia, and there is no relationship between the presence of HIV and the development of myelopathy [13,16–19]. There is also no correlation between CSF levels of HIV-1 RNA and TNF-α and the presence or severity of myelopathy [20].

There is histopathological resemblance between subacute combined degeneration in vitamin B12 (cobalamin) deficiency and HIV myelopathy [2]. This led to speculations of a possible role of vitamin B12 deficiency in the pathogenesis of HIV-VM. However, patients with HIV myelopathy generally have normal serum vitamin B12 levels. Vitamin B12 supplements do not alter the course of the disease. An alternative indirect mechanism may involve impaired methylation, secondary to a deficiency of *S*-adenosylmethionine (SAM), a major methyl group donor in the nervous system. SAM is converted from methionine by methionine synthase. During this process of conversion, vitamin B12 acts as a vital coenzyme [21]. SAM-dependent methylation is essential in myelin formation, stabilization, and repair and in the metabolism of nucleic acids and neurotransmitters. There is evidence linking impaired methylation in the nervous system and neurological complications of AIDS in adults and children [22–24]. The pathogenesis of VM may therefore be related to a complex chain of events that is initiated by the viral infection with consequent macrophage activation and cytokine release. These events may ultimately lead to transmethylation impairment and myelin vacuolization and destruction [21,24].

1.6 Diagnosis

Vacuolar myelopathy is a clinical diagnosis. The diagnosis is largely based on the slow progression of the symptoms, the typical symptoms and findings on neurological examination, and the exclusion of other causes of spinal cord disease. The differential diagnosis of VM is extensive and includes other infections such as human T-lymphotropic virus type I or II, cytomegalovirus, herpes simplex virus type 2, toxoplasma, tuberculosis, syphilis, and metabolic (B12 deficiency) and neoplastic diseases, particularly lymphoma, and the workup should include spinal MR, serological studies, and CSF studies. Although the clinical presentation and the evolution of VM are fairly typical, the diagnosis may be difficult in patients with advanced AIDS and severe systemic disease or coexisting peripheral neuropathy and dementia [2,25,26].

A rapidly progressing myelopathy over days or weeks, the presence of a discrete sensory level, CSF pleocytosis greater than 30 cells/mL and back pain are all evidence against the diagnosis of vacuolar myelopathy and should lead to diagnostic investigations to disclose the cause of spinal cord disease.

1.7 Treatment

There is no definitive treatment for VM. Supplementation with vitamin B12 has proven ineffective in improving the symptoms or delaying the progression of VM [27]. Nevertheless, the vitamin is commonly administered to patients with clinical evidence of VM. Corticosteroids and intravenous gamma globulin have also been ineffective in uncontrolled clinical experience [28]. One study showed a beneficial response to zidovudine (ZDV) (10 mg/kg) [29], whereas another reported no benefit from this antiretroviral agent [30].

To date, there has been no controlled study describing the effect of highly active antiretroviral therapy (HAART) on the clinical manifestations or electrophysiological measures of VM. A case report described clinical improvement of myelopathy after the introduction of HAART in one patient [31]. Although HAART reduced plasma viral load, there was no report of CSF viral load measurement and electrophysiological results before and after treatment. Although it is possible that HAART had a specific effect on myelopathy, it is also possible that the symptomatic improvement may have been related to general improvement of health without a specific effect on the spinal cord [32]. These cases further illustrate the role of electrophysiological studies to monitor progression of the disease and therapeutic effect.

The effect of combination antiretroviral drugs and HAART, in improving the symptoms or slowing the progression of VM is therefore not known. Geraci and Di Roco [33] reported that the use of different combinations of antiretroviral drugs was ineffective in preventing the onset of myelopathy. It is possible, however, that the medications were able to delay the progression or alter the course of the disease. It is also not known whether the introduction of these drugs has decreased the incidence of VM.

Recently there have been attempts to treat the possibly underlying metabolic disorder. A pilot study using high doses of oral ell L-methionine led to improvement in clinical and electrophysiological features of the disease in an open label clinical trial [24], and a double-blind, placebo-controlled study using methionine to treat AIDS-associated VM is near completion. Symptomatic treatment is obviously indicated for patients with spasticity and urinary dysfunction.

2 HIV PERIPHERAL NEUROPATHY

2.1 Distal Symmetrical Polyneuropathy

Distal symmetrical polyneuropathy (DSP) has been noted to be a common complication of HIV infection since early on in the epidemic [26,34–42]. The reported incidence of peripheral neuropathy varies based on CD4 cell count and HIV disease stage. Advanced HIV infection and low CD4 cell counts [43] (especially below 100 cells/μL) increase the risk of peripheral neuropathy [44].

Clinical Features. The major presenting symptoms of patients with HIV DSP are paresthesias, dysesthesias, and numbness that begin in the toes and soles of the feet. This progresses to involve the extremities in a stocking and glove distribution. A complaint of "burning feet" is common and is elicited in 23–100% of cases [34,35,38,42]. Symptoms are often severe at night. Contact sensitivity may be evident, and even the lightest contact with bed sheets, socks, or the examiner's touch may result in extreme pain. Pain has a marked effect on the patient's quality of life. In exceptionally severe cases, patients may be suicidal. Muscle weakness rarely occurs except in advanced cases.

Neurological examination reveals depressed or absent ankle reflexes or reduced ankle reflexes relative to the knees. In a setting of concomitant myelopathy, increased knee reflexes with relatively normal ankle reflexes are present. Vibratory threshold is increased, and pinprick and temperature sensations are reduced in a stocking and glove distribution. Proprioception is usually relatively preserved [39] unless a concurrent myelopathy is present. Weakness, if present, is generally limited to the intrinsic foot muscles.

Electrophysiological Data Nerve conduction studies usually reveal significant reduction or absence of sural nerve amplitude [26,35,45]. However, DSP may be associated with normal sural nerve conduction studies [26,35,38,39]. Less commonly, reduced sensory or motor amplitude of the median or ulnar nerves is found. Late responses are delayed in a pattern consistent with axonal neuropathy. Electromyography may show active or chronic denervation with reinervation.

Laboratory Data. Cerebrospinal fluid analysis is usually not necessary unless there is suspicion of an inflammatory or alternative infectious cause of peripheral neuropathy. Although it is prudent to look for an underlying reversible cause of neuropathy, exhaustive laboratory investigations are usually uninformative. Nerve biopsy is rarely necessary except in cases with atypical features

Pathology. Nerve histology in DSP reveals degeneration of myelinated and unmyelinated axons [34,36,41,46,47]. Mild perivascular mononuclear inflammation is present in the epineurium and endoneurium in up to two-thirds of specimens [36,41,46,47]. The inflammatory cells consist mainly of T lymphocytes and macrophages with predominance of suppressor/cytotoxic cells (CD8) endoneurally, and a more equal ratio with helper/inducer (CD4) cells epineurially and perineurially.

Pathogenesis. The pathogenesis of DSP is unknown. Early attempts to demonstrate the presence of HIV in dorsal root ganglion cells, nerve roots, or peripheral nerves were largely unsuccessful [36,41,48,49], making direct HIV infection of peripheral nerves an unlikely cause of DSP. However, a role of the glycoprotein 120 (gp120) subunit of HIV has been proposed as a cofactor in the pathogenesis of DSP [50].

The lack of evidence for significant in vivo infection of neurons, together with widespread functional and pathological damage to the peripheral nervous system, leads to the possibility that indirect mechanisms such as cytokine dysregulation and immunological dysfunction are etiological factors. Elevated levels of tumor necrosis factor-alpha, interleukin-1, and interleukin-6 have been identified in peripheral nerve and dorsal root ganglia of patients with AIDS [34–36,40–42,51–53], and their expression may be induced via low levels of HIV-1 replication acting on cytotoxic T lymphocytes, resulting in degeneration of sensory neurons within the dorsal root ganglion [53].

In some cases of DSP, paraproteinemia has been implicated as an etiology of neuropathy. Investigators have identified IgM and IgG reactivity against myelin proteins in a majority of tested samples from HIV-infected individuals [54]. There are also high titers of antisulfatide antibodies in some of these patients [55], and these antibodies show immunofluorescence at the paranodal regions of nerves [56]. However, these data have not been causally linked to the development of neuropathy in patients with AIDS.

Reports of identification of CMV inclusions or antigens in peripheral nerve specimens, nerve roots, or dorsal root ganglion cells in some DSP patients [36,41,48,49] and the presence of active systemic CMV infection in patients with painful neuropathy [57] have led other investigators to suspect a possible role of CMV in DSP pathogenesis. However, there is no compelling evidence that direct CMV infection causes DSP.

Antiretroviral Drug–Associated Neuropathy

Nucleoside analog–associated DSP is clinically and electrophysiologically indistinguishable from HIV-associated DSP. Toxic neuropathy tends to be more painful, abrupt in onset, and rapidly progressive [58]. Most important, it bears a temporal relationship to the offending drug. However, in clinical practice, with retrospective histories, it is often difficult to determine the precise date of onset of symptoms relative to the frequent changes in a patient's antiretroviral drug regimen. Among the 15 currently approved antiretroviral agents used for HIV infection, only the dideoxynucleoside analogs (d-drugs) are clearly neurotoxic. One of the first nucleoside analogs found to cause peripheral neuropathy was adenosine arabinoside (vidarabine) for hepatitis B [59]. The neurotoxic antiretroviral d-drugs used in patients with HIV infection include didanosine, zalcitabine and stavudine [60–64].

Patients with advanced HIV infection and severe immunosuppression are at greatest risk of neurotoxicity due to antiretroviral agents. Early studies of nucleoside analog–associated peripheral neuropathy employed dosage schedules that are above the presently approved dosages. Thus, in the analysis of incidence data, it is important to take into consideration the dosages of the drug administered and the stage of disease of the patient cohort under study. Early hypotheses held that ddC, which contains a cytosine base, interfered with the production of sphingomyelin via the formation of a ddC-diphosphocholine metabolite [65]. However, a similar toxicity is observed in d4T and ddI, which do not have a cytosine base. In vitro and animal experiments have suggested inhibition of mitochondrial DNA synthesis, particularly the potent inhibition of DNA polymerase gamma, as the mechanism for neurotoxicity [66–71].

The nucleoside analogs, when phosphorylated to their active triphosphate form, are incorporated into target viral DNA, resulting in inhibition of the reverse transcriptase or chain termination. This is the mechanism responsible for its antiviral activity. However, the same phosphorylated analogs may be used by mammalian mitochondrial DNA polymerase gamma leading to mitochondrial toxicity. In vitro studies and experience with a similar nucleoside analog, fialuridine, which causes a painful sensory neuropathy, have shown evidence of associated mitochondrial toxicity. Cases of biopsy-proven mitochondrial damage in patients with nucleoside analog–associated peripheral neuropathy have been reported [72]. Delayed onset of symptoms may be due to the gradual decline of mitochondrial DNA levels in the cell, and the coasting phenomenon may result from abnormal signaling that may occur as mitochondrial DNA is gradually restored during recovery [58].

Other investigators have suggested that neurotoxicity of nucleoside analogs is due to depletion of acetyl-L-carnitine. Acetyl-L-carnitine may promote peripheral nerve regeneration following injury by causing the release of nerve growth factor [73]. Carnitine depletion causes a disruption of mitochondrial metabolism with resultant toxic accumulation of fatty acids [74]. Lower levels of acetyl-L-carnitine were found in HIV-infected patients taking dideoxynucleosides with peripheral neuropathy than in those without neuropathy [75]. However, data from ACTG 291 do not support this hypothesis [75a]. There was no difference in carnitine levels in those with differing degrees of severity of peripheral neuropathy.

Detailed discussion of other neurotoxic neuropathies is beyond the scope of this chapter. A list of medications that can cause toxic neuropathy is presented in Table 1.

Treatment. The mainstay in the management of DSP is symptomatic treatment with adequate pain control. Effective analgesia can usually be attained by following the World Health Organization's (WHO) suggested analgesic ladder of stepwise pain control (see Fig. 2). In step 1, patients receive nonopioid analgesics such as acetaminophen or a nonsteroidal anti-inflammatory drug (NSAID). If pain remains uncontrolled, the patient is advanced to step 2, where a weak opioid such as codeine is used. In step 3, patients whose pain is not controlled by step 1 or 2 drugs are treated with a strong opioid such as morphine. In each step, adjuvant medications may be employed together with the primary analgesics. The adjuvant agents consist of tricyclic antidepressants (e.g., amytriptyline, nortriptyline) and anticonvulsants (e.g., lamotrigine, gabapentin, diphenylhydantoin, carbamazepine). Desipramine and amitriptyline at doses of up to 150 mg/day can be used [76,77]. Carbamazepine and phenytoin have been reported to reduce neuropathic pain [76,77]. Lamotrigine provided significantly greater reduction in average pain than placebo

Table 1 Drugs That Can Cause Toxic Neuropathy in HIV

Drug	Use in HIV patients
Thalidomide	Aphthous ulcers, wasting syndrome, cachexia, diarrhea
Isoniazid	Mycobacterial infections
Ethambutol	Mycobacterial infections
Pyridoxine	With INH
Metronidazole	Infections caused by anaerobes and protozoans
Dapsone	PCP prophylaxis in cases of sulfonamide intolerance
Vincristine	Lymphoma
Chloramphenicol	Salmonella infections
Etoposide	Kaposi's sarcoma

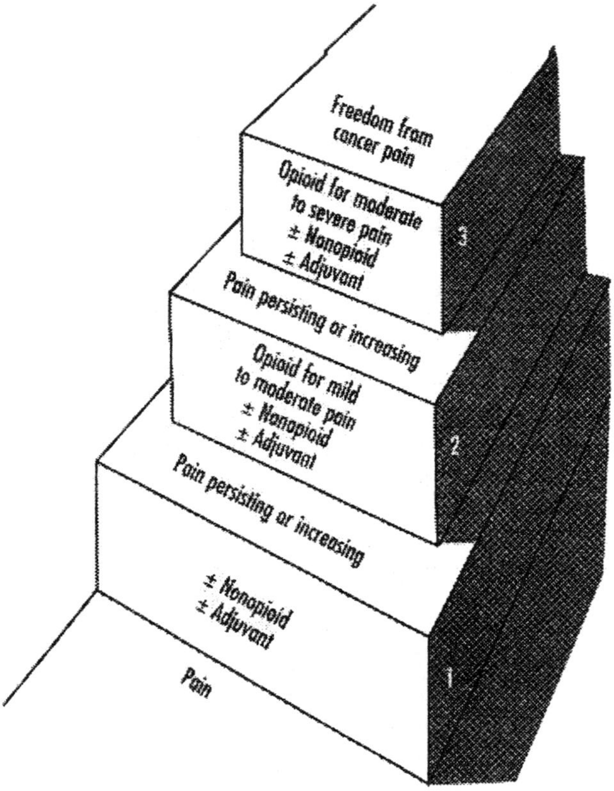

Figure 2 The WHO analgesic ladder. (Reproduced from Ref. 81 with permission.)

in 42 patients [78] with HIV-associated DSP. Gabapentin is an attractive therapy for painful neuropathy due to its favorable pharmacokinetic profile and sparse drug-drug interactions [79,80], although it has not been studied in controlled trials in HIV neuropathy. Topical agents are also effective adjuncts. Lidocaine, now available as a patch formulation, may provide benefit in DSP, although controlled studies are awaited.

An important step in the management of neurotoxic neuropathy is identification of the offending drug and, if possible, its discontinuation. However, the decision to discontinue a drug, especially an antiretroviral drug, should not be automatic. It should follow only after careful weighing of benefits versus risks. At times, dose reduction without sacrificing virological control is enough to alleviate symptoms. In cases where not enough choice of antiretroviral agents is available and substitution is not possible without jeopardizing virological control, symptomatic treatment with adequate pain control is necessary.

2.2 Inflammatory Demyelinating Polyneuropathy

In HIV-seropositive individuals, as in seronegative individuals, two major forms of inflammatory demyelinating polyneuropathy (IDP) may occur. The acute type, with rapidly progressive ascending weakness, minor sensory symptoms, and generalized areflexia, is referred to as acute inflammatory demyelinating polyneuropathy or AIDP (also called

Table 2 Common Drugs Used in the Treatment of DSP

Class/drug	Dosage
Nonprescription analgesics	
Nonsteroidal anti-inflammatory drugs	Variable, depending on the agent
Acetaminophen	500–1000 mg every 4–6 h
Tricyclic antidepressants	
Amitryptiline or desipramine	Less than or equal to 150 mg/day at bedtime
Anticonvulsants	
Gabapentin	300–1200 mg tid
Lamotrigine	200–250 mg bid (after slow escalation)
Topical analgesics	
Lidocaine patch	One patch to affected areas bid
Capsaicin 0.075%	Apply on affected areas bid
Narcotic analgesics	
Oxycodone	1–2 tablets every 4–6 h
Morphine	Variable, depending on the pain intensity
Fentanyl patch	Variable, usually start at 25 μg/h q72h

Guillain-Barré syndrome). The chronic, more slowly progressive, monophasic or relapsing form is chronic inflammatory demyelinating polyneuropathy or CIDP. A lymphocytic pleocytosis (10–50 cells/mm^3) in the cerebrospinal fluid in HIV-seropositive patients with IDP [82] helps to distinguish HIV-infected from seronegative individuals. The presence of more than five per cubic millimeter CSF lymphocytes in this setting should raise suspicion of undiagnosed HIV infection.

Electrophysiological findings in HIV-related IDP are similar to those in HIV-uninfected patients. These include markedly decreased motor and sensory nerve conduction velocity, conduction block, prolonged distal latencies, reduced compound muscle action potential (CMAP) amplitudes, reduced sensory nerve action potential (SNAP) amplitudes, and reduced motor unit recruitment proportional to the degree of weakness.

The pathogenesis of IDP is likely immune-mediated, particularly in patients early in the course of HIV infection, with relatively high CD4 lymphocyte counts. In advanced patients, with CD4 counts <100 cells/mm^3, CMV infection may cause IDP. Other laboratory abnormalities in IDP include increased CSF levels of soluble CD8 and neopterin [83] and anti-peripheral nerve myelin antibody titers that parallel the course of the disease [84].

Nerve biopsy in patients with CIDP reveal inflammatory cell infiltrates and internodal demyelination [82]. In mildly affected nerves, there is moderate subperineurial edema and small numbers of lymphocytes near endoneurial vessels. Within the endoneurial space, there is significant demyelination. In more severe cases, there is phagocyte-mediated myelin stripping [82] and axonal degeneration. In two cases of AIDS associated IDP, CMV inclusions were seen detected in Schwann cells, implicating a direct role of this virus [85].

Case series have shown a positive response of HIV-related IDP to immunomodulating therapy, such as corticosteroids, high-dose intravenous immunoglobulins (0.5–1.0 g/kg for 2 days) and plasmapheresis (four to five exchanges). In advanced HIV infection, antiviral therapy against CMV may be indicated.

2.3 Mononeuropathy Multiplex

Mononeuropathy multiplex (MM) is an infrequent neurological manifestation of HIV infection. MM is characterized clinically by multifocal nerve abnormalities. Distinguishing features include asymmetrical distribution, with involvement of cutaneous nerves, mixed nerves, nerve roots and cranial nerves. Tendon reflexes are preserved in uninvolved areas. The incidence of MM is bimodal. The first peak occurs early in the course of HIV infection, when CD4 cell counts are above 200 cells/mm^3, with a limited distribution of deficits. Facial nerve palsy is a common presentation of this form of MM. These deficits in early MM commonly resolve within several months with or without immunomodulating treatment. Pathological findings reveal axonal degeneration with epineural and endoneural perivascular inflammatory infiltrates.

Mononeuropathy, multiplex may also occur in advanced immunosuppression, with CD4 counts of 50 cells/mm^3 or less. Extensive nerve involvement may rapidly progress to include multiple cranial nerves. This form of MM may appear similar to IDP or progressive polyradiculopathy (PP). Nerve biopsy reveals numerous polymorphonuclear infiltrates, with mixed axonal and demyelinative lesions. In several patients with HIV-associated MM, CMV was identified in peripheral nerve specimens with various assays [86]. Nerve biopsy revealed evidence for necrotizing arteritis [87] and vasculitis in some patients with HIV-associated MM.

As in IDP, an autoimmune phenomenon is the proposed pathogenesis underlying MM that occurs early in HIV infection. Additional mechanisms may be involved in the late onset of MM, including herpes zoster [88], lymphoma [89], cryoglobulinemia [90], cryptococcal meningitis (particularly in cranial neuropathies) [88], and CMV infection [91].

Early MM may resolve spontaneously within several months [42,92]. In cases of delayed or incomplete recovery, corticosteroids, plasmapheresis, or intravenous immunoglobulins may be indicated. In late onset MM occurring in advanced HIV infection, empirical therapy for CMV should be considered.

2.4. Progressive Polyradiculopathy

Progressive polyradiculopathy (PP) occurs most commonly in advanced HIV infection when CD4 cell counts drop below 50 cells/mm^3. PP is characterized clinically by the rapid onset of radicular pain and paresthesias in a cauda equina distribution, followed by signs of progressive involvement of multiple nerve roots, usually lumbar and sacral. Clinical signs include flaccid paraparesis, sphincter dysfunction, and lower extremity areflexia. Some patients have associated myelopathic features, leading to its occasional description as polyradiculomyelopathy or myeloradiculitis [93,94]. Although PP is usually a late manifestation of AIDS, it may be the presenting manifestation of AIDS in rare cases [94].

Cerebrospinal fluid findings in PP include marked polymorphonuclear pleocytosis (mean $651 \pm 1053 \times 10^6$/L), elevated protein (mean 2.28 ± 1.78 g/dL), a and hypoglycorrhachia (mean CSF glucose ratio 0.48 ± 0.17) [94].* Electromyography reveals a reduced number of motor units and abnormal spontaneous activity in weak muscles. Nerve conduction velocities are only mildly abnormal. The severe and widespread proximal axonal pathology in lumbar nerve root segments helps to differentiate PP from MM or IDP. MRI

* The mean values given are based on CMV PP, which accounts for the majority of cases of PP.

of the lumbosacral spine may show enhancement of the roots, in some cases also including the conus medullaris (see Fig. 3).

Progressive polyradiculopathy is an inflammatory, necrotizing polyradiculitis most commonly involving the lumbosacral ventral and dorsal nerve roots, with less frequent thoracic cervical and cranial nerve pathology. Most reported autopsied cases of patients with PP reveal nuclear and cytoplasmic cytomegalic inclusions within endothelial, Schwann and ependymal cells that stain positively for CMV by immunohistochemistry and in situ hybridization [95–97]. CMV detection by PCR analysis of CSF in PP has a sensitivity of 92% and specificity of 94% [98–100].

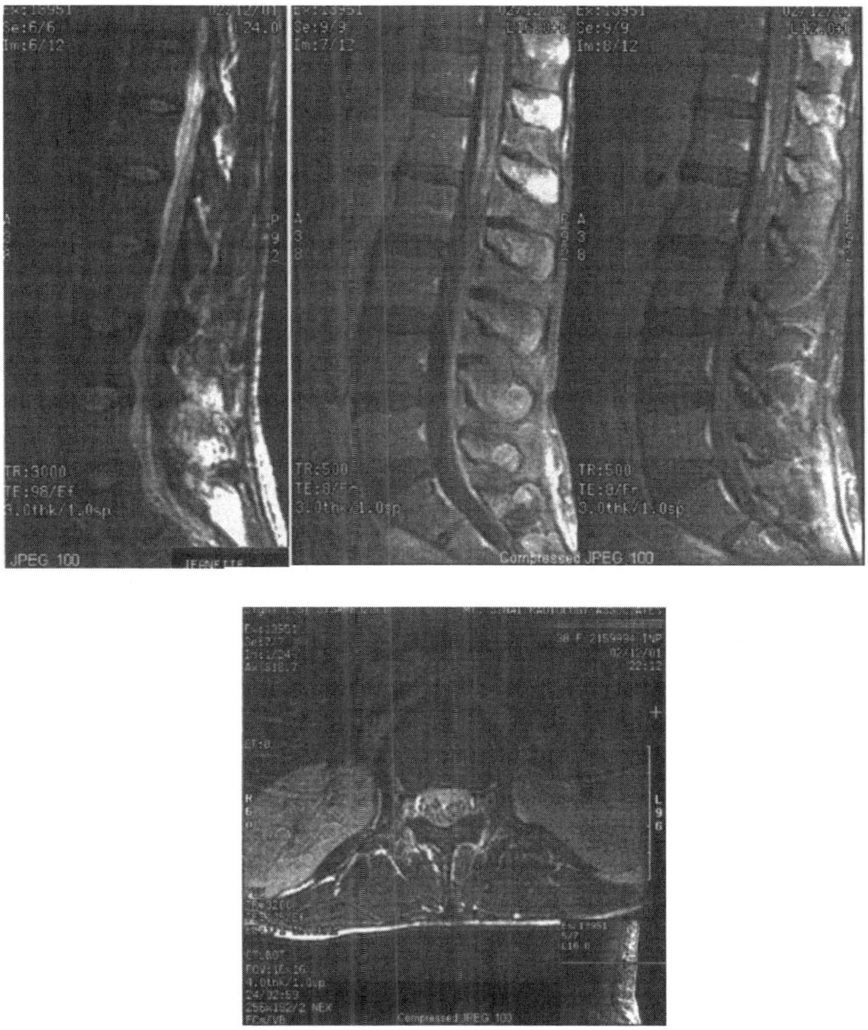

Figure 3 Saggital and axial MRI of the spinal cord of an HIV-seropositive patient with CMV myeloradiculopathy showing increased T2 signal and enhancement at the conus with clumping of the roots. LP revealed WBC of2400, 90% poly morphonuclear cells; CSF CMV PCR was detected.

Cytomegalovirus infection is the major cause of PP in patients with AIDS; less common causes include neurosyphilis [101] and lymphomatous meningitis [102]. Case series have reported improvement of CMV-associated PP with antiviral therapy, including the use of ganciclovir, foscarnet, and cidofovir. A prospective study evaluating treatment of AIDS-associated PP was suspended due to the marked reduction in incidence of CMV disease in the current highly active antiretroviral therapy (HAART) era.

2.5 Autonomic Neuropathy

Early case studies reported clinical and laboratory evidence of autonomic impairment in HIV-infected patients [103,104]. Clinical findings include syncope, dizziness, orthostatic hypotension, diarrhea, decreased sweating, impotence, bladder dysfunction, resting tachycardia, and fatal arrythmias [105]. Autonomic neuropathy has rarely been seen since the introduction of HAART, and many of the earlier cases may have been partly due to the debilitating effects of late-stage HIV infection.

Autonomic neuropathy may have numerous causes, including drugs such as tricyclic antidepressants, vincristine, and pentamidine. Malnutrition and dehydration may be contributory in later stages of disease. Rarely, patients with severe DSP may develop autonomic dysfunction.

Pathological examination reveals abnormalities in several sites of the autonomic nervous system. Autonomic axons are depleted in small bowel mucosa of HIV-infected patients [106]. T-lymphocyte inflammatory infiltrates, ganglion nerve cell loss, and macrophages were noted in the sympathetic ganglia of autopsied cases [107]. A reduction in the number of oxytocin neurons in the paraventricular nucleus of the hypothalamus has also been described, although the clinical significance of this is unclear [108].

Management of autonomic neuropathy is mainly supportive. It includes recognition and discontinuation of offending medications, correction of fluid and electrolyte imbalance, use of compressive stockings and abdominal binders, liberal salt intake, and reconditioning exercises as well as maneuvers such as squatting and standing with crossed legs. In severe cases, pharmacological agents such as fludricortisone, midodrine, and antiarrhythmic agents may be needed.

HIV MYOPATHY

The incidence of HIV myopathy has not been established in prospective studies, although it appears to be relatively uncommon. One series reported two cases of polymyositis out of 101 patients with HIV [109]. In another report, 11 of 15 AIDS patients had positive electromyographic evidence of myopathy; four out of seven muscle biopsies were positive for myopathic changes. In AIDS Clinical Trials Group (ACTG) 016, a retrospective analysis of a primary antiretroviral protocol, the incidence of myopathy was 0.4% in the placebo group ($n=351$) [110].

3.1 Clinical Features

HIV myopathy usually presents with slowly progressive weakness of proximal muscles. Typical complaints include difficulty in lifting arms above the head, difficulty in rising up from a chair, and difficulty climbing stairs. Myalgias are common and may be present in 25–50% of cases.

Neurological examination reveals symmetrical weakness of proximal muscles of the extremities. Neck flexors may be involved as well. Functional tests of muscle strength include having the patient rise unassisted from a seated position and a squat, and having the patient sustain arm extension above the head for 15 s. In pure myopathy cases, deep tendon reflexes are normal or preserved. However, in HIV disease, more than one neurological complication may coexist, such as myelopathy or neuropathy, giving a combination of signs. In cases like these, ancillary tests may be necessary to distinguish the different neurological conditions.

3.2 Laboratory Data

Serum creatine phosphokinase (CPK) levels are usually elevated to a moderate degree, with a median level of approximately 500 IU/L [111]. Levels greater than 1500 IU/L have been reported with resultant rhabodmyolysis [112]. CPK is not a specific marker of myopathy in HIV-positive patients. In ACTG 016, a majority of the patients with elevated CPK did not have clinical evidence of weakness or myopathy [110]. An isolated CPK elevation is not sufficient to make a diagnosis of myopathy without the accompanying clinical features.

3.3 Electrophysiological data

Electromyography (EMG) is sensitive and specific in the diagnosis of myopathy. In a series of 50 patients with myopathy, 94% had myopathic EMG findings [111] consisting of short, brief motor unit action potentials, recruiting with an early and full interference pattern, with or without irritative activity.

3.4 Pathology and Pathogenesis

From early reports of HIV myopathy prior to the availability of antiretroviral agents, two histological patterns were described. The first was HIV-related polymyositis, characterized by the presence of myofiber necrosis and phagocytosis and by mononuclear cell inflammation in the interstitial and interfascicular areas. Immunohistochemistry revealed HIV in the CD4 positive cells [113] and macrophages [114] within the infiltrate. CD8 positive cells made up the majority of the cells in the infiltrate [114]. Because of the nature of the inflammation, some investigators have proposed an autoimmune mechanism as the pathogenesis. The second pattern was a structural myopathy characterized by the presence of abnormal myofiber structure with rod and cytoplasmic bodies and basophilic granular material [115,116]. Inflammatory cell infiltrates varied from severe to absent. HIV could not be demonstrated by immunohistochemistry or in situ hybridization. With the advent of reverse transcriptase polymerase chain reaction (RT-PCR), HIV was demonstrated within endomyseal macrophages and myocyte nuclei [117]. The latter finding raises the possibility of direct infection of myocytes by HIV in zidovudine (ZDV/AZT)-naive HIV patients with myopathy.

The role of AZT in HIV myopathy is controversial. AZT belongs to the family of nucleoside analogs, which, when phosphorylated to their active triphosphate forms, compete with the triphosphatases as substrates for DNA polymerases. They are then incorporated into the target viral DNA, resulting in inhibition of the reverse transcriptase or chain termination of the virus. However, the same phosphorylated analogs have been shown in vitro to be used by mammalian mitochondrial DNA polymerase gamma, thereby inhibiting it.

In 1988, Bessen et al. [118] reported polymyositis in four HIV-seropositive patients treated with AZT, three of whom had improvement after AZT withdrawal. Several case reports followed describing myopathies in AZT-treated patients whose condition improved after drug withdrawal [119–121]. Muscle biopsy from AZT-treated patients with myopathy revealed ragged red fibers (RRF), the percentage of which was reported to correlate with the severity of clinical myopathy. Other features of inflammatory myopathy were present. Electron microscopy (EM) showed proliferation and enlargement of mitochondria with paracrystalline inclusions. These mitochondrial abnormalities were not seen in the AZT-naive group. Another nucleoside analog, stavudine (d4T), has recently been implicated in the same entity, with similar histopathological features [122,123].

However, there is controversy surrounding the role of AZT in the production of AZT-induced mitochondrial myopathy. Analysis of 26 muscle biopsies failed to distinguish AZT-treated from AZT-naive patients [111]. Mitochondrial abnormalities correlated with the degree of myofiber degeneration in HIV-positive patients regardless of AZT exposure history in several studies [124–126].

3.5 Treatment

Some patients improve with AZT withdrawal [118,120,127,128], but others do not [111,129–131]. AZT rechallenge in some patients has not reproduced their myopathic symptoms [121,128].

Corticosteroids may provide benefit in HIV myopathy [127,130,132–134]. However, they should be used with caution because of the immunosuppressant effects. Although intravenous immunoglobulins may be an alternative option without risk of immunosuppression, there is only limited reported experience in HIV myopathy.

REFERENCES

1. Snider, W.; Simpson, D.M.; Nielsen, S.; Metroka, C.; Posner, J. Neurological complications of acquired immunodeficiency syndrome: analysis of 50 patients. Ann Neurol. 1983, 14, 403–418.
2. Petito, C.K.; Navia, B.A.; Cho, E.S.; Jordan, B.D.; George, D.C.; Price, R.W. Vacuolar myelopathy pathologically resembling subacute combined degeneration in patients with the acquired immunodeficiency syndrome. N Engl J Med. 1985, 312(14), 874–829.
3. Artigas, J.; Grosse, G.; Niedobitek, F. Vacuolar myelopathy in AIDS: a morphological analysis. Pathol Res Pract. 1990, 186, 228–237.
4. Dal Pan, G.J.; Glass, J.D.; McArthur, J.C. Clinicopathologic correlations of HIV-1-associated vacuolar myelopathy: an autopsy-based case-control study. Neurology. 1994, 44(11), 2159–2164.
5. Lang, W.; Miklossy, J.; Deruaz, J.; Pizzolato, G.; Probst, A.; Schaffner, T. Neuropathology of the acquired immunodeficiency syndrome: a report of 135 consecutive autopsy cases from Switzerland. Acta Neuropathol (Berl). 1989, 77, 379–390.
6. Chong, J.; Di Rocco, A.; Tagliati, M.; Danisi, F.; Simpson, D.M.; Atlas, S.W. MR findings in AIDS-associated myelopathy. AJNR Am J Neuroradiol. 1999, 20(8), 1412–1416.
7. Overhage, J.; Greist, A.; Brown, D. Conus medullaris syndrome. Am J Med. 1990, 89, 814–815.
8. Tucker, T.; Dix, R.; Katzen, C.; Davis, R.; Schmidley, J. CMV and HSV ascending myelitis in a patient with AIDS. Ann of Neurol. 1985, 18, 74–79.

9. Tagliati, M.; Di Rocco, A.; Danisi, F.; Simpson, D. The role of somatosensory evoked potentials in the diagnosis of AIDS-associated myelopathy. Neurology. 2000, 54, 1477–1482.
10. Jakobsen, J.; Smith, T.; Gaub, J.; Helweg-Larsen, S.; Trojaborg, W. Progressive neurological dysfunction during latent HIV infection. Br Med J. 1989, 299, 225–228.
11. Pierelli, F.; Garrubba, C.; Tilia, G.; Farisi, L.; Fattapposta, F.; Pozzessere, G.; Soldati, G.; Stanzione, P.; D'Offizi, G.; Mezzaroma, I. Multimodal evoked potentials in HIV-1-seropositive patients: relationship between the immune impairment and the neurophysiological function. Acta Neurol Scand. 1996, 93(4), 266–271.
12. Simpson, D.; Tagliati, M. Neurological manifestations of HIV infection. Ann Intern Med. 1994, 121, 769–785.
13. Tan, S.; Guiloff, R.; Scaravilli, F. AIDS-associated vacuolar myelopathy. A morphometric study. Brain. 1995, 118, 1247–1261.
14. Goldstick, L.; Mandybur, T.; Bode, R. Spinal cord degeneration in AIDS. Neurology. 1985, 35, 103–106.
15. Budka, H. Human immunodeficiency virus (HIV) envelope and core proteins in CNS tissues of patients with the acquired immune deficiency syndrome (AIDS). Acta Neuropathol (Berlin). 1990, 79, 611–619.
16. Eilbott, D.; Peress, N.; Burger, H. Human immunodeficiency virus type 1 in spinal cords of acquired immunodeficiency syndrome patients with myelopathy: expression and replication in macrophages. Proc Natl Acad Sci USA. 1989, 86, 3337–3341.
17. Kure, K.; Llena, J.; Lyman, W.; Soreiro, R.; Weidenheim, K.; Hirano, A.; Dickson, D. Human immunodeficiency virus-1 infection of the nervous system: an autopsy study of 268 adult, pediatric and fetal brains. Hum Pathol. 1991, 22, 700–710.
18. Petito, C.K.; Vecchio, D.; Chen, Y.T. HIV antigen and DNA in AIDS spinal cords correlate with macrophage infiltration but not with vacuolar myelopathy. J Neuropathol Exp Neurol. 1994, 53(1), 86–94.
19. Rosenblum, M.; Scheck, A.C.; Cronin, K.; Brew, B.J.; Khan, A.; Paul, M.; Price, R.W. Dissociation of AIDS-related vacuolar myelopathy and productive HIV-1 infection of the spinal cord. Neurology. 1989, 39(7), 892–896.
20. Geraci, A.; Di Rocco, A.; Liu, M.; Werner, P.; Tagliati, M.; Godbold, J.; Simpson, D.; Morgello, S. AIDS myelopathy is not associated with elevated HIV viral load in cerebrospinal fluid. Neurology. 2000, 55(3), 440–442.
21. Tan, S.V.; Guiloff, R.J. Hypothesis on the pathogenesis of vacuolar myelopathy, dementia, and peripheral neuropathy in AIDS [see comments]. J Neurol Neurosurg Psychiatry. 1998, 65(1), 23–28.
22. Surtees, R.; Hyland, K.; Smith, I. Central nervous system methyl group metabolism in children with neurological complications of HIV infection. Lancet. 1990, 335(1), 619–621.
23. Castagna, A.; Le Grazie, C.; Accordini, A.; Giulidori, P.; Cavalli, G.; Bottiglieri, T.; Lazzarin, A. Cerebrospinal fluid S-adenosylmethionine (SAMe) and glutathione concentrations in HIV infection: effect of parenteral treatment with SAMe. Neurology. 1995, 45(9), 1678-1683.
24. DiRocco, A.; Tagliati, M.; Danisi, F.; Dorfman, D.; Moise, J.; Simpson, D. L-Methionine for AIDS-associated vacuolar myelopathy. Neurology. 1998, 51, 266–268.
25. Di Rocco, A. Diseases of the spinal cord in human immunodeficiency virus infection. Semin Neurol. 1999, 19(2), 151–155.
26. Snider, W.; Simpson, D.; Nielsen, S. Neurologic complications of acquired immunodeficiency syndrome; analysis of 50 patients. Ann Neurol. 1983, 14, 403–418.
27. Kieburtz, K.; Giang, D.; Schiffer, R.; Vakil, N. Abnormal vitamin B12 metabolism in human immunodeficiency virus infection: association with neurologic dysfunction. Arch Neural. 1991, 48(3), 312–314.
28. Dal Pan, G.; Berger, J. Spinal cord disease in human immunodeficiency virus infection. In:; Berger, J., Levy, R., Eds. AIDS and the Nervous System; Lippincott-Raven: Philadelphia, 1997, 173–187.

29. Oksenhendler, E.; Ferchal, F.; Cadranel, J.; Sauvageon-Martre, H.; Clauvel, J. Zidovudine for HIV-related myelopathy. Am J Med. 1990, 88(5N), 65N–66N.
30. Yarchoan, R.; Berg, G.; Brouwers, P. Response of immunodeficiency virus-associated neurological disease to 3′-azido-3′-deoxythymidine. Lancet. 1987, 1, 132–135.
31. Staudinger, R.; Henry, K. Remission of HIV myelopathy after highly active antiretroviral therapy. Neurology. 2000, 54, 267–268.
32. Di Rocco, A.; Geraci, A.; Tagliati, M. Remission of HIV myelopathy after highly active antiretroviral therapy [letter]. Neurology. 2000, 55, 456.
33. Geraci, A.; Di Rocco, A. Anti-HIV therapy. AIDS. 2000, 14(13), 2059–2061.
34. Bailey, R.O.; Baltch, A.L.; Venkatesh, R.; Singh, J.K.; Bishop, M.B. Sensory motor neuropathy associated with AIDS. Neurology. 1988, 38(6), 886–891.
35. Cornblath, D.R.; McArthur, J.C. Predominantly sensory neuropathy in patients with AIDS and AIDS-related complex. Neurology. 1988, 38(5), 794–796.
36. de la Monte, S.M.; Gabuzda, D.H.; Ho, D.D.; Brown, R.H., Jr; Hedley-Whyte, E.T.; Schooley, R.T.; Hirsch, M.S.; Bhan, A.K. Peripheral neuropathy in the acquired immunodeficiency syndrome. Ann Neurol. 1988, 23(5), 485–492.
37. Janssen, R.S.; Saykin, A.J.; Kaplan, J.E.; Spira, T.J.; Pinsky, P.F.; Sprehn, G.C.; Hoffman, J.C.; Mayer, W.B.; Schonberger, L.B. Neurological complications of human immunodeficiency virus infection in patients with lymphadenopathy syndrome. Ann Neurol. 1988, 23(1), 49–55.
38. Lange, D.J.; Britton, C.B.; Younger, D.S.; Hays, A.P. The neuromuscular manifestations of human immunodeficiency virus infections. Arch Neurol. 1988, 45(10), 1084–1088.
39. Leger, J.M.; Bouche, P.; Bolgert, F.; Chaunu, M.P.; Rosenheim, M.; Cathala, H.P.; Gentilini, M.; Hauw, J.J.; Brunet, P. The spectrum of polyneuropathies in patients infected with HIV. J Neurol Neurosurg Psychiatry. 1989, 52(12), 1369–1374.
40. Levy, R.; Bredesen, D.; Rosenblum, M. Neurological manifestations of the acquired immunodeficiency syndrome: experience at UCSF and review of the literature. J Neurosurg. 1985, 62, 475–495.
41. Mah, V.; Vartavarian, L.M.; Akers, M.A.; Vinters, H.V. Abnormalities of peripheral nerve in patients with human immunodeficiency virus infection. Ann Neurol. 1988, 24(6), 713–717.
42. So, Y.T.; Holtzman, D.M.; Abrams, D.I.; Olney, R.K. Peripheral neuropathy associated with acquired immunodeficiency syndrome. Prevalence and clinical features from a population-based survey. Arch Neurol. 1988, 45(9), 945–948.
43. Bacellar, H.; Munoz, A.; Miller, E.N.; Cohen, B.A.; Besley, D.; Selnes, O.A.; Becker, J.T.; McArthur, J.C. Temporal trends in the incidence of HIV-1-related neurologic diseases: Multicenter AIDS Cohort Study, 1985–1992. Neurology. 1994, 44(10), 1892–1900.
44. Tagliati, M.; Grinnell, J.; Godbold, J.; Simpson, D.M. Peripheral nerve function in HIV infection: clinical, electrophysiologic, and laboratory findings. Arch Neurol. 1999, 56(1), 84–89.
45. Simpson, D.; Cohen, J.; Sivak, M. Neuromuscular complications in association with acquired immunodeficiency syndrome. Ann Neurol. 1985, 18, 160.
46. Griffin, J.W.; Crawford, T.O.; Tyor, W.R., et al. Predominantly sensory neuropathy in AIDS: Distal axonal degeneration and unmyelinated fiber loss. Neurology. 1991, 41(suppl 1), 374.
47. Lipkin, W.I.; Parry, G.; Kiprov, D.; Abrams, D. Inflammatory neuropathy in homosexual men with lymphadenopathy. Neurology. 1985, 35(10), 1479–1483.
48. Grafe, M.; Wiley, C. Spinal cord and peripheral nerve pathology in AIDS: the roles of cytomegalovirus and human immunodeficiency virus. Ann of Neurol. 1989, 25, 561–566.
49. Rance, N.E.; McArthur, J.C.; Cornblath, D.R.; Landstrom, D.L.; Griffin, J.W.; Price, D.L. Gracile tract degeneration in patients with sensory neuropathy and AIDS. Neurology. 1988, 38(2), 265–271.
50. Apostolski, S.; McAlarney, T.; Quattrini, A.; Levison, S.W.; Rosoklija, G.; Lugaressi, A.; Corbo, M.; Sadiq, S.A.; Lederman, S.; Hays, A.P. The gp120 glycoprotein of human immunodeficiency virus type 1 binds to sensory ganglion neurons. Ann Neurol. 1993, 34(6), 855–863.

51. Wesselingh, S.; Glass, J.; McArthur, J.; Griffin, J.; Griffin, D. Cytokine dysregulation in HIV-associated neurological disease. Adv Neuroimmunol. 1994, 4(3), 199–206.
52. Tyor, W.R.; Wesselingh, S.L.; Griffin, J.W.; McArthur, J.C.; Griffin, D.E. Unifying hypothesis for the pathogenesis of HIV-associated dementia complex, vacuolar myelopathy, and sensory neuropathy. J Acquir Immune Defic Syndr Hum Retrovirol. 1995, 9(4), 379–388.
53. Yoshioka, M.; Shapshak, P.; Srivastava, A.K.; Stewart, R.V.; Nelson, S.J.; Bradley, W.G.; Berger, J.R.; Rhodes, R.H.; Sun, N.C.; Nakamura, S. Expression of HIV-1 and interleukin-6 in lumbosacral dorsal root ganglia of patients with AIDS [published erratum appears in. Neurology]. Neurology. 1994, 44(6), 1120–1130.
54. Petratos, S.; Turnbull, V.; Papadopoulos, R. Antibodies against peripheral myelin glycolipids in people with HIV infection. Immunol Cell Biol. 1998, 76, 535–541.
55. Petratos, S.; Turnbull, V.; Papadopoulos, R. High titer anti-sulfatide antibodies in HIV-infected individuals. Neuroreport. 1999, 10, 2557–2562.
56. Petratos, S.; Turnbull, V.; Papadopoulos, R. Peripheral nerve binding patterns of anti-sulphatide antibodies in HIV- infected individuals. Neuroreport. 1999, 10, 1659–1664.
57. Fuller, G.N.; Jacobs, J.M.; Guiloff, R.J. Association of painful peripheral neuropathy in AIDS with cytomegalovirus infection [see comments]. Lancet. 1989, 2(8669), 937–941.
58. Moyle, G.J.; Sadler, M. Peripheral neuropathy with nucleoside antiretrovirals: risk factors, incidence and management. Drug Saf. 1998, 19(6), 481–494.
59. Garcia, G.; Smith, C. Adenosine arabinoside monophosphate (vidarabine phosphate) in combination with human leukocyte interferon in the treatment of chronic hepatitis B: a randomized, double-blinded, placebo-controlled trial. Ann Intern Med. 1987, 107, 278–285.
60. Yarchoan, R.; Perno, C. Phase I studies of 2′,3′-deoxycytidine in severe human immunodeficiency virus infection as a single agent and alternating with zidovudine. Lancet. 1988, 1, 76–81.
61. Merigan, T.; Skowron, G. Circulating p24 antigen levels and responses to dideoxycytidine in human immunodeficiency virus (HIV) infections. Ann Intern Med. 1989, 110, 189–194.
62. Cooley, T.; Kunches, L. Once-daily administration of 2′3′-dideoxyinosine (ddI) in patients with the acquired immunodeficiency syndrome or AIDS-related complex. N Engl J Med. 1990, 322, 1340–1345.
63. Lambert, J.; Seidlin, M. Dideoxyinosine (ddI) in patients with the acquired immunodeficiency syndrome or AIDS-related complex: results of phase I trial. N Engl J Med. 1990, 322, 1333–1340.
64. Browne, M.J.; Mayer, K.H.; Chafee, S.B.; Dudley, M.N.; Posner, M.R.; Steinberg, S.M.; Graham, K.K.; Geletko, S.M.; Zinner, S.H.; Denman, S.L. 2′,3′-Didehydro-3′-deoxythymidine (d4T) in patients with AIDS or AIDS- related complex: a phase I trial. J Infect Dis. 1993, 167(1), 21–29.
65. Cooney, D.; Dalal, M. Initial studies on the cellular pharmacology of 2′3′-dideoxycytidine, an inhibitor of HTLV-III infectivity. Biochem Pharmacol. 1986, 35, 2065–2068.
66. Pezeshkpour, G.; Krarup, C. Peripheral neuropathy in mitochondrial disease. J Neurol Sci. 1987, 77, 285–304.
67. Balzarini, J.; Herdewlin, P. Differential patterns of intracellular metabolism of 2′3′-dideoxythymidine and 3′-azido-2′3′-dideoxythymidine, two potent anti-human immunodeficiency virus compounds. J Biol Chem. 1989, 264, 6127–6133.
68. Chen, C.H.; Vazquez-Padua, M.; Cheng, Y.C. Effect of anti-human immunodeficiency virus nucleoside analogs on mitochondrial DNA and its implication for delayed toxicity. Mol Pharmacol. 1991, 39(5), 625–628.
69. Chen, C.; Cheng, Y. Delayed cytotoxicity and selective loss of mitochondrial DNA in cells treated with the anti-human immunodeficiency virus compound 2′3′-dideoxycytidine. J Biol Chem. 1989, 264, 11934–11937.
70. Martin, J.; Brown, C. Effects of antiviral nucleoside analogs on human DNA polymerases and mitochondrial DNA synthesis. Antimicrob Agents Chemother. 1994, 38, 2743–2749.

71. Cui, L.; Locatelli, L. Effect of nucleoside analogs on neurite regeneration and mitochondrial DNA synthesis in PC-12 cells. J Pharmacol Exp Ther. 1997, 280, 1228–1234.
72. Gasnault, J.; Pinganaud, C.; Kousignian, P. Subacute ascendant polyneuropathy revealing a nucleoside-induced mitochondrial cytopathy–: Lisbon: Portugal, 1999.
73. Angelucci, L.; Ramacci, M.; Taglialatela, G. Nerve growth factor binding in aged rat central nervous system: effect of acetyl-L-carnitine. J Neurosci Res. 1988, 20, 491–496.
74. Colucci, W.; Gandour, R. Carnitine acetyltransferase: a review of its biology, enzymology and bioorganic chemistry. Bioorg Chem. 1988, 16, 307–334.
75. Famularo, G.; Moretti, S.; Marcellini, S.; Trinchieri, V.; Tzantzoglou, S.; Santini, G.; Longo, A.; De Simone, C. Acetyl-carnitine deficiency in AIDS patients with neurotoxicity on treatment with antiretroviral nucleoside analogues. AIDS. 1997, 11(2), 185–190.
75a. Simpson, D.; Katzenstein, D.; Haidichi, B., et al. and the AIDS Clinical Trials Group Protocol 291/860 Study Team. Plasma carnitine in HIV-associated neuropathy. AIDS. 2001, 15(16), 2207–2208.
76. Simpson, D.M.; Tagliati, M.; Ramcharitar, S. Neurologic complications of AIDS: new concepts and treatments. Mt Sinai J Med. 1994, 61(6), 484–491.
77. Simpson, D.; Tagliati, M. Nucleoside analogue-associated peripheral neuropathy in human immunodeficiency virus infection. J AIDS. 1995, 9, 153–161.
78. Simpson, D.M.; Olney, R.; McArthur, J.C.; Khan, A.; Godbold, J.; Ebel-Frommer, K. A placebo-controlled trial of lamotrigine for painful HIV-associated neuropathy [see comments]. Neurology. 2000, 54(11), 2115–2119.
79. Mellick, G.; Mellick, L. Reflex sympathetic dystrophy treated with gabapentin. Arch Phys Med Rehab. 1997, 78, 89–105.
80. Rosner, H.; Rubin, L.; Kestenbaum, A. Gabapentin adjunctive therapy in neuropathic pain states. Clin J Pain. 1996, 12(1), 56–8.
81. Simpson, D. Neurologic complications of HIV: Diagnosing and Managing Peripheral Neuropathies; Bristol-Myers Squibb Monograph: New York, 2000, p. 12.
82. Cornblath, D.R.; McArthur, J.C.; Kennedy, P.G.; Witte, A.S.; Griffin, J.W. Inflammatory demyelinating peripheral neuropathies associated with human T-cell lymphotropic virus type III infection. Ann Neurol. 1987, 21(1), 32–40.
83. Griffin, D.; McArthur, J.C. Soluble interleukin-2 receptor and soluble CD8 in serum and cerebrospinal fluid during human immunodeficiency virus-associated neurologic disease. J Neuroimmunol. 1990, 28, 97–109.
84. Misha, B.; Sommers, W.; Koski, C. Acute inflammatory demyelinating polyneuropathy in the acquired immunodeficiency syndrome. Ann Neurol. 1985, 18, 131–132.
85. Morgello, S.; Simpson, D.M. Multifocal cytomegalovirus demyelinative polyneuropathy associated with AIDS. Muscle Nerve. 1994, 17(2), 176–182.
86. Said, C. Multifocal neuropathy in HIV infection (abst.). In: Proceedings of the Seventh International Conference of HIV Infection; Padova: Italy, 1991, p 89.
87. herardi, G.R.; Lebargy, F.; Gaulard, P.; Mhiri, C.; Bernaudin, J.F.; Gray, F. Necrotizing vasculitis and HIV replication in peripheral nerves [letter]. N Engl J Med. 1989, 321(10), 685–686.
88. Engstrom, J.; Lewis, E.; McGuire, D. Cranial neuropathy and the acquired immunodeficiency syndrome. Neurology. 1991, 41(suppl 1), 374.
89. Berger, J.; Flaster, M.; Schatz, N. Cranial neuropathy heralding otherwise occult AIDS-related large cell lymphoma. J Clin Neurophysiol. 1993, 13, 113–118.
90. Stricker, R.; Sanders, K.; Owen, W. Mononeuritis multiplex associated with cryoglobulinemia in HIV infection. Neurology. 1992, 42, 2103–2105.
91. Roullet, E.; Asseurus, J.; Gozlan, J. Cytomegalovirus multifocal neuropathy in AIDS: analysis of consecutive patients. Neurology. 1994, 44, 2174–2182.
92. So, Y.; Olney, R. The natural history of mononeuritis multiplex and simplex in HIV infection. Neurology. 1991, 41(suppl 1), 375.

93. Cohen, B.J.; McArthur, J.C.; Grohman, S. Neurologic prognosis of cytomegalovirus polyradiculomyelopathy in AIDS. Neurology. 1993, 43, 493–499.
94. Anders, H.; Goebel, F. Cytomegalovirus polyradiculopathy in patients with AIDS. Clin Infect Dis. 1998, 27, 345–352.
95. Eidelberg, D.; Sotrel, A.; Vogel, H.; Walker, P.; Kleefield, J.; Crumpacker, C.S. Progressive polyradiculopathy in acquired immune deficiency syndrome. Neurology. 1986, 36(7), 912–916.
96. Miller, R.; Storey, J.; Greco, C. Ganciclovir in the treament of progressive AIDS-related polyradiculopathy. Neurology. 1990, 40, 569–574.
97. de, J.; J, G.a.n.s.; Portegies, P.; Tiessens, G. Therapy for cytomegalovirus polyradiculopathy in patients with AIDS. Treatment with ganciclovir. AIDS. 1990, 4, 421–425.
98. Gozlan, J.; El, M.; Baudrimont, M. A prospective evaluation of clinical criteria and polymerase chain reaction assay of cerebrospinal fluid for the diagnosis of cytomegalovirus-related neurological diseases during AIDS. AIDS. 1995, 9, 253–260.
99. Shinkai, M.; Spector, S. Quantitation of human cytomegalovirus (HCMV) DNA in cerebrospinal fluid by competitive PCR in AIDS patients with different HCMV central nervous system diseases. Scan J Infect Dis. 1995, 27, 559–561.
100. Vogel, J.; Cinatl, J.; Lux, A. New PCR assay for rapid and quantitative detection of human cytomeglovirus in cerebrospinal fluid. J Clin Microbiol. 1996, 34, 482–483.
101. Lanska, M.; Lanska, D.J.; Schmidley, J. Syphilitic polyradiculopathy in an HIV-positive man. Neurology. 1988, 38, 1297–1301.
102. Leger, J.M.; Henin, D.; Belec, L.; Mercier, B.; Cohen, L.; Bouche, P.; Hauw, J.J.; Brunet, P. Lymphoma induced polyradiculopathy in AIDS: 2 cases. J Neurol. 1992, 239, 132–134.
103. Freeman, R.; Roberts, M.; Friedman, L. Autonomic function and human immunodeficiency virus infection. Neurology. 1990, 40, 575–580.
104. Shahmanesh, M.; Bradbeer, C.; Edwards, A. Autonomic dysfunction in patients with human immunodeficiency virus infection. Int J STD AIDS. 1991, 2, 419–423.
105. Craddock, C.; Pasvol, G.; Bull, R. Cardiorespiratory arrest and autonomic neuropathy in acquired immunodeficiency syndrome. Lancet. 1987, 2, 16–18.
106. Batman, P.A.; Miller, A.R.; Sedgwick, P.M.; Griffin, G.E. Autonomic denervation in jejunal mucosa of homosexual men infected with HIV. AIDS. 1991, 5(10), 1247–1252.
107. Chimelli, L.; Scaravilli, F. Morphological changes in the autonomic nervous system of patients with AIDS: Padova: Italy, 1991, 89.
108. Purba, J.S.; Hofman, M.A.; Portegies, P.; Troost, D.; Swaab, D.F. Decreased number of oxytocin neurons in the paraventricular nucleus of the human hypothalamus in AIDS. Brain. 1993, 116(4), 795–809.
109. Berman, A.; Espinoza, L.; Diaz, J. Rheumatic manifestations of human immunodeficiency virus infection. Am J Med. 1988, 85, 59–64.
110. Simpson, D.M.; Slasor, P.; Dafni, U.; Berger, J.; Fischl, M.A.; Hall, C. Analysis of myopathy in a placebo-controlled zidovudine trial. Muscle Nerve. 1997, 20(3), 382–385.
111. Simpson, D.M.; Citak, K.A.; Godfrey, E.; Godbold, J.; Wolfe, D.E. Myopathies associated with human immunodeficiency virus and zidovudine: can their effects be distinguished? [see comments]. Neurology. 1993, 43(5), 971–976.
112. Chariot, P.; Ruet, E.; Authier, F.; Levy, Y.; Gherardi, R. Acute rhabdomyolysis in patients infected with human immunodeficiency virus. Neurology. 1994, 44, 1692–1696.
113. Dalakas, M.; Pezeshkpour, G. Gravell, M.; Sever, J. Polymyositis associated with AIDS retrovirus. JAMA. 1986, 256, 2381–2383.
114. Iiia, I.; Nath, A.; Dalakas, M. Immunocytochemical and virological characteristics of HIV-associated inflammatory myopathies: similarities with seronegative patients. Ann Neurol. 1991, 29, 474–481.
115. Simpson, D.M.; Bender, A.N. Human immunodeficiency virus-associated myopathy: analysis of 11 patients. Ann Neurol. 1988, 24(1), 79–84.

116. Gonzalez, M.; Olney, R.; So, Y. Subacute structural myopathy associated with human immunodeficiency virus infection. Arch Neurol. 1988, 45, 585–587.
117. Seidman, R.; Peress, N.; Nuovo, G. In situ detection of polymerase chain reaction-amplified HIV-1 nucleic acids in skeletal muscle in patients with myopathy. Mod Pathol. 1994, 7, 369–375.
118. Bessen, L.J.; Greene, J.B.; Louie, E.; Seitzman, P.; Weinberg, H. Severe polymyositis-like syndrome associated with zidovudine therapy of AIDS and ARC. N Engl J Med. 1988, 318(11), 708.
119. Helbert, M.; Fletcher, T.; Peddle, B.; Harris, J.; Pinching, A. Zidovudine-associated myopathy. Lancet. 1988, 1, 689–690.
120. Gorard, D.; Henry, K.; Guiloff, R. Necrotizing myopathy and zidovudine. Lancet. 1988, 1, 1050–1051.
121. Gertner, E.; Thurn, J.; Williams, D. Zidovudine-associated myopathy. Am J Med. 1989, 6, 814–818.
122. Miller, K.D.; Cameron, M.; Wood, L.V.; Dalakas, M.C.; Kovacs, J.A. Lactic acidosis and hepatic steatosis associated with use of stavudine: report of four cases. Ann Intern Med. 2000, 133(3), 192–196.
123. Mokrzycki, M.H.; Harris, C.; May, H.; Laut, J.; Palmisano, J. Lactic acidosis associated with stavudine administration: a report of five cases. Clin Infect Dis. 2000, 30(1), 198–200.
124. Kuncl, R.; George, E. Toxic neuropathies and myopathies. Curr Opin Neurol. 1993, 6, 695–704.
125. Lane, R.; McLean, K.; Moss, J.; Woodrow, D. Myopathy in HIV infection: the role of zidovudine and the significance of tubuloreticular inclusions. Neuropathol Appl Neurobiol. 1993, 19, 406–413.
126. Morgello, S.; Wolfe, D.; Godfrey, E.; Feinstein, R.; Tagliati, M.; Simpson, D.M. Mitochondrial abnormalities in human immunodeficiency virus-associated myopathy. Acta Neuropathol. 1995, 90(4), 366–374.
127. Dalakas, M.C.; Illa, I.; Pezeshkpour, G.H.; Laukaitis, J.P.; Cohen, B.; Griffin, J.L. Mitochondrial myopathy caused by long-term zidovudine therapy [see comments]. N Engl J Med. 1990, 322(16), 1098–1105.
128. Panegyres, P.; Tan, M.; Kakulas, B. Necrotising myopathy and zidovudine. Lancet. 1988, 1, 1050–1051.
129. Espinoza, L.; Aguilar, J.; Espinoza, C. Characteristics and pathogenesis of myositis in human immunodeficiency virus infection. Distinction from azidothymidine-induced myopathy. Rheum Dis Clin North Am. 1991, 107, 598–599.
130. Manji, H.; Harrison, M.; Round, J.; Jones, D.A.; Connolly, S.; Fowler, C.J.; Williams, I.; Weller, I.V. Muscle disease, HIV and zidovudine: the spectrum of muscle disease in HIV-infected individuals treated with zidovudine. J Neurol. 1993, 240(8), 479-488.
131. Till, M.; MacDonnell, K. Myopathy with human immunodeficiency virus type 1 (HIV-1) infection: HIB-1 or zidovudine. Ann Intern Med. 1990, 113, 492–494.
132. Mhiri, C.; Baudrimont, M.; Bonne, G.; Geny, C.; Degoul, F.; Marsac, C.; Roullet, E.; Gherardi, R. Zidovudine myopathy: a distinctive disorder associated with mitochondrial dysfunction. Ann Neurol. 1991, 29(6), 606–614.
133. Chalmers, A.C.; Greco, C.M.; Miller, R.G. Prognosis in AZT myopathy [see comments]. Neurology. 1991, 41(8), 1181–1184.
134. Simpson, D.; Godbold, J.; Hassett, J. HIV-associated myopathy, and the effects of zidovudine and prednisone. Clin Neuropathol. 1993, 12(suppl 1), S20.

13

HTLV-I and HTLV-II

Mitsuhiro Osame
Kagoshima University Faculty of Medicine
Kagoshima, Japan

I. INTRODUCTION

Human T-cell lymphotropic virus type I (HTLV-I) is known as the causative agent for adult T-cell leukemia (ATL). This same virus was found to be related to another human disease, a progressive spastic paraparesis, found independently in two areas of the world, the Caribbean basin and Japan. In the Caribbean basin, 59% of patients with tropical spastic paraparesis (TSP) had antibodies to HTLV-I [1]. In Japan, a high prevalence of primary lateral sclerosis or spinal spastic paraparesis was found in South Kyushu [2]. A follow-up study of this disorder established the existence of a new disease associated with HTLV-I, which was named HTLV-I associated myelopathy (HAM) [3–5]. The disease is now known under the acronym HAM/TSP [6,7].

2 HTLV-I

HTLV-I is a type C retrovirus, subfamily *Oncoviridae*. In contrast to the human immunodeficiency virus (HIV), HTLV-I causes disease in only about 5% of infected people. HTLV-I is estimated to infect approximately 10 million people worldwide. There are large endemic areas in southern Japan, Central and West Africa, the Caribbean, Central and South America, and the Middle East and smaller foci in the aboriginal populations of Australia, Papua New Guinea, and northern Japan. In Europe and North America the virus is found chiefly in immigrants from these endemic areas and in some communities of intravenous drug users.

Within the endemic areas, the seroprevalence varies between 1% and 20%. The numbers of patients with ATL and HAM/TSP are estimated to be more than 3000 and 5000, respectively. HTLV-I has been shown to be associated not only with HAM/TSP but also with T-lymphocytic alveolitis, polymyositis, arthritis, and sicca syndrome. There are also less certain associations with chronic infective dermatitis, Behçet disease, pseudohypoparathyroidism, and systemic lupus erythematosus [8].

There are three important modes of transmission: parental and neonatal infection from a seropositive mother, in which breastfeeding is a significant factor; sexual transmission, particularly from males to females; and transmission by infected blood, either by transfusion or by sharing of needles among drug users. Transmission of the virus depends on transfer of cells from infected people. Blood for transfusion is now routinely screened for HTLV-I in several countries, including Japan [9], the United States, and Brazil. In Japan, the screening could prevent the occurrence of new HAM/TSP patients derived from blood transfusion [9].

HTLV-I is known as a complex retrovirus. In addition to the three genes present in other typical replication-competent exogenous retroviruses (*gag*, *pol*, and *env*), it encodes at least two other proteins: Tax, which stimulates transcription of the proviral genome, and Rex, which controls the splicing of HTLV-I mRNA.

Although it can infect a wide variety of cell types in vitro, HTLV-I appears to replicate efficiently mainly in $CD4^+$ (helper) T cells; these are the cells that are transformed in ATL. $CD8^+$ T cells are an additional viral reservoir in vivo for HTLV-I [10,11]. Antibodies against the Gag protein are the first to appear after infection, and they predominate in the first 2 months. Thereafter, anti-envelope antibodies predominate, and about half of infected individuals subsequently produce antibodies to the Tax protein. Diagnosis of HTLV-I infection depends on the detection of specific antibodies by particle agglutination or ELISA assays and confirmation by PCR or Western blot assay.

3 ADULT T-CELL LEUKEMIA (ATL)

A person infected with HTLV-I has about a 5% lifetime risk of developing ATL. The main features of ATL are (1) age of onset from about 20 to 70 years, the mean age at onset of ATL being about 60 years in Japan and 40 years in the Caribbean and Brazil, the reason for this difference unknown; (2) lymphadenopathy and hepatosplenomegaly; (3) skin lesions similar to mycosis fungoides, or Sezary syndrome; (4) presence of proliferated T-cell lymphocytes with flowerlike nuclei; (5) hypercalcemia; and (6) poor prognosis with a mean survival of 10 months. ATL clinically presents in four different ways: (1) acute ATL, whose clinical manifestations are those listed above; (2) chronic ATL; (3) smoldering ATL; and (4) lymphoma-type ATL.

Southern blot analysis indicates the presence of oligoclonal or monoclonal proliferation of $CD4^+$ cells that carry the HTLV-I provirus in the cellular DNA.

The mean survival times (untreated) for acute, lymphomatous, and chronic (smoldering) ATL in Japan are 6.2, 10.2, and 24.3 months, respectively. The disease often responds initially to standard chemotherapeutic regimens, but early relapse is common, and the disease typically becomes refractory to further chemotherapy after 2–6 months. Recently, significant progress has been made with the discovery that a combination of extensive chemotherapeutic regimens and bone marrow transplantation can lead to complete remission for some patients [12].

More than half of ATL cases will have neurological complications during the course of the disease. These manifestations mainly consist of (1) altered consciousness derived mainly from hypercalcemia, (2) dementia, (3) seizures, (4) hemiparesis and pyramidal tract signs, (5) cranial nerve deficits, (6) meningeal irritation, and (7) polyneuropathy.

Except for the altered consciousness, these complications have been associated with direct tumor cell invasion in most cases [13].

4 HTLV-I ASSOCIATED MYELOPATHY (HAM/TSP)

4.1 Clinical Features of HAM/TSP

HAM/TSP is characterized by a spastic paraparesis that is slowly progressive, or in some cases static after initial progression, and anti-HTLV-I antibody positivity in serum and cerebrospinal fluid [3,4]. Almost all patients show spasticity and/or hyperreflexia of the lower extremities, initially presenting as gait and urinary disturbances. Many patients manifest with low back pain, weakness of the lower extremities, and a poorly defined (mild) sensory affect. Rarely, the disease presents as cerebellar ataxia. Patients with younger age of onset (<15 years old) tend to have short height and slow progression of the disease, whereas patients with older age of onset (>61 years old) show faster progression regardless of the mode of transmission [14].

The clinical and laboratory guidelines for the diagnosis of HAM/TSP are summarized in Table 1, based on the recommendations of a 1988 WHO meeting [6,7].

Patients with HAM/TSP have high antibody titers to HTLV-I both in serum and in CSF [4]. Aside from HTLV-I antibody positivity, other essential laboratory findings include lymphocytic pleocytosis in the CSF and higher than normal CSF neopterin levels [14]. High signals of T2-weighted magnetic resonance images are observed in the white matter of the brain similar to those found in multiple sclerosis. Swelling or atrophy of the spinal cord has been reported in a few cases of HAM/TSP.

4.2 Treatment of HAM/TSP

There is no definitive treatment yet established for HAM/TSP. Several clinical trials have shown transient beneficial effects of corticosteroids, α-interferon, azathioprine, high-dose vitamin C, pentoxifylline, danazol, and plasmapheresis [15]. A double-blind, multicenter study on the therapeutic effect of treatment with natural interferon-α showed significantly effective results [16]. There is recent evidence that treatment with the nucleotide analog lamivudine can reduce the provirus load of HTLV-I [17], but the clinical impact of such treatment is not yet known.

4.3 Histopathological Features of HAM/TSP

Pathological analysis indicates that the disease affects the spinal cord, predominantly at the thoracic level (Fig. 1). There is degeneration of the lateral corticospinal tract as well as of the spinocerebellar or spinothalamic tract of the lateral column [18] (Fig. 2). These lesions are associated with perivascular and parenchymal lymphocytic infiltration with the presence of foamy macrophages, proliferation of astrocytes, and fibrillary gliosis [19] (Fig. 3). There is also widespread loss of myelin and axons, particularly in the corticospinal tracts of the spinal cord. Damage is most severe in the middle to lower thoracic regions of the spinal cord. These findings are consistent with a patient's neurological symptoms such as paraparesis, spasticity, hyperreflexia, and Babinski's sign [19].

A nonrandom distribution of affected regions was suggested by an autopsy study that showed that the regions mainly affected are the so-called watershed zones of the spinal cord in patients with HAM/TSP [18]. Similar findings were observed in the brain,

Table 1 Diagnostic Guidelines for HAM/TSP

I. **Criteria.** The florid picture of chronic spastic paraparesis is not always seen when the patient first presents. A single symptom or physical sign may be the only evidence of early HAM/TSP.

   A. *Age and sex incidence.* Mostly sporadic and adult but sometimes familial, occasionally seen in childhood; females predominant.

   B. *Onset.* This is usually insidious but may be sudden.

   C. *Main neurological manifestations.*

      1. Chronic spastic paraparesis that usually progresses slowly, sometimes remains static after initial progression.
      2. Weakness of the lower limbs, more marked proximally.
      3. Bladder disturbance usually an early feature, constipation usually occurs later, and impotence or decreased libido is common.
      4. Sensory symptoms such as tingling, "pins and needles," and burning are more prominent than objective physical signs.
      5. Low lumbar pain with radiation to the legs is common.
      6. Vibration sense is frequently impaired, proprioception less often affected.
      7. Hyperreflexia of the lower limbs, often with clonus and Babinski's sign.
      8. Hyperreflexia of upper limbs, and positive Hoffmann's and Trömner signs frequent. Weakness may be absent.
      9. Exaggerated jaw jerk in some patients.

   D. *Less frequent neurological findings.* Cerebellar signs; optic atrophy; deafness nystagmus other cranial nerve deficits; hand tremor; absent or depressed ankle jerk. Convulsions, cognitive impairment, dementia, and impaired consciousness are rare.

   E. *Other neurological manifestations that may be associated with HAM/TSP.* Muscular atrophy; fasciculations (rare); polymyositis; peripheral neuropathy; polyradiculopathy; cranial neuropathy; meningitis; encephalopathy.

   F. *Systemic non-neurological manifestations that may be associated with HAM/TSP.* Pulmonary alveolitis; uveitis; Sjögren's syndrome; arthropathy; vasculitis; ichthyosis; cryoglobulinemia; monoclonal gammopathy; adult T-cell leukemia/lymphoma.

II. **Laboratory diagnosis**

   A. HTLV-I antibodies or antigens are present in blood and cerebrospinal fluid (CSF).

   B. CSF may show mild lymphocyte pleocytosis.

   C. Lobulated lymphocyte may be present in blood and/or CSF.

   D. Mild to moderate increase of protein may be present in CSF.

   E. Viral isolation is possible from blood and/or CSF.

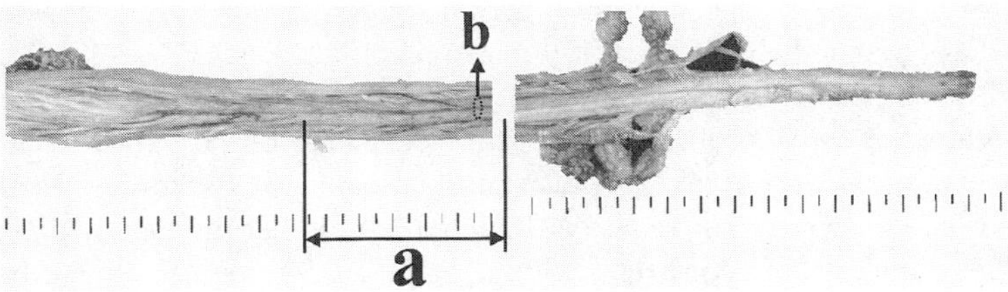

Figure 1 Macroscopic view of a spinal cord of an autopsied patient with HAM/TSP. (a) The middle to lower regions of the spinal cord, where atrophy is prominent. (b) Marked atrophy of the spinal cord.

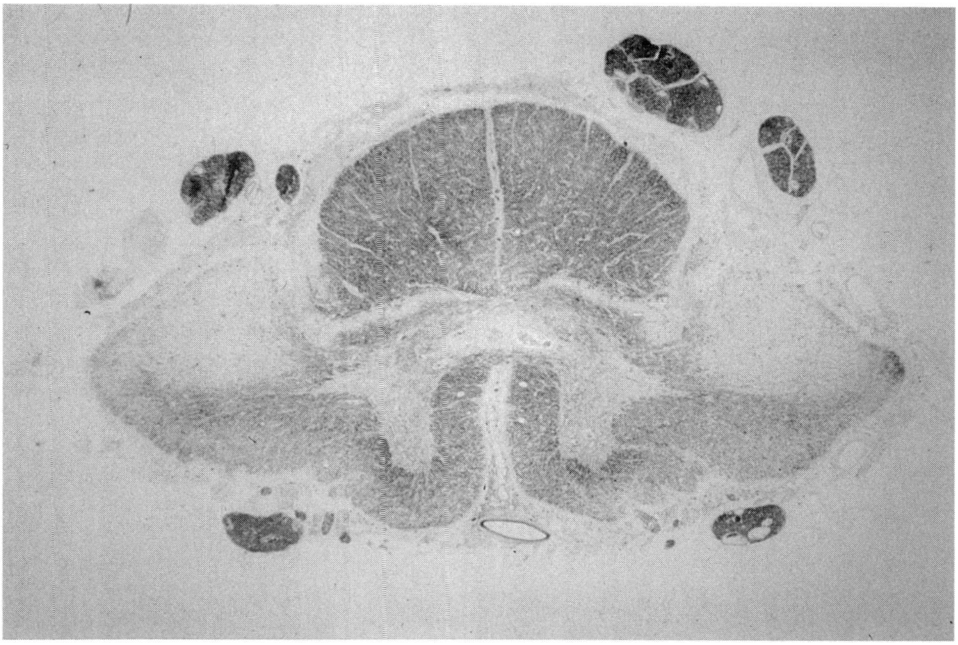

Figure 2 Neuropathological findings in an autopsied patient with HAM/TSP. Midthoracic spinal cord showing symmetrical degeneration in the lateral colums (Klüver-Barrera stain).

although to a lesser degree [20]. These results suggest that inflammatory changes occurred simultaneously in the spinal cord and in the brain, with the distribution of inflamed vessels closely correlated with the characteristic vascular architecture of the brain and the spinal cord, which led to a slowing of blood flow. Inflammatory changes are inversely correlated with the duration of the disease, and significant cellular inflammation with expression of inflammatory cytokines and major histocompatibility complexes (MHC) (classes I and II) is observed in cases of short duration [19,21].

In presumably early lesions, the axons are relatively preserved. A predominance of $CD8^+$ cells has been observed in cases of longer disease duration [19]. Both HTLV-I proviral DNA and HTLV-I *tax* gene expression were observed mostly in the $CD4^+$ infiltrating mononuclear cells and not in other tissue cells in the spinal cord lesions [22,23]. Activity of inflammation corresponded well with the amount of HTLV-I proviral DNA in situ and with the presence of apoptosis of $CD4^+$ T cells [24]. Metalloproteinases (MMP)-2 (gelatinase A) and MMP-9 (gelatinase B), which are implicated in extracellular matrix degradation, were also detected in infiltrating cells [25]. MMP-9 and tissue inhibitors of metalloproteinase-3 (TIMP-3) levels in CSF of HAM/TSP patients were higher than that of HTLV-I carriers without neurological symptoms [26]. Early axonal damage was also demonstrated using immunoreactivity for β-amyloid precursor protein (APP), which was a sensitive marker for impairment of fast axonal transport. These axonal damages were more intensively expressed in areas of active inflammatory lesions than in areas of inactive chronic lesions [27].

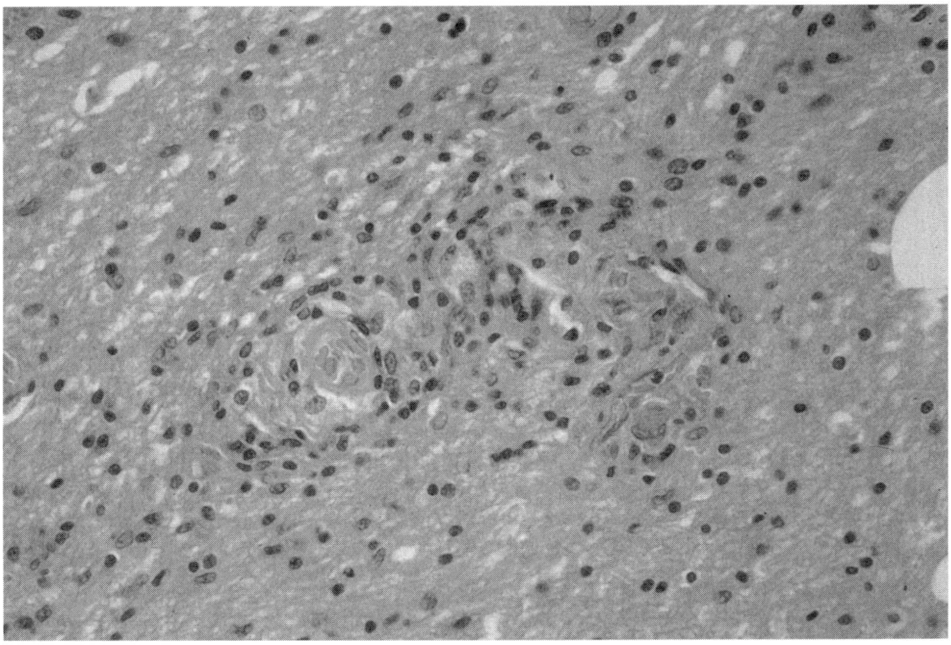

Figure 3 Microscopic findings in an autopsied patient with HAM/TSP. Perivascular and parenchymal mononuclear infiltration in the lower thoracic spinal cord (hematoxylin-eosin stain).

4.4 Risk Factors for HAM/TSP

The prevalence of HAM/TSP is between 0.1% and 2% of HTLV-I-infected individuals. The lifetime risk of developing this disease among carriers is estimated to be 0.23% in Japan [28]. About two-thirds of patients are female [14]. Other known risk factors for HAM/TSP include a high proviral load of HTLV-I [29] and a certain HTLV-I subgroup [30]. Most people infected with HTLV-I mount a strong cytotoxic T-lymphocyte (CTL) response to the virus [31]. This strong CTL response protects against the development of HAM/TSP by reducing the proviral load [31]. However, when the proviral load exceeds a threshold level, HTLV-I-specific CTL could contribute to inflammation [29,32]. The immune response to HTLV-I is now closer to being understood [33].

4.5 Pathological Mechanisms

Why does HTLV-I cause HAM/TSP in only less than 5% of infected people? To answer this question, many risk factors have been identified as mentioned above. Individuals who possess more risk factors may have a greater tendency to develop the disease. $CD8^+$ CTL interactions with target cells (i.e., HTLV-I infected $CD4^+$ T cells) may play an important role in the CNS inflammatory process in HAM/TSP. Recently, in addition to $CD4^+$ cells, $CD8^+$ cells were also found to be a viral reservoir in vivo for HTLV-I [10,11]. Therefore the $CD8^+$ T cells in the CNS may also be infected by HTLV-I, although there has been no direct evidence for it. HTLV-I Tax will induce upregulation of various molecules such as adhesion molecules, MMP-9, and inflammatory cytokines (TNF-α, IL1-β, INF-γ), which result in infiltration of HTLV-I-infected T cells into the CNS. After trafficking into the CNS, HTLV-I-infected T cells exhibit significant viral antigen expression, and the interaction of HTLV-I-infected CD4$^+$ T cells and HTLV-I-specific CD8$^+$ CTLs may lead to secretion of cytokines, MMPs, and Fas ligand in the CNS. In addition, the spontaneous secretion of IFN-γ from HTLV-I-infected CD4$^+$ T cells could activate macrophage/microglias, which could also secret IFN-γ. These cytokines would damage bystander neural tissue (mechanism I, Fig. 4).

A recent study supports an additional hypothesis that antibodies would identify a CNS autoantigen in HAM/TSP [34]. Immunoglobulin G isolated from HAM/TSP patients identified heterogeneous nuclear ribonuclear protein-A1 (hnRNP-A1) as the autoantigen. Antibodies to hnRNP-A1 cross-reacted with HTLV-I Tax. Immunoglobulin G specifically stained human Betz cells. Infusion of autoantibodies in brain sections inhibited neuronal firing [34]. These data suggest the importance of molecular mimicry between HTLV-I and hnRNP-A1 in the pathogenesis of HAM/TSP (mechanism II, Fig. 4).

Which of these two mechanisms might play a more important role in developing HAM/TSP? The distribution of the affected lesions in the CNS of HAM/TSP [18,20] and the histopathological view described in this review support the greater importance of mechanism I. Immunological studies show the important role of T cells [35], again indicat-

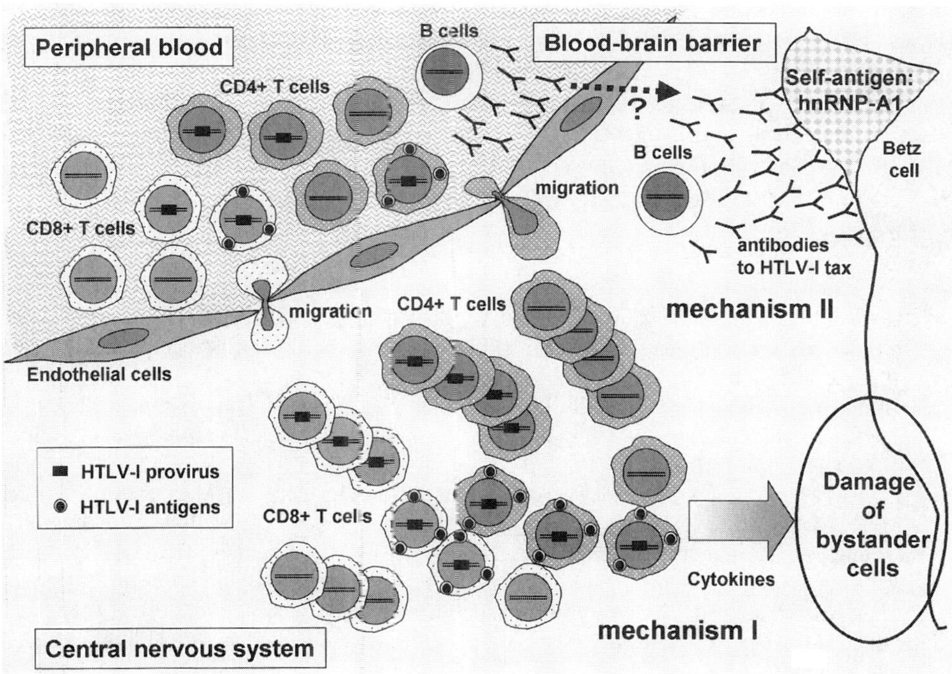

Figure 4 Pathological mechanisms of developing HAM/TSP. Mechanism I: CD8$^+$ CTL interaction with target cells (i.e., HTLV-I infected CD4$^+$ or CD8$^+$ T cells) would produce cytokines including IFN-γ, which would damage bystander neural tissue. Mechanism II: Immunoglobulin G specific to HTLV-I tax would cross-react with heterogeneous nuclear ribonuclear protein-Al (hnRNP-Al) expressed in a Betz cell and damage the cell.

ing the importance of mechanism I. HAM/TSP might be caused mainly by mechanism I, and mechanism II might be playing an additional role in the development of HAM/TSP (Fig. 4).

3 HTLV-II

A virus related to HTLV-I, HTLV-II is also a type C retrovirus in the *Oncoviridae* subfamily. The epidemiology of this virus is somewhat different from that of HTLV-I. The populations principally affected are Americans and intravenous drug abusers [36]. Three genetic serotypes have been identified. The mode of transmission of HTLV-II parallels those of HTLV-I and HIV. Breastfeeding, sexual transmission, contaminated blood products, and intravenous drug abuse are all established modes of transmission [36].

A spectrum of neurological diseases have been reported in association with HTLV-II infection but with a much lower frequency than is observed with HTLV-I. These disorders include a myelopathy that is seemingly identical to HAM/TSP [37–43]. In some patients, HTLV-II RNA has been amplified from the spinal cord. A spinocerebellar syndrome has been described in which ataxia, oculomotor abnormalities, and dysarthria precede by several years the development of spinal cord manifestations [44]. Cerebellar atrophy predominated in the superior vermis, and pathological findings included axonal neuropathy, dorsal column involvement, and inflammatory myopathy [44]. However, Peters and colleagues [42] urge caution in attributing a link between HTLV-II and myelopathy or other neurological disease and suggest the need for exhaustive study. Berger and colleagues [37] reported an HAM/TSP illness occurring in a patient coinfected with HIV and HTLV-II.

The rarity of these neurological disorders accompanying HTLV-II has precluded meaningful analysis of treatment options. However, the same therapies as those employed in HAM/TSP have been generally recommended.

REFERENCES

1. Gessain, A.; Barin, F.; Vernant, J.C.; Gout, O.; Maurs, L.; Calender, A.; de The, G.. Antibodies to human T-lymphotropic virus type-I in patients with tropical spastic paraparesis. Lancet. 1985, 2, 407–410.
2. Osame, M.; Arima, H.; Norimatsu, K.; Kawahira, M.; Okatsu, Y.; Nagamatsu, K.; Igata, A. Epidemiostatistical studies of muscular atrophy in southern Kyushu (Kagoshima and Okinawa prefectures). Jpn J Med. 1975, 14, 230–231.
3. Osame, M.; Usuku, K.; Izumo, S.; Ijichi, N.; Amitani, H.; Igata, A.; Matsumoto, M.; Tara, M. HTLV-I associated myelopathy, a new clinical entity [letter]. Lancet. 1986, 1, 1031–1032.
4. Osame, M.; Matsumoto, M.; Usuku, K.; Izumo, S.; Ijichi, N.; Amitani, H.; Tara, M.; Igata, A. Chronic progressive myelopathy associated with elevated antibodies to human T-lymphotropic virus type I and adult T-cell leukemia like cells. Ann Neurol. 1987, 21, 117–122.
5. Osame, M.; Igata, A. The history of discovery and clinico-epidemiology of HTLV-I-associated myelopathy (HAM). Jpn J Med. 1989, 28, 412–414.
6. World Health Organization. Report of the Scientific Group on HTLV-I and Associated Diseases, Kagoshima, Japan, December 1988. Available from Manila, Philippines: World Health Organization, March 1989. Virus Diseases. Human T-lymphotropic virus type I, HTLV-1. WHO Weekly Epidemiol Rec 49 382–383 1989.
7. Osame, M. Review of WHO Kagoshima meeting and diagnostic guidelines for HAM/TSP. In Human Retrovirology: HTLV; Blattner, W.A., Ed.; Raven Press: New York, 1990, 191–197.

8. Kubota, R.; Osame, M.; Jacobson, S. Retrovirus: human T-cell lymphotropic virus type I-associated diseases and immune dysfunction. In Effects of Microbes on the Immune System; Cunningham, M.W., Fujinami, R.S., Eds.; Lippincott Williams & Wilkins: Philadelphia, 2000, 349–371.
9. Osame, M.; Janssen, R.; Kubota, H.; Nishitani, H.; Igata, A.; Nagataki, S.; Mori, M.; Goto, I.; Shimabukuro, H.; Khabbaz, R.; Kaplan, J. Nationwide survey of HTLV-I-associated myelopathy in Japan: association with blood transfusion. Ann Neurol. 1990, 28, 50–56.
10. Hanon, E.; Stinchcombe, J.C.; Saito, M.; Asquith, B.E.; Taylor, G.P.; Tanaka, Y.; Weber, J.N.; Griffiths, G.M.; Bangham, C.R. Fratricide among CD8(+) T lymphocytes naturally infected with human T cell lymphotropic virus type I. Immunity. 2000, 13, 657–664.
11. Nagai, M.; Brennan, M.B.; Sakai, J.A.; Mora, C.A.; Jacobson, S. CD8+ T cells are an in vivo reservoir for human T-cell lymphotropic virus type I. Blood. 2001, 98, 1858–1861.
12. Obama, K.; Tara, M.; Sao, H.; Taji, H.; Morishima, Y.; Mougi, H.; Maruyama, Y.; Osame, M. Allogeneic bone marrow transplantation as a treatment for adult T-cell leukemia. Int J Hematol. 1999, 69, 203–205.
13. Tara, M.; Tokunaga, M.; Osame, M.; Niina, K. Neurological complications of adult T-cell leukemia/lymphoma. In HTLV-I and the Nervous System; Roman, G.C., Vernant, J.-C., Osame, M., Eds.. 1989, 73–82.
14. Nakagawa, M.; Izumo, S.; Ijichi, S.; Kubota, H.; Arimura, K.; Kawabata, M.; Osame, M. HTLV-I-associated myelopathy: analysis of 213 patients based on clinical features and laboratory findings. J Neurovirol. 1995, 1, 50–61.
15. Nakagawa, M.; Nakahara, K.; Maruyama, Y.; Kawabata, M.; Higuchi, I.; Kubota, H.; Izumo, S.; Arimura, K.; Osame, M. Therapeutic trials in 200 patients with HTLV-I-associated myelopathy/tropical spastic paraparesis. J Neurovirol. 1996, 2, 345–355.
16. Izumo, S.; Goto, I.; Itoyama, Y.; Okajima, T.; Watanabe, S.; Kuroda, Y.; Araki, S.; Mori, M.; Nagataki, S.; Matsukura, S.; Akamine, T.; Nakagawa, M.; Yamamoto, I.; Osame, M. Interferon-alpha is effective in HTLV-I-associated myelopathy: a multicenter randomized, double-blind, controlled trial. Neurology. 1996, 46, 1016–1021.
17. Taylor, G.P.; Hall, S.E.; Navarrete, S.; Michie, C.A.; Davis, R.; Witkover, A.D.; Rossor, M.; Nowak, M.A.; Rudge, P.; Matutes, E.; Bangham, C.R.; Weber, J.N. Effect of lamivudine on HTLV-I DNA copy number, T-cell phenotype and anti-Tax cytotoxic T-cell frequency in patients with HTLV-I-associated myelopathy. J Virol. 1999, 73, 10289–10295.
18. Izumo, S.; Ijichi, T.; Higuchi, I.; Tashiro, A.; Takahashi, K.; Osame, M. Neuropathology of HTLV-I-associated myelopathy: a report of two autopsy cases. Acta Paediatr Jpn. 1992, 34, 358–364.
19. Umehara, F.; Izumo, S.; Nakagawa, M.; Ronquillo, A.T.; Takahashi, K.; Matsumuro, K.; Sato, E.; Osame, M. Immunocytochemical analysis of the cellular infiltrate in the spinal cord lesions in HTLV-I-associated myelopathy. J Neuropathol Exp Neurol. 1993, 52, 424–430.
20. Moe Moe Aye; Matsuoka, E.; Moritoyo, T.; Umehara, F.; Suehara, M.; Hokezu, Y.; Yamanaka, H.; Isashiki, Y.; Osame, M.; Izumo, S. Histopathological analysis of four autopsy cases of HTLV-I-associated myelopathy/tropical spastic paraparesis: inflammatory changes occur simultaneously in the entire central nervous system. Acta Neuropathol. 2000, 100, 245–252.
21. Umehara, F.; Izumo, S.; Ronquillo, A.T.; Matsumuro, K.; Sato, E.; Osame, M. Cytokine expression in the spinal cord lesions in HTLV-I-associated myelopathy. J Neuropathol Exp Neurol. 1994, 53, 72–77.
22. Moritoyo, T.; Reinhart, T.A.; Moritoyo, H.; Sato, E.; Izumo, S.; Osame, M.; Haase, A.T. Human T-lymphotropic virus type I-associated myelopathy and tax gene expression in CD4+ T lymphocytes. Ann Neurol. 1996, 40, 84–90.
23. Matsuoka, E.; Usuku, K.; Jonosono, M.; Takenouchi, N.; Izumo, S.; Osame, M. CD44 splice variant involvement in the chronic inflammatory disease of the spinal cord: HAM/TSP. J Neuroimmunol. 2000, 102, 1–7.

24. Umehara, F.; Nakamura, A.; Izumo, S.; Kubota, R.; Ijichi, S.; Kashio, N.; Hashimoto, K.; Usuku, K.; Sato, E.; Osame, M. Apoptosis of T lymphocytes in the spinal cord lesions in HTLV-I-associated myelopathy: a possible mechanism to control viral infection in the central nervous system. J Neuropathol Exp Neurol. 1994, 53, 617–624.
25. Umehara, F.; Okada, Y.; Fujimoto, N.; Abe, M.; Izumo, S.; Osame, M. Expression of matrix metalloproteinases and tissue inhibitors of metalloproteinases in HTLV-I-associated myelopathy. J Neuropathol Exp Neurol. 1998, 57, 839–849.
26. Lezin, A.; Buart, S.; Smadja, D.; Akaoka, H.; Bourdonne, O.; Perret-Liaudet, A.; Cesaire, R.; Belin, M.F.; Giraudon, P. Tissue inhibitor of metalloproteinase 3, matrix metalloproteinase 9, and neopterin in the cerebrospinal fluid: preferential presence in HTLV type I-infected neurologic patients versus healthy virus carriers. AIDS Res Hum Retroviruses. 2000, 16, 965–972.
27. Umehara, F.; Abe, M.; Koreeda, Y.; Izumo, S.; Osame, M. Axonal damage revealed by accumulation of β-amyloid precursor protein in HTLV-I-associated myelopathy. J Neurol Sci. 2000, 176, 95–101.
28. Kaplan, J.E.; Osame, M.; Kubota, H.; Igata, A.; Nishitani, H.; Maeda, Y.; Khabbaz, R.F.; Janssen, R.S. The risk of development of HTLV-I-associated myelopathy/tropical spastic paraparesis among persons infected with HTLV-I. J Acquir Immune Defic Syndr. 1990, 3, 1096–1101.
29. Nagai, M.; Usuku, K.; Matsumoto, W.; Kodama, D.; Takenouchi, N.; Moritoyo, T.; Hashiguchi, S.; Ichinose, M.; Bangham, C.R.; Izumo, S.; Osame, M. Analysis of HTLV-I proviral load in 202 HAM/TSP patients and 243 asymptomatic HTLV-I carriers: high proviral load strongly predisposes to HAM/TSP. J Neuroviral. 1998, 4, 586–593.
30. Furukawa, Y.; Yamashita, M.; Usuku, K.; Izumo, S.; Nakagawa, M.; Osame, M. Phylogenetic subgroups of human T cell lymphotropic virus (HTLV) type I in the tax gene and their association with different risks for HTLV-I-associated myelopathy/tropical spastic paraparesis. J Infect Dis. 2000, 182, 1343–1349.
31. Jacobson, S.; Shida, H.; McFarlin, D.E.; Fauci, A.S.; Koenig, S. Circulating CD8 + cytotoxic T lymphocytes specific for HTLV-I pX in patients with HTLV-I associated neurological disease. Nature. 1990, 348, 245–248.
32. Jeffery, K.J.; Siddiqui, A.A.; Bunce, M.; Lloyd, A.L.; Vine, A.M.; Witkover, A.D.; Izumo, S.; Usuku, K.; Welsh, K.I.; Osame, M. CRM Bangham. The influence of HLA class I alleles and heterozygosity on the outcome of human T cell lymphotropic virus type I infection. J Immunol. 2000, 165, 7278–7284.
33. Bangham, C.R.M. The immune response to HTLV-I. Curr Opin Immunol. 2000, 12, 397–402.
34. Levin, M.C.; Lee, S.M.; Kalume, F.; Morcos, Y.; Dohan, F.C., Jr.; Hasty, K.A.; Callaway, J.C.; Zunt, J.; Desiderio, D.M.; Stuart, J.M. Autoimmunity due to molecular mimicry as a cause of neurological disease. Nat Med. 2002, 8, 509–513.
35. Nagai, M.; Jacobson, S. Immunopathogenesis of human T cell lymphotropic virus type I-associated myelopathy. Curr Opin Neurol. 2001, 14, 381–386.
36. Lowis, G.W.; Sheremata et al., W.A. Epidemiologic features of HTLV-II: serologic and molecular evidence. Ann Epidemiol. 2002, 12(1), 46–66.
37. Berger, J.R.; Svenningsson et al., A. Tropical spastic paraparesis-like illness occurring in a patient dually infected with HIV-I and HTLV-II. Neurology. 1991, 41(1), 85–87.
38. Harrington, W.J., Jr.; Sheremata et al., W. Spastic ataxia associated with human T-cell lymphotropic virus type II infection. Ann Neurol. 1993, 33(4), 411–414.
39. Jacobson, S.; Lehky et al., T. Isolation of HTLV-II from a patient with chronic, progressive neurological disease clinically indistinguishable from HTLV-I-associated myelopathy/tropical spastic paraparesis. Ann Neurol. 1993, 33(4), 392–396.
40. Sheremata, W.A.; Harrington, W.J., Jr. Association of ''(tropical) ataxic neuropathy'' with HTLV-II. Virus Res. 1993, 29(1), 71–77.
41. Lehky, T.J.; Flerlage et al., N. Human T-cell lymphotropic virus type II-associated myelopathy: clinical and immunologic profiles. Ann Neurol. 1996, 40(5), 714–723.

42. Peters, A.A.; Oger et al., J.J. An apparent case of human T-cell lymphotropic virus type II (HTLV-II)-associated neurological disease: a clinical molecular, and phylogenetic characterisation. J Clin Virol. 1999, 14(1), 37–50.
43. Silva, E.A.; Otsuki et al., K. HTLV-II infection associated with a chronic neurodegenerative disease: clinical and molecular analysis. J Med Virol. 2002, 66(2), 253–257.
44. Castillo, L.C.; Gracia et al., F. Spinocerebellar syndrome in patients infected with human T-lymphotropic virus types I and II (HTLV-I/HTLV-II): report of 3 cases from Panama. Acta Neurol Scand. 2000, 101(6), 405–412.

14

Rabies

Erawady Mitrabhakdi and Thiravat Hemachudha
Chulalongkorn University Hospital
Bangkok, Thailand

Henry Wilde
Queen Saovabha Memorial Institute, Thai Red Cross Society, and
Chulalongkorn University Hospital
Bangkok, Thailand

1 INTRODUCTION

Rabies is one of the most dramatic infections of the nervous system owing to its horrifying clinical presentations and almost invariably fatal outcome. It is also a complex disease, and the mechanism for the diverse presentations has not yet been clarified. The clinical symptomatology can vary considerably, particularly in cases associated with bats. Atypical clinical presentations resembling other viral encephalitides have been increasingly observed in Thailand since 1997, after exposure to canine rabies variants (crv). Differences in cellular tropism either at the inoculation site or in the nervous system, differences in routes of spread or host response, or differences in viral strains may account for this diversity.

With rapid movement of people among continents, human and animal cases can appear in regions where rabies was once eradicated or where it has never been recorded. It is crucial that healthcare providers be able to diagnose and differentiate rabies from other neurological disease and know how to provide postexposure prophylaxis.

2 HISTORY

The disease has been associated with the supernatural and the dog since antiquity. An early record appeared in the pre-Mosaic Eshnunna Code of Mesopotamia about 2300 BCE [1]. The name "rabies" has its roots in Sanskrit, rabhas, which refers to the god of death and his dog, from the Vedic period of India (thirtieth century BCE). In ancient Egypt, the god Sirius was pictured as a furious dog. *Rabhas*, *Lyssa* or *Lytta* (Greek), and *rage* (French) refer to the cause of violence or madness. The term ''hydrophobia'' was coined by Cornelius Celsus, a Roman of the first century CE who provided a classic clinical description of this disease. The infectivity of saliva and urine from rabid dogs was suspected as early as Roman times, and the disease was attributed to a poison (*virus* in Latin). In 1804, transmission of infection by saliva was demonstrated. Burning and cupping of the wounds of those bitten by rabid dogs had been practiced since antiquity.

Girolamo Fracastoro (1478–1553) an Italian scientist, accurately determined the incubation period of rabies in humans. Virus advance to the central nervous system (CNS) via the nerves was postulated in 1769 by Giovanni Battista Morgagni on the basis of symptoms of paresthesia at the bite site. Pasteur demonstrated the CNS as a prime target and proved that the disease mainly affected the brainstem [2]. He also showed that nervous tissue was infectious as well as saliva. Moreover, the virus can be attenuated by serial passages. On July 6,1885, he used his rabbit spinal cord rabies vaccine in two patients who had been exposed to rabies [1,2]. The development of this vaccine involved over 90 serial intracerebral passages of rabies virus in rabbits, followed by air drying, which resulted in loss of infectivity. Although Pasteur's treatment saved the lives of countless victims, it carried some serious neuroparalytic risks, an immune-mediated encephalitis and neuritis [3–5]. The discovery of ''endocellular Negri bodies'' was made by Adelchi Negri in 1903. The diagnostic value of this was demonstrated in 1913 by Negri's wife, Lina Negri-Luzzani [6].

Modern cell culture rabies vaccines have proved to be close to ideal immunogens because of their efficacy, the few injections required, and their relative lack of side effects. Human diploid cell rabies vaccine and antirabies serum produced in mules protected all but one of 45 persons who were severely bitten by rabid dogs and wolves in northern Iran in 1975 [7]. Other equally safe and effective vaccines include purified Vero cell, chick embryo cell, and duck embryo rabies vaccines [5]. Experimental vaccines being developed include oral and parenteral poxvirus- and adenovirus-vectored recombinant and parenteral plasmid rabies vaccines and edible vaccine, a purified virus protein from tobacco plants infected with recombinant alfalfa mosaic virus displaying rabies glyco- and nucleoproteins [8].

It has become clear that aggressive wound care and the administration of purified rabies immunoglobulin (RIG) plus vaccine have saved more severely exposed patients than vaccine alone [5].

3 EPIDEMIOLOGY

Rabies is a zoonosis of domestic and wild mammals. About 50,000 people die of rabies every year. However, with deaths in India alone reaching 30,000 annually, this is likely to be a gross underestimate. Underestimation is undoubtedly a contributory factor to rabies being ranked low on the priority lists for disease control programs of the World Health Organization (WHO) and developing countries [9]. The domestic dog is the principal but not exclusive reservoir host in Asia, Africa, the Pacific islands, and South America. Wildlife such as the mongoose, jackal, and meerkat of South Africa; the pariah dog and jackal of Southeast Asia; the vampire bat of South America; and the fox of eastern Europe also play a significant role in epizootic transmission [8,9].

In North America, there is an epizootic in raccoons in the Mid-Atlantic and northeastern states. Other reported rabid animals include skunks, foxes, insectivorous bats, cats, horses, and dogs [8]. Transmission from bats was the most common cause of human cases of rabies. In the United States and Canada, during 1980 and 2000, bat rabies variants (brv) were identified in 27 of 42 patients [8,10]. Twenty of these 27 cases had evidence of infection with a variant found primarily in the silver-haired bat (*Lasionycteris noctivagans*) or eastern pipistrelle bat (*Pipistrellus subflavus*). Only two gave a definite history of bat bite. Australia, a previously "rabies-free" continent, had become a lyssavirus-endemic area by 1996 [8,9]. This new variant pteropid lyssavirus or Ballina virus (for Ballina, New South Wales, Australia, where the first human infection was contracted) has been found in fruit bats (flying foxes, genus *Pteropus*) and has also been identified in other bats, including insectivorous species. Since November 1996, two human cases of rabies-like illness have occurred. Virtually all of Asia's human deaths from rabies were of people who did not receive postexposure treatment. Of 13 treatment failure cases between 1992 and 2000 in Thailand, all but three had treatment flaws or deviation from WHO recommendations [9,11; personal experience (TH)].

Although legally enforced canine rabies immunization, a practice of strict quarantine, and rigid control of stray dogs have virtually eliminated canine rabies in the western hemisphere, Australasia, and Japan, rabies has shifted to wild or sylvan carnivorous animals in some of these areas. In many Asian countries, cultural and religious customs still prevent reduction of the stray dog population.

4 VIROLOGY

Rabies virus with its distinct bullet shape belongs to the *Lyssavirus* genus of the *Rhabdoviridae* family. The classical rabies virus, isolated from terrestrial mammals including dogs and hematophagous and insectivorous bats, is of sero- and genotype 1 which is the most prevalent worldwide [8]. Comparison of the viral nucleoprotein gene (N) allowed delineation of seven genotypes. Serotype classification is based on differences in viral antigens and antibodies produced by the host (Table 1). The rabies virus contains a single-stranded, antisense, nonsegmented RNA molecule of 11,932 nucleotides of negative polarity. It measures approximately 180 mm × 75 nm and has regularly spaced knoblike spikes on a cylindrical envelope, except for the flat end of the "bullet." The rabies genome contains a leader of 50 nucleotides, followed by genes that encode five proteins: nucleoprotein (N), phosphoprotein (P), matrix protein (M), glycoprotein (G), and polymerase (L). All rhabdoviruses have two major structural components: a helical ribonucleoprotein core (RNP) and a surrounding envelope. In the RNP, the genome RNA is tightly encased by the N protein. Two other viral proteins, P and L, are associated with the RNP. Transcription and replication of the virus are ensured by the RNP complexes of these N, P, and L proteins. The glycoprotein forms approximately 400 trimeric spikes, which are tightly arranged on the surface of the virus. Changes in the G protein amino acid sequence have a strong influence on viral virulence. The M protein is associated with both the envelope and the RNP and may be the central protein of rhabdoviral assembly. The arrangement of these proteins and the RNA genome determine the structure of the rabies virus.

5 PATHOPHYSIOLOGY

Human rabies is almost always attributed to a rabid animal bite. The risk of rabies due to bite is about 50 times that of scratches (5–80% versus 0.1–1%) [9]. Besides severity of the bite, efficient transmission also depends on the number of acetylcholine receptors at the site and the amount of virus in the saliva of the rabid animal [1,9]

Table 1 Seven Putative Genotypes in the Genus *Lyssavirus*

Genotype[a]	Serotype	Virus[b]
1	1	Classical rabies virus
2	2	Lagos bat
3	3	Mokola
4	4	Duvenhage
5	5	EBL-1
6	5	EBL-2
7		Australian bat lyssavirus (Ballina virus)

[a] Genotypes 1 and 2 belong to phylogroup 2; the remainder belong to phylogroup 1.
[b] EBL = European bat lyssavirus.

Although rabies usually follows bite exposures, it can be acquired via inhalation from aerosolized virus in caves inhabited by rabid bats and in laboratory accidents with infected aerosolized tissues [8,9]. Transmission of rabies is also associated with handling and skinning of infected carcasses and exposure of the conjunctiva, oral mucous membranes, genitalia, and skin abrasions to the saliva of rabid animals. Human-to-human transmission other than by corneal transplantation has not been well documented [1,8,9]. Transplacental transmission has been rarely reported in humans, and infants born to mothers with rabies encephalitis were found to be healthy [9].

Binding of viral glycoprotein to the alpha subunit of nicotinic AchR (nAchR) leads to multiplication in the muscle cells [9]. Following primary infection, the virus undergoes an ''eclipse'' phase. This silent phase is variable and may be explained by localization of virus within the muscle, which in turn provides an opportunity for host immune clearance and for postexposure treatment. Rabies antigen and genome may exist for as long as 2 months after inoculation into the muscle [12]. It is not known which factors control the length of this silent delay period.

After budding from plasma membranes of the muscle cells, virus or its genome is taken up into unmyelinated nerve endings at the neuromuscular junctions or at the muscle spindles. Rabies virus is then transported to the CNS via retrograde axoplasmic flow. The virus then infects and replicates again in the dorsal root ganglia and anterior horn cells [13,14]. At the dorsal root ganglia, it is presumed that viral replication can then be recognized and attacked by immune effectors, giving rise to clinical prodromal symptoms at the bite site [9]. Direct viral entry into the nerves without prior replication in the muscles may explain the short incubation period of less than 7 days demonstrated in a patient who had bite injury to the brachial plexus [15].

In the case of cryptic bat rabies, where a history of exposure is rarely obtained, the epidermis and dermis, rather than muscle, may serve as portals of entry [16]. Local sensory prodromes reflecting ganglioneuritis are present in 30% of crv cases (versus 70% in bat-related cases). This suggests that in brv cases, a sensory pathway may be preferential [9].

Travel time from the peripheral nerve to the CNS is relatively constant at a rate of 8–20 mm/day and depends on the proximity of the inoculation site to the CNS [14]. Studies of a preparation of challenge virus standard (CVS) in chick nerve–muscle cocultures have shown that the neuromuscular junction is the major site of entry to neurons. Colocalization of virus and endosome tracers within the nerve terminals, along with progressive accumulation of virus and tracers in axons and nerve cells, indicated retrograde transport of endocytosed virus from motor nerve terminals [17]. Because nAchR is not present on all categories of neurons susceptible to rabies virus, it is unlikely to be the only receptor that mediates viral entry into neurons. Rabies virus may also use carbohydrates, phospholipid, gangliosides, neural cell adhesion molecule (NCAM or CD 56), and low-affinity nerve growth factor receptor-P75 neurotrophin receptor (P75NTR) to gain entry into the cells [9].

Once rabies reaches the CNS, rapid amplification occurs. Virus disseminates via plasma membrane budding and direct cell-to-cell transmission or by trans-synaptic propagation [14]. The G protein is required for attachment to neuronal receptors and trans-synaptic spread [18]. Following stereotactic inoculation into the striatum, rabies virus has been shown to travel by retrograde fast axonal transport. This transport involves an interaction between viral capsid P protein and microtubule dynein [19,20]. Neurons are the CNS cells selectively and dominantly involved. However, infection of astrocytes and glial cells in animals and humans has also been reported [1]. Negri bodies, a pathological hallmark of rabies, result from excessive accumulation of ribonucleoprotein (RNP) in the cytoplasm [8]. In human rabies, clinical and laboratory evidence suggests that differential response at various CNS regions may contribute to the diversity of clinical manifestations [9,15]. Regional CNS rabies antigen distribution as well as magnetic resonance imaging (MRI) of the brain are similar in both encephalitic and paralytic forms [9,21] (Figs. 1A, 1B). Brainstem, thalamus, basal ganglia, and spinal cord are preferential sites in both forms. Minimal or absence of rabies viral antigen was found in the hippocampus and neocortex. Despite a similar MRI localization in brainstem in both brv and crv rabies patients, brainstem signs and myoclonus are usually lacking in the latter [9,22]. Inflammatory reactions are usually scant and when they are present do not correlate with clinical manifestations. The presence of virus in the CNS does not determine the clinical severity of the disease. High titers of virus in the brain and spinal cord can be found in animals long before clinical signs appear [1]. The degree of muscarinic AchR function modification in the hippocampus of rabid dogs was not dependent on the viral load [23]. Rabies virus antigen was readily demonstrable by immunofluorescence in the frontal area of one paralytic patient who remained fully conscious and had quadriplegia and respiratory failure requiring ventilatory support [1].

The animal host and viral strain may not be the major determinants of clinical manifestations, although rabies after a vampire bat bite is almost always of the paralytic form [1]. A recent outbreak of human rabies in the Peruvian jungle that was transmitted by vampire bats, however, presented as the furious (encephalitic) form [24]. Furthermore, the same dog that transmitted paralytic rabies to one patient also caused classical encephalitic rabies in another [9]. Incomplete immunization was not associated with any certain clinical form of human rabies [1].

It remains uncertain why rabies in patients with intact cellular immunity to rabies virus tends to manifest as encephalitic rabies. In theory, rabies infection of the CNS, particularly the brainstem, leads to the production of cytokines and proinflammatory molecules such as IL-1, alpha/beta, IL-6, IL-10, tumor necrosis factor alpha (TNF-α), interferons (IFN), and NO and to the secretion of chemokines [9]. These cytokines can activate the TNF-α p55 kDa receptor, resulting in the recruitment of T and B cells. This may therefore lead to the promotion of immune recognition against rabies virus at such an ''immune-privileged'' site. In addition, these cytokines can modify hippocampus and other limbic system functions, including the electrical cortical and HPA axis activities and serotonin metabolism. Immune activation also leads to further cytokine production, thus accentuating limbic symptomatology. Delayed mortality was observed in mice deficient

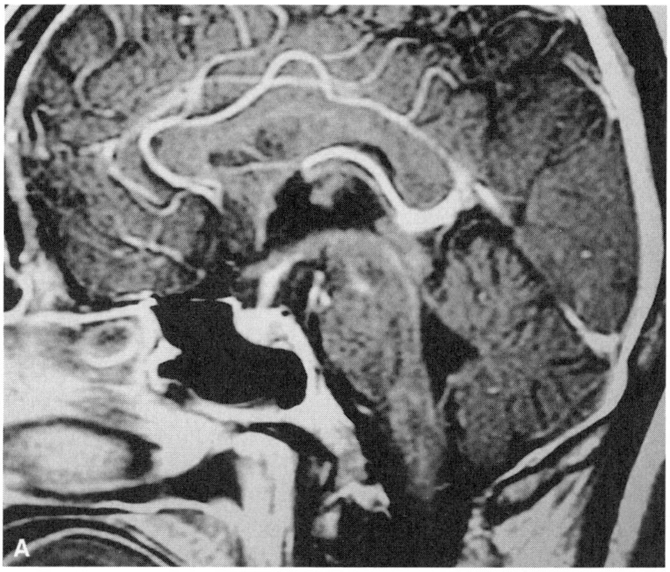

Figure 1 (A) Sagittal T1-weighted MRI with gadolinium. Enhanced lesions at the brainstem in a paralytic rabies patient.

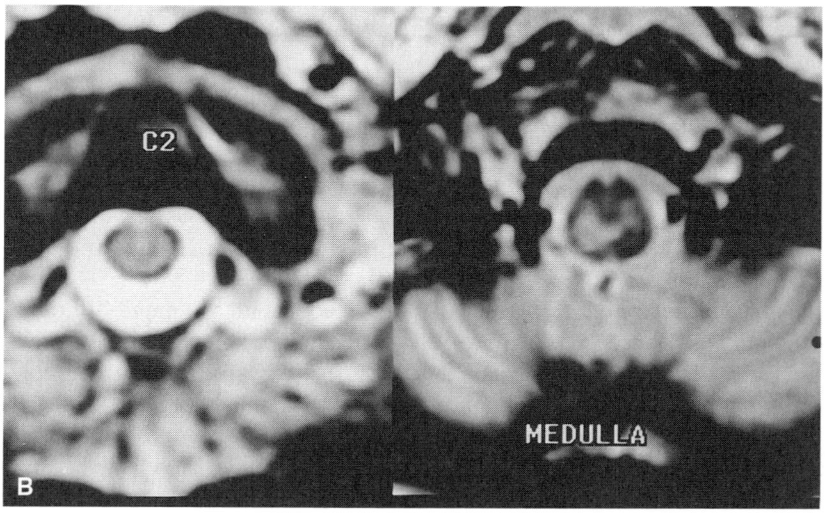

Figure 1 (B) Axial T1-weighted MRI with gadolinium. Enhanced lesions at the level of cervical cord and medulla in a paralytic rabies patient. (Courtesy of Dr. Jiraporn Laothamatas, Ramathibodi Hospital, Bangkok, Thailand.)

for the p55 kDa TNF-α receptor as a result of an increase in IFN-γ and IL-10 concentration and a reduction in inflammatory cell infiltration in the CNS. This may indicate that cytokines, which signal via p55 kDa TNF-α receptor, have a negative effect on survival of the host [9]. Furious rabies patients die faster, on the average in 5 days (versus 13 days in dumb rabies). In paralytic rabies, the exaggeration of cytokine production might not occur due to the lack of immune recognition of rabies virus in the brainstem. This possibly explains the relative paucity of limbic dysfunction and the absence of cellular immune activity to rabies virus in the paralytic patients.

The nature of weakness in paralytic rabies has not yet been fully explained. Although spinal cord motor neurons are likely to be involved at some point, it is still not known whether anterior horn cells or peripheral nerve dysfunction contribute to such clinical weakness, particularly in the early phase of clinical illness. There is also no conclusive information regarding the neuropathic pain at the bitten region, which is thought to be related to sensory ganglioneuropathy. Our preliminary studies of nerve conduction do not differentiate paralytic rabies from Guillain-Barré syndrome (GBS).

Eventual centrifugal spread from the CNS along neural pathways to the heart, skin, and other organs, especially salivary and serous glands of the tongue, is an important component of the complete rabies cycle [8]. Peripheral tissues such as the nape of the neck and cornea can serve as diagnostic tools [5]. Because all neural and non-neural organs, except for blood, may contain viable virus, transplant organs obtained from patients with an unexplained neurological disease may transmit rabies [5].

6 CLINICAL FEATURES

Not every rabid animal bite results in clinical rabies. Rabies mortality after untreated bites by rabid dogs varies from 35% to 57%, depending on the severity and location of the wound and presumed virus concentration in the saliva [8,9,15]. Transdermal bites, particularly with bleeding, on the head, face, neck, and hand carry the highest risk and are usually associated with a shorter incubation period. Nevertheless, all bites should be treated with the same urgency. In the case of rabid bats, the risk is present even from a scratch, owing to the unique ability of these agents to replicate in the skin.

The clinical features of rabies can be classified as classical and nonclassical. The classical encephalitic (furious) and paralytic (dumb) forms are almost always attributed to the canine rabies variant (crv) of genotype 1. Nonclassical rabies is found in patients exposed to bats (genotype 1, 5, 6, or 7) and has recently been found in Thai crv rabies – infected patients [9]. Atypical presentations were also seen in rare rabies survivors [1,9] (see Sec. 6.5).

The clinical features of classical rabies can be divided into five stages: the incubation period, the prodrome, the acute neurological phase, coma, and death. The first four are discussed briefly in Secs. 6.1–6.4.

6.1 Incubation period

The incubation period for rabies is usually 1–2 months but may range from less than 7 days to more than 6 years [9,25,26]. The cases with unusually long incubation periods (27 months; 4 and 6 years) were associated with Australian bat lyssavirus and have been found in immigrants to the United States from southeast Asia. An incubation period of less than a week has been seen with direct inoculation of the virus into nervous tissue, as in patients with brachial plexus injury from dog bites [15]. Rabies cannot be excluded by the absence of a history of rabid animal bite, particularly in rabies-endemic areas where unrecognized exposures are common.

6.2 Prodrome

The prodromal stage begins when the virus enters the dorsal root ganglia and subsequently the CNS. At this stage, symptoms are vague and nonspecific such as fever, generalized muscle aches, and gastrointestinal disturbances. However, approximately one-third of patients with dog-related infections (regardless of clinical types) and three-fourths of those with bat-related disease experience local neuropathic symptoms at the bite site. The intense local prodrome of burning, itching, or piloerection, which starts at the wound and gradually spreads to the entire limb or the same side of the face in a nonradicular pattern, is a reliable indicator of rabies [1]. Rarely, these symptoms occur at locations remote from the bite site. For instance, two patients developed severe itching on the ears as a prodrome after having been bitten on their toes [1,15].

6.3 Acute Neurological Phase

Within hours or days after the prodrome, rabies patients enter the acute neurological phase. Two-thirds suffer from an encephalitic (furious) form, and the remainder present with paralysis resembling GBS [1,15]. The average length of survival is 5–7 days in furious cases and 13 days in paralytic cases. Mental dysfunction can be seen in furious cases as well as in some paralytic patients but to a much greater degree.

Encephalitic (Furious) Rabies

The earliest neurological feature resembles an intense anxiety reaction or acute psychosis. This can be aggravated by thirst, fear, bright light, or loud noise. Fever is a constant finding and may have started after the prodromal phase. There are three major cardinal signs of furious rabies: (1) fluctuating consciousness, in which the mental state alternates between periods of severe agitation and periods of normality or depression; (2) phobic and inspiratory spasms, which result from spasms of the accessory respiratory muscles of the neck and diaphragm followed by neck flexion or extension and end with perception of dyspnea; and (3) signs of autonomic dysfunction, with hypersalivation, pupillary abnormality, piloerection, excessive sweating, priapism, repeated ejaculation, and neurogenic pulmonary edema.

In classic furious rabies, there are usually no cranial nerve deficits or hemitract signs (i.e., hemiparesis or hemianesthesia). Seizures are rare but may occasionally be seen in fully developed rabies or in the preterminal phase. As the disease progresses, fluctuation of mental state is no longer observed, and the period of irritability is followed by deterioration of consciousness and coma.

Aero- and hydrophobia can be incited by blowing or fanning air on the face or chest wall of the patient and by encouraging the patient to swallow or by merely offering a cup of water. During the induced spasms, patients are extremely aroused and may display a fearful facial expression. Pharyngeal spasms may not necessarily be present, but when evident the patients may spit abundant saliva. This creates another characteristic and terrifying image of rabies. The first attack of hydrophobia may occur suddenly without prior swallowing difficulties or pain on swallowing. This argues against this being a conditioned reflex. One patient had his first experience of hydrophobia while taking a bath. Soft palate and pharyngeal sensation remains intact, but there is a hyperactive gag reflex. Once the patients lapse into coma, phobic spasms are replaced by intermittent inspiratory spasms that occur spontaneously every few minutes. These inspiratory spasms may actually present in the early stage but may not be identified because of their nonintense character and infrequent occurrence.

Paralytic (Dumb) Rabies

Paralytic (dumb) rabies can be difficult to diagnose due to a lack of aggression and relative sparing of consciousness. The major cardinal signs of rabies appear late or may be mild in paralytic cases. Phobic spasms occur in only half of the patients, whereas inspiratory spasms occur in all cases during the preterminal phase and may not be recognized as such. Weakness usually starts in the bitten extremity and progresses to all limbs and bulbar and respiratory muscles. Facial diaparesis is as common as in sporadic GBS. In the case of facial bites, weakness may initially involve ipsilateral facial and oculomotor muscles. However, no correlation between the development of paralytic rabies and the site of the bite has been found [21]. Presentations mimicking ascending myelitis with fasciculation, loss of joint position sense, or hypoesthesia to pinprick up to the thoracic level have been rarely observed [1].

The following features suggest paralytic rabies and serve to differentiate this disorder from GBS: persistent fever from the onset of limb weakness, intact modalities of sensory functions except at the bitten region, percussion myoedema, and bladder dysfunction [1,9]. It is unknown why percussion myoedema is present in paralytic rabies but not in encephalitic rabies or in neuroparalytic accidents following neural tissue vaccination. Myoedema has also been found in extreme cachexia, hyponatremia, hypothyroidism, renal failure, and syndrome of inappropriate secretion of antidiuretic hormone (SIADH) and therefore needs to be interpreted with caution.

Clinical manifestations in rabies patients after exposure to virus of insectivorous or frugivorous bat origin (genotype 1, 5, 6, or 7) differ in many ways from the classic forms of crv furious and dumb rabies. Local prodromes are much more common, as reported in 30 out of 46 brv-rabies patients documented between 1951 and 2000 [9,15]. There have been reports of radicular pain, objective sensory or motor deficits, and choreiform movements of the bitten extremity during the prodromal phase. Both focal brainstem signs and myoclonus are common. These brv-rabies manifestations correlate with MRI findings of abnormalities in the brainstem [22]. It is intriguing that despite the similar MRI abnormalities in crvrabies, obvious brainstem signs are usually not noted (Fig. 1). Other patients have been described as having hemiparesis or hemisensory loss, ataxia, vertigo, or Horner's syndrome. Convulsive and nonconvulsive seizures and hallucinations are frequent. Phobic spasms were described in only one out of six brv-rabies patients during 1997–2000 [1,9,10].

In crv rabies, weakness of the bitten extremity was usually observed only in patients who subsequently developed dumb rabies. Myoclonus, tremor, oculomotor abnormalities, and cerebellar signs were not present. Neither hemisensory loss nor hemiparesis was observed in crv patients. Horner's syndrome or loss of sweat on one side of the face and trunk was noted in one crv patient. Nonconvulsive seizures during the early neurological phase were seen in one patient [1]. Hallucinations were seen in only two crv patients in our experience.

Nonclassic presentations have been noted in at least six patients with dog-related rabies since 1997 at Chulalongkorn University Hospital alone [1]. One patient presented with ocular myoclonus and hemichorea. The other had spontaneous repeated pleasurable ejaculations and did not exhibit cardinal signs of rabies until the preterminal phase. Other manifestations included paraparesis, facial and bulbar weakness with preserved arm strength, or bilateral arm weakness. No patients had phobic spasm or autonomic hyperactivity.

6.4 Coma

It is extremely difficult to diagnose rabies at the coma stage. The two forms of rabies are indistinguishable once the patients become comatose. Inspiratory spasms are useful in diagnosis at this stage but are difficult to detect in paralytic rabies due to weakness. In encephalitic rabies, abnormal breathing patterns and depression of consciousness appear simultaneously. Regular breathing interspersed with inspiratory spasms is replaced by tachypnea, apneustic respiration, and, finally, ataxic respiration. These patterns are not observed in paralytic rabies, in which alveolar hypoventilation and ventilatory failure develop before the patients become obtunded. Sinus tachycardia, disproportionate to the fever, is evident in most cases even when adequate hydration is maintained. Coma precedes circulatory insufficiency, the prime cause of death in most cases. Hematemesis is seen in 30–60% of patients 6–12 h before death [1,9].

6.5 Recovery

Seven patients with atypical rabies presentations have been reported to survive [9]. None of them had phobic spasms or other cardinal features of rabies. The first patient (1972), who was bitten by a bat, had unsteady gait, dysarthria, and hemiparesis. The second patient (1976), bitten by a rabid dog, had quadriparesis and generalized myoclonus at the early stage and later developed cerebellar signs (ataxia, dysmetria, and dysdiadochokinesia), frontal lobe dysfunction, and bibrachial weakness. The third patient (1977) had aerosol exposure to a highly concentrated fixed rabies virus strain. Of the remaining four patients, three were bitten by rabid dogs and one by a vampire bat. Each patient received prophylaxis promptly with cell culture vaccine but not rabies immunoglobulin. All the children developed encephalitis within a month of exposure, with high concentrations of neutralizing antibodies to rabies virus detected in the CSF. Acute signs persisted for months, and there were chronic sequelae.

7 LABORATORY FINDINGS

Routine laboratory studies are nondiagnostic. Complete blood counts are usually normal or show mild leukocytosis with neutrophilia. Hyponatremia is present in approximately one-third of the patients regardless of the clinical type or stage of the disease [1]. This can be explained by inadequate intake from dysphagia and hydrophobia or SIADH. Hypernatremia with polyuria is rare. CSF examination is normal in most cases. However, mild CSF pleocytosis (less than 30 cells/dL) with lymphocytic predominance and slightly elevated protein level (less than 100 mg/dL) in GBS-like patients who are HIV-seronegative should alert the clinician, particularly when fever, hyponatremia, and bladder dysfunction occur early in the course of illness. A pleocytosis of over 100 cell/dL (110–950) is rare and suggests other diagnosis. Magnetic resonance imaging can be helpful in antemortem diagnosis of rabies [9]. Paralytic and encephalitic rabies patients had similar distributions of abnormal, ill-defined, mildly hypersignal T2 images involving the brainstem, hippocampus, hypothalamus, deep and subcortical white matter, and deep and cortical gray matter with varying degrees of severity depending on the stage of disease. Gadolinium enhancement is clearly shown only in later stages (Figs. 1 and 2). Although a seemingly similar involvement of thalamus and midbrain on magnetic resonance images can also be seen in Japanese encephalitis, more prominent hypersignal T2 changes and foci of hemorrhages are observed in this arboviral encephalitis than in rabies [9].

8 DIFFERENTIAL DIAGNOSIS

Differential diagnosis includes encephalitis caused by arboviruses such as Japanese, eastern equine, and West Nile viruses and enterovirus 71 and Nipah virus infections [9]. Diffuse flaccid paralysis was found in 10% of patients with West Nile virus encephalitis, with no discernible ascending pattern of progression [27]. Asymmetrical weakness in an unimmunized patient in an epidemic setting suggests paralytic poliomyelitis or atypical forms of Japanese encephalitis [9]. The Trinidad outbreak of paralytic rabies was initially thought to be poliomyelitis.

Acute hepatic porphyria with neuropsychiatric disturbances can be confused with rabies. Phobic and inspiratory spasms are seen only in rabies. Other conditions mimicking rabies include intoxication by a variety of substances such as atropine-like compounds and cannabis, alcohol withdrawal (delirium tremens), and acute serotonin syndrome from taking serotonin reuptake inhibitors [1]. Tetanus resembles rabies only in the form of reflex spasms [1]. All tetanus patients have a clear sensorium. Rabies patients do not have persistent rigidity or sustained contraction of axial musculature such as the jaw, neck, back, and abdomen as in tetanus. Spasms in rabies predominatly involve accessory respira-

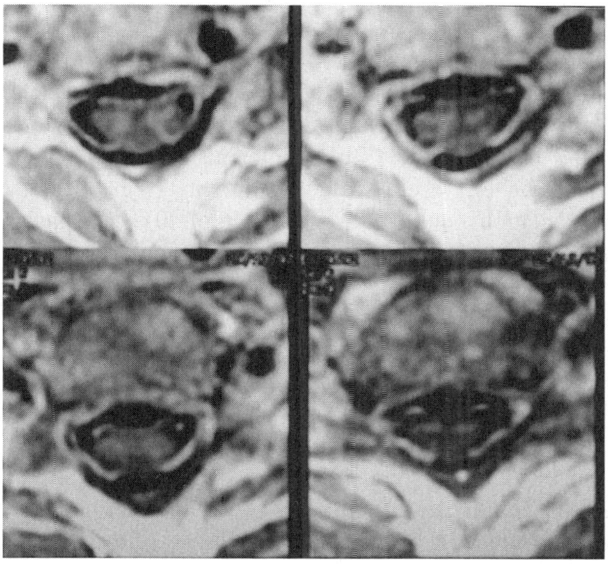

Figure 2 Axial T1-weighted MRI with gadolinium. Enhanced lesions in the gray matter of cervical cord and nerve roots in a paralytic rabies patient. (Courtesy of Dr. Jiraporn Laothamatas, Ramathibodi Hospital, Bangkok, Thailand.)

tory muscles and the diaphragm, whereas in tetanus spasms occur in axial muscles. Opisthotonos is present extremely rarely if ever in rabies.

In some parts of the world where nervous tissue rabies vaccine is still widely used, allergic encephalomyelitis must be included in the differential diagnosis. These vaccine-induced ''accidents'' develop in as many as 1 in 400 Semple vaccine–treated patients and less often in subjects who received mouse brain vaccine [3,4]. Neither phobic spasms, paresthesias at bite sites, nor fluctuating consciousness are present in these postvaccination reactions.

Acute motor axonal neuropathy (AMAN), an axonal form of GBS, shares many clinical features with paralytic rabies [28,29]. AMAN following *Campylobacter jejuni* infection may have preceding diarrhea that may be mistakenly diagnosed as a prodromal symptom of rabies. Areflexic quadriparesis and bilateral facial weakness without sensory deficits are observed in both conditions. Urinary incontinence is a common early symptom in paralytic rabies but rare in GBS. Inspiratory spasms with abnormal behavior may appear late in the clinical course and may be masked by generalized paralysis and superimposed metabolic disturbances that may occur in both conditions. Acute inflammatory demyelinating GBS (AIDP) or acute motor sensory axonal neuropathy (AMSAN) and GBS-like syndrome following nervous tissue rabies vaccine may exhibit some degree of sensory deficit. These are usually absent in paralytic rabies. Furthermore, the presence of local prodromal symptoms, even without a history of bite exposure and early autonomic dysfunctions (especially hypersalivation, abnormal pupils, and piloerection) suggest paralytic rabies.

9 ESTABLISHING THE DIAGNOSIS

An early antemortem diagnosis of rabies is extremely important. Delays in diagnosis result in much anxiety, potential spread of contamination, and an unnecessary expensive postexposure prophylaxis.

The diagnosis of furious rabies can be made with confidence when three classic or major cardinal signs are present together: fluctuating consciousness, phobic spasms, and autonomic dysfunction. However, in areas where bats are the principal vector of rabies to humans, these clinical expressions may be variable [9,10]. Phobic spasms were found in only half (10 of 20) of the cases. However, either phobic spasms alone or the presence of three or more of the following—agitation, confusion, seizures (16 of 20) or dysphagia (7 of 20), hypersalivation (10 of 20), limb pain, paresthesias (9 of 20), limb weakness, and paralysis or ataxia (3 of 20)—were significantly associated with antemortem diagnosis [18]. Local prodromal symptoms alone have to be interpreted with great caution because they may be modified by the patient's fear of rabies. A definite history of a bite, although commonly found in crv cases, is not helpful in cases associated with brv. Of the six brv cases reported between 1998 and 2000 in North America, only one had a definite history of a bat bite.

Serological testing in the serum and CSF may produce variable results [1,10,30,31]. Only 20% (6 of 31) of nonvaccinated Thai rabies patients had CSF rabies antibody (by rapid fluorescent focus inhibition test) within 1–26 days after the onset of the disease. All antibody-positive serum samples were obtained within 9 days after onset (3/6 within the first 3 days). None of the 27 Thai nonvaccinated crv-rabies patients were CSF antibody–positive to rabies virus [1,30]. Results obtained from the analysis of 102 samples from 39 cases in the United States and 16 in France since 1960 showed that serum antibody usually developed if the patients survived more than 8 days (6 of 43 between days 1 and 8 versus 34 of 59 from day 9). Antibody in the CSF appeared later (0 of 19 between days 1 and 8 versus 10 of 28 on day 9) [10,31].

Rabies virus may be isolated in mouse neuroblastoma cells from saliva specimens [10]. This cell culture isolation is sensitive and specific, and results are known within 4–5 days. However, all samples to be tested must be maintained frozen after collection with no preservatives. The success rate also depends on the status of rabies antibody (13 of 15 were positive in antibody-negative patients, compared to 0 of 17 in antibody-positive patients). The intermittence of rabies virus shedding in the saliva also confuses these findings. One must also understand that false negative results may be obtained from decomposed brains.

Rabies viral antigen may be detected by fluorescent antibody technique performed on frozen sections of nuchal skin samples [8–10,31]. An examination of at least 20 sections is required to detect the rabies nucleocapsid inclusions around the base of hair follicles. The result is unrelated to the presence or absence of antibody. Earlier studies showed that the proportion of positive results tends to increase as the disease progresses [1]. However, in another 26 rabies patients, antigen could be detected in as many as five of six (82%) within 4 days after onset. This number dropped to six of 10 (60%) between days 5 and 8 and to seven of 10 (70%) from day 9 [31]. Detection of rabies viral antigen in corneal and salivary impression smears may yield false positive or negative results [32].

Brain biopsy with antigen detection yielded the highest sensitivity in two series [32]. False negative results may occur when brain biopsy is performed during the first few days of the clinical illness. This may be due to a relative lack of viral antigen in the frontotemporal region and can be overcome by RT-PCR. Our experience with the nested PCR in 500 dog and five human brain samples showed 100% sensitivity without positive or false negative results [30,33].

In addition to CNS tissue, saliva, CSF, tears, skin biopsy samples, and urine may be sources for detection of rabies virus RNA by RT-PCR or nucleic acid sequence based amplification (NASBA) [8,9,33–35. Serial samples should be tested, because not all are positive, owing to intermittent shedding of virus. Sensitivity was not affected by antibody status of the patients. The success rate also depends on primer selection and surveillance of genetic variation among different reservoirs in a certain geographical location.

Where postmortem diagnosis from a complete brain autopsy is not possible, brain tissue can be obtained by Trucut needle aspiration through a transorbital approach [9]. It should be noted that only a minimal amount of brain tissue can be obtained by this technique and antigen detection by fluorescent antibody test may give a false negative result [personal experience].

10 MANAGEMENT

There is no specific treatment for rabies once clinical signs develop. Treatment is purely symptomatic, aiming to lessen the degree of agitation and to comfort the patient and family as much as possible. Preexposure vaccination is recommended for attending nurses and physicians who routinely care for patients with rabies. Only those who had true exposure despite precautions should receive postexposure treatment. Treatment with interferon and antiviral drugs such as ribavirin, adenine arabinoside, acyclovir, and inosine pronobex, intrathecal or systemic administration of rabies polyclonal antibody, and immunosuppresive therapy such as high-dose steroids or antithymocyte globulin did not alter the course of the disease in humans [9,30].

11 PREVENTION

Successful postexposure prophylaxis (PEP) relies on prompt start of treatment whenever the biting animal cannot be killed and its brain examined immediately by fluorescence microscopy [5,8] (see Table 2). Whether or not a dog in an endemic area has attacked without provocation should not be considered in the treatment decision. Treatment may be discontinued if the dog or cat, but no other mammal, remains healthy throughout an observation period of 10 days. Treatment must be initiated if the dog or cat develops abnormal signs during this observation period. Treatment should be started in an exposed person regardless of the time interval that elapsed since exposure but is usually not administered after a time interval longer than one year. Pregnancy is not a contraindication for PEP.

Postexposure prophylaxis consists of local wound care (thorough cleansing with soap and water, followed by application of 70% ethanol or a solution of iodine), use of modern tissue culture rabies vaccine, and wound infiltration with human or purified rabies immunoglobulin (RIG) in the case of single or multiple transdermal bites or scratches or licks over mucous membranes [8] (Table 2). RIG serves to neutralize some, if not all, of the virus inoculum at the bite site and closes the time gap until neutralizing antibody elicited

Table 2 Guide for Postexposure Treatment

Category	Type of contact with a suspect or confirmed domestic or wild[a] animal, or animal unavailable for observation	Recommended treatment
I	Touching or hand feeding of animals; licks on intact skin	None, if reliable case history is available.
II	Nibbling of uncovered skin; minor scratches or abrasions without bleeding; licks on broken skin	Administer vaccine immediately.[b] Stop treatment if animal remains healthy throughout an observation period[c] of 10 days or if animal is euthanized and found to be negative for rabies by appropriate laboratory techniques.
III	Single or multiple transdermal bites or scratches; contamination of mucous membrane with saliva (i.e., licks)	Administer rabies immunoglobulin and vaccine immediately.[b] Stop treatment if animal remains healthy throughout an observation period[c] of 10 days or if animal is killed humanely and found to be negative for rabies by appropriate laboratory techniques.

[a] Exposure to rodents, rabbits, and hares seldom, if ever, requires specific anti-rabies treatment.

[b] If an apparently healthy dog or cat in or from a low-risk area is placed under observation, it may be justified to delay specific treatment.

[c] This observation period applies only to dogs and cats. Except in the case of threatened or endangered species, other domestic and wild animals suspected of being rabid should be euthanized and their tissues examined using appropriate laboratory techniques.

by active immunization appears. The wound should not be sutured. As much as possible of the RIG [human (HRIG) 20 IU/kg; equine (ERIG) 40 IU/kg] should be injected in and around the wounds. We discourage the practice of multiple punctures during this process because this, like wound suturing, may cause additional injuries to nerves [11]. If the entire volume cannot be administered at the wound, the remainder can be administered at a distant site, such as the deltoid opposite the vaccine dose or the anterior thigh. In the case of multiple severe wounds, where RIG is insufficient in volume for infiltration of all wounds, dilution with saline solution to an adequate volume is recommended. RIG can be administered with a delay of up to 7 days after the initiation of vaccine treatment. A skin test prior to ERIG administration is required. ERIG is approximately 80% less expensive than HRIG in Thailand. Unfortunately, a shortage of ERIG supply may occur in the near future [36]. All tissue culture rabies vaccines such as human diploid cell (HDCV), purified Vero cell (PVRV), and purified chick embryo cell (PCECV) rabies vaccines are equally safe and effective. These vaccines can be given by the intramuscular (IM) route (at deltoid or anterolateral thigh muscles in children) on days 0, 3, 7, 14, and 28 or 30. The economical Thai Red Cross intradermal (ID) multisite regimen (2–2–2–0–1–1) consists of 0.1 mL of any potent tissue culture vaccine injected at two different sites on days 0, 3, and 7 and at one site on days 30 and 90. The Oxford intradermal multisite regimen (8–0–4–0–1–1) consists of 0.1 mL of any potent tissue culture vaccine injected intradermally at eight sites on day 0 (both sides of deltoid, lateral thigh, suprascapular region, and lower quadrant of the abdomen), at four sites on day 7, and at one site on days 28 and 90 [8]. Both intradermal regimens have been approved by WHO.

Persons who have been previously vaccinated with either pre- or postexposure regimens using a tissue culture or avian origin vaccine should receive two boosters on days 0 and 3 after a potential reexposure. No RIG is needed [8].

To date, there is no approved PEP for HIV-infected individuals. These patients have shown a low or even absence of antibody response after rabies vaccination [8].

Preexposure vaccination is recommended for subjects who are at a continuing risk of exposure (e.g., laboratory technicians and veterinarians.) [8]. The recommended preexposure immunization schedule consists of three intramuscular doses or three intradermal (0.1 mL) injections on days 0, 7, and 28 at the deltoid area. Also, neutralizing antibody titers should be checked every 6 months. If the value is less than 0.5 IU/mL, a booster dose of vaccine should be given using either the intramuscular or intradermal route.

For an individual who is on antimalarial chemoprophylaxis or is immunocompromised, IM injections are preferable.

ACKNOWLEDGMENTS

We are indebted to the medical residents and nurses and technicians of the Chulalongkorn University Hospital, Prommitr Hospital, and Queen Saovabha Memorial Institute and Neuroimaging Center, Ramathibodi Hospital, and to Professors Athasit Vejjajiva, Prida Phuapradit, Phrao Nivatvongs, Piyarat Ratanavanich, and Jiraporn Laothamatas for helpful advice. Clinical and research work was supported in part by grants from the Thai Red Cross Society and the Thailand Research Fund, General Prayudh Charumani, Mr, Cherdchai Wilailak, and the Phraya Athakraweesunthorn and Khunying Foundation.

REFERENCES

1. Hemachudha, T.; Mitrabhakdi, E. Rabies. In Infectious Diseases of the Nervous System; Davis, L.E., Kennedy, P.G.E., Eds.; Butterworth-Heineman: Oxford: UK, 2000, 401–444.
2. Baer, G.M.; Bellini, W.J.; Fishbein, D.B. Rhabdoviruses. In Virology; Fields, B.N., Knipe, D.M., Eds.; Academic Press: New York, 1990, 883–930.
3. Hemachudha, T.; Griffin, D.E.; Giffels, J.J.; Johnson, R.T.; Moser, A.B.; Phanupak, P. Myelin basic protein as an encephalitogen in encephalomyelitis and polyneuritis following rabies vaccination. N Engl J Med. 1987, 316, 369–374.
4. Hemachudha, T.; Phanuphak, P.; Johnson, R.T.; Griffin, D.E.; Ranatavongsiri, J.; Siriprasomsap, W. Neurologic complications of Semple-type rabies vaccine: clinical and immunologic studies. Neurology. 1987, 37, 550–556.
5. WHO. Eighth Report of the WHO Expert Committee on Rabies: Geneva, 1992.
6. Kristensson, K.; Dastur, D.K.; Manghani, D.K.; Tsiang, H.; Bentivoglio, M. Rabies: interactions between neurons and viruses. A review of the history of Negri inclusion bodies. Neuropathol Appl Neurobiol. 1996, 22, 179–187.
7. Bahmanyar, M.; Fayaz, A.; Nour-Saleski, S.; Mohammadi, M.; Koprowski, K. Successful protection of humans exposed to rabies infection. JAMA. 1976, 236, 2751–2754.
8. Rupprecht, C.E.; Hanlon, C.A.; Hemachudha, T. Rabies re-examined. Lancet Infect Dis. 2002, 2, 337–353.
9. Hemachudha, T.; Laothamatas, J.; Rupprecht, C.E. Human rabies: a disease of complex neuropathogenetic mechanisms and diagnostic challenges. Lancet Neurol. 2002, 1, 101–109.
10. Noah, D.L.; Drenzek, C.L.; Smith, J.S.; Krebs, J.W.; Orciari, L.; Shaddock, J.; Sanderlin, D.; Whitfield, S.; Fekadu, M.; Olson, J.G.; Rupprecht, C.E.; Childs, J.E. Epidemiology of human rabies in the United States, 1980 to 1996. Ann Intern Med. 1998, 128, 922–930.
11. Hemachudha, T.; Mitrabhakdi, E.; Wilde, H.; Vejabhuti, A.; Siripataravanit, S.; Kingnate, D. Additional reports of failure to respond to treatment after rabies exposure in Thailand. Clin Infect Dis. 1999, 28, 143–144.
12. Charlton, K.M.; Nadin-Davis, S.; Casey, G.A.; Wandeler, A.I. The long incubation period in rabies: delayed progression of infection in muscle at the site of exposure. Acta Neuropathol. 1997, 94, 73–77.
13. Tsiang, H. Interaction of rabies virus and host cells. In Rabies; Campbell, J.B., Charlton, K.M., Eds.; Kluwer: Boston, 1998, 67–100.
14. Tsiang, H. Pathophysiology of rabies virus infection of the nervous system. Adv Virus Res. 1993, 42, 375–412.
15. Hemachudha, T.; Phuapradit, P. Rabies. Curr Opin Neurol. 1997, 10, 260–267.
16. Morimoto, K.; Patel, M.; Corisdeo, S.; Hooper, D.C.; Fu, Z.F.; Rupprecht, C.E.; Koprowski, H.; Dietzschold, B. Characterization of a unique variant of bat rabies virus responsible for newly emerging human cases in North America. Proc Natl Acad Sci USA. 1996, 93, 5653–5658.
17. Lewis, P.; Fu, Y.; Lentz, T.L. Rabies virus entry at the neuromuscular junction in nerve-muscle cocultures. Muscle Nerve. 2000, 23, 720–730.
18. Etessami, R.; Conzelmann, K.K.; Fadai-Ghotbi, B.; Natelson, B.; Tsiang, H.; Cecaldi, P.E. Spread and pathogenic characteristic of a G-deficient rabies virus recombinant: an in vitro and in vivo study. J Gen Virol. 2000, 81, 2147–2153.
19. Raux, H.; Flamand, H.; Blondel, H. Interaction of the rabies virus P protein with the LC8 dynein light chain. J Virol. 2000, 74, 10212–10216.
20. Jacob, Y.; Badrane, H.; Ceccaldi, P.E.; Tordo, N. Cytoplasmic dynein LC8 interacts with lyssavirus phosphoprotein. J Virol. 2000, 74, 10217–10222.
21. Tirawatnpong, S.; Hemachudha, T.; Manutsathit, S.; Shuagnshoti, S.; Phanthumchinda, K.; Phanuphak, P. Regional distribution of rabies viral antigen in the central nervous system of human encephalitic and paralytic rabies. J Neurol Sci. 1989, 92, 91–99.

22. Pleasure, S.J.; Fischbein, N.J. Correlation of clinical and neuroimaging findings in a case of rabies encephalitis. Arch Neurol. 2000, 57, 1765–1769.
23. Dumrongphol, H.; Srikiatkhachorn, A.; Hemachudha, T.; Kotchabhakdi, N.; Govitrapong, P. Alteration of muscarinic acetylcholine receptors in rabies viral-infected dog brains. J Neurol Sci. 1996, 137, 1–6.
24. Warner, C.K.; Zaki, S.R.; Shieh, W.J.; Whitfield, S.G.; Smith, J.S.; Orciari, L.A.; Shaddock, J.H.; Niezgoda, M.I.; Wright, C.W.; Goldsmith, C.S.; Sanderlin, D.W.; Yager, P.A.; Rupprecht, C.E. Laboratory investigation of human deaths from vampire bat rabies in Peru. Am J Trop Med Hyg. 1999, 60, 502–507.
25. Smith, J.S.; Fishbein, D.B.; Rupprecht, C.E.; Clark, K. Unexplained rabies in three immigrants in the United States: a virologic investigation. N Engl J Med. 1991, 324, 205–211.
26. Hanna, J.N.; Carney, I.K.; Smith, G.A.; Tannenberg, A.E.; Deverrill, J.E.; Botha, J.A.; Serafin, I.L.; Harrower, B.J.; Fitzpatrick, P.F.; Searle, J.W. Australian bat lyssavirus infection: a second human case, with a long incubation period. Med J Aust. 2000, 72, 597–599.
27. Nash, D.; Mostashari, F.; Fine, A. The outbreak of West Nile virus infection in the New York City area in 1999. N Engl J Med 2001, 344, 1807–1814.
28. McKhann, G.M.; Cornblath, D.R.; Griffin, J.W.; Ho, T.W.; Li, C.Y.; Jiang, Z.; Wu, H.S.; Zhaori, G.; Liu, Y.; Jou, L.P. Acute motor axonal neuropathy: a frequent cause of acute flaccid paralysis in China. Ann Neurol. 1993, 33, 333–342.
29. Hafer-Macko, C.; Hsieh, S.T.; Li, C.Y.; Ho, T.W.; Sheikh, K.; Cornblath, D.R.; McKhann, G.M.; Asbury, A.K.; Griffin, J.W. Acute motor axonal neuropathy: an antibody-mediated attack on axolemma. Ann Neurol. 1996, 40, 635–644.
30. Hemachudha, T. Human rabies: clinical aspects, pathogenesis and potential therapy. In Lyssaviruses; Rupprecht, C.E., Dietzschold, B., Koprowski, H., Eds.; Springer-Verlag: New York, 1994, 121–143.
31. Crepin, P.; Audry, L.; Rotivel, Y.; Gaccin, A.; Caroff, C.; Bourhy, H. Intravitam diagnosis of human rabies by PCR using saliva and cerebrospinal fluid. J Clin Microbiol. 1998, 36, 1117–1121.
32. Hemachudha, T.; Phanuphak, P.; Sriwanthana, B.; Manutsathit, S.; Phanthumchinda, K.; Siriprasomsup, W.; Ukachoke, C.; Rasameechan, S.; Kaoroptham, S. Immunologic study of human encephalitic and paralytic rabies. Preliminary report of 16 patients. Am J Med. 1988, 84, 673–677.
33. Kamolvarin, N.; Tirawatnpong, T.; Rattanasiwamoke, R.; Tirawatnpong, S.; Panpanich, T.; Hemachudha, T. Diagnosis of rabies by polymerase chain reaction with nested primers. J Infect Dis. 1993, 167, 207–210.
34. Wacharapluesadee, S.; Hemachudha, T. Nucleic acid sequence based amplification in the rapid diagnosis of rabies. Lancet. 2001, 358, 892–893.
35. Wacharapluesadee, S.; Hemachudha, T. Urine as a source of rabies RNA detection in the diagnosis of human rabies. Clin Infect Dis. 2002, 34, 874–875.
36. Wilde, H.; Khawplod, P.; Hemachudha, T. Sitprija, V. Postexposure treatment of rabies infection: can it be done without immunoglobulin. Clin Infect Dis. 2002, 34, 477–480.

15

Arthropod-Borne Virus Encephalitis

John Booss
Veterans Affairs (VA) Connecticut Healthcare System
West Haven, and
Yale University School of Medicine
New Haven, Connecticut, U.S.A.

Nick Karabatsos
Centers for Disease Control and Prevention
Fort Collins, Colorado, U.S.A.

1 INTRODUCTION

The biological success of the arboviruses seems to be as improbable as it is undeniable. Because they are dependent on separate replication cycles in insect and vertebrate hosts and transmission to susceptible vertebrate hosts by the insect vector (arthropod-borne, or arbo), there would appear to be several vulnerable points in the life cycle of arboviruses. There are, in fact, marked ecological and atmospheric constraints on the abundance of the vector. Some of these factors can be tracked by indicators including remote sensing devices on satellites [1]. Despite the apparent fragility of the life cycle, it has been adopted by over 500 viruses in six virus families that blanket the world [2]. Furthermore, arboviruses are ecological opportunists, expanding or jumping into new territories when facilitating factors allow. The appearance of West Nile virus in New York City in 1999 and its spread in succeeding years, including a massive outbreak crossing the United States in 2002, and the movement of Rift Valley fever virus into the Arabian Peninsula in 2000 are but two recent examples of arboviruses behaving as emerging infectious diseases.

Arboviral infection of humans is not fundamental to the virus life cycle. It often results from infection with a different vector than the one that maintains the virus cycle in small mammals or birds. Humans are usually dead-end hosts because insufficient viremia is achieved to allow transmission by the vector. The effect in humans is highly variable, with some arboviruses responsible for a high ratio of inapparent to apparent infections and/or producing only mild undifferentiated febrile illness, whereas others may produce highly lethal hemorrhagic disease or encephalitis.

2 THE VIRUSES

The term arbovirus has remained useful, emphasizing as it does vector specificity and hence a means of public health control. Phylogenetic studies of the flaviviruses, an arbovirus genus of the family *Flaviviridae*, suggest that they may have evolved first as non-vector-borne viral clusters and that tick-borne and mosquito-borne viral clusters evolved subsequently [3]. Based on physical-chemical relationships, the arboviruses fall into six families, of which three contain particularly important agents causing encephalitis (Table 1). All three contain single-stranded RNA viruses, ranging in size from the 40–60 nm of the *Flaviviridae* family to 80–120 nm of the *Bunyaviridae*. Alpha and flaviviruses contain one RNA molecule of about 10–12 kDA, whereas the genome of viruses of the *Bunyaviridae* family consists of three segments of RNA.

The arboviruses were traditionally isolated and grown in suckling mice. A variety of cell culture systems such as Vero or BHK21 and mosquito cell cultures can also be used for virus isolation. Traditional diagnosis also relied heavily on serological changes between acute and convalescent samples, and results were often unavailable during the acute illness. Current more rapid diagnostic technologies such as IgM capture ELISA, antigen testing, and PCR allow diagnoses during the acute illness [4,5]. This is of great benefit to the clinician in the acute differential diagnostic process and will be crucial as specific antiviral agents are developed.

The histories of some of the arbovirus encephalidites bear a relationship to that of von Economo's encephalitis lethargica. That mysterious illness, which appeared during

Table 1 Arboviral Encephalitides Discussed in this Chapter[a]

Virus	Virus family	Geographic location
North America		
Eastern equine encephalitis	*Togaviridae*	Atlantic and Gulf coasts of U.S. and South America.
Western equine encephalitis	*Togaviridae*	West of Mississippi River in U.S. and adjacent regions in Canada.
St. Louis encephalitis	*Flaviviridae*	Most of U.S.
West Nile virus	*Flaviviridae*	Eastern U.S., but the principal areas are Africa, southern Europe, Asia, and Australia.
La Crosse virus	*Bunyaviridae*	Upper Midwest to West Virginia.
Powassan virus	*Flaviviridae*	Canada, eastern U.S., former Soviet Union.
Colorado tick fever	*Reoviridae*	Much of western U.S.
International		
Venezuelan equine encephalitis	*Togaviridae*	South and Central America and Mexico.
Tick-borne encephalitis	*Flaviviridae*	Central Europe, eastern former Soviet Union.
Japanese encephalitis	*Flaviviridae*	China, India, Japan, Southeast Asia.
Murray Valley encephalitis	*Flaviviridae*	Australia, Papua New Guinea.
Rift Valley fever	*Bunyaviridae*	Africa, Arabian Peninsula.

[a] Dengue virus is discussed in Chapter 22 in this volume.

World War I and the etiology of which has never been determined, stimulated great interest in viral encephalitis. Australian X disease, which was almost certainly Murray Valley fever [6], and Japanese encephalitis, also termed Japanese B encephalitis, were each named in distinction to encephalitis lethargica. Vector and host control methods coupled with national vaccination programs have led to remarkable reductions of Japanese encephalitis viral disease in Japan itself and in Korea. Those successes are at once heartening but simultaneously and paradoxically disheartening. Japanese encephalitis virus (JEV) annually kills and maims tens of thousands of persons, and the territory in which it is established is continuing to grow.

A major theme is the association of human arbovirus encephalitis with epizootics. Disease, mortality, and virus isolation were reported in horses for the eastern, western, and Venezuelan varieties of equine encephalitis in the 1930s before the corresponding achievements in the human disease counterparts [7]. More recently, the spread of West Nile virus in Africa and from there to the Mediterranean Basin and to Europe may have been dependent on migratory birds [8]. The cause of the jump of the disease to New York City in 1999, in the form of the Israeli goose viral strain, remains unknown. Crows have been particularly sensitive natural sentinels in the United States. The spread of Rift Valley fever virus (RVFV), which moved into the Arabian Peninsula in 2000, has been attributed in previous territorial enlargements to the movements of domestic animals. Thus, Hoogstraal et al. [9] implicated camels as a mechanism of the spread of RVF virus into southern Egypt. As the facilitators of emerging viral diseases continue, such as human encroachment on previously isolated habitats, global warming producing shifts in ecosystems, exposure of immunologically vulnerable populations, and inadvertent rapid global transportation of vectors and hosts, arbovirus encephalitides are likely to continue to appear in previously naive settings. Nonetheless, location, season, vector, host, and certain facilitating factors are crucial to the understanding of arbovirus ecology and are tabulated in each of the virus discussions that follow.

The presentation of arboviral encephalitidies in this chapter is organized by North American and international arboviral diseases. Such a division is necessarily arbitrary. Thus, eastern equine encephalitis, which has North American and South American subtypes, is discussed with the North American viruses. Conversely, Venezuelan equine encephalitis is discussed in the international category because of its importance to Mexico and Central and South America. Disease emergence and reemergence play an important role in the discussions that follow. Thus, prior to 1999, West Nile virus encephalitis might not have been discussed, let alone considered, with North American arboviruses. It emerged in New York City, reemerged in Israel, caused a major outbreak in the former Soviet Union, and afflicted horses in the Camargue in the south of France as well as horses in Italy. Its discussion with the North American arboviral diseases is admittedly arbitrary but emphasizes the emergent nature of the disease. Rift Valley fever virus too enlarged its territory by moving into the Arabian Peninsula in 2000. This arbovirus of humans and animals seems a prime candidate to continue to enlarge its territory because it can be spread not only by numerous mosquito vectors but also by direct contact of humans with dead or sick animals and aborted animal fetuses.

3 NORTH AMERICAN ARBOVIRAL ENCEPHALITIDES

According to the Centers For Disease Control and Prevention (CDC), the most common arbovirus encephalitis in the United States over the period 1964–1999 was the California serogroup, mainly the La Crosse virus (Sec. 3.5) with a median of 66 cases annually. St. Louis encephalitis virus (Sec. 3.3) produced a median number of 26 cases in the same period, eastern equine encephalitis (Sec. 3.1) a median of four cases, and western equine encephalitis (Sec. 3.2) a median of three cases [10]. West Nile virus (Sec. 3.4) emerged in New York City in 1999 when 62 cases were identified by intense surveillance. By 2001, human cases ranged up and down the Atlantic seaboard. The year 2002 saw a massive epidemic with cases from coast to coast. However, despite the media attention focused on West Nile fever virus and the resulting public anxiety, it is neither the most lethal nor the most feared of the North American arboviruses. That distinction goes to eastern equine encephalitis virus.

3.1 Eastern Equine Encephalitis

An alphavirus in the *Togaviridae* family, eastern equine encephalitis virus (EEEV) was isolated from horses in 1933 and from an outbreak of 34 cases of severe encephalitis in Massachusetts [7]. The virus circulates in swamp habitats between *Culiseta melanura* mosquitoes and passerine birds (Table 2). Other species of mosquitoes serve as bridge vectors to humans and horses. In the north the virus can be active from July to October, whereas year round cases can occur in Florida. The locations for EEEV tend to be stable, reflecting specific ecosystems, offering some advantage to public health officials. Cases have tended to occur sporadically. The ratio of inapparent to apparent infection has been low, ranging from 8:1 to 50:1 [11]. Attack rates and fatality rates are highest in the elderly and in children [12].

Neurological signs reflect the severity of the pathological changes in the brain [11,13]. On gross examination, the brain appears to be congested and swollen and to have flattened gyri. Microscopically, neutrophils are much more prominent than in most other types of encephalitis. These appear with mononuclear cells in the meninges, in perivascular infiltrates, and in the parenchyma. Vascular thrombosis is found, as are foci of necrosis. The basal ganglia, substantia nigra, cerebral cortex, and hippocampus are the most severely afflicted. White matter is affected only when it is near involved gray matter [11].

The incubation period has been estimated to be 3–10 days. A prodromal phase may be encountered with fever, headache and abdominal distress [14]. In other cases, however, the onset may be explosive, with high fever, impairment of consciousness, vomiting, focal weakness and seizures. Infants may demonstrate a bulging fontanelle. Evolution of the illness is rapid, close to 70% of patients becoming stuporous or comatose within 2 days of hospital admission [14]. The duration of coma in cases with a favorable outcome has

Table 2 Eastern Equine Encephalitis in the United States

Location	Atlantic seaboard and Gulf Coast
Principal vector	*Culiseta melanura* to birds, other mosquito species as bridge vectors to humans
Animal hosts	Birds
Season	Florida, throughout the year; northern seaboard, July–October
Facilitating factors	Increased rainfall, which facilitates multiplication of the principal and bridge vectors

Source: Refs. 11 and 12.

been found to be no more than 5 days. Early studies emphasized the high mortality rates, e.g., 68% in Massachusetts, and severe neurological sequelae [7]. The review of U.S. cases between 1988 and 1994 by Deresiewicz et al. [14] revealed a mortality rate of 36% and moderate or severe disability in 35% of the survivors.

Fever, an elevated white count with elevated neutrophils, and hyponatremia are likely to be found. Spinal fluid examination has shown elevated white cells at an average of 370 cells/mm^3 on first exam, a predominance of neutrophils, protein levels at approximately 100 mg/dL, and no reduction in sugar levels [14]. EEGs tend to show bilateral slowing, although findings mimicking herpes simplex encephalitis (HSE), such as lateralized periodic epileptiform discharges, can be observed [11,14]. Neuroimaging, both CT scans and MRI, show involvement of basal ganglia and thalami in the majority of patients studied. Cortex and brainstem were less commonly abnormal. Twenty-one of 32 CT scans and 13 of 14 MRIs were abnormal. The authors emphasized the usefulness of neuroimaging studies to differentiate EEEV from HSE [14].

Viral isolation from CSF or blood is not likely during life. Virus has been visualized by electron microscope (EM) on brain biopsy [15]. A presumptive diagnosis can be made by a positive serum IgM capture ELISA. Confirmation requires a fourfold rise in serum hemagglutination inhibition, complement fixation, immunofluorescence, or neutralization titers between acute and convalescent sera. Positive IgM capture ELISA on a spinal fluid sample is also considered to be a confirmed positive [14].

There are no proven antiviral drugs with which to treat EEEV infections. The rapid impairment of consciousness requires aggressive medical management in the setting of an intensive care unit (ICU). Seizure control, intracranial pressure management, fever reduction, electrolyte correction prevention of aspiration, skin erosion, and the development of contractures all require the constant attention available in an ICU. The neuropathology of fatal cases demonstrates the aggressiveness of the process, with neutrophilic perivascular infiltrates and vessel wall infiltration sometimes mimicking arteritis [16].

Although a vaccine is available for animals and for exposed personnel such as laboratory workers, it is not available to the general public. Hence, public health measures to control the vector and personal actions to avoid the vector are indicated. These include the use of screening and household air conditioning, insect repellents, and long sleeves and trousers and avoidance of outdoor activities at times of heavy mosquito activity.

3.2 Western Equine Encephalitis

First isolated from a 1931 epizootic in the San Joaquin Valley in California, western equine encephalitis virus (WEEV) was suspected as the cause of six human cases in Minnesota in the late summer of 1937 [17]. WEEV has been responsible for massive outbreaks of encephalitis, yet cases have virtually disappeared in horses and humans in recent years. Eklund and Blumstein [17] reported that of 737,000 horses in Minnesota in the summer of 1937, over 41,000 became sick and 9200 died. Eklund [18] reported that almost 800 human cases of encephalitis occurred in Minnesota in the summer of 1941, and almost 500 human cases were reported from the Saskatchewan epidemic in 1965 [19]. Adjacent U.S. and Canadian areas were often involved—Manitoba, North Dakota, and Minnesota in 1941 and 1975, for example [20]. Mysteriously, there has been an absence of epidemic WEEV activity in recent years. The CDC reported only three isolated cases in the United States in the decade 1990–1999 [10], and seroprevalence rates have declined in the western United States. Tsai and Monath [12] suggest that a reduced rural population and changes in land use patterns and lifestyles may have contributed to a reduced risk of infection.

Whereas humans and horses represent dead end hosts, mounting an insufficient viremia to sustain a cycle of infection, the *Culex tarsalis* mosquito and passerine birds are the principal components of the normal cycle of viral replication and transmission (Table 3). Both older people and infants are at increased risk of infection, with a wide range of the inapparent to apparent infection ratios by age group.

The original descriptions of the encephalitis of WEEV noted the rapid onset of fever, headache, vomiting and depressed level of consciousness [17,18]. The incubation period has been cited as between 1 and 2 weeks. Study of the 1975 epidemic in North Dakota and western Minnesota by Leech et al. [21] demonstrated significantly more cases of febrile headache and aseptic meningitis than of encephalitis. Stiff neck as part of the whole spectrum of illness has been reported in several case series. Tremors may occur. The acute course of the encephalitis has tended to be 10 days to 2 weeks. When coma occurred, it usually reversed within 4 days. Although there is a wide variation in the reported mortality, probably reflecting the different populations studied, overall it has been under 5% [22].

Sequelae in adults have been reported to be related to the duration of coma [23] and to be most severe in infants; these include cognitive, behavioral, and motor defects and seizures [24]. Finley et al. [25] emphasized the effect of WEEV infection during the perinatal period on ontogenesis, which thus produced continuing and delayed sequelae. Neuropathological examination of autopsied cases revealed vascular congestion, with perivascular infiltrates and focal areas of inflammation, necrosis, and gliosis noted upon microscopic exam [11,26].

General laboratory studies and neurodiagnostic tests do little to explicitly identify the illness. There may be an elevation of the peripheral white cell count. The CSF shows the typical profile of viral infection with a mononuculear pleocytosis after the first 2 days, normal or elevated protein, and a normal sugar value. Whereas the EEG findings are usually nonspecific, lateralized findings suggesting HSE were reported by Bia et al. [27]. The paucity of cases in recent years has resulted in no significant case series reports of CT or MRI. Diagnosis is dependent on serological changes. These are crucially important because the territories and season of St. Louis encephalitis virus and those of WEEV overlap and their diseases can present as a mixed epidemic [28].

Management of acute cases of encephalitis caused by WEEV is supportive in nature because there are no antiviral agents demonstrated to be clinically effective. The neuropathological literature does not suggest that massive increase in intracranial pressure will occur, and the clinical literature does not suggest that seizure control is a common acute management problem. Further, the literature reports that coma is brief in duration, 3–4 days, before reversal. Hence, one gains the impression that should epidemics of WEEV disease reappear, careful clinical care will be rewarded with good outcomes.

Table 3 Western Equine Encephalitis

Location	West of Mississippi River in U.S., adjacent regions of Canada
Vector	*Culex tarsalis*
Animal hosts	Wild birds
Season	Summer and fall

Source: Refs. 11 and 12.

A vaccine is available for horses and for persons whose occupations put them at special risk. It is anticipated that vector control would be useful in the event of a reemergence of a WEEV epidemic.

3.3 St. Louis Encephalitis

Thoroughly studied in St. Louis in 1933 when the disease was characterized and the virus isolated [29], St. Louis encephalitis had been the most widespread of the arbovirus encephalitides in the United States. Thus, a 1975 epidemic of 2800 cases occurred in 31 states [30]. It is found from coast to coast and from north to south. Only New England as a region has escaped. A flavivirus of the *Flaviviridae* family, it had also been the most common arbovirus to strike the United States in epidemic form [10]. However, it has both an epidemic form in eastern and midwestern states and an endemic form in western states (Table 4). In 2001, an outbreak of over 60 cases was the largest outbreak ever recorded in Louisiana. Different mosquito vectors transmit the virus to birds in different regions of the country. Cases occur from June through October with a peak in August and September.

The incubation period is highly variable, from 4 to 21 days, as is the inapparent to apparent infection ratio, from 16:1 to 425:1 [30]. Disease and mortality are significantly more common in the elderly. A prodrome of fever, headache, malaise, and myalgia may precede the onset of frank parenchymal signs of encephalitis. There may be a variety of systemic signs including gastrointestinal (GI) disturbances, sore throat and cough, and dysuria [31]. Evolution to stiff neck may evolve, along with confusion and tremor. In the 1966 epidemic in Dallas, mental aberration was found in 79%, nuchal rigidity in 65%, and tremor in 53% of patients [32]. Of a total of 95 cases, 11 were classified as aseptic meningitis, four as febrile headache, one as nonspecific illness, and 79 as encephalitis. The preponderance of encephalitis reflects a hospital based population. Less severe forms of the illness would likely be found more commonly in a community based serosurvey. The case fatality rate in the Dallas study was 17%. Risk factors for acquiring the disease reflect exposure to the vector. These have been found to include inadequate screening, lack of air conditioning, and sitting outside a residence [33].

Laboratory studies include CSF typical of viral infection with a mononuclear pleocytosis and a moderate protein elevation [32]. A peripheral leukocytosis is more common than leukopenia, although a normal WBC may be found. White cells may be found in the urine, in the absence of bacteriuria, with an elevated BUN and/or creatinine. Electromyographic abnormalities may be found compatible with lower motor neuron dysfunction [32]. A study of 11 patients from the 1995 Dallas epidemic found abnormal EEGs in the nine patients who had EEGs performed, including diffuse background slowing in seven

Table 4 St. Louis Encephalitis

	Urban (epidemic)	Rural (endemic)
Location	Midwest, eastern states	Western states
Vector	*Culex pipiens*	*Culex tarsalis*
Animal hosts	Birds	Birds
Season	Summer, fall	Summer, fall

Source: Refs. 11 and 31.

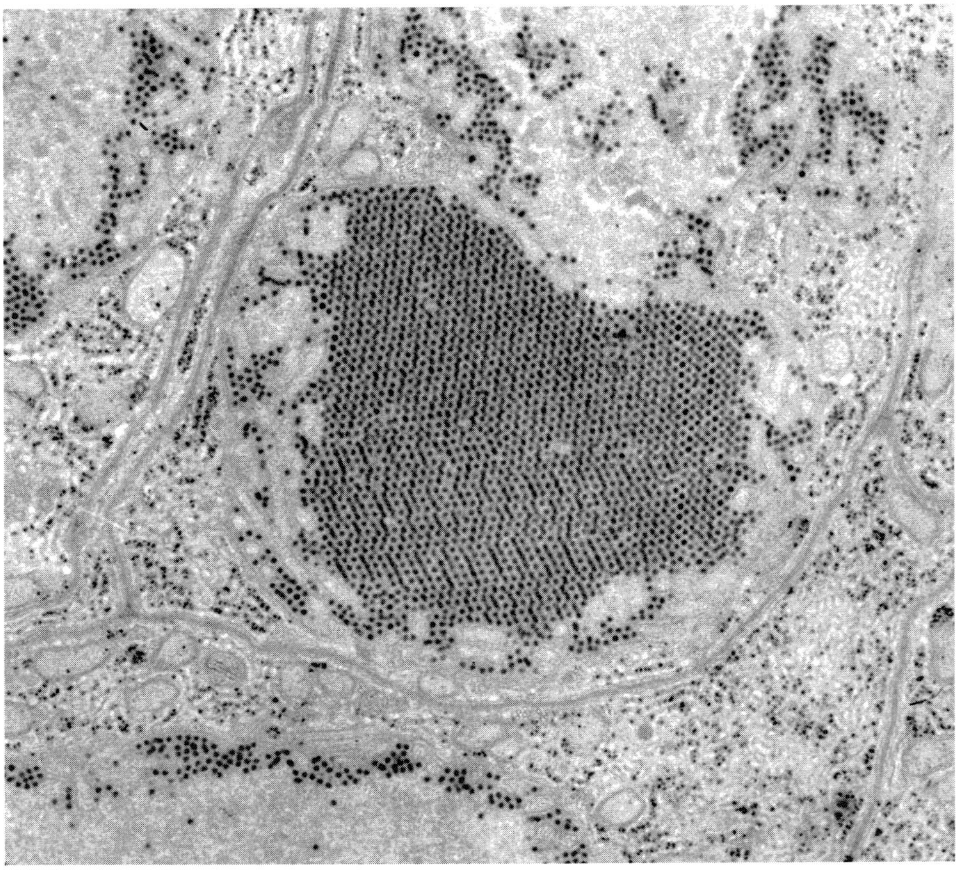

Figure 1 Photomicrograph of St. Louis encephalitis virus. Paracrystalline array of virions in salivary gland of *Culex pipiens* mosquito. Magnification of original about 30,000×. (Courtesy of F. A. Murphy, School of Veterinary Medicine, University of California, Davis.)

patients, status epilepticus (one patient), and bilateral periodic lateralized epileptiform discharges (one patient). CT scans revealed nonspecific findings such as atrophy. However, review of five MRI scans showed T2-weighted hyperintensity of the substantia nigra [34]. Presumptive viral diagnosis during the acute illness relies on IgM capture ELISA. However, verification requires assays such as serum neutralization because of cross-reactivity with closely related viruses [31].

There is no antiviral therapy available, hence management of the acute phase of the illness requires vigorous supportive treatment. Six patients in the 1995 Dallas epidemic required mechanical ventilation, four had tonic clonic seizures, and one required pentobarbital anesthesia for status epilepticus [34]. The average duration of stay in the hospital was 17 days. It has been commented that reversal of impaired consciousness may be rapid with neurological improvement after several days. However, convalescence from symptoms such as anxiety, forgetfulness, tremor, headache, and unsteadiness can take months, or even years, in some patients [31].

No vaccine is available, so vector control and avoidance are the crucial aspects of prevention. A review of the 1986 outbreak in Baytown and Houston, Texas emphasized the importance of mosquito surveillance to facilitate vector control [33]. It also suggested that assistance in repairing window screens might be an important preventive measure in impoverished neighborhoods, where the risk of infection is highest.

3.4 West Nile Virus Encephalitis

West Nile virus encephalitis reached North America in 1999 when 62 cases were identified in and around New York City. The cases were concentrated in a northern area of the Borough of Queens [35]. The means of transfer to North America has not been determined. Many mechanisms have been suspected, including migratory birds, imported exotic birds, a viremic human traveler, inadvertent transport of infected mosquitoes, and the suggestion of bioterrorism. The New York West Nile virus strain is most closely related to a West Nile virus isolated in Israel from a goose [36]. That relationship is telling, because the 1999 New York human outbreak was also associated with an outbreak in exotic and domestic birds [37]. Crows have been a particular target, with a die-off observed before the human cases in 1999 and isolations of the virus from crows occurring early in the season, as natural sentinels, in subsequent years. The territory of the virus expanded from 1999 to 2002, and viremic birds seem the likely cause. Additionally, the virus appears to be capable of overwintering in mosquitoes. In 2002, a major human outbreak crossed the United States, with over 3,700 cases [37a].

The virus was originally isolated from a woman in the West Nile district of Uganda during a yellow fever survey [39]. Subsequently, it has disseminated widely (Table 5). Cases have been reported from Egypt, Israel, South Africa, Romania, and the former Soviet Union, among other locations. In 1999 when the encephalitis made itself apparent in New York City, the virus caused a massive outbreak in the Volgograd region of Russia [40]. It reemerged, causing human cases in Israel, in 2000 and was the cause of equine encephalomyelitis in the Camargue in France as well as in Italy in 1998. Kunjin virus, closely related to West Nile virus, is found in Southeast Asia and Australia [38].

Culicine mosquitoes are the usual vector. As household mosquitoes, they have been associated with flooded domestic basements, for example, in the Bucharest outbreak in 1996 [41]. Outdoor pools, birdbaths, and old tires containing water are peridomestic sites for mosquito breeding. The usual hosts are birds, although the massive die offs of crows experienced in North America are a unique feature. Migrating birds are suspected of being the mechanism to spread the virus to new regions. Phylogenetic analysis of viral isolates support the hypothesis that migrating birds transported the virus from sub-Saharan Africa

Table 5 West Nile Virus Encephalitis

Locations	Africa, Middle East, former Soviet Union, Europe, India, Indonesia, and United States
Vector	*Culex* and other species
Animal hosts	Birds
Season	Summer to early fall
Facilitating factors	Peridomestic flooding

Source: Refs. 35, 38, and 44.

to northern Africa and thence to southern Europe [42]. During the 2002 U.S. epidemic, transmission mechanisms included intrauterine and breast milk [42a], organ transplantation [42b], and blood transfusion [42c].

In the 1999 human outbreak in New York City, the virus particularly targeted the medulla [43]. Cerebral swelling and necrosis were not found. However, the targeting of the medulla by the virus and the clinical observations of an axonopathy and anterior horn cells indicate that clinicians need to be vigilant of the need to institute assisted respiration.

A significant amount of clinical information was gathered in the 1996 Bucharest outbreak [44]. The inapparent to apparent disease ratio varied from 1:140 to 1:320. Three manifestations—meningitis, meningoencephalitis, and encephalitis—were encountered. The onset was noted to be abrupt as it had been in other outbreaks, with fever, headache, neck stiffness, and vomiting. Lymph node swelling and rash have been variably found in different outbreaks. In the 1996 Bucharest outbreak, disorientation, confusion, a reduced level of consciousness, and weakness were observed. Mortality was 4.3%, affecting those over 50 years of age.

The spectrum of illness in the 1999 New York City outbreak was investigated by household-based cluster sampling [45]. In the outbreak's epicenter, about 2.6% of individuals over 5 years of age were found to have been infected. Of these, about 20% had a mild illness that included headache, muscle and joint pains, and fatigue. The ratio of asymptomatic or mildly symptomatic cases to diagnosed meningoencephalitis was 140:1 at the epicenter.

In the prodrome to the meningoencephalitis, a large proportion of patients had gastrointestinal symptoms [46]. Flaccid weakness, resembling Guillain Barré syndrome, was a common finding. Seven of 62 afflicted individuals in the New York outbreak, or 11%, died. Of the survivors in New York City one year later, fewer than 40% had made full recoveries [47]. A polio-like illness was found in 2002 [47a], confirming earlier findings [47b].

Cerebrospinal fluid and EEG data are available as well as some neuroimaging results. The CSF reveals a mononuclear pleocytosis following a polymorphonuclear pleocytosis and protein elevation. The blood can reveal a lymphopenia. Electroencephalogram findings include high-voltage slowing. Head CT scans were normal in 43 patients studied, but enhancement in the leptomeninges and/or in periventricular areas was found in 31% of 16 patients studied with MRI [35].

Rapid viral diagnosis employs an IgM capture ELISA. Because there is cross-reactivity, such as with St. Louis encephalitis virus, presumptive positives must be confirmed with other assays such as fourfold changes in neutralizing antibody. Virus isolation from CSF, identification of virus in tissue, and PCR have not yet proven clinically useful. Thus, IgM capture methodology for presumptive diagnosis followed by acute and convalescent serum antibody determinations for confirmation are the diagnostic methods of choice.

Management must include vigilance for the need for respiratory support due to flaccid weakness from motor unit and/or brainstem involvement. Ribavirin [48,49] and interferon alpha-2b [49] have been reported to reduce replication of West Nile virus in vitro. In August 2002, the U.S. Food and Drug Administration (FDA) approved a national clinical trial of alpha-interferon [49a]. Recalcitrant seizures and significant cerebral swelling do not appear to have been clinical management issues.

A vaccine for horses has been released for use. Public health efforts in the United States have focused on surveillance, mosquito larvicide and adulticide, distribution of diagnostic reagents, and public information efforts [50].

3.5 La Crosse Virus Encephalitis

La Crosse virus is the most common cause of endemic encephalitis in children in the United States. An average of 74 cases were reported annually from 1964 to 1999 in the United States [10]. A member of the California serogroup bunyaviruses of the *Bunyaviridae* family, La Crosse is a single stranded RNA virus with a negative sense genome of three segments [51]. First isolated from the brain of a four-year-old girl in La Crosse, Wisconsin in 1960 [52], it was not the first virus in the serogroup to be isolated. California encephalitis virus was first isolated from mosquitoes in Kern County at the southern end of the San Joaquin Valley in California in 1943 [53]. It was associated with three cases of encephalitis two years later, but no further cases were reported until that of a 65-year-old man who fully recovered from encephalitis in Marin County, California in 1996 [54]. In contrast, another member of the California serogroup, Jamestown Canyon virus, appears to be significantly more common than previously appreciated. Mayo et al [55] recently found seroprevalence rates between 3.9% and 10.1% in Connecticut, citing similar prevalence rates in Massachusetts.

The virus is found almost entirely east of the Mississippi, a swath of states from Wisconsin and Minnesota to Ohio and West Virginia accounting for the vast majority of cases [56]. It is transmitted by a cycle between its mosquito vector, *Aedes triseriatus*, and small mammals such as chipmunks and squirrels [57]. It is also transmitted from chronically infected female *Aedes triseriatus* mosquitoes to their offspring and can overwinter in infected eggs. (See Table 6.) The vector favors hardwood forests, laying its eggs in tree holes, but other sites such as old tires will serve. Most infections occur in or in proximity to woodlands, with a predominance in boys, from July to September. The vast majority of cases are in children below age 16.

There is not an extensive neuropathological literature on La Crosse virus encephalitis, which is somewhat surprising in light of its frequency. Kalfayan [58] reported two cases in which gross inspection revealed diffuse congestion, edema, flattening of the convolutions, and uncal herniation. Microscopic exam revealed focal areas of inflammation, mild leptomeningitis, and vasculitis. Kalfayan speculated that the vasculitis may play a key role in the pathogenesis of the process. Fatal human infections with La Crosse virus may be associated with a given genotype of the virus that has a highly conserved M segment RNA both by nucleotide and deduced amino acid sequence [59].

There are a wide range of estimates of the ratio of inapparent or mild to significant illness, from as low as 2:1 to as high as 1500:1 [57]. However, there is a high seropositivity rate in endemic areas, 20% at La Crosse, Wisconsin, for example, with even higher rates in some exposed woodland workers [57]. The incubation period has been variously estimated to be from 3 to 7 or up to 15 days [51,57]. From the earliest large clinical studies,

Table 6 La Crosse Virus Encephalitis

Locations	East of Mississippi, upper Midwest to West Virginia
Vector	*Aedes triseriatus*
Animal hosts	Small mammals, chipmunks, squirrels
Season	July to September

Source: Refs. 11, 51, and 57.

it was recognized that the disease could be relatively mild and self-limited or severe and life threatening [60].

McJunkin, et al. [61] recently reviewed their clinical experience in 127 patients with La Crosse virus encephalitis in West Virginia. Headache, fever, and vomiting were found in the majority, and disorientation and seizures were each found in over 40%. Important differential diagnostic considerations included enteroviral aseptic meningitis, plus HSE for those with temporal lobe localization.

Somewhat over half of the West Virginia series required pediatric intensive care unit admission [61]. Observed complications were seizures, including status epilepticus, raised intracranial pressure, including herniation, inappropriate ADH secretion, and the need for mechanical respiration. Predictors of deterioration included a low Glascow Coma Scale score on admission, often 12 or less, and a rise in temperature, often above 38.5°C. Although a history of vomiting and/or seizures was often present, they were also found commonly in those who did not deteriorate. The mean duration of hospitalization was 6.2 days.

Examination of the spinal fluid usually reveals normal glucose and protein values, although the latter may be moderately elevated, and a moderate pleocytosis with a predominance of mononuclear cells. The peripheral white cell count is usually elevated. Hyponatremia and findings of inappropriate ADH secretion are common. EEGs usually reveal slowing but can also show focal abnormalities or periodic lateralizing epileptiform discharges suggesting HSE. McJunkin, et al. [61] reported that only 12% of CT scans demonstrated abnormalities with edema the most common finding. [61] Four of the MRIs were abnormal, with note made of focal cortical gadolinium enhancement.

The diagnosis relies on serological changes rather than virus isolation or PCR. Currently employed antibody studies include indirect immunofluorescence [60] and IgM capture immunoassay [62]. In vitro studies of ribavirin are promising [63], and a clinical trial of ribavirin is reported to be under way [61]. The need to treat cases compatible with HSE with acyclovir is important until an alternative diagnosis is established.

Mortality is low in cases of La Crosse virus encephalitis, on the order of 0.5% [57]. The duration of illness is usually 2 weeks or less. Sequelae include epilepsy in up to 28% of cases, motor abnormalities in under 3%, and behavioral abnormalities during recovery [57]. McJunkin et al. [61] found impaired mean I.Q. in a group of 28 children studied 10–18 months after hospitalization. Fifteen of 25 children had test results supporting a diagnosis of attention deficit hyperactivity disorder (ADHD).

No commercial vaccine is yet available. In light of the high levels of seropositivity in endemic areas noted above, a vaccine could be targeted to certain geographic areas for children. Meanwhile, reduction of exposure to the breeding areas of the daytime feeding vector is a key protective strategy. Elimination of breeding sites, such as old tires where water can pool, should be undertaken.

3.6 Powassan Virus Encephalitis

Powassan virus (POWV) was first isolated from the brain of a five-year-old boy from Powassan, Ontario, Canada, who died of acute encephalitis in September 1958 [64]. By 1998, 27 cases of POW encephalitis had been reported from the United States and Canada, seven of which were from Ontario and 10 from New York State [65]. With increased arbovirus surveillance occasioned by the incursion of West Nile fever, four cases were discovered in northern New England, of which three were in Maine and one in Vermont

[66]. Transmitted by four tick species in North America, the principal vectors are *Ixodes* ticks. The primary hosts are medium-sized mammals, particularly woodchucks, but household pets can also carry Powassan-infected ticks. Infections have occurred from May to December but peak from June to September [66]. Several strains of POWV exist, including a recently described deer tick virus; they separate into two phylogenetic lineages, each of which can cause disease in humans [67].

Powassan virus appears to produce a multifocal and diffuse gray matter infection of all levels of the central nervous system, so no one presentation is characteristic. The incubation period ranges from 8 to 34 days. Fever may be absent at the beginning but is a consistent finding thereafter. Signs of CNS involvement can be diffuse, with headache, obtundation, and vomiting. Focal or generalized seizures can occur. Motor weakness can be manifest as a monoparesis, hemiparesis, or quadriparesis and be of the upper or lower motor type. Gait ataxia can be found. POWV shares with its TBE counterparts the capacity to produce acute spinal cord clinical disorders, followed by muscular atrophy. A case has been reported that resembled herpes simplex encephalitis [68]. Ophthalmoplegia has been documented [66].

As pleomorphic as the clinical picture is, there are no strongly suggestive laboratory features. There may be an elevated peripheral white cell count including increased neutrophils. The CSF too can reveal a significant component of neutrophils, although mononuclear cells usually predominate, with some elevation of CSF protein. EEGs may show diffuse or lateralized slowing of mild or marked degree. Neuroimaging, CT scanning, or MRI has not been diagnostically useful [65,66]. Virus isolation has been achieved at autopsy [64,67] but not in life. Diagnosis is dependent on antibody changes, including virus specific IgM and neutralizing antibody in CSF and serum. Positive identification has been achieved in this way as early as 3 days after hospitalization [66]. No antiviral therapy is available. Acute care may require ICU support of respiration and therapy of seizures, which can evolve to status epilepticus. Prognosis includes mortality of 10–15% [66] with sequelae in over half of the survivors [65]. No vaccine is available, so the only means of protection is avoidance of the vector and small and medium-sized mammals such as woodchucks and skunks [66].

3.7 Colorado Tick Fever

Colorado tick fever (CTF) is usually a self-limiting febrile illness, but it can be complicated by neurological involvement [68a]. Caused by a virus in the *Coltivirus* genus of the *Reoviridae* family, it is transmitted by *Dermacentor andersoni* ticks [68b]. Often found in persons who work or have recreation in the Rocky Mountains, it has for years been confused with Rocky Mountain spotted fever. However, CTF is often distinguished by a biphasic, or saddlebacked, fever pattern and leukopenia. Rash, although found occasionally, is infrequent. The onset is abrupt, with fever, headache, and muscle aches. Duration of illness is 7–10 days, punctuated often by a 2–3 day afebrile period midway in the course. Neurological complications include meningitis, meningoencephalitis, and encephalitis [68a,68c]. Virus can be isolated from the blood throughout the febrile course.

4 INTERNATIONAL ARBOVIRAL ENCEPHALIDITES

Japanese encephalitis is undoubtedly the most widespread destructive arbovirus encephalitis, afflicting tens of thousands of people annually in Asia. However, we start this section with Venezuelan equine encephalitis, which can also afflict tens of thousands of people as well as horses in South and Central America and Mexico, but in outbreaks separated by several years or even decades. A very large number of people are afflicted by the Central European, or milder, form of tick-borne encephalitis, and an unknown number in the former Soviet Union by the severe Far Eastern form of Russian spring-summer tick-borne encephalitis. We next consider Murray Valley encephalitis of Australia, before concluding with a consideration of Rift Valley fever, which has recently moved into the Arabian Peninsula from Africa.

4.1 Venezuelan Equine Encephalitis

First isolated from the brains of horses during an epizootic in Venezuela in 1936, Venezuelan equine encephalitis virus (VEEV) was the third equine encephalitis virus to be identified [69,70]. However, outbreaks had been recorded since the 1920s. Since that time, VEEV disease has presented dramatic contrasts: it can be manifested as a brief dengue-like illness or as a severe encephalitis; it has seasons in which it produces massive epizootics and epidemics in which tens of thousands of horses and people become ill followed by years in which no epizootic/epidemic activity is found [71].

There are seven or eight sets or subsets of antigenically related VEEV that produce two patterns of epidemiological behavior [12]. The epizootic pattern is associated with large outbreaks in horses and humans followed by years of apparent inactivity. In contrast, the sylvatic pattern is associated with endemic virus infection in forests to which humans are exposed sporadically and in which equine infection is rare.

As might be expected, the sylvatic and epizootic patterns not only have different ecological and clinical behaviors but are also associated with different hosts and vectors (Table 7). Sylvatic varieties are sustained in a cycle between birds and small mammals and the *Culex melaconion* mosquito. Transmission to humans is by *Aedes taeniorhynchus* as well as *Culex melaconion* mosquitoes. Epizootic varieties have numerous mosquito vectors and infect horses, donkeys, and burros.

A massive epizootic/epidemic of IC VEEV in Venezuela and Columbia in 1995 was estimated to have involved 75,000–100,000 people [71]. It moved with great rapidity and was brought under control by interdiction of animal transport, vector control by spraying

Table 7 Venezuelan Equine Encephalitis

	Sylvatic cycle	Epizootic cycle
Behavior	Endemic, sporadic	Epizootic, epidemic
Varieties and serotypes	ID, IE, II–VI	IAB, IC
Location	South and Central America and Florida Everglades	South and Central America, reaching South Texas
Vectors	To humans: *Aedes taeniorhynchus* To birds, small mammals, and humans: *Culex melaconion*	Numerous species
Animal hosts	Birds and small mammals	Horses, donkeys, and mules
Season	Year-round in tropical and subtropical sites	Rainy season

Source: Ref. 12.

of insecticides, and vaccination of horses. The 1995 outbreak bore a striking resemblance to an outbreak in 1962–1964: the same IC variety, the same geographic region, and both outbreaks followed unusually heavy rains that allowed multiplication of the vector. The authors concluded that the epizootic strain may have emerged from an enzootic cycle. Human–mosquito–human infection may have occurred, in light of the rapid spread of the infection within communities, and human-to-human spread might also have occurred, resulting from the presence of virus in the throat.

The clinical manifestations of the IC VEEV outbreak in 1995 consisted of fever, chills, headache, myalgia, prostration, and vomiting [71]. The illness lasted about 3–4 days. Complications included seizures, abortions, and stillborn births. The case fatality rate was 0.7. Reporting from the La Guajira state in Colombia on the 1995 epidemic, Rivas et al. [72] found that neurological symptoms started an average of 4 days after the onset of the illness. [72] Seizures were common among the neurological complications, hemiparesis or behavioral changes were each seen in 11%, and stupor and coma in 3%.

A low inapparent to apparent infection ratio of 10:1 is reported [12], as is a brief incubation period of 1–4 days [60–73]. Neurological complications are more common in children. In the southern Texas series reported by Bowen et al. [73], seizures, temporary paralysis, or coma were seen in almost one-fourth of the patients under 17 years of age. The percentage of patients with long-term neurological residues appears to be low, but that observation is not based on a large systematic follow-up study. The case fatality rate among those with encephalitis is noted to be between 10% and 25%, and overall it is 1% or less [9,12].

Laboratory studies are nonspecific. Leukopenia is commonly found [73]. The CSF contains a few to hundreds of lymphocytes, with moderate depression of glucose and elevation of protein [73]. There is a paucity of neurophysiological and neuroimaging data. A case report of VEEV meningitis-encephalitis described slowing and disorganization on EEG that cleared after discharge and a radioisotope brain scan that showed no abnormalities [74]. Viral diagnoses can employ virus isolation from blood in the early stages of the illness. Serological evidence includes IgM capture ELISA and diagnostic fourfold changes in hemagglutination inhibition titers [71].

Because the acute systemic illness is usually brief and self-limited, general supportive measures are necessary. No specific antiviral compounds are available. Neurological complications should also be managed with supportive measures. Although some cases have suggested a delayed onset of CNS complications, there are insufficient data to support an immune-mediated process and immune intervention such as steroids. In addition, these could be contraindicated in light of the leukopenia commonly observed.

Prevention relies on vaccination of horses, insecticide spraying, and prohibition of transport of horses, donkeys, and mules. A vaccine that is not licensed for general use should be given to exposed laboratory workers because of numerous reports of laboratory infection.

4.2 Tick-Borne Encephalitis

From 1932, various theories of causation were proposed for an acute CNS disease with a high mortality found in the eastern regions of the then Soviet Union [75]. Originally classified as toxic influenza, an etiology of Japanese summer encephalitis was next entertained. In 1937, a special expedition to study the disease successfully isolated the agent, defined the vector, documented the neuropathology, and described the clinical manifesta-

Table 8 Selected Tick-Borne Encephalitis Serocomplex Agents

Agent	Locations	Reference
Tick-borne encephalitis: Central European subtype, Russian spring-summer subtype	See Table 9.	See text.
Powassan virus	Eastern U.S., Canada, former Soviet Union	See text.
Louping ill	Scotland, England, Wales, Ireland	77
Kyasanur Forest disease	India	78

tions [75]. Over the following decades it became apparent that a related illness with a lower mortality was found throughout much of Europe. Two antigenic variants of tick-borne encephalitis (TBE) virus have been found—a European subtype, usually called Central European encephalitis virus, and a Far Eastern subtype, often called Russian spring-summer encephalitis virus [31,76]. The viruses are part of a tick-borne encephalitis virus group of the *Flaviviridae* family. Selected members of this serocomplex are noted in Table 8.

Tick-borne encephalitis is an endemic disease with the distribution of cases reflecting the density of tick infestation. Hundreds of cases are observed annually in European countries and 5,000–10,000 cases per year are reported from Russia [79]. There are slightly different periods of transmissibility to humans of the two TBE viral subtypes (Table 9). In general this is spring and summer, but the Central European cases extend into the fall. Outdoor work and recreational activities in endemic areas by forestry workers and picnickers, respectively, are associated with transmission. Seroprevalence in endemic areas can range up to 30% [31]. Implementation of vaccine programs has resulted in a remarkable reduction of cases. In Austria, for example, vaccination has been associated with a reduction of the numbers of cases from several hundred per year to under 100 cases annually [76]. Ingestion of raw (unpasteurized) goat's milk has also been a means of human infection, with health education reducing this risk [31].

Table 9 Tick-Borne Encephalitis (TBE): Central European and Russian Spring-Summer

	Central European encephalitis	Russian spring–summer encephalitis
Locations	Europe, former Soviet Union	Far eastern former Soviet Union, Mongolia (China)
Season	Late spring to autumn	Spring and early summer
Vector	*Ixodes ricinus*	*Ixodes persulcatus*
Animal hosts	Mammals, livestock, rodents	Mammals, livestock, rodents
Transmission	Tick bite, raw goat's milk	Tick bite, raw goat's milk
Facilitating factors	Incursion into vector habitats, heavy vector infestation	Incursion into vector habitats, heavy vector infestation

Source: Refs. 11 and 31.

Osetowska [80] summarized neuropathological data indicating that there are only slight neuropathological differences between the subtypes of TBE virus. Gray matter lesions are found at several levels of the nervous system, including the cerebral cortex, basal nuclei, medulla, cerebellum, and spinal cord. The white matter appears to be spared. Inflammation consists of perivascular infiltrates, microglial nodules, and microglia in areas of spongiosis and necrosis. The targeting of gray matter and the anatomic distribution of lesions make the distinction from polio difficult at times. There is a clinical corollary in that TBE can present as a polio-like illness [81]. It also results in the commonly observed atrophy and weakness of the shoulder girdle, neck, and arms as sequelae to TBE. Multifocal targeting of gray matter, ''spotty accumulation'' [80], may underlie several of the clinical observations, including intractable focal seizures (Kozhevnikov's epilepsy). Chronic forms of TBE have been documented [82].

The incubation period from exposure to clinical expression of disease is variably reported but ranges from 4 to 20 days [11]. A large prospective study from Sweden documented the clinical course and outcomes of the Central European viral subtype [83]. Meningeal (55%) and encephalitic (45%) forms were observed. Eighty-seven percent of the cases had a biphasic course. The first phase, which was characterized by fever and headache, lasted from 1 to 8 days, and the latency interval after the first phase before neurological involvement had a range of 1–33 days with a median of 8 days. Ataxia, altered consciousness, and irritability were common findings in the neurological phase. Spinal and/or radicular findings were observed in slightly over 10% of cases. Median duration of hospitalization in the acute phase was about 1 week. No mortality was observed in this study. Clinical observations may vary from region to region. For example, a report from eastern Croatia found a predominance of the monophasic course; a lower percentage, approximately 10%, of the meningitic form; and a mortality of 3.3% [84]. There are relatively few clinical data in the recent English language literature on the Russian spring-summer viral subtype. Mortality of 20–30%, a prolonged convalescence, and residual disabilities of 20–30% were noted by Silber and Soloviev in their 1946 treatise [75].

Follow-up evaluations at 1 year in the large Swedish prospective study of Günther et al. [83] found sequelae in 33 of the 80 patients evaluated following meningoencephalitis. The most common manifestations were impaired memory and concentration, ataxia, dysphasia, and headache. Tetraparesis and bilateral shoulder paralysis were also observed at 1 year. In those patients in whom spinal nerve paralysis resolved, it took between 14 and 197 days (median of 96.5 days). Follow-up of the eastern Croatian group by Anic et al. [84] found that convalescence lasted up to 2 months with complaints of headache, insomnia, vertigo, tinnitis and hands tremor. Residual abnormalities apparently occur with a higher frequency in the Far Eastern subtype [11]. These include atrophy of neck, shoulders, and arms and occasional cases of continuous focal seizures (Kozhevnikov's epilepsy).

Neither the prospective Swedish study [83] nor the eastern Croatian study [84] reported neuroimaging data. EEGs performed during the acute phase of the illness revealed diffuse nonspecific changes in slightly under two thirds of the patients [84]. Pleocytosis in the spinal fluid has been found to be moderate, with an increasing majority of mononuclear cells over the first 9 days of the illness [83]. CSF albumin climbed over the course of 6 weeks, reflecting damage to the blood brain barrier. Two thirds of the patients in the Swedish study demonstrated intrathecal IgG synthesis at the sixth week of illness [83]. None of the CSF findings correlated with the clinical outcome at 6 weeks or 1 year.

Specific viral diagnosis can be made by isolating virus from the blood during the first phase of the illness. More commonly, however, a presumptive diagnosis can be made with an IgM capture ELISA [85] and confirmed by complement fixing and/or increases in IgG ELISA antibody between acute and convalescent serum samples [83].

Acute treatment strategies are supportive in nature, because as there are no specific antiviral compounds. Most intense support will be necessary for the Far Eastern (Russian spring-summer) viral subtype disease. The convalescent period is more prolonged than in non-TBE cases [83]. The development of neck, shoulder, and limb atrophy would call for physical rehabilitative intervention. Vaccination has been highly successful where systematically employed [76]. It is unclear whether vector control with pesticides has materially reduced the incidence of infection.

4.3 Japanese Encephalitis

As recounted by Clarke and Casals [86], recognition of a disease compatible with that caused by JEV reaches back to 1871 in Japan. It was originally named Japanese B encephalitis to distinguish it from type A encephalitis, or von Economo's disease. Determination of the clinical features and isolation of a filterable agent that was pathogenic for rabbits occurred as a result of an epidemic in Japan in 1924 [86]. In the following decade, suspicion that the disease was mosquito-borne was confirmed by isolation of the virus from culicine species.

Although originally identified in Japan, JEV blankets an enormous swath of Asia, expanding its territory from Pakistan in the west to Japan in the east and from China in the north to Australia in the south [87] (Table 10). It encompasses some of the world's most populous countries, including China and India, and is endemic to areas in which nearly 3 billion people live [88]. Its annual incidence in these areas ranges from 10 to 100 per 100,000, with an estimated 50,000 cases annually worldwide. It is among the most damaging of the encephalitides, killing an estimated 10,000 people per year and producing a high percentage of neurological, psychiatric, and cognitive abnormalities in survivors. JEV disease–associated mortality and sequelae are most common in children, leaving in their wake an enormous human loss and public health burden.

The ecological and public health factors that bring about the infection are well characterized (Table 10). Transmitted by culicine mosquitoes, principally *Culex tritaeniorhychus*, certain agricultural practices and climatic conditions such as paddies, irrigation, and heavy rainy seasons favor mosquito multiplication. Wading birds serve as hosts, but domestic pigs are instrumental in human infection. Human and pig habitats often intermin-

Table 10 Japanese Encephalitis

Locations	Expanding territory in Asia and Oceania; principally China, Japan, Southeast Asia, and India.
Season	May to October in temperate regions, following the rainy season in tropical areas, and year-round where irrigation supports mosquito breeding.
Vector	Several Culicine species, but particularly *Culex tritaeniorhychus*.
Hosts	Wading birds and pigs are most important.
Facilitating factors	Still water farming practices, e.g., paddies; physical proximity of human habitation with domestic pigs.

Source: Refs. 31 and 88.

gle. High levels of viremia, in the absence of visible illness in domestic pigs, facilitates transmission to humans by the vector. Centralized pig production, modern farming practices, mosquito eradication, and widespread immunization have markedly reduced the incidence of JEV disease in certain countries such as Japan and South Korea.

The pathogenesis of JEV reflects intracellular infection of neurons. Desai and coworkers [87] found persistence of viral antigen in CSF in the presence of immune complexes and IgM and neutralizing antibodies. This suggested to the investigators that the intracellular spread of the virus precluded limitation of infection in the extracellular space by humoral immunity [87]. This factor and infection of white cells may underlie recurrence or persistence of infection in some cases [31]. Viral antigen was found in the thalamus, hippocampus, substantia nigra, and medulla oblongata [89]. It has been speculated that the anatomical localization of viral antigen explained the acute impairment of consciousness and respiration and the long-term sequela of cognitive impairment. Of note, 38% of autopsied cases in the series also demonstrated neurocysticercosis [89]. The role of the dual infection is unclear.

There has been a very wide range in the estimated apparent to inapparent infection ratio. A figure of 1:270 was reported from an endemic region in India after a 3-year prospective study [90]. The onset of overt illness follows an incubation period of 4–14 days. The onset of the first phase of the illness includes fever and chills, headache, and GI symptoms, particularly in children. An undetermined percentage of cases have an uneventful recovery at this stage [88]. Involvement of the CNS evolves with clouding of consciousness, seizures, particularly in children, motor loss, and abnormal movements. Meningeal signs can be observed, particularly in adults. In a study of U.S. military personnel in Korea in 1950, Dickerson et al. [91] characterized the following findings as pathognomonic: altered sensorium, mask-like facies, thick retarded speech, coarse ocular tremor, symmetrical paresis, and hyperactive deep tendon reflexes. The acute encephalitic phase lasts 1–2 weeks, with death, should it occur, frequently in the third to eighth days [92]. Estimates of fatal outcome range from 10% to 40%, with children under 10 years of age the most susceptible. Other indicators of poor prognosis include depth and duration of impaired consciousness, reflex changes, and severity of CSF and EEG abnormalities [87,93]. Ravi et al. [87] observed that only one-third of patients fully recover from the encephalitis. The gamut of sequelae includes motor paresis, movement disorders, behavioral abnormalities, mental retardation, and learning disability.

Although various patterns of EEG abnormalities have been described, none are pathognomonic. Kalita and Misra [94] described patterns of diffuse delta waves, delta waves with spikes, and alpha pattern coma. In contrast, neuroimaging studies have been more specific [93,94]. CT scans show thalamic abnormalities in most patients and abnormaities of the basal ganglia, midbrain, and pons in some patients [93]. MRI has greater sensitivity and shows additional areas. Pradhan, et al. [95] correlated involvement of the substantia nigra on MRI with Parkinsonian sequelae. The CSF white cell count is usually in the range of 20–400 cells/mm^3, with an average of 185 [91]. Polymorphonuclear cells can be prominent in the first 4 days, but their number drops off thereafter. Protein elevation is moderate. Ravi et al. [87] noted that a normal CSF cell count can be observed in some patients. A peripheral leukocytosis of 10,000–19,000 can be found with a predominance of polymorphonuclear cells. An elevated sedimentation rate is also described as uniform [91].

The presumptive diagnostic test of choice is IgM capture ELISA, with very high sensitivity from 1 to 2 weeks after onset in serum and CSF [96]. However, cross-reactions with other serologically closely related viruses require confirmation by diagnostic fourfold rises between acute and convalescent serum samples tested by other assays such as neutralization or complement fixation. Virus can be isolated on some occasions, particularly early in the illness, from CSF and serum. However, the time required to achieve and characterize an isolate renders isolation technologies much less useful in acute disease management than for epidemiological studies.

Management in the first 8 days is crucial to survival. There are no proven specific antiviral compounds for JEV. Controlled study of interferon alpha-2A revealed no benefit [97]. Although high-dose steroids are commonly considered in the management of edema associated with encephalitis, Hoke et al. [98] failed to find significant benefit in the treatment of encephalitis caused by JEV. In an adult population in which mortality was 10%, Dickerson, et al. [91] emphasized the vital role of an adequate airway. Ravi, et al. [87] suggested that since cerebral edema may be the most important cause of death, measures to control raised intracranial pressure are recommended. Fluid and electrolyte management, often complicated by inappropriate ADH secretion, is key. Control of recurrent seizures and treatment of superimposed pulmonary infections are also required.

Preventive measures of proven efficacy include national vaccination programs such as those in Japan and South Korea [99]. Other measures include mosquito eradication programs, modern agricultural methods to reduce still water breeding grounds for the vectors, insecticides, and modern centralized animal husbandry to remove pigs from close contact with human living quarters. Insecticide-impregnated bed nets have been advocated and have proven protective as an emergency response to outbreaks of JEV disease [100]. Although vaccination has been broadly successful in national programs, vaccination is not routinely recommended for travelers because of vaccine associated adverse events [31]. It is recommended for travelers only if there is to be prolonged exposure in endemic areas.

4.4 Murray Valley Encephalitis

In the years between 1917 and 1925, Australia experienced several outbreaks of a highly lethal encephalitis [6]. It was called Australian X disease to distinguish it from von Economo's disease, which was raging worldwide during those years. Virus was isolated but lost in subsequent years [11]. In 1951 the disease reappeared in the Murray River and Darling River areas and was called Murray Valley encephalitis. Virus was isolated and characterized as a group B arbovirus [6]. Because the isolate of Australian X disease had been lost, identity of the agents could not be achieved, but clinical and epidemiological features led to the conclusion that they were in fact identical [101]. Following an outbreak in 1974, the illness was called Australian encephalitis [6]. However, Australian encephalitis also includes cases due to Kunjin virus [102], a flavivirus related to Murray Valley encephalitis virus (MVEV) but most closely related to West Nile virus.

Surveillance by seroconversion in sentinel chicken flocks and virus isolation from trapped mosquitoes have been used to study the appearance of virus activity [103]. The principal cycle is between the *Culex annulirostris* mosquito and waterfowl (Table 11). Proliferation of the vector and extended migration of infected waterfowl occur following heavy wet season rainfall and flooding. Broom et al. [103] provide data in support of reintroduction of virus into arid regions by both viremic vertebrate hosts and vertical transmission in mosquito eggs.

Table 11 Murray Valley Encephalitis

Locations	Australia, Papua New Guinea
Season	Summer, January to May
Vector	*Culex annulirostris*
Hosts	Waterbirds
Facilitating factors	Heavy rainfall and flooding, facilitating movement of waterbirds and breeding of vectors

Source: Ref. 102.

The pathogenesis of the encephalitis reflects direct acute attack on neurons themselves [104]. Evolving lesions may become necrotic and can be found in white matter, thalamus, and cerebellum. The histopathology reveals a microglial response to be the first response and prolonged and to be associated with histiocytic activity, lymphocytic infiltration, and astrocytic proliferation. In the chronic stage, widespread necrosis may be found.

It has been estimated that only one in 1000 infections is manifested by clinical illness [102]. The incubation period is judged to be between 1 and 4 weeks [105]. The prodrome consists of fever, headache, nausea, and vomiting for 2–5 days followed by signs of encephalitis [105]. Respiratory tract signs have also been reported in the prodromal stage [106]. Encephalitis is manifest as impaired consciousness, from obtundation to coma; altered mentation, from confusion to delirium; seizures; neck stiffness; tremor; motor weakness; and ataxic gait [105,106]. A plethora of neurological findings, including brainstem and spinal cord findings, make it exceptionally difficult to distinguish disease caused by MVEV from other types of encephalitis on clinical grounds alone [105]. Of great management importance, however, is respiratory failure. The course of the acute illness is approximately 2 weeks. Of 18 cases of encephalitis caused by MVEV reported by Mackenzie et al. [106], 10 were classified as moderate, of which four had residual, five as severe, of which all had residual, and three that were fatal. Burrow et al. [107] reported mortality of 31% and residual neurological disability in 25%.

Electroencephalographic studies do not reveal etiologically suggestive features. Diffuse changes compatible with encephalitis and the absence of focal abnormalities have been reported from several studies [105–107]. Similarly, CAT scans of the brain are either normal or show diffuse abnormalities [106,107]. The CSF, too, is without characteristic features, described as a ''non purulent meningitis'' [105], with a mononuclear pleocytosis.

For diagnostic purposes, virus isolation has not been successful in live patients; hence serological changes have been the cornerstone of viral diagnosis. Care must be taken to exclude closely related viruses [31]. IgM antibody specific for MVEV has been used to exclude cross-reacting viruses [108].

Management follows procedures for other types of severe encephalitis in which no specific antiviral medication is available. Bennett [105] emphasized the importance of artificial respiration. Management of seizures, raised intracranial pressure, prevention of bacterial superinfection, maintenance of fluid and electrolyte balance, frequent repositioning, and passive range-of-motion exercises are each indicated.

Prevention is dependent on serological surveillance of sentinel chicken flocks and virus isolation in trapped mosquitoes to alert exposed populations and initiate vector control. No vaccine is available.

4.5 Rift Valley Fever

Starting in July, 1930, Daubney, Hudson and colleagues investigated an outbreak of illness causing the deaths of newborn lambs in the Rift Valley in Kenya [109]. The report of those investigations the following year in *The Journal of Pathology and Bacteriology* was remarkably complete. The disease, which was found to affect sheep, young lambs, cattle, goats, and humans, was thought to have been present in the Rift Valley for some years. The researchers found it to be transmitted by a filterable agent that they grouped with dengue and yellow fever viruses. The disease, which caused hepatic necrosis in domesticated animals, was also responsible for livestock abortions. Transmission by mosquitoes was strongly suspected. Remarkably, all four Europeans engaged in the investigation developed a febrile illness that was characterized as dengue-like. It was associated with malaise, rigors, headache, fever, and joint pains and cleared within 4 days. One of the individuals had a biphasic illness in which the second period was associated with ''headache and defective vision for some weeks afterward.'' This is suggestive of either the ocular or CNS complication that were later described in humans [109].

Virus isolation from human blood was achieved during a South African epizootic in 1950–1951, and a serous retinopathy was described (reviewed in Ref. 110). Van Velden et al. [111] reported an enlarged spectrum of human disease during a 1975 epizootic in South Africa. These included involvement of the central nervous system and a hemorrhagic form. Transmission occurred during handling of tissues or carcasses of dead animals, the incubation period was 4–6 days, and a biphasic pattern was observed. The illness was confined to sub-Saharan Africa until a 1977 epizootic in Egypt [112]. In the fall of 2000 an outbreak of Rift Valley fever viral disease in humans in association with animal deaths was observed for the first time outside of Africa on the Arabian Peninsula at the southwestern border of Saudi Arabia and Yemen [113]. Concern arises that the enzootic/epidemic disease may spread to still other areas in Europe and Asia [114].

The epidemiology of Rift Valley fever virus (RVFV) is remarkable for the wide spectrum of vertebrate hosts and mosquito vectors and also for the several mechanisms by which humans can be infected (Table 12). These include the bite of an infected vector; handling of carcasses, tissues, or abortuses of infected animals; and the aerosol route. Torrentially flooded agricultural plains and irrigation canals promote breeding sites for vectors. RVFV is one of the infectious agents for which satellite imaging is useful to determine locations that would support viral transmission. Several such areas were identified by aerial surveys and satellite data in the 2000 outbreak in Yemen [115].

Table 12 Rift Valley Fever

Locations	Africa and Arabian Peninsula (Saudi Arabia and Yemen)
Season	Variable, dependent on region and mechanism of transmission
Vector	Numerous species of mosquitoes
Transmission to humans	Vector, aerosol, handling of infected animal carcasses or abortuses
Animal hosts	Numerous, including domestic sheep, cattle, goats, and camels and many other amplifying hosts
Facilitating factors	Heavy rainfall, flooded agricultural fields, flood plains, dam construction

Source: Refs. 9 and 114.

Whereas certain other arboviral replication systems are restricted with respect to vector and host, RVFV is promiscuous. Numerous mosquito species have been implicated in transmission and as potential interepidemic reservoirs [116]. Similarly, although deaths in young domesticated animals such as sheep and goats and animal abortions draw attention to epizootics, a wide range of vertebrates serve as hosts and vehicles for dissemination. Hoogstraal et al. [9] speculated that camels, long used for slaughter, draft animals, or mounts, may have been the mechanism to introduce RVFV into Egypt. So promiscuous is RVFV that editorial comment on the Arabian Peninsula outbreak of 2000 noted that ''this virus may be able to establish itself almost anywhere in the world based on the availability of potentially permissive vectors and animal reservoirs'' [116]. Although currently restricted to Africa and the Arabian Peninsula, the potential for widespread dissemination exists.

The clinical forms and the veterinary pathology [109] suggest a wide tissue tropism. However, the human neuropathology has not yet been systematically reported. Van Velden et al. [111] did include a brief description of neuropathological findings in three of their cases from South Africa. The findings included focal areas of necrosis in one case and perivascular cuffing in two other cases. The relative infrequency of meningoencephalitis (under 1%) and the delay in onset following systemic signs of infection (1–3 weeks) [114] in addition to the paucity of systematic description of the human neuropathology leave open the mechanism of injury to the CNS.

Transmission of infection to humans is by vector, exposure to sick or dead animals or abortuses, or the aerosol route. The incubation period is brief; 2–6 days. The onset of illness in humans is sudden, with malaise, fever, rigor, headache, myalgia, and GI symptoms of anorexia, nausea, and vomiting [110]. The period of systemic illness is brief, with defervescence of fever within 4 days and full recovery within 2 weeks [110]. In previous epidemics, the vast majority of infections were of the self limited and inocuous type, with mortality under 1%. Death, when it occurs, is most commonly associated with the hemorrhagic form. In the Saudi Arabian outbreak of 2000, 17% of hospitalized patients suspected of having severe RVFV disease died. Cases that were originally identified as severe illness with hemorrhagic fever were associated with acute renal failure [116].

Ocular and neurological forms are infrequent, usually under 2% and 1%, respectively, and death is uncommon [114]. Ocular complications include bilateral macular, paramacular, and extramacular vasculitis, retinitis, and vascular occlusion [117]. In the cases reported by Siam et al. [117], permanent loss of vision occurred in 40–50% of patients after resolution of the ocular lesions. Neurological symptoms in humans have been recognized from the earliest report of disease caused by RVFV [109] but, as noted, are infrequent and hitherto not commonly lethal. Of 348 patients with RVFV disease observed during an epidemic in southern Mauritania, 17 had encephalitis [118]. Van Velden [111] reported that neurological symptoms followed resolution of the acute phase by 3–6 days. Findings included meningeal irritation, confusion, impaired consciousness, bruxism, and hypersalivation. Less commonly found abnormalities (one patient each) included visual hallucinations, locked-in syndrome, and upper limb choreiform movements. The clinical course, management challenges, and neurological and psychiatric sequelae have not been systematically reported. Of five patients with meningoencephalitis described by Laughlin et al. [110] one who had exhibited decerebrate posturing died 2 months after onset of the illness and two had residual hemiparesis. In light of the generally low mortality reported for RVFV-produced meningoencephalitis, this early experience may reflect a selection bias.

The CSF has been reported to show a pleocytosis, predominantly lymphocytic, from 20 to 600 cells/mm^3 [110]. CSF protein levels have been reported as normal [110] or elevated [111]. EEG and neuroimaging studies have not been systematically reported.

The diagnosis should be suspected in an appropriate region in the presence of an epizootic in sheep and goats. For purposes of defining the Saudi Arabian epidemic of 2000, the CDC used antigen detection and IgM antibody by ELISA, PCR, virus isolation, and immunohistochemistry [113]. ELISA IgG assays are as sensitive as or more sensitive than traditional serodiagnosis such as hemagglutination inhibition, complement fixation; and plaque reduction neutralization assays [119].

Intravenous ribavirin has been considered for treatment of RVFV infections [113]. Management of neurological complications has not been defined. Vaccines have not been licensed for human use but have been investigated in high risk individuals [114]. Prediction of epizootics and epidemics can make use of telesatellite surveillance for permissive ecological features. Prevention and interruption of outbreaks depends on livestock vaccination, control of animal movement, vector control, insecticide impregnated bednets, and protective gloves and clothing when working with sick or dead animals and abortuses during an epizootic. Because of the potential for infection by direct human-to-human contact, universal precautions and barrier nursing techniques should be employed by healthcare workers [114].

5 CONCLUDING REMARKS: REPRISE, RESOURCES, AND NEEDS

Arboviruses are agents of location, season, particular ecological conditions, and epizootics. These factors are of considerable benefit to the clinician in narrowing the diagnostic possibilities in evaluating cases of encephalitis. In some instances, such as La Crosse virus encephalitis in the United States or tick-borne encephalitis in Central Europe, the diseases are endemic and recur with seasonal regularity in particular regions. In other instances, the diseases may appear at varying intervals in epidemic proportions. However, clues are found in the locations of the outbreaks such as epidemics caused by Venezuelan equine encephalitis virus originating in the Guajira Peninsula shared by northern Colombia and Venezuela. The association with epizootics is diagnostically useful. The equine encephalitdes spring to mind, as does the vulnerability of domestic livestock to Rift Valley fever virus and of crows to West Nile virus in the United States. Heavy rains and flooding or an intense tropical rainy season are the necessary ecological precedents for vector proliferation for many arboviral agents including Murray Valley encephalitis or Japanese encephalitis viruses.

In addition to the four factors of location, season, ecological conditions, and epizootics, the clinician should stay informed as to current and anticipated virological activity. The principal means is regular contact with the state department of health laboratories or, in larger medical centers, with the diagnostic virology laboratories. Other resources (Table 13) include the *ProMED Digest*, a daily web-based independent surveillance service, *Morbidity and Mortality Weekly Report* (*MMWR*) of the CDC, the *Weekly Epidemiological Record* (*WER*) of WHO, and *Emerging Infectious Diseases*, a journal published by the CDC.

Despite the capacity of laboratory and public health officials to track infectious diseases, emerging infectious diseases will continue to pose clinical challenges. The 2000 epidemic/epizootic of Rift Valley fever virus in Saudi Arabia and Yemen and the epidemic of West Nile virus encephalitis in southern Russia in 1999 are but two recent examples.

Table 13 Information Resources Concerning Infectious Diseases for the Practicing Clinician

Source	Contact information
ProMED Digest	Online: www.promedmail.org/
Morbidity and Mortality Weekly Report (MMWR)	Online: www.cdc.gov/mmwr/ Subscription for paper copies: Massachusetts Medical Society P.O. Box 9120 Waltham, MA 02454–9120
WHO Weekly Epidemiological Record (WER)	Online: www.who.int/wer/
Emerging Infectious Diseases	Online: www.cdc.gov/eid/ Subscription for paper copies: National Center for Infectious Diseases, Centers for Disease Control and Prevention (CDC) 1600 Clifton Rd. Mailstop D-61, Atlanta, GA 30333

Facilitating factors include intrusion of susceptible hosts into endemic regions, disruption of previously stable ecosystems, modern transportation systems, and shifts in migratory patterns of birds. The widespread ecological shifts and displacements anticipated in the wake of global warming will require international tracking and communication systems. The increased incidence of TBE in association with milder winters and earlier springs in Sweden may be an early example of global warming changing the ecology of disease [120]. Vaccine development, production, and distribution capacities will likely increase in importance. The proven usefulness of vaccines against Japanese encephalitis virus in Japan and Korea and the virus of TBE in Austria and the role of immunization of horses to halt the progress of Venezuelan equine encephalitis virus should convince funding bodies, international agencies, and voluntary organizations of the efficacy of vaccination programs against arboviral infections. Antiviral drug development programs along the lines of therapy trials for encephalitis caused by human herpesvirus 1 and HIV treatment would ideally receive public funding to stimulate pharmacological development.

ACKNOWLEDGMENTS

We are grateful to Drs. M. M. Esiri, D. Mayo, and A. Tselis for critical reading of the manuscript and constructive criticism. Any mistakes or omissions, of course, remain our responsibility. We are also grateful to Ms. Marybeth Fernandez for careful manuscript preparation.

REFERENCES

1. Linthicum, K.J.; Anyamba, A.; Tucker, C.J.; Kelley, P.W.; Myers, M.F.; Peters, C.J. Climate and satellite indicators to forecast Rift Valley fever epidemics in Kenya. Science. 1999, 285, 397–400.
2. Karabatsos, N. International Catalogue of Arboviruses Including Certain Other Viruses of Vertebrates; Am Soc Trop Med Hyg: San Antonio, TX, 1985.

3. Kuno, G.; Chang, G.-J.J.; Tsuchiya, K.R.; Karabatsos, N.; Cropp, C.B. Phylogeny of the genus *Flavivirus*. J Virol. 1998, 72, 73–83.
4. Calisher, C.H.; Beaty, B.J.; Chandler, L.J. Arboviruses. In: Laboratory Diagnosis of Viral Infections, 3rd ed.; Lennette, E.H., Smith, T.F., Eds.; Marcel Dekker: New York, 1999, 305–332.
5. Roehrig, J.T. Arboviruses. In: Clinical Virology Manual, 3rd ed.; Specter, S., Hodinka, R.L., Young, S.A., Eds.; ASM Press: Washington, DC, 2000, 356–373.
6. Mackenzie, J.S.; Broom, A.K. Australian X disease, Murray Valley encephalitis and the French connection. Vet Microbiol. 1995, 46, 79–90.
7. Feemster, R.F. Equine encephalitis in Massachusetts. N Engl J Med. 1957, 257, 701–704.
8. Rappole, J.H.; Derrickson, S.R.; Hubalek, Z. Migratory birds and spread of west Nile virus In the western hemisphere. Emerg Infect Dis. 2000, 6, 319–328.
9. Hoogstraal, H.; Meegan, J.M.; Khalil, G.M. The Rift Valley fever epizootic in Egypt 1977–78. 2. Ecological and entomological studies. Trans Roy Soc Trop Med Hyg. 1979, 73, 624–629.
10. Centers for Disease Control and Prevention. Summary of Notifiable Diseases, United States 1999. MMWR. 2001, 48, 1–101.
11. Booss, J.; Esiri, M.M. Viral Encephalitis. Pathology, Diagnosis and Management; Blackwell Sci: Oxford: UK, 1986, 152–184.
12. Tsai, T.F.; Monath, T.P. Alphaviruses. In Clinical Virology; Richman, D.D., Whitley, R.J., Hayden, F.G., Eds.; Churchill Livingstone: New York, 1997, 1217–1255.
13. Farber, S.; Hill, A.; Connerly, M.L.; Dingle, J.H. Encephalitis in infants and children caused by the virus of the eastern variety of equine encephalitis. JAMA. 1940, 114, 1725–1731.
14. Deresiewicz, R.L.; Thaler, S.J.; Hsu, L.; Zamani, A.A. Clinical and neuroradiographic manifestations of eastern equine encephalitis. N Engl J Med. 1997, 336, 1867–1874.
15. Kim, J.H.; Booss, J.; Manuelidis, E.E.; Duncan, C.C. Human eastern equine encephalitis. Electron microscopic study of a brain biopsy. Am J Clin Pathol. 1985, 84, 223–227.
16. Wesselhoeft, C.; Smith, E.C.; Branch, C.F. Human encephalitis. Eight fatal cases, with four due to the virus of equine encephalomyelitis. JAMA. 1938, 111, 1735–1741.
17. Eklund, C.M.; Blumstein, A. The relation of human encephalitis to encephalomyelitis in horses. JAMA. 1938, 111, 1734–1735.
18. Eklund, C.M. Human encephalitis of the western equine type in Minnesota in 1941. Clinical and epidemiological study of serologically positive cases. Am J Hyg. 1946, 43, 171–193.
19. Rozdilsky, B.; Robertson, H.E.; Chorney, J. Western encephalitis: report of eight fatal cases. Saskatchewan epidemic, 1965. Can Med Assoc J. 1968, 98, 79–86.
20. Waters, J.R.V. An epidemic of western encephalomyelitis in humans—Manitoba. Can J Pub Health. 1976, 67(suppl 1), 28–32.
21. Leech, R.W.; Harris, J.C.; Johnson, R.M. 1975 encephalitis epidemic in North Dakota and western Minnesota. An epidemiologic, clinical, and neuropathologic study. Minn Med. 1981, 545–548.
22. McGowan, J.E.; Bryan, J.A.; Gregg, M.B. Surveillance of arboviral encephalitis in the United States, 1955–1971. Am J Epidemiol. 1973, 97, 199–207.
23. Mulder, D.W.; Parrott, M.; Thaler, M. Sequelae of western equine encephalitis. Neurology. 1951, 1, 318–327.
24. Earnest, M.P.; Goolishian, H.A.; Calverley, J.R.; Hayes, R.O.; Hill, H.R. Neurologic, intellectual, and psychologic sequelae following western encephalitis. A follow-up study of 35 cases. Neurology. 1971, 21, 969–974.
25. Finley, K.H.; Fitzgerald, L.H.; Richter, R.W.; Riggs, N.; Shelton, J.T. Western encephalitis and cerebral ontogenesis. Arch Neurol. 1967, 16, 140–164.
26. Baker, A.B.; Noran, H.H. Western variety of equine encephalitis in man. Arch Neurol Psych. 1942, 47, 565–587.
27. Bia, F.J.; Thornton, G.F.; Main, A.J.; Fong, C.K.Y.; Hsiung, G.D. Western equine encephalitis mimicking herpes simplex encephalitis. JAMA. 1980, 244, 367–369.

28. Kokernot, R.H.; Shinefield, H.R.; Longshore, W.A., Jr. The 1952 outbreak of encephalitis in California. Differential diagnosis. Calif Med. 1953, 79, 73–77.
29. Public Health Bull 214, Report on St. Louis Outbreak of Encephalitis; US Gov Printing Office: Washington, DC, 1935.
30. Monath, T.P. Epidemiology. In St. Louis Encephalitis; Monath, T.P., Ed.; Am Publ Health Assoc: Washington, DC, 1980, 239–312.
31. Monath, T.P.; Tsai, T.F. Flaviviruses. In Clinical Virology; Richman, D.D., Whitley, R.J., Hayden, F.G., Eds.; Churchill Livingstone: New York, 1997, 1133–1185.
32. Southern, P.M., Jr; Smith, J.W.; Luby, J.P.; Barnett, J.A.; Sanford, J.P. Clinical and laboratory features of epidemic St. Louis encephalitis. Ann Intern Med. 1969, 71, 681–689.
33. Centers for Disease Control and Prevention. St. Louis encephalitis—Baytown and Houston, Texas. MMWR. 1986, 35, 693–695.
34. Wasay, M.; Diaz-Arrastia, R.; Suss, R.A.; Kojan, S.; Haq, A.; Burns, D.; Van Ness, P. St. Louis encephalitis. A review of 11 cases in a 1995 Dallas, Tex. epidemic. Arch Neurol. 2000, 57, 114–118.
35. Nash, D.; Mostashari, F.; Fine, A.; Miller, J.; O'Leary, D.; Murray, K.; Huang, A.; Rosenberg, A.; Greenberg, A.; Sherman, M.; Wong, S. M Layton for the 1999 West Nile Outbreak Response Working Group. The outbreak of West Nile virus infection in the New York City area in 1999. N Engl J Med. 2001, 344, 1807–1814.
36. Lanciotti, R.S.; Roehrig, J.T.; Deubel, V.; Smith, J.; Parker, M.; Steele, K.; Crise, B.; Volpe, K.E.; Crabtree, M.B.; Scherret, J.H.; Hall, R.A.; MacKenzie, J.S.; Cropp, C.B.; Panigrahy, B.; Ostlund, E.; Schmitt, B.; Malkinson, M.; Banet, C.; Weissman, J.; Komar, N.; Savage, H.M.; Stone, W.; McNamara, T.; Gubler, D.J. Origin of the West Nile virus responsible for an outbreak of encephalitis in the northeastern United States. Science. 1999, 286, 2333–2337.
37. Steele, K.E.; Linn, M.J.; Schoepp, R.J.; Komar, N.; Geisbert, T.W.; Manduca, R.M.; Calle, P.P.; Raphael, B.L.; Clippinger, T.L.; Larsen, T.; Smith, J.; Lanciotti, R.S.; Panella, N.A.; McNamara, T.S. Pathology of fatal West Nile virus infections in native and exotic birds during the 1999 outbreak in New York City, New York. Vet Pathol. 2000, 37, 208–224.
37a. Centers for Disease Control and Prevention. West Nile virus activity—United States, Nov. 21–26, 2002. MMWR. 2002, 51 1072–1073.
38. Hubalek, Z.; Halouzka, J. West Nile fever—a reemerging mosquito-borne viral disease In Europe. Emerg Infect Dis. 1999, 5, 643–650.
39. Smithburn, K.C.; Hughes, T.P.; Burke, A.W.; Paul, J.H. A neurotropic virus isolated from the blood of a native of Uganda. Am J Trop Med. 1940, 20, 471–492.
40. Platonov, A.E.; Shipulin, G.A.; Shipulina, O.Y.; Tyutyunnik, E.N.; Frolochkina, T.I.; Lanciotti, R.S.; Yazyshina, S.; Platanova, O.V.; Obukhov, I.L.; Zhukov, A.N.; Vengerov, Y.Y.; Pokrovskii, V.I. Outbreak of West Nile virus infection, Volgograd region, Russia, 1999. Emerg Infect Dis. 2001, 7, 128–132.
41. Han, L.L.; Popovici, F.; Alexander, J.P.; Laurentia, V.; Tengelsen, L.A.; Cernescu, C.; Gary, H.E.; Ion-Nedelcu, N.; Campbell, G.L.; Tsai, T.F. Risk factors for West Nile virus infection and meningoencephalitis, Romania, 1996. J Infect Dis. 1999, 179, 230–233.
42. Savage, H.M.; Ceianu, C.; Nicolescu, G.; Karabatsos, N.; Lanciotti, R.; Vladimirescu, A.; Laiv, L.; Ungureanu, A.; Romanca, C.; Tsai, T.F. Entomologic and avian investigations of an epidemic of West Nile fever in Romania in 1996, with serologic and molecular characterization of a virus isolate from mosquitoes. Am J Trop Med Hyg. 1999, 61, 600–611.
42a. Centers for Disease Control and Prevention. Intrauterine West Nile Virus infection, New York, 2002. MMWR. 2002, 51, 1135–1136.
42b. Centers for Disease Control and Prevention. West Nile infection in organ donor and transplant recipients—Georgia and Florida. MMWR. 2002, 51, 790.
42c. Couzin, J West Nile virus. Blood banks in a 'Race against the mosquitoes.' Science. 2003, 299, 1824.

43. Sampson, B.A.; Ambrosi, C.; Charlot, A.; Reiber, K.; Veress, J.F.; Armbrustmacher, V. The pathology of human West Nile virus infection. Hum Pathol. 2000, 31, 527–531.
44. Tsai, T.F.; Popovici, F.; Cernescu, C.; Campbell, G.L.; Nedelcu, N.I. West Nile encephalitis epidemic in southeastern Romania. Lancet. 1998, 352, 767–771.
45. Mostashari, F.; Bunning, M.L.; Kitsutani, P.T.; Singer, D.A.; Nash, D.; Cooper, M.J.; Katz, N.; Liljebjelke, K.A.; Biggerstaff, B.J.; Fine, A.D.; Layton, M.C.; Mullin, S.A.; Johnson, A.J.; Martin, D.A.; Hayes, E.B.; Campbell, G.L. Epidemic West Nile encephalitis, New York, 1999: results of a household-based seroepidemiological survey. Lancet. 2001, 358, 261–264.
46. Asnis, D.S.; Conetta, R.; Teixeira, A.A.; Waldman, G.; Sampson, B.A. The West Nile virus outbreak of 1999 in New York: the Flushing Hospital experience. Clin Infect Dis. 2000, 30, 413–418.
47. Nash, D.; Labowitz, A.; Maldin, B.; Martin, D.; Mostashari, F.; Fine, A.; Roehrig, J.T.; Campbell, G.; Layton, M. A follow-up study of persons infected with west Nile virus during a 1999 outbreak in the New York City area, Abstracts of the IDSA 39th Annual Meeting, 2001, 25.
47a. Centers for Disease Control. Acute flaccid paralysis syndrome associated with West Nile virus infection—Mississippi and Louisiana, July–Aug 2002. MMWR. 2002, 51, 825–828.
47b. Gadot, N.; Weitzman, S.; Lehmann, E.E. Acute anterior myelitis complicating West Nile fever. Arch. Neurol. 1979, 36, 172–173.
48. Jordan, I.; Briese, T.; Fischer, N.; Yiu-Nam Lau, J.; Lipkin, W.I. Ribavirin inhibits West Nile virus replication and cytopathic effect in neural cells. J Infect Dis. 2000, 182, 1214–1217.
49. Anderson, J.F.; Rahal, J.J. Efficacy of interferon alpha-2b and ribavirin against West Nile virus in vitro. Emerg Infect Dis. 2002, 8, 107–108.
49a. Altman, L.A. Agency approves trial of interferon to treat west nile virus; New York Times, 21 Aug 2002, Sect A, p. 13.
50. Centers for Communicable Diseases. Epidemic/epizootic West Nile virus in the United States: revised guidelines for surveillance, prevention and control. Workshop. Charlotte, NC. Jan 31–Feb 4, 2001, 1–104.
51. Mertz, G.J. Bunyaviridae: bunyaviruses, phleboviruses, nairoviruses, and hantaviruses. In Clinical Virology; Richman, D.D., Whitley, R.J., Hayden, F.G., Eds.; Churchill Livingstone: New York, 1997, 943–971.
52. Thompson, W.H.; Kalfayan, B.; Anslow, R.O. Isolation of California encephalitis group virus from a fatal human illness. Am J Epidemiol. 1965, 81, 245–253.
53. Hammon, W.M.; Reeves, W.C.; Galindo, P. Epidemiologic studies of encephalitis in the San Joaquin Valley of California, 1943, with the isolation of viruses from mosquitoes. Am J Hyg. 1945, 42, 299–306.
54. Eldridge, B.F.; Glaser, C.; Pedrin, R.E.; Chiles, R.E. The first reported case of California encephalitis in more than fifty years. Emerg Infect Dis. 2001, 7, 451–452.
55. Mayo, D.R.; Karabatsos, N.; Scarano, F.J.; Brennan, T.; Buck, D.; Fiorentino, T.; Mennone, J.; Tran, S. Jamestown Canyon virus: seroprevalence in Connecticut. Emerg Infect Dis. 2001, 7, 911–912.
56. McJunkin, J.E.; Khan, R.R.; Tsai, T.F. California-La Crosse encephalitis. Infect Dis Clin North Am. 1998, 12, 83–93.
57. Rust, R.S.; Thompson, W.H.; Matthews, C.G.; Beaty, B.J.; Chun, R.W.M. La Crosse and other forms of California encephalitis. J Child Neurol. 1999, 14, 1–14.
58. Kalfayan, B. California Serogroup Viruses; Alan R Liss: New York, 1983, 179–186.
59. Huang, C.; Thompson, W.H.; Karabatsos, N.; Grady, L.; Campbell, W.P. Evidence that fatal human infections with La Crosse virus may be associated with a narrow range of genotypes. Virus Res. 1997, 481, 143–148.
60. Balfour, H.H.; Siem, R.A.; Bauer, H.; Quie, P.G. California arbovirus (La Crosse) infections. I. Clinical and laboratory findings in 66 children with meningoencephalitis. Pediatrics. 1973, 52, 680–691.

61. McJunkin, J.E.; De Los Reyes, E.C ; Irazuzta, J.E.; Caceres, M.J.; Khan, R.R.; Minnich, L.L.; Fu, K.D.; Lovett, G.D.; Tsai, T.; Thompson, A. La Crosse encephalitis in children. N Engl J Med. 2001, 344, 801–807.
62. Dykers, T.I.; Brown, K.L.; Gundersen. C.B.; Beaty, B.J. Rapid diagnosis of La Crosse encephalitis: detection of specific immunoglobulin M in cerebrospinal fluid. J Clin Microbiol. 1985, 22, 740–744.
63. Cassidy, L.F.; Patterson, J.L. Mechanism of La Crosse virus inhibition by ribavirin. Antimicrob Agents Chemother. 1989, 33, 2009–2011.
64. McLean, D.M.; Donohue, W.L. Powassan virus: isolation of virus from a fatal case of encephalitis. Can Med Assoc J. 1959, 80, 708–711.
65. Gholam, B.I.A.; Puksa, S.; Provias, J.P Powassan encephalitis: a case report with neuropathology and literature review. Can Med Assoc. 1999, 161, 1419–1422.
66. Centers for Communicable Diseases. Outbreak of Powassan encephalitis—Maine and Vermont, 1999–2001. MMWR. 2001, 50, 761–764.
67. Kuno, G.; Artsob, H.; Karabatsos, N.; Tsuchiya, K.R.; Chang, G.J.J. Genomic sequencing of deep tick virus and the phylogeny of the Powassan-related viruses of North America. Am J Trop Med Hyg. 2001, 65, 671–676.
68. Embil, J.A.; Camfield, P.; Artsob, H.; Chase, D.P. Powassan virus encephalitis resembling herpes simplex encephalitis. Arch Intern Med. 1983, 143, 341–343.
68a. Spruance, S.L.; Bailey, A. Colorado tick fever. A review of 115 laboratory confirmed cases. Arch Intern Med. 1973, 131, 288–293
68b. Tsai, T.F. Coltivirus (Colorado tick fever) In Principles and Practice of Pediatric Infectious Diseases; Long, S.S., Pickering, L.K., Prober, C.G., Eds.; Churchill Livingstone: New York, 1997, 1209–1211.
68c. Eklund, C. Colorado tick fever. Neurology. 1958, 8, 889.
69. Kubes, V.; Rios, F.A. The causative agent of infectious equine encephalomyelitis in Venezuela. Science. 1939, 90, 20–21.
70. Beck, C.E.; Wyckoff, R.W.G. Venezuelan equine encephalomyelitis. Science. 1938, 88, 530.
71. Weaver, S.C.; Salas, R.; Rico-Hesse, R.; Ludwig, G.V.; Oberste, M.S.; Boshell, J.; Tesh, R.B. Reemergence of epidemic Venezuelan equine encephalomyelitis in South America. Lancet. 1996, 348, 436–440.
72. Rivas, F.; Diaz, L.A.; Cardenas, V.M.; Daza, E.; Bruzon, L.; Alcala, A.; De la Hoz, O.; Caceres, F.M.; Aristizabal, G.; Martinez, J.W.; Revelo, D.; De la Hoz, F.; Boshell, J.; Camacho, T.; Calderon, L.; Olano, V.A.; Villarreal, L.I.; Roselli, D.; Alvarez, G.; Ludwig, G.; Tsai, T. Epidemic Venezuelan equine encephalitis in La Guajira, Colombia, 1995. J Infect Dis. 1997, 175, 828–832.
73. Bowen, G.S.; Fashinell, T.R.; Dean, P.B.; Gregg, M.B. Clinical aspects of human Venezuelan equine encephalitis in Texas. PAHO Bull. 1976, X, 46–57.
74. Ehrenkranz, N.J.; Sinclair, M.C.; Buff, E.; Lyman, D.O. The natural occurrence of Venezuelan equine encephalitis in the United States. First case and epidemiologic investigations. N Engl J Med. 1970, 282, 298–302.
75. Silber, L.A.; Soloviev, V.D. Far Eastern tick-borne spring-summer (spring) encephalitis. Am Rev Sov Med Spec. 1946(Suppl), 1–80.
76. Heinz, F.X.; Mandl, C.W. The molecular biology of tick-borne encephalitis virus. APMIS. 1993, 101, 735–745.
77. Davidson, M.M.; Williams, H ; Macleod, J.A.J. Louping ill in man: a forgotten disease. J Infect. 1991, 23, 241–249.
78. Pavri, K. Clinical, clinicopathologic, and hematologic features of Kyasanur Forest disease. Rev Infect Dis. 1989, 11(Suppl. 4), S854–S859.
79. Netesov, S.V.; Conrad, J.L. Emerging infectious diseases in Russia, 1990–1999. Emerg Infect Dis. 2001, 7, 1–5.

80. Osetowska, E. Tick-borne encephalitides. In: Clinical Virology. The Evaluation and Management of Human Viral Infections; Debré, R., Celers, J., Eds.; WB Saunders: Philadelphia, 1970, 182–193.
81. Schellinger, P.D.; Schmutzhard, E.; Fiebach, J.B.; Pfausler, B.; Maier, H.; Schwab, S. Poliomyelitic-like illness in Central European encephalitis. Neurology. 2000, 55, 299–302.
82. Ogawa, M.; Okubo, H.; Tsuji, Y.; Yasui, N.; Someda, K. Chronic progressive encephalitis occurring 13 years after Russian spring-summer encephalitis. J Neurol Sci. 1973, 19, 363–373.
83. Gunther, G.; Haglund, M.; Lindquist, L.; Forsgren, M.; Skoldenberg, B. Tick-borne encephalitis in Sweden in relation to aseptic meningo-encephalitis of other etiology: a prospective study of clinical course and outcome. J Neurol. 1997, 244, 230–238.
84. Anic, K.; Soldo, I.; Peric, L.; Karner, I.; Barac, B. Tick-borne encephalitis in eastern Croatia. Scand J Infect Dis. 1998, 30, 509–512.
85. Heinz, F.X.; Roggendorf, M.; Hofmann, H.; Kunz, C.; Deinhardt, F. Comparison of two different enzyme immunoassays for detection of immunoglobulin M antibodies against tick-borne encephalitis virus in serum and cerebrospinal fluid. J Clin Microbiol. 1981, 14, 141–146.
86. Clarke, D.H.; Casals, J. Japanese encephalitis virus. In Viral and Rickettsial Infections of Man, 4th ed; Horsfall, F.L., Jr, Tamm, I., Eds.; Lippincott: Philadelphia, 1965, 626–631.
87. Ravi, V.; Desai, A.; Shankar, S.K.; Gourie-Devi, M. Japanese Encephalitis. In Infectious Diseases of the Nervous System; Davis, L.E., Kennedy, P.G.E., Eds.. 2000, 231–257.
88. World Health Organization. Japanese Encephalitis Vaccines. WHO Position Paper Weekly. Epidemiol Rec. 1998, 73, 337–344.
89. Desai, A.; Shankar, S.K.; Ravi, V.; Chandramuki, A.; Gourie-Devi, M. Japanese encephalitis virus antigen in the human brain and its topographic distribution. Acta Neuropathol. 1995, 89, 368–373.
90. Gajanana, A.; Thenmozhi, V.; Samuel, P.P.; Reuben, R. A community-based study of subclinical flavivirus infections in children in an area of Tamil Nadu, India, where Japanese encephalitis is endemic. Bull WHO. 1995, 73, 237–244.
91. Dickerson, R.B.; Newton, J.R.; Hansen, J.E. Diagnosis and immediate prognosis of japanese B encephalitis. Am J Med. 1952, 12, 277–288.
92. Hinh, L.D. Clinical aspects of Japanese B encephalitis in North Vietnam. Clin Neurol Neurosurg. 1986, 88, 189–192.
93. Misra, U.K.; Kalita, J.; Srivastava, M. Prognosis of Japanese encephalitis: a multivariate analysis. J Neurol Sci. 1998, 161, 143–147.
94. Kalita, J.; Misra, U.K. EEG in Japanese encephalitis: a clinico-radiological correlation. Electroenceph Clin Neurophysiol. 1998, 106, 238–243.
95. Pradham, S.; Pandey, N.; Shashank, S.; Gupta, R.K.; Mathur, A. Parkinsonism due to predominant involvement of substantia nigra in Japanese encephalitis. Neurology. 1999, 53, 1781–1786.
96. Burke, D.S.; Nisalak, A.; Ussery, M.A.; Laorakpongse, T.; Chantavibul, S. Kinetics of IgM and IgG response to Japanese encephalitis virus in human serum and cerebrospinal fluid. J Infect Dis. 1985, 151, 1093–1099.
97. Solomon, T.; Dung, N.M.; Wills, B.; Kneen, R.; Gainsborough, M.; Diet, T.V.; Thuy, T.N.; Loan, H.T.; Khanh, V.C.; Vaughn, D.W.; White, N.J.; Farrar, J.J. Interferon alfa-2a in Japanese encephalitis: A randomised double-blind placebo-controlled trial. Lancet, 361(9360), 821–826.
98. Hoke, C.H., Jr; Vaughn, D.W.; Nisalak, A.; Intralawan, P.; Poolsuppasit, S.; Jongsawas, V.; Titsyakorn, U.; Johnson, R.T. Effect of high-dose dexamethasone on the outcome of acute encephalitis due to Japanese encephalitis virus. J Infect Dis. 1992, 165, 631–637.
99. Sohn, Y.M. Japanese encephalitis immunization in South Korea: past, present, and future. Emerg Infect Dis. 2000, 6, 17–24.

100. Dapeng, L.; Konghua, Z.; Jinduo, S.; Renguo, Y.; Hongru, H.; Baoxiu, L.; Yong, L.; Ze, W. The protective effect of bed nets impregnated with pyrethroid insecticide and vaccination against Japanese encephalitis. Trans Roy Soc Trop Med Hyg. 1994, 88, 632–634.
101. Anderson, S.G. Murray Valley encephalitis and Australian X disease. J Hyg. 1954, 52, 447–468.
102. Mackenzie, J.S.; Lindsay, M.D.; Coelen, R.J.; Broom, A.K.; Hall, R.A.; Smith, D.W. Arboviruses causing human disease in the Australasian zoogeographic region. Arch Virol. 1994, 136, 447–467.
103. Broom, A.K.; Lindsay, M.D.A.; Johansen, C.A.; Wright, A.E. (T.); Mackenzie, J.S. Two possible mechanisms for survival and initiation of Murray Valley encephalitis virus activity in the Kimberley region of Western Australia. Am J Trop Med Hyg. 1995, 53, 95–99.
104. Robertson, E.G. Murray Valley encephalitis: pathological aspects. Med J Aust. 1952, 1, 107–110.
105. Bennett, N.M. Murray Valley encephalitis, 1974. Med J Aust. 1976, 2, 446–450.
106. Mackenzie, J.S.; Smith, D.W.; Broom, A.K.; Bucens, M.R. Australian encephalitis in western Australia, 1978–1991. Med J Aust. 1993, 158, 591–595.
107. Burrow, J.N.; Whelan, P.I.; Kilburn, C.J.; Fisher, D.A.; Currie, B.J.; Smith, D.W. Australian encephalitis in the Northern Territory: clinical and epidemiologic features, 1987–1996. Aust NZJ Med. 1998, 28, 590–596.
108. Wiemers, M.A.; Stallman, N.D. Immunoglobulin M in Murray Valley encephalitis. Pathology. 1975, 7, 187–191.
109. Daubney, R.; Hudson, J.R.; Garnham, P.C. Enzootic hepatitis or Rift Valley fever. An undescribed virus disease of sheep, cattle and man from East Africa. J Pathol Bacteriol. 1931, 34, 545–579.
110. Laughlin, L.W.; Meegan, J.M.; Strausbaugh, L.J.; Morens, D.M.; Watten, R.H. Epidemic Rift Valley fever In Egypt: observations of the spectrum of human illness. Trans Roy Soc Trop Med Hyg. 1979, 73, 630–633.
111. Van Velden, D.J.J.; Meyer, J.D.; Olivier, J.; Gear, J.H.S.; McIntosh, B. Rift Valley fever affecting humans in South Africa. A clinicopathological study. South African Med J. 1977, 51, 867–871.
112. Meegan, J.M.; Hoogstraal, H.; Mousa, M.I. An epizootic of Rift Valley fever in Egypt in 1977. Vet Rec. 1979, 105, 124–125.
113. Centers for Disease Control and Prevention. Outbreak of Rift Valley fever—Saudi Arabia, August–October, 2000. MMWR. 2000, 49, 905–908.
114. World Health Organization. Rift Valley Fever. Fact Sheet 207. Revised September 2000. Available at http://www.who.int/inf-fs/en/fact207.html.
115. Centers for Disease Control and Prevention. Outbreak of Rift Valley fever—Yemen, August–October, 2000. MMWR. 2000, 49, 1065–1066.
116. Centers for Disease Control and Prevention. Update: Outbreak of Rift Valley Fever—Saudi Arabia, August–November, 2000. MMWR. 2000, 49, 982–985.
117. Siam, A.L.; Meegan, J.M.; Gharbawi, K.F. Rift Valley fever ocular manifestations: observations during the 1977 epidemic in Egypt. Br J Ophthalmol. 1980, 64, 366–374.
118. Riou, O.; Phillippe, B.; Jouan, A; Coulibaly, I.; Mondo, M.; Digoutte, J.P. Neurologic and neurosensory forms of Rift Valley fever in Mauritania (in French). Bull Soc Pathol Exot Ses Fils. 1989, 82, 605–610.
119. Niklasson, B.; Peters, C.J.; Grandien, M.; Wood, O. Detection of human immunoglobulins G and M antibodies to Rift Valley fever virus by enzyme-linked immunosorbent assay. J Clin Microbiol. 1984, 19, 225–229.
120. Lindgren, E.; Gustafson, R. Tick-borne encephalitis in Sweden and climate change. Lancet. 2001, 358, 16–18.

16

Enteroviruses

Stacie L. Ropka and Burk Jubelt
State University of New York (SUNY) Upstate Medical University
Syracuse, New York, U.S.A.

1 DESCRIPTION OF THE VIRUSES

1.1 Classification

Enteroviruses are one of the five subfamilies (genera) in the family *Picornaviridae*. They are found in humans (human enteroviruses) and many animals and are species specific. The *Picornaviridae* are small RNA viruses, thus the term ''picornavirus'' was derived from ''pico,'' meaning very small, and ''RNA'' for the type of genomic nucleic acid. There are at least 67 recognized enterovirus serotypes specific for humans (Table 1). The enteroviruses share a number of characteristics. They replicate at 37°C, lack a lipid envelope, and are stable at acid pH, which allows them to survive and replicate in the gastrointestinal tract. The virion is composed of a positive single strand of RNA of approximately 7400 nucleotides and a 3′ poly-A tail [1]. The polyprotein is translated into one long single protein, which is then cleaved to form all the individual viral proteins [2]. The capsid is an icosahedron (spheroidal) that is 22–30 nm in diameter and composed of four proteins. Three of them, VP1, VP2, and VP3, are each repeated 60 times and compose the external surface of the capsid. Once the virus completes the replication cycle it is generally released from the host cell via cell lysis, thus killing the infected cell. However, recent studies suggest that a solely lytic infection is not always the case in that the virus or at least viral RNA may persist for months or years after the acute infection [3].

Originally enteroviruses were categorized by host range. However, as new knowledge has surfaced the original definitions have blurred, and new enteroviruses are not classified but are just assigned a number. For example, the 70th and 71st enteroviruses discovered are designated by number and known as enterovirus (EV) 70 and EV71, respectively.

Table 1 Genus *Enterovirus*

Poliovirus	3 (types 1–3)
Coxsackievirus group A	23 (types A1–A22, A24)[a]
Coxsackievirus group B	6 (types B1–B6)
Echovirus	31 (types 1–9, 11–27, 29–33)[b]
Enterovirus (human)	4 (types 68–71)
Enterovirus (non-human)	At least 34 types

[a] Coxsackievirus A23 is the same virus previously identified as echovirus 9.
[b] Echovirus 10 has been reclassified as reovirus, echovirus 28 as rhinovirus 1A, and echovirus 34 as coxsackievirus A24.
Source: Ref. 9.

1.2 History of the Viruses

Presumed cases of poliomyelitis were probably first reported during the eighteenth Egyptian dynasty (1580–1350 BCE) [4]. Although cases of poliomyelitis were reported in 1840, it was not until 1908 that the viral cause of poliomyelitis was recognized [5]. Over the next five decades numerous epidemics of poliomyelitis were recorded. The poliomyelitis epidemics were eventually controlled with killed vaccine in the 1950s and later with the live oral polivirus in the 1960s. The coxsackie viruses and echoviruses (ECHO) were identified around 1950. In the 1970s diseases associated with two new enteroviruses, EV70 and EV71, were characterized [6,7]. Recently some enterovirus strains that were isolated in the 1950s but not classified have been restudied. Antigenic tests together with molecular characterizations revealed that some of these isolates are human enteroviruses. Currently, no disease specific to the most recently isolated enterovirus (EV73) has been described [8].

1.3 Epidemiology

Human enteroviruses are endemic worldwide [9]. However, they have been referred to as ''summer viruses'' because in temperate northern climates the majority of infections occur epidemically and are seen from May through October, with roughly 70% of illness reported in July through mid-December [10]. In tropical climates enteroviral disease is endemic and present year round [5].

Enteroviruses are primarily spread from host to host by fecal–oral or fecal–hand–oral transmission. Infection is acquired orally, and virus replicates in the oropharynx and lower gastrointestinal tract. Virus is shed for about a week in the oral secretions but sometimes for several months (usually 2–4 weeks) in the feces [11]. Thus the usual source of infection is fecal contamination (fingers, utensils, food, etc.), although less frequently transmission may also occur via oral secretions. Respiratory spread by airborne aerosols is very rare. The viruses are frequently found in water (both salt and fresh). Thus contact with contaminated water during either recreational activities or land irrigation occasionally serves as the mode of infection. Good sewage practices have been attributed to keeping this source of infection from people who take part in water recreational activities.

The distribution and transmission of enteroviral infections depend on a number of factors, including environment (sanitation and standards of hygiene, crowded conditions), geography, season, and host characteristics (especially age). Thus, transmission is easily achieved in group settings such as day care facilities, families with young children, sports teams, or hospital wards. Enteroviruses are spread horizontally in the community and are usually introduced into the household by young children. Viral cultures of the skin often reveal enteric flora, which epidemiologically is referred to as the "fecal veneer." This fecal veneer is more abundant on and more effectively spread by children.

Worldwide the enteroviruses are responsible for the majority of viral infections. In the United States the enteroviruses are responsible for 10–15 million (3–5% of the population) symptomatic infections per year [12,13]. Routine surveillance in the United States reveals that during nonepidemic periods enteroviruses are recovered from the environment in the following proportions: 68.8% ECHO, 2.8% Cox B, 9% Cox A, 13.1% other [14] (Table 2).

1.4 Systemic Pathogenesis

Enteroviruses replicate in the mucosa throughout most of the gastrointestinal tract. After replication in the mucosa there is local spread and replication in the lymphatic tissues, lamina propria, and tonsils in the oropharynx and Peyer's patches in the ileum. Within 24–48 hours, virus is released and can be detected in the throat and stool. Virus also may spread through lymphatics to lymph nodes and from there to the bloodstream. This initial or primary viremia does not usually result in viral invasion of the central nervous system (CNS) but does result in further amplification of viral titer in non-neural tissues. A secondary viremia of greater magnitude occurs next and is more likely to result in CNS invasion.

Most enteroviral infections are inapparent (subclinical) [11]. Occasionally gastrointestinal infection may result in nausea, vomiting, abdominal discomfort, and loose stools. If viremia occurs, systemic manifestations may be seen. The most common (>50%) systemic illness of enteroviral infection is nonspecific febrile illness. The fever is manifested at 38.5–40°C and lasts an average of 3 days. Occasionally there is a biphasic pattern with fever for 1 day followed by 2–3 days of normal temperature, then fever again for 2–4 more days. Other symptoms that might present are malaise, myalgia, headache, sore throat, and mild conjunctivitis. Generally after these symptoms clear the patient makes a complete and uneventful recovery. Because up to 50% of enteroviral infections are asymptomatic or cause only a mild febrile illness it is believed that many cases go undiagnosed or unreported. A seroepidemiological study conducted in New York found that 26% of the adult population had been exposed to an enterovirus [15] supporting the hypothesis that enteroviral infections are far more common than the reported number of cases would suggest.

Table 2 Prevalence of Non-Polio Enterovirus

	United Kingdom 1975–1994[a]	United States 1997–1999[b]
ECHO	61%	68.8%
Cox group A	10%	2.8%
Cox group B	29%	9.0%
Unknown	n/a	13.1%

[a] Data from 40,366 isolates. From Ref. 10.
[b] Data from 1672 isolates. From Ref. 14.

Although the vast majority of enteroviral infections are asymptomatic and the symptomatic diseases are generally mild, some enteroviral infections result in more obvious and enterovirus-specific systemic disease, such as carditis, acute hemorrhagic conjunctivitis (AHC), herpangina, pleurodynia, and hand-foot-and-mouth disease (HFMD) (Table 3). All of these systemic manifestations can occur sporadically or in epidemics. However, both EV70 AHC and EV71 HFMD have resulted in dramatic epidemics. EV70 spreads by hand-to-eye contact. The aberrant mode of spread (hand-to-eye) probably explains the explosive epidemics, and the rapid spread probably exhausts the population of nonimmune hosts, resulting in the disappearance of virus until sufficient mutations accumulates to initiate another epidemic [16]. EV70 AHC has at times progressed to severe neurological complications, especially acute flaccid paralysis [17,18]. However, in other epidemics no neurological disease was seen [19,20], suggesting that some strains of EV70 are more neurotropic. Although most cases of HFMD are relatively uneventful, sometimes when EV71 is the causative agent, neurological complications can be frequent, quite variable, and severe (Tables 4 and 5) [21].

A high-risk group for neurological complications during enteroviral infection are neonates [22]. In general, ECHO and Cox B account for the majority of neonatal enteroviral infections. The infection in neonates is usually acquired during birth or by postnatal contact with an infected caretaker. Transplacental transmission, though rare, has been reported in the literature. More often the mother passes the infection to the child at delivery [22,23]. Infection occurs in 20–50% of infants in whom there is a maternal history of illness in the week preceding delivery [22]. Severity of neonatal illness is related to the severity of maternal illness at the time of delivery as well as the age of the infant at time of infection. Infants younger than 10 days of age are at higher risk for severe disease because of their relative inability to mount a significant immune response and their lack of serotype-specific maternal antibody. Again, as with most enteroviral infections the infection in neonates is asymptomatic. However, unlike adults, where only 5% of the infections are symptomatic, up to 20% of infected infants will develop symptomatic disease [22]. The most common neurological manifestations are meningoencephalitis, aseptic meningitis, and encephalitis.

Table 3 Non-Neurological Syndromes Caused by Enteroviruses

Rashes
Coxsackieviruses groups A and B
Echovirus
Hand-foot-and-mouth disease
Coxsackieviruses group A
EV 71
Herpangina
Coxsackieviruses group A
Pleurodynia (epidemic myalgia, Bornholm disease)
Coxsackieviruses Group B
Myocarditis/pericarditis
Coxsackieviruses Group B
Conjunctivitis
EV 70 (AHC)
Coxsackieviruses A24

Table 4 Neurological Complications Caused by Genus *Enterovirus*

Poliovirus	Flaccid paralysis
	Encephalitis
	Cranial nerve palsies
	Opsoclonus-myoclonus
	Aseptic meningitis
Enterovirus 71[a]	Aseptic meningitis
	Encephalitis
	Flaccid paralysis
	Transverse myelitis
	Cerebellar ataxia
	Opsoclonus-myoclonus
	Febrile convulsion
	Guillain-Barré syndrome
Enterovirus 70	Flaccid paralysis
	Cranial nerve palsies
ECHO virus[b]	Aseptic meningitis
	Encephalitis
	Paralysis
Coxsackievirus group A[c]	Flaccid paralysis
	Rhabdomyolysis
	Aseptic meningitis
	Encephalitis
Coxsackievirus group B[d]	Aseptic meningitis
	Encephalitis
	Flaccid paralysis

[a] EV 71 (together with coxsackieviruses A7 and A9) is the most common cause of flaccid paralysis since the widespread use of polio vaccine.
[b] ECHO viruses 30, 11, 9, and 7 are the most common ECHO virus serotypes associated with epidemic aseptic meningitis. However, starting in 2000 there have been an increasing number of reports from Europe and the United States of aseptic meningitis due to ECHO virus 13.
[c] Coxsackievirus A9 is the most common group A virus associated with neurological disease.
[d] Coxsackieviruses B5, B3, B4, and B6 are the most common group B viruses associated with neurological disease.

Table 5 Recent EV71 Epidemics/Outbreaks with Neurological Complications

Virus	Country (year)	Primary disease	Neurological manifestations
EV 71	Australia (1999)	HFMD	Aseptic meningitis, flaccid paralysis, Guillian-Barré syndrome, cerebellar ataxia, opso-myoclonus
EV 71	Taiwan (1998)	HFMD	Aseptic meningitis, rhombencephalitis, flaccid paralysis
EV 71	Malaysia (1997)	HFMD	Aseptic meningitis, rhombencephalitis, flaccid paralysis

More rare is acute non-polio flaccid paralysis. Death occurs in up to 10% of cases, usually due to the systemic complications (cardiac, hepatic, pulmonary) [24] and rarely due to CNS disease. CNS deficits may be severe and permanent [22]. Another common source of infection is the caretaker. Thus, nosocomial transmission has accounted for multiple outbreaks in nurseries.

2 ILLNESSES: NEUROLOGICAL COMPLICATIONS OF ENTEROVIRAL INFECTION

Enteroviral CNS disease syndromes are diverse, and each syndrome can be caused by many different enteroviruses. The syndromes associated with specific infecting viruses shown in Table 4 are meant to be only general guidelines for analyzing clinical cases.

2.1 Viral (Aseptic) Meningitis

History of the Illness

Coxsackievirus and echovirus are the main causes of enteroviral meningitis. EV71 and poliovirus (PV) are less frequent causes. Cox A was isolated from a patient with paralysis in Coxsackie, New York, in 1948. Subsequent isolates have primarily occurred from cases of viral meningitis rather than paralysis, which is rarely caused by coxsackievirus. Cox B was isolated from a patient with viral meningitis in 1949. The echoviruses were isolated in 1951 but could initially not be related to any disease, thus the name ''ECHO'' for ''enteric cytopathogenic human orphan'' viruses. EV71 was isolated from patients with CNS disease in California in 1969 through 1973 [7].

Epidemiology

Of the reported cases of viral meningitis, enteroviruses are the most common cause. Enteroviral meningitis may occur in either epidemic (Table 6) or sporadic fashion. Over 10,000 cases of sporadic aseptic meningitis are reported to the CDC yearly [25]. However, due to underreporting, the actual yearly incidence is believed to be closer to 75,000 cases [26]. Several enteroviral species can cause aseptic meningitis (Tables 4–6). Viral meningitis can be sporadic, a neurological complication during systemic epidemic disease, or the only disease manifestation. Whether sporadic or epidemic, echovirus is most often isolated. Over the past 35 years ECHO 3, 4, 7, 11, 18, 19, and 30 have been the cause of viral meningitis outbreaks, but more than 50% of the cases of viral meningitis are due to ECHO 30. However, in 2000 (in Europe) and 2001 (United States) ECHO 13 [27] was associated with a number of outbreaks of viral meningitis (Table 6). This is believed to be the first time that ECHO 13 has been associated with viral meningitis. If this signals the first time for ECHO 13 infections, then the population is at risk owing to the lack of previous exposure. This could result in an increase in viral meningitis due to ECHO 13 [27]. In addition to ECHO, coxsackieviruses, particularly Cox B5, have been associated with outbreaks of viral meningitis [28]. However, when viral meningitis is a neurological complication of another enteroviral disease, ECHO and coxsackie viruses are not the predominant enteroviruses recovered. For example, during HFMD, viral meningitis is a complication and the HFMD is due to EV71 [29].

Pathophysiology

As with other causes of viral meningitis, enteroviruses probably replicate in meningeal and ependymal cells and spread via cerebrospinal fluid (CSF) pathways.

Table 6 Recent Epidemics/Outbreaks of Aseptic Meningitis

Predominant enterovirus serotype	Country (year)
ECHO 13	United States (2001)
ECHO 13	Europe (2000)
ECHO 30	Romania (1999)
ECHO 30	Turkey (1999)
ECHO 30	France (1999)
ECHO 30	Japan (1991)

Clinical Manifestation

Viral meningitis presents with headache, nausea, photophobia, fever, a stiff neck, and general irritability. Onset can be abrupt. Headaches may be intense and so severe as to require narcotics for relief of pain [30]. Although photophobia, nausea, and vomiting are common, the presentation of viral meningitis is strongly influenced by age. In older children there is frequently a short prodromal period of 2–3 days before medical attention is sought for severe headache. Upon presentation these children frequently have fever, and up to 33% of them will have positive Kernig and/or Brudzinski signs. In younger children signs are much less specific and include increased irritability and nonspecific rash. Nuchal rigidity is not always apparent and in infants is rarely seen. Fifty percent of children older than 1 will develop nuchal rigidity [22]. Generally children less than 3 years of age are most susceptible to viral meningitis, whether sporadically, as a primary epidemic, or as the neurological complication of another enteroviral epidemic. For all age groups disease can progress to involve additional organ systems (e.g., renal failure, carditis). A mild maculopapular rash associated with viral meningitis may resemble the petechial rash of meningococcemia. Additional differential diagnoses of viral meningitis include Rocky Mountain spotted fever, Lyme disease, ehrlichiosis, and other viral infections (arbovirus, herpes simplex 2, mumps, varicella, LCM) (Table 7). Viral meningitis is generally benign and self-limiting, and the patients make a complete recovery. However, some studies suggest that children less than 3 months of age may suffer from speech and language delay [22]. The duration of illness is 3–7 days, although adolescents and adults may be symptomatic for several weeks [26]. Although many cases of viral meningitis result in hospitalization, this is primarily due to the time it takes to perform the differential diagnosis. It is important to distinguish viral meningitis due to enteroviral infection from aseptic meningitis due to herpesvirus (treatable with acyclovir) or bacterial meningitis (life-threatening without antibiotic treatment). Although viral meningitis, has an excellent prognosis, it is important to obtain a definite diagnosis to rule out other less benign and treatable diseases with similar presentations.

Radiographic and Neurophysiological Findings

Computerized tomography (CT), magnetic resonance imaging (MRI), and electroencephalography (EEG) are usually normal. Occasionally the EEG may reveal diffuse slowing without clinical encephalitis.

CSF Findings

The CSF profile in viral meningitis usually consists of a mononuclear cell pleocytosis with a normal glucose. However, virus has been recovered from the CSF with normal leukocyte counts [31]. Occasionally during the first 24–48 h of the infection, polymorpho-

Table 7 Relative Frequency of Meningitis and Encephalitis of Known Viral Etiology

Viral agent	Viral meningitis 1976[a]	Viral encephalitis 1976[b,c]	Viral encephalitis 1981[c,d]
Enterovirus	324 (83%)	13 (2%)	82 (23%)
Mumps	28 (7%)	71 (10%)	7 (2%)
Arbovirus	6 (2%)	424 (60%)	107 (30%)
Herpes Simplex	15 (4%)	69 (19%)	97 (27%)
Measles	3 (1%)	44 (8%)	1 (0.3%)
Varicella	5 (1%)	58 (8%)	30 (8%)

[a] Data from Ref. 90. There were 2534 cases of indeterminate etiology.
[b] Data from Ref. 91. There were 1121 cases of indeterminate etiology.
[c] Includes both primary and postinfectious encephalitis. Almost all cases caused by measles and varicella are postinfectious.
[d] Data from Ref. 92. There were 1156 cases of indeterminate etiology.

nuclear cells may be seen, mimicking bacterial meningitis. Rarely the CSF glucose may be low, as in fungal and tuberculous meningitis [5].

Diagnostic Strategies

The neurological manifestations of viral meningitis are not diagnostically useful for differentiating the cause. Systemic manifestations (Table 3) as well as epidemic disease and household disease may provide clues. Except for PV and EV70, virus may be cultured from the CSF, usually from the stool, and occasionally from the throat. PCR techniques are also available.

Treatment

Although current treatment of viral meningitis is mostly supportive, specific agents are under study. Pleconaril is an antiviral agent specific for enteroviruses. Pleconaril has been shown to be effective at interfering with viral replication. Early studies indicate that pleconaril is effective in treating enteroviral meningitis (38–50% improvement in drug group vs. placebo) [30].

2.2 Encephalitis

History of the Illness

It has long been known that PV can cause encephalitis that primarily involves the motor cortex and brainstem [5]. During the 1960s it was recognized that echoviruses, and coxsackieviruses could cause a mild encephalitis [32]. During the next decade it was recognized that some cases of enteroviral encephalitis were due to EV71 [33].

Epidemiology

Enteroviral encephalitis is an acute infection of the brain parenchyma. The most common causes of virus induced encephalitis are herpes simplex virus and arboviruses. However, enteroviruses have frequently been associated with encephalitis [34]. There are roughly 20,000 cases of encephalitis reported every year in the United States. Generally encephalitis due to enteroviruses is found either as a neurological complication during enteroviral epidemics or as the result of enteroviral infection in immunocompromised people.

Pathophysiology

The encephalitis is caused by enteroviral invasion of the brain parenchyma, where these viruses infect neurons. The viruses spread interstitially from cell to cell. However, some enteroviruses (PV, EV70, EV71) can spread axonally [5,35].

Clinical Manifestation

As with aseptic meningitis, viral encephalitis may present with a febrile illness and headache. However, unlike patients with viral meningitis, encephalitic patients experience signs of brain involvement with altered consciousness, alteration in behavior and language, and sometimes focal signs. A few generalized seizures may occur.

The differential diagnosis of enteroviral encephalitis includes many of the same viruses that are responsible for aseptic meningitis [mumps, varicella, arboviruses (toga- and bunyaviruses, and herpes simplex] and other viruses, e.g., adenovirus, cytomegalovirus, Epstein-Barr virus, herpes zoster [36–38]. It is important to distinguish those viral encephalitides that are treatable with specific agents, e.g., herpes simplex, varicella-zoster. A wide variety of nonviral infections are included in the differential: mycoplasma, Legionnaires' disease, Lyme disease, syphilis, brucellosis, subacute bacterial endocarditis, brain abscess, Rocky Mountain spotted fever, toxoplasmosis, cysticercosis, amebiosis, schistosomiasis, malaria, trypanosomiasis, echinococcosis, and trichinosis [39–43]. Noninfectious diseases to be considered include vasculitis, sarcoidosis, and gliomatosis cerebrii [40,41].

Usually the encephalitis is mild and there is good recovery, although during epidemics of EV71 HFMD there have been reports of fatal cases of encephalitis. Severe encephalitis may also occur as part of the severe systemic coxsackievirus infections in neonates.

Radiographs and Neurophysiological Findings

The CT and MRI are usually normal. Rarely, focal parenchymal lesions may be seen [44]. The EEG usually reveals generalized slowing but focal, slowing or sharp waves may be seen.

CSF findings

The CSF profile for encephalitis resembles that of aseptic meningitis. Thus there is a lymphocytic pleocytosis, mildly to moderately elevated protein, and normal glucose.

Diagnostic Strategies

The neurological manifestations of viral encephalitis are not diagnostically useful for differentiating the etiology. Systemic manifestations, the occurrence of epidemic disease, and disease in family members may provide diagnostic clues. Coxsackievirus, ECHO, and EV71 may be isolated from the CSF, stool, or throat. PV is not isolated from the CSF but is found in the stool and throat.

Treatment

Treatment for enteroviral encephalitis is supportive. Although there is no specific treatment for encephalitis due to enteroviral infection, it is important to distinguish it from herpes simplex, for which treatment with acyclovir is effective. Usually the encephalitis is mild and there is good recovery, although during epidemics of EV71 HFMD there have been reports of fatal cases of encephalitis.

2.3 Rhombencephalitis

History of the Illness

Rhombencephalitis (brainstem encephalitis) is a more serious neurological complication of EV71 infection, which appeared during epidemics of EV71 HFMD in Taiwan and

Malaysia in the last 5 years. Unlike the other neurological complications, rhombencephalitis is often fatal. During the recent outbreaks of HFMD, rhombencephalitis occurred with a fatality rate of 14% [21–45,46].

Epidemiology

To date the only cases of rhombencephalitis have been reported in association with EV71 HFMD epidemics in Asia. In these epidemics the vast majority of cases that progressed to rhombencephalitis occurred in children, generally under 5 years of age, and all were infected with EV71 [47].

Pathophysiology

Rhombencephalitis is an encephalitis of the brainstem due to direct viral infection. In the fatal cases EV71 was detected in the CNS by both tissue culture isolation and RT-PCR amplification.

Clinical Manifestations

For rhombencephalitis, initial symptoms are myoclonic jerks, tremor, and ataxia. With progressive and more severe disease, cranial nerve palsies (ocular and bulbar) and respiratory failure are seen. During these epidemics a few cases (<10%) of aseptic meningitis and acute flaccid paralysis were also seen, sometimes in the same patients.

Radiographs and Neuropathological Findings

Magnetic resonance imaging T-2 scans reveal hyperintensities in the brainstem, usually in the pontine tegmentum. Occasionally lesions extended to the thalamus and the cervical cord. Reports from autopsies confirmed extensive inflammation of the meninges and the parenchyma, with the brainstem and spinal cord most heavily involved [21]. In animal studies cellular infiltrates consisted of mononuclear and polymorphynuclear cells and probably macrophages [48,49].

CSF Findings

Routine CSF findings (mononuclear pleocytosis, normal glucose) are typical of a CNS viral syndrome.

Diagnostic Strategies

During the 1998 HFMD outbreak in Taiwan both EV71 and Cox A16 enteroviruses were circulating but the neurological complications occurred only in those infected with EV71. Likewise, an outbreak of HFMD in Malaysia in 1997 also proved to be the result of several different strains of enterovirus (ECHO 1, Cox A9, EV71), but again those progressing to rhombencephalitis were associated with EV71. HFMD due to EV71 is indistinguishable from that caused by ECHO or coxsackieviruses. However, the neurotropic nature of EV71 underscores the need to elucidate the enteroviral strain involved so healthcare workers can be alerted to the possibility of serious neurological complications. Virus can be isolated from the throat, stool, and CSF [50].

Treatment

Treatment is supportive.

2.4 Paralytic Disease

History of the Illness

Epidemics of poliomyelitis first occurred in the second half of the nineteenth century and peaked in the 1950s. Once poliovirus epidemics were reduced it was found that infection with other enteroviruses could result in acute flaccid paralysis (Table 4) [51]. EV71 has caused flaccid paralysis as one of the neurological complications during HFMD epidemics [52]. In addition, there have been reports of EV71 outbreaks in which the only symptom was flaccid paralysis. EV70 has also caused severe flaccid paralysis during epidemics of AHC [51]. The paralytic disease caused by coxsackievirus and ECHO is indistinguishable from poliomyelitis but usually of a mild degree.

Epidemiology

The World Health Organization has a vigorous campaign under way that was aimed at eradicating poliomyelitis from the globe by the end of 2002 but has been extended to 2005. In the western hemisphere, the last known case of poliomyelitis due to an indigenous wild-type strain occurred in Peru in 1991 [53]. More recent cases in the Dominican Republic and Haiti appear to have been caused by mutants of the oral polio vaccine. Paralytic diseases caused by coxsackievirus and ECHO are isolated and rare. In the period 1976–1979, 52 cases of paralytic disease caused by enteroviruses were recorded in the United States; 25 were caused by poliovirus, 18 by echoviruses, seven by coxsackieviruses, and two by EV71 [54]. Paralysis from EV70 occurs during epidemics of AHC. Approximately 1 in 10,000–15,000 cases of AHC are complicated by neurological involvement, which primarily affects adults. EV71 has quite variable manifestations. It has resulted in both epidemic and isolated cases throughout the world.

Pathophysiology

Paralysis is due to infection of lower motor neurons in the spinal cord anterior horns and brainstem.

Clinical Manifestations

Acute flaccid paralysis usually begins with fever, flu-like symptoms, and sometimes muscle cramps followed by muscle weakness in one or more limbs. The prodromal symptoms may not occur with EV70. Paralysis is usually asymmetrical, flaccid, more proximal than distal, and often patchy [55]. The reflexes are lost as paralysis progresses. Over the next several days, paralysis may develop in other extremities and bulbar involvement with impaired respiration may occur. Extension of paralysis is unlikely to occur after the fifth or sixth day. Paralysis caused by coxsackievirus and ECHO are usually mild compared to that seen with PV, EV70, and EV71. Generally the weakened muscles regain some strength. The differential diagnosis of paralytic disease should thus include all of the enteroviruses that can cause paralysis (Table 4). In addition, several other viruses, e.g., rabies [56] and herpes zoster [57], can occasionally cause acute lower motor neuron paralysis. Other entities in the differential include acute inflammatory polyradiculitis (Guillain-Barré syndrome), botulism, acute toxic neuropathies, acute intermittent porphyria, acute transverse myelitis, and acute spinal cord compression from epidural abscess [58–60].

Radiographic and Neurophysiological and Neuropathological Findings

Radiographic studies are almost always normal but there have been a few reports of spinal cord T2 hyperintensity in the anterior horns on MRI [61].

Neurophysiological studies during acute poliomyelitis are relatively sparse because many of these techniques were developed after the elimination of poliomyelitis epidemics. During the first week of paresis, stimulation of motor nerves may reveal reduced to absent compound muscle action potential (CMAP) amplitudes, depending upon the severity of weakness [62,63]. F-wave responses are often unobtainable [64]. Electromyographic (EMG) needle testing reveals decreased to absent motor unit potentials (MUPs) depending on the degree of weakness and reduced recruitment in weak muscles [62–65]. Diffuse fibrillation potentials begin to be seen toward the end of week 1 [62,63] and increase dramatically in subsequent weeks [64,65]. By 4–6 weeks after the onset of weakness, signs of denervation, fibrillation, and positive sharp waves are profound, while CMAPs and MUP's remain similar to those of week 1, i.e., reduced to absent CMAPs, decreased to absent MUPs, and decreased recruitment [62,65]. In MUPs that are seen, the mean duration may begin to increase [66]. By 3 months after the onset of weakness, additional CAMPs can be elicited and abnormal MUPs may appear (polyphasic, long and short duration, low to normal amplitudes [62,65]. For months to years, CMAPs remain reduced in weak muscles [62], and signs of acute denervation may persist [65,67]. Many MUPs with increased amplitudes and duration are now seen and may remain polyphasic [65]. Recruitment of MUPs is usually increased compared to the onset of acute weakness but usually does not return to a normal level [65]. Sensory nerve action potentials (SNAPs) and sensory and motor conduction velocities are usually normal [62–64,68,69], although a few contradictory studies have been reported [67,70].

Neuropathological studies have revealed meningeal and parenchymal inflammation. Meningeal and perivascular inflammation consists primarily of lymphocytes and macrophages with fewer polymorphonuclear leukocytes (PMNs). Occasionally in the first 24 h PMNs may predominate. The parenchymal inflammation includes lymphocytes, microglial cells, fewer PMNs, and macrophages. The parenchymal inflammation is maximal in the anterior horns, where degenerating motor neurons are seen [48,49].

CSF Findings

The findings are similar to those seen in aseptic meningitis and encephalitis.

Diagnostic Strategies

Poliovirus, coxsackievirus, ECHO, and EV71 can be isolated from the stool and throat but rarely from the CSF when the presentation is paralytic. EV70 is not found in the stool or throat and is usually not isolated from the CSF.

Treatment

Treatment for paralytic disease is supportive except for PV, for which there is vaccination. Widespread epidemics due to poliovirus were virtually eliminated by the use of vaccine. Currently two vaccines are available. The one that is generally credited with eliminating poliomyelitis from the United States is the live attenuated Sabin vaccine, which is given orally (OPV). Other countries (notably Sweden) relied on the killed Salk vaccine, which is injected (IPV). Until recently most developed countries were considered to be free of endemic wild-type poliovirus. Recently poliovirus serotype 1 was recovered in the waters around the Florida Keys (reported in the Miami *Herald*, June 21, 2000), and poliovirus serotype 1 was recovered in Greece in 1996, 14 years after the last reported poliovirus. It has been suggested that poliovirus may be endemic in some populations and that the virus can be carried from an endemic area into so-called clean areas [71]. The recovery of wild-type poliovirus underscores the difficulty in eliminating poliovirus from the environment and the need to continue vaccinating children. The goal of the World Health Organization (WHO) is to eliminate poliovirus from the environment so that vaccination will be unnecessary. However, there is some debate about which vaccine protocol (OPV vs. IPV) will accomplish this goal [72].

The live oral vaccine offers many benefits. Even though it is given as a series of three doses, 50% of first-time vaccinees have sufficient antibody (Ab) levels after just one does and over 95% will have an adequate Ab response after the third dose [73]. It is believed that the protection is lifelong [73]. However, the oral vaccine has some drawbacks. When the live oral vaccine was used in the United States, 8–10 cases of vaccine-associated poliomyelitis were reported annually [73]. Some of these cases were due to the vaccination of immunodeficient children, and persistent disease of the vaccinee usually resulted in death. Other cases of vaccine-associated disease were due to a reversion of the vaccine strain to a more virulent strain during replication in the gut of the vaccinee [74]. The revertant strain could then infect unprotected caregivers. Because of the possible chance of getting polio from the live vaccine, in June 2000 the CDC issued a new vaccination policy. Now all children in the United States receive four doses of IPV.

With IPV, the virus has been killed so it is not possible to become infected and paralyzed with poliovirus from the vaccine. However, there are some negatives associated with the Salk vaccine. Some reports have indicated that after the first does of IPV only 36% of the patients will have developed a sufficient Ab response: thus it is more imperative to complete the series when the killed vaccine is used [75]. Repeat vaccination has been a problem in countries with migratory populations. Second, killed vaccine does not promote the ''herd'' effect whereby a vaccinated person can vaccinate another in the population via contact. Third, the killed vaccine is not as efficient at promoting gut immunity [76,77]. Therefore, those people vaccinated only with killed vaccine may still be able to host wild-type poliovirus replication in their gut. Should wild-type poliovirus be reintroduced into the United States the IPV alone will prevent disease but will be insufficient to eradicate polivirus from the environment [73]. Considering the latest findings of wild-type virus in the coastal waters surrounding the United States, it is conceivable that those receiving the killed vaccine will serve as hosts and allow for the repopulation of the environment with wild-type poliovirus. OPV remains the vaccine of choice for controlling poliovirus outbreaks [78].

2.5 Persistent Infections

Persistent infections of the CNS due to enteroviruses primarily occur in immunodeficient children with agammaglobulinemia. The viruses that cause these infections are of low virulence. Most are caused by ECHO [79]. Vaccine-like PVs have caused several dozen cases [5,80]. Coxsackieviruses have also caused several cases [79]. Persistent ECHO infections primarily involve the brain with alterations in behavioral and mental status, headaches, seizures, pyramidal tract involvement, ataxia, and tremors [79]. About one-half of the patients with chronic ECHO infections develop a dermatomyositis-like syndrome, presumably from viral invasion of the muscle [81], although virus has been isolated from muscle in only two instances. In the PV cases, there is a prolonged incubation period of several months from the time of vaccination until the onset of neurological disease [80]. Some cases begin with lower motor neuron paralysis but the persistent CNS infection continues to cause progressive intellectual and cerebral dysfunction with paralysis occurring later [82].

In spite of the humoral immunodeficient state, there is usually an intense inflammatory reaction in the CNS with a mononuclear cell meningeal and perivascular reaction, mononuclear and microglial rod cell parenchymal inflammation, neuronophagia, and microglial nodules (reviewed in Ref. 5). Examination of the CSF reveals a mononuclear pleocytosis. ECHO but not poliovirus has been frequently isolated from the CSF. Either virus can at times be isolated from the stool because of intestinal infection.

The use of intravenous immunoglobulin for the treatment of agammaglobulinemia has markedly reduced the incidence of these infections [83]. Oral poliovaccine should not be given to these patients. Unfortunately, in many cases the immunodeficiency was not obvious at the time of vaccination.

2.6 Other CNS Disease

Uncommon acute neurological syndromes reported to be due to enteroviruses have included acute cerebellar ataxia, acute infantile hemiplegia, opsiclonus-myoclonus, isolated cranial nerve palsies (especially facial), parkinsonism, hemichorea, transverse myelitis, and Guillain-Barré syndrome [5,63]. These disease syndromes are very rare, and a cause-and-effect relationship is not always clear.

In the 1980s people who had apparently recovered from acute flaccid paralysis due to poliovirus infection reported experiencing resieived weakening and new muscle weakness. This phenomenon is now known as post-polio syndrome (PPS), and although it is known that the initial cause was paralytic disease due to poliovirus infection, it is not clear if the virus plays an active role in the new weakness [84,85].

Because the enteroviruses are so prevalent and there are a number of neurological diseases associated with enteroviral infections, studies looking into neurological disease of unknown etiology are being reexamined with respect to the possibility of an enteroviral etiology. An enteroviral etiology has been postulated for amyotrophic lateral sclerosis (ALS) [86].

3 DIAGNOSIS

Historically, most enteroviral infections were diagnosed clinically and the virus was not isolated and identified. However, a positive identification of enterovirus even in sporadic cases can alert healthcare workers to the presence of enteroviral strains that are more closely associated with high morbidity or mortality and those most likely to cause epidemics. This information can allow for the expectation that more aggressive supportive care may be necessary. This is especially important during epidemics of EV70 and EV71, both of which have been associated with a large percentage of serious complications and death. In addition, a positive diagnosis of enterovirus-induced disease will allow the physician to rule out other more serious and treatable diseases and will also promote the responsible use of antibiotics that will have no effect on the viral infection. There are several factors that aid in the clinical diagnosis. seasonality, especially in northern temperate zones, exposure history (family members, work contacts, community exposure); and characteristic physical findings.

There are three methods for detecting enterovirus in clinical samples: serology, viral culture, and nucleic acid amplification via the polymerase chain reaction (PCR) [87]. Serology is of limited value because it is slow, requires both acute and convalescent blood samples, and is difficult owing to the large number of antigenically distinct serotypes (at least 60) that need to be included in the assay. Thus serology is not very useful for routine diagnosis of enteroviral infections. Viral culture has been the gold standard for detecting enterovirus [22]. During infection, enterovirus is shed from the intestine and oropharynx; thus stool samples and oropharyngeal swabs can be grown in tissue culture and the identification made based on cytopathic effect. Although viral cultures take 4–8 days (up to 14 days depending on the strain), they are useful for the isolation and identification of serotypes. However, viral culture is limited because of low relative sensitivity ($<75\%$ for diagnosis of aseptic meningitis) and the fact that enteroviruses can be shed for weeks after exposure and may be part of an asymptomatic infection and not the disease at hand [30]. Commercial PCR methods (e.g., Amplicon®) specific for enteroviruses have been developed and are becoming more readily available in the clinical setting. PCR requires a small amount of clinical material and is rapid (12–24 hs), very sensitive, and highly specific. As kits become available, this method of diagnosis will become the gold standard [24].

4 TREATMENT

Treatment during enteroviral epidemics consists mostly of controlling the epidemic with good hygiene practices and giving support to those with severe neurological complications. Poliovirus epidemics may be controlled through prevention by aggressive vaccination programs. Currently, vaccination is available only for the polioviruses. Pleconaril is the only antiviral agent that has efficacy against enteroviral infections [30]. However, pleconaril has yet to receive approval from the U.S. Food and Drug Administration (FDA). Thus, there is virtually no effective antiviral treatment for enteroviruses other than supportive therapy for the symptoms. Intravenous immunoglobulin has been used in neonates and immunodeficient children with mixed results. In some cases in the neonates it has been associated with faster cessation of viremia and viruria [88], but in agammaglobulinemic children the results are inconsistent. Early reports of combination therapy (high dose immunoglobulin coupled with pleconaril) have shown some promise [89].

REFERENCES

1. Pevear, D.C.; Oh, C.K.; Cunningham, L.L.; Calenoff, M.; Jubelt, B. Localization of genomic regions specific for the attenuated, mouse-adapted poliovirus type 2 strain W-2. J Gen Virol. 1990, 71, 43–52.
2. Kitamura, N.; Semler, B.; Rothberg, P.G. Primary structure, gene organization and polypeptide expression of poliovirus RNA. Nature. 1981, 291, 547–553.
3. Girard, S.; Gosselin, A.S.; Pelletier, I.; Colbere-Garapin, F.; Couderc, T.; Blondel, B. Restriction of poliovirus RNA replication in persistently infected nerve cells. J Gen Virol. 2002, 83, 1087–1093.
4. Paul, J.R. Epidemiology of Poliomyelitis. In: Poliomyelitis, WHO Monograph Ser No. 26; WHO: Geneva, 1955, 9–29.
5. Jubelt, B.; Lipton, H.L. Enterovirus infections. In ed.; McKendall, R.R. Handbook of Clinical Neurology, Vol 12; Elsevier Science: Amsterdam, 1989, 307–347.
6. Kono, R. On the causative agent of accute haemorrhagic conjunctivitis and its epidemiology. Nippon Ganka Gakkai Zasshi—Acta Soc Ophthalmol Jpn. 1974, 78, 333–341.
7. Schmidt, N.J.; Lennette, E.H.; Ho, H.H. An apparently new enterovirus isolated from patients with disease of the central nervous system. J Infect Dis. 1974, 129, 304–309.
8. Oberste, M.S.; Schnurr, D.; Maher, K.; al-Busaidy, S.; Pallansch, M.A. Molecular identification of new picornaviruses and characterization of a proposed enterovirus 73 serotype. J Gen Virol. 2001, 82, 409–416.
9. Melnick, J.L. Poliovirus and other enteroviruses. In Viral Infections of Humans; Evans, A.S., Kaslow, R.A., Eds.; Plenum: New York, 1997, 583–663.
10. Atkinson, P.J.; Sharland, M.; Maguire, H. Predominate enteroviral serotypes causing meningitis. Arch Dis Child. 1998, 78, 373–374.
11. Horstmann, D.M. Epidemiology of poliomyelitis and allied diseases—1963. Yale J Biol Med. 1963, 36, 5–26.

12. Spigland, I.; Fox, J.P.; Elveback, L.R. The Virus Watch program: a continuing surveillance of viral infections in metropolitan New York families. II. Laboratory methods and preliminary report on infections revealed by virus isolation. J Epidemiol. 1966, 83, 413–435.
13. Strikas, R.A.; Anderson, L.J.; Parker, R.A. Temporal and geographic patterns of isolates of nonpolio enterovirus in the United States, 1970–1983. J Infect Dis. 1986, 153, 346–351.
14. Centers for Disease Control. Enterovirus surveillance—United States, 1997–1999. MMWR. 2000, 49, 913–916.
15. Deibel, R.; Gross, L.L.; Collins, D.N. Isolation of a new enterovirus. Proc Soc Exp Biol Med. 1975, 148, 203–207.
16. Johnson, R.T. Viral Infections of the Nervous System, 2nd ed. 1998.
17. Katiyar, B.C.; Misra, S.; Singh, R.B.; Singh, A.K.; Gupta, S.; Gulati, A.K. Adult polio-like syndrome following enterovirus 70 conjunctivitis (natural) history of the disease. Acta Neurol Scand. 1983, 67, 263–274.
18. Wadia, N.H.; Katrak, S.M.; Misra, V.P. Polio-like motor paralysis associated with acute hemorrhagic conjunctivitis in an outbreak in 1981 in Bombay, India: clinical and serologic studies. J Infect Dis. 1983, 147, 660–668.
19. Patriarca, P.A.; Onorato, I.M.; Sklar, V.E. Acute hemorrhagic conjunctivitis: investigation of a large-scale community outbreak in Dade County, Fla. JAMA. 1983, 249, 1283–1289.
20. Bern, C.; Pallansch, M.A.; Gary, H.E., Jr Acute hemorrhagic comjunctivitis due to enterovirus 70 in American Samoa: serum-neutralizing antibodies and sex-specifc protection. Am J Epidemiol. 1992, 136, 1502–1506.
21. Huang, C.C.; Liu, C.C.; Chang, Y.C.; Chen, C.Y.; Wang, S.T.; Yeh, T.F. Neurologic complications in children with enterovirus 71 infection. N Engl J Med. 1999, 341, 936–942.
22. Zaoutis, T.; Klein, J.D. Enterovirus infections. Pediatr Rev. 1998, 19, 183–191.
23. Modlin, J.F. Perinatal echovirus and group B coxsackievirus infections. Clin Perinatol. 1988, 15, 233–246.
24. Kaplan, M.H.; Klein, S.W.; McPhee, J.; Harper, R.G. Group B coxsackievirus infections in infants younger than three months of age: a serious childhood illness. Rev Infect Dis. 1983, 5, 1019–1032.
25. Centers for Disease Control. Deaths among children during an outbreak of hand, foot, and mouth disease—Taiwan, Republic of China, April – July 1998. MMWR. 1998, 47, 629–632.
26. Rotbart, H.A.; McCracken, G.H.; Whitley, R.J. Clinical significance of enteroviruses in serious summer febrile illnesses of children. Pediatr Infect Dis J. 1999, 18, 869–874.
27. Centers for Disease Control. Echovirus type 13 — United States, 2001. MMWR. 2001, 50, 777–780.
28. Goldwater, P.N. Immunoglobulin M capture immunoassay in investigation of coxsackievirus B5 and B6 outbreaks in South Australia. J Clin Microbiol. 1995, 33, 1628–1631.
29. Chang, L.Y.; Lin, T.Y.; Huang, Y.C. Comparison of enterovirus 71 and coxsackievirus A16 clinical illnesses during the Taiwan enterovirus epidemic, 1998. Pediatr Infect Dis J. 1999, 18, 1092–1096.
30. Sawyer, M.H. Enterovirus infections: diagnosis and treatment. Pediatr Infect Dis J. 1999, 18, 1033–1040.
31. Wenner, H.A.; Abel, D.; Olson, L.C.; Burry, V.F. A mixed epidemic associated with echovirus types 6 and 11. Am J Epidemiol. 1981, 114, 369–378.
32. Horstmann, D.M.; Yamada, N. Enterovirus infections of the central nervous system. In ed; Zimmerman, H.M. Association for Research in Nervous and Mental Disease: Infections of the Nervous System; Proceedings of the Association, 1964 New York, NY; Williams & Wilkins: Baltimore, 1968, 236–253.
33. Chonmaitree, T.; Menegus, M.A.; Schervish-Swierkosz, E.M.; Schwalenstocker, E. Enterovirus 71 infection: report of an outbreak with two cases of paralysis and a review of the literature. Pediatrics. 1981, 67, 489–493.
34. Pruit, A.A. Infections of the nervous system. Neurol Clinics North Am. 1998, 16, 419–447.

35. Hashimoto, I.; Hagiwara, A. Studies on the pathogenesis of and propagation of enterovirus 71 in poliomyelitis-like disease in monkeys. Acta Neuropathol. 1982, 58, 125–132.
36. Centers for Disease Control. Poliomyelitis—Pennsylvania, Maryland. MMWR. 1979, 28, 49–50.
37. Koskiniemi, M.; Vaheri, A. Acute encephalitis of viral origin. Scand J Infect Dis. 1982, 14, 181–187.
38. Beghi, E.; Nicolosi, A.; Mulder, D.W.; Hauser, W.A.; Shuster, L. Encephalitis and aseptic meningitis, Olmsted County, Minnesota, 1950–1981: I. Epidemiology. Ann Neurol. 1984, 16, 283–294.
39. Ho, M. Acute viral encephalitis. In Handbook of Clinical Neurology, Vol. 34. Infections of the Nervous System, Part II; Vinken, P.J., Bruyn, G.W., Klawans, H.L., Eds.: Amsterdam: North-Holland; 1978, 63–82.
40. Johnson, R.T. Viral Infections of the Nervous System; Raven Press: New York, 1998.
41. Tyler, K.L. Diagnosis and management of acute viral encephalitis. Semin Neurol. 1984, 4, 25–33.
42. Jubelt, B.; Miller, J.R. Viral infections. In ed; Rowland, L.P. Merritt's Textbook of Neurology, 8th ed; Lea and Febiger: Philadelphia, 1989, 96–136.
43. Miller, J.R.; Jubelt, B. Parasitic infections. In ed; Rowland, L.P. Merritt's Textbook of Neurology, 8th ed; Lea and Febiger: Philadelphia, 1989, 164–175.
44. Modlin, J.F.; Dagan, R.; Berlin, L.E.; Virshup, D.M.; Yolken, R.H.; Menegus, M. Focal encephalitis with enterovirus infections. Pediatrics. 1991, 88, 841–845.
45. Lum, L.C.; Wong, K.T.; Lam, S.K. Fatal enterovirus 71 encephalomyelitis. J Pediatr. 1998, 133, 795–798.
46. Hsueh, C.; Jung, S.M.; Shih, S.R. Acute encephalomyelitis during an outbreak of enterovirus 71 infection in Tawain: report of an autopsy case with pathologic, immunofluorescence, and molecular studies. Mod Pathol. 2000, 13, 1200–1205.
47. Dolin, R. Enterovirus 71—emerging infections and emerging questions. N Engl J Med. 1999, 341, 984–985.
48. Bodian, D. Poliomyelitis. In ed; Minckler, J. Pathology of the Nervous System; McGraw-Hill: New York, 1972, 2323–2394.
49. Ford, D.J.; Ropka, S.L.; Collins, G.H.; Jubelt, B. The neuropathology observed in wild-type mice inoculated with human poliovirus mirros human paralytic poliomyelitis. Microb Pathol. 2002, 33, 97–107.
50. Ho, M.; Chen, E.R.; Hsu, K.W. An epidemic of enterovirus 71 infection in Taiwan. N Engl J Med. 1999, 341, 929–935.
51. Chopra, J.S.; Sawhney, I.M.S., Dhand, U.K.; Prabhakar, S.; Naik, S.; Sehgal, S. Neurological complications of acute haemorrhagic conjunctivitis. J Neurol Sci. 1986, 73, 177–191.
52. Gilbert, G.L.; Dickson, K.E.; Waters, M.J.; Kennett, M.L.; Land, S.A.; Sneddon, M. Outbreak of enterovirus 71 infection in Victoria, Australia, with a high incidence of neurologic involvement. Pediatr Infect Dis J. 1988, 7, 484–488.
53. Centers for Disease Control. Progress toward global eradication of poliomyelitis, 1988–1991. MMWR. 1993, 42, 486–496.
54. Centers for Disease Control. Enterovirus surveillance report, 1970–1979. MMWR. 1981.
55. Howe, H.A.; Wilson, J.L. Poliomyelitis. In Viral and Rickettsial Infections of Man, 3rd ed; Rivers, T.M., Horsfall, F.L., Eds.; Lippincott: Philadelphia, 1957, 432–478.
56. Chopra, J.S.; Banerjee, A.K.; Murthy, J.M.; Pal, S.R. Paralytic rabies: a clinico-pathological study. Brain. 1980, 103, 789–802.
57. Johnson, R.T. Herpesvirus Infection. In Viral Infections of the Nervous System, 2nd ed.; Lippincott: Philadelphia, 1998, 133–168.
58. Price, R.W.; Plum, F. Poliomyelitis. In Handbook of Clinical Neurology, Vol 34. Infections of the Nervous System, Part II; Vinken, P.J., Bruyn, G.W., Klawans, H.L., Eds.: Amsterdam: North-Holland; 1978, 93–132.

59. Asbury, A.K. Diagnostic considerations in Guillain-Barre syndrome. Ann Neurol. 1981, 9(suppl), 1–5.
60. Gear, J.H. Nonpolio causes of polio-like paralytic syndromes. Rev Infect Dis. 1984, 6(suppl. 2), S379–S384.
61. Rao, D.G.; Bareman, D.E. Hyperintensities of the anterior horn cells on MRI due to poliomyelitis. J Neurol Neurosurg Psych. 1997, 63, 720.
62. David, W.S.; Doyle, J.J. Acute infantile weakness: a case of vaccine-associated poliomyelitis. Muscle Nerve. 1997, 20, 747–749.
63. Edwards, E.A.; Grant, C.C.; Huang, Q.S.; Powell, K.F.H.; Croxson, M.C. A case of vaccine-associated paralytic poliomyelitis. J Paediatr Child Health. 2000, 36, 408–411.
64. So, Y.T.; Olney, R.K. AAEM case report #23: acute paralytic poliomyelitis. Muscle Nerve. 1991, 14, 1159–1164.
65. Wiechers, D.O. New concepts of the reinnervated motor unit revealed by vaccine-associated poliomyelitis. Muscle Nerve. 1988, 11, 356–364.
66. Pinelli, P.; Buchthal, F. Duration, amplitude, and shape of muscle action potentials in poliomyelitis. Electroencephalogr Clin Neurophysiol. 1951, 3, 497–504.
67. Keren, O.; Grosswasser, Z.; Heller, L.; Ring, H. Electrophysiological findings in poliomyelitis patients at the subacute phase. Electromyoger Clin Neurophysiol. 1992, 32, 547–554.
68. Johnson, E.W.; Guyton, J.D.; Olsen, K.J. Motor nerve conduction velocity studies in poliomyelitis. Arch Phys Med Rehab. 1960, 41, 185–190.
69. Bhaskar, S.H.; Bhatia, B.D.; Mahadevan, S.; Thombre, D.P.; Krishnamurthy, N. Electrophysiological studies in children with paralytic poliomyelitis. Electromyogr Clin Neurophysiol. 1997, 37, 33–37.
70. Valli, G.; Jann, S.; Colturani, G.; Scarlato, G. Neurophysiological problems in poliomyelitis diagnosis: a case report. Electromyogr Clin neurophysiol. 1987, 27, 73–75.
71. Fiore, L.; Genovese, D.; Diamanti, E. Antigenic and molecular characterization of wild type 1 poliovirus causing outbreaks of poliomyelitis in Albania and neighboring countries in 1996. J Clin Microbiol. 1998, 36, 1912–1918.
72. Henderson, D.A. Developed countries should not use inactivated polio vaccine for the prevention of poliomyelitis. Rev Med Virol. 1997, 7, 83–86.
73. CDC. Poliomyelitis prevention in the United States: inroduction of a sequential vaccination schedule of inactivated poliovirus vaccine followed by oral poliovirus vaccine. MMWR. 1997, 46, 1–25.
74. Kew, O.M.; Nottay, B.K.; Hatch, M.H.; Nakano, J.H.; Obijeski, J.F. Multiple genetic changes can occur in the oral poliovaccines upon replication in humans. J Gen Virol. 1981, 56, 337–347.
75. Robertson, S.E.; Traverso, H.P.; Drucker, J.A. Clinical efficacy of a new, enhanced-potency, inactivated poliovirus vaccine. Lancet. 1988, 1, 897–899.
76. Onorato, I.M.; Modlin, J.F.; McBean, A.M.; Thomas, M.L.; Losonsky, G.A.; Bernier, R.H. Mucosal immunity induced by enhanced-potency inactivated and oral polio vaccines. J Infect Dis. 1991, 163, 1–6.
77. Herremans, T.M.P.T.; Reimerink, J.H.J.; Busiman, A.M.; Kimman, T.G.; Koopmans, P.G. Induction of mucosal immunity by inactivated polivirus vaccine is dependent on previous mucosal contact with live virus. J Immunol. 1999, 162, 5011–5018.
78. Prevots, D.R.; Ciofe, M.; Sallabanda, A. Outbreak of paralytic poliomyelitis in Albania, 1966: high attack rate among adults and apparent interruption of transmission following a nationwide mass vaccination. Clin Infect Dis. 1998, 26, 419–425.
79. McKinney, R.E., Jr; Katz, S.L.; Wilfert, C.M. Chronic enteroviral meningoencephalitis in agammaglobulinemic patients. Rev Infect Dis. 1987, 9, 334–356.
80. Wyatt, H.V. Poliomyelitis in hypogammaglobulinemics. J Infect Dis. 1973, 128, 802–806.
81. Hays, A.P.; Gamboa, E.T. Acute viral myositis. In Myology, Vol 2; England, A.G., Banker, B.Q., Eds.; McGraw-Hill: New York, 1986, 1439–1466.

82. Davis, L.E.; Bodian, D.; Price, D.; Butler, I.; Vickers, J. Chronic progressive poliomyelitis secondary to vaccination of an immunodeficient child. N Engl J Med. 1977, 297, 241–245.
83. Misbah, S.A.; Spickett, G.P.; Ryba, P.C J. Chronic enteroviral meningitis in agammaglobulinemia: case report and literature review. J Clin Immunol. 1992, 12, 266–270.
84. Jubelt, B.; Drucker, J. Poliomyelitis and the post-polio syndrome. In ed; Younger, D.S. Motor Disorders; Lippincott Williams & Wilkins: Philadelphia, 1999, 381–395.
85. Jubelt, B.; Agre, J.C. Characteristics and management of post-polio syndrome. JAMA. 2000, 284, 412–414.
86. Jubelt, B.; Berger, J.R. Does viral disease underlie ALS? Lessons from the ALS pandemic. Neurology. 2001, 57, 945–946.
87. Hamilton, M.S.; Jackson, M.A.; Abel, D. Clinical utility of polymerase chain reaction testing for enteroviral meningitis. Pediatr Infect Dis J. 1999, 18, 533–537.
88. Nagington, J.; Gandy, G.; Walker, J.; Gray, J.J. Use of normal immunoglobulin in an echovirus 11 outbreak in a special-care baby unit Lancet. 1983, 2, 443–446.
89. Quartier, P.; Foray, S.; Casanova, J.L.; Hau-Rainsard, I.; Blanche, S.; Fischer, A. Enteroviral meningoencephalitis in X-linked agammaglobulinemia: intensive immunoglobulin therapy and sequential viral detection in cerebrospinal fluid by polymerase chain reaction. Pediatr Infect Dis J. 2000, 19, 1106–1108.
90. Centers for Disease Control. Aseptic Meningitis Surveillance, Annual Summary 1976. MMWR. 1979, 28, 22–24.
91. Centers for Disease Control. Annual Summary 1977. MMWR. 1978, 26, 13.
92. Centers for Disease Control. Annual Summary 1981: Reported Morbidity and Mortality in the United States. MMWR. 1982, 30, 34.

17

Adenoviruses

Flor M. Munoz
Baylor College of Medicine
Houston, Texas, U.S.A.

Robert J. Baumann
University of Kentucky College of Medicine
Lexington, Kentucky, U.S.A.

1 INTRODUCTION

Adenoviruses are a common cause of febrile illnesses in young children. They are most frequently associated with upper respiratory tract infections such as pharyngitis or coryza, but they are also the cause of pneumonia and gastrointestinal, ophthalmological, genitourinary, and central nervous system (CNS) diseases [1,2]. Although most adenoviral illnesses are self-limiting, severe, and even fatal, infections can occur in immunocompromised hosts and occasionally in immunocompetent children and adults. In addition to their clinical importance, adenoviruses are being studied intensively as vectors to deliver foreign genes for gene therapy and for immunization against other pathogens.

2 DESCRIPTION AND CLASSIFICATION OF ADENOVIRUSES

Adenoviruses that cause human disease belong to the family *Adenoviridae* and the genus *Mastadenovirus*. They are classified into six subgroups or species (formerly called subgenera), A–F, on the basis of their physiochemical, biological, and genetic properties (Table 1) [3,4]. Adenovirus subgroups B and C are usually implicated in cases of CNS disease. Subgroups are further classified into 49 distinct adenovirus (Ad) serotypes or subspecies based upon antigenic determinants detected by viral neutralization assay [5]. Two new candidate serotypes, Ad50 and Ad51, belonging to species B1 and D, respectively, were described in 1999 [6]. Serotypes within each subgroup are closely related at the DNA level and frequently share similar biological properties. DNA analysis using restriction endonucleases allowed the identification of subspecies that appear intermediate between established serotypes, suggesting the occurrence of viral recombination in nature and resulting in the classification of adenoviruses into several new genotypes (e.g., Ad7a and Ad7b) [7]. More rapid and sensitive polymerase chain reaction (PCR) assays for the identification of adenovirus serotypes are under active development [8]. Although typing of adenoviruses into subgroups and serotypes is not routinely performed in most clinical laboratories, specific identification can be of clinical and epidemiological importance [9].

The 60–90 nm adenovirus virion contains a double-stranded DNA genome of approximately 35 kb surrounded by a nonenveloped icosahedral protein capsid containing 252 capsomeres and fiberlike projections from each of 12 pentagonal vertices. The fiber protein is attached noncovalently to the icosahedron by a pentameric polypeptide called the penton base. The fiber protein mediates attachment to cells and is probably an important determinant of tissue tropism. A cellular receptor known as ''coxsackie B and adenovirus receptor'' or CAR because it also binds coxsackie B virus has been identified for the subgroup C adenovirus types 2 and 5 [10]. The major surface protein of the virion is the trimeric polypeptide hexon. Group-reactive antigenic determinants are present on the hexon proteins from all human adenoviruses. Type-specific neutralizing epitopes are present on both the fiber and hexon proteins, with minor sites on the penton base. In addition, many adenoviruses hemagglutinate rat or rhesus monkey red blood cells; this property is related to the fiber proteins and is used to classify adenoviruses into four hemagglutination groups (Table 1).

Adenoviruses are highly stable in adverse physical conditions of pH and temperature and resist many chemical agents, including lipid solvents [1]. These properties account for their ability to spread and survive outside host cells. Adenoviruses can survive freezing with minimal loss of infectivity. Maximal infectivity occurs at pH ranges of 6.0–9.5 at 24°C. The virus can be inactivated by heat at 56°C for 30 min, by 0.25% sodium dodecyl sulfate, which causes a disruption of the capsid, and by formaldehyde.

3 DESCRIPTION OF ILLNESSES CAUSED BY ADENOVIRUS

Adenoviruses cause a variety of symptoms and well-described illness syndromes that occur as epidemics or endemic and sporadic disease. Adenoviruses (usually subgroups B, C, and E) are a common cause of febrile upper and lower respiratory tract illness in infants and young children year-round. Symptoms frequently associated with adenoviral infection include conjunctivitis, tonsillitis, laryngotracheobronchitis (croup), and pneumonia. Severe pneumonia has been associated with concurrent infection with measles and *B. pertussis* [11,12] and with adenovirus serotypes 3 and 7 in otherwise healthy children [13,14]. Disseminated adenoviral disease with multiorgan involvement has been described in association with these and other adenovirus serotypes in immunocompromised patients [13]. Adenoviruses have also been implicated in outbreaks of febrile respiratory disease in summer camps and swimming pools [15]. The epidemic syndrome of conjunctivitis with pharyngeal symptoms, pharyngoconjunctival fever (PCF), is characteristic of adenovirus. Adenoviruses (usually subgroup D) are the most common cause of epidemic keratoconjunctivitis (EKC), a syndrome characterized by eye pain and inflammation, fever, and preauricular lymphadenopathy. Adenovirus serotype 4 (the only member of subgroup E) has also been associated with outbreaks of conjunctivitis [4]. Enteric adenoviruses (subgroup A or subgroup F types 40 and 41) can cause outbreaks of diarrheal illness in

Table 1 Adenovirus Classification and Characteristic Disease Syndromes[a]

Subgroup or species	Serotypes or subspecies	Percent DNA homology within species/(G + C)[b]	Hemagglutination pattern[c]	Oncogenic potential: In hamsters	Oncogenic potential: In tissue	Characteristic disease syndrome
A	**12**, 18, 31	48–69/48–49	IV	High	Mod	Gastroenteritis in infants
B1	**3**, **7**, 16, 21, 50	89–94/50–52	I	Mod	Mod	Respiratory disease,[d] pneumonia, PCF,[e] acute hemorrhagic cystitis
B2	**11**, 14, 34, 35			Mod	Mod	
C	1, **2**, **5**, 6	99–100/57–59	III	Low–none	Low	Respiratory disease, gastroenteritis, and intussuception
D	8, 19, 37	94–99/57–61	II	Low–none	Mod	EKC[f]
	9, 10, 13, 15, 17, 19, 20, 22–30, **32**, 33, 36, 38, 39, 42, 43–47, 51					Infection in immunocompromised hosts
E	4	N.A./57–59	III	Low–none	Low	Respiratory disease, PCF
F	40, 41	62/57–59	Unknown	Unknown	Unknown	Gastroenteritis

[a] Adenovirus types associated with CNS disease are in **boldface** type.
[b] Percent guanine plus cytosine content in DNA.
[c] Hemagglutination patterns: I, complete agglutination of monkey erythrocytes; II, complete agglutination of rat erythrocytes; III, partial agglutination of rat erythrocytes; IV, little or no agglutination.
[d] Respiratory diseases include coryza, pharyngitis, tonsillitis, laryngotracheobronchitis, bronchiolitis, and pneumonia.
[e] PCF-pharyngoconjunctival fever.
[f] EKC-epidemic keratoconjunctivitis.
Source: Refs. 18, 32, 62, 64, 68, and 70.

infants, typically in group care settings such as day care centers. Symptoms can be prolonged and similar to those caused by rotavirus.

Central nervous system manifestations associated with adenoviral infection include meningoencephalitis, aseptic meningitis, febrile seizures, and Reye's syndrome–like illness [16–21]. These have been described more frequently in young children and in immunocompromised hosts, often in association with multiorgan involvement, particularly pneumonia and hepatitis, and during disseminated adenoviral disease. Adenovirus type 7 is most commonly associated with CNS symptoms and is the serotype that has been most frequently isolated from the spinal fluid. Other adenovirus types isolated from the CSF of patients with neurological manifestations include types 2, 3, 5, 6, 9, 11, 12, and 32 [18,22,23].

4 HISTORY OF THE ILLNESS

In 1956, the name adenovirus was given to an agent that was first isolated in tissue cultures of surgically removed human adenoids by Rowe and collaborators in 1953 [24]. The characteristic cytopathic changes produced by this agent were also described by Hilleman and Werner in 1954 [25] when they cultured throat washings of military recruits with febrile acute respiratory illness. Adenovirus was then identified as the cause of acute respiratory disease in military recruits [26], and shortly thereafter the association between adenovirus and epidemic keratoconjunctivitis and pharyngoconjunctival fever were recognized. In fact, these epidemic syndromes caused by adenovirus had been described in Europe, Asia, and the United States since the nineteenth century and throughout the first half of the twentieth century [1,26]. Because adenovirus was an important cause of acute respiratory disease in military recruits, intensive research was carried out that resulted in the identification and classification of dozens of adenovirus types and in the development of vaccines against adenovirus types 4 and 7 to control epidemics in the military [27,28]. A suspension in the production of these vaccines in 1994 resulted in the reemergence of adenovirus epidemics in military trainees [29]. Enteric adenoviruses (types 40 and 41), first reported in 1975, were initially found by electron microscopy and subsequently shown to be a cause of diarrheal illness in children [30].

The first association of adenovirus and CNS disease appeared in the literature in 1956 in a report of five French children who developed neurological symptoms during an outbreak of respiratory disease caused by adenovirus [31]. Among them, adenovirus type 7 was isolated from brain tissue of a child who developed fatal encephalitis. Several similar cases were reported in the following decades, describing an association between CNS manifestations and adenoviral disease in otherwise healthy children with sporadic infection and during outbreaks of adenovirus types 7, 3, and 5 [16,32–34]. Some children had concomitant or recent viral infections such as varicella [22,35,36]. Although isolation of adenovirus from brain tissue or spinal fluid was not always performed, adenovirus infection was documented in patients who developed neurological manifestations without other identified etiology [21,37–39]. Eight cases of meningoencephalitis in children and infants were reported in association with an adenovirus type 7 epidemic in northern Finland in 1970 [40]. CSF cultures were not obtained, but adenovirus was isolated from the nasopharynx in six children and from the lung in one of two patients who died. More recently, adenovirus subgenus B species were identified in CSF samples of children dying of cardiomyopathy and encephalitis during an epidemic of enterovirus 71 hand, foot, and mouth disease [41].

These reports suggest that CNS involvement can occur in primary adenoviral infections or in association with concomitant infection with other pathogens. It has been postulated that adenovirus serotypes frequently associated with CNS disease may share common neurotropic characteristics that are not yet well understood. Host factors are also likely to play a role, as suggested by the substantial number of reports in the last two decades of cases of multisystemic adenoviral infections with CNS involvement in immunocompromised patients, particularly transplant recipients [13,42–47].

5 EPIDEMIOLOGY

Adenoviruses have a worldwide distribution. Infections may occur throughout the year, but the incidence of adenovirus-associated respiratory disease is usually higher in the late winter, spring, and early summer. Adenoviral infections are common in households with young children and in institutions such as day care centers and the military [1,26,48,49]. Nosocomial outbreaks have also been documented [35,50]. Transmission may occur by direct contact, via aerosol particles, by the fecal–oral route, or by contact with contaminated fomites or water [1,2]. In addition, vertical transmission may occur from exposure to cervical secretions at birth [51]. Fatal neonatal infections have been described with subgroup B types 11 and 35, isolates that are common genitourinary pathogens [52,53].

Patients with deficiencies in cell-mediated immune responses are at higher risk for severe adenoviral infection. Susceptible populations include infants with congenital immune deficiencies, bone marrow and solid organ transplant recipients, and patients with acquired immunodeficiency syndrome (AIDS) [19,54,55]. Severe infection has also been described immediately following infection with measles and *B. pertussis* [11,12].

Adenoviruses can cause persistent infection with prolonged viral shedding in the feces that may last months to years after the initial acute infection [56]. Serological evidence that adenoviruses can be transmitted from kidney and liver transplants also suggests that these organs can occasionally harbor adenoviruses in a latent form [57]. Reactivation of endogenous virus may play a role in adenoviral diseases in immunocompromised patients. The major reservoirs for persistent adenoviral infection and the mechanisms of viral persistence are unknown.

Most individuals have serological evidence of prior adenoviral infection by 10 years of age. The most common isolates are the subgroup C types 1, 2, 5, and 6, which are associated primarily with upper respiratory tract illnesses in children under 2 years of age and are endemic in most areas of the world [14]. Infections with adenoviruses types 4, 7, 11, 14, and 21 typically occur later in life in the form of epidemic acute respiratory disease. Epidemic syndromes are associated with specific serotypes. Pharyngoconjuctival fever (PCF) is most commonly caused by Ad3 or Ad7 and occasionally by Ad1, Ad4, and Ad14. Keratoconjunctivitis (EKC) is usually caused by adenovirus types 8, 19, and 37, but types 4, 7, 10, 11, 14, and 15 have also been implicated. Many adenoviruses can cause gastroenteritis, particularly those belonging to subgroups A, B, and C, but Ad40 and Ad41, subgroup F, cause prolonged watery diarrhea in infants. Acute hemorrhagic cystitis is associated with Ad11 and Ad21.

The epidemiology of central nervous system infection by the adenoviruses is not well defined. There are multiple sources of information, each having its own strengths and weaknesses. The interpretation of the results of these epidemiological studies is complicated by a lack of characteristic clinical features and the commonness of adenoviral infections; it is quite possible for a non-neurological adenoviral infection and an unrelated neurological disorder to occur simultaneously. In addition, adenoviral infections are not generally reportable, many infections occur without clinical symptomatology, and some infected individuals can shed virus after clinical signs of infection have disappeared. There have been studies of adenoviral infection in large populations, but these studies did not include mandatory reporting of patients with suspected meningitis or encephalitis. Most studies report community and family epidemics and outbreaks in special populations, especially military recruits and immunosuppressed persons.

The Virus Watch program investigated the natural occurrence and consequences of common viral infections in two U.S. communities during the 1960s [56,58]. Surveillance did not rely upon reporting by healthcare providers or on clinical symptomatology. Specimens were collected on a biweekly schedule from index family members (usually a child, newborn to 9 years old) and examined for viruses while sera collected periodically from all family members were tested for antibodies. The New York study followed 25 families on Shelter Island for 560 family-months of observation and 153 more urban families for 2076 family-months of observation. Adenoviruses were found to be the most frequently encountered major virus group. Infections showed seasonal variability, and the overall incidence was inversely correlated with age. Two patterns of infection were noted: one of brief excretion with little or no illness, and one of persistent excretion frequently associated with illness. The second group had prolonged and intermittent (or recrudescent) excretion and demonstrated prolonged intrafamilial spread. Fifty-five percent of infections were subclinical; 82% of clinical illness was respiratory, predominantly upper respiratory tract. Except for an association between Ad2 and pharyngitis, there was no clear relationship between virus type, clinical symptoms, and age. No cases of meningitis or encephalitis were observed.

An analysis of data reported to the World Health Organization from 1967 to 1975 was similar in that approximately 88% of all adenoviral infections involved children. Twelve percent (2023/16,458) had CNS manifestations, but the rate of isolation of virus from cerebral spinal fluid or brain was not reported for children. Among adults, 2% (65/2753) reportedly had adenovirus isolated from their spinal fluid [14]. Data from Virus Watch and from WHO indicate that types 2, 1, 7, 3, and 5 account for most symptomatic CNS infections [56–59].

In its surveillance of aseptic meningitis in 1971 and of encephalitis in 1973, the Centers for Disease Control and Prevention reported that adenoviral infection accounted for 0.3% (2/633) of aseptic meningitis and 1.8% (4/520) of encephalitis cases, with one reported death from encephalitis in 1973 [60].

6 PATHOPHYSIOLOGY

Adenovirus enters the body through the mucosal surface of the respiratory tract, conjunctivae, or gastrointestinal tract. Under natural circumstances, the incubation for respiratory syndromes is on average 6–9 days; it is longer for EKC (3–22 days) and shorter for enteric adenoviruses (2–10 days). Usually the infectious cycle is initiated by binding of the virion fiber to a 46 kDa protein receptor (CAR) in the surface of cells [10]. An additional interaction between cell surface integrins ($\alpha V\beta 3$ and $\alpha V\beta 5$) and the penton base protein in which the fiber is embedded has been shown to be important for internalization of many adenoviruses into the cell [61]. Viremia may occur early in the course of disease and can be associated with a maculopapular rash. Adenovirus can be recovered from respiratory secretions from most body organs in disseminated disease, and from spinal fluid in patients with meningoencephalitis. A particular neurotropism has been attributed to some adenovirus serotypes (Ad3, 5, and 7) because they seem to be recovered from the spinal fluid more often than others. However, there are not enough data to better define the pathogenesis of adenovirus in the central nervous system. No characteristic CSF or histopathological findings have been described in patients with neurological manifestations during adenoviral disease or in patients in whom adenovirus was isolated from spinal fluid (Table 2). Inflammatory and/or immunological mechanisms may play a role in CNS disease associated with adenovirus. Diffuse encephalitis, necrosis, hemorrhage, and glial nodules have been described in brain autopsies of fatal cases [16,17,23,40,41,43,60,62–64].

Adenoviruses may remain latent intracellularly in lymphadenoid tissue, lymphocytes, and kidney after an acute infection. Species-specific neutralization antibody has been associated with prevention of symptomatic disease but not necessarily of infection. Maternally derived IgG antibody is likely protective against serious adenoviral infections early in life, because infection rates are lower in infants under 6 months of age. Both humoral and cellular immune mechanisms are important for protection, as evidenced by the increased severity of illness in patients with impaired IgA responses and the increased severity of disease and risk of dissemination in patients with altered cell-mediated immune mechanisms [19,42,45,65,66].

7 NEUROLOGICAL MANIFESTATIONS OF ADENOVIRAL INFECTION

There are no pathognomonic features indicative of adenoviral infection of the central nervous system. Almost all reported cases have occurred in children and only four in adults (Table 2). One of the adults was a military recruit [67]. Outbreaks of Ad7 are well recognized among new recruits. Pneumonia and pharyngitis are common manifestations, but CNS disease is uncommon even during outbreaks [63]. Two adult patients were immunosuppressed, one with malignant lymphoma and one with AIDS [62,64]. The fourth reported adult case was atypical and difficult to evaluate [68]. It involved a patient who had had a previous episode of hearing loss. With a second episode of hearing loss, which was both more severe and more persistent, the patient was also symptomatic with an upper respiratory infection, and Ad3 was isolated from the pharynx. Especially considering the previous episode of hearing loss, it is difficult to accept the Ad3 as causative and ignore the possibility that it represented an unrelated episode of pharyngitis in a patient with a Ménière's type illness [69].

Forty-seven children with adenovirus-associated CNS disease are described in Table 2. Although the completeness and availability of clinical data vary considerably from case to case, it is possible to make a rough tabulation of symptoms. Alterations in consciousness and seizures were the most common neurological manifestations. A few patients were mentioned as having corticospinal tract signs, and one patient was ataxic. With the exclusion of the adult patient with hearing loss mentioned above, only one patient was reported to have a cranial nerve palsy [70]. Although 15 patients apparently died of their adenovirus infection and two others from their underlying disease, it should be kept in mind that patients who have died are likely to be overrepresented in the published literature. CNS manifestations occurred in both immunocompromised and immunocompetent patients, with different degrees of severity of disease, usually in association with respiratory tract symptoms but also in the context of disseminated adenoviral disease [13]. All surviving patients seemed to have recovered completely, except for one child described as having a persisting EEG abnormality [32].

Table 2 Reported Cases of Adenovirus-Associated Central Nervous System Disease

Case	Age (yr)	Epidemiology	Clinical non-CNS	Clinical CNS; outcome	Imaging	EEG	CSF	Diagnostic test	Ref.
1	42	Immunosuppressed; immunosuppressive therapy	Malignant lymphoma	Seizures, fever, confusion, coma; death	Radionuclide scan normal	Diffuse generalized slowing	7 WBC (M), Pr 162 mg%, Glu 86 mg%	Ad32 isolated from brain. Autopsy: focal encephalitis. EM: viral particles. Immunofluorescent staining: positive.	62
2	38	Community infection	URI	Sudden unilateral hearing loss with tinnitus (similar episode 1 yr earlier resolved); partial resolution with limited follow-up	N/A	N/A	Normal	Ad3 isolated from pharynx.	68
3	36	Immunosuppressed (AIDS)	N/A	N/A; death	N/A	N/A	N/A	Ad31 and Ad49 isolated from brain.	64
4	22	Military recruit; Ad7 prevalent in recruits in the same group	URI	Status epilepticus, respiratory failure; death	CT: sinusitis	N/A	Elevated protein	Autopsy: histological changes consistent with viral encephalitis; Ad PCR positive, lung and brain.	63
5	12	Community infection	Fever, headache, vomiting	Headache, slight rigidity of the back; full recovery	N/A	N/A	367 WBC, 87% N	CSF and stool negative for Ad; Ad antibody titer rose from 16 to 64.	70
6	10	6 weeks s/p resection of cerebellar astrocytoma	Fever and neck pain	Full recovery	N/A	N/A	1019 WBC, 79% N, 21% M, Pr 253 mg%	Ad12 isolated from CSF.	18

7	9	Community infection	Fever, pneumonia	Stupor, hallucinations; clinically normal, abnormal EEG	N/A	Diffuse slowing	Normal	Ad7 isolated from stool.	32
8	8	Community infection	Fever, vomiting	Headache, stiff neck, delirium; recovered	N/A	N/A	110 WBC, Pr 60 mg%	Ad7 isolated from stool.	39
9	6	Acute lymphocytic leukemia (ALL) in relapse	Fever, vomiting, *P. aeruginosa* skin infection	Progressive lethargy; died in ALL relapse	N/A	N/A	172 WBC (mostly M), Pr 50 mg%	Ad7a isolated from CSF.	18
10	6	Congenital; immunosuppression; sibling of case 25	Panhypogammaglobulinemia	Fever, seizures lasting 6 wk; died	CT: multiple areas of low attenuation	High-amplitude slow waves; multiple bizarre discharges	170 WBC	Ad and ECHO 11 in stool and nasopharynx but not in CSF.	72
11–19[a]	1–7	Community infection	Fever (8), hepatomegaly (8), vomiting (6), pneumonia (6)	Stupor (7), full fontanel (4), convulsions (3), loss of consciousness (3); death (3)	N/A	Normal (1), abnormal (5)	WBC normal (5), elevated (2); protein; normal (5), elevated (2)	Ad7 isolated from pharynx or stool; 3 deaths—brain showed only edema and cellular changes in two, lymphocyte infiltration in one.	40
20	5	Community infection	Fever, cough, anorexia; pneumonia on X-ray	Severe headache, lethargy, slight nuchal rigidity; full recovery	N/A	N/A	370 WBC, 93% N, 7% L, Pr 52 mg%	Ad12 isolated from CSF.	18
21	4	Community infection	Vomiting and fever for 3 days	Tremors (focal seizures?), papilledema, generalized seizures for 4 days, coma; full recovery	N/A	N/A	No cells	CSF produced cytopathogenic effect (CPE); Ad7 neutralization positive.	16
			Headache,						

(*Continued*)

Table 2 *Continued*

Case	Age (yr)	Epidemiology	Clinical non-CNS	Clinical CNS; outcome	Imaging	EEG	CSF	Diagnostic test	Ref.
22	4	Community infection	fatigue, vomiting, fever	Right facial paresis; full recovery including normal EEG	N/A	Suggested encephalitis in the left cerebral lobe	No cells	Ad7 isolated from stool.	70
23	4	Community infection	Exudative pharyngitis, anorexia, fever, headache over 9 days	Lethargy, hyperactive reflexes, sustained ankle clonus, bladder distention; full recovery	N/A	N/A	23 WBC, 90% L, Pr 39 mg%	Ad6 isolated from CSF.	18
24	3.6	Community infection	High fever, cough for 3 wk, pneumonia	Apathy, stiff neck, hypotonia, coma for 6 days; normal on discharge	N/A	N/A	1st CSF no cells; 2 days later, 46 WBC, Pr 120 mg%	Ad3 cultured from stool and pharynx.	73
25	3	Congenital, immuno-suppression; sibling of case 10	Panhypogamma-globulinemia; fever, vomiting, abnormal liver function tests	Headache; normal outcome after intraventricular immunoglobulin (100 mg/day × 3)	CT: marked ventricular dilatation with peri-ventricular lucency	N/A	180 cells	Ad and ECHO 11 isolated from CSF.	72
26	3	Family epidemic	Fever, URI, otitis, pneumonia	Stupor; full recovery	N/A	Diffuse slowing	Normal CSF	Ad7 isolated from stool and pharynx.	38
27	3	Community infection after herpes zoster	No fever, URI	Irritable, without energy; outcome not reported	N/A	Normal	No cells; Pr 16 mg%	Ad5 isolated from stool and CSF.	22
28	3	Community infection after chickenpox	Fever	Ataxia, long tract signs, stupor; outcome not reported	N/A	Normal	55 WBC, Pr 36 mg%	Ad5 isolated from CSF.	22
			Fever, hepatitis,						

29	2.5	Orthotopic liver transplant for cirrhosis due to biliary atresia	pneumonia with effusion, DIC, multiorgan involvement	Seizures; death from disseminated Ad disease	N/A	N/A	N/A	Ad2 isolated from CSF, blood, urine, pharynx, lung, liver, spleen. Autopsy: subdural and subarachnoid hemorrhages.	13
30	2.4	CHARGE syndrome with congenital heart disease	Fever, pneumonia, bronchiolitis	Seizures; cardiac arrest and death	N/A	N/A	N/A	Ad3 isolated from lung, heart. Autopsy: acute encephalitis, no viral inclusions, negative in situ hybridization.	13
31	2	Community infection; sibling of case 36	URI, fever, diarrhea, pneumonia, hepatocellular liver injury	Status epilepticus as other findings were resolving; death, pulmonary hemorrhage	N/A	Pseudoperiodic discharges and burst suppression pattern	No cells, Pr 80 mg%	Ad7 from lung but not CSF, penton antigen in serum.	17
32	1.8	Family epidemic	Fever, URI, stomatitis, otitis, vomiting	Irritable, stiff neck, focal seizures, stupor; full recovery	N/A	Diffuse slowing	No cells, Pr 40 mg%	Ad7 isolated from stool and pharynx,	38
33	1.8	Community infection	High fever, rhinitis, cough for 2 wk	Increased tone, generalized seizure; death	N/A	N/A	8000 WBC, 3000 RBC, Pr 117 mg%	Ad3 isolated from CSF, stool, and throat. Autopsy: brain edema without inflammatory infiltrates; Ad3 isolated from lung but not from brain.	73

(Continued)

Table 2 *Continued*

Case	Age (yr)	Epidemiology	Clinical non-CNS	Clinical CNS; outcome	Imaging	EEG	CSF	Diagnostic test	Ref.
34	1.7	Community infection	pneumonia with effusion, diarrhea, hepatitis, pancreatitis	Stupor, posturing, hypertonia, hyperreflexia, no seizures; full recovery	CT: normal	Diffuse slowing, consistent with encephalo-pathy	2 WBC, Pr 36 mg%	Ad7 isolated from blood, tracheal secretions, stool.	13
35	1.5	Community infection	Skin rash, pulmonary congestion, fever for 5 days	Found dead in bed	N/A	N/A	N/A	Ad5 isolated from brain. Autopsy: cerebral edema.	33
36	1	Community infection; sibling of case 31	URI, fever, pneumonia, rash, hepatocellular liver injury	Status epilepticus as other findings were resolving; death, respiratory failure	N/A	Pseudoperiodic discharges and burst suppression pattern	21 L, Pr 133 mg%	Ad7 from lung but not liver; penton antigen in serum.	17
37	1	Community infection	Cough, fever, diarrhea, URI, pneumonia	Stupor and death	N/A	N/A	69 cells, Pr 35 mg%	Autopsy: Ad isolated from brain, lung, and CSF.	31
38	1	Orphanage epidemic	Pneumonia	N/A death	N/A	N/A	N/A	Ad7 isolated from CSF, and from lung and brain at autopsy.	93
39	1	Community infection	Fever, URI, pneumonia, hepatomegaly and liver failure	Seizures, stupor; death	N/A; death	Diffuse slowing	No cells, Pr 40 mg%	Ad7 isolated from stool but not CSF. Autopsy: cerebral edema without inflammation.	32
			Pneumonia						

40	1	Community infection	URI, fever, otitis, diarrhea, lethargy, pneumonia	Stiff neck; no other details given	N/A	N/A	Normal	Ad7 isolated from stool and pharynx.	32
41	0.6	Community infection; lived in same town as cases 31 and 36		Status epilepticus as other findings were resolving; death, myocardial dysfunction	N/A	Pseudoperiodic discharges and burst suppression pattern	5 L, Pr 18 mg%	Ad7 from lung, penton antigen in serum. Autopsy: brain and other organs negative for Ad7.	17
42	NB	Community infection	Cyanosis, hypothermia, pneumonia	Hemiconvulsions; died day 21	N/A	N/A	No cells	Ad11 isolated from brain, lung. Brain showed edema, perivascular cuffing by mononuclear cells, small foci of cerebellar hemorrhage.	52
43–51	?	Routine culture of stool specimens 1949–1954	Massachusetts residents suspected of having polio	Three had fever and meningismus but no cells in CSF; outcomes not reported	N/A	N/A	Six had CSF pleocytosis	Ad3, Ad9, Ad11, Ad12. One patient also had paralytic polio with type 1 poliovirus.	37

Abbreviations: Ad, adenovirus; CNS, central nervous system; EEG, electroencephalogram; CSF, cerebrospinal fluid; CT, computed tomography; URI, upper respiratory tract infection; WBC, white blood cells; M, monocytes; N, neutrophils; L, lymphocytes; Pr, protein; Glu, glucose; NB, newborn; RBC, red blood cells.

[a]The number in parentheses is the number of patients with the indicated finding.

Ladisch et al. [17] reported three patients with a Reye's syndrome–like illness. The children presented from the same geographic area within a short time of each other. All had pneumonia due to Ad7. As the pneumonia seemed to be improving, the children developed seizures and became comatose. Laboratory studies showed evidence of hepatocellular dysfunction and disseminated intravascular coagulation in two of the children. Interestingly enough, these patients did not have vomiting, elevated ammonia levels, or papilledema, usually seen in Reye's syndrome. It is not known whether these children received aspirin. All three children died. Pathologically, the brains did not show evidence of encephalitis but rather showed cortical neuron depletion and necrosis. Their livers were enlarged and demonstrated microvesicular and univacuolar fatty infiltration and irregular Kupffer cell hyperplasia. On electron microscopy the affected livers showed dilated rough endoplastic reticulum and did not show the abnormal mitochondria and smooth endoplasmic reticulum reported in Reye's syndrome. Additionally, all the children had evidence of vascular injury in multiple other tissues. Ad7 was isolated from the lung in each case but not from any of the other tissues. Serum from each patient was positive for an adenovirus penton antigen. The authors postulate that it was this antigen that caused the extrapulmonary tissue injury that led to death. Possible mechanisms of action include direct toxicity or an immunopathological injury caused by circulating antigen–antibody complexes.

Other reported cases might also fit this Reye's syndrome–like illness. After personal communication, Ladisch et al. decided that the case reported by Escobar and Sodhi [71] was similar to theirs. Also similar is the case of the 1-year-old girl reported by Steen-Johnsen et al. [32] who developed an upper respiratory infection and then pneumonia. A week later she worsened and developed peripheral cyanosis, seizures, and coma. There were thrombocytopenia and derangement of her coagulation factors. With intensive supportive treatment she began to improve, but after 3 weeks of slow improvement she suddenly deteriorated with marked respiratory distress, cardiac failure, fever, and leukocytosis and died. At autopsy her brain showed cerebral edema without inflammation, and her liver showed fatty changes. Ad7 was isolated from feces but not CSF.

Two of the patients reported by Simila et al. [40] also seem to fit this pattern. They were part of an Ad7 outbreak, and Ad7 was isolated from pharynx or feces. Two of the three patients who died showed brain edema and cellular changes without evidence of inflammation. Two of the fatal cases had "severe thrombocytopenic haemorrhagic diathesis," although which two was not stated. Simila et al. [40] report that hepatomegaly and edema were inconstant parts of the clinical picture but do not link these findings to specific patients.

Relatively few reports included brain imaging or electroencephalography (EEG) in patients with adenoviral infection. Each of two siblings with panhypogammaglobulinemia had either periventricular lucencies or multiple areas of low attenuation on computed tomographic (CT) scanning [72], suggesting diffuse injury apparently unique to them. The reported EEGs did not demonstrate any pathognomonic features. Pseudo-periodic discharges and burst suppression [17] are indicative of diffuse and severe cortical involvement, as is the presence of high-amplitude slow waves with multiple bizarre discharges [72].

It is interesting that some CSF examinations did not demonstrate any cells. A few but not all of these patients were immunosuppressed. When cells were present there were less than 1000 cells/mm^3, except for a patient with a preceding surgical removal of a cerebellar tumor [18] and one with red and white blood cells in the fluid [73].

Adenovirus type 7 was the adenovirus by far the most commonly associated with CNS disease. Two or three cases have been reported in association with adenovirus types 3, 5, 11, and 12, and single reports involve types 6, 9, and 32. One patient was positive for both types 31 and 49. Ad7 has also been the adenovirus serotype reported in association with the Reye's syndrome–like illness.

8 DIAGNOSTIC STRATEGIES

Several diagnostic approaches are available for the specific diagnosis of adenoviral infection. In most cases, adenoviruses can be readily recovered from clinical samples early in the course of the disease, and viral isolation is the diagnostic method of choice. Viral isolation also provides the opportunity to determine the species and subspecies (serotyping), which is important for epidemiological studies. Appropriate samples for viral culture include nasopharyngeal swabs or aspirates, throat swabs or washes, conjunctival swabs or scrapings, stool or rectal swabs, urine, cerebrospinal fluid (CSF), and tissue samples. Other methods of diagnosis include antigen detection assays, monoclonal antibody tests, nucleic acid amplification tests, and serological assays.

8.1 Viral Culture

Viral culture is the most sensitive and most specific method for detecting most adenoviruses. All serotypes, except types 40 and 41 (enteric adenoviruses), cause a characteristic cytopathic effect (CPE) in human epithelial cell lines such as HeLa, A549, or HEp2 and in primary human embryonic kidney (HEK) cells. The CPE—rounding of cells and clumping in grapelike clusters (Fig. 1)—generally occurs within 2–7 days with the common lower serotypes, but with others, especially subgroup D serotypes, it can require up to 28 days. Centrifugation of the sample and inoculation in shell vials may hasten detection. Isolation of enteric adenoviruses requires special cell lines such as an HEK cell line transformed by adenovirus type 5 and Chang conjunctival cells.

Several adenovirus serotypes have been isolated in spinal fluid or brain tissue of patients with symptomatic neurological disease, particularly children. However, isolation of adenovirus from CSF is rare. A presumptive diagnosis of CNS involvement during adenoviral infection is usually possible in patients with neurological manifestations unexplained by other pathologies and isolation of adenovirus from respiratory secretions, stool, or tissue from other affected organs.

8.2 Serotyping of Isolates

Serotype analysis can be performed in a reference virology laboratory. Use of group-specific (antihexon) antibody identifies the isolate as a member of the adenovirus group. Serotyping is traditionally performed by hemagglutination inhibition and/or serum neutralization assays using a selected panel of type-specific rabbit antisera.

8.3 Viral Antigen Assays

Direct detection of adenovirus antigens in clinical samples may be performed by adenovirus-specific enzyme-linked immunosorbent assay (ELISA), immunofluorescence assay (IFA), or radio and enzyme immunoassay (RIA or EIA). Commercially available assays use group-specific monoclonal antibodies that react with common antigenic determinants

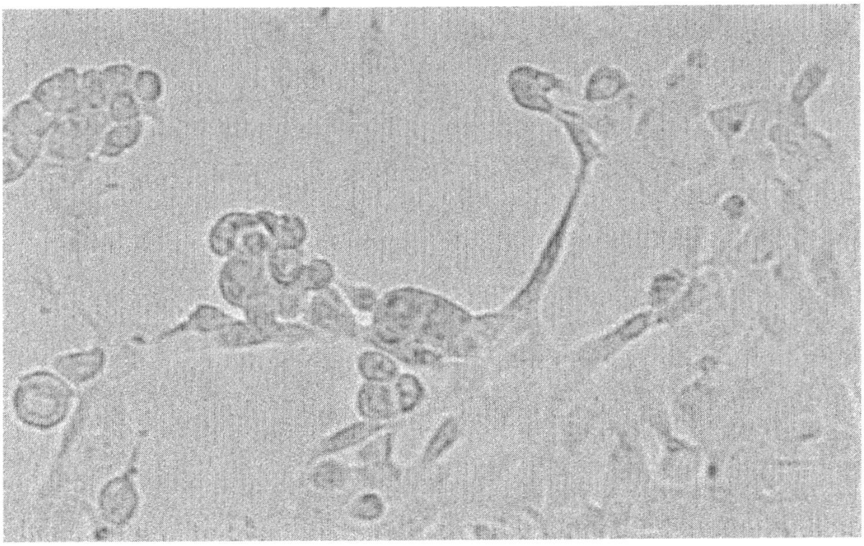

Figure 1 Adenovirus cytopathic effect (CPE) in tissue culture cells. Rounding of cells and clumping in grapelike clusters is observed. (Photograph courtesy of Gail Demmler, MD and the Diagnostic Virology Laboratory at Texas Children's Hospital, Houston, Texas.)

on all serotypes. Although these assays can provide a rapid diagnosis, they are not specific enough for epidemiological studies and are not as sensitive as viral culture for the detection of most serotypes.

Antigen assays are particularly useful for the detection of the fastidious adenovirus types 40 and 41 in stool samples. Another potential application is in the rapid diagnosis of epidemic diseases such as keratoconjunctivitis (EKC). Direct adenovirus antigen assays can also be used to screen cell cultures before the development of CPE as well as to confirm the presence of adenovirus in cell cultures positive for CPE.

8.4 Restriction Endonuclease Analysis

Restriction endonuclease (RE) analysis can distinguish between different clinical isolates of the same serotype. RE analysis is useful for epidemiological studies, particularly during outbreaks of adenoviral infection.

8.5 Polymerase Chain Reaction

Polymerase chain reaction (PCR) is a highly sensitive and specific assay that can be used to detect adenoviral DNA from a variety of clinical specimens including fixed tissues. Because different adenovirus serotypes are very heterogeneous at the DNA level, PCR primers may be selected to detect specific serotypes or related serotypes. Other primers have been designed to detect the majority of serotypes, although most "universal" primers do not perform equally well against all serotypes.

The experience with PCR detection of adenovirus in CSF is limited, and data are not available on sensitivity or specificity. Nested primers are used to increase the probability of detection of adenoviruses in the CSF. In one study [74], adenovirus was detected by PCR in one culture-positive clinical sample of CSF. In three prospective studies that included PCR assays for the detection of adenovirus in CSF samples of patients with clinical neurological symptoms of aseptic meningitis or encephalitis, 0 of 27, 383, and 2333 adult and child specimens, respectively, were positive for adenovirus by PCR [75–77]. PCR techniques allowed the identification of subgenus B adenovirus in the CSF of 10 children who died during an epidemic of enterovirus-71 hand-foot-and-mouth disease in Malaysia in 1997, raising the possibility that coinfection with adenovirus contributed to the CNS manifestations and mortality in these children [41].

Polymerase chain reaction has been used to diagnose adenoviral myocarditis, an entity difficult to diagnose by culture or histopathology alone [78,79]. PCR has also been used to look for persistent or latent adenoviral infection in tissue samples and peripheral blood mononuclear cells (PBMCs) [80,81]. In HIV-infected patients, PCR has been shown to be a sensitive and specific tool for detecting adenovirus in various samples during acute infection and to determine the presence of asymptomatic viral shedding [82]. Investigational multiplex PCR assays allow the detection of several viruses in a single test and the possibility of a rapid diagnosis of adenovirus in children with acute respiratory infections [83,84]. Multiplex PCR has been found to be more sensitive than viral culture for the diagnosis of adenoviral keratoconjunctivitis [85], but the experience with CSF specimens is limited.

8.6 Histopathology

Histopathology may be nonspecific, especially in the early stages of infection. Adenoviruses can cause characteristic intranuclear inclusions in infected organs [86]. Early postinfection cells may display small eosinophilic inclusions. During the later stages of infection, basophilic inclusions appear, which initially may be surrounded by a clear halo within the nucleus. When these intranuclear inclusions enlarge and obscure the nuclear membrane, the cells are referred to as "smudge" cells. However, these findings have not been described in the brain. Adenoviruses cause neither intracytoplasmic inclusions nor multinucleated cells [87].

If routine histopathology is nondiagnostic and viral culture of tissue is negative (or not done), more specialized tests may be performed on tissue samples. Electron microscopy can be used to detect the characteristic icosahedral virions that typically form large paracrystalline aggregates with the nuclei of infected cells [88]. Adenovirus-specific immunohistochemical assays and in situ DNA assays are also available.

8.7 Serology

Recent infection may be documented by assay of paired acute and convalescent sera for anti-adenovirus antibodies [2]. It is important to document a fourfold or greater rise in antibody titer, because there is a high prevalence of anti-adenovirus antibodies in the general population, and there are numerous cross-reactions between related serotypes.

The most widely available assay measures complement fixation antibodies that react against all adenovirus serotypes. EIA is much more sensitive than complement fixation and is being increasingly used in diagnostic laboratories. Both of these assays measure group-specific anti-hexon antibodies and do not provide information about the serotype. In contrast, detection of hemagglutination inhibition antibodies or neutralizing antibodies is more sensitive and serotype-specific. These assays are primarily performed in reference laboratories and are best interpreted when the patient's sera are tested against the patient's own isolate.

9 TREATMENT

There is no specific treatment for adenoviral infection. Most illnesses caused by adenoviruses are self-limited, and patients with severe disease can recover with adequate supportive therapy. Antiviral drugs such as ribavirin, ganciclovir, vidarabin, and cidofovir have been used alone or in combination with intravenous immunoglobulin in isolated cases of hemorrhagic cystitis, severe pneumonia, and disseminated disease, particularly in immunocompromised patients and neonates [55,89,90]. Although these drugs may have in vitro activity against certain adenovirus serotypes [91,92], the clinical effect of antiviral therapy has not been studied systematically and is difficult to ascertain in cases reported to date where the final outcome tends to vary according to the host's underlying immune status and the extent of adenoviral disease.

REFERENCES

1. Foy, H.M. Adenoviruses. In: Viral Infections of Humans: Epidemiology and Control; Evans, A.S., Kaslow, R.A., Eds.; Plenum Medical: New York, 1997, 119–138.
2. Horwitz, M.S. Adenoviruses. In: Fields Virology; Fields, B.N., Knipe, D.M., Howley, P.M., Eds.; Lippincott-Raven: Philadelphia, 1996, 2160.
3. Benko, M.; Harrach, B.; Russell, W.C. Family Adenoviridae. In: Seventh Report of the International Committee on Taxonomy of Viruses. Virus Taxonomy; Van Regenmortel, M.H.V., Fauquet, C.M., Bishop, D.K.L., Carstens, E.B., Estes, M.K., Lemon, S.M., et al., Eds.; Academic Press: New York, 1999, 227–238.
4. Wadell, G. Molecular epidemiology of human adenoviruses. Curr Top Microbiol Immunol. 1984, 110, 191–220.
5. Hierholzer, J.C. Adenoviruses. In: Diagnostic Procedures for Viral, Rickettsial, and Chlamydial Infections; Lennette, E.H., Lennette, D.A., Lennette, E.T., Eds.; Am Public Health Assoc: Washington, DC, 1995, 169–188.
6. De Jong, J.C.; Wermenbol, A.G.; Verweij-Uijterwaal, M.W.; Slaterus, K.W.; Wertheim-Van Dillen, P.; GJ Van Doornum, G.J. Adenoviruses from human immunodeficiency virus-infected individuals, including two strains that represent new candidate serotypes Ad50 and Ad51 of species B1 and D, respectively. J Clin Microbiol. 1999, 37(12), 3940–3945.
7. Adrian, T.; Wadell, G.; Hierholzer, J.C.; Wigand, R. DNA restriction analysis of adenovirus prototypes 1 to 41. Arch Virol. 1986, 91(3–4), 277–290.
8. Elnifro, E.M.; Cooper, R.J.; Klapper, P.E.; Bailey, A.S. PCR and restriction endonuclease analysis for rapid identification of human adenovirus subgenera. J Clin Microbiol. 2000, 38(6), 2055–2061.
9. Kajon, A.; Wadell, G. Genome analysis of South American adenovirus strains of serotype 7 collected over a 7-year period. J Clin Microbiol. 1994, 32(9), 2321–2323.
10. Bergelson, J.M.; Cunningham, J.A.; Droguett, G.; Kurt-Jones, E.A.; Krithivas, A.; Hong, J.S. Isolation of a common receptor for coxsackie B viruses and adenoviruses 2 and 5. Science. 1997, 275(5304), 1320–1323.
11. Arditi, M.; Ostrove, J.M.; Shulman, S.T. Acute measles complicated by fatal adenovirus pneumonia with bone marrow failure: lack of evidence for direct viral invasion of the bone marrow. Pediatr Infect Dis J. 1992, 11(4), 327–330.

12. Severien, C.; Teig, N.; Riedel, F.; Hohendahl, J.; Rieger, C. Severe pneumonia and chronic lung disease in a young child with adenovirus and Bordetella pertussis infection. Pediatr Infect Dis J. 1995, 14(5), 400–401.
13. Munoz, F.M.; Piedra, P.A.; Demmler, G.J. Disseminated adenovirus disease in immunocompromised and immunocompetent children. Clin Infect Dis. 1998, 27(5), 1194–1200.
14. Schmitz, H.; Wigand, R.; Heinrich, W. Worldwide epidemiology of human adenovirus infections. Am J Epidemiol. 1983, 117(4), 455–466.
15. D'Angelo, L.J.; Hierholzer, J.C.; Keenlyside, R.A.; Anderson, L.J.; Martone, W.J. Pharyngoconjunctival fever caused by adenovirus type 4: report of a swimming pool-related outbreak with recovery of virus from pool water. J Infect Dis. 1979, 140(1), 42–47.
16. Gabrielson, M.O.; Joseph, C.; Hsiung, G.D. Encephalitis associated with adenovirus type 7 occurring in a family outbreak. J Pediatr. 1966, 68(1), 142–144.
17. Ladisch, S.; Lovejoy, F.H.; Hierholzer, J.C.; Oxman, M.N.; Strieder, D.; Vawter, G.F. Extrapulmonary manifestations of adenovirus type 7 pneumonia simulating Reye syndrome and the possible role of an adenovirus toxin. J Pediatr. 1979, 95(3), 348–355.
18. Kelsey, D.S. Adenovirus meningoencephalitis. Pediatrics. 1978, 61(2), 291–293.
19. Hierholzer, J.C. Adenoviruses in the immunocompromised host. Clin Microbiol Rev. 1992, 5(3), 262–274.
20. Koskiniemi, M.; Rantalaiho, T.; Piiparinen, H.; von Bonsdorff, C.H.; Farkkila, M.; Jarvinen, A. Infections of the central nervous system of suspected viral origin: a collaborative study from Finland. J Neurovirol. 2001, 7(5), 400–408.
21. Ruuskanen, O.; Meurman, O.; Sarkkinen, H. Adenoviral diseases in children: a study of 105 hospital cases. Pediatrics. 1985, 76(1), 79–83.
22. Faulkner, R.; VanRooyen, C.E. Adenovirus types 3 and 5 isolated from the cerebrospinal fluid of children. Can Med Assoc J. 1962, 87, 1123.
23. Roos, R.; Chou, S.M.; Rogers, N.G.; Basnight, M.; Gajdusek, D.C. Isolation of an adenovirus 32 strain from human brain in a case of subacute encephalitis. Proc Soc Exp Biol Med. 1972, 139(2), 636–640.
24. Rowe, W.P.; Huebner, R.J.; Gilmore, L.K.; Parrot, R.H.; Ward, T.G. Isolation of a cytopathogenic agent from human adenoids undergoing spontaneous degeneration in tissue culture. Proc Soc Exp Biol Med. 1953, 84, 570–573.
25. Hilleman, M.R.; Werner, J.H. Recovery of new agents from patients with acute respiratory illness. Proc Soc Exp Biol. 1954, 85, 183–188.
26. Dingle, J.; Langmuir, A.D. Epidemiology of acute respiratory disease in military recruits. Am Rev Respir Dis. 1968, 97, 1–65.
27. Top, F.H., Jr; Grossman, R.A.; Bartelloni, P.J.; Segal, H.E.; Dudding, B.A.; Russell, P.K. Immunization with live types 7 and 4 adenovirus vaccines. I. Safety, infectivity, antigenicity, and potency of adenovirus type 7 vaccine in humans. J Infect Dis. 1971, 124(2), 148–154.
28. Top, F.H., Jr; Dudding, B.A.; Russell, P.K.; Buescher, E.L. Control of respiratory disease in recruits with types 4 and 7 adenovirus vaccines. Am J Epidemiol. 1971, 94(2), 142–146.
29. Barraza, E.M.; Ludwig, S.L.; Gaydos, J.C.; Brundage, J.F. Reemergence of adenovirus type 4 acute respiratory disease in military trainees: report of an outbreak during a lapse in vaccination. J Infect Dis. 1999, 179(6), 1531–1533.
30. De Jong, J.C.; Bijlsma, K.; Wermenbol, A.G.; Verweij-Uijterwaal, M.W.; Van Der Avoort, H.G.A.M.; Wood, D.J. Detection, typing, and subtyping of enteric adenoviruses 40 and 41 from fecal samples and observation of changing incidences of infections with these types and subtypes. J Clin Microbiol. 1993, 31, 1562–1569.
31. LeLong, M.; Lepine, P.; Alison, F. La pneumonie a virus du groupe APC chez le nourrison isolement du virus: les lesions anatomohistologiques. Arch Fr Pediatr. 1956, 13, 1092.
32. Steen-Johnsen, J.; Orstavik, I.; Attramadal, A. Severe illnesses due to adenovirus type 7 in children. Acta Paediatr Scand. 1969, 58(2), 157–163.

33. Chatterjee, N.K.; Samsonoff, W.A.; Balasubramaniam, N.; Rush-Wilson, K.; Spargo, W.; Church, T.M. Isolation and characterization of adenovirus 5 from the brain of an infant with fatal cerebral edema. Clin Infect Dis. 2000, 31(3), 830–833.
34. Sutton, R.N.P.; Pullen, H.J.M.; Blackledge, P.; Brown, E.H.; Sinclair, L.; Swift, P.N. Adenovirus type 7; 1971–74. Lancet. 1976, Nov 6, 2(7993), 987–991.
35. Barr, J.; Kjellen, L.; Svedmyr, A. Hospital outbreak of adenovirus type 3 infections. Acta Paediatr. 1958, 47, 365–382.
36. Levy, Y.; Nitzan, M.; Beharab, A.; Zeharia, A.; Schoenfeld, T.; Nutman, J. Adenovirus type 3 infection with systemic manifestation in apparently normal children. Isr J Med Sci. 1986, 22(11), 774–778.
37. Kibrick, S.; Melendez, L.; Enders, J.F. Clinical associations of enteric viruses with particular reference to agents exhibiting properties of the ECHO group. Ann NY Acad Sci. 1957, 67, 311–325.
38. Bebe, M. Epidemie familiale de maladie a adenovirus type 7. Arch Franc Pediatr. 1964, 21, 87–100.
39. Pereira, M.; MacCallum, F. Infection with adenovirus type 12. Lancet. 1964, 1, 198.
40. Simila, S.; Jouppila, R.; Salmi, A.; Pohjonen, R. Encephalomeningitis associated with an adenovirus type 7 epidemic. Acta Paediatr Scand. 1970, 59, 310.
41. Cardosa, M.J.; Krishnan, S.; Tio, P.H.; Perera, D.; Wong, S.C. Isolation of subgenus B adenovirus during a fatal outbreak of enterovirus 71-associated hand, foot, and mouth disease in Sibu, Sarawak. Lancet. 1999, 354(9183), 987–991.
42. Blanke, C.; Clark, C.; Broun, E.R.; Tricot, G.; Cunningham, I.; Cornetta, K. Evolving pathogens in allogeneic bone marrow transplantation: increased fatal adenoviral infections. Am J Med. 1995, 99(3), 326–328.
43. Flomenberg, P.; Babbitt, J.; Drobyski, W.R.; Ash, R.C.; Carrigan, D.R.; Sedmak, G.V. Increasing incidence of adenovirus disease in bone marrow transplant recipients. J Infect Dis. 1994, 169(4), 775–781.
44. Krilov, L.R.; Kaplan, M.H.; Frogel, M.; Rubin, L.G. Fatal adenovirus disease and human immunodeficiency virus infection. Pediatr Infect Dis J. 1990, 9(10), 753.
45. Krilov, L.R.; Rubin, L.G.; Frogel, M.; Gloster, E.; Ni, K.; Kaplan, M. Disseminated adenovirus infection with hepatic necrosis in patients with human immunodeficiency virus infection and other immunodeficiency states. Rev Infect Dis. 1990, 12(2), 303–307.
46. Michaels, M.G.; Green, M.; Wald, E.R.; Starzl, T.E. Adenovirus infection in pediatric liver transplant recipients. J Infect Dis. 1992, 165(1), 170–174.
47. Myerowitz, R.L.; Stalder, H.; Oxman, M.N.; Levin, M.J.; Moore, M.; Leith, J.D. Fatal disseminated adenovirus infection in a renal transplant recipient. Am J Med. 1975, 59(4), 591–598.
48. Edwards, K.M.; Thompson, J.; Paolini, J.; Wright, P.F. Adenovirus infections in young children. Pediatrics. 1985, 76(3), 420–424.
49. Pacini, D.L.; Collier, A.M.; Henderson, F.W. Adenovirus infections and respiratory illnesses in children in group day care. J Infect Dis. 1987, 156(6), 920–927.
50. Wesley, A.G.; Pather, M.; Tait, D. Nosocomial adenovirus infection in a paediatric respiratory unit. J Hosp Infect. 1993, 25(3), 183–190.
51. Montone, K.T.; Furth, E.E.; Pietra, G.G.; Gupta, P.K. Neonatal adenovirus infection: a case report with in situ hybridization confirmation of ascending intrauterine infection. Diagn Cytopathol. 1995, 12(4), 341–344.
52. Osamura, T.; Mizuta, R.; Yoshioka, H.; Fushiki, S. Isolation of adenovirus type 11 from the brain of a neonate with pneumonia and encephalitis. Eur J Pediatr. 1993, 152(6), 496–499.
53. Pinto, A.; Beck, R.; Jadavji, T. Fatal neonatal pneumonia caused by adenovirus type 35. Report of one case and review of the literature. Arch Pathol Lab Med. 1992, 116(1), 95–99.
54. Abzug, M.J.; Levin, M.J. Neonatal adenovirus infection: four patients and review of the literature. Pediatrics. 1991, 87(6), 890–896.

55. Gavin, P.J.; Katz, B.Z. Intravenous ribavirin treatment for severe adenovirus disease in immunocompromised children. Pediatrics. 2002, 110(1 Pt 1), E9–E9.
56. Fox, J.P.; Hall, C.E.; Cooney, M.K. The Seattle Virus Watch. VII. Observations of adenovirus infections. Am J Epidemiol. 1977, 105(4), 362–386.
57. Koneru, B.; Atchison, R.; Jaffe, R.; Cassavilla, A.; Van Thiel, D.H.; Starzl, T.E. Serological studies of adenoviral hepatitis following pediatric liver transplantation. Transplant Proc. 1990, 22(4), 1547–1548.
58. Fox, J.P.; Brandt, C.D.; Wassermann, F.E.; Hall, C.E.; Spigland, I.; Kogon, A. The Virus Watch program: a continuing surveillance of viral infections in metropolitan New York families. VI. Observations of adenovirus infections: virus excretion patterns, antibody response, efficiency of surveillance, patterns of infections, and relation to illness. Am J Epidemiol. 1969, 89(1), 25–50.
59. Assad, F.; Cockburn, W.C. A seven year study of WHO virus laboratory reports on respiratory viruses. Bull WHO. 1974, 51, 437–445.
60. Centers for Disease Control. Neurotropic viral diseases surveillance. Encephalitis. 2; Dep. Health, Educ Welfare: Atlanta, 1973.
61. Wickham, T.J.; Mathias, P.; Cheresh, D.A.; Nemerow, G.R. Integrins alpha v beta 3 and alpha v beta 5 promote adenovirus internalization but not virus attachment. Cell. 1993, 73(2), 309–319.
62. Chou, S.M.; Roos, R.; Burrell, R.; Gutmann, L.; Harley, J.B. Subacute focal adenovirus encephalitis. J Neuropathol Exp Neurol. 1973, 32(1), 34–50.
63. Ryan, M.A.K.; Gray, G.C.; Malasig, M.D. Two fatal cases of adenovirus related illness in previously healthy young adults—Illinois 2000. MMWR. 2001, 50(26), 553–555.
64. Schnurr, D.; Bollen, A.; Crawford-Miksza, L.; Dondero, M.E.; Yagi, S. Adenovirus mixture isolated from the brain of an AIDS patient with encephalitis. J Med Virol. 1995, 47(2), 168–171.
65. Dagan, R.; Schwartz, R.H.; Insel, R.A.; Menegus, M.A. Severe diffuse adenovirus 7a pneumonia in a child with combined immunodeficiency: possible therapeutic effect of human immune serum globulin containing specific neutralizing antibody. Pediatr Infect Dis. 1984, 3(3), 246–251.
66. Hromas, R.; Clark, C.; Blanke, C.; Tricot, G.; Cornetta, K.; Hedderman, A. Failure of ribavirin to clear adenovirus infections in T cell-depleted allogeneic bone marrow transplantation. Bone Marrow Transplant. 1994, 14(4), 663–664.
67. Ryan, M.A.; Gray, G.C.; Smith, B.; McKeehan, J.A.; Hawksworth, A.W.; Malasig, M.D. Large epidemic of respiratory illness due to adenovirus types 7 and 3 in healthy young adults. Clin Infect Dis. 2002, 34(5), 577–582.
68. Jaffe, B.F.; Maassab, H.F. Sudden deafness associated with adenovirus infection. N Engl J Med. 1967, 276(25), 1406–1409.
69. Thai-Van, H.; Bounaix, M.J.; Fraysse, B. Meniere's disease: pathophysiology and treatment. Drugs. 2001, 61(8), 1089–1102.
70. Janson, E. Epidemic occurrence of adenovirus type 7 infection in Helsinki. Ann Paediatr Fenn. 1962, 8, 24–34.
71. Escobar, M.R.; Sodhi, H. Diagnostic problems in adenoviral disease. VA Med Mon (1918). 1973, 100(3), 258–260.
72. Lau, Y.L.; Levinsky, R.J.; Morgan, G.; Strobel, S. Dual meningoencephalitis with echovirus type 11 and adenovirus in combined (common variable) immunodeficiency. Pediatr Infect Dis J. 1988, 7(12), 873–876.
73. Sattelkau, V.G. Der Nachweis einer Adenovirusinfektion (Typ 3) bei schwer bzw. todlich verlaufener Meningoenzephalitis. Arch Kinderheilk. 1965, 170, 174–181.
74. Avellon, A.; Perez, P.; Aguilar, J.C.; Lejarazu, R.; Echevarria, J.E. Rapid and sensitive diagnosis of human adenovirus infections by a generic polymerase chain reaction. J Virol Methods. 2001, 92(2), 113–120.

75. Studahl, M.; Bergstrom, T.; Hagberg, L. Acute viral encephalitis in adults—a prospective study. Scand J Infect Dis. 1998, 30(3), 215–220.
76. Chesky, M.; Scalco, R.; Failace, L.; Read, S.; Jobim, L.F. Polymerase chain reaction for the laboratory diagnosis of aseptic meningitis and encephalitis. Arq Neuropsiquiatr. 2000, 58(3B), 836–842.
77. Read, S.J.; Jeffery, K.J.; Bangham, C.R. Aseptic meningitis and encephalitis: the role of PCR in the diagnostic laboratory. J Clin Microbiol. 1997, 35(3), 691–696.
78. Martin, A.B.; Webber, S.; Fricker, F.J.; Jaffe, R.; Demmler, G.; Kearney, D. Acute myocarditis. Rapid diagnosis by PCR in children. Circulation. 1994, 90(1), 330–339.
79. Towbin, J.A.; Griffin, L.D.; Martin, A.B.; Nelson, S.; Siu, B.; Ayres, N.A. Intrauterine adenoviral myocarditis presenting as nonimmune hydrops fetalis: diagnosis by polymerase chain reaction. Pediatr Infect Dis J. 1994, 13(2), 144–150.
80. Flomenberg, P.; Gutierrez, E.; Piaskowski, V.; Casper, J.T. Detection of adenovirus DNA in peripheral blood mononuclear cells by polymerase chain reaction assay. J Med Virol. 1997, 51(3), 182–188.
81. Horvath, J.; Palkonyay, L.; Weber, J. Group C adenovirus DNA sequences in human lymphoid cells. J Virol. 1986, 59(1), 189–192.
82. Echavarria, M.; Forman, M.; Ticehurst, J.; Dumler, J.S.; Charache, P. PCR method for detection of adenovirus in urine of healthy and human immunodeficiency virus-infected individuals. J Clin Microbiol. 1998, 36(11), 3323–3326.
83. Grondahl, B.; Puppe, W.; Hoppe, A.; Kuhne, I.; Weigl, J.A.; Schmitt, H.J. Rapid identification of nine microorganisms causing acute respiratory tract infections by single-tube multiplex reverse transcription-PCR: feasibility study. J Clin Microbiol. 1999, 37(1), 1–7.
84. Na, B.K.; Kim, J.H.; Shin, G.C.; Lee, J.Y.; Lee, J.S. Detection and typing of respiratory adenoviruses in a single-tube multiplex polymerase chain reaction. J Med Virol. 2002, 66(4), 512–517.
85. Elnifro, E.M.; Cooper, R.J.; Klapper, P.E.; Yeo, A.C.; Tullo, A.B. Multiplex polymerase chain reaction for diagnosis of viral and chlamydial keratoconjunctivitis. Invest Ophthalmol Vis Sci. 2000, 41(7), 1818–1822.
86. Becroft, D.M. Histopathology of fatal adenovirus infection of the respiratory tract in young children. J Clin Pathol. 1967, 20(4), 561–569.
87. Landry, M.L.; Fong, C.K.; Neddermann, K.; Solomon, L.; Hsiung, G.D. Disseminated adenovirus infection in an immunocompromised host. Pitfalls in diagnosis. Am J Med. 1987, 83(3), 555–559.
88. Pinkerton, H.; Carroll, S. Fatal adenovirus pneumonia in infants. Correlation of histologic and electron microscopic observations. Am J Pathol. 1971, 65(3), 543–548.
89. Bordigoni, P.; Carret, A.S.; Venard, V.; Witz, F.; Le Faou, A. Treatment of adenovirus infections in patients undergoing allogeneic hematopoietic stem cell transplantation. Clin Infect Dis. 2001, 32(9), 1290–1297.
90. Aebi, C.; Headrick, C.L.; McCracken, G.H.; Lindsay, C.A. Intravenous ribavirin therapy in a neonate with disseminated adenovirus infection undergoing extracorporeal membrane oxygenation: pharmacokinetics and clearance by hemofiltration. J Pediatr. 1997, 130(4), 612–615.
91. Sidwell, R.W. In vitro and in vivo inhibition of DNA viruses by ribavirin. In: Clinical Applications of Ribavirin; Smith, R.A., Knight, V., Smith, J.A.D., Eds.; Academic Press: New York, 1984, 19–32.
92. de Oliveira, C.B.; Stevenson, D.; LaBree, L.; McDonnell, P.J.; Trousdale, M.D. Evaluation of cidofovir (HPMPC, GS-504) against adenovirus type 5 infection in vitro and in a New Zealand rabbit ocular model. Antiviral Res. 1996, 31(3), 165–172.
93. Chany, C.; Lepine, P.; LeLong, M. Severe fatal pneumonia in infants and young children associated with adenovirus infections. Am J Hyg. 1958, 67, 637.

18

Measles and Its Neurological Complications

Benedikt Weissbrich, Jürgen Schneider-Schaulies, and Volker ter Meulen
University of Würzburg
Würzburg, Germany

1 INTRODUCTION

Although the incidence of measles has declined substantially since the introduction of live attenuated vaccines, it is still a disease of worldwide importance [1]. With about 900,000 deaths per year, measles remains among the three most common causes of childhood mortality and is a major cause of neurological deficits in countries where its incidence is high. Acute measles can be accompanied by early or late central nervous system (CNS) complications [2]. These include acute postinfectious measles encephalomyelitis (APME), which usually develops within 1 week after the onset of the rash, late complications measles inclusion body encephalitis (MIBE) affecting immunocompromised patients, and subacute sclerosing panencephalitis (SSPE), which presents months to years after the initial measles infection. These three diseases can be distinguished clinically, and they have distinct histopathological characteristics. Moreover there are specific laboratory tests that help to differentiate these conditions. Yet their pathogenesis is not fully understood.

APME is the most common CNS complication and occurs in about 0.1% of natural measles infections [3]. Its mortality ranges between 10% and 40%, and most of the survivors are left with long-term sequelae. SSPE and MIBE are rare complications but are almost always fatal. Currently, neither acute measles nor its neurological complications are susceptible to any specific therapy. Treatment is primarily symptomatic, and its effectiveness is only palliative if that. Prevention of the primary disease through vaccination is the only effective means of preventing its severe neurological complications.

2 VIRUS FAMILY, STRUCTURE, AND REPLICATION

Measles virus (MV) is a member of the newly introduced *Mononegavirales* group, which comprises the *Rhabdo-*, *Filo-*, and *Paramyxoviridae*. As a paramyxovirus, MV possesses structural and biochemical features associated with this group; however, it lacks a detectable virion-associated neuraminidase activity. Therefore it has been grouped into a separate genus, the morbilliviruses. Other members of this group include rinderpest virus, which infects cattle; peste des petit ruminants, which infects sheep and goats; canine distemper virus, which infects various carnivores; phocine distemper virus; dolphin morbillivirus; and porpoise morbillivirus. All these viruses exhibit antigenic similarities and produce similar diseases in their host species, but their neuroinvasiveness differs considerably. Whereas canine distemper virus causes neurological disease in approximately 50% of dogs, MV causes encephalomyelitis in about 0.1% of cases.

Measles virus virions are highly pleomorphic, with an average size of 120–250 nm, and both filamentous and irregular forms are known (Fig. 1). MV particles consist of a lipid envelope surrounding the viral ribonucleoprotein (RNP) complex (Fig. 2). Both viral transmembrane proteins, the fusion (F) protein composed of F1 (41 kDa) and F2 (22 kDa) and the hemagglutinin (H) protein (80 kDa), are present on the envelope surface and appear as projections from the particle. The cytoplasmic domains of both are believed to interact with the matrix (M) protein (37 kDa), which in turn links the envelope to the RNP core structure. The viral genomic RNA is fully encapsidated by nucleocapsid (N) protein (60 kDa) to form the RNP core structure that resists RNAse degradation. Because the viral genome cannot serve as mRNA, the viral polymerase complex consisting of the phosphoprotein (P; 70 kDa) and the large protein (L; 220 kDa) is part of the RNP core complex.

The viral genome is a nonsegmented RNA molecule of negative polarity that is about 16 kb in length [4]. The genome encodes six structural genes for which the reading frames are arranged linearly and without overlap in the order 3′-N-P-M-F-H-L-5′ (Fig. 2). The genome is flanked by noncoding 3′-leader and 5′-trailer sequences that are thought to contain specific encapsidation signals and the viral promoters used for viral transcription and/or replication. From the P gene, three nonstructural proteins, C (20 kDa), V (46 kDa), and R (46 kDa), can be expressed [5].

Replication of MV is confined to the cytoplasmic compartment. Following delivery of the viral RNP complex into the cytoplasm of a susceptible host cell, viral transcription is initiated after specific attachment of the polymerase complex to the promoter located within the 3′ end of the genome and progresses to the 5′ end by transcribing mono- and bicistronic mRNAs. At each gene boundary, the polymerase complex resumes transcription of the distal gene or, controlled by unknown factors, leaves the template to reinitiate at the promoter region [6]. As a consequence, a polar gradient is established for the frequency of viral mRNAs, with the N-specific mRNA being the most abundant and the L-specific mRNA the least represented. At the 3′ end of each gene, poly(A) tracts are added to the mRNA transcripts, most probably by a polymerase stuttering mechanism at the termination signals. In addition, bi- and polycistronic polyadenylated transcripts spanning two or more adjacent genes are produced.

One of the most important characteristics determining viral tropism is the use of specific receptors on the surface of susceptible target cells that allow viral attachment and penetration. MV is highly species-specific in that it does not naturally replicate in nonprimate hosts. In vivo it reveals a pronounced tropism for epithelial cells and cells of hemato-

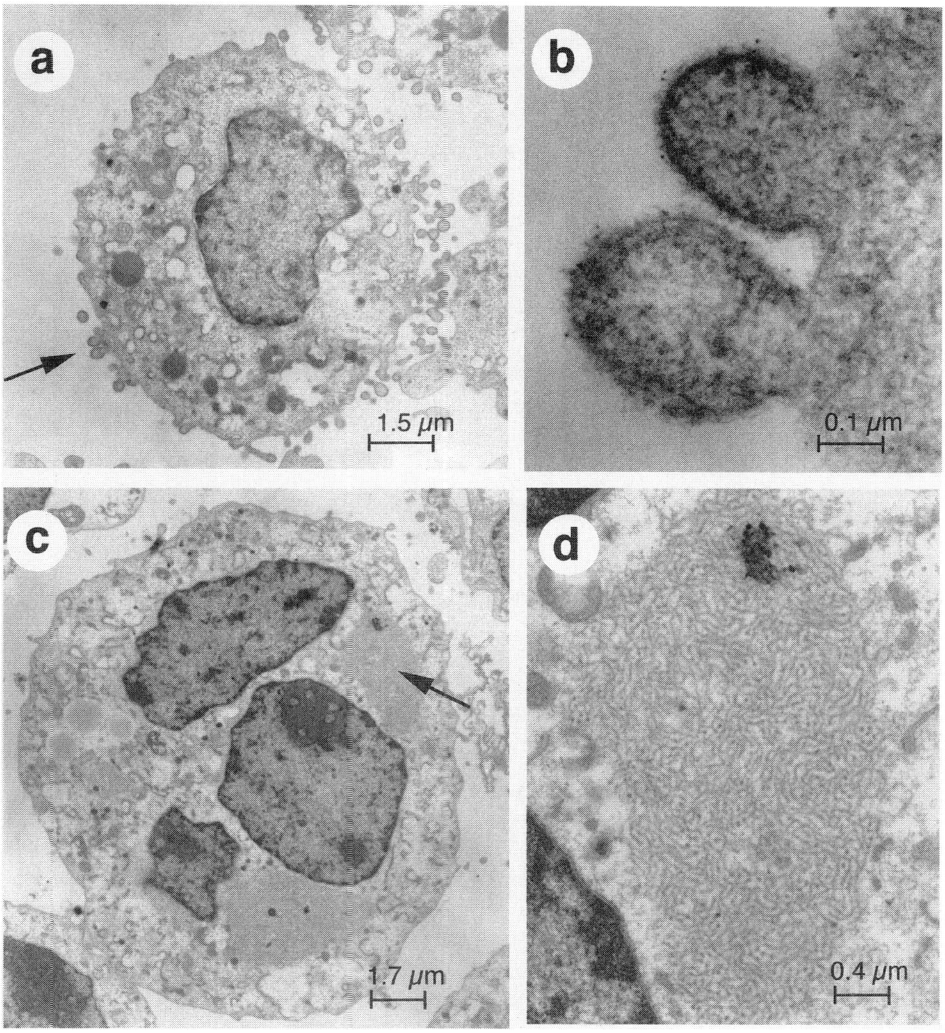

Figure 1 Electron microscopic pictures of monocytes (U937 cells) persistently infected with measles virus. (a) Budding of newly assembled viruses at the cell surface. Characteristic alterations of the plasma membrane in various stages of budding are detected around the cell. (b) Enlargement from panel a (arrow) showing two almost completely budded viruses. (c) Large aggregates of nucleocapsids are present within another persistently infected U937 cell, from the surface of which almost no budding occurs. (d) Enlargement from panel c (arrow). (Photo courtesy of Dr. R. Firsching, Institute for Virology and Immunobiology, University of Würzburg, Germany.)

poietic lineage. Multinucleate giant cells are formed in lymph nodes; they are pathognomonic for the measles infection. However, endothelial cells and neural cells such as neurons, astrocytes, and microglial cells can also be infected. As cellular receptors for MV, the widely expressed transmembrane protein CD46 [7,8] and the lymphoid cell specific signaling lymphocytic activation molecule (SLAM, CD150) [9,10] have been identi-

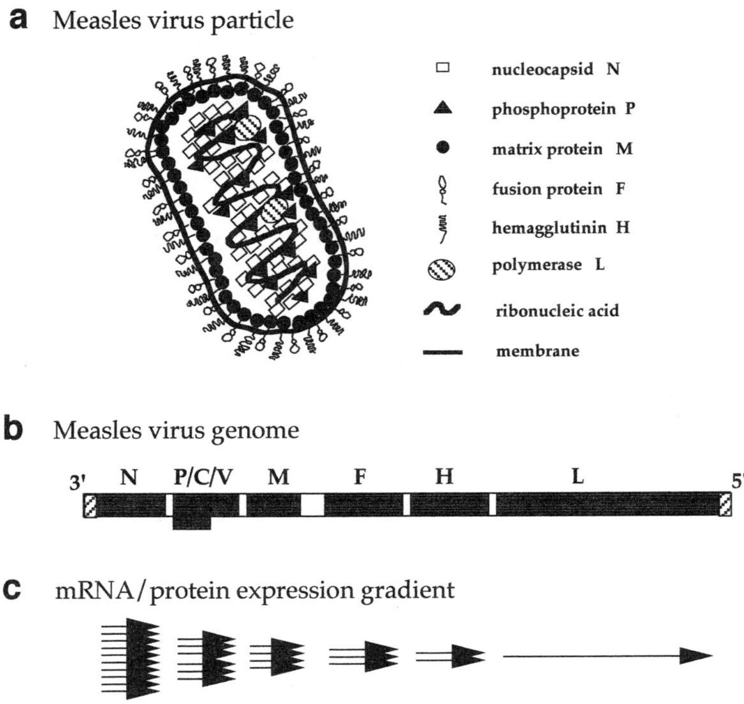

Figure 2 (a) Model of the measles virus particle including structural proteins. The envelope proteins M, F, and H are not required for the intracellular multiplication of ribonucleocapsid protein (RNP) complexes consisting of the viral RNA and N, P, and L (polymerase) proteins. (b) The MV single-stranded RNA genome of negative polarity contains 15,894 nucleotides. (c) The genome is transcribed with decreasing efficiency from its 3′ end to the 5′ end with highest relative frequencies of the N transcripts resulting in a steep expression gradient.

fied. In contrast to CD46, all MV strains and isolates tested so far interact with SLAM on the surface of activated B and T cells, monocytes, and dendritic and memory cells. Interaction of the virus with SLAM may induce a signaling event, and infection leads to the downregulation of SLAM from the cell surface [10]. The question as to whether SLAM involvement during measles contributes to the virally induced immunosuppression and has pathogenetic consequences requires further investigation.

Measles virus is a monotypic virus, and infection with one circulating genotype or vaccine virus induces immunity to all other MV strains. MV isolates have been obtained from many different locations and from patients with different clinical conditions. Although much effort has been invested in attempts to distinguish viruses that might be predisposed to the production of encephalitis or to the production of SSPE, the existence of such strains leading to certain variants of measles could not be demonstrated [11]. Based on sequence analysis of vaccine and wild-type MV strains and of SSPE isolates, the relationship of these various MVs has been established by assigning them to lineage groups [12]. Within the 151 COOH-terminal amino acids of the N protein, approximately 10% divergence between the most unrelated strains has been found. MV strains fall into eight different clades comprising 20 genotypes, some of which are extinct (i.e., have not been isolated recently) [13,14]. In populations where mass vaccination campaigns have been undertaken it is important to define whether measles cases are caused by an imported virus or represent endemic circulating virus due to inadequate vaccine coverage or vaccine failure. Using routine investigation and molecular epidemiological methods, it was possible to demonstrate that after 1993, indigenous transmission of MV was interrupted in the United States [15].

3 ACUTE MEASLES AND ACUTE POSTINFECTIOUS MEASLES ENCEPHALOMYELITIS

3.1 History

The first description of measles has been attributed to the Persian physician Al-Rhazes and dates back to the tenth century [16]. In 1911, Goldberger and Anderson demonstrated that measles was a viral disease [17], but it was only in 1954 that Enders and Peebles were able to isolate and propagate MV in cell culture [18]. With this achievement, studies of the virus structure and replication strategy as well as the preparation of a live attenuated vaccine became possible.

3.2 Epidemiology

Measles virus is a highly contagious agent that causes a well-characterized childhood disease. In the pre-vaccine era in industrialized countries, maximum incidence of measles occurred in children between 5 and 9 years of age. By the age of 20 years, nearly 99% of the population had been exposed to the virus. With the introduction of measles vaccination, the age incidence and percentage of measles cases in different age groups has changed markedly. In countries with optimal vaccine utilization, wild-type measles infections have shifted to the teenage group [19]. In contrast, in the developing countries with ineffective vaccine programs, measles has its greatest incidence in children under 2 years of age. Here the disease is a serious problem with a high rate of mortality (up to 10%). It has been found that the severity of acute measles and mortality correlate in general with the severity of malnutrition. Therefore, the pattern of epidemiology observed differs markedly in different parts of the world, and a thorough understanding of this is essential for the development of successful vaccination programs.

Natural immunity to measles is known to last for at least 65 years. In 1781 measles disappeared from the Faroe Islands following an epidemic. It was not reintroduced until 1846. Individuals old enough to have experienced the disease 65 years previously were still protected. This unusual persistence of immunity even in the absence of reexposure from the environment has suggested that MV may normally persist inside the body, possibly in lymphocytes, and thus restimulate immunity from within.

The efficient spread of the virus is mediated by aerosol droplets and respiratory secretions, which can remain infectious for several hours. The disease incidence in the northern hemisphere tends to rise in winter and spring, when lowered relative humidity favors this form of transmission. In equatorial regions epidemics of measles are less marked but can occur in the hot dry seasons.

Acute encephalomyelitis is the most common neurological complication of measles. It is observed at a frequency of about 1 per 1000 cases of acute measles. In children under the age of 2, the frequency is lower. Encephalitis is not considered to be a complication of vaccination with the live attenuated vaccine strains in use today. Only one case of encephalitis per million children was reported after immunization, which is actually less than the estimated background of encephalitis cases in children [3]. However, some clustering of the cases was observed during days 6–15 after vaccination [20].

3.3 Clinical Manifestations

Acute Measles

Even before the onset of clinical symptoms a pronounced lymphopenia and a defect in cell-mediated immunity are observed, as demonstrated in tuberculin-positive individuals who become tuberculin-negative. These immunosuppressive effects are the result of virus interactions with cells of the lymphoid system (see below). After 10–11 days, the patient enters the prodromal phase, which lasts 2–4 days. The initial symptoms consist of fever, malaise, sneezing, rhinitis, congestion, conjunctivitis, and cough. The severity of these symptoms increases over the next few days. At the beginning of the prodromal stage, a transitory rash is sometimes observed that has an urticarial or macular appearance but disappears prior to the onset of the typical exanthema. Giant cells are present in the sputum, nasopharyngeal secretions, and urinary sediment cells. Virus is present in blood and respiratory secretions, and the patient is highly infectious. During the prodromal stage, Koplik's spots, the pathognomonic enanthema of measles, appear on the buccal and lower labial mucosa opposite the lower molars. These raised spots with white centers are characteristic of measles and begin to fade some 2–4 days after the onset of the prodromal phase as the rash develops.

The distinctive maculopapular rash appears about 14 days after exposure and starts behind the ears and on the forehead. From there the exanthema spreads within 3 days and involves the face, neck, trunk, and upper and lower extremities. Once the entire body is covered, the rash fades on the third or fourth day and a brownish discoloration occurs, sometimes accompanied by a fine desquamation. Concurrently the fever usually falls, and the conjunctivitis as well as the respiratory symptoms begin to subside. Virus shedding decreases, and the patient is in general no longer infectious 4 days after appearance of the rash. Normally, the patient shows rapid improvement. Continuation of clinical symptoms of the respiratory tract or fever suggests complications.

Complications of acute measles are not uncommon and result mainly from opportunistic secondary infections of MV-infected necrotic surfaces such as those in the respiratory tract or the intestines. Invasion of bacteria or other viruses can result in otitis media, bronchitis, pneumonia, or diarrhea. The most frequent complication affecting the CNS is acute postinfections measles encephalomyelitis (APME) (see below). Giant cell pneumonia is a frequently fatal complication of acute measles in immunocompromised patients. Other unusual manifestations that may complicate acute measles include myocarditis, pericarditis, hepatitis, appendicitis, mesenteric lymphadenitis, and ileocolitis. MIBE and SSPE are caused by persistent MV in the CNS (see below).

Modified measles (white measles) occurs in partially immunized children. These may be infants with residual maternal antibodies or individuals who have received immune serum globulin for protection. Occasionally, this form of measles has also been seen in the course of live vaccine failure. In general, the illness is mild and follows the regular sequence of events seen in acute measles but with reduced symptoms.

Black measles is a very rare form of hemorrhagic measles. Because infection of endothelial cells during acute measles is normally associated with the exanthema, an extension to confluent hemorrhagic skin eruptions can occur. Patients with this illness have signs of encephalitis or encephalopathy, and pneumonia. Extensive bleeding from the mouth, nose, and bowels can be associated with disseminated intravascular coagulation.

Atypical measles is a form of measles no longer seen. It occurred in persons who had been vaccinated with formalin-inactivated measles vaccine used in the 1960s. Protection by these vaccines was short-lived, and infection with the wild-type MV caused a centripetal rash and pneumonia. A similar disease characterized by reduced cytotoxic T-cell responses can be induced experimentally in rhesus macaques by formalin-inactivated MV [21].

Acute Postinfectious Measles Encephalomyelitis

It is likely that CNS involvement is common even in uncomplicated measles. Transient abnormality of the electroencephalogram (EEG) is detected in about 50% of measles patients, and CSF pleocytosis is observed in 30% [22–24].

Acute encephalomyelitis occurs in approximately 0.1% of wild-type measles infections. It usually develops within a period of 8 days after the onset of rash, usually when the rash has started to fade and defervescence has occurred. Occasionally APME may take place during the prodromal stage or as late as 3 weeks after the onset of rash [25]. Appearance of neurological signs is usually sudden and is accompanied by resurgence of fever. Initial symptoms include headache, depressed level of consciousness, ranging from mild obtundation to profound coma, and focal or generalized seizures. The course of APME is quite variable. Motor deficits of varying severity in the form of paraparesis, hemiparesis, or cranial nerve palsies are seen. Cerebellar ataxia, choreoathetosis, and myoclonus are also observed. Improvement can occur 2–3 days after the onset of neurological symptoms, but persistence of a comatose state for weeks is not uncommon. About 10–40% of the APME cases are fatal, and most of the survivors are left with lasting neurological sequelae. These include behavioral abnormalities, mental retardation, recurrent seizures, deafness, and persistent motor deficits [26,27].

3.4 Pathophysiology

Acute Measles

Measles virus (MV) first gains entry into the body through the upper respiratory tract, the nose, and possibly the conjunctivae. Primary replication is assumed to occur at the site of entry; however, the first sign of infection is viral replication in the draining lymph nodes, syncytium formation, and destruction of lymphoid tissue. The virus then spreads to the rest of the reticuloendothelial system and respiratory tract through the blood (primary viremia). Giant cells containing inclusion bodies (Warthin-Finkeldy cells) are formed in lymphoid tissue and also on the epithelial surfaces of the trachea and bronchi. About 5 days after the initial infection the virus spreads from the compartments in which it has been replicating to infect the skin and viscera, kidney, and bladder (secondary viremia) (Fig. 3). Infectious virus is shed by the upper respiratory tract, and cell-bound virus can be isolated from urine.

Unlike other viruses, MV causes infection characterized by lymphoid hyperplasia and inflammatory mononuclear cell infiltrates in all the infected organs. The epithelia of the respiratory tract and conjunctiva are relatively thin, with about one or two cell layers. These soon begin to break down, and the inflammatory reaction leads to the symptoms observed at the beginning of the prodromal phase: runny nose, conjunctivitis, malaise, and fever. The thicker mucosal surfaces of the buccal cavity are then affected, and Koplik spots appear about 11 days after infection. The appearance of these spots marks the com-

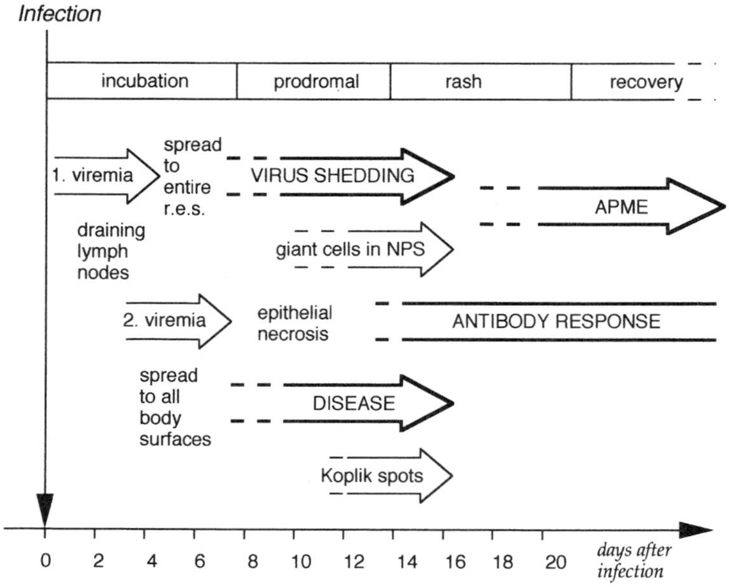

Figure 3 Schematic time course of acute measles. Abbreviations: r.e.s., reticuloendothelial system; NPS, nasopharyngeal secretion; APME, acute postinfectious measles encephalitis.

mencement of a delayed type hypersensitivity reaction (Fig. 4). The spots usually fade by day 3 after their appearance as the rash itself develops. The mechanisms underlying the production of both the spots and the rash are thought to be the same. Unlike other sites of replication, virus antigen is absent from the lesions themselves. Virus antigen can be detected in the skin, but it is concentrated near the blood vessels and in the endothelial cells of the dermal capillaries themselves. The rash is characterized by vascular congestion, edema, epithelial necrosis, and round cell infiltration, but giant cells are absent. Viral replication does not break through the skin, and virus is not shed from this surface. The containment of infection in the skin is thought to be due to the development of cytotoxic T cells that destroy infected tissue and due to interferon (IFN) production, which acts to promote cellular resistance to infection. The rash itself results from the accumulated damage to the vascular walls caused by this delayed type hypersensitivity reaction and is thus usually not seen in the immunosuppressed.

Although antibody titers normally rise in the exanthematous stage of the illness, they are not thought to be the major factor in promoting recovery. MV-infected cells are lysed inefficiently by the classic pathway of the complement activation, although more so by the alternative pathway. Furthermore, patients with agammaglobulinemia handle MV infection normally and recover. However, those with T-cell deficiencies usually do not develop the rash and can be severely ill. Evidently, measles infection in the immunocompetent host triggers an efficient virus-specific immune response that leads to the clearance of the virus from the peripheral blood and the establishment of lifelong immunity against reinfection. Paradoxically, at the same time a general suppression of responsiveness to other pathogens is established, which is the major reason for the high morbidity and mortality rate associated worldwide with acute measles.

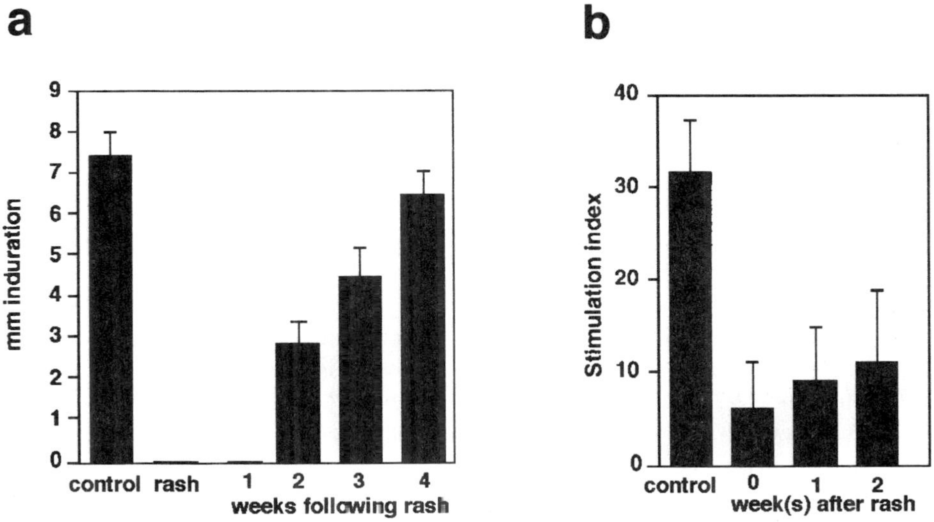

Figure 4 In the course of and following measles, both delayed-type hypersensitivity (DTH) reactions as measured by (a) tubulin test in size (mm) of skin induration and (b) in vitro proliferative responses of lymphocytes to mitogen stimulation, are suppressed. (Data from Refs. 142 and 143.)

Initially, the patients reveal a marked lymphopenia that affects both B and T cells and most likely results from a general loss of lymphocytes due to viral infection. In addition to lymphocytes, MV-infected monocytes/macrophages have also been found, predominantly at later stages of the disease. Immunosuppression, however, continues for several weeks after the onset of the rash until the lymphocyte counts have returned to normal and MV-infected cells are present with only low frequency or are no longer detectable (Fig. 4). Key features of MV-induced immunosuppression are inhibition of delayed type hypersensitivity responses and a restricted ability of lymphocytes to proliferate in response to recall antigens as well as allogenic and mitogenic stimulation [28–31].

Because only a few infected cells are usually detected, several hypotheses have been put forward to explain these findings, which include the production of inhibitory factors by infected cells that have not yet been identified [32], interference with the production of stimulatory cytokines by monocytes/macrophages [33], and induction of a cell cycle arrest in uninfected lymphocytes [34]. Interestingly, professional antigen-presenting cells such as dendritic cells (DCs) have also been found to impair rather than stimulate the activation of T cells in vitro once they express viral glycoproteins on their surface [35,36]. Thus, MV-infected dendritic cells could play a central role in the induction of widespread immune suppression [37].

Acute Postinfectious Measles Encephalomyelitis

The most frequent, but also the least understood, neurological complication of measles is acute postinfectious measles encephalomyelitis (APME). In common with postinfectious encephalitis induced by other viruses, its neuropathology is characterized by demyelination, perivascular cuffing, gliosis, and the appearance of fat-laden macrophages near the blood vessel walls. Petechial hemorrhages may be present, and in some cases inclusion bodies have been observed in brain cells.

As pointed out above, CNS involvement in acute measles is thought to be common, but the questions of whether and how MV reaches the CNS in the course of the acute infection are still a matter of controversy. MV is highly lymphotropic and could be carried into the CNS even in cases where encephalitis has not been recognized. However, only in exceptional cases can MV be isolated from the brain tissue of APME patients. In the majority of APME cases studied, neither MV antigen nor RNA has been found in the CNS [38,39]. Therefore, current theories favor an autoimmune reaction as the possible cause of CNS damage, because APME patients may exhibit a proliferative T-lymphocyte response to myelin basic protein (MBP) [40]. In addition, in CSF specimens of such patients MBP was detected as a consequence of myelin breakdown. Such MBP-specific lymphoproliferative responses have been seen not only after measles but also in patients with postinfectious encephalomyelitis following rubella or varicella infection or after rabies immunization [27]. The latter disorder is probably the human equivalent of experimental allergic encephalitis (EAE), because such patients received rabies vaccine prepared in brain tissue. Because APME is characterized by demyelinating lesions in association with blood vessels as in EAE, the finding of an MBP-specific lymphoproliferative response in measles infection is considered to be of pathogenic importance [40]. How MV leads to a T-cell-mediated autoimmune response is still unknown. At present, the possibilities of molecular mimicry or a deregulation of autoreactive cells occurring secondary to viral infections of lymphocytes are being considered.

3.5 Radiographic and Neurophysiological Findings

In APME, the EEG is nonspecific and shows diffuse or focal slowing, which may persist for weeks after clinical recovery. Magnetic resonance imaging (MRI) of the brain is characterized by diffuse white matter disease.

3.6 CSF Findings

The CSF in APME patients can be normal, but in two-thirds of cases there is a lymphocytic pleocytosis in the range of 10–500 cells/μL. About half the patients show a moderate to pronounced elevation of protein levels. With rare exceptions, there is no intrathecal synthesis of measles antibodies [27].

3.6 Diagnostic Strategies

Because its symptoms are typical, acute measles can be diagnosed clinically by experienced physicians. Criteria for clinical diagnosis of measles, as defined by the CDC, are listed in Table 1. Laboratory confirmation is possible by demonstration of MV-specific IgM antibodies and/or by demonstration of a seroconversion or rise of IgG antibodies, which are frequently negative in blood samples obtained during the acute phase. Confirmation of acute measles by serological tests or virus isolation should be obtained regularly in countries with low measles incidence. Diagnostic tests are also indicated in immunosuppressed patients, in whom rash may be missing or be atypical.

Table 1 CDC Criteria for the Clinical Diagnosis of Measles

Typical rash $\geq$ 3 days
Temperature $\geq 38.3\,^{\circ}C$
At least one of the triad of cough, conjunctivitis, coryza

Diagnosis of APME is based on the close temporal relationship between acute measles and the onset of encephalomyelitis. Virological studies of the CSF are rarely helpful in confirming the diagnosis, because MV RNA and an intrathecal synthesis of MV-specific antibodies are usually absent in APME.

3.7 Treatment and Prevention

Treatment of Acute Measles

Currently, there is no specific antiviral therapy for measles. Symptomatic treatment depends on the clinical manifestations and may include antipyretics and cough suppressants. Antibiotics are indicated when there is bacterial superinfection.

Treatment of Acute Postinfectious Measles Encephalomyelitis

Treatment of APME is also only symptomatic and supportive and does not differ from that of any other postinfectious encephalitis. Seizure control requires anticonvulsive drugs. The effectiveness of corticosteroids is controversial, because of the variable and unpredictable course of APME and because of the lack of randomized placebo-controlled studies. A retrospective study in Norway revealed a higher mortality in comatose patients treated with corticosteroids than in those who were not so treated. Hyperimmune gamma globulin is not effective [41].

Prevention

Attenuated live vaccines prevent measles. Because they can be neutralized by transplacentally transmitted maternal antibodies that persist during the first year of life [42], initial administration of the vaccine in the industrialized countries is recommended at the age of 12–15 months [43,44]. In some of the developing countries where measles is still rampant and where infection does occur during the first year of life, earlier immunization has been recommended and practiced. Usually, measles vaccine is given in combination with mumps and rubella vaccine. One dose of the vaccine leads to seroconversion in more than 90% of the recipients. A second dose of measles vaccine is recommended for all children under the age of 6 years to protect nonresponders to the first vaccination. The second dose may be given as early as 4 weeks after the first. The measles vaccines in use today have an excellent safety record. Some recipients develop a transient rash or low-grade fever 5–12 days after the vaccination, but they remain otherwise asymptomatic; complications are extremely rare. Acute encephalitis after vaccination has been reported with a frequency of less than one per million doses administered compared to one per 1000 children with natural measles. Vaccination also significantly reduced the occurrence of SSPE and other complications associated with measles [3,45,46].

In unvaccinated persons exposed to measles, symptomatic disease can be prevented by the administration of live vaccine within 3 days after exposure. For immunocompromised patients or children with chronic diseases, passive immunization with human immunoglobulin is recommended. If given within the first 3 days after exposure, it may prevent the development of disease. Its effectiveness is diminished if it is given 4–6 days after exposure; beyond this time it is ineffective.

Because there is no MV animal reservoir, eradication of measles is possible, and it has been the aim of a WHO campaign that started in 1984 [47]. Considerable success has been made toward this goal, as exemplified by the elimination of endemic measles from the United States and from some parts of Europe [48].

4 SUBACUTE SCLEROSING PANENCEPHALITIS

4.1 History

Subacute sclerosing panencephalitis (SSPE) is a rare, slowly progressing, fatal degenerative disease of the brain. The earliest reports of its clinical and pathological characteristics were in the 1930s by Dawson, who described a subacute encephalitis with intranuclear inclusion bodies and postulated its viral origin [49,50]. Crediting his contribution, the disease was also referred to as "Dawson's inclusion body encephalitis." Pette and Döring in 1939 [51] and Van Bogaert in 1945 [52] described similar clinical diseases as nodular panencephalitis and subacute sclerosing leukoencephalitis, respectively. In 1950 it was realized by Greenfield [53] that all three descriptions were of the same disease, and it has been called SSPE since then. Ultrastructural studies by Bouteille et al. in 1965 [54] revealed that the inclusion bodies in SSPE brains contained nucleocapsids resembling those of a paramyxovirus. Involvement of MV in this disease was finally established by serological demonstration of elevated measles antibodies in the CSF and of MV antigen in the brain tissue by immunofluorescence [55,56]. In 1969, MV was isolated from SSPE brains by cocultivation techniques [57,58].

4.2 Epidemiology

Subacute sclerosing panencephalitis follows infection with wild-type measles virus after an interval of 6–8 years. However, some SSPE cases have occurred as early as 2 years and as late as 20–30 years after the primary infection. The average age of onset is between 9 and 13 years [59,60], but the range of ages may vary considerably. Because the average age of measles infection has increased in immunized populations, the age of SSPE onset may possibly rise as well. However, so far adult onset SSPE is rare [61].

The incidence in unvaccinated populations is one per 100,000–500,000 cases of acute measles. Males are slightly more likely than females to develop SSPE, and about half of SSPE patients contracted measles before the age of 2 years. It is important to note that acute measles infection occurring in the first year of life carries an extra risk of subsequent SSPE. However, in that age group, when the infection occurs in the presence of maternal antibodies, clinical measles may not be recognized for what it is because of the attenuation of the symptoms. No unusual features of the acute measles infection or other factors predisposing to the development of SSPE have ever been demonstrated. In cases of children who developed SSPE after having been vaccinated against measles, wild-type measles infection usually occurred before the vaccination.

In the wake of routine vaccination programs against measles, SSPE has become rare in industrialized countries, where cases of SSPE are mostly seen in immigrant children or in unimmunized adolescents. However, in parts of the world where measles vaccination of infants is not universal, SSPE remains one of the most common acquired neurological disorders. This is true for parts of eastern Europe, Africa, and Asia [62].

4.3 Clinical Manifestations

The course of SSPE is quite variable, and various staging schemes have been described [63–66]. The disease usually starts insidiously with psychological disturbances and/or a generalized intellectual deterioration that may become apparent by declining school performance, emotional lability, personality change, forgetfulness, poor attention span, or difficulty sleeping (stage I according to Ref. 64). This stage may last for weeks or months and may not be recognized as illness until more definite neurological signs appear (stage II). These may take the form of myoclonic jerks — which are characteristic of SSPE — dyspraxia, generalized convulsions, aphasia, and visual disturbances. Myoclonic jerks occur at a frequency of 5–15 min^{-1} and tend to be synchronous with the EEG spike and wave complexes (Fig. 5). Invasion of the retina leads to chorioretinitis (Fig. 6) in 75% of the SSPE cases. This may ultimately lead to blindness. Retinal depigmentation, optic atrophy, optic neuritis, and papilledema are also observed. Progressive cerebral degeneration heralds stage III, which is characterized by decorticate or decerebrate posturing, loss of brainstem function, and coma. Approximately two-thirds of the patients die during stage III. In stage IV, the patient no longer has cortical functions.

However, progression of the disease is highly variable. Remissions are common, and some stages overlap one another so that progression of symptoms may be atypical. The illness usually lasts 1–3 years. Much more rapid forms that lead to death within a few months as well as prolonged courses with a duration of more than 20 years have been described [67]. Periods with spontaneous improvement or disease stagnation are not unusual and occur in 20% and 50% of the patients, respectively [65].

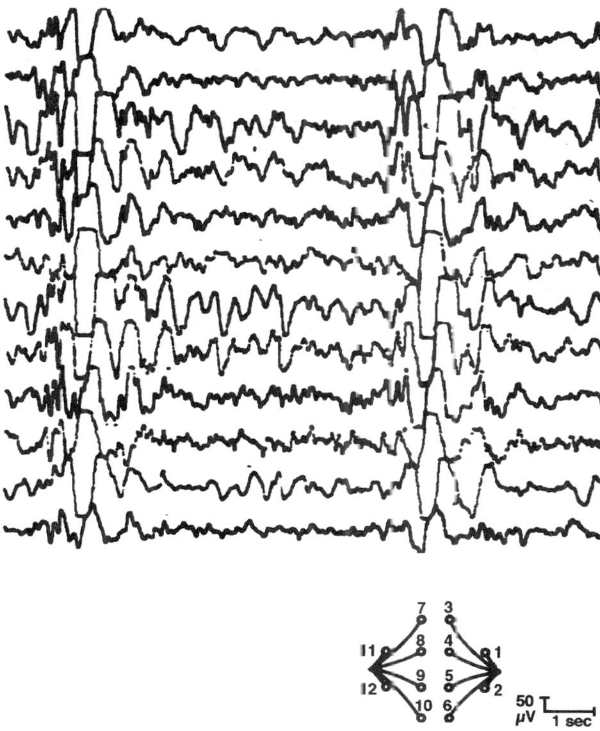

Figure 5 Typical electroencephalogram (EEG) of an SSPE patient with Radermecker complexes every 5–15 s.

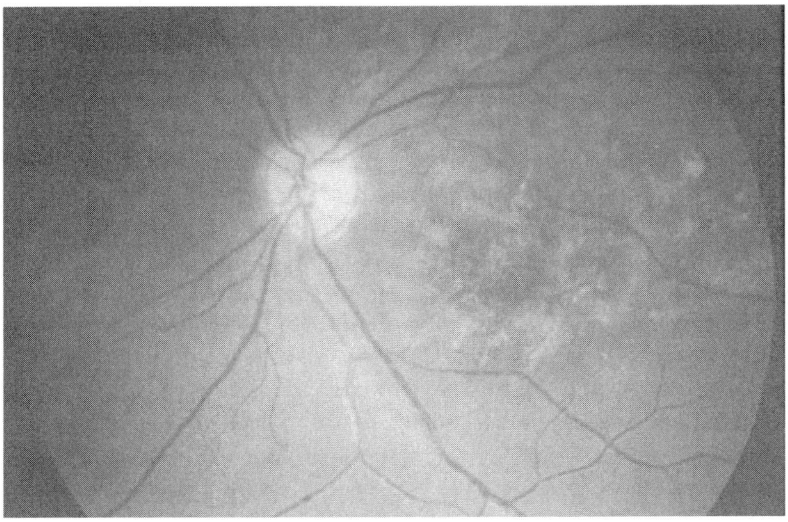

Figure 6 Chorioretinitis with a central location in SSPE. Patchy edema of the superficial neuronal layers grouped around the macula can be seen.

4.4 Pathophysiology

Subacute sclerosing panencephalitis affects the cerebral cortex and brainstem; the cerebellum and spinal cord are less involved. Neuropathological findings include diffuse encephalitis affecting both the gray and white matter, characterized by perivascular cuffing and diffuse lymphocytic infiltration. Gliosis is usual. Brain endothelial cells, neurons, fibrous astrocytes, oligodendroglia, and macrophages or microglial cells contain large aggregates of intranuclear inclusion bodies consisting of MV nucleocapsid structures. Giant cell formations or membrane changes consistent with viral maturation have not been detected [68–71].

Since the first implication of MV in the etiology of SSPE in 1967, extensive studies of the association of MV and SSPE have been carried out by many laboratories (reviewed in Refs. 72 and 73). However, the manner in which the persistent infection is first established in the brain and the exact steps leading to the onset of the disease are still largely unknown. PCR-based studies have even suggested that MV may in fact frequently reach the brain and establish lifelong persistent infections that remain asymptomatic and do not lead to the development of SSPE [74,75]. The virus is thought to gain entry to the CNS during viremia in acute measles or by infected lymphocytes, but once there replication proceeds only slowly and a widespread encephalitis is not established. It is also not known to what extent viral replication per se is responsible for the development of lesions or what part is played by the immune system.

Virological Aspects

Important clues have come from the study of MV replication. The virus normally replicates by the production of giant cells and release of infectious progeny. In SSPE, however, free infectious virus has never been isolated either from the brain or from CSF, and histopathological examinations have consistently failed to reveal the morphological changes associated with virus maturation [72,76]. As with MIBE, giant cells and thickening of the plasma membrane at points of budding have never been observed, suggesting the absence of viral glycoproteins. Viral nucleocapsids present in the cytoplasm are randomly scattered and show no sign of regular alignment beneath the plasma membrane. Inasmuch as budding of viral particles and infectious virus have not been detected, it is likely that the infection spreads slowly in a cell-associated manner [77].

In SSPE brain sections studied by immunohistochemistry, the expression of the MV N and P proteins was consistently detected in infected cells. In contrast, expression of the envelope proteins was never detected. Molecular biological studies on SSPE brain tissue revealed extensive transcriptional and translational alterations affecting mainly MV M, F, and H genes [78–81]. As in MIBE, the envelope protein–specific mRNAs were detected in only low copy numbers and were highly impaired in directing the synthesis of the corresponding gene products in vitro [82]. Intracellular factors induced by type 1 IFN, such as MxA, contribute to the low level of viral gene expression [83]. Sequence analyses revealed a high rate of mutations located all over the MV genome, although different genes were affected at different levels. The highest number of alterations were found in the M gene, followed by F, H, P, and N genes, which were mutated to about the same extent, whereas the L gene was most conserved [84]. Those introduced mutations either were point mutations, most probably accumulating because of the infidelity of the viral polymerase, or appeared as clustered transitions that are thought to result from the activity of a cellular enzyme complex that actively modifies viral genetic information. As a result of either of these events, viral protein synthesis is abolished or truncated or it generates unstable MV proteins (Fig. 7).

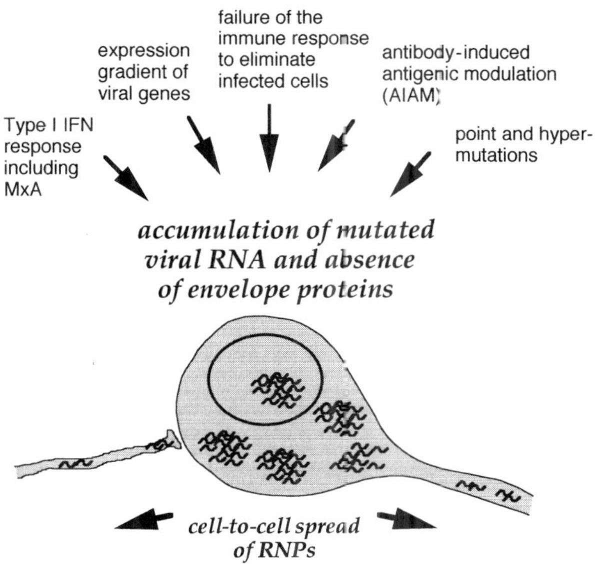

Figure 7 Several factors influence the establishment of a persistent MV infection in the brain: the IFN response, the neural cell type–specific steep expression gradient, antibody-induced antigenic modulation, and hypermutations allow low-level intracellular MV replication and prevent expression of viral envelope proteins. The virus spreads from cell to cell as RNP when the cell-mediated immune response cannot eliminate the intracellular pathogen from the CNS.

These molecular biological data explain the absence of infectious MV particles and the lack of budding and cell fusion in infected SSPE brain tissue. Infectious virus can be rescued only occasionally from brain tissue by cocultivation [57,58,84a]. So-called SSPE isolates can be of two different types: cytolytic budding or cell-associated viruses, the latter spreading through the culture with a gradually enlarging area of cytopathic effect (CPE). In the second type of isolate, mRNAs for the MV envelope proteins are detectable in infected cells, although their function is impaired and they are not synthesized in infected cells or in in vitro translation experiments. As revealed by recent sequence analyses, some of the SSPE isolates are likely contaminations and are in fact ordinary laboratory MV strains. It is still unclear whether those that are true SSPE isolates represent a small subpopulation of replication/maturation-competent viruses or revertants that have been selected by the isolation procedure and thus may not be representative of the dominant virus population in the infected brain.

On the basis of these MV sequence analyses of brain tissues from SSPE patients, it was possible to identify the wild-type MVs that circulated in the population years ago when these SSPE patients were infected with MV. These findings resulted in two important conclusions: first, that SSPE develops after infection with a wild-type virus and not following measles vaccination, and second, that SSPE is caused by circulating normal wild-type viruses and not by some neurotropic strains.

Immunological Aspects

One of the immunological hallmarks in SSPE is the hyperimmune response to MV antigens, which includes neutralizing antibodies in serum and CSF [76]. Yet this immune response fails to control viral infection. This phenomenon has led to the proposition that measles antibodies support persistence rather than interfere with it. Because complement-mediated lysis is of low efficiency in the brain as a result of low complement concentration in this compartment, it is likely that the major effect of antibody in the brain tissue is promotion of clearance of antigen from the cell surfaces by capping or shedding rather than lysis of the infected cells. This process could explain the lack of membrane glycoproteins on the surface of brain cells but cannot explain the lack of intracellular envelope proteins. Additional evidence has been obtained in tissue culture experiments that suggests that the capping process interferes with viral protein synthesis. It has been shown that antibodies directed against the hemagglutinin of MV can exert an inhibitory effect on the viral mRNA and protein expression such as is found in SSPE brains [85,86]. Thus in SSPE the host immune response could contribute to the production of this fatal disease.

Cell-mediated immunity (CMI) is more important in the control of MV infection than the humoral immune response. Cellular immune responses are indicated by the infiltration of mononuclear cells into areas of infection in the CNS. Most of these infiltrating cells are B cells and CD4 positive T cells, and there is local production of TNF-α and IFN-γ [87,88]. CSF has increased levels of α_2-microglobulin, soluble IL-2 receptor, and soluble CD8, all of which are indicative of a local cellular immune reaction [89]. Thus, in general, there is no evidence of a defect of CMI in these patients. It is possible that the specific response to MV antigens is impaired, because anamnestic skin tests with measles antigens are often negative. There is a disparity in results obtained by using MV antigens in assays for cytotoxic T cells. Depending on the test system employed and on the potency and purity of the virus antigens used, either a minor inhibition of CMI or normal reactions in comparison to controls can be detected in SSPE patients [90,91]. Studies of the T helper (Th) 1 and Th2 cytokine production of peripheral blood mononuclear cells in response to measles virus have shown a decreased virus-specific production of IFN-γ in some SSPE patients, suggesting a Th1/Th2 imbalance [92]. The pathological relevance of this finding is unclear.

4.5 Radiographic and Neurophysiological Findings

Neuropathological lesions underlie the characteristic EEG changes, which consist of periodic high-amplitude slow wave complexes that are synchronous with the myoclonic jerks recurring at 3.5–20 s intervals (Fig. 5) [93]. These periodic complexes (Radermecker complexes) are remarkably stereotyped in that their form is identical in all leads. They are bilateral, usually synchronous and symmetrical. Moreover, they usually consist of two or more delta waves and are biphasic or polyphasic. The pathophysiology of this abnormal EEG pattern is poorly understood, but most investigators regard these complexes in SSPE as characteristic and even pathognomonic. It is noteworthy that this EEG pattern is variable within the course of the disease and from one patient to another [94]. Moreover, these typical complexes can disappear as the disease progresses and a low-voltage background dominates the EEG.

Magnetic resonance images may be normal in the first weeks or months after the onset of mental deterioration. The first changes observed are mostly focal increased T2 signals in the white matter of the corona radiata. With disease progression, increased T2 signal intensity is also observed elsewhere in the cerebral white matter and the brainstem. Brain atrophy becomes evident with thinning of the cortex and dilatation of the ventricles (Fig. 8). In advanced disease, there is almost complete loss of the white matter [95]. In this stage, involvement of the basal ganglia and/or the cortical gray matter is seen in about one-third of the patients. Although the progress of MRI abnormalities in SSPE patients often follows a constant pattern, correlation between the clinical status and the initial MRI findings is poor. MRI is not specific and is not necessary or even helpful in making the diagnosis. There is also doubt as to whether MRI can be helpful in monitoring progression of the disease. Clinical improvement has been noted while the MRI abnormality progressed; the opposite has also been reported [96]. In contrast to the MRI studies, computed tomographic (CT) scans frequently appear normal until the advanced disease stages, mainly because the white matter changes are difficult to detect in CT examinations.

4.6 CSF Findings

In contrast to APME and MIBE, SSPE is characterized by an excessive intrathecal synthesis of MV-specific IgG. Oligoclonal bands are always positive, and the IgG index is always elevated. The intrathecal synthesis of MV-specific IgG can be identified by calculation of the MV-specific antibody index [97] or by isoelectric focusing with MV-specific immunoblotting [98,99]. Serum titers of MV-specific IgG are also usually substantially elevated. Because the MV antibody titers in the CSF and serum are extremely high, it is important to test these fluids in high dilutions in order to obtain correct titer values for the calculation of the MV-specific antibody index. Testing CSF and serum in the standard dilutions may give incorrectly low values and result in falsely normal antibody indices.

Because MV replication occurs only within the brain cells and viral particles are not released, MV-specific sequences in the CSF samples cannot usually be amplified by RT-PCR.

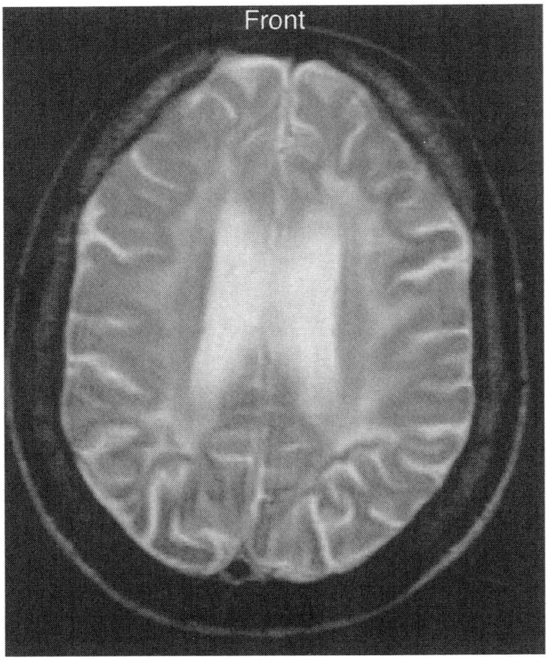

Figure 8 T2-weighted magnetic resonance (MR) image of an SSPE patient (33 years of age) with demyelination in the subcortical white matter. (Photo courtesy of Prof. Dr. L. Solymosi, Department of Neuroradiology, University of Würzburg, Germany.)

4.7 Diagnostic Strategies

Usually the diagnosis of SSPE begins with clinical observation of symptoms and signs characteristic of this disease. These, however, are subject to large variability in both the course of the disease and age of onset. The demonstration of intrathecally produced MV IgG is necessary to confirm the diagnosis of SSPE. Further support is provided by a typical EEG pattern with periodic slow wave complexes (Radermecker) (Fig. 5). Oligoclonal bands in the CSF are always present. Negative oligoclonal bands and a normal total IgG index virtually rule out the diagnosis.

Although the techniques used for the examination of intrathecal measles antibody production vary widely among the laboratories, demonstration of pathological measles antibody titers is an essential noninvasive confirmatory test of suspected SSPE. Brain biopsy revealing MV proteins by immunocytochemical tests or MV RNA by RT-PCR in tissue samples is the alternative confirmatory method, but this invasive procedure is rarely necessary. Availability of brain tissue samples makes it possible to determine the genotype of the MV and detect SSPE-specific mutation patterns. The finding of extremely elevated MV-specific IgG antibodies in the serum supports the diagnosis of SSPE but is not specific enough to confirm it. With rare exceptions, serum measles IgM is negative in SSPE patients.

There can be an overlap of diagnoses revealing similar laboratory abnormalities. For example, intrathecal production of MV-specific IgG antibodies is also seen in about

80% of patients with multiple sclerosis and in some patients with other CNS diseases of autoimmune nature [100]. However, in such cases differential diagnosis usually presents no difficulty.

4.8 Treatment and Prevention

A variety of approaches to treatment of SSPE have been attempted, but it has been extremely difficult to evaluate their effectiveness, SSPE is a rare disease, and therefore clinical trials are frequently based on a very small number of patients, which makes interpretation difficult. Moreover, the course of SSPE is highly variable and spontaneous remissions are common. The aim of any therapy must be to prevent the spread of the virus within the CNS and guide the immune system toward elimination of the virus. Ideally, therapy should be started as early as possible to maximize the potential benefit of any of the regimens described below.

Because of the lack of antiviral agents with a proven specific effect against MV, many efforts to treat SSPE have been directed toward the activation or potentiation of the immune defense in the brain. Several antiviral agents and immunomodulators have been found to be ineffective for the treatment of SSPE, although the reports have been anecdotal [101]. Conflicting results have been reported about the use of amantadine [102,103]. Inosiplex (Isoprinosine®), an inosine derivative with immunomodulating properties that augments IFN production and activates natural killer cells [104], was the first drug that appeared to have a beneficial effect on long-term survival in some reports [105–107]. This therapy appeared to be more effective in patients with slowly progressing forms of SSPE than in those with rapid progression [108–110]. The data from the inosiplex monotherapy studies were reviewed by Taylor et al. [101].

With the availability of IFN-α, several groups explored its use for the treatment of SSPE by subcutaneous, intravenous, intrathecal, and intraventricular administration with varying success [111–113]. Trials using a combination of inosiplex and intraventricular IFN-α showed a positive effect in about 50% of the SSPE patients if started in the early stages of the disease [66,114]. This combination therapy represents a significant improvement over historical controls. However, neither the optimal doses nor the optimal length of treatment have been established, and the beneficial effect on the survival rate appears to be only temporary [115,116]. An international randomized clinical trial comparing treatment with inosiplex alone and the combination with intraventricular IFN-α was recently completed. Preliminary results revealed no evidence of a difference in survival between the two groups at week 30 [117]. Another treatment option currently being investigated is the combination of ribavirin and intraventricular IFN-α. Data from an SSPE hamster model suggested a synergistic effect of the two agents [118]. A few case reports suggest clinical improvement in patients undergoing this combination therapy [119,119a], but it is too early to draw any conclusions about the possible benefits of this therapeutic regimen. Recent studies suggest that magnetic resonance spectroscopy may be used for monitoring effects of treatment [119b].

Despite many efforts at the establishment of an effective therapy of SSPE, it has not been possible to cure the disease. At best, palliation or some temporary remission of its course can be achieved in some patients by prolonged treatment. Vaccination against measles at the earliest recommended age remains the most effective means of preventing SSPE.

5 MEASLES INCLUSION BODY ENCEPHALITIS

5.1 History

Measles inclusion body encephalitis (MIBE) was first described in 1973 [120]. In different reports it has been referred to as subacute measles encephalitis with immunosuppression, immunosuppressive measles encephalitis, acute measles encephalitis of the delayed type, atypical measles encephalitis, and measles inclusion body encephalitis.

5.2 Epidemiology

Giant cell pneumonia and MIBE are two severe complications of measles affecting children and adults with defective cell-mediated immune responses. MIBE is therefore probably best regarded as an opportunistic MV infection. A literature review in 1993 summarized the data of 33 cases reported by then [121]. A few cases have been added subsequently [122–128]. Currently MIBE is most common in children receiving immunosuppressive drugs or axial radiation therapy for the treatment of malignant diseases. Acute lymphoblastic leukemia as the underlying disease has been reported in 70% of the cases and neuroblastoma in 9% [121]. Human immunodeficiency virus (HIV) infection is another important predisposing cause [121,124–126]. In contrast, children with hypogammaglobulinemia recover normally from acute measles and are not at increased risk of MIBE.

Of the 33 patients reviewed, about one-half had a history of classical measles preceding the onset of neurological symptoms by 1–7 months [121]. Some patients had atypical measles. The rash may be missing, because it develops through cellular immune response. Some patients had a known history of exposure to measles but did not develop clinical disease. Eighteen percent had neither clinical measles nor a known exposure. Prior measles immunization is not fully protective against the development of MIBE in an immunosuppressed child [121–123,125]. There has been one report of an MIBE caused by an MV vaccine strain in a child with a profoundly depressed CD8 count and dysgammablobulinemia, which was not suspected at the time of vaccination [127].

5.3 Clinical Manifestations

Typically the neurological disease begins with seizures that are often severe and refractory to treatment. They may take the form of epilepsia partialis continua. Only one patient has been reported to have had fever. As the disease evolves the patients develop neurological deficits such as hemiparesis, ataxia, aphasia, and visual symptoms. Eventually all patients exhibit altered levels of consciousness, which can progress to stupor and coma. The course of the disease is usually rapid, and death follows in 75% of the patients within days to a few months. Some patients die of the underlying disease. The surviving patients are left with permanent neurological deficits [121].

5.4 Pathophysiology

Histopathological studies of brain tissue are characterized by massive eosinophilic inclusion bodies consisting of viral nucleocapsids in the nucleus and/or the cytoplasm of neurons and glial cells. There is also glial cell proliferation, focal necrosis, perivascular cuffing by lymphocytes and/or plasma cells, and myelin loss. However, any of these features can be missing [121].

Immunohistological and molecular biological studies of the brain of one patient suggested defects in viral replication. Of the five major structural proteins of MV, only

N and P proteins were consistently detected in infected brain cells, whereas the envelope proteins were missing. In contrast, the mRNAs specific for the five viral proteins were detectable in brain-derived total RNA samples by Northern blot analyses, although the mRNAs for the envelope proteins were underrepresented in comparison with lytically infected cells. In vitro, N and P proteins were efficiently synthesized from their corresponding mRNAs, indicating a restriction of the expression of the MV envelope proteins in MIBE [129]. This restriction has been partially explained by sequence analyses that revealed a high rate of mutations distributed over the entire MV genome. For the MV M gene, mutations have eliminated the initiation codon, which explains the failure of MV M protein synthesis in infected MIBE brain tissue [84]. Defects in MV mRNA transcription and envelope protein synthesis apparently do not significantly affect the activity of the RNP complex, which spreads to different areas of the patient's brain. Inasmuch as infectious virus particles are not likely to form, because of the restriction of the envelope proteins required for assembly, budding and giant cell formation have never been observed. Virus is thought to spread by microfusion events.

5.5 Radiographic and Neurophysiological Findings

Computed tomographic scans of the brain have been normal in patients with MIBE [121]. In a few patients, areas of reduced density have been reported [122,124]. MRI can show focal increased signal intensities on T2-weighted images, but only a few MRI results have been reported so far [121,123,127,128,130]. EEG findings are generally abnormal and include various nonspecific changes such as diffuse slowing and spike wave activity [121].

5.6 CSF Findings

The CSF findings are nonspecific. There is usually no protein elevation or pleocytosis. Elevation in measles IgG antibodies has been found in about 50% of the MIBE patients [121,122]. In many of the old reports, insensitive antibody assays such as complement fixation were used. The percentage of patients with intrathecal synthesis of MV IgG may be higher when more sensitive tests such as enzyme immunoassays are employed. In contrast to SSPE, however, an excessive intrathecal antibody synthesis is not a hallmark of MIBE. There have been no reports on RT-PCR analysis of CSF for MV RNA.

5.7 Diagnostic Strategies

Although brain biopsy remains the definitive basis of confirmation of MIBE, it has been suggested that the combination of the following findings is highly suggestive of MIBE in an immunocompromised patient [121]: refractory focal seizures, history of measles or exposure to measles within 7 months of presentation, absence of fever, and normal CSF analysis (protein, cell count).

Serum MV IgM should be tested because it is usually positive for several weeks to several months after a primary measles infection. CSF analysis for MV antibodies and MV RNA by RT-PCR should be performed as well. Demonstration of MV RNA in the CSF or intrathecal synthesis of MV IgG supports the suspected diagnosis. However, negative results do not rule out MIBE. The examination of follow-up CSF samples may reveal increases in antibody titer.

A definitive diagnosis of MIBE can be obtained by brain biopsy [121,126]. The neuropathological changes were described in Sec. 5.4. In addition, in most patients electron microscopy reveals tubular structures typical of paramyxovirus nucleocapsids. Measles etiology should be further confirmed by immunohistochemistry and/or by RT-PCR for MV RNA from brain tissue.

MIBE is frequently confused with SSPE. However, the interval between acute measles and the onset of neurological disease, the disease course, the serological CSF findings, and the immunological predisposition clearly distinguish MIBE from SSPE as well as from APME.

5.8 Treatment and Prevention

There is no effective standard therapy for MIBE. Because the number of cases is small, it is impossible to carry out controlled studies. Therapy with IFN-α has shown no clear benefit [131,132]. A few patients have been treated with ribavirin [121,127]. The results are conflicting. There appeared to be a favorable effect in one case when ribavirin was started early in the disease course.

As is true of APME and SSPE, vaccination of children at the earliest recommended age is the best means of preventing MIBE. To prevent exposure of immunosuppressed patients to measles, vaccination of non-immune household contacts is essential.

6 UNCONFIRMED ASSOCIATIONS OF MEASLES VIRUS TO OTHER CNS DISEASES

6.1 Multiple Sclerosis

An infectious etiology of multiple sclerosis has been suggested for a long time because of a number of epidemiological observations and the pathological characteristics of this disease. Although many different pathological agents have been suspected of being etiological agents, so far none has been identified unequivocally [133].

One of the first viruses linked to the pathogenesis of multiple sclerosis by assessing virus antibodies in sera and CSF, was measles virus [134,135]. In these studies, a slightly higher titer was found in patients with MS than in controls. These early observations were followed by numerous reports from other groups that basically confirmed the original observations, although there were negative reports as well [136]. Moreover, using molecular biological techniques, measles virus–specific RNA was detected in scattered cells in brain tissue of MS patients but also in that of controls [137,138]. Other investigators failed to detect any measles virus genetic information in MS brains [139–141]. Obviously, measles virus can persist in human brain, but it is not etiologically linked to multiple sclerosis.

ACKNOWLEDGMENTS

We thank S. Schneider-Schaulies, F. Grehn, and L. Solymosi for helpful comments and M. Katz for critical reading of the manuscript.

REFERENCES

1. ter Meulen, V.; Billeter, M.A. Measles Virus. Volume 191 of Current Topics in Microbiology and Immunology; Capron, A., Compans, R.W., Cooper, M., Koprowski, H., McConnell, I.,

Melchers, F., Oldstone, M., Olsnes, S., Potter, M., Saedler, H., Vogt, P.K., Wagner, H., Wilson, I., Eds.; Springer-Verlag: New York, 1995.
2. Schneider-Schaulies, S.; ter Meulen, V. Measles. In Principles and Practice of Clinical Virology; Zuckerman, A.J., Banatvala, J.E., Pattison, J.R., Eds.; Wiley: Chichester, 2000, 357–386.
3. Duclos, P.; Ward, B.J. Measles vaccines: a review of adverse events. Drug Saf. 1998, 19, 435–454.
4. Radecke, F.; Billeter, M.A. Measles virus antigenome and protein consensus sequences. In Measles Virus; Billeter, M.A., ter Meulen, V., Eds.; Springer-Verlag: New York, 1995, 181–192.
5. Liston, P.; Briedis, D.J. Ribosomal frameshifting during translation of measles virus P protein mRNA is capable of directing synthesis of a unique protein. J Virol. 1995, 69, 6742–6750.
6. Parks, C.L.; Lerch, R.A.; Walpita, P.; Wang, H.P.; Sidhu, M.S.; Udem, S.A. Analysis of the noncoding regions of measles virus strains in the Edmonston vaccine lineage. J Virol. 2001, 75, 921–933.
7. Dörig, R.E.; Marcil, A.; Chopra, A.; Richardson, C.D. The human CD46 molecule is a receptor for measles virus (Edmonston strain). Cell. 1993, 75, 295–305.
8. Naniche, D.; Varior-Krishnan, G.; Cervoni, F.; Wild, T.F.; Rossi, B.; Rabourdin-Combe, C.; Gerlier, D. Human membrane cofactor protein (CD46) acts as a cellular receptor for measles virus. J Virol. 1993, 67, 6025–6032.
9. Tatsuo, H.; Ono, N.; Tanaka, K.; Yanagi, Y. SLAM (CDw150) is a cellular receptor for measles virus. Nature. 2000, 405, 893–897.
10. Erlenhoefer, C.; Wurzer, W.J.; Löffler, S.; Schneider-Schauliesxy, S.; ter Meulen, V.; Schneider-Schaulies, J. CD150 (SLAM) is a receptor for measles virus but is not involved in viral contact-mediated proliferation inhibition. J Virol. 2001, 75, 4499–4505.
11. Rota, J.S.; Hummel, K.B.; Rota, P.A.; Bellini, W.J. Genetic variability of the glycoprotein genes of current wild-type measles isolates. Virology. 1992, 188, 135–142.
12. Rima, B.K.; Earle, J.A.; Baczko, K.; ter Meulen, V.; Liebert, U.G.; Carstens, C.; Carabana, J.; Caballero, M.; Celma, M.L.; Fernandez-Munoz, R. Sequence divergence of measles virus haemagglutinin during natural evolution and adaptation to cell culture. J Gen Virol. 1997, 78, 97–106.
13. World Health Organization. Nomenclature for describing the genetic characteristics of wild-type measles viruses (update). Part I. Wkly Epidemiol Rec. 2001, 76, 242–247.
14. World Health Organization. Nomenclature for describing the genetic characteristics of wild-type measles viruses (update). Wkly Epidemiol Rec. 2001, 76, 249–251.
15. Bellini, W.J.; Rota, P.A. Genetic diversity of wild-type measles viruses: implications for global measles elimination programs. Emerg Infect Dis. 1998, 4, 29–35.
16. Greenhill, W.A. A Treatise on the Smallpox and Measles. (Translated from the original Arabic); Sydenham Society: London, 1848.
17. Goldberger, J.; Anderson, J.F. An experimental demonstration of the presence of the virus of measles in the mixed buccal and nasal secretions. JAMA. 1911, 57, 476–478.
18. Enders, J.F.; Peebles, T.C. Propagation in tissue cultures of cytopathogenic agents from patients with measles. Proc Soc Exp Biol Med. 1954, 36, 277–286.
19. Centers for Disease Control and Prevention. Global measles control and regional elimination, 1998–1999. MMWR; 1999; Vol. 48, 1124–1130.
20. Landrigan, P.J.; Witte, J.J. Neurologic disorders following live measles-virus vaccination. JAMA. 1973, 223, 1459–1462.
21. Polack, F.P.; Auwaerter, P.G.; Lee, S.H.; Nousari, H.C.; Valsamakis, A.; Leiferman, K.M.; Diwan, A.; Adams, R.J.; Griffin, D.E. Production of atypical measles in rhesus macaques: evidence for disease mediated by immune complex formation and eosinophils in the presence of fusion-inhibiting antibody. Nat Med. 1999, 5, 629–634.
22. Ojala, A. On changes in the cerebrospinal fluid during measles. Ann Med Intern Fenn. 1947, 36, 321–331.

23. Gibbs, F.A.; Gibbs, E.L.; Carpenter, P.R.; Spies, H.W. Electroencephalographic abnormality in ''uncomplicated'' childhood diseases. JAMA. 1959, 171, 1050–1055.
24. Hanninen, P.; Arstila, P.; Lang, H.; Salmi, A.; Panelius, M. Involvement of the central nervous system in acute, uncomplicated measles virus infection. J Clin Microbiol. 1980, 11, 610–613.
25. Griffin, D.E.; Hemachudha, T.; Johnson, R.T. Postinfectious and postvaccinal encephalomyelitis. In Clinical and Molecular Aspects of Neurotropic Virus Infection; Gilden, D.H., Lipton, H.L., Eds.; Kluwer Academic: Boston, 1989, 501–527.
26. Meyer, E.; Byers, R.K. Measles encephalitis. Am J Dis Child. 1952, 84, 543–579.
27. Johnson, R.T.; Griffin, D.E.; Hirsch, R.L.; Wolinsky, J.S.; Roedenbeck, S.; Lindo de Soriano, I.; Vaisberg, A. Measles encephalomyelitis—clinical and immunologic studies. N Engl J Med. 1984, 310, 137–141.
28. von Pirquet, C. Das Verhalten der kutanen Tuberkulin-Reaktion während der Masern. Dtsch Med Wochenschr. 1908, 34, 1297–1300.
29. Sanchez-Lanier, M.; Guerin, P.; McLaren, L.C.; Bankhurst, A.D. Measles virus-induced suppression of lymphocyte proliferation. Cell Immunol. 1988, 116, 367–381.
30. Borrow, P.; Oldstone, M.B.A. Measles virus–mononuclear cell interactions. In: Measles Virus; Billeter, M.A., ter Meulen, V., Eds.; Springer-Verlag: New York, 1995, 85–100.
31. Schlender, J.; Schnorr, J.J.; Spielhoffer, P.; Cathomen, T.; Cattaneo, R.; Billeter, M.A.; ter Meulen, V.; Schneider-Schaulies, S. Interaction of measles virus glycoproteins with the surface of uninfected peripheral blood lymphocytes induces immunosuppression in vitro. Proc Natl Acad Sci USA. 1996, 93, 13194–13199.
32. Fujinami, R.S.; Sun, X.; Howell, J.M.; Jenkin, J.C.; Burns, J.B. Modulation of immune system function by measles virus infection: role of soluble factor and direct infection. J Virol. 1998, 72, 9421–9427.
33. Karp, C.L.; Wysocka, M.; Wahl, L.M.; Ahearn, J.M.; Cuomo, P.J.; Sherry, B.; Trinchieri, G.; Griffin, D.E. Mechanism of suppression of cell-mediated immunity by measles virus. Science. 1996, 273(5303), 228–231.
34. Avota, E.; Avots, A.; Niewiesk, S.; Kane, L.P.; Bommhardt, U.; ter Meulen, V.; Schneider-Schaulies, S. Disruption of Akt kinase activation is important for immunosuppression induced by measles virus. Nat Med. 2001, 7, 725–731.
35. Schnorr, J.J.; Seufert, M.; Schlender, J.; Borst, J.; Johnston, I.C.; ter Meulen, V.; Schneider-Schaulies, S. Cell cycle arrest rather than apoptosis is associated with measles virus contact-mediated immunosuppression in vitro. J Gen Virol. 1997, 78, 3217–3226.
36. Steineur, M.P.; Grosjean, I.; Bella, C.; Kaiserlian, D. Langerhans cells are susceptible to measles virus infection and actively suppress T cell proliferation. Eur J Dermatol. 1998, 8, 413–420.
37. Klagge, I.M.; Schneider-Schaulies, S. Virus interactions with dendritic cells. J Gen Virol. 1999, 80, 823–833.
38. Gendelman, H.E.; Wolinsky, J.S.; Johnson, R.T.; Pressman, N.J.; Pezeshkpour, G.H.; Boisset, G.F. Measles encephalomyelitis: lack of evidence of viral invasion of the central nervous system and quantitative study of the nature of demyelination. Ann Neurol. 1984, 15, 353–360.
39. Moench, T.R.; Griffin, D.E.; Obriecht, C.R.; Vaisberg, A.J.; Johnson, R.T. Acute measles in patients with and without neurological involvement: distribution of measles virus antigen and RNA. J Infect Dis. 1988, 158, 433–442.
40. Johnson, R.T. The pathogenesis of acute viral encephalitis and postinfectious encephalomyelitis. J Infect Dis. 1987, 155, 359–364.
41. Moench, T.R.; Johnson, R.T. Measles. In Clinical and Molecular Aspects of Neurotropic Virus Infection; Gilden, D.H., Lipton, H.L., Eds.; Kluwer Academic: Boston, 1989, 201–229.
42. Caceres, V.M.; Strebel, P.M.; Sutter, R.W. Factors determining prevalence of maternal antibody to measles virus throughout infancy: a review. Clin Infect Dis. 2000, 31, 110–119.
43. Albrecht, P.; Ennis, F.A.; Saltzman, E.J.; Krugman, S. Persistence of maternal antibody in infants beyond 12 months: mechanism of measles vaccine failure. J Pediatr. 1977, 91, 715–718.

44. Schlereth, B.; Rose, J.K.; Buonocore, L.; ter Meulen, V.; Niewiesk, S. Successful vaccine-induced seroconversion by single-dose immunization in the presence of measles virus-specific maternal antibodies. J Virol. 2000, 74, 4652–4657.
45. Davidkin, I.; Valle, M. Vaccine-induced measles virus antibodies after two doses of combined measles, mumps and rubella vaccine: a 12-year follow-up in two cohorts. Vaccine. 1998, 16, 2052–2057.
46. Davidkin, I.; Valle, M.; Peltola, H.; Hovi, T.; Paunio, M.; Roivainen, M.; Linnavuori, K.; Jokinen, S.; Leinikki, P. Etiology of measles- and rubella-like illnesses in measles, mumps, and rubella-vaccinated children. J Infect Dis. 1998, 178, 1567–1570.
47. Mitchell, C.D.; Balfour, H.H., Jr. Measles control: so near and yet so far. Prog Med Virol. 1985, 31, 1–42.
48. Orenstein, W.A.; Strebel, P.M.; Papania, M.; Sutter, R.W.; Bellini, W.J.; Cochi, S.L. Measles eradication: is it in our future?. Am J Public Health. 2000, 90, 1521–1525.
49. Dawson, J.R.J. Cellular inclusions in cerebral lesions of lethargic encephalitis. Am J Pathol. 1933, 9, 7.
50. Dawson, J.R.J. Cellular inclusions in cerebral lesions of epidemic encephalitis. Arch Neurol Psychiatry. 1934, 31, 685–700.
51. Pette, H.; Döring, G. Über einheimische Panencephalomyelitis vom Charakter der Encephalitis japonica. Dtsch Z Nervenheilkd. 1939, 149, 7–44.
52. van Bogaert, L. Une leuco-encéphálite sclérosante suaigü. J Neurol Neurosurg Psychiatry. 1945, 8, 101.
53. Greenfield, J.G. Encephalitis and encephalomyelitis in England and Wales during the last decade. Brain. 1950, 73, 141–166.
54. Bouteille, M.; Fontaine, C.; Vedrenne, C.L.; Dalarue, J. Sur un cas de encéphalite suaigue à inclusions: étude anatomoclinique et ultrastructurale. Rev Neurol (Paris). 1965, 118, 454–458.
55. Connolly, J.H.; Allen, I.V.; Hurwitz, L.J.; Millar, J.H. Measles-virus antibody and antigen in subacute sclerosing panencephalitis. Lancet. 1967, 1, 542–544.
56. Freeman, J.M.; Magoffin, R.L.; Lennette, E.H.; Herndon, R.M. Additional evidence of the relation between subacute inclusion-body encephalitis and measles virus. Lancet. 1967, 2, 129–131.
57. Horta-Barbosa, L.; Fuccillo, D.A.; Sever, J.L.; Zeman, W. Subacute sclerosing panencephalitis: isolation of measles virus from a brain biopsy. Nature. 1969, 221, 974.
58. Payne, F.E.; Baublis, J.V.; Itabashi, H.H. Isolation of measles virus from cell cultures of brain from a patient with subacute sclerosing panencephalitis. N Engl J Med. 1969, 281, 585–589.
59. Haddad, F.S.; Risk, W.S.; Jabbour, J.T. Subacute sclerosing panencephalitis in the Middle East: report of 99 cases. Ann Neurol. 1977, 1, 211–217.
60. Dyken, P.R. Subacute sclerosing panencephalitis. Current status. Neurol Clin. 1985, 3, 179–196.
61. Singer, C.; Lang, A.E.; Suchowersky, O. Adult-onset subacute sclerosing panencephalitis: case reports and review of the literature. Mov Disord. 1997, 12, 342–353.
62. Gascon, G.G. Subacute sclerosing panencephalitis. Semin Pediatr Neurol. 1996, 3, 260–269.
63. Freeman, J.M. The clinical spectrum and early diagnosis of Dawson's encephalitis, with preliminary notes on treatment. J Pediatr. 1969, 75, 590–603.
64. Jabbour, J.T.; Garcia, J.H.; Lemmi, H.; Ragland, J.; Duenas, D.A.; Sever, J.L. Subacute sclerosing panencephalitis. A multidisciplinary study of eight cases. JAMA. 1969, 207, 2248–2254.
65. Risk, W.S.; Haddad, F.S. The variable natural history of subacute sclerosing panencephalitis: a study of 118 cases from the Middle East. Arch Neurol. 1979, 36, 610–614.
66. Gascon, G.; Yamani, S.; Crowell, J.; Stigsby, B.; Nester, M.; Kanaan, I.; Jallu, A. Combined oral isoprinosine-intraventricular alpha-interferon therapy for subacute sclerosing panencephalitis. Brain Dev. 1993, 15, 346–355.

67. Grunewald, T.; Lampe, J.; Weissbrich, B.; Reichmann, H. A 35-year-old bricklayer with hemimyoclonic jerks. Lancet. 1998, 351, 1926.
68. Esiri, M.M.; Oppenheimer, D.R.; Brownell, B.; Haire, M. Distribution of measles antigen and immunoglobulin-containing cells in the CNS in subacute sclerosing panencephalitis (SSPE) and atypical measles encephalitis. J Neurol Sci. 1982, 53, 29–43.
69. Kirk, J.; Zhou, A.L.; McQuaid, S.; Cosby, S.L.; Allen, I.V. Cerebral endothelial cell infection by measles virus in subacute sclerosing panencephalitis: ultrastructural and in situ hybridization evidence. Neuropathol Appl Neurobiol. 1991, 17, 289–297.
70. Allen, I.V.; McQuaid, S.; McMahon, J.; Kirk, J.; McConnell, R. The significance of measles virus antigen and genome distribution in the CNS in SSPE for mechanisms of viral spread and demyelination. J Neuropathol Exp Neurol. 1996, 55, 471–480.
71. Mesquita, R.; Castanos-Velez, E.; Biberfeld, P.; Troian, R.M.; de Siqueira, M.M. Measles virus antigen in macrophage/microglial cells and astrocytes of subacute sclerosing panencephalitis. APMIS. 1998, 106, 553–561.
72. ter Meulen, V.; Hall, W.W. Slow virus infections of the nervous system: virological, immunological and pathogenetic considerations. J Gen Virol. 1978, 41, 1–25.
73. ter Meulen, V.; Carter, M.J. Measles virus persistency and disease. Prog Med Virol. 1984, 30, 44–61.
74. Nakayama, T.; Mori, T.; Yamaguchi, S.; Sonoda, S.; Asamura, S.; Yamashita, R.; Takeuchi, Y.; Urano, T. Detection of measles virus genome directly from clinical samples by reverse transcriptase-polymerase chain reaction and genetic variability. Virus Res. 1995, 35, 1–16.
75. Katayama, Y.; Kohso, K.; Nishimura, A.; Tatsuno, Y.; Homma, M.; Hotta, H. Detection of measles virus mRNA from autopsied human tissues. J Clin Microbiol. 1998, 36, 299–301.
76. ter Meulen, V.; Stephenson, J.R.; Kreth, H.W. Subacute sclerosing panencephalitis. In Comprehensive Virology; Fraenkel-Conrat, H., Wagner, R.R., Eds.; Plenum Press: New York, 1983, 105–159.
77. Schneider-Schaulies, J.; Niewiesk, S.; Schneider-Schaulies, S.; ter Meulen, V. Measles virus in the CNS: the role of viral and host factors for the establishment and maintenance of a persistent infection. J Neurovirol. 1999, 5, 613–622.
78. Carter, M.J.; Willcocks, M.M.; ter Meulen, V. Defective translation of measles virus matrix protein in a subacute sclerosing panencephalitis cell line. Nature. 1983, 305, 153–155.
79. Baczko, K.; Liebert, U.G.; Billeter, M.; Cattaneo, R.; Budka, H.; ter Meulen, V. Expression of defective measles virus genes in brain tissues of patients with subacute sclerosing panencephalitis. J Virol. 1986, 59, 472–478.
80. Cattaneo, R.; Schmid, A.; Rebmann, G.; Baczko, K.; ter Meulen, V.; Bellini, W.J.; Rozenblatt, S.; Billeter, M.A. Accumulated measles virus mutations in a case of subacute sclerosing panencephalitis: interrupted matrix protein reading frame and transcription alteration. Virology. 1986, 154, 97–107.
81. Liebert, U.G.; Baczko, K.; Budka, H.; ter Meulen, V. Restricted expression of measles virus proteins in brains from cases of subacute sclerosing panencephalitis. J Gen Virol. 1986, 67, 2435–2444.
82. Cattaneo, R.; Rebmann, G.; Baczko, K.; ter Meulen, V.; Billeter, M.A. Altered ratios of measles virus transcripts in diseased human brains. Virology. 1987, 160, 523–526.
83. Schneider-Schaulies, S.; Schneider-Schaulies, J.; Schuster, A.; Bayer, M.; Pavlovic, J.; ter Meulen, V. Cell type-specific MxA-mediated inhibition of measles virus transcription in human brain cells. J Virol. 1994, 68, 6910–6917.
84. Cattaneo, R.; Schmid, A.; Eschle, D.; Baczko, K.; ter Meulen, V.; Billeter, M.A. Biased hypermutation and other genetic changes in defective measles viruses in human brain infections. Cell. 1988, 55, 255–265.
84a. Ogura, H.; Ayata, M.; Hayashi, K.; Seto, T.; Matsuoka, O.; Hattori, H.; Tanaka, K.; Tanaka, K.; Takano, Y.; Murata, R. Efficient isolation of subacute sclerosing panencephalitis virus

from patient brains by reference to magnetic resonance and computed tomographic images. J Neurovirol. 1997, 3, 304–309.
85. Fujinami, R.S.; Oldstone, M.B. Antiviral antibody reacting on the plasma membrane alters measles virus expression inside the cell. Nature. 1979, 279, 529–530.
86. Schneider-Schaulies, S.; Liebert, U.G.; Segev, Y.; Rager-Zisman, B.; Wolfson, M.; ter Meulen, V. Antibody-dependent transcriptional regulation of measles virus in persistently infected neural cells. J Virol. 1992, 66, 5534–5541.
87. Hofman, F.M.; Hinton, D.R.; Baemayr, J.; Weil, M.; Merrill, J.E. Lymphokines and immunoregulatory molecules in subacute sclerosing panencephalitis. Clin Immunol Immunopathol. 1991, 58, 331–342.
88. Nagano, I.; Nakamura, S.; Yoshioka, M.; Kogure, K. Immunocytochemical analysis of the cellular infiltrate in brain lesions in subacute sclerosing panencephalitis. Neurology. 1991, 41, 1639–1642.
89. Mehta, P.D.; Kulczycki, J.; Mehta, S.P. Sobczyk, W.; Coyle, P.K.; Sersen, E.A.; Wisniewski, H.M. Increased levels of beta 2-microglobulin, soluble interleukin-2 receptor, and soluble CD8 in patients with subacute sclerosing panencephalitis. Clin Immunol Immunopathol. 1992, 65, 53–59.
90. Dhib-Jalbut, S.; Jacobson, S.; McFarlin, D.E.; McFarland, H.F. Impaired human leukocyte antigen-restricted measles virus-specific cytotoxic T-cell response in subacute sclerosing panencephalitis. Ann Neurol. 1989, 25, 272–280.
91. Griffin, D.E. Immune responses during measles virus infection. In Measles Virus; Billeter, M.A., ter Meulen, V., Eds.; Springer-Verlag: New York, 1995, 117–134.
92. Hara, T.; Yamashita, S.; Aiba, H.; Nihei, K.; Koide, N.; Good, R.A.; Takeshita, K. Measles virus-specific T helper 1/T helper 2-cytokine production in subacute sclerosing panencephalitis. J Neurovirol. 2000, 6, 121–126.
93. Radermecker, J. Aspects électroéncephalographiques dans trois cas d'encéphalite subaigue. Acta Neurol Psychiatr Belg. 1949, 49, 222–232.
94. Dogulu, C.F.; Ciger, A.; Saygi, S.; Renda, Y.; Yalaz, K. Atypical EEG findings in subacute sclerosing panencephalitis. Clin Electroencephalogr. 1995, 26, 193–199.
95. Bohlega, S.; al-Kawi, M.Z. Subacute sclerosing panencephalitis. Imaging and clinical correlation. J Neuroimaging. 1994, 4, 71–76.
96. Brismar, J.; Gascon, G.G.; von Steyern, K.V.; Bohlega, S. Subacute sclerosing panencephalitis: evaluation with CT and MR. Am J Neuroradiol. 1996, 17, 761–772.
97. Reiber, H.; Lange, P. Quantification of virus-specific antibodies in cerebrospinal fluid and serum: sensitive and specific detection of antibody synthesis in brain. Clin Chem. 1991, 37, 1153–1160.
98. Dörries, R.; ter Meulen, V. Detection and identification of virus-specific, oligoclonal IgG in unconcentrated cerebrospinal fluid by immunoblot technique. J Neuroimmunol. 1984, 7, 77–89.
99. Pohl-Koppe, A.; Kaiser, R.; ter Meulen, V.; Liebert, U.G. Antibody reactivity to individual structural proteins of measles virus in the CSF of SSPE and MS patients. Clin Diagn Virol. 1995, 4, 135–147.
100. Felgenhauer, K.; Reiber, H. The diagnostic significance of antibody specificity indices in multiple sclerosis and herpes virus induced diseases of the nervous system. Clin Investig. 1992, 70, 28–37.
101. Taylor, W.J.; DuRant, R.H.; Dyken, P.R. Treatment of subacute sclerosing panencephalitis: an overview. Drug Intell Clin Pharm. 1984, 18, 375–381.
102. Haslam, R.H.; McQuillen, M.P.; Clark, D.B. Amantadine therapy in subacute sclerosing panencephalitis. A preliminary report. Neurology. 1969, 19, 1080–1086.
103. Robertson, W.C., Jr; Clark, D.B.; Markesbery, W.R. Review of 38 cases of subacute sclerosing panencephalitis: effect of amantadine on the natural course of the disease. Ann Neurol. 1980, 8, 422–425.

104. Gadoth, N.; Kott, E.; Levin, S.; Hahn, T. The interferon system in subacute sclerosing panencephalitis and its response to isoprinosine. Brain Dev. 1989, 11, 308–312.
105. Dyken, P.R.; Swift, A.; DuRant, R.H. Long-term follow-up of patients with subacute sclerosing panencephalitis treated with inosiplex. Ann Neurol. 1982, 11, 359–364.
106. Jones, C.E.; Dyken, P.R.; Huttenlocher, P.R.; Jabbour, J.T.; Maxwell, K.W. Inosiplex therapy in subacute sclerosing panencephalitis. A multicentre, non-randomised study in 98 patients. Lancet. 1982, 1, 1034–1037.
107. Fukuyama, Y.; Nihei, K.; Matsumoto, S.; Ebina, T.; Kamoshita, S.; Sato, T.; Arima, M.; Yabuuchi, H.; Ueda, S.; Ohtahara, S. Clinical effects of MND-19 (inosiplex) on subacute sclerosing panencephalitis—a multi-institutional collaborative study—The Inosiplex-SSPE Research Committee. Brain Dev. 1987, 9, 270–282.
108. Huttenlocher, P.R.; Mattson, R.H. Isoprinosine in subacute sclerosing panencephalitis. Neurology. 1979, 29, 763–771.
109. DuRant, R.H.; Dyken, P.R.; Swift, A.V. The influence of inosiplex treatment on the neurological disability of patients with subacute sclerosing panencephalitis. J Pediatr. 1982, 101, 288–293.
110. DuRant, R.H.; Dyken, P.R. The effect of inosiplex on the survival of subacute sclerosing panencephalitis. Neurology. 1983, 33, 1053–1055.
111. Bye, A.; Balkwill, F.; Brigden, D.; Wilson, J. Use of interferon in the management of patients with subacute sclerosing panencephalitis. Dev Med Child Neurol. 1985, 27, 170–175.
112. Huttenlocher, P.R.; Picchietti, D.L.; Roos, R.P.; Cashman, N.R.; Horowitz, B.; Horowitz, M.S. Intrathecal interferon in subacute sclerosing panencephalitis. Ann Neurol. 1986, 19, 303–305.
113. Panitch, H.S.; Gomez-Plascencia, J.; Norris, F.H.; Cantell, K.; Smith, R.A. Subacute sclerosing panencephalitis: remission after treatment with intraventricular interferon. Neurology. 1986, 36, 562–566.
114. Yalaz, K.; Anlar, B.; Oktem, F.; Aysun, S.; Ustacelebi, S.; Gurcay, O.; Gucuyener, K.; Renda, Y. Intraventricular interferon and oral inosiplex in the treatment of subacute sclerosing panencephalitis. Neurology. 1992, 42, 488–491.
115. Anlar, B.; Yalaz, K.; Oktem, F.; Kose, G. Long-term follow-up of patients with subacute sclerosing panencephalitis treated with intraventricular alpha-interferon. Neurology. 1997, 48, 526–528.
116. Cianchetti, C.; Marrosu, M.G.; Muntoni, F.; Fratta, A.; Zuddas, A. Intraventricular alpha-interferon in subacute sclerosing panencephalitis. Neurology. 1998, 50, 315–316.
117. Gascon, G.G.; Anlar, B.A.; Fojas, M.; Udani, V. Randomized treatment study of inosiplex versus combined inosiplex and intraventricular alpha interferon in subacute sclerosing panencephalitis (SSPE) [abstract FO-14-5]. Brain Dev. 2002, 24, 409.
118. Takahashi, T.; Hosoya, M.; Kimura, K.; Ohno, K.; Mori, S.; Takahashi, K.; Shigeta, S. The cooperative effect of interferon-alpha and ribavirin on subacute sclerosing panencephalitis (SSPE) virus infections, in vitro and in vivo. Antiviral Res. 1998, 37, 29–35.
119. Tomoda, A.; Shiraishi, S.; Hosoya, M.; Hamada, A.; Miike, T. Combined treatment with interferon-alpha and ribavirin for subacute sclerosing panencephalitis. Pediatr Neurol. 2001, 24, 54–59.
119a. Solomon, T.; Hart, C.A.; Vinjamuri, S.; Beeching, N.J.; Malucci, C.; Humphery, P. Treatment of subacute sclerosing panencephalitis with interferon-alpha, ribavirin and inosiplex. J Child Neurol. 2002, 9, 703–705.
119b. Takashima, H.; Eriguchi, M.; Nakamura, T.; Satoh, J.; Kuroda, Y.; Udono, H.; Uchino, A. Interferon therapy-responsive brain metabolic abnormalities in a case of adult-onset subacute sclerosing panencephalitis evaluated by 1H MRS analysis. J Neurol Sci. 2003, 207, 59–63.
120. Breitfeld, V.; Hashida, Y.; Sherman, F.E.; Odagiri, K.; Yunis, E.J. Fatal measles infection in children with leukemia. Lab Invest. 1973, 28, 279–291.

121. Mustafa, M.M.; Weitman, S.D. Winick, N.J.; Bellini, W.J.; Timmons, C.F.; Siegel, J.D. Subacute measles encephalitis in the young immunocompromised host: report of two cases diagnosed by polymerase chain reaction and treated with ribavirin and review of the literature. Clin Infect Dis. 1993, 16, 654–660.
122. Hughes, I.; Jenney, M.E.; Newton, R.W.; Morris, D.J.; Klapper, P.E. Measles encephalitis during immunosuppressive treatment for acute lymphoblastic leukaemia. Arch Dis Child. 1993, 68, 775–778.
123. Chen, R.E.; Ramsay, D.A.; deVeber, L.L.; Assis, L.J.; Levin, S.D. Immunosuppressive measles encephalitis. Pediatr Neurol. 1994. 10, 325–327.
124. Vigliano, P.; Boffi, P.; Giordana, M.T.; Tovo, P.A.; Palomba, E.; Rigardetto, R. Subacute measles encephalitis in a boy with perinatal HIV-1 infection. Dev Med Child Neurol. 1995, 37, 1117–1119.
125. Budka, H.; Urbanits, S.; Liberski, P.P.; Eichinger, S.; Popow-Kraupp, T. Subacute measles virus encephalitis: a new and fatal opportunistic infection in a patient with AIDS. Neurology. 1996, 46, 586–587.
126. Poon, T.P.; Tchertkoff, V.; Win, H. Subacute measles encephalitis with AIDS diagnosed by fine needle aspiration biopsy A case report. Acta Cytol. 1998, 42, 729–733.
127. Bitnun, A.; Shannon, P.; Durward, A.; Rota, P.A.; Bellini, W.J.; Graham, C.; Wang, E.; Ford-Jones, E.L.; Cox, P.; Becker, L.; Fearon, M.; Petric, M.; Tellier, R. Measles inclusion-body encephalitis caused by the vaccine strain of measles virus. Clin Infect Dis. 1999, 29, 855–861.
128. Gazzola, P.; Cocito, L.; Capello, E.; Roccatagliata, L.; Canepa, M.; Mancardi, G.L. Subacute measles encephalitis in a young man immunosuppressed for ankylosing spondylitis. Neurology. 1999, 52, 1074–1077.
129. Baczko, K.; Liebert, U.G.; Cattaneo, R.; Billeter, M.A.; Roos, R.P.; ter Meulen, V. Restriction of measles virus gene expression in measles inclusion body encephalitis. J Infect Dis. 1988, 158, 144–150.
130. Barthez Carpentier, M.A.; Billard, C.; Maheut, J.; Jourdan, M.L.; Degenne, D.; Ruchoux, M.M.; Goudeau, A.; Santini, J.J. Acute measles encephalitis of the delayed type: neuroradiological and immunological findings. Eur Neurol. 1992, 32, 235–237.
131. Olding-Stenkvist, E.; Forsgren, M.; Henley, D.; Kreuger, A.; Lundmark, K.M.; Nilsson, A.; Wadell, G. Measles encephalopathy during immunosuppression: failure of interferon treatment. Scand J Infect Dis. 1982, 14, 1–4.
132. Simpson, R.; Eden, O.B. Possible interferon response in a child with measles encephalitis during immunosuppression. Scand J Infect Dis. 1984, 16, 315–319.
133. ter Meulen, V.; Katz, M. The proposed viral etiology of multiple sclerosis and related demyelinating diseases. In ; Raine, C.S., McFarland, H.F., Tourtellotte, W.W., Eds. Multiple Sclerosis. Clinical and Pathogenetic Basis; Chapman and Hall: London, 1997, 287–305.
134. Adams, J.M.; Imagawa, D.T Measles antibodies in multiple sclerosis. Proc Soc Exp Biol Med. 1962, 111, 562–566.
135. Johnson, R.T. The possible viral etiology of multiple sclerosis. Adv Neurol. 1975, 13, 1–46.
136. Norrby, E. Viral antibodies in multiple sclerosis. Prog Med Virol. 1978, 24, 1–39.
137. Haase, A.T.; Ventura, P.; Gibbs, C.J.; Tourtellotte, W.W. Measles virus nucleotide sequences: detection by hybridization in situ. Science. 1981, 212, 672–675.
138. Cosby, S.L.; McQuaid, S.; Taylor, M.J.; Bailey, M.; Rima, B.K.; Martin, S.J.; Allen, I.V. Examination of eight cases of multiple sclerosis and 56 neurological and non-neurological controls for genomic sequences of measles virus, canine distemper virus, simian virus 5 and rubella virus. J Gen Virol. 1989 70, 2027–2036.
139. Stevens, J.G.; Bastone, V.B.; Ellison, G.W.; Myers, L.W. No measles virus genetic information detected in multiple sclerosis–derived brains. Ann Neurol. 1980, 8, 625–627.
140. Hall, W.W.; Choppin, P.W. Failure to detect measles virus proteins in brain tissue of patients with multiple sclerosis. Lancet. 1982, 1, 957.

141. Godec, M.S.; Asher, D.M.; Murray, R.S.; Shin, M.L.; Greenham, L.W.; Gibbs, C.J., Jr; Gajdusek, D.C. Absence of measles, mumps, and rubella viral genomic sequences from multiple sclerosis brain tissue by polymerase chain reaction. Ann Neurol. 1992, 32, 401–404.
142. Tamashiro, V.G.; Perez, H.H.; Griffin, D.E. Prospective study of the magnitude and duration of changes in tuberculin reactivity during uncomplicated and complicated measles. Pediatr Infect Dis J. 1987, 6, 451–454.
143. Hirsch, R.L.; Griffin, D.E.; Johnson, R.T.; Cooper, S.J.; Lindo de Soriano, I.; Roedenbeck, S.; Vaisberg, A. Cellular immune responses during complicated and uncomplicated measles virus infections of man. Clin Immunol Immunopathol. 1984, 31, 1–12.

19

Mumps Virus

Steven A. Rubin and Kathryn M. Carbone
U.S. Food and Drug Administration
Bethesda, Maryland, U.S.A.

1 INTRODUCTION

Mumps, once commonly referred to as epidemic parotitis, is an acute communicable viral disease of childhood. Mumps virus is transmitted by oropharyngeal secretions, with primary replication in the respiratory mucosa. Following the development of viremia, mumps virus can infect a number of organ systems, producing a variety of acute inflammatory reactions. Clinical manifestations of mumps were first reported in the fifth century BCE by Hippocrates in his *Book of Epidemics*, where symptoms of parotitis and orchitis in an epidemic illness were described. In 1790, the Royal Society of Edinburgh published a paper by Hamilton describing CNS involvement in mumps [1]. The predilection of the virus for the CNS was made evident in a study by Bang and Bang in 1943 [2] demonstrating CSF pleocytosis in 65% of mumps cases. Later epidemiological studies would reveal mumps virus as the leading cause of virus-induced aseptic meningitis and encephalitis in the United States and Europe. Although usually benign and self-limited, mumps virus infection of the CNS can lead to permanent neurological damage including deafness. Due to the success of vaccination programs, mumps is now mostly restricted to unvaccinated populations; the incidence of mumps in the United States is currently less than one case per 100,000 population [3].

2 BIOLOGICAL, PHYSICAL, AND CHEMICAL CHARACTERISTICS OF THE AGENT

2.1 Classification

Mumps virus is a member of the *Paramyxoviridae* family, subfamily *Paramyxovirinae*, genus *Rubulavirus*. Other members of this genus include parainfluenza viruses 2 and 4,

Newcastle disease virus and Simian virus 5 [4]. Of these, only mumps virus and parainfluenza viruses 2 and 4 are human pathogens.

2.2 Virion Structure and Genome

Mumps virions are pleomorphic particles ranging from 100 to 600 nm in size and consisting of a helical ribonucleocapsid core surrounded by a host cell–derived lipid envelope. The virus genome is a nonsegmented, single-stranded RNA macromolecule of negative polarity containing 15,384 nucleotides. The gene order is 3′ N-P-M-F-SH-HN-L 5′, representing the genes for the nucleoprotein, phosphoprotein, matrix protein, fusion protein, small hydrophobic protein, hemagglutinin-neuraminidase protein, and polymerase, respectively [5]. Each gene is translated into a single protein except for the P gene, from which mRNAs are transcribed that encode the phosphoprotein and two nonstructural proteins, V and I [6,7].

2.3 Viral Proteins and Function

Many investigators have contributed to the identification of the viral proteins and their function (see review by Carbone and Wolinsky [8]). The N, P, and L proteins associate with the viral RNA genome forming the ribonucleocapsid, of which the L protein is the RNA-dependent RNA polymerase [9–11]. Surrounding the ribonucleocapsid and forming the luminal side of the viral envelope is the M protein, which is believed to be responsible for the alignment of the ribonucleocapsid during virus assembly [12]. The HN protein, present on the outer surface of the viral envelope, is a disulfide-bonded oligomer responsible for virus binding and release [13]. The hemagglutinin activity of the protein is responsible for the binding of the virus to its cellular receptor sialic acid (*N*-acetyl neuraminic acid), a sugar group common to many glycosylated molecules, whereas the neuraminidase activity of the protein is responsible for the cleavage of sialic acid from sugar side chains, thereby releasing progeny virions from the cell surface [13,14]. The neuraminidase also functions to elute virus from inappropriately bound cells or debris and to remove sialic acid from nascent virions to prevent self-association. The F protein, also located on the outer surface of the viral envelope, is composed of two disulfide-linked gylcopeptides, F_1 and F_2 , products of host cell proteolytic cleavage of the immature protein F_0 [15,16]. The hydrophobic amino terminus of the F_1 protein triggers virus-to-cell or cell-to-cell fusion, an event mediated by the HN protein by an unknown mechanism [17]. Only antibodies generated against the HN and F surface proteins appear to be capable of neutralizing the virus [18]. The functions of the SH, V, and I proteins are less clear. The SH protein has been identified only in infected cells and may or may not be present in the intact virion [19]. The V and I proteins are, by inference from other members of the *Paramyxoviridae* family, likely involved in the regulation of gene transcription and translation [20,21].

2.4 Viral Replication

The replicative cycle of paramyxoviruses has been described in detail [4]. Replication begins with the fusion of the viral and host cell lipid membranes. Both the HN and F glycoproteins are required for this event, although the specific mechanism is not clear [22,23]. Following fusion, the viral ribonucleocapsid enters the host cell cytoplasm, where the genomic RNA is transcribed by the viral RNA–dependent RNA polymerase into monocistronic mRNAs, which are translated by the host cell. As intracellular levels of viral proteins reach a threshold, the activity of the viral polymerase switches to synthesizing full-length antisense genomic RNA, which is transcribed into negative sense progeny RNA. Virion assembly begins in the cell cytoplasm with the binding of the N proteins with the genomic RNA followed by the binding of the P and L protein complex [24]. This encapsidation process is believed to be initiated by specific leader and trailer sequences of the viral genome [25]. The ribonucleocapsid is then brought to the cellular cytoplasmic membrane, an event presumably mediated by the M protein [12], whereby the virion picks up its lipid envelope containing the HN and F viral glycoproteins as it exits the cell, usually resulting in cell lysis.

3 EPIDEMIOLOGY

3.1 Epidemics

Mumps is endemic worldwide, occurring epidemically every 2–5 years in unvaccinated populations [26,27]. Historically, mumps was most commonly seen from January to May in temperate zones [26,27] and mostly in children, with the highest incidence rates reported between 5 and 9 years of age, although the disease can be contracted at any age. Over the past 10 years there has been a gradual shift in the typical age of infection toward the 10–19-year-old group, and seasonality is no longer evident [28,29]. Prior to recent immunization programs, greater than 90% of most populations had serological evidence of exposure to mumps virus by age 15 [30,31]. Mumps is rarely seen in children younger than 1 year of age, most likely due to acquisition of immunity by placental transfer of maternal antibody.

3.2 Incidence

Vaccine use is responsible for a significant decline in the incidence of mumps worldwide. In the years following licensure of the live virus mumps vaccine in 1967, the number of cases of mumps in the United States decreased by more than 99%, reaching an all-time low of less than one case per 100,000 population in 1998 [3] (Fig. 1). The success of national vaccine programs can, in large part, be attributed to the high rate of seroconversion following vaccination (>90%) and the requirement of very low levels of antibody for protection; neutralizing antibody titers as low as 1:2 have been shown to be protective [32]. The effectiveness of the humoral immune response was demonstrated in one study wherein seropositive patients challenged intravenously with 10^9 plaque-forming units of mumps virus failed to develop symptoms of reinfection [33]. However, there is some evidence that symptomatic reinfections can occur [34,35].

3.3 Morbidity and Mortality

Although a hallmark sign of mumps virus infection, parotitis is not invariably seen. Serious mumps morbidity is primarily due to complications of meningitis, encephalitis, and orchitis. Prior to 1975, mumps was the most commonly diagnosed cause of encephalitis in the United States [36] and can account for up to half of all reported encephalitis cases in unvaccinated populations. Meningoencephalitis is most common among children between the ages of 5 and 9, with a 3:1 or higher male/female ratio [37,38]. The case fatality rate for mumps is between 1.6 and 3.8 per 10,000 [39]. Mumps virus infection may also involve other organs (see below), although serious complications are rare.

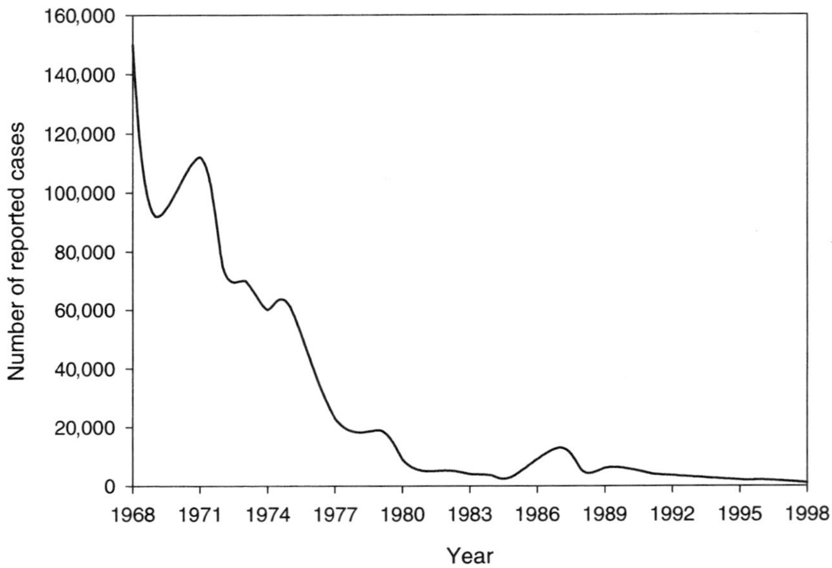

Figure 1 Number of reported cases of mumps in the United States, 1968–1998. (From National Notifiable Diseases Surveillance System, Centers for Disease Control and Prevention, Atlanta, Georgia, 1999.)

3.4 Transmission

Humans are the only natural host for mumps virus infection; there is no known animal reservoir. Mumps virus is readily transmitted by oropharyngeal secretions via direct contact, respiratory droplet, or fomites entering the nose or mouth. Virus is shed in the saliva over a period spanning 3–5 days before and after the appearance of clinical symptoms. The incubation period is typically 16–18 days with a range of 2–4 weeks. Individuals with inapparent infection are also contagious. Virus can also be present in breast milk and urine, although transmission via these routes has not been established. Mumps is less contagious than measles or varicella, with secondary attack rates of approximately 30% [40].

4 PATHOGENESIS

Primary replication occurs in the nasal mucosa or upper respiratory mucosal epithelium and, following infection of regional lymph nodes, widely disseminates throughout the body. Viremia develops 2–3 weeks after the primary infection and lasts for 3–5 days. Disappearance of viremia is coincident with the development of virus-specific IgA. Mumps virus preferentially infects T lymphocytes, a fact that may explain the ability of the virus to spread despite the rapid development of a humoral immune response [41–43]. Although parotid tissue is the most common site of viral infection following viremia, parotid involvement is not an obligate step in the infection. Mumps virus also has a particularly high tropism for the CNS; pleocytosis of the spinal fluid is seen in more than half of all clinically diagnosed cases of mumps [2]. Studies in animals have suggested that virus invades the CNS via infection of the choroid plexus. Progeny virions released from the infected choroid plexus enter the CSF circulation, mostly infecting meningeal and ventricular ependymal cells [42,44]. Infrequently, though, virus can penetrate deeper into the brain parenchyma, causing encephalitis and numerous neurological complications. Most other organs also become infected following viremia, causing a variety of acute inflammatory reactions (see Table 1).

Table 1 Reported CNS and Non-CNS Complications of Mumps Virus Infection

CNS	Non-CNS
Ataxia	Albuminuria
Blindness	Appendicitis
Cerebellitis	Arthritis
Cerebral diplegia	Basophilism
Choreoathetosis	Diabetes mellitus
Choroiditis	Endocardial fibroelastosis
Coma	Endocarditis
Congenital fetal defects	Epididymoorchitis
Death	Glaucoma
Deafness	Hepatitis
Encephalitis	Hypertension
Facial paralysis	Keratitis
Guillain-Barré syndrome	Laryngeal edema
Hemiplegia	Mastitis
Hydrocephalus	Myocarditis
Labyrinthitis	Nephritis
Landry's paralysis	Oophoritis
Meningitis	Orchitis
Mental disorder	Pancreatitis
Mesencephalic syndrome	Parotitis
Myelitis/neuromyelitis	Pericarditis
Ocular paralysis	Phlebitis
Optic atrophy	Pleurisy
Optic neuritis	Pneumonia
Papilledema	Presternal edema
Polyneuritis	Prostatitis
Retinitis	Pulmonary infarction
Subarachnoid hemorrhage	Scleritis
	Serositis
	Sialadenitis
	Sublinguitis
	Submandibular adenitis
	Thrombocytopenia purpura
	Thyroiditis
	Uveitis

Source: Adapted from Ref. 109.

5 CLINICAL MANIFESTATIONS

Prodromal symptoms are nonspecific and include low-grade fever, myalgias, malaise, anorexia, abdominal pain, and headache. Although mumps virus commonly infects glandular tissue and the CNS, virtually all tissues and organ systems can be infected and symptomatic. Despite the invasiveness of the virus, up to one-third of susceptible individuals have subclinical infection [45,46] and an additional 20–40% may have only nonspecific or primarily respiratory symptoms.

5.1 Glandular Involvement

Parotitis

Salivary gland swelling, typically parotitis, is the hallmark of mumps virus infection, occurring in up to 95% of all symptomatic cases [47]. Parotitis is usually bilateral, with inflammation of one gland preceding that of the other by several days. Unilateral involvement occurs in 25% of cases. Parotid swelling usually develops within a day following prodromal symptoms and may be initially noted as an earache or tenderness on palpation of the angle of the jaw. The gland reaches maximum size over the next 2–3 days, and swelling and tenderness can persist for more than a week. The submaxillary and sublingual salivary glands can also be involved. In rare cases, sialectasia can result as a complication of chronic sialadenitis [48]. The incidence of parotitis is similar for both males and females.

Gonadal Infection

Epididymoorchitis is relatively common in postpubescent males, occurring in approximately 35% of cases, and it usually develops during the first 2 weeks following the onset of parotitis but can occur in its absence [47,49]. Orchitis, typically unilateral, is characterized by painful swelling due to profuse edema in the interstitial connective tissue. Inflammation can be severe, with swelling up to four times normal size, but resolves within 4–10 days following symptomatic treatment.

Impotence is not a sequela, despite the fact that testicular atrophy can be detected in 50% of cases months to years later [50]. In rare cases, sterility [51] and testicular malignancy [52] due to mumps orchitis have been reported. Oophoritis is less common, developing in 5% of postpubertal women with mumps. Infection of the ovary is often accompanied with pelvic pain and, in exceedingly rare cases, can cause fertility impairment and premature menopause [53].

5.2 Central Nervous System

Mumps virus CNS invasion, as demonstrated by CSF pleocytosis, occurs in more than half of patients presenting with clinical mumps [2]; however, symptomatic CNS disease is apparent in only 1–15% of cases [54,55]. Serious neurological complications are rare, occurring at a rate of one per 6000 cases of mumps [56]. Neurological manifestations appear with a 3:1 or greater male/female ratio [37,38] and are generally preceded by parotitis by 4–5 days but can occur prior to or in the total absence of detectable salivary gland swelling. Thus, mumps virus should not be excluded as a possible etiology of cases presenting with CNS signs solely because of the absence of parotitis.

Cerebrospinal Fluid Abnormalities

Pleocytosis of the spinal fluid is often the only evidence of mumps virus CNS infection [2,57,58]. Lymphocytes are the predominant cell type found in the CSF, with white blood cell counts averaging 250 cells/mm^3 , with a typical range of 10–500 cells/mm^3 [59,60]. CSF pressure [61] and glucose and protein levels are generally within the normal range; however, glucose concentrations under 40 mg/dL (hypoglycorrhachia) and protein levels in excess of 300 mg/dL have been reported [60,62–64]. Intrathecal production of mumps virus–specific IgM and IgG has been reported in 40% of patients with symptomatic CNS infection [65]. Pleocytosis and intrathecal antibody synthesis, although typically transient, have been found to persist for months to years following mumps CNS infection [66–69], suggesting continued antigenic stimulation and persistence of mumps virus in the CNS.

Meningitis

Clinical meningitis is the most common neurological manifestation of mumps, diagnosed in up to 15% of cases [2,55,70]. Meningeal symptoms include headache, vomiting, fever, and nuchal rigidity, usually appearing 3–5 days after the onset of parotitis, with a range of 1 week before to 2 weeks after [2,71]. Meningeal involvement is also suggested by positive Brudzinski's and Kernig's signs and can be differentiated from encephalitis by a normal EEG. In many cases the diagnosis can be confirmed by an elevated CSF/serum antibody ratio [43,72]. Mumps meningitis is generally a benign condition, with complete recovery in 3–4 days.

Encephalitis

Encephalitis is one of the most serious complications of mumps virus infection, occurring in less than 0.2% of apparent mumps infections [37,56], and is responsible for the majority of fatal cases. In the prevaccine era, mumps was the leading cause of viral encephalitis in the United States, accounting for approximately one-third of the cases. Mumps encephalitis is now rarely seen. Although the encephalitis is typically the result of direct viral invasion, cases of postinfectious encephalitis, an autoimmune process likely triggered by the neural breakdown products of the primary infection, has been known to occur. Symptoms of primary encephalitis appear before or during the development of parotitis, and those of postinfectious encephalitis appear 1–3 weeks after the onset of parotitis and are almost always associated with demyelination [73,74]. Encephalitis is readily distinguished from meningitis by abnormal EEG and clinical findings suggestive of supratentorial involvement (e.g., decreased mental alertness).

Clinical Features. Clinical features of encephalitis include ataxia, exaggerated or diminished tendon reflexes, muscle weakness, paresis, paralysis of one or more limbs, spasticity, cranial nerve palsies, transient cortical blindness, personality changes, and impaired consciousness28,37,62,75–77. Coma and respiratory failure may also occur and are unfavorable prognostic signs. Death occurs in 0.5–2.3% of cases [78]. The incidence of complications is low, and the prognosis for complete recovery is excellent, usually resolving over a period of 1–2 weeks.

Syndromes. Neurological syndromes associated with encephalitis include cerebellar ataxia [79], transverse myelitis [80,81], seizures [76], ascending polyradiculitis [82], a poliomyelitis-like syndrome [83,84], and aqueductal stenosis and hydrocephalus [70,85]. Many of these likely reflect outcomes of postinfectious encephalitis.

Pathology. Changes typical of encephalitis have been reported upon autopsy and include edema and congestion throughout the brain with hemorrhages, lymphocytic infiltration, perivascular gliosis, and demyelination. Findings in the spinal cord include early degenerative changes in the anterior horn cells and perineuronal edema [76].

Deafness

Deafness is a well-known neurological complication of mumps. Although typically seen in patients presenting with encephalitis, there is no known etiological association between the two CNS manifestations [86]. Deafness is believed to be the result of direct damage to the cochlea and cochlear neurons, which become infected by the perilymph, which freely communicates with the CSF [86–89]. Transient, high-frequency deafness is the most common form, occurring in 4–5% of cases of mumps. Permanent deafness is rare (one per 20,000 cases of mumps) and is usually unilateral. Males and females are involved equally.

5.3 Other Organ Systems

As a systemic disease, a number of other organs and/or tissues are involved in mumps, although clinical manifestations, with the exception of the kidney (hematuria, proteinuria, nephritis, viruria), are infrequent. These organs include the ovary (oophoritis), prostate (prostatitis), liver (hepatitis), pancreas (pancreatitis), thyroid (thyroiditis), heart (ECG changes, myocarditis, endocardial fibroelastosis), mammary glands (mastitis), lungs (upper respiratory symptoms), and joints (polyarthralgia, arthritis). Mumps virus infection during the first trimester of pregnancy has also been associated with fetal death [90]. Additional clinical manifestations are listed in Table 1.

6 DIAGNOSIS

A clinical diagnosis of mumps is made on the basis of fever and constitutional symptoms such as parotitis, usually developing within 3 weeks after known exposure. White blood cell and differential counts are usually normal, although leukocytosis is not uncommon in cases presenting with meningitis, orchitis, or pancreatitis. Serum amylase levels are elevated in cases of parotitis or pancreatitis, the latter being differentiable from parotitis by isoenzyme analysis or serum pancreatic lipase determinations [74]. CSF findings are described above. Laboratory confirmation is often required because a number of infectious agents, drugs, and conditions can cause mumps-like symptoms (see Sec. 7). The laboratory diagnosis is typically based on isolation of infectious virus, serological confirmation, or the identification of viral genome by reverse transcriptase polymerase chain reaction (RT-PCR).

6.1 Viral Culture

Because mumps virus replication is transient, there is a relatively small window of opportunity for successful virus isolation. Virus can be readily isolated from swabs of the opening of the duct of Stensen or from saliva 2 days before to 6 days after the onset of parotitis, from the CSF within 6–9 days of meningeal symptoms, and from urine (preferably the first voided morning urine) during the first 2 weeks of illness59,91–93. Virus has only rarely been isolated from blood [94,95], despite the apparent frequency of viremia. Specimens should be collected as early in the disease as possible and, due to the friable nature of the virus, be kept on ice until use or stored at −70°C or below. Mumps virus will infect a wide range of mammalian cell lines, both transformed and primary. Vero cells (African green monkey kidney) are highly permissive to mumps virus infection and thus the cell line of choice for isolating clinical specimens. Conventionally, cell cultures are observed for characteristic cytopathic effects of mumps virus, i.e., syncytial formation followed by lysis (Fig. 2) appearing between 5 and 10 days post-inoculation. However, cytopathic effects per se are not necessarily diagnostic, because some mumps virus strains are noncytopathic and many of the viruses in the differential diagnosis of mumps cause

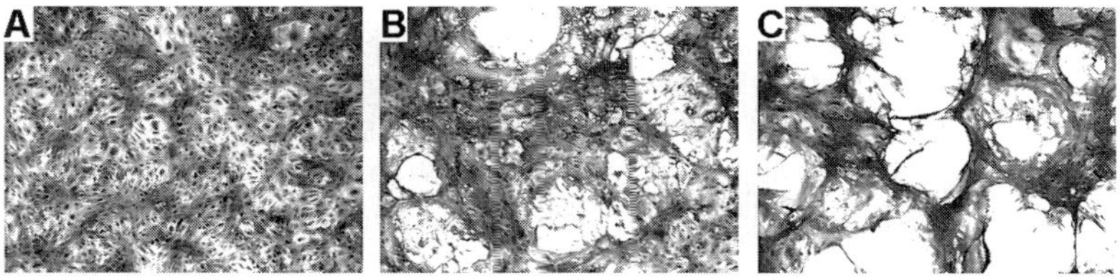

Figure 2 Phase contrast image showing the progression of cytopathic effects of a mumps virus clinical isolate incubated on Vero cells on days 1 (A), 4 (B), and 7 (C) after plating. The classic cytopathic effect of cell-to-cell fusion (syncytia formation) can be seen in panels B and C. By day 9, the entire monolayer was lysed.

cellular pathology indistinguishable from that induced by mumps virus. A definitive diagnosis therefore requires the use of mumps virus–specific antiserum, typically using immunofluorescence assays. Alternatively, cultures can be tested by RT-PCR for the identification of mumps virus RNA (see Sec. 6.3).

6.2 Serology

Because most people have been either vaccinated or naturally infected with mumps virus, simple detection of mumps virus antibodies in patient serum is not sufficient for a diagnosis of acute infection. Rather, a fourfold or higher rise in antibody titer needs to be demonstrated in convalescent serum drawn 2–4 weeks after the onset of clinical symptoms relative to that in serum drawn prior to or at the onset of symptoms. Alternatively, mumps virus IgM and IgG levels can be compared when only acute-phase serum is available. The mumps virus–specific IgM response will exceed the IgG response early in the infection before diminishing over the next few weeks to months [43,96]. The IgG response persists indefinitely. Thus, acute infection can be distinguished from a previous exposure based on the presence and levels of serum IgM and IgG, most commonly measured by enzyme-linked immunosorbent assay (ELISA). Other serological tests used to detect mumps virus antibodies include complement fixation, hemagglutination inhibition, and virus neutralization. Although all tests are valid means of diagnosing mumps infection, the ELISA and virus neutralization tests are preferred for assessing mumps immunity. Of these, the virus neutralization assay is the most accurate, but the technical features of this assay make routine use in clinical settings unwieldy. The complement fixation and hemagglutination inhibition tests only detect the presence of antibody directed against certain mumps virus proteins and thus are not a reliable means of determining mumps immunity. Further, the hemagglutination inhibition test is affected by cross-reacting heterologous antibodies induced by parainfluenza 3 virus

6.3 Polymerase Chain Reaction

Clinical samples can be tested by RT-PCR for the presence of mumps virus RNA either directly or following an intervening in vitro tissue culture step. The RT-PCR assay is highly sensitive and specific and can detect mumps virus in a significant percentage of clinical samples testing negative by the tissue culture method [97,98]. In cases presenting with parotitis, saliva and throat swab specimens are the preferred clinical samples. RT-

PCR testing of CSF in cases of suspected mumps meningitis is highly successful. In a recent study, mumps virus RNA was detected in 96% of patients with a clinical diagnosis of viral CNS disease, whereas only 39% of the patients were mumps virus positive by culture [97]. Although viruria is common in mumps, RT-PCR-based virus detection rates in urine are low [98]. RT-PCR detection of mumps virus RNA additionally provides for the identification of the specific strain of mumps virus involved, a feature key to epidemiological studies.

7 DIFFERENTIAL DIAGNOSIS

In the presence of salivary gland swelling, a clinical diagnosis of mumps is usually straightforward, although a number of other infectious agents should be considered. These include parainfluenza virus types 1 and 3, influenza A virus, coxsackievirus, lymphocytic choriomeningitis virus, HIV, and suppurative infections including *Staphylococcus aureus* and atypical mycobacteria. All can be easily differentiated from mumps virus by serology or culture. Parotitis can also be caused by starch ingestion or drugs such as phenylbutazone, thiouracil, iodides, and phenothiazines and by metabolic disorders including diabetes mellitus, cirrhosis, uremia, and malnutrition. Rare conditions such as Mikulicz's, Parinaud's, and Sjogren's syndromes can also be confused with mumps [74]. Other possible causes of parotid swelling include tumors, cysts, and salivary stones. In the absence of parotitis, laboratory confirmation of the diagnosis is required.

8 TREATMENT AND CONTROL

8.1 Treatment

No specific treatment for mumps is available. There is some evidence that administration of immunoglobulin could be helpful in selected cases [99]; however, it has not been shown to be effective during an epidemic [100], and the product is no longer available. Treatment of mumps is thus symptomatic and supportive.

8.2 Control

The apparent ineffectiveness of passive protection and the near impossibility of preventing virus spread by case isolation (considering that virus is shed prior to the appearance of clinical symptoms and a significant portion of infected individuals are asymptomatically infected) leaves vaccination as the only practical control measure. Fortunately, mumps vaccines are highly effective, being solely responsible for the extraordinary decline in the incidence of mumps witnessed over the past 25 years. All mumps vaccines currently in use are composed of live attenuated virus. Mumps vaccine is typically administered in combination with measles and rubella vaccines (MMR) to children 12–15 months of age; a second dose is recommended one month later or beyond. A single dose of the Jeryl Lynn mumps virus strain, the only strain used in the United States, induces protective levels of virus-neutralizing antibodies in more than 95% of recipients [101]. The duration of vaccine-induced immunity is believed to be lifelong. Similar rates of protection have been reported following vaccination with other mumps virus vaccines, although the protective efficacy of others, such as the Rubini strain used in many European countries, has been reported to be as low as 6% under field conditions [102–104]. The safety of mumps vaccines, like their efficacy, is strain-dependent. For example, although few serious adverse events have been reported in association with vaccination with the Jeryl Lynn strain [105,106], the Urabe vaccine strain, used throughout Europe, England, Canada, and Japan, has been causally associated with aseptic meningitis in one in 11,000 vaccinees [107,108]. Most countries have replaced the Urabe strain with the Jeryl Lynn strain. A causal association of CNS clinical symptoms following the administration of the Jeryl Lynn mumps virus vaccine strain has not been documented [106].

REFERENCES

1. Hamilton, R. An account of a distemper by the common people of England vulgarly called the mumps. London Med J. 1790, 11, 190–211.
2. Bang, H.O.; Bang, J. Involvement of the central nervous system in mumps. Acta Med Scand. 1943, 113, 487–505.
3. Centers for Disease Control and Prevention. Summary of Notifiable Diseases, United States, 1998. MMWR. 1999, 47, ii–92.
4. Lamb, R.A.; Kolakofsky, D. Paramyxoviridae: the viruses and their replication. In Fields Virology; Fields, B.N., Ed.; Lippincott-Raven: Philadelphia, 1996, 1177–1204.
5. Elango, N.; Varsanyi, T.M.; Kovamees, J.; Norrby, E. Molecular cloning and characterization of six genes, determination of gene order and intergenic sequences and leader sequence of mumps virus. J Gen Virol. 1988, 69, 2893–2900.
6. Elliott, G.D.; Yeo, R.P.; Afzal, M.A.; Simpson, E.J.; Curran, J.A.; Rima, B.K. Strain-variable editing during transcription of the P gene of mumps virus may lead to the generation of non-structural proteins NS1 (V) and NS2. J Gen Virol. 1990, 71, 1555–1560.
7. Paterson, R.G.; Lamb, R.A. RNA editing by G-nucleotide insertion in mumps virus P-gene mRNA transcripts. J Virol. 1990, 64, 4137–4145.
8. Carbone, K.M.; Wolinsky, J.S. Mumps virus. In Fields Virology; Knipe, D.M., Howley, P.M., Eds.; Lippincott-Raven: Philadelphia, 2001, 1381–1400.
9. Orvell, C. Structural polypeptides of mumps virus. J Gen Virol. 1978, 41, 527–539.
10. McCarthy, M.; Johnson, R.T. A comparison of the structural polypeptides of five strains of mumps virus. J Gen Virol. 1980, 46, 15–27.
11. Okazaki, K.; Tanabayashi, K.; Takeuchi, K.; Hishiyama, M.; Yamada, A. Molecular cloning and sequence analysis of the mumps virus gene encoding the L protein and the trailer sequence. Virology. 1992, 188, 926–930.
12. Matsumoto, T. Assembly of paramyxoviruses. Microbiol Immunol. 1982, 26, 285–320.
13. Orvell, C. Immunological properties of purified mumps virus glycoproteins. J Gen Virol. 1978, 41, 517–526.
14. Jensik, S.C.; Silver, S. Polypeptides of mumps virus. J Virol. 1976, 17, 363–373.
15. Merz, D.C.; Server, A.C.; Waxham, M.N.; Wolinsky, J.S. Biosynthesis of mumps virus F glycoprotein: non-fusing strains efficiently cleave the F glycoprotein precursor. J Gen Virol. 1983, 64, 1457–1467.
16. Naruse, H.; Nagai, Y.; Yoshida, T.; Hamaguchi, M.; Matsumoto, T.; Isomura, S.; Suzuki, S. The polypeptides of mumps virus and their synthesis in infected chick embryo cells. Virology. 1981, 112, 119–130.
17. Elango, N.; Varsanyi, T.M.; Kovamees, J.; Norrby, E. The mumps virus fusion protein mRNA sequence and homology among the paramyxoviridae proteins. J Gen Virol. 1989, 70, 801–807.
18. Houard, S.; Varsanyi, T.M.; Milican, F.; Norrby, E.; Bollen, A. Protection of hamsters against mumps virus (MuV) infection by antibodies raised against the MuV surface glycoproteins expressed from recombinant vaccinia virus vectors. J Gen Virol. 1995, 76, 421–423.

19. Takeuchi, K.; Tanabayashi, K.; Hishiyama, M.; Yamada, A. The mumps virus SH protein is a membrane protein and not essential for virus growth. Virology. 1996, 225, 156–162.
20. Curran, J.; Marq, J.B.; Kolakofsky, D. The Sendai virus nonstructural C proteins specifically inhibit viral mRNA synthesis. Virology. 1992, 189, 647–656.
21. Horikami, S.M.; Hector, R.E.; Smallwood, S.; Moyer, S.A. The Sendai virus C protein binds the L polymerase protein to inhibit viral RNA synthesis. Virology. 1997, 235, 261–270.
22. Tanabayashi, K.; Takeuchi, K.; Okazaki, K.; Hishiyama, M.; Yamada, A. Expression of mumps virus glycoproteins in mammalian cells from cloned cDNAs: both F and HN proteins are required for cell fusion. Virology. 1992, 187, 801–804.
23. Lamb, R.A. Paramyxovirus fusion: a hypothesis for changes. Virology. 1993, 197, 1–11.
24. Kingsbury, D.W.; Hsu, C.H.; Murti, K.G. Intracellular metabolism of Sendai virus nucleocapside. Virology. 1978, 91, 86–94.
25. Blumberg, B.M.; Kolakofsky, D. Intracellular vesicular stomatitis virus leader RNAs are found in nucleocapsid structures. J Virol. 1981, 40, 568–576.
26. Centers for Disease Control and Prevention. Mumps surveillance 1973. MMWR. 1974, 23, 431.
27. Centers for Disease Control and Prevention. Summary of notifiable diseases, United States, 1991. MMWR. 1991, 40, 3.
28. Centers for Disease Control and Prevention. Summary of notifiable diseases, United States, 1996. MMWR. 1996, 45, 1–88.
29. Holmes, S.J.; Sirotkin, B.I.; Williams, W.W.; Cochi, S.L.; Hadler, S.C.; Lindegren, M.L. Mumps surveillance—United States, 1988–1993. MMWR. 1995, 44, 1–14.
30. Centers for Disease Control and Prevention. Mumps—United States, 1984–1985. MMWR. 1986, 35, 216–219.
31. Mortimore, P.P. Mumps prophylaxis in the light of a new test for antibody. Br Med J. 1978, 2, 1523–1524.
32. Weibel, R.E.; Buynak, E.B.; Whitman, J.E.; Leagus, M.B.; Stokes, J.; Hilleman, M.R. Jeryl Lynn strain live mumps virus vaccine. JAMA. 1969, 207, 1667–1670.
33. Okuno, Y.; Asada, T.; Yamanishi, K.; Otsuka, T.; Takahashi, M.; Tanioka, T.; Aoyama, H.; Fukui, O.; Matsumoto, K.; Uemura, F.; Wada, A. Studies on the use of mumps virus for treatment of human cancer. Biken J. 1978, 21, 37–49.
34. Gut, J.P.; Lablache, C.; Behr, S.; Kirn, A. Symptomatic mumps virus reinfections. J Med Virol. 1995, 45, 17–23.
35. Nojd, J.; Tecle, T.; Samuelsson, A.; Orvell, C. Mumps virus neutralizing antibodies do not protect against reinfection with a heterologous mumps virus genotype. Vaccine. 2001, 19, 1727–1731.
36. Centers for Disease Control and Prevention. Mumps—United States. MMWR. 1978, 27, 379–381.
37. Koskiniemi, M.; Donner, M.; Pettay, O. Clinical appearance and outcome in mumps encephalitis in children. Acta Paediatr Scand. 1983, 72, 603–609.
38. Leboreiro, A. Leboreiro-Fernandez, M. Moura-Ribeiro, I; Sawan, F.; Ventura, A.; Barbosa, K. Mumps meningoencephalitis, an epidemiological approach. Arq Neuropsiquiatr. 1997, 55, 12–15.
39. Centers for Disease Control and Prevention. Mumps surveillance report. MMWR. 1972, 21, 1–12.
40. Hope-Simpson, R.E. Infectiousness of communicable diseases in the household (measles, chickenpox, and mumps). Lancet. 1952, 2, 549–554.
41. Fleischer, B.; Kreth, H.W. Mumps virus replication in human lymphoid cell lines and in peripheral blood lymphocytes: preference for T cells. Infect Immun. 1982, 35, 25–31.
42. Wolinsky, J.S.; Klassen, T.; Baringer, J.R. Persistence of neuroadapted mumps virus in brains of newborn hamsters after intraperitoneal inoculation. J Infect Dis. 1976, 133, 260–267.

43. Ukkonen, P.; Granstrom, M.L.; Penttinen, K. Mumps-specific immunoglobulin M and G antibodies in natural mumps infection as measured by enzyme-linked immunosorbent assay. J Med Virol. 1981, 8, 131–142.
44. Wolinsky, J.S.; Baringer, J.R.; Margolis, G.; Kilham, L. Ultrastructure of mumps virus replication in newborn hamster central nervous system. Lab Invest. 1974, 31, 403–412.
45. Levitt, L.P.; Mahoney, D.H.; Casey, H.L.; Bond, J.O. Mumps in a general population: a seroepidemiologic study. Am J Dis Child. 1970, 120, 134–138.
46. Maris, E.P.; Enders, J.F.; Stokes, J.; Kane, L.W. Immunity in mumps: IV. The correlation of the presence. J Exp Med. 1946, 84, 323–339.
47. Philip, R.N.; Reinhard, K.P.; Lachman, D.B. Observations on a mumps epidemic in a ''virgin'' population. Am J Hyg. 1959, 69, 91–111.
48. Travis, L.W.; Hecht, D.W. Acute and chronic inflammatory diseases of the salivary glands, diagnosis and management. Otolarying Clin North Am. 1977, 10, 329–338.
49. Candel, S. Epididymitis in mumps, including orchitis: further clinical studies and comment. Ann Intern Med. 1951, 34, 20.
50. Beard, C.M.; Benson, R.C.; Kelalis, P.P.; Elveback, L.R.; Kurland, L.T. The incidence and outcome of mumps orchitis in Rochester, Minnesota, 1935 to 1974. Mayo Clin Proc. 1977, 52, 3–7.
51. Shulman, A.; Shohat, B.; Gillis, D.; Yavetz, H.; Homonnai, Z.T.; Paz, G. Mumps orchitis among soldiers: frequency, effect on sperm quality, and sperm antibodies. Fertil Steril. 1992, 57, 1344–1346.
52. Kaufman, J.J.; Bruce, P.T. Testicular atrophy following mumps, a case of testis tumor?. J Urol. 1963, 35, 67–69.
53. Morrison, J.C.; Givens, J.R.; Wiser, W.L.; Fish, S.A. Mumps oophoritis: a cause of premature menopause. Fertil Steril. 1975, 26, 655–659.
54. Hammer, S.M.; Connolly, K.J. Viral aseptic meningitis in the United States: clinical features, viral etiologies, and differential diagnosis. Curr Clin Top Infect Dis. 1992, 12, 1–25.
55. Immunization Practices Advisory Committee. Mumps prevention. MMWR. 1989, 38, 397–400.
56. Russell, R.R.; Donald, J.C. The neurological complications of mumps. Br Med J. 1958, 2, 27–35.
57. Brown, J.W.; Kirkland, H.B.; Hein, G.E. Central nervous system involvement during mumps. Am J Med Sci. 1948, 215, 434–441.
58. Finkelstein, H. Meningo-encephalitis in mumps. JAMA. 1938, 3, 17–19.
59. Wolontis, S.; Bjorvatin, B. Mumps meningoencephalitis in Stockholm November 1964–July 1971. II. Isolation attempts from the cerebrospinal fluid in a hospitalized study group. Scand J Infect Dis. 1973, 5, 261–271.
60. Kilham, L. Mumps meningoencephalitis with and without parotitis. Am J Dis Child. 1949, 78, 324–333.
61. Bruyn, H.B.; Sexton, H.M.; Brainerd, H.D. Mumps meningoencephalitis. A clinical review of 119 cases with one death. Calif Med. 1957, 86, 153–160.
62. Koyama, S.; Morita, K.; Yamaguchi, S.; Fujikane, T.; Sasaki, N.; Aizawa, H.; Kikuchi, K. An adult case of mumps brain stem encephalitis. Intern Med. 2000, 39, 499–502.
63. Wilfert, C.M. Mumps meningoencephalitis with low cerebrospinal-fluid glucose, prolonged pleocytosis and elevation of protein. N Engl J Med. 1969, 280, 855–859.
64. Johnstone, J.A.; Ross, C.A.C.; Dunn, M. Meningitis and encephalitis associated with mumps infection. Arch Dis Child. 1972, 47, 647–651.
65. Vandvik, B.; Nilsen, R.E.; Vartdal, F.; Norrby, E. Mumps meningitis: specific and nonspecific antibody responses in the central nervous system. Acta Neurol Scand. 1982, 65, 468–487.
66. Vandvik, B.; Norrby, E.; Steen-Johnson, J.; Stensvold, K. Mumps meningitis: prolonged pleocytosis and occurrence of mumps virus-specific oligoclonal IgG in the cerebrospinal fluid. Eur Neurol. 1978, 17, 13–22.

67. Link, H.; Laurenzi, M.A.; Fryden, A. Viral antibodies in oligoclonal and polyclonal IgG synthesized within the central nervous system over the course of mumps meningitis. J Neuroimmunol. 1981, 1, 287–298.
68. Fryden, A.; Link, H.; Moller, E. Demonstration of CSF lymphocytes sensitized against mumps virus antigens in mumps meningitis. Acta Neurol Scand. 1978, 57, 396–404.
69. Julkunen, I.; Lehtokoski-Lehtiniemi, E.; Koskiniemi, M.; Vaheri, A. Elevated mumps antibody titers in the cerebrospinal fluid suggesting chronic mumps virus infection in the central nervous system. Pediatr Infect Dis. 1985, 4, 99.
70. Timmons, G.D.; Johnson, K.P. Aqueductal stenosis and hydrocephalus after mumps encephalitis. N Engl J Med. 1970, 283, 1505–1507.
71. Ritter, B.S. Mumps meningoencephalitis in children. J Pediatr. 1958, 52, 424–433.
72. Morishima, T.; Miyazu, M.; Ozaki, T.; Isomura, S.; Suzuki, S. Local immunity in mumps meningitis. Am J Dis Child. 1980, 134, 1060–1064.
73. Taylor, F.B.; Toreson, W.E. Primary mumps meningoencephalitis. Arch Intern Med. 1963, 112, 114–119.
74. Baum, S.G.; Litman, N. Mumps virus. In Principles and Practice of Infectious Diseases; Mandell, G.L.; Churchill Livingstone: Philadelphia, 2000, 1776–1781.
75. Nussinovitch, M.; Volovitz, B.; Varasano, I. Complications of mumps requiring hospitalization in children. Eur J Pediatr. 1995, 154, 732–734.
76. Miller, H.G.; Stanton, J.B.; Gibbons, J.L. Para-infectious encephalomyelitis and related syndromes. A critical review of the neurological complications of certain specific fevers. Quart J Med. 1956, 25, 427–505.
77. Leonberg, S.C. Nervous system affections caused by the mumps virus. Neurol Neurocir Psiquiatr. 1977, 18, 485–493.
78. Modlin, J.F.; Orenstein, W.A.; Brandling-Bennett, A.D. Current status of mumps in the United States. J Infect Dis. 1975, 132, 106–109.
79. Cohen, H.A.; Ashkenazi, A.; Nussinovitch, M.; Amir, J.; Hart, J.; Frydman, M. Mumps-associated acute cerebellar ataxia. Am J Dis Child. 1992, 146, 930–931.
80. Nussinovitch, M.; Brand, N.; Frydman, M.; Varsano, I. Transverse myelitis following mumps in children. Acta Paediatr. 1992, 81, 183–184.
81. Venketasubramanian, N. Transverse myelitis following mumps in an adult—a case report with MRI correlation. Acta Neurol Scand. 1997, 96, 328–330.
82. Ghosh, S. Guillain-Barre syndrome complicating mumps. Lancet. 1967, 1, 895.
83. Lennette, E.H.; Caplan, G.E.; Magoffin, R.L. Mumps virus infection simulating paralytic poliomyelitis. A report of 11 cases. Paediatrics. 1960, 25, 788–797.
84. Kilham, L.; Levens, J.; Enders, J.F. Nonparalytic poliomyelitis and mumps meningoencephalitis. Differential diagnosis. JAMA. 1949, 140, 934–936.
85. Ogata, H.; Oka, K.; Mitsudome, A. Hydrocephalus due to acute aqueductal stenosis following mumps infection: report of a case and review of the literature. Brain Dev. 1992, 14, 417–419.
86. McKenna, M. Measles, mumps, and sensorineural hearing loss. Ann NY Acad Sci. 1997, 830, 291–298.
87. Hall, R.; Richards, H. Hearing loss due to mumps. Arch Dis Child. 1987, 62, 189–191.
88. Smith, G.A.; Guessen, R. Inner ear pathologic features following mumps infection. Arch Otolaryngol. 1976, 102, 108–111.
89. Westmore, G.A.; Pickard, B.H.; Stern, H. Isolation of mumps virus from the inner ear after sudden deafness. Br Med J. 1979, 1, 14–15.
90. Siegel, M.; Fuerst, H.T. Low birth weight and maternal virus diseases. A prospective study of rubella, measles, mumps, chickenpox, and hepatitis. JAMA. 1966, 197, 680–684.
91. Chiba, Y.; Tsutsumi, H.; Nakao, T.; Wakisaka, A.; Aizawa, M. Human leukocyte antigen-linked genetic controls for T-cell mediated cytotoxicity response to mumps virus in humans. Infect Immun. 1982, 35, 600–604.

92. Utz, J.P.; Kasel, J.A.; Cramblett, H.G.; Szwed, C.F.; Parrott, R.H. Clinical and laboratory studies of mumps. I. Laboratory diagnosis by tissue-culture techniques. N Engl J Med. 1957, 257, 497–502.
93. McLean, D.M.; Bach, R.D. Larke, R.P.; McNaughton, G.A. Mumps meningoencephalitis Toronto, 1963. Can Med Assoc J. 1964, 90, 458–462.
94. Overman, J.R. Viremia in human mumps virus infections. Arch Intern Med. 1958, 102, 354–356.
95. Kilham, L. Isolation of mumps virus from the blood of a patient. Proc Soc Exp Biol Med. 1948, 69, 99–100.
96. Bjorvatn, B.; Wolontis, S. Mumps meningoencephalitis in Stockholm November 1964–July 1971. I. Analysis of a hospitalized study group. Questions of selection and representativity. Scand J Infect Dis. 1973, 5, 253–260.
97. Poggio, G.P.; Rodriguez, C; Cisterna, D.; Freire, M.C.; Cello, J. Nested PCR for rapid detection of mumps virus in cerebrospinal fluid from patients with neurological diseases. J Clin Microbiol. 2000, 38, 274–278.
98. Cusi, M.G.; Bianchi, S.; Valassina, M.; Santini, L.; Arnetoli, M.; Valensin, P.E. Rapid detection and typing of circulating mumps virus by reverse transcription/polymerase chain reaction. Res Virol. 1996, 147, 227–232.
99. Gellis, S.S.; McGuiness, A.C.; Peters, M. A study of the prevention of mumps orchitis by gammaglobulin. Am J Med Sci. 1945, 210, 661–664.
100. Reed, D.; Brown, G.; Merrick, R.; Sever, J.; Feltz, E. A mumps epidemic on St. George Island, Alaska. JAMA. 1967, 199, 113–117.
101. Buynak, E.B.; Hilleman, M.R Live attenuated mumps virus vaccine. 1. Vaccine development. Proc Soc Exp Biol Med. 1966, 123, 768–775.
102. German, D.; Ströhle, A.; Eggenberger, K.; Steiner, C.E.; Matter, L. An outbreak of mumps in a population partially vaccinated with the Rubini strain. Scand J Infect Dis. 1996, 28, 235–238.
103. Toscani, L.; Batou, M.; Bouvier, P.; Schlaepfer, A. Comparison of the efficacy of various strains of mumps vaccine: a school survey. Soz Praventivmed. 1996, 41, 341–347.
104. Strohle, A.; Eggenberger, K.; Steiner, C.A.; Matter, L.; Germann, D. Mumps epidemic in vaccinated children in West Switzerland. Schweiz Med Wochenschr. 1997, 127, 1124–1133.
105. Hilleman, M.R. Past, present, and future of measles, mumps, and rubella virus vaccines. Pediatrics. 1992, 90, 149–153.
106. Balraj, V.; Miller, E. Complications of mumps vaccines. Rev Med Virol. 1995, 5, 219–227.
107. Colville, A.; Pugh, M. Mumps meningitis and measles, mumps and rubella vaccine. Lancet. 1992, 340, 786.
108. Miller, E.; Goldacre, M.; Pugh, S.; Colville, A.; Farrington, P.; Flower, A.; Nash, J.; MacFarlane, L.; Tettmar, R. Risk of aseptic meningitis after measles mumps and rubella vaccine in UK children. Lancet. 1993, 341, 979–982.
109. Hyatt, H.W. Complications of mumps. GP. 1962, 25, 124–126.

20

Rubella

Avindra Nath
The Johns Hopkins University School of Medicine
Baltimore, Maryland, U.S.A.

1 INTRODUCTION

Rubella virus is the causative agent of the disease known more popularly as German measles. Rubella is predominantly a childhood disease and is endemic throughout the world. It most often causes a benign illness characterized by fever and rash that first appears on the face and spreads from head to toe, lasting about 3 days. Other symptoms include lymphadenopathy, frequently behind the ears and on the back of the neck, malaise, and/or conjunctivitis. Arthralgia and arthritis occur in ≤70% of infected adult and adolescent females. The incubation period for rubella is 12–23 days, and 20–50% of rubella infections are asymptomatic. Persons with rubella are most infectious when rash is erupting but can shed virus from 7 days before to 5–7 days after rash onset (i.e., the infectious period). However, occasionally neurological involvement may occur. Sequelae of rubella virus infection include three distinct neurological syndromes: a postinfectious encephalitis following acute infection, a spectrum of neurological manifestations following congenital infection, and an extremely rare neurodegenerative disorder, progressive rubella panencephalitis (PRP), that can follow either congenital or postnatal infection.

2 DESCRIPTION OF THE VIRUS

Rubella virus infects only humans. There is no known animal reservoir. This virus is a single-stranded, plus-sense RNA virus belonging to the *Togaviridae* family. It replicates in the cytoplasm. Surface projections of the envelope are distinct glycoprotein spikes composed of two virus proteins forming heterodimers. Its nucleocapsid has icosahedral symmetry. There are three structural virion proteins: the capsid protein, envelope glycoprotein E1, and envelope glycoprotein E2 [1]. Rubella virus is relatively unstable and is inactivated by lipid solvents, trypsin, formalin, ultraviolet light, extremes of pH and heat, and amantadine.

3 EPIDEMIOLOGY

Since the licensure of the rubella vaccine in 1969, the number of cases of rubella in the United States has decreased 99%, from 57,686 cases in 1969 to 271 cases in 1999. Seventy-three percent of persons with rubella in 1969 were Hispanic, compared with 4% in 1991 [2]. Most of these patients were recent immigrants from countries with suboptimal rubella immunization programs. High levels of seronegativity to rubella among women aged 15–19 years are reported in France (12%), Italy (10%), Germany (8%), and Taiwan (6%) and 1–3% in the United Kingdom, Netherlands, and Finland [3]. Many Asian and African countries do not have rubella vaccination programs, and these schemes have only recently been introduced in much of Latin America and the Caribbean. No accurate data exist for seroprevalence of rubella virus antibody or incidence of congenital rubella syndrome (CRS) in many of these countries. Neurological complications with rubella infections are rare, varying from one in 6000 to one in 24,000. An estimated incidence of 44–275 cases of congenital rubella syndrome per 100,000 live births is quoted [4].

4 PATHOGENESIS

Following respiratory transmission of rubella virus, replication of the virus is thought to occur in the nasopharynx and regional lymph nodes. A viremia occurs 5–7 days after exposure, with spread of the virus throughout the body. Transplacental infection of the fetus occurs during viremia. Fetal damage occurs through destruction of cells as well as mitotic arrest. The pathogenesis of the neurological syndromes associated with rubella virus is not well understood. Viral invasion and replication in the brain have been definitively demonstrated in CRS and appear to account for the majority of neurological lesions observed in this disease. Patients with CRS are unable to mount an immune response against the E2 envelope glycoprotein or the core antigen, which may explain in part the reason for viral perisistence [5]. The pathogenesis of rubella encephalitis following acute infection has not been determined [6]; however, the presence of virus in the cerebrospinal fluid (CSF) of these patients suggests that direct invasion by the virus may occur [7].

Immune-mediated pathology is particularly evident in progessive rubella panencephalitis (PRP) and may be autoimmune in nature, possibly triggered by molecular mimicry between viral and host epitopes, considering the apparent lack of virus in the brain [8]. Molecular mimicry may occur between rubella antigens and retinal or myelin antigens [9,10].

5 CLINICAL MANIFESTATIONS

5.1 Congenital Rubella

Intrauterine rubella infection can cause miscarriage, stillbirth, or abortion. However, 25% of babies whose mothers contract rubella during the first trimester of pregnancy can develop a spectrum of congenital defects seen in the newborn, known as congenital rubella syndrome (CRS). A rubella epidemic in the United States in 1964–1965 resulted in 12.5

million cases of rubella infection with about 20,000 newborns having CRS. The teratogenic effects are rare if infection occurs after 16 weeks of gestation. The stigmata of CRS are widespread, encompassing neural, ocular, and systemic development, and manifest as a spectrum of involvement [11,12]. Deafness is the most common and often the sole manifestation of congenital rubella infection, especially after the fourth month of gestation [13]. Eye defects, including cataracts, glaucoma, pigmentary retinopathy, and microphthalmia, may occur [14]. Cardiac defects such as patent ductus arteriosus, ventricular septal defect, pulmonic stenosis, and coarctation of the aorta are possible [15]. Neurological abnormalities include microcephaly, mental retardation, and meningoencephalitis. Other abnormalities that may occur include osteoporosis, splenomegaly, hepatitis, and thrombocytopenia with purpura. Some infected babies appear normal at birth and during infancy. However, all babies whose mothers had rubella during pregnancy should be monitored carefully, because problems with vision, hearing, learning, and behavior may first become noticeable during childhood.

Patients with neurological involvement on magnetic resonance imaging show basal ganglia and white matter hyperintensities characterized as bilateral T2 signal hyperintensities in periventricular and subcortical regions, punctate or linear in shape; they have been observed predominantly in the parietal lobes [16,17]. Basal ganglia calcification may also be found and is best demonstrated by computerized tomography [18]. Although serological testing remains the most available laboratory method for confirmation, CRS can also be confirmed by reverse transcriptase polymerase chain reaction (RT-PCR) assays, which detect rubella virus [19]. Any infant infected with rubella in utero can shed virus for about a year, sometimes longer. Rubella virus can be isolated from nasal, blood, throat, urine, or cerebrospinal fluid specimens (best results come from throat swabs). Serological testing requires demonstration of rubella-specific IgM antibody or an infant's rubella antibody level that persists above and beyond the expected from passive transfer of maternal antibody (i.e., rubella titer that does not drop at the expected rate of a twofold dilution per month). False positive serum rubella IgM tests have occurred in persons with parvovirus infections, with a positive heterophile test for infectious mononucleosis, or with a positive rheumatoid factor.

A 60-year follow-up study showed an increase in prevalence of diabetes (22%), thyroid disorders (19%), early menopause (73%), and osteoporosis (12.5%) in CRS patients [20]. Congenital rubella is probably underdiagnosed, especially among affected children who have sensorineural hearing loss as the only defect. Awareness of congenital rubella is currently low, few young health professionals will ever have seen a case, and by the time hearing loss has been recognized and investigated, many children already have rubella antibodies from the measles, mumps, and rubella (MMR) vaccination.

5.2 Rubella Encephalitis

Encephalitis occurs in one in 5000 cases of acute rubella infection, more frequently in adults (especially in females) than in children. Mortality estimates vary from 0% to 50%. Some patients may recover without sequelae [21]. Early diagnosis can be established by PCR on CSF [22]. Only a few cases have come to autopsy. In one patient who died several years after acute rubella encephalitis neuropathological findings were predominantly in the cerebral cortex with widespread necrosis and gliosis. Some neuronal cell loss was also noted in the basal ganglia, whereas the white matter appeared normal [23].

5.3 Progressive Rubella Panencephalitis

Progressive rubella panencephalitis (PRP) is a slow viral infection of the central nervous system. PRP was first reported in 1974 [24], and approximately 20 cases have been reported since then. All patients were male and between the ages of 8 and 21 years at onset, and most had signs of congenital rubella syndrome [25], although some cases of PRP occurred following rubella infections. Although PRP may exhibit clinical features resembling subacute sclerosing panencephalitis (SSPE) associated with measles virus, the age at onset is much older and the clinical course is more benign, extending over several years with periods of stabilization and remission [26]. Some patients may develop myoclonus, but this is not as prominent as that seen with SSPE. The main neurological features of PRP are dementia, cerebellar ataxia, and seizures. CSF may be acellular or have a mild pleocytosis, with elevation of protein. It is characterized by high levels of immunoglobulins, oligoclonal bands, and high CSF/serum rubella antibody titer ratios, suggesting intrathecal synthesis of rubella specific antibodies [27]. Diffuse atrophy of the brain, particularly the cerebellum, with ventricular dilatation and high signal intensity lesions in white matter may be found on MRI. EEG recordings often show slowing, although high-voltage burst suppression and epileptiform abnormalities have been described [28]. The prominent pathological findings in the brain included diffuse destruction of white matter with perivascular inflammatory cells and gliosis, moderate neuronal loss, numerous amorphous vascular deposits in the white matter, and severe generalized cerebellar atrophy [29]).

Rarely, rubella infection in adults may lead to pericarditis and myocarditis [30]. Aseptic arthritis or juvenile rheumatoid arthritis has also been associated with rubella infection [31] as well as with the rubella vaccine [32,33]. There is no specific treatment available for any of the rubella-related symdromes. Complications of rubella vaccine are discussed in Chapter 27.

REFERENCES

1. Waxham, M.N.; Wolinsky, J.S. A model of the structural organization of rubella virions. Rev Infect Dis. 1985, 7(suppl 1), S133–S139.
2. CDC. Control and prevention of rubella: evaluation and management of suspected outbreaks, rubella in pregnant women, and surveillance for congenital rubella syndrome. Morb Mortal Wkly Rep. 2001, 50, 1–23.
3. Nardone, A.; Gay, N.J.; Edmunds, W.J. Congenital rubella: down but not out. Lancet. 2002, 360, 804.
4. Cutts, F.T.; Vynnycky, E. Modelling the incidence of congenital rubella syndrome in developing countries. Int J Epidemiol. 1999, 28(6), 1176–1184.
5. de Mazancourt, A.; Waxham, M.N.; Nicolas, J.C.; Wolinsky, J.S. Antibody response to the rubella virus structural proteins in infants with the congenital rubella syndrome. J Med Virol. 1986, 19(2), 111–122.
6. Frey, T.K. Neurological aspects of rubella virus infection. Intervirology. 1997, 40(2–3), 167–175.
7. Squadrini, F.; Taparelli, F.; De Rienzo, B.; Giovannini, G.; Pagani, C. Rubella virus isolation from cerebrospinal fluid in postnatal rubella encephalitis. Br Med J. 1977, 2, 1329–1330.
8. Coyle, P.K.; Wolinsky, J.S. Characterization of immune complexes in progressive rubella panencephalitis. Ann Neurol. 1981, 9(6), 557–562.
9. Williams, L.L.; Lew, H.M.; Davidorf, F.H.; Pelok, S.G.; Singley, C.T.; Wolinsky, J.S. Altered membrane fatty acids of cultured human retinal pigment epithelium persistently infected with rubella virus may affect secondary cellular function. Arch Virol. 1994, 134(3–4), 379–392.

10. Nath, A.; Wolinsky, J.S. Antibody response to rubella virus structural proteins in multiple sclerosis. Ann Neurol. 1990, 27(5), 533–536.
11. Cooper, L.Z. Rubella in pregnancy. Postgrad Med. 1969, 46(6), 106–107.
12. Miller, E.; Cradock, J.E.-; Pollock, T.M. Consequences of confirmed maternal rubella at successive stages of pregnancy. Lancet. 1982, 2, 781–784.
13. Trybus, R.J.; Karchmer, M.A.; Kerstetter, P.P.; Hicks, W. The demographics of deafness resulting from maternal rubella. Am Ann Deaf. 1980, 125(8), 977–984.
14. Arnold, J. Ocular manifestations of congenital rubella. Curr Opin Ophthalmol. 1995, 6(3), 45–50.
15. Thanopoulos, B.D.; Rokas, S. Frimas, C.A.; Mantagos, S.P.; Beratis, N.G. Cardiac involvement in postnatal rubella. Acta Paediatr Scand. 1989, 78(1), 141–144.
16. Lane, B.; Sullivan, E.V.; Lim, K.O.; Beal, D.M.; Harvey, Jr, R.L.; Meyers, T. White matter MR hyperintensities in adult patients with congenital rubella. Am J Neuroradiol. 1996, 17(1), 99–103.
17. Yamashita, Y.; Matsuishi, T.; Murakami, Y.; Shoji, H.; Hashimoto, T.; Utsunomiya, H. Neuroimaging findings (ultrasonography, CT, MRI) in 3 infants with congenital rubella syndrome. Pediatr Radiol. 1991, 21(8), 547–549.
18. Chang, Y.C.; Huang, C.C.; Liu, C.C. Frequency of linear hyperechogenicity over the basal ganglia in young infants with congenital rubella syndrome. Clin Infect Dis. 1996, 22(3), 569–571.
19. Tanemura, M.; Suzumori, K.; Yagami, Y.; Katow, S. Diagnosis of fetal rubella infection with reverse transcription and nested polymerase chain reaction: a study of 34 cases diagnosed in fetuses. Am J Obstet Gynecol. 1996, 174(2), 578–582.
20. Sullivan, E.M.; Burgess, M.A.; Forrest, J.M. The epidemiology of rubella and congenital rubella in Australia, 1992 to 1997. Commun Dis Intell. 1999, 23(8), 209–214.
21. Merrer, J.; Perin-Dureau, F.; Appere, C.; Palmer, P.; Santoli, F.; De Jonghe, B. Severe forms of rubella encephalitis: arguments for a better vaccination policy. Presse Med. 1999, 28(8), 395–397.
22. Date, M.; Gondoh, M.; Kato, S.; Fukushima, M.; Nakamoto, N.; Kobayashi, M. A case of rubella encephalitis: rubella virus genome was detected in the cerebrospinal fluid by polymerase chain reaction. No To Hattatsu. 1995, 27(4), 286–290.
23. Warzok, R.; Wockel, W.; Scholtze, P. Late neuropathological findings after acute rubella encephalitis. Zentralbl Allg Pathol. 1979, 123(4), 301–306.
24. Lebon, P.; Lyon, G. Letter: Non-congenital rubella encephalitis (Letter). Lancet. 1974, 2, 468.
25. Townsend, J.J.; Baringer, J.R.; Wolinsky, J.S.; Malamud, N.; Mednick, J.P.; Panitch, H.S. Progressive rubella panencephalitis. Late onset after congenital rubella. N Engl J Med. 1975, 292(19), 990–993.
26. Wolinsky, J.S.; Berg, B.O.; Maitland, C.H. Progressive rubella panencephalitis. Arch Neurol. 1976, 33(10), 722–723.
27. Wolinsky, J.S.; Waxham, M.N.; Hess, J.L.; Townsend, J.J.; Baringer, J.R. Immunochemical features of a case of progressive rubella panencephalitis. Clin Exp Immunol. 1982, 48(2), 359–366.
28. Guizzaro, A.; Volpe, E.; Lus, G.; Bravaccio, F.; Cotrufo, R.; Paolozzi, C. Progressive rubella panencephalitis. Follow-up EEG study of a case. Acta Neurol (Napoli). 1992, 14(4–6), 485–492.
29. Townsend, J.J.; Wolinsky, J.S.; Baringer, J.R. The neuropathology of progressive rubella panencephalitis of late onset. Brain. 1976, 99(1), 81–90.
30. Harada, T.; Ohtaki, E.; Tobaru, T.; Kitahara, K.; Sumiyoshi, T.; Hosoda, S. Rubella-associated perimyocarditis—a case report. Angiology. 2002, 53(6), 727-732.
31. Bosma, T.J.; Etherington, J.; O'Shea, S.; Corbett, K.; Cottam, F.; Holt, L. Rubella virus and chronic joint disease: is there an association?. J Clin Microbiol. 1998, 36(12), 3524–3526.

32. Geier, D.A.; Geier, M.R. Rubella vaccine and arthritic adverse reactions: an analysis of the Vaccine Adverse Events Reporting System (VAERS) database from 1991 through 1998. Clin Exp Rheumatol. 2001, 19(6), 724–726.
33. Tingle, A.J.; Mitchell, L.A.; Grace, M.; Middleton, P.; Mathias, R.; MacWilliam, L. Randomised double-blind placebo-controlled study on adverse effects of rubella immunisation in seronegative women. Lancet. 1997, 349, 1277–1281.

21

Influenza and CNS Complications

Marie Studahl
Sahlgrenska University Hospital/Östra, Göteborg University
Göteborg, Sweden

Annika Linde
Swedish Institute for Infectious Disease Control
Solna, and
Karolinska Institute
Stockholm, Sweden

1 DESCRIPTION OF THE VIRUS

Influenza viruses (A, B, and C) are enveloped RNA viruses and belong to the family *Orthomyxoviridae*. The envelope is covered with glycoprotein spikes: the hemagglutinin (HA), neuraminidase (NA), and membrane channel (M2) proteins (influenza A). Identified in influenza A are 15–16 HA and nine NA subtypes. H1–H3 in combination with N1 or N2 are known to have caused epidemics in humans. HA is the ligand, attaching the virus to its sialic acid receptor, and it is the primary target for neutralizing antibodies (Fig. 1). HA and M2 are also active in uncoating of the virus, and NA is a receptor-destroying enzyme that releases newly produced viruses from the infected cell. The genome is multipartite, which facilitates reassortment of genomic segments from different influenza viruses infecting the same cell. This may result in antigenic shifts, if one or both surface glycoproteins are exchanged, and introduction of novel influenza subtypes to which the population lacks immunity. Minor changes in antigenicity of influenza virus by amino acid changes in the surface antigen (HA and NA) due to point mutations of the genome, known as antigen drift, occur yearly or every few years. Antigenic changes are frequent in influenza A and less frequent in influenza B and C [1]. Each subtype of influenza A is designed by its HA and NA [2].

2 ILLNESS

Influenza virus infection is a highly contagious respiratory illness. It is transmitted from an infected to a susceptible person by transfer of virus-containing respiratory droplets.

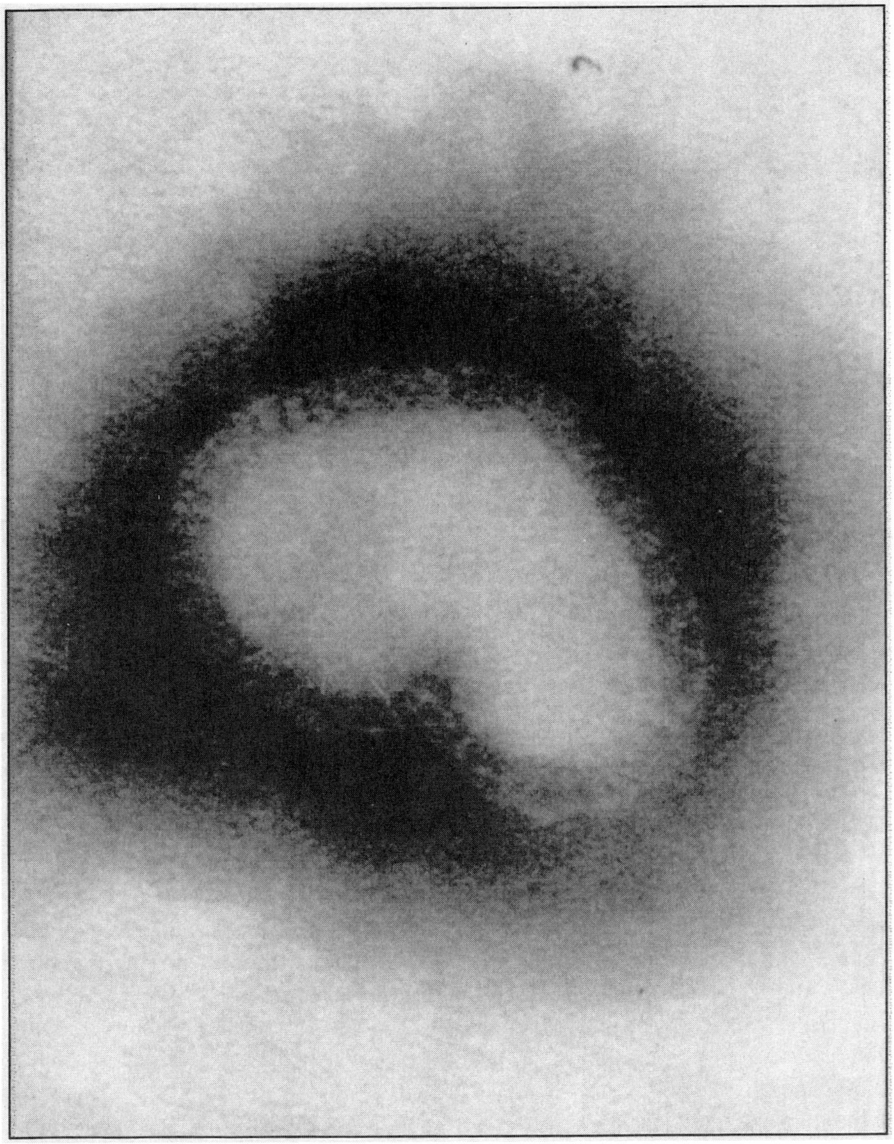

Figure 1 Electromicrosic image of influenza virus. (Virus preparation, Camilla Kolmskog; photo courtesy of Kjell-Olof Hedlund.)

The clinical outcome may range from inapparent infection, pharyngitis and/or tracheobronchitis, to systemic disease (''flu'') with a risk of complications in inner organs such as the lung, heart, brain, liver, kidneys, and muscles. The most common causes of death are respiratory complications or cardiovascular diseases. After an incubation period of 1–5 days, the onset of illness is generally abrupt [1], but it may be insidious and atypical, especially in the elderly. The initial symptoms are nonproductive cough, high fever

(38–40°C), chills, headache, anorexia, myalgia affecting the limbs and back, and dizziness. Additional symptoms are nasal discharge or obstruction, sneezing, painful eye movements, and, less frequently, productive cough, hoarseness, substernal soreness, diarrhea, and vomiting. The duration of the fever is between 1 and 5 days, although cough and fatigue may persist for several weeks after the fever has gone down [2]. During a known influenza epidemic the specificity of the clinical diagnosis is 60–80% [3].

In children, the dominating symptoms are fever, headache, cough, and sore throat [4]. Children often present with febrile convulsions [5]. Otis media, croup, and gastrointestinal manifestations are common, especially in infants. In neonates, the symptoms are even more nonspecific, with lethargy, poor feeding, apnea, and bronchiolitis [6].

Influenza A H3N2 has been suggested to produce more severe illness than influenza A H1N1, and influenza B is intermediate in severity [7,8]. Influenza C causes only mild respiratory infections that mainly affect children or the elderly [9]. The patients who are at risk of developing serious morbidity and mortality are those with chronic heart and lung disease, the elderly, patients with other debilitating conditions, transplant recipients, smokers, and pregnant women.

3 EPIDEMIOLOGY

Influenza virus infection is a common disease worldwide. In temperate climates, influenza virus epidemics occur almost exclusively during the coldest months of the year, because transmission is facilitated by indoor crowding, low temperature, and low humidity. In contrast, influenza can be isolated year-round in tropical countries, with peaks during monsoon and wet seasons [10]. Both influenza A and B are associated with high mortality and morbidity, although, owing to its antigenic stability, influenza B does not cause pandemics [11].

The natural reservoirs of influenza A are infected pigs and aquatic birds [12]. New pandemic strains to which the population lacks immunity emerge either by genetic reassortment in cells dually infected by a human and an animal influenza A virus or by direct transmission of an animal strain to humans. China and/or south-east Asia are regarded as a ''mixing vessels'' for this interchange, and most pandemics have originated from this area [11].

4 HISTORY OF THE ILLNESS

Epidemic fevers have been described in the medical literature since the time of Hippocrates (460–357 B.C.) [13]. However, because there are no pathognomonic symptoms in influenza disease, it was after the isolations of influenza A, B, and C viruses [14–17] that a more detailed picture of the infection and course of epidemics was first discernible [2]. Neurological complications associated with influenza were mentioned in the prediagnostic era [18], and encephalitis was reported during the ''Spanish influenza'' epidemic in 1918–1919 [19]. This large pandemic was caused by a swine influenza virus (H1N1sw) [20], and 20–40 million people died. Encephalitis lethargica (von Economo's disease) has been associated with the 1918–1919 influenza pandemics [21], but proof of a causal relationship is lacking [2]. During the following decades, sporadic reports were published on neurological complications during epidemics of influenza [22,23]. During the Asian influenza in 1957–1958, several authors reported cases of encephalitis and encephalopathy in association with influenza virus infection diagnosed virologically [24–28].

5 PATHOPHYSIOLOGY

The precise pathogenesis of the various neurological syndromes related to influenza infection is unknown. Influenza virus isolates from CSF and/or brain have previously been regarded as rare, but several cases are documented (Table 1). Influenza RNA has repeatedly been detected by PCR in the CSF or brain [29–34], indicating the presence of virus in the CNS (Table 2). Influenza virus may spread to the brain or CSF either by the hematogenous route via the choroid plexus or through endothelial cells [35] and/or neuronally through the olfactory and trigeminal nerve system [36,37]. The latter pathway to the brain, as shown in animal models [36,38–41], may be facilitated by the fact that free nerve endings of the olfactory nerves near the epithelium of the upper respiratory tract are destroyed by the influenza infection [37]. A prerequisite for hematogenous spread is viremia, which is sparsely reported in influenza [42–46]. Viremia is present mainly during the incubation period [43,44], which explains the failure to isolate virus from the blood in patients with influenza A infection [47–49] and to detect RNA from peripheral blood mononuclear cells (PBMC) and plasma [50]. However, a few reports of RNA detected by PCR in PBMC [30,51] and findings of viral antigen in CD8 positive T lymphocytes [52] have been published. Depending on the anatomical site of viral entry into the CNS, different populations of cells may become infected [36], and the immune response in the brain may also differ [53]. A vascular endothelial infection has been suggested as part of the pathogenesis of acute encephalopathy [30]. This is supported by a postmortem study of immunosuppressed influenza patients in which influenza viruses were isolated from brains in which ependymal cells were positive for virus antigen by immune staining [54]. In mice models, wild-type influenza strains replicate primarily in ependymal cells in the CNS, whereas neuro-adapted strains may replicate in neurons and glial cells [55,56]. The neuro-adapted genotype is associated with changes in the neuraminidase gene [56–59]. Virus antigen positivity was shown in Purkinje cells in the cerebellum and in several neurons in the pontine nuclei in a child who died of influenza encephalopathy [33]. A limited direct invasion of neurons cannot be excluded. Postmortem studies of brains from patients with CNS complications associated with influenza are generally lacking. Selective fatal cases show congestion and hyperemia of the brain [26–28,33,60] and lack of inflammatory reaction [24,27,33,54,61]. There are also rare reports of demyelination [24,27] and neuron degeneration [28].

6 CLINICAL MANIFESTATIONS

A wide spectrum of signs of CNS involvement during influenza A infection have been observed. Children are more often afflicted than adults [2]. However, many case descriptions in the older literature provide insufficient virological support for the diagnosis, making the association between the neurological syndromes and influenza virus difficult to interpret. The two most commonly described syndromes are the clinical entities encephalitis and encephalopathy, which are not always distinguishable from each other [62]. Encephalitis is more often associated with (1) focal neurological signs and/or (2) focal abnormalities on CT/MRI, plus (3) fever and (4) inflammatory changes in the CSF. Encephalopathy, on the other hand, is more often reflected by (1) diffuse CNS dysfunction with loss of consciousness, (2) usually a normal CT/MRI, (3) absent or low-grade fever, and (4) lack

Table 1 Virologically Verified Influenza Infections of the CNS Described in the Literature

Author/year [Ref. no.]	No. of cases	Sex/age (years)	Clinical description of symptoms/onset after flu (days)	Type of infl.	CSF findings Leukocytes/ mono/μL	CSF findings Protein (g/L)	CT/MRI	Outcome	EEG	Method of virological diagnosis
Flewett 1958 [27]	1	N.D./3 1/2	Convulsions "at the height of attack of influenza"	Infl. A	N.D.	N.D.	N.D.	Death	N.D.	Virus isolation from brain + lungs
Kapila 1958 [28]	1	M/N.D.	Sudden weakness of respiratory muscles/0	Infl. A	N.D.	N.D.	N.D.	Death	N.D.	Virus isolation from brain
Price 1976 [60]	1	M/3	Congenital hydrocephalus + encephalopathy/N.D.	Infl. A_2	N.D.	N.D.	N.D.	Death	N.D.	Virus isolation from brain and lungs
Salonen 1978 [67]	1	F/62	Myelitis/N.D.	Infl. A	17	2.4	MRI; contrast enhancement of the cervical medulla	Permanent sequelae	N.D.	Intrathecal antibodies in CSF
Paisley 1978	1	F/15 mo	Fever and irritability/N.D.	Infl. A_2	Normal	Normal	N.D.	Recovered	N.D.	Virus isolated from CSF and NPA and throat
Rose 1982 [82]	3	F/18 F/35 M/6	Encephalopathy/7 Encephalopathy/N.D. Transverse myelitis and arachnoid cyst/N.D.	Infl. A H3N2 H1N1 H1N1	35/35 0 0	0.76 0.85 0.14	Normal CT Normal CT N.D.	Recovered Quadriparesis Sequelae	N.D.	Virus isolation from CSF
Rantala 1990	1	N.D.	Febrile convulsions/N.D.	Infl. B	N.D.	N.D.	N.D.	N.D.	N.D.	Virus isolation from CSF
Studahl 1998 [99]	1	M/40	Encephalitis/0	Infl. A	4/4	0.46	CT: low attenuation occipitally left	Recovered	Pathological	Virus isolation from CSF
Okabe 2000 [32]	13	N.D.	Encephalitis, encephalopathy, Reyes' syndrome/N.D.	Infl. A H3N2	N.D.	N.D.	N.D.	N.D.	N.D.	Virus isolation from CSF

(*Continued*)

Table 1 *continued*

Author/year [Ref. no.]	No. of cases	Sex/age (years)	Clinical description of symptoms/onset after flu (days)	Type of infl	CSF findings: Leukocytes/ mono/μL	CSF findings: Protein (g/L)	CT/MRI	Outcome	EEG	Method of virological diagnosis
Hakoda 2000 [88]	1	F/32	Encephalopathy/5	Infl. A H3N2	Normal	Normal	Normal CT, MRI	Persistent vegetative state	Normal	Virus isolation from CSF + significant titer rise (HI)
Studahl 2000 [100]	1	F/25	Meningomyelitis/14	Infl. A	224/222	0.5	Normal CT; MRI spine: high signals within arachnoidea	Decreased sensibility, no patellar reflexes	N.D.	Virus isolation from CSF
Fujimoto 2000 [87]	1	M/7	Encephalopathy/1	Infl. A H1N1	1	1.3	MRI: symmetrical thalamic lesions	Recovered	N.D.	Intrathecal antibodies in CSF

ND: not done, no data.
HI: Hemagglutination inhibition.
NPA: nasopharyngeal aspirate.

Table 2 Cases with Influenza Virus–Induced Encephalitis/Encephalopathy from whom CSF Samples were Positive for Influenza Virus RNA

Author/year [Ref. no.]	No. of cases	Clinical description	Type of influenza virus	Onset of neurological symptoms after onset of "flu" (days)
Fujimoto 1998 [29]	5	Encephalitis/encephalopathy	Infl. A H3N2	1 (mean)
Ito 1999 [30]	1	Encephalopathy	Infl. A H3N2	N.D.
McCullers 1999 [31]	1	Encephalitis	Infl. B	2
Togashi 2000 [34]	9	Encephalitis/encephalopathy	Infl. A H3	N.D.
Takahashi 2000 [33]	1	Encephalopathy	Infl. A H3N2	1
Okabe 2000 [32]	14	Encephalitis/encephalopathy	Infl. A H3N2	N.D.

N.D.: No data

of pleocytosis in the CSF. A special form of encephalopathy is Reye's syndrome, a metabolic and neurological disorder due to mitochondrial damage that affects mostly children. The disease is characterized by acute onset of lowered consciousness, convulsions, and fatty degeneration of the liver [63]. It may be preceded by influenza A or, more often, influenza B infection [64,65] as well as by other viral infections [66] in combination with treatment with acetylsalicylic acid.

Medullary involvement is rare in influenza of the CNS, but a case with MRI-verified cervical myelitis diagnosed by intrathecally produced antibodies against influenza was recently reported [67].

In a few cases, influenza virus has been described as one of the viral agents preceding acute disseminated encephalomyelitis (ADEM) [19,68,69], a demyelinating autoimmune disease probably mediated by T cells [70]. Similarly, influenza virus has not been proven to be a common antecedent infection in Guillain-Barré syndrome, although it may occur [71].

The onset of neurological symptoms of encephalitis or encephalopathy after the first signs of influenza infection is usually within 8 days, ranging from 0 to 21 days [2]. Febrile convulsions have frequently been observed during influenza epidemics in children [5]. Fever, seizures, and decreased consciousness are the most common signs in encephalitis or encephalopathy associated with influenza infection, irrespective of whether the patient presents with pathological or normal neuroimaging. Additional symptoms, more sparsely reported, are focal neurological symptoms such as paresis, aphasia, cranial nerve palsies, and choreoathetosis [72–74].

During the last decade, outbreaks of influenza-associated (A and B) encephalopathy with rapid progression and high fatality rate have been described among young Japanese children [75–79]. An estimation of the incidence of influenza encephalitis/encephalopathy was about 100–200 in 1997–1998 in Japan [32]. The children present with convulsions at an early stage and lose consciousness and become comatose within 24 h. Characteristic findings are symmetrical thalamic lesions, brain edema, and lesions in the periventricular region and medullary substance of cerebellum visualized on CT and/or MRI. Whether this clinical entity is absent, underdiagnosed, and/or not reported in the rest of the world is unknown.

7 RADIOGRAPHIC AND NEUROPHYSIOLOGICAL FINDINGS

A review was performed of the English language literature from 1979 to 2000 that describes MRI and/or CT scans in patients with encephalitis/encephalopathy associated with influenza infection [31,35,37,72–74,76,77,79–89]. The reports from Japan dominate, where influenza virus–associated encephalopathy, especially among children, seems to have increased [32]. The clinical data (number of cases, age, CSF leukocytes, onset of neurological symptoms in days, prognosis) of the patients with normal versus pathological CT/MR are shown in Table 3. Discrimination between encephalitis and encephalopathy is not entirely distinct. The two groups did not differ concerning days with influenza symptoms before the appearance of neurological symptoms: 3.6 and 4.3 days, respectively. The main difference between the groups was that patients with pathological CT/MRI were significantly younger and had more severe sequelae than patients with normal CT/MRI. A patient with initially normal radiographic findings may within just a few days develop pathological changes detectable by CT or MRI [85,89]. Most patients with normal CT/MRI recovered or had mild sequelae (Table 4) [72,76,81–83,89]. This is similar to other encephalitides [90], although severe sequelae such as choreoathetosis, altered personality, spastic quadriparesis, and persistent vegetative state may occur [72,82,88] (Table 4). Abnormal neuroimaging findings in 21 patients with influenza virus–associated encephalitis/encephalopathy (Table 5) varied widely but could be divided into three main groups [76]:

Group 1. Symmetrical lesions in the thalami and/or brainstem, basal ganglia, and cerebellum with or without severe brain edema. MRI, T2-weighted: high signal intensity lesions in the thalami [72–74,77,79,80,85,87,89]. Only a few patients recovered [85]. Often the outcome was either fatal [79,89] or patients suffered severe brain damage [73,77].

Group 2. Lesions in thalami only (Group 2A). Hyperintensity in the gray matter and subcortical white matter with various localization (Group 2B) [35,72,84,85], on T2-weighted MRI with outcome varying between recovery [84,85] and more severe neurological sequelae [72,76].

Group 3. Diffuse severe brain edema [37] resulting in severe brain damage or death.

Electroencephalogram (EEG) is usually nonspecific with pathological, diffuse slowing of brain waves consistent with encephalitis [27,73,74,80–82].

Table 3 Comparison Between Patients with Influenza Virus–Associated Encephalitis/Encephalopathy and Normal Versus Pathological Radiographic Findings

	Normal CT/MRI	Pathological CT/MRI
Number of cases	14	21
Age, mean (median, range) (years)	21.4 (24.5, 2–50)	5.4 (3, 1–14)
CSF leukocytes/μL, mean (median, range)	16 (2, 0–43)	45 (9, 0–318)
Onset of neurological symptoms after "flu" (days) (range)	3.6 (0–7)	4.3 (0–20)
Prognosis	7 recovered, 6 sequelae, 1 N.D.	3 recovered, 9 sequele, 4 deaths, 5 N.D.
Type of influenza	A: 10, B: 4	A: 20, B: 1

N.D.: no data.
Source: Data abstracted from Ref. 31, 35, 37, 72–74, 76–77, and 79–89.

Table 4 Cases of Influenza Virus–Associated Encephalitis/Encephalopathy with Normal Brain CT/MRI Findings

Author/year [Ref. no.]	Sex/age (years)	Onset after "flu" (days)	Type of infl.	CSF findings: Leukocytes/ mono/(μL)	CSF findings: Protein (g/L)	Outcome	Method of virological diagnosis
Sulkava 1981 [81]	M/50	7	Infl. A H3N2	22/22	0.7	Recovered	Significant[a] rise in titer (HI)
Sulkava 1981 [81]	M/36	7	Infl. A H3N2	19/19	0.7	Recovered	Significant rise in titer (HI)
Rose 1982 [82]	F/18	7	Infl. A H3N2	34/34	0.76	Recovered	Virus isolation from CSF
Rose 1982 [82]	F/35	5	Infl A H1N1	0	0.85	Spastic quadriparesis	Virus isolation from CSF
Hawkins 1987 [83]	F/37	0	Infl. B	43/43	0.32	Recovered	Significant rise in titer (CF)
Hawkins 1987 [83]	F/38	2	Infl. B	Normal	Normal	Recovered	Significant rise in titer (CF)
Hayase 1997 [86]	F/31	Gradually	Infl. B	Normal	Normal	N.D.	PCR positive in CSF
Kimura 1998 [76]	M/2	3	Infl. A	Normal	N.D.	Recovered	Significant rise in titer (HI)
Kimura 1998 [76]	F/7	3	Infl. A	Normal	N.D.	Epilepsy	Significant rise in titer (HI)
Ryan 1999 [72]	F/34 mo	2	Infl. A	5/5	0.25	Choreoathetosis, slight hemiparesis, altered personality	Virus isolation from NPA[d]
Ryan 1999 [72]	M/4	2	Infl. A	0/0	0.3	Tremor, ataxia, recovered after 6 months	Virus isolation from NPA
Mc Cullers 1999 [31]	F/6	2	Infl. B	22/22	0.28	Ataxia, difficulties in school	Virus isolation from NPA; RT-PCR positive in CSF
Hakoda 2000 [88]	F/32	5	Infl. A H3N2	Normal	Normal	Persistent vegetative state	Virus isolation from CSF; significant rise in titer (HI)
Sugaya 2000 [89]	M/2	1	Infl. A H1N1	1	Normal	Recovered	Virus isolation from throat; significant rise in titer (HI)

[a] Four-fold or greater rise.
HI: Hemagglutin inhibition.
CF: Complement fixation.
NPA: Nasopharyngeal aspirate.

8 CEREBROSPINAL FLUID FINDINGS

Cerebrospinal Fluid (CSF) from patients with influenza virus–associated encephalopathy or encephalitis often has a normal cell count and normal content of protein and glucose, but a mild pleocytosis and a slightly increased protein level may be present [24–26,91]. More seldom there is a high cell count [85] or high protein content [73]. Cytokines, such as soluble tumor necrosis factor receptor-1 and interleukin-6, have been measured in CSF

Table 5 Cases of Influenza Virus–Associated Encephalitis/Encephalopathy with Pathological Brain CT/MRI Findings

CT/MRI	Author/year [Ref. no.]	Sex/ age (years)	Onset after "flu" (days)	Type of influenza	CSF findings: Leukocytes mono/ μL	CSF findings: Protein (g/L)	Outcome	Method of virological diagnosis
Group 1: Symmetrical lesions in the thalami and/or brainstem, basal ganglia, cerebellum, with or without brain edema	Delorme 1979 [80]	M/13	2	A	N.D.	N.D.	Difficulties with fine motor coordination	Virus isolated from NPA + significant titer rise (CF and HI)
	Hattori 1983 [74]	F/2	5	A	2	0.72	Slight dumpiness, mild mental deficit	Significant titer rise (CF)
Group 2A: MRI T_2-weighted: high-signal-intensity lesions in the thalami	Protheroe 1991 [73]	M/4	1	A	Normal	0.3	Cranial nerve paresis	Significant titer rise (CF)
	Protheroe 1991 [73]	F/11	3	A	0	1.2	Severe neurological sequele	Significant titer rise (CF)
	Nagai 1993 [77]	M/11 mo	3	A H3N2	9	0.32	Vegetative state	Significant titer rise (HI)
	Nagai 1993 [77]	M/27 mo	2	A H3N2	9	0.32	Spastic quadriplegia	Significant titer rise (HI)
	Kimura 1995 [85]	M/14	5	A or B	317/190	0.54	Recovered	Significant titer rise (HI)
	Shinjho 2000 [79]	F/3	2	A H3N2	0	N.D.	Death	Virus isolated from throat + significant titer rise (HI)
	Shinjoh 2000 [79]	M/1	0	A H3N2	1	N.D.	Death	Virus isolated from throat + significant titer rise (HI)
	Fujimoto 2000 [87]	M/7	2	A H1N1	1	1.3	Recovered	Significant titer rise (HI) + intrathecal antibodies
	Sugaya 2000 [89]	F/3	1	A H3N2	0	N.D.	Death	Virus isolated from throat + significant titer rise
Group 2B: MRI T_2-weighted: hyperintensity in the gray matter and subcortical white matter with various localization	Fuji 1992 [84]	M/22 mo	1	A	1	0.13	Recovered	Significant titer rise (CF)
	Kimura 1995 [85]	F/13	20	B	318/238	0.44	Mild epilepsy	Significant titer rise (HI)
	Kimura 1998 [76]	M/2	3	A	N.D.	N.D.	Severe brain damage	Significant titer rise (HI)
	Tsuchiya 2000 [35]	F/1.5	3	B	N.D.	N.D.	N.D.	Significant titer rise (HI)
	Tsuchiya 2000 [35]	F/1.6	12	A H3N2	27	N.D.	N.D.	Significant titer rise (HI)
	Tsuchiya 2000 [35]	F/3	4	A H3N2	N.D.	N.D.	N.D.	Significant titer rise (HI)
	Tsuchiya 2000 [35]	M/9	5	A H3N2	N.D.	N.D.	N.D.	Significant titer rise (HI)
	Tsuchiya 2000 [35]	M/13	11	A H3N2	41	N.D.	N.D.	Significant titer rise (HI)
	Ryan 1999 [72]	21 mo	0	A	1/1	0.11	Left hemiparesis, dystonia, epilepsy, altered personality	Virus isolated from NPA
Group 3CT: Severe brain edema	Yokota 2000 [37]	M/5	0	A	Normal	N.D.	Death	Virus isolated from NPA

ND: Not done, no data.
NPA: nasopharyngeal aspirate.
CF: Complement fixation.

from children with influenza virus–associated encephalopathy but were not elevated in the majority of the patients [30]

9 DIAGNOSTIC STRATEGIES

Virus isolation is the gold standard for laboratory diagnosis of influenza [92], and attempts to isolate the virus from the CSF are warranted (Table 1). Detection of influenza virus A and B–specific RNA fragments in CSF or brain tissue by RT-PCR has been reported in cases with influenza virus–associated encephalitis/encephalopathy with a rapid onset after influenza symptoms [29–34]. This indicates that virus is present in the CNS in at least some of the patients with CNS involvement (Table 2). Complementary diagnostics are virus isolation or antigen detection from nasopharyngeal aspirates and/or throat or bronchial washings, which is a sign of ongoing infection [93] indicating that the CNS manifestations may be elicited by influenza. Rapid antigen tests have varying sensitivity but are useful in the acute stage of illness. Earlier diagnosis of CNS manifestations associated with influenza virus was based mostly on serology and clinical descriptions [73]. The serological methods available are hemagglutination inhibition, enzyme immunoassay (EIA), complement fixation, or neutralization tests. A serologically verified infection [fourfold or greater rise in titer of specific IgG using acute and convalescent sera (10–14 days later)] may indicate that the virus is the cause of the concurrent CNS disease. However, the need for paired serum samples for serological diagnostics limits its usefulness during acute illness. No study on the diagnostic usefulness of intrathecally produced antibodies against influenza has been performed, although case reports have been published on myelitis using EIA [67] and on encephalopathy using hemagglutination inhibition [87].

10 TREATMENT

The therapy against CNS complications of influenza is usually symptomatic, with supervision of vital functions in the intensive care unit, antiepileptic treatment against seizures, and corticosteroid treatment against brain edema. Mild hypothermia therapy has been used for suppressing brain edema in encephalopathy associated with influenza [37]. Because the pathogenesis is largely unknown in influenza CNS complications, the role of antiviral treatment is unclear. No controlled trials on antiviral treatment have been made of CNS complications associated with influenza A infection. Amantadine, which has therapeutic and prophylactic effect against influenza A infection, has been used in single cases [89]. The CSF concentrations of amantadine are approximately half of the concurrent plasma levels [94,95], and the bioavailability is complete after oral intake [96]. In cases with influenza virus present in the CNS, where treatment can be started early amantadine may be effective, but at the present stage of knowledge amantadine therapy is purely experimental. Prevention with drugs (amantadine, rimantadine) or immunizations is associated with decreased numbers of deaths from influenza-related complications [97], but the protective role in CNS syndromes is unknown. The new neuraminidase inhibitors against influenza (zanamivir, oseltamivir) are registered for therapeutic use and shorten the duration of symptoms of influenza A and B infections if administered soon after the onset of disease [98]. Their role as protectors against CNS syndromes is uncertain. Because the most devastating CNS complications associated with influenza affect children, mostly in Japan, immunization programs have been proposed for small children [33,34]. The role of influenza in causing neurological disease is, however, incompletely examined, and to increase the knowledge of the pathogenesis in humans pathological and virological studies in fatal cases should be undertaken as well as controlled treament studies with antiviral agents.

REFERENCES

1. Hayden, F.G.; Palese, P. Influenza virus. In: Clinical Virology; Richman, D.D., Whitley, R.J., Hayden, F.G., Eds.; Churchill Livingstone: London, 1997, 911–942.
2. Nicholson, K.G. Human influenza. In: Textbook of Influenza; Nicholson, K.G., Hay, A.J., Webster, R.G., Eds.; Blackwell Sci: Boston, 1998, 219–264.
3. Monto, A.S.; Gravenstein, S.; Elliott, M.; Colopy, M.; Schweinle, J. Clinical signs and symptoms predicting influenza infection. Arch Intern Med. 2000, 160, 3243–3247.
4. Podosin, R.L.; Felton, W.L. The clinical picture of Far East influenza occurring at the Fourth National Boy Scout Jamboree. N Engl J Med. 1958, 258, 778–782.
5. Brocklebank, J.T.; Court, S.D.M.; Mcquillin, J.; Gardner, P.S. Influenza-A infection in children. Lancet. 1972, ii, 497–500.
6. Meibalane, R.; Sedmak, G.V.; Sasidharan, P.; Garg, P.; Grausz, J.P. Outbreak of influenza in a neonatal intensive care unit. J Pediatr. 1977, 91, 974–976.
7. Monto, A.S.; Koopman, J.S.; Longini, I.M. Tecumseh study of illness. XIII. Influenza infection and disease, 1976–1981. Am J Epidemiol. 1985, 121, 811–822.
8. Wright, P.F.; Thompson, J.; Karzon, D.T. Differing virulence of H1N1 and H3N2 influenza strains. Am J Epidemiol. 1980, 112, 814–819.
9. Dykes, R.C.; Cherry, J.D.; Nolan, C.E. A clinical, epidemiologic, serologic, and virologic study of influenza C virus infection. Arch Intern Med. 1980, 140, 1295–1298.
10. Hope-Simpson, R.E. The role of season in the epidemiology of influenza. J Hyg Cambridge. 1981, 86, 35–47.
11. Nguyen-Van-Tam, J.S. Epidemiology of influenza. In: Textbook of Influenza; Nicholson, K.G., Hay, A.J., Webster, R.G., Eds.; Blackwell Sci: Boston, 1998, 181–206.
12. Webster, R.G.; Bean, W.J.; Gorman, O.T.; Chambers, T.M.; Kawaoka, Y. Evolution and ecology of influenza A viruses. Microbiol Rev. 1992, 56, 152–179.
13. Potter, C.W. Chronicle of influenza pandemics. In Textbook of Influenza; Nicholson, K.G., Hay, A.J., Webster, R.G., Eds.; Blackwell Sci: Boston, 1998, 3–18.
14. Smith, W.; Andrewes, C.H.; Laidlaw, P.P. A virus obtained from influenza patients. Lancet. 1933, i, 66–68.
15. Francis, T. A new type of virus from epidemic influenza. Science. 1940, 92, 405–406.
16. Magill, T.P. A virus from cases of influenza-like upper-respiratory infection. Proc Soc Exp Biol (NY). 1940, 45, 162–164.
17. Taylor, R.M. Studies on survival of influenza virus between epidemics and antigenic variants of the virus. Am J Pub Health. 1949, 39, 171–178.
18. Bury, J.S. A discussion on influenza as it effects the nervous system. Br Med J. 1900, 2, 877–884.
19. Crookshank, F.G. Epidemic encephalomyelitis and influenza. Lancet. 1919, i, 79–80.
20. Taubenberger, J.K.; Reid, A.H.; Krafft, A.E.; Bijwaard, K.E.; Fanning, T.G. Initial genetic characterization of the 1918 ''Spanish'' influenza virus. Science. 1997, 275, 1793–1796.
21. Ravenholt, R.T.; Foege, W.H. Before our time. 1918 influenza, encephalitis lethargica, Parkinsonism. Lancet. 1982, ii, 860–864.
22. Leigh, A.D. Infections of the nervous system occurring during an epidemic of influenza B. Br Med J. 1946, 2, 936–938.
23. Jennings, G.H. The clinical features of the pneumonias undergoing virus tests. Br Med J. 1952, 1, 123–129.
24. Horner, F.A. Neurologic disorders after Asian influenza. N Engl J Med. 1958, 258, 983–985.

25. Dubowitz, V. Influenzal encephalitis. Lancet. 1958, 1, 140–141.
26. Dunbar, J.M.; Jamieson, W.M.; Langlands, J.H.M. Encephalitis and influenza. Br Med J. 1958, 1, 913–915.
27. Flewett, T.H.; Hoult, J.G. Influenzal encephalopathy and postinfluenzal encephalitis. Lancet. 1958, 2, 11–15.
28. Kapila, C.C.; Kaul, S.; Kalayaram, T.S.; Banerjee, D. Neurological and hepatic disorders associated with influenza. Br Med J. 1958, 2, 1311–1314.
29. Fujimoto, S.; Kobayashi, M.; Uemura. O.; Iwasa, M.; Ando, T.; Katoh, T.; Nakamura, C.; Maki, N.; Togari, H.; Wada, Y. PCR on cerebrospinal fluid to show influenza- associated acute encephalopathy or encephalitis. Lancet. 1998, ii, 873–875.
30. Ito, Y.; Ichiyama, T.; Kimura, H.; Shibata, M.; Ishiwada, N.; Kuroki, H.; Furukawa, S.; Morishima, T. Detection of influenza virus RNA by reverse transcription-PCR and proinflammatory cytokines in influenza-virus-associated encephalopathy. J Med Virol. 1999, 58, 420–425.
31. McCullers, J.A.; Facchini, S.; Chesney, P.J.; Webster, R.G. Influenza B virus encephalitis. Clin Infect Dis. 1999, 28, 898–900.
32. Okabe, N.; Yamashita, K.; Taniguchi, K.; Inouye, S. Influenza surveillance system of Japan and acute encephalitis and encephalopathy in the influenza season. Pediatr Int. 2000, 42, 187–191.
33. Takahashi, M.; Yamada, T.; Nakashita, Y.; Saikusa, H.; Deguchi, M.; Kida, H.; Tashiro, M.; Toyoda, T. Inflenza virus-induced encephalopathy: clinicopathologic study of an autopsied case. Pediatr Int. 2000, 42, 204–214.
34. Togashi, T.; Matsuzono, Y.; Narita, M. Epidemiology of influenza-associated encephalitis-encephalopathy in Hokkaido, the northernmost island of Japan. Pediatr Int. 2000, 42, 192–196.
35. Tsuchiya, K.; Katase, S.; Yoshino, A.; Hachiya, J. MRI of influenza encephalopathy in children: value of diffusion-weighted imaging. J Comput Assist Tomogr. 2000, 24, 303–307.
36. Reinacher, M.; Bonin, J.; Narayan, O.; Scholtissek, C. Pathogenesis of neurovirulent influenza A virus infection in mice. Lab Invest. 1983, 49, 686–692.
37. Yokota, S.; Imagawa, T.; Miyamae, T.; Ito, S.; Nakajima, S.; Nezu, A.; Mori, M. Hypothetical pathophysiology of acute encephalopathy and encephalitis related to influenza virus infection and hypothermia therapy. Pediatr Int. 2000, 42, 197–203.
38. Shinya, K.; Silvano, F.D.; Morita, T.; Shimada, A.; Nakajima, M.; Ito, T.; Otsuki, K.; Umemura, T. Encephalitis in mice inoculated intransally with an influenza virus strain originated from a water bird. J Vet Med Sci. 1998, 60, 627–629.
39. Mori, I.; Diehl, A.D.; Chauhan, A.; ljunggren, H.G.; Kristensson, K. Selective targeting of habenular, thalamic midline and monoaminergic brainstem neurons by neurotropic influenza A virus in mice. J Neurovirol. 1999, 5, 355–362.
40. Mori, I.; Kimura, Y. Apoptotic neurodegeneration induced by influenza A virus infection in the mouse brain. Microbes Infect. 2000, 2, 1329–1334.
41. Shinya, K.; Shimada, A.; Ito, T.; Otsuki, K.; Morita, T.; Tanaka, H.; Takada, A.; Kida, H.; Umemura, T. Avian influenza virus intranasally inoculated infects the central nervous system of mice through the general visceral afferent nerve. Arch Virol. 2000, 145, 187–195.
42. Naficy, K. Human influenza infection with proved viremia. N Engl J Med. 1963, 269, 964–966.
43. Stanley, E.D.; Jackson, G.G.E.E. Viremia in Asian influenza. Trans Assoc Am Phys. 1969, 79, 376–387.
44. Khakpour, M.; Saidi, A.; Naficy, K. Proved viremia in Asian influenza (Hong Kong variant) during incubation period. Br Med J. 1969, 4, 208–209.
45. Lehmann, N.I.; Gust, I.D. Viremia in influenza: a report of two cases. Med J Aust. 1971, 2, 1166–1169.
46. Xu, H.; Yasui, O.; Tsuruoka, H.; Kuroda, K.; Hayashi, K.; Yamada, A.; Ishizaki, T.; Yamada, Y.; Watanabe, T.; Hosaka, Y. Isolation of type B influenza virus from the blood of children. Clin Infect Dis. 1998, 27, 654–655.

47. Kilbourne, E.D. Studies on influenza in the pandemic of 1957–1958. III. Isolation of influenza A (Asian strain) viruses from influenza patients with pulmonary complications. Details of virus isolation and characterization of isolates, with quantitative comparison of isolation methods. J Clin Invest. 1959, 38, 266–274.
48. Minuse, E.; Willis, P.W.; Davenport, F.M.; Francis, T. An attempt to demonstrate viremia in cases of Asian influenza. J Lab Clin Med. 1962, 59, 1016–1019.
49. Morris, J.A.; Kasel, J.A.; Saglam, M.; Knight, V.; Loda, F.A. Immunity to influenza as related to antibody levels. N Engl J Med. 1966, 274, 527–535.
50. Mori, I.; Nagafuji, H.; Matsumoto, K.; Kimura, Y. Use of the polymerase chain reaction for demonstration of influenza virus dissemination in children. Clin Infect Dis. 1997, 24, 736–737.
51. Tsuruoka, H.; Xu, H.; Kuroda, K.; Hayashi, K.; Yasui, O.; Yamada, A.; Ishizaki, T.; Yamada, Y.; Watanabe, T.; Hosaka, Y. Detection of influenza virus RNA in peripheral blood mononuclear cells of influenza patients. Jpn J Med Sci Biol. 1997, 50, 27–34.
52. Takahashi, M.; Yamada, T. Viral etiology for Parkinson's disease—a possible role of influenza A virus infection. Jpn J Infect Dis. 1999, 52, 89–98.
53. Stevenson, P.G.; Hawke, S.; Sloan, D.J.; Bangham, C.R. The immunogenicity of intracerebral virus infection depends on anatomical site. J Virol. 1997, 71, 145–151.
54. Frankova, V.; Jirasek, A.; Tumova, B. Type A influenza: postmortem virus isolations from different organs in human lethal cases. Arch Virol. 1977, 53, 265–268.
55. Johnson, R.T.; Mims, C.A. Pathogeneisis of viral infections of the nervous system. N Engl J Med. 1968, 278, 23–30, 84–92.
56. Ward, A.C. Neurovirulence of influenza A. J Neurovirol. 1996, 2, 139–151.
57. Sugiura, A.; Ueda, M. Neurovirulence of influenza virus in mice. I. Neurovirulence of recombinants betwen virulent and avirulent virus strains. Virology. 1980, 101, 440–449.
58. Nakajima, S.; Sugiura, A. Neurovirulence of influenza virus in mice. II. Mechanism of virulence as studied in a neuroblastoma cell line. Virology. 1980, 101, 450–457.
59. Maurizi, C.P. Why was the 1918 influenza pandemic so lethal? The possible role of a neurovirulent neuraminidase. Med Hypotheses. 1985, 16, 1–5.
60. Price, D.A.; Postlethwaite, R.J.; Longson, M. Influenzavirus A2 infections presenting with febrile convulsions and gastrointestinal symptoms in young children. Clin Pediatr. 1976, 15, 361–367.
61. Oseasohn, R.; Adelson, L.; Kaji, M. Clinicopathologic study of thirty-three fatal cases of Asian influenza. N Engl J Med. 1959, 260, 509–518.
62. Davis, L.E. Diagnosis and treatment of acute encephalitis. Neurologist. 2000, 6, 145–159.
63. Reye, R.D.K.; Morgan, G.; Baral, J. Encephalopathy and fatty degeneration of the viscera—a disease entity in childhood. Lancet. 1963, ii, 749–752.
64. Norman, M.G.; Lowden, J.A.; Hill, D.E.; Bannatyne, R.M. Encephalopathy and fatty degeneration of the viscera in childhood: II. Report of a case with isolation of influenza B virus. Can Med Assoc J. 1968, 99, 549–554.
65. Hall, B.D.; Hughes, W.T.; Kmetz, D. Reye's syndrome: an association with influenza A infection. J KY Med Assoc. 1969, 67, 269–271.
66. Corey, L.; Rubin, R.J.; Hattwick, M.A.W.; Nobel, G.R.; Cassidy, E. A nationwide outbreak of Reye's syndrome. Its epidemiologic relationship to influenza B. Am J Med. 1976, 61, 615–625.
67. Salonen, O.; Koskiniemi, M.; Saari, A.; Myllylä, V.; Pyhälä, R.; Airaksinen, L.; Vaheri, A. Myelitis associated with influenza A virus infection. J Neurovirol. 1997, 3, 83–85.
68. Okuno, T.; Takao, T.; Ito, M.; Mirakawa, H.; Nakano, Y. Contrast enhanced hypodense area in a case of acute disseminated encephalitis following influenza A virus. Comput Radiol. 1982, 6, 215–217.
69. Johnson, R.T. Acute encephalitis. Clin Infect Dis. 1996, 23, 219–226.

70. Tselis, A.C.; Lisak, R.P. Acute disseminated encephalomyelitis. In Clinical Neuroimmunology; Antel, J., Ed.; Blackwell Sci: Oxford: UK, 1998, 116–147.
71. Jacobs, B.C.; Rothbarth, P.H.; van der Meche, F.G.; Herbrink, P.; Schmitz, P.I.; de Klerk, M.A.; van Doorn, P.A. The spectrum of antecedent infections in Guillain-Barre syndrome: a case-control study. Neurology. 1998, 51, 1110–1115.
72. Ryan, M.M.; Procopis, P.G.; Ouvrier, R.A. Influenza A encephalitis with movement disorder. Pediatr Neurol. 1999, 21, 669–673.
73. Protheroe, S.M.; Mellor, D.H. Imaging in influenza A encephalitis. Arch Dis Child. 1991, 66, 702–705.
74. Hattori, H.; Kawamori, J.; Takao, T.; Ito, M.; Nakano, S.; Okuno, T.; Mikawa, H. Computed tomography in postinfluenzal encephalitis. Brain Dev. 1983, 5, 564–567.
75. Mizuguchi, M.; Abe, J.; Mikkaichi, K.; Yoshida, K.; Yamanaka, T.; Kamoshita, S. Acute necrotizing encephalopathy of childhood: a new syndrome presenting with multifocal, symmetric lesions. J Neurol Neurosurg Psych. 1995, 58, 555–561.
76. Kimura, S.; Ohturi, N.; Nezu, A.; Tanaka, M.; Takeshita, W.S. Clinical and radiological variability of influenza-related encephalopathy or encephalitis. Acta Paediatr Jpn. 1998, 40, 264–270.
77. Nagai, T.; Yagashita, A.; Tsuchiya, Y.; Asamura, S.; Kurokawa, H.; Matsuo, N. Symmetrical thalamic lesion on CT in influenza A virus infection presenting with or without Reye syndrome. Brain Dev. 1993, 15, 67–74.
78. Sugaya, N.; Miura, M. Amantadine therapy for influenza type A-associated encephalopathy. Pediatr Infect Dis. 1999, 18, 734.
79. Shinjoh, M.; Bamba, M.; Jozaki, K.; Takahashi, E.; Koinuma, G.; Sugaya, N. Influenza A-associated encephalopathy with bilateral thalamic necrosis in Japan. Clin Infect Dis. 2000, 32, 611–613.
80. Delorme, L.; Middleton, P.J. Influenza A virus associated with acute encephalopathy. Am J Dis Child. 1979, 133, 822–824.
81. Sulkava, R.; Rissanen, A.; Pyhälä, R. Post-influenzal encephalitis during the influenza A outbreak in 1979/1980. J Neurol Neurosurg Psych. 1981, 44, 161–163.
82. Rose, E.; Prabhakar, P. Influenza A virus associated neurological disorders in Jamaica. West Indian Med J. 1982, 31, 29–33.
83. Hawkins, S.A.; Lyttle, J.A.; Connolly, J.H. Two cases of influenza B encephalitis. J Neurol Neurosurg Psych. 1987, 50, 1236–1237.
84. Fujii, Y.; Kuriyama, M.; Konishi, Y.; Sudo, M. MRI and SPECT in influenzal encephalitis. Pediatr Neurol. 1992, 18, 133–136.
85. Kimura, S.; Kobayashi, T.; Osaka, H.; Shimizu, C.; Uehara, S.; Ohtuki, N. Serial magnetic resonance imaging in post-infectious focal encephalitis due to influenza virus. J Neurol Sci. 1995, 131, 74–77.
86. Hayase, Y.; Tobita, K. Probable post-influenza cerebellitis. Intern Med. 1997, 36, 747–749.
87. Fujimoto, Y.; Shibata, M.; Tsuyuki, M.; Okada, M.; Tsuzuki, K. Influenza A virus encephalopathy with symmetrical thalamic lesions. Eur J Pediatr. 2000, 159, 319–321.
88. Hakoda, S.; Nakatani, T. A pregnant woman with influenza A encephalopathy in whom influenza A/Hong Kong virus (H3) was isolated from cerebrospinal fluid. Arch Intern Med. 2000, 160, 1041, 1045.
89. Sugaya, N. Influenza-associated encephalopathy in Japan: pathogenesis and treatment. Pediatr Int. 2000, 42, 215–218.
90. Buttner, T.; Dorndorf, W. Prognostic value of computed tomography and cerebrospinal fluid analysis in viral encephalitis. J Neuroimmunol. 1988, 20, 163–164.
91. McConkey, B.; Daws, R.A. Neurological disorders associated with Asian influenza. Lancet. 1958, ii, 15–17.
92. Cox, N.J.; Subbarao, K. Influenza. Lancet. 1999, 9, 1277–1282.

93. Frank, A.L.; Taber, L.H.; Wells, C.R.; Wells, J.M.; Glezen, W.P.; Paredes, A. Patterns of shedding of myxoviruses and paramyxoviruses in children. J Infect Dis. 1981, 144, 433–441.
94. Brenner, M.; Haass, A.; Jacobi, P.; Schimrigk, K. Amantadine sulfate in treating Parkinson's disease: clinical effects, psychometric tests, and serum concentrations. J Neurol. 1989, 236, 153–156.
95. Kornhuber, J.; Quack, G.; Dansz, W. Therapeutic brain concentrations of the NMDA receptor antagonist amantadine. Neuropharmacology. 1995, 34, 713–721.
96. Tominack, R.L.; Hayden, F.G. Rimantadine hydrochloride and amantadine hydrochloride use in influenza A virus infections. Infect Dis Clin North Am. 1989, 1, 459–478.
97. Anon, F.G. Prevention and control of influenza: recommendations of the Advisory Committee on Immunization Practices (ACIP), Centers for Disease Control and Prevention. MMWR. 1998, 47, 1–26.
98. Long, J.K.; Mossad, S.B.; Goldman, M.P. Antiviral agents for treating influenza. Cleve Clin J Med. 2000, 67, 92–95.
99. Studahl, M.; Bergström, T.; Hagberg, L. Acute viral encephalitis in adults—a prospective study. Scand J Infect Dis. 1998, 30, 215–220.
100. Studahl, M.; Hagberg, L.; Rekabdar, E.; Bergström, T. Herpesvirus DNA detection in cerebrospinal fluid: differences in clinical presentations between alpha-, beta-, and gamma-herpesviruses. Scand J Infect. 2000, 32, 237–248.

22

Dengue

Tom Solomon
University of Liverpool
Liverpool, United Kingdom

Alan D. T. Barrett
The University of Texas Medical Branch at Galveston
Galveston, Texas, U.S.A

1 INTRODUCTION

Dengue virus is numerically the most important arbovirus (arthropod-borne virus) of humans, with an estimated 100 million cases per year and 2.5 billion people at risk of contracting the disease [1]. The geographical area affected by dengue is expanding, and almost every country between the tropics of Capricorn and Cancer is now affected (Fig. 1). The disease has been recognized by the World Health Organization (WHO) as one of the most important emerging diseases of humans. There is no antiviral treatment, and no vaccines are available. Dengue viruses are best known for causing a fever–arthralgia–rash syndrome (dengue fever), which has probably existed for hundreds of years, and a hemorrhagic syndrome [dengue hemorrhagic fever (DHF)], which was first recognized in the 1950s. The question of neurological manifestations has been a long-standing controversy that has only recently received prominent attention. For this reason it is considered separately in this volume, rather than along with the other arboviruses, for which central nervous system (CNS) disease is the most important manifestation (see Chap. 15).

There has been considerable debate about the relative importance or even existence of the neurological manifestations of dengue and the mechanism by which they occur. In this chapter we consider the classical and neurological manifestations of dengue infection, focusing on the latter, reexamine some of the existing epidemiological data, and look at the extent to which recent studies are beginning to unravel the many unanswered questions about this disease.

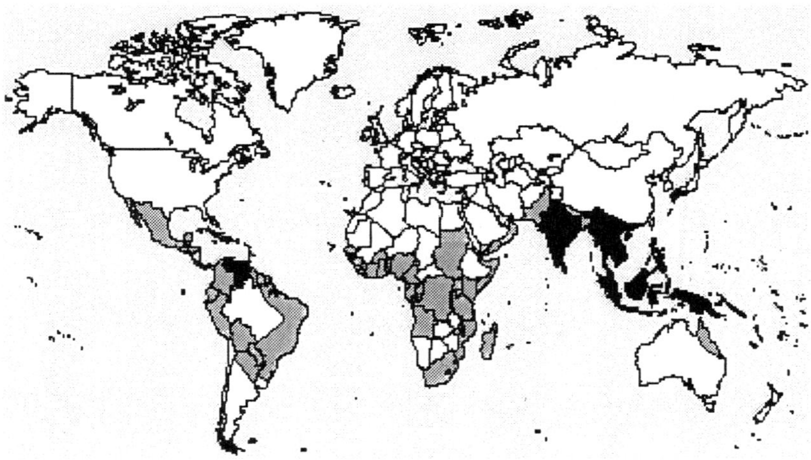

Figure 1 Global distribution of dengue. Countries where dengue fever (gray) and dengue hemorrhagic fever (black) have been reported in the past 25 years.

2 DENGUE VIRUS

Dengue is a member of the *Flavivirus* genus within the *Flaviviridae* family, named after the prototype yellow fever virus (Latin *flavus*, yellow) [2]. The genus contains approximately 70 members, many of which are human pathogens, including Japanese encephalitis virus, West Nile virus, Murray Valley encephalitis virus, tick-borne encephalitis virus, Omsk hemorrhagic fever virus, louping ill virus and Kyasanur forest disease virus [3].

Like all flaviviruses, dengue viruses are small (50 nm) spherical enveloped particles and comprise approximately 11 kilobases (kb) of single-stranded positive sense RNA wrapped in a nucleocapsid core protein, within a membrane that contains the envelope and membrane proteins (Fig. 2). The genome codes for three structural proteins—envelope (E), premembrane (PrM), and core (C)—and seven nonstructural proteins in a single reading frame flanked by a short 5′ and a longer 3′ untranslated region (UTR) [4]. The dengue serocomplex of the *Flavivirus* genus contains four serologically related viruses named dengue-1, dengue-2, dengue-3, and dengue-4 viruses. Although these viruses are called dengue ''serotypes'', they are in fact distinct viruses.

2.1 Viral Attachment and Entry

The envelope (E) protein is the largest structural protein, consisting of nearly 500 amino acids with six disulfide bridges and two potential glycosylation sites, one, two, or neither of which may be used, depending on the virus strain. The six disulfide bridges are shared by all flavivirus E proteins, so the overall structure of the E protein is thought to be very similar for all flaviviruses. The E protein is the major target of the humoral immune response and, as described below, is thought to be important in viral entry into host cells [5]. Studies with monoclonal antibodies suggested three antigenic domains [6,7]. The three-dimensional structure of the ectodomain (the external portion) of Central European tick-borne encephalitis virus has been determined by X-ray crystallography [8]. This has

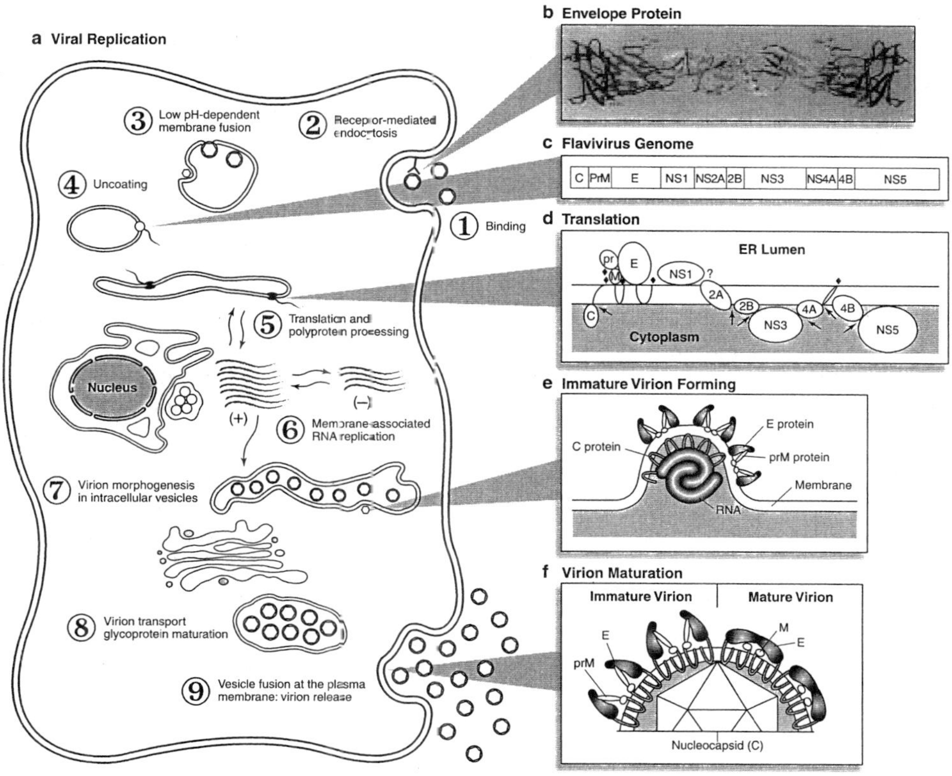

Figure 2 Flavivirus replication. (a) Overview of replication steps 1–9. (b–f) Key elements in this process. At (1) the virus binds to the putative host cell receptor, using the E (envelope) protein (b). Receptor-mediated endocytosis occurs (2), and following low-pH-dependent fusion of the viral and host endosomal membranes (3) the virion uncoats (4) and the nucleocapsid is released. The single-stranded positive sense 11 kb genome (c) is translated into a 3000 amino acid polyprotein that is processed co- and post-translationally by viral and host proteases (5) into the three structural proteins C (core) PrM (premembrane), and E (envelope), and seven nonstructural proteins (d). RNA replication occurs via a negative sense intermediate (6). The positive sense RNA so formed is associated with C protein (e), and immature virions are thought to form by budding into the endoplasmic reticulum (7). Immature virions are transported in vesicles to the host cell surface (8). The PrM protein is cleaved to produce the mature virion (f), as release occurs by exocytosis (9). (Modified from Refs. 8, 183, and 184.)

shown that the E protein forms dimers that are parallel to the surface of the virion membrane (Fig. 2b). In addition, the three antigenic domains suggested by studies with monoclonal antibodies were confirmed by structural studies and termed domains II, III, and I, respectively. Domain III is the putative receptor binding domain (by which virions attach to the yet-to-be-identified host cell receptor), domain II is the dimerization domain and contains a putative fusion sequence (which facilitates fusion of viral and host cell membrane), and domain I has a central beta barrel and is the hinge domain that links the other two domains. Recently the structure of the dengue-2 virion was determined with cryoelectron microscopy to 24 Å resolution [9]. This showed that the virion has a well-organized outer protein shell, a lipid bilayer, and a less well defined inner nucleocapsid core. The 3-D structure of E protein, determined by X-ray crystallography, was fitted into the cryoelectron microscopic reconstruction and showed that the icosohedral scaffold consists of 90 E dimers. The membrane (M) protein was thought to be located in a hole between the E dimers [9].

A series of studies have shown that amino acid changes at the putative receptor binding site on the exposed outer surface of domain III critically affect attachment of flaviviruses to host cells [10,11]. The host cell receptor molecule has not been identified, but Fc receptors, glycoproteins, and glycosaminoglycans have all been implicated as involved in binding of virus to cells. These molecules are not mutually exclusive, and it is possible that dengue viruses use different molecules to attach to different cell types or that the molecules act cooperatively. For example, the highly sulfated heparan sulfate is a glycosaminoglycan molecule postulated to be a cell-binding site [12]. It is possible that glycosaminoglycans act as initial cell-binding sites and that proteinacious molecules are the true receptors. After attachment, the virion is taken up by endocytosis (Fig. 2a). The endocytotic vesicle so formed becomes more acidic, which is thought to trigger a pH dependent conformational change in the envelope protein from a dimer to a trimer, around the ''hinge'' region of domain I, bringing the fusion region of domain II up against the host cell membrane. This fusion region comprises 14 amino acids (residues 98–111) and includes a classic motif made up of three amino acids, glycine–X–phenylalanine, where X can be any amino acid [13,14]. Fusion of the virion membrane and host membrane allows release of the virion nucleocapsid into the cytoplasm of the cell (Fig. 2a).

2.2 Viral Replication and Release

The nucleocapsid uncoats, and viral replication begins on the surface of the endoplasmic reticulum. First the viral RNA is translated into a single 3000 amino acid polyprotein, which is co- and post-translationally cleaved by host and viral proteases into the three structural and seven nonstructural proteins that remain closely associated with the membrane of the endoplasmic reticulum [4]. The PrM and E proteins project into the luminal side of the endoplasmic reticulum, whereas the C protein projects into the cytoplasmic side (Fig. 2d). The seven nonstructural proteins include NS1, a membrane-associated glycoprotein thought to be involved in viral replication; NS3, which has protease (in combination with NS2B), helicase, and RNA triphosphatase activities; and NS5, which includes the RNA-dependent RNA polymerase and methyltransferase activities. This is found in all flaviviruses (i.e., conserved).

Next, RNA replication begins on membranes close to the nucleus. A replication complex on the membranes comprises viral RNA, nonstructural proteins (NS2A, NS3, NS4A, and NS5), and cellular proteins. The best characterized cellular protein is translation elongation factor 1α (EF-1α), a protein whose usual function is the binding of amino acid–tRNA complexes to the ribosome during translation [15]. Computer predictions of the secondary structure of the 5′ UTR show that it includes a stem–loop structure that is conserved. The 3′ UTR has a similar conserved stem–loop structure as well as other lengths of conserved sequence. These observations are consistent with the UTRs' suggested function as RNA polymerase recognition sites that are preserved during evolution of flaviviruses and are important in viral replication [16]. The positive-sense RNA is transcribed into a negative-sense intermediate, which acts as the template for the production of multiple positive-sense RNA molecules that are progeny genomes (Fig. 2a). Newly formed progeny viral RNA then interacts with multiple copies of the C protein to form the nucleocapsid.

The central portion of the C protein is embedded in the membrane of the endoplasmic reticulum, but its amino and carboxy termini project into the cytoplasm and have many basic amino acids that combine with the acidic RNA. The viral PrM and E proteins are translocated to the lumen of the endoplasmic reticulum, where PrM/E heterodimers are formed. The nucleocapsid is then thought to bud through the endoplasmic reticulum membrane, becoming enveloped in membrane containing E and PrM proteins and thus forming immature virions inside cytoplasmic vesicles (Fig. 2e). Most flaviviruses assemble in the cisternae of the rough endoplasmic reticulum, are transferred to the Golgi apparatus, and are then transported via secretory pathways to the cell surface. During this process the PrM/E heterodimers are further modified by addition of N-linked glycans and glycan trimming. Immediately prior to release, the PrM protein is cleaved by a furin-like protease to its mature M protein form (Fig. 2f), thus allowing the formation of E protein homodimers and ''activating'' the E protein for the pH-dependent conformational changes that occur during subsequent attachment and entry into cells [17]. Virions are released at the cell surface by exocytosis—trans-type maturation [18]. However, for dengue-2, cis-type maturation has also been demonstrated in mosquito cells: free nucleocapsid-like structures are detected in the cytoplasm, and virions bud at the plasma membrane [19].

2.3 Dengue Serotypes

Four antigenically distinct dengue viruses (dengue-1, -2, -3, and -4) have been identified by cross-neutralization tests using polyclonal antibodies. Primary infection refers to an individual's first infection with any one of the four dengue viruses, and secondary infection is a second infection with a different dengue virus. Infection with one dengue virus confers lifelong immunity to that virus but does not protect against secondary infection with a different type. The E protein is the main determinant of serological cross-reactivity. Comparisons of the amino acid sequences of the E protein from the four dengue viruses show 62–78% homology [20]. Dengue viruses 1 and 3 are most closely related, and dengue-4 virus is the most distant. Comparison of the complete genomes confirms these relationships. Phylogenetic studies indicate that dengue-4 virus derived first from the common dengue ancestor, followed by dengue-2 and then dengue-1 and dengue-3 viruses [21], although some reports suggest that dengue-1 virus may have derived after dengue-3 virus. For each of the four dengue viruses, genetic variants have been identified, initially as topotypes on the basis of RNA oligonucleotide fingerprinting [22] and subsequently as genotypes on the basis of nucleotide sequencing. This sequencing has identified at least five genotypes of each dengue virus [20,23,24]. These genotypes generally correlate with geographical areas. Within an area, changes in the virus population occur slowly over time as the virus evolves. In addition, new genotypes may be introduced from other areas. Genotype classification is thus helpful in determining the origins and spread of epidemics. For example, dengue-2 genotype I has persisted in Asia since it was first identified there in 1944, and was introduced into the Americas in 1981, causing the first epidemic of dengue hemorrhagic fever (DHF) in Cuba [25].

Like most other positive-sense RNA viruses, dengue has a high rate of mutation during replication, because it lacks mechanisms for proofreading and nucleic acid repair [26]. Thus, any single isolate of dengue may actually represent a quasi-species with a mixture of slightly different viruses, which may have slightly different properties [27]. Quasi-species for the E gene of dengue-3 have recently been reported in plasma samples of six patients [28], which may have implications for the disease pattern.

3 GEOGRAPHICAL DISTRIBUTION

During the twentieth century, the geographical area affected by dengue increased. It is now the most widely distributed mosquito-borne virus of humans, occurring in virtually every country between the tropics of Capricorn and Cancer (Fig. 1) [29]. Although dengue fever was recognized in many tropical countries during the twentieth century, dengue hemorrhagic fever (DHF) was largely confined to southeast Asia until the 1980s. Since then, however, it has reemerged in the Indian subcontinent and occurred for the first time in China, Tahiti and Cuba, the Caribbean, the Pacific Islands, Venezuela, and Brazil [30]. There are now an estimated 250,000–500,000 cases of DHF globally each year [1]. The spread of dengue since World War II has been linked to a worldwide resurgence of *Aedes aegypti* following poor vector control, overcrowding of refugee and urban populations, and increasing human travel [31,32].

4 NATURAL CYCLE

Unlike many other arboviruses, for which a zoonotic (animal) cycle occurs and human infections are coincidental (see Chap. 15), dengue is primarily a virus of humans, transmitted between them by *Aedes* mosquitoes, especially *Aedes aegypti*. Dengue also exists in a sylvatic (forest) cycle, where non-human primates are the host. Sylvatic dengue occurs in parts of tropical southeast Asia and in West Africa, and recent genetic evidence suggests that each of the four serotypes of dengue virus may have spread to humans from these sylvatic cycles [33]. Forest cycles have been documented for many mosquito species of three subgenera (*Stegomyia*, *Finlaya*, and *Diceromyia*) of the genus *Aedes*. However, for human transmission *Aedes aegypti* (subgenus *Stegomyia*, genus *Aedes*) is the most important.

4.1 Dengue Vectors

Aedes aegypti is a ''domestic'' mosquito that is anthrophilic (i.e, feeds on humans) and breeds in peridomestic collections of clean water (storage jars, containers etc.). Only the females seek blood meals; they feed principally during the day and feed repeatedly on various hosts, enhancing their role as vectors. The infected mosquito remains infectious for its entire life. Transovarial transmission of dengue (i.e., from the mosquito into its eggs) has been documented. *Aedes aegypti* eggs can survive for long periods in dry conditions, which may explain the occurrence of spontaneous dengue outbreaks. *Aedes aegypti* is thought to have originated in the forests of Africa and has now spread across the globe, being found worldwide between latitudes 35°N and 35°S. It was eradicated from most of Central and South America during a Pan American Health Organization program against yellow fever in the 1950s and 1960s [34]. However, it reinfested much of the Americas after the program ended in the 1970s [1]. Since the 1980s, dengue virus has also been reintroduced into the Caribbean, the Pacific, and Australia.

Aedes albopictus also transmits dengue viruses. It is indigenous to southeast Asia and may have been responsible for the postulated spread of dengue viruses from their forest cycle among lower primates to humans. In the late 1980s and 1990s, *Aedes albopictus* spread to much of the Americas, including the United States, and to southern Europe, largely as a result of intercontinental transport of used car tires containing the mosquito's eggs [35]. Because of its better survival at cooler temperatures it has been suggested that this mosquito could be responsible for a major European outbreak [36]. Other species implicated as important vectors include *Ae. scutellaris hebrideus* in New Guinea, *Ae. polynesiensis* in Tahiti, *Ae. albopictus* in the Americas, and *Ae. cooki* in Niue.

5 HISTORY

The name dengue is thought to derive from a Swahili term to describe a dengue-like illness on the east coast of Africa in the mid-nineteenth century: Ki-Dinga pepo, ''*a disease characterized by a sudden cramp-like seizure, caused by an evil spirit*'' [37]. This was shortened to ''denga'' or ''dyenga'' and is thought to have become ''dengue'' as a Spanish derivative of the African term when the slave trade brought the disease to the West Indies [37].

5.1 Early History

The earliest description of a disease compatible with dengue fever comes from a Chinese encyclopedia of disease symptoms and remedies edited during the Northern Sung dynasty (A.D. 992) but first published during the Chin dynasty (A.D. 265–420) [37]. The disease was characterized by rash, fever, eye pain, arthralgias, myalgias, and hemorrhagic manifestations and was called ''water poison'' because of its apparent connection to flying insects associated with water. Epidemics of febrile disease attributed to dengue also occurred in the French West Indies in 1635 and in Panama in 1699. However, Benjamin Rush's classic description of ''breakbone fever,'' an epidemic that struck Philadelphia in 1780 during a pandemic that also affected Cairo, Egypt, and Batavia (Jakarta) Indonesia, is generally accepted as the first accurate clinical description of dengue fever (Fig. 3) [38]. In Cairo the disease was called *mal de genoux* (knee trouble), and in Indonesia, *knockelkoorts* (bone fever). Rush's account [38] described fever; severe pain in the head, back, and limbs; rash; and bleeding ranging from ''a few spoonfuls'' to ''profuse hæmorrhage'' and interestingly appeared to recognize neurological presentations of the disease [38] (Fig. 3).

Numerous epidemics of disease compatible with dengue fever occurred in Asia, Africa, and America in the nineteenth and twentieth centuries. These included outbreaks in Zanzibar (1823 and 1870), Calcutta (1824, 1853, 1871, and 1905), the West Indies (1827), Hong Kong (1901), Greece (1927–1928), Australia (1925–1926, 1942), the United States (1922), and Japan (1942–1945) [39]. Retrospective serological surveys have been possible for epidemics in the first half of the twentieth century. These suggest that dengue-like illnesses in the United States, Greece, Australia, and Japan were indeed due to dengue virus [30]. Dengue was the second human disease, after yellow fever, attributed to a ''filterable virus,'' and the mosquito's role in transmission was demonstrated in volunteer experiments in 1903. In the 1940s the search for a vaccine led to the distinction of more than one immunological type (serotypes). The prototype dengue-1 strain (Hawaii) and the prototype dengue-2 strain (New Guinea C) were recovered by inoculating the serum of dengue patients from those locations into volunteers in the United States. Dengue-3 (prototype strain H-87) and dengue-4 (prototype strain H-241) were isolated in the Philippines when the first epidemics of DHF occurred there in the 1950s (see below).

Neurological manifestations were a recognized complication of dengue fever in many of the early epidemics. For example, Cleland et al. recognized neurological complications during an outbreak in the Pacific in 1918 [40]. During the 1940s the Japanese described many cases in Taiwan, Okinawa, and Japan [41]. Many neurological complications, including some fatal cases, were seen during a large epidemic of dengue-like illness in Greece in 1928 [41], which subsequent serological studies indicate was due to dengue

Figure 3 Breakbone fever. (Left) William Rush, whose classical description of ''breakbone fever,'' ''An Account of the Bilious Remitting Fever'', (right), included observations on patients who presented with coma, convulsions, and delirium.

[42]. However, as described below, from the 1950s, with the emergence of a new hemorrhagic fever syndrome caused by dengue viruses, neurological manifestations tended to either be forgotten or ignored.

5.2 The Emergence of Dengue Hemorrhagic Fever

In the 1950s outbreaks of an apparently new hemorrhagic fever occurred in the Philippines and Thailand and were soon shown to be caused by dengue viruses [43]. Following further massive epidemics of dengue hemorrhagic fever (DHF) in Thailand, WHO adopted clinical and laboratory criteria for diagnosing and treating dengue fever and DHF [44,45] (see Sec. 10). These definitions included a severity grading, which, although undoubtedly helpful in defining DHF, caused controversy because there was no mention of other severe manifestations.

5.3 The Dengue Encephalopathy Controversy

Of prime concern among the other severe manifestations was neurological disease. At a meeting of SEAMEO-TROPMED in Bangkok in 1976, three separate presentations from Thailand, Burma, and Indonesia drew attention to neurological manifestations of dengue [46–48]. With the publication of further case series over the next decade [47,49], it gradually became accepted that patients with severe DHF could develop encephalopathy secondary to the many complications of severe disease [50–54]. These included hepatic dysfunction (sometimes as part of a Reye's-like syndrome), hyponatremia, hypoxia, cerebral edema, or hemorrhage. But whether dengue viruses could cross the blood-brain barrier to cause a true viral encephalitis was not certain [55].

Compounding the controversy was the fact that many publications consisted of isolated case reports or retrospective case series, some of which appeared to describe the same patients. Neurological disease was sometimes grouped together with other "unusual manifestations" such as hepatic, renal, or cardiac dysfunction. Studies tended to look for neurological complications in patients presenting with dengue fever and DHF rather than looking for evidence of dengue infection among patients presenting with central nervous system (CNS) disease. Definitions of encephalopathy and encephalitis were vague, and capabilities for precisely diagnosing dengue infection and excluding other potential causes of CNS disease were variable. In recent decades the improved ability to diagnose dengue infection allowed the detection of anti-dengue antibody and even virus in the cerebrospinal fluid (CSF) [56,57]. Although these reports were highly suggestive of CNS infection, without detailed prospective clinical and epidemiological data their significance remained unclear: could these have been contaminants from a traumatic lumbar puncture or coincidental infections that were not the cause of the CNS disease? Recent detailed prospective studies appear to have answered some of these questions [58–60]. As discussed below, these have shown that in some settings dengue is a relatively important cause of neurological disease. Moreover, in combination with recent immunohistochemical studies, these data provide evidence that in some instances dengue viruses do indeed cross the blood-brain barrier to replicate in the CNS [61,62].

6 EPIDEMIOLOGY OF DHF

Since it was first described in the Philippines and Thailand in the 1950s, epidemic DHF has spread across all of southeast Asia (Fig. 1). Subsequently, its range has extended even further. The first epidemic of DHF in China occurred on Hainan Island in 1985–1986 with a morbidity of 1913 per 100,000 residents [63]. India experienced its first outbreak in 1988 (24 patients and eight deaths in 2 months) [64], Taihiti and New Caledonia in 1989, and Sri Lanka in 1990 (935 cases, 54 deaths) [39]. In Central and South America there has been long-standing dengue virus activity, but only since the 1980s have massive epidemics of dengue fever and DHF been reported. In Cuba in 1981 there were 24,000 cases of DHF with 158 deaths in 3 months [65]. In Brazil there were an estimated million cases of dengue fever in 1986–1987. In Iquitos, Peru, one-fourth of the population of 300,000 had dengue fever during 1990. In the same year Venezuela reported 3108 DHF cases with 78 deaths. Between 1986 and 1990 1.5 million cases of DHF with 15,940 deaths were reported to WHO, giving approximately 300,000 cases and 3000 deaths annually [39].

7 EPIDEMIOLOGY OF NEUROLOGICAL DENGUE

Neurological manifestations of dengue infection have been described in almost all areas where dengue occurs, including Asia, the Pacific, Australia, Africa, the Americas, and Europe [41]. Many of the more recent epidemiological and clinical studies have come from Asia. Data from several studies that provided basic epidemiological descriptions are reevaluated below to identify risk factors for neurological disease [49,51,53, 58,59,64,66–70] (Table 1). Differences in the setting and the population studied, the methods used for diagnosing dengue infection, and what constitutes a neurological manifestation make comparison between studies difficult. Despite these differences a comparison of some of the relevant studies shows that approximately 1–5% of patients with a clinically apparent dengue infection have neurological manifestations (Table 1). Higher percentages have been found in Indonesia, which may reflect a broader definition of CNS disease (for example, including lethargy and apathy) [49,53]. Interestingly the percentage of neurological manifestations is broadly similar whether the group being studied comprises patients with dengue fever, DHF, or both. Neurological manifestations may also be important among returning travelers with dengue [71].

Fewer studies have looked for dengue infection among patients presenting with neurological disease. In Vietnam, where only 1% of dengue admissions had neurological manifestations, dengue accounted for 16 (4.2%) of 378 patients with suspected CNS infection seen over one year [58]. A similar study in Bangkok, Thailand, found that dengue was responsible for 8 (20%) of 40 children admitted with suspected encephalitis [60]. Dengue was more important than any other single agent, including Japanese encephalitis virus, another flavivirus. In northern Thailand, four (10%) of 44 patients with suspected encephalitis were infected with dengue [72]. In a study of febrile convulsions in Nigeria, dengue accounted for 3% of children, or 20% of those from whom a virus was isolated [73]. After coxsackieviruses, it was the most common cause of febrile convulsions.

7.1 Dengue Virus Types

All four dengue viruses have been associated with neurological dengue, though dengue-2 and dengue-3 viruses have been most frequently implicated. In southeast Asia, where more than one strain cocirculates, several strains have been isolated from neurological patients during the same study, and occasionally even from the same patient [58]. In an Indian study, dengue-2 virus was most frequently isolated [52]; in Thailand, dengue-1, dengue-2, and dengue-3 viruses were detected in equal numbers [60]; in Vietnam, dengue-3 virus was detected in five patients, dengue-2 virus in four, and dengue-1 virus in one [58]. In Indonesia, where the incidence of neurological dengue appears to be especially high, dengue-3 virus has been most frequently isolated [49]. To determine if there is a significant association with neurological disease, it would be necessary to compare the frequency of dengue-3 virus isolation in neurological cases with the frequency in non-neurological cases. Unfortunately these data are not available, but an association can be deduced from data that have been published: Kho et al. isolated dengue virus from 140 dengue patients, 41 of whom had neurological features [49]. Although the serotypes are not given for all 41 isolates, they are given for the 29 patients who died. From these data we can derive that dengue-3 virus was isolated from 20 (69%) of 29 neurological patients who died, compared with 44 (40%) of 111 patients who did not have fatal neurological manifestations [odds ratio (95% confidence interval [C.I.]) 3.38 (1.31–8.91); Fisher's exact test $p=0.005$]. A related paper described 30 fatal DHF cases, 21 of whom had encephalitic signs [74]. The data for the encephalitis patients are not given separately, but 21 (70%) of the patients who died were infected with dengue-3 virus, compared with 73 (48%) of those who survived [odds ratio 2.06 (1.04–6.56), $\chi^2 = 5$, $p=0.02$]. These data suggest dengue-3 virus may indeed be associated with a greater risk of neurological disease.

7.2 Other Risk Factors for Neurological Dengue

Neurological manifestations of dengue have been described in all age groups. However, young children and adults appear to be at greater risk of neurological disease than older

Table 1 Summary of Some Key Papers on the Epidemiology of Neurological Dengue[a]

Reference	Location, year	Neurological dengue patients/all patients in study	Details of patient group studied	Diagnostic methods [no. of neurological dengue patients]	Comments
Studies looking for neurological manifestations among patients with DF/DHF					
Kaplan and Lindgren [66]	Central Pacific, 1944	13/1488 (1%)	Service personnel, ages 19–31 yr, with DF	Clinical diagnosis	Large outbreak, not clear in how many patients dengue was confirmed.
Kuberski et al. [186]	Fiji, 1975	3/65 (5%)	Children and adults (5–50 yr) with DF and DHF	Serology (HI and PRNT)	Two with neck stiffness, one semicomatose.
Kho et al. [49]	Jakarta, Indonesia, 1975–1977	119/672 (18%)	Children (8 mo–14 yr) with DHF	Serology (HI) and virus isolation from serum [41]	Retrospective study in three hospitals. Included 35 patients with lethargy; excluding these, 84 (13%) had neurological disease.
Srivastava et al. [64]	Delhi, India, 1988	3/24 (13%)	Children with DHF	Serology (HI and CFT)	Retrospective study.
Hendarto and Hadinegro [53]	Jakarta, Indonesia, 1975–1976, 1985–1986, 1988–1989	33/358 (9%), 21/745 (3%), 98/1329 (7%)	Children with DHF	Serology (HI)	Retrospective study during three time periods. Included patients with apathy.
Thisyakorn and Thisyakorn [51]	Bangkok, Thailand, 1987	12/505 (2.4%)	Children with DF and DHF	Serology and/or isolation	Prospective study.
Brown et al. [71]	Travelers Clinic, UK, 1982	1/13 (7.7%)	Adults (returning travelers), 22–63 yrs with DF	Serology (IgM ELISA) and PCR of serum	Returning travelers from India (9 patients), Thailand (3), Guyana (1, who was encephalopathic).
Row et al. [68]	Australia, 1993	2/210 (1%)	Adults with DF, plus one with DHF	Serology (IgM ELISA)	
Thisyakorn et al. [69]	Bangkok and Sonkhla, Thailand, 1987–1994	30/2975 (1%)	Children, mostly DHF?	Antibody in serum [30]; virus isolation in serum [2]	Prospective study.

(*Continued*)

Table 1 *continued*

Reference	Location, year	Neurological dengue patients/all patients in study	Details of patient group studied	Diagnostic methods [no. of neurological dengue patients]	Comments
Solomon et al. [58]	Southern Vietnam 1994–1995	16/1691 (1%)	Children and adults, mostly DHF	Serology (IgM and IgG ELISA) [15; 2 with Ab in CSF]; isolation [5; 2 in CSF]; PCR [4; 2 in CSF]	Prospective study.
Pancharoen and Thisyakorn [70]	Bangkok, Thailand, 1987–1998	80/1493 (5.4%)	Children with DF and DHF	Serology	Included 35 children with febrile seizures.
Cam et al. [59]	Ho Chi Minh City, Vietnam, 1997–1999	27/5400 (0.5%)	Children with DHF	Serology (HI and IgM ELISA) [14 with Ab in CSF], PCR of CSF [1]	
Studies looking for dengue infection among patients presenting with CNS disease					
Familusi et al. [73]	Nigeria, 1968	3/105 (3%)	Children with febrile convulsions	Virus isolation from serum [3]	Patients presented with signs of URTI, then a convulsion.
Burke et al. [72]	Northern Thailand	4/44 (10%)	Children with encephalitis	Serology (IgM ELISA)	Inadvertently identified during a study of Japanese encephalitis.
Solomon et al. [58]	Southern Vietnam 1994–1995	16/378 (4%)	Children and adults with suspected CNS infections	Serology, isolation, and PCR (see above)	On admission, 6 had DHF, 2 had DF; 1 developed DHF subsequently.
Chokephaibulkit et al. [60]	Bangkok, Thailand, 1996–1998	8/40 (20%)	Children with encephalitis	Serology [8], isolation/PCR of blood [6], PCR of CSF [1]	Seven developed DHF.

CSF = cerebrospinal fluid, DF = dengue fever, DHF = dengue hemorrhagic fever, HI = hemagglutination inhibition, CFT = complement fixation test, PRNT = plaque reduction neutralization test, ELISA = enzyme-linked immunosorbent assay, Ab = antibody.

[a] Definitions of neurological disease and diagnostic methods vary between studies. Also, data from some of the original publications have been modified to provide a consistent format for inclusion in this table.

children. For example, a reanalysis of data from Thailand [51] shows that 7 (5.5%) of 141 dengue virus–infected children less than 5 years old presented with neurological disease, compared with just 5 (1.4%) of 371 older children [odds ratio (95% C.I.) 4.03 (1.1–16.4), $p=0.019$]. This may be partly because younger children have an increased risk of febrile convulsions. Pancharoen and Thisyakorn [75] compared the features of 77 dengue-infected children less than 2 years old with the same number of children between 2 and 15 years old. Younger children were more likely to have seizures (26% vs. 1.3%, $p<0.0001$), a rash, and diarrhea but were less likely to have a positive tourniquet test. Adults are also more likely to present with neurological dengue. In Vietnam [58], 10 (3.8%) of 280 adults hospitalized with dengue infection had neurological disease compared with six (0.4%) of 1411 children [odds ratio (95% confidence interval) 8.67 (2.8–29.2), $p<0.0001$]. Of course, the observation that older children are less likely to a present to a hospital with neurological dengue may just be a reflection of the fact that they are especially likely to present with DHF.

As is clear from Table 1, neurological manifestations have been associated with dengue fever and all four severity grades of DHF. However, it appears that patients with severe DHF (grades III and IV, also known collectively as dengue shock syndrome) are especially likely to have neurological disease. Hendarto et al. found that 76 (75%) of 98 dengue patients with encephalopathy in Indonesia had DHF grades III or IV [53]. A reanalysis of data from Vietnam [58] shows that five (0.36%) of 301 patients with DHF grades III and IV had neurological disease, compared with three (0.21%) of 1382 with milder disease (dengue fever or DHF grades 1 to 2) [odds ratio 1.6 (1.5–50.2), $p=0.004$]. This supports the contention that severity of DHF is an important contributor to the development of neurological disease.

Where primary and secondary dengue infection have been distinguished, it is clear that both are associated with neurological disease [57,58,60,69]. Secondary infection appears to be associated with encephalopathic DHF, whereas primary infection tends to be associated with convulsions (especially in younger children) [73,76] and peripheral neuropathies (especially in adults) [66,77].

8 PATHOPHYSIOLOGY OF DHF

8.1 Antibody-Dependent Enhancement

Dengue hemorrhagic fever is characterized by increased permeability of the blood vessels (allowing leakage of plasma from the vessels), hemorrhagic manifestations (which may be mild), and thrombocytopenia [78] (Fig. 4). Few questions in arbovirology have excited as much interest as the pathogenesis of DHF. Although DHF can occur in a primary dengue infection, epidemiological evidence shows that it is more likely when infection with one dengue virus is followed by a secondary infection with a different virus [79]. For example, in 1981 when a massive outbreak of DHF in Cuba was caused by dengue-2 virus infecting a population previously exposed to dengue-1, 95% of DHF patients had a secondary infection [65]. In Santiago de Cuba, no dengue had occurred since a dengue-1 outbreak in 1977–1979 until an outbreak of dengue-2 virus in 1997. During this outbreak all symptomatic patients were adults born before the 1977–1979 outbreak, and 98% of them had secondary dengue infections. In contrast, almost all of those who had asymptomatic infections had a primary dengue infection [80,81].

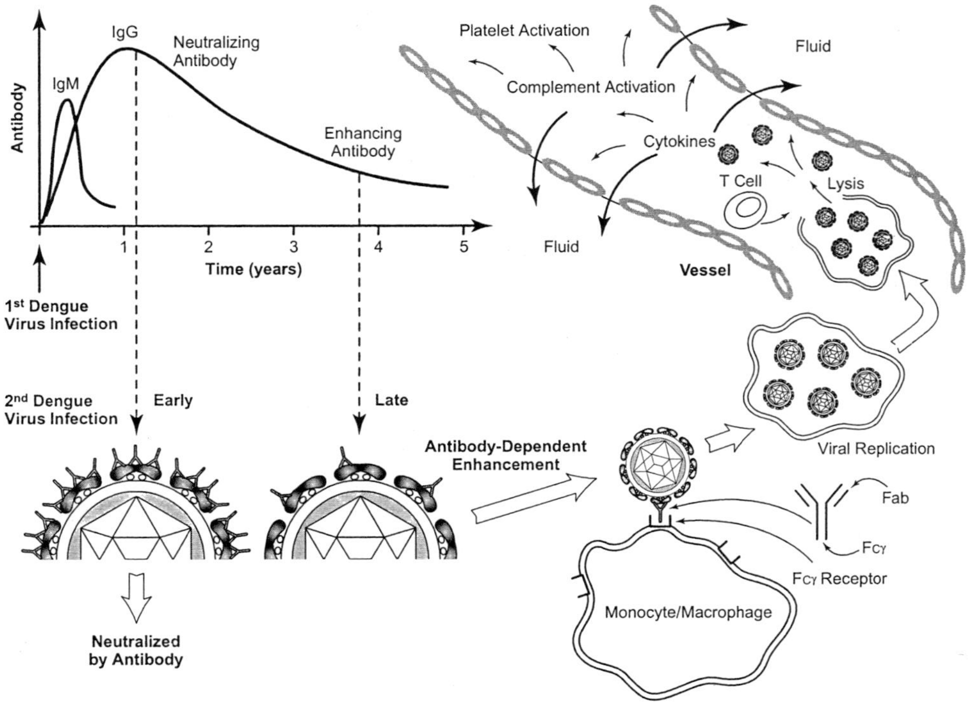

Figure 4 The postulated role of antibody-dependent enhancement in the pathogenesis of DHF. In this schematic representation, antibody levels following a first dengue virus infection are shown in the top left panel. If there is ''early'' infection with a second serotype, then there is sufficient cross-reactive IgG antibody to neutralize the second infection. However, if a second infection occurs in later years, neutralizing IgG antibodies against critical epitopes have decreased, but IgG antibodies against noncritical sites result in viable virus–antibody complexes. Using Fcγ receptors, the IgG antibodies enhance the entry of virions into monocytes, where viral replication occurs. The resulting T-cell-orchestrated immune response results in high levels of cytokines and complement and platelet activation, which in turn lead to increased vascular permeability and vascular leakage.

The presence of circulating antibodies from either previous dengue infection with a different virus type or passively acquired maternal immunoglobulin (Ig) G has consistently been reported as a risk factor for DHF [79]. Laboratory experiments indicate that this may be because IgG antibody to the first virus cross-reacts with and binds to the second virus without neutralizing it (Fig. 4). This antibody, by binding with Fc gamma receptors on the surface of macrophages is postulated to enhance the entry of the second virus into macrophages. This ''antibody dependent enhancement,'' by allowing more virus into the macrophages, is thought to lead to increased intracellular multiplication and more severe disease [82,83]. All four dengue viruses are thought to have been around for approximately 10,000 years, and disease consistent with dengue fever has been described since antiquity. Why epidemic DHF appears to have emerged only in the last 50–100 years is not certain, but the increased intensity with which different dengue viruses cocirculate is one possible explanation [31].

8.2 Strain Determinants

Evidence is accumulating that the specific dengue virus types and the order of infection may also be important. For example, data from Thailand and Cuba suggest that most primary infections with dengue-2 and dengue-4 appear to be asymptomatic [80,81,84]. In Rayong, Thailand, sequential infection with dengue-1 and then dengue-2 virus seemed to be more likely to lead to dengue shock syndrome than other combinations [85]. More recently it has become clear that the genotype of dengue virus may also be relevant. In the Americas, dengue-2 virus has been present for many years, but epidemics of DHF appeared only after a southeast Asian dengue-2 genotype was introduced in 1981 [86–88]. Sequencing analysis of dengue-2 strains from dengue fever and DHF patients suggested specific amino acid changes, particularly in the E protein 5′ and 3′ untranslated regions (UTRs), may be important determinants of pathogenicity—the E protein by altering virion binding to host cells, and the UTRs by altering secondary structures at the ends of the viral genome and thus affecting viral translation and replication [89].

Whatever the relative contributions of viral strain and antibody dependent enhancement, increased intracellular viral replication is thought to be the critical common pathway in the development of the plasma leakage that characterizes DHF (Fig. 4). Using serial dilution, Vaughn et al. [84] recently demonstrated higher dengue-2 peak viremias in DHF patients than in dengue fever patients. Pathologically there is no evidence of necrosis or inflammation of the blood vessels in DHF. The increased capillary permeability may be due to gaps in the endothelium [90]. High levels of cytokines and other markers of activated T cells (TNF-α, IFN-γ, sTNF receptors, sIL-2 receptor, IL-1β, IL-6) support a role for cytokines in increasing capillary permeability [91–93].

8.3 Genetic Susceptibility

Dengue hemorrhagic fever is more common in southeast Asia than in Africa and America. This may be because of the greater intensity with which the four serotypes of dengue viruses circulate in southeast Asia, or it may be because of differences in genetic susceptibility. The observation that during the 1981 Cuban outbreak black individuals were less likely to present with DHF is consistent with the hypothesis that resistance and susceptibility genes exist for DHF. Recently HLA haplotypes associated with susceptibility to DHF were identified in Vietnamese and Thai children [93,94].

9 PATHOPHYSIOLOGY OF NEUROLOGICAL COMPLICATIONS

The pathophysiological mechanisms underlying neurological dengue have also been the subject of some controversy. Although several DHF autopsy series were published in the 1960s and 1970s, there have been few series of patients with neurological disease. Nevertheless, as discussed below, the evidence from a variety of clinical and pathological studies suggests that a range of complications of severe DHF may contribute to reduced consciousness in some patients. These include hepatic dysfunction, which may be part of a Reye's-like syndrome, hyponatremia, renal failure, metabolic acidosis, intracranial hemorrhage (which may be frank or microvascular), disseminated intravascular coagulation, hypoxia, and raised intracranial pressure. However, in other patients who do not appear to have any of these complications, the evidence suggests that on occasion dengue virus crosses the blood-brain barrier to enter the CNS and cause encephalitis.

9.1 Hepatic Dysfunction and Reye's Syndrome

Hepatomegaly and elevated liver transaminases are common in dengue patients, both those with and those without neurological manifestations. In one study approximately one-half of DHF patients had hepatomegaly, detected on ultrasound; in addition, one-third had edematous thickening of the gallbladder wall [95]. Some elevation of transaminases was seen in 90% of Taiwanese patients infected with dengue [96] and 72% of Indian children with neurological dengue [52]. In most cases this elevation is moderate, transient, and without consequences [96]; however, in about 10% of cases aspartate aminotransferase or alanine aminotransferase is considerably elevated, usually in association with jaundice and prolonged prothrombin times. For example, in one series, five (23%) of 21 patients with neurological dengue had transaminases greater than 10 times normal, and three of them were also jaundiced [58]. In addition to occurring in DHF, severe hepatic dysfunction is also seen in patients with dengue fever and occasionally in patients with no features of either DHF or dengue fever, suggesting that severe dengue virus–mediated liver injury can occur independently of the mechanisms of plasma leakage in DHF [30]. In some encephalopathic patients the elevated transaminases have been associated with hypoglycemia and raised ammonia levels suggesting a Reye's-like syndrome [49,97]. In addition to direct virally induced liver damage, other factors that may contribute to hepatic dysfunction include ingestion of salicylate, paracetamol, and other drugs and toxins.

In histopathological studies, two patterns of damage are seen. In some cases, the livers of dengue patients are similar to those of yellow fever patients, with necrosis of the intermediate zone of liver lobules and the presence of cytoplasmic change resembling Councilman bodies [98]. Councilman bodies are now thought to be hepatocytes that have undergone apoptosis, as is seen when dengue viruses replicate in hepatocytes in vitro [99]. In other patients there is microvascular fatty infiltration of hepatocytes, reminiscent of that seen in Reye's syndrome [46]. Dengue virus has been isolated from liver tissue at autopsy and also detected by immunohistochemistry and by the reverse transcriptase polymerase chain reaction [100,101].

9.2 Other Metabolic Abnormalities

Mild hyponatremia is common in patients with neurological dengue, particularly during treatment for DHF. In Vietnam, five of 21 patients had a sodium level of between 130 and 135 mmol/L; in one patient it was 128 mmol/L. More severe hyponatremia (<125 mmol/L) was reported for two patients in a Thai series [55]. Although such severe hyponatremia may contribute to coma in some patients, it is unlikely that milder derangements of sodium concentration are important. Renal function is typically well preserved in dengue. When impairment occurs it is usually a consequence of hypovolemic shock and responds to appropriate therapy. Other metabolic derangements that may be seen, particularly in dengue shock syndrome, include hypoglycemia, acidosis, and hypoxia. Disseminated intravascular coagulation, complement activation, deposition of antigen–antibody complexes, and release of ''toxins'' such as histamine, serotonin, bradykinin, and slow-reacting substance A have also all been postulated as possible causes of coma [49,54]. Any explanation of coma in DHF must account for the fact that although all these metabolic processes occur in many patients with DHF, only a small proportion of patients develop coma.

9.3 Cerebral Edema

A series of studies have shown that cerebral edema occurs frequently in fatal dengue infection, whether or not there were neurological features. In Thailand, Bhamarapravati et al. [98] performed a limited examination of the brains of 42 patients who had died of DHF. The pia-arachnoid was edematous in most cases, with prominence of the Virchow-Robin space (the extension of the subarachnoid space around the cerebral arterioles as they penetrate the brain) [98]. Burke [102] studied 12 fatal DHF cases in Singapore. All had heavy brains with congested edematous leptomeninges. Nimmannitya et al. [55] noted edema in three of 10 fatal DHF cases with encephalopathy. Edema was also seen in three of five cases in Brazil [54].

With the advent of cranial imaging techniques these findings have been confirmed in nonfatal DHF cases. Lum et al. performed CT scans in three of six patients with coma and DHF and found all had cerebral edema [57]. In Vietnam, MRIs were performed on 18 of 27 DHF patients with neurological presentations: 12 had cerebral edema [59]. Cerebral edema on CT has also been described in dengue fever and in a patient with no features of dengue fever or DHF [58]. A case report by Janssen et al. [103] describing fatal cerebral edema associated with primary dengue infection in a returning traveler was particularly instructive. Histological examination showed diffuse endothelial thickening of capillaries in the white matter and focal extravasation of erythrocytes, without infiltration of mononuclear cells [103]. Immunological staining showed IgM, C1q, and the terminal complement complex C5b-9 in the cerebral capillaries. These changes suggest that in this patient, who did not have DHF, pathophysiological processes similar to DHF including complement activation may have led to loss of integrity of the vascular endothelium, resulting in leakage of plasma and cerebral edema. Increased permeability of the blood-brain barrier secondary to cytokine activation has also been seen in a mouse model of neurological dengue infection [90].

9.4 Cerebral Hemorrhage

Histopathological studies of fatal DHF cases suggest that both gross hemorrhage and petechial hemorrhage contribute to encephalopathy in some patients. Mild focal subarachnoid hemorrhage was seen in the brains of approximately 10% of 42 DHF patients who had limited examinations [98]. In another series, six of 10 encephalopathic DHF patients had intracranial hemorrhage at autopsy [55]. Burke et al. examined 12 DHF cases in Singapore and found intracranial hemorrhage in four (one subarachnoid, one subdural, two intracerebral); petechial hemorrhage was seen frequently in the white matter [102]. Similar findings have recently been shown in nonfatal cases using cerebral imaging. In the series from Vietnam described above, one patient had intracranial hemorrhage in addition to edema [59]. Patey et al. [104] reported a patient with dengue fever who developed a focal subarachnoid hemorrhage demonstrated by >1000 red cells in the CSF and computer tomographic (CT) and magnetic resonance imaging (MRI) abnormalities. And in the fatal case report of Janssen et al. described above [103], in addition to edema there was a pontine hemorrhage, which was believed to be the cause of the patient's sudden deterioration.

9.5 Viral Invasion Across the Blood-Brain Barrier and Encephalitis

Although a range of complications of dengue fever and DHF may lead to encephalopathy, several authors reported patients who had neurological symptoms before the other features of dengue were apparent [47,57]. And more recently neurological dengue patients have been identified who never had any features of dengue fever or DHF [58]. Could such patients have a true viral encephalitis?

Strictly speaking, encephalitis is a pathological diagnosis that should be made only with histological confirmation at either autopsy or brain biopsy [105]. Recent histopathological evidence suggests that dengue infection is occasionally associated with inflammation and other changes at autopsy. Such a patient was reported in Thailand [51,69,70]. This child, who to date has not been fully written up, is reported to have had elevated IgM in the CSF and cerebral edema on CT scan [30]; at autopsy the brain was grossly swollen with meningoencephalitis and perivascular infiltration of mononuclear cells in sections of the cerebrum. In addition there was massive necrosis of the liver, and periadrenal hemorrhage [30].

In Bhamarapravati et al.'s [98] autopsy series of DHF patients from Thailand, occasional neurons showed acidophilia, and shrinkage of the cytoplasm was seen in 12 (25%) of 42 brains. Chimelli et al. [54] examined the neuropathological findings in five fatal DHF cases from Brazil that had neurological features. Four of the five patients had nonspecific changes similar to those described in earlier series: edema, vascular congestion, focal hemorrhages, and perivenous lymphocytic infiltration. However, one patient who presented with a delayed onset of neurological symptoms (20 days into the illness) and had 1275 leukocytes in the CSF had frequent foci of perivenous inflammatory mononuclear cell infiltrate and glial reaction, at autopsy, with several foci of perivenous demyelination. This perivenous leukoencephalitis was thought to represent an immunological mechanism similar to acute disseminated encephalomyelitis [54]. Dengue viral antigens were subsequently demonstrated in the brains of three of these five cases [61]. More detailed immunolabeling in the case with perivascular encephalitis showed a moderate reactive astrocytosis in the Virchow-Robin space and suggested that CD68+ macrophages were infected with dengue virus both within and outside small veins. Macrophages expressing viral antigen were found in both the white and gray matter. In the gray matter they were often juxtaposed to neurons that appeared to display cytopathic effects. It was suggested that CD68+ macrophages might carry dengue viruses across the blood-brain barrier in a "Trojan horse" mechanism [106] and that neurons might be injured by contact with these dengue-infected macrophages, similar to what has been proposed for HIV [107]. Rosen et al. attempted to amplify by the polymerase chain reaction (PCR) dengue virus from the brain tissue of 15 children who died of DHF, including one from whose midbrain dengue virus had been isolated previously [101]. No viral RNA was detected in any samples. However, more recently, Ramos et al. [62] reported the autopsy findings of a 17-year-old who presented with fever, myalgia, convulsions, and coma followed by DHF during an outbreak in Mexico. Immunohistochemical staining showed dengue-4 virus in the neurons, astrocytes, microglia, and endothelial cells of the inferior olivary nucleus in the medulla and the granular layer of the cerebellum. Dengue-4 virus RNA was amplified from the same tissue using reverse transcriptase PCR. Interestingly, even though neurons were infected with dengue virus, the characteristic features of viral encephalitis—inflammatory perivascular cuffing, neuronal death with neurophagia, and activated microglial cells—were again not observed in this case. The question remains, should their be called encephalitis if the viral infection has not precipitated an inflammatory response in the brain? For several viruses, including Japanese encephalitis virus, a patient may die with immunohistochemical evidence of virus in neurons without any evidence of a cellular inflammatory response. It could be argued that proof of a viral CNS infection (be that immunohistochemical proof, virus isolation from the CSF, or antibody detection in the CSF), in the context of an appropriate clinical picture, is a more appropriate gold standard than a histopathological description of an inflammatory response that can be seen only at autopsy or on brain biopsy.

In summary, then, in the majority of dengue-infected encephalopathic patients, neurological changes can be accounted for by the complications of severe DHF—edema, hemorrhage, and metabolic derangement. However, when patients present with encephalopathy and no other features of dengue infection, it is unclear whether the same pathophysiological processes—for example, subtle increases in vascular permeability and microvascular hemorrhage—are responsible or whether some other mechanisms may be involved. Clinical virology studies suggest that dengue viruses can cross the blood-brain barrier to enter and replicate in the CNS. The means by which this occurs is uncertain, though carriage by cells of the monocyte/macrophage lineage has been suggested. Proof of encephalitis, as defined by immunohistochemistry, may not be evident, but recent immunohistochemical and nucleic acid hybridization studies have confirmed that dengue viruses can infect neurons [61,62].

9.6 Dengue Neuropathies and Myelopathy

There are even fewer data to explain how dengue causes peripheral disease. A post- or parainfectious etiology is suggested by the fact that symptoms often occur toward the end of or after the acute infection [77]. Where they have been performed, nerve conduction studies have shown decreased motor conduction velocity [77,108], which would suggest a demyelinating process. The cause is uncertain, though deposition of immune complexes has also been suggested.

9.7 Virulence Determinants

The factors that determine whether a particular infection will result in hemorrhagic disease, neurological disease, or both are not completely understood. As discussed above for hemorrhagic disease, the evidence suggests that both the host immune response and the viral determinants may be important. For some flaviviruses their introduction to a new geographical area is associated with changes in the clinical epidemiology, such as when West Nile virus caused encephalitis epidemics in Romania and New York [109]. To examine this possibility, dengue-2 isolates from Vietnamese patients with neurological disease were compared with other dengue-2 strains to determine their geographical origin [58]. A limited phylogenetic analysis based on a 240 base-pair fragment encoding the junction between the E and NS1 genes showed that they were closely related to other strains from Vietnam and were representative of locally circulating dengue-2 viruses rather than being imported from elsewhere [58].

Investigations of neurovirulence determinants in animal models of neurological dengue have focused on the E protein. A comparison of a mouse adapted neurovirulent strain of dengue-4 with its non-neurovirulent parental strain identified two amino acid substitutions in the envelope (E) protein (Fig. 2b) that appear to be associated with neurovirulence [110]. One of these ablated a glycosylation site within domain I (the hinge domain); the other was in the postulated stem anchor region adjacent to domain III (the putative receptor binding domain). A separate comparison of a mouse-adapted neurovirulent dengue-2 strain and its parental virus identified two different amino acid changes [111]. One was a conservative change in domain I, but the other was a change from glycine (a negatively charged amino acid) to lysine (a positively charged amino acid) in domain II (the fusion domain).

Thus a series of changes in the critical regions of the E protein may be responsible for neurovirulence of dengue in mouse. Whether similar changes are important in human dengue infections is not known. In a preliminary study, the E protein of a dengue-2 strain isolated from the serum of a child with DHF and encephalopathy was compared with other dengue-2 strains from the same epidemic season. The only change was a conservative change from an alanine to a valine at position 173 in domain I [112].

10 CLINICAL FEATURES

10.1 Dengue Fever

Symptoms develop after an incubation period of 4–7 days. In young children dengue infection often causes an undifferentiated febrile illness. Thirty percent of children diagnosed with an upper respiratory tract infection in a Bangkok outpatient clinic were shown by virus isolation or serology to have acute dengue infections [113]. In older children and adults, dengue fever presents as a classical fever–arthralgia–rash syndrome [114]. There is an abrupt onset of high fever, vomiting, muscle and joint aches (which may be severe), often with retro-orbital pain, photophobia, and lymphadenopathy. Nausea, abdominal pain, and a complaint of a metallic taste of food are also often reported. Skin eruptions occur in 50% of patients and include a transient mottling, flushing, or maculopapular rash. Petechiae and other bleeding manifestations such as gum, nose, or gastrointestinal hemorrhage are not uncommon in dengue fever but do not in themselves make it into DHF (see below). In some patients there is no acute rash, but during convalescence a fine maculopapular rash is seen on the limbs (a ''recovery rash''). The tourniquet test—a measure of capillary fragility—is traditionally recommended [78]: A blood pressure cuff inflated to half-way between systolic and diastolic pressure for 5 min produces 20 or more petechiae in a 2.5 cm square on the forearm (Fig. 5). However, recent studies from Vietnam and Thailand show that although it is highly specific, the sensitivity of the tourniquet test is low, being positive in approximately 30–35% of patients with dengue fever and 45% of patients with DHF [115,116]. In most patients it provides little additional diagnostic information because there is already obvious evidence of hemorrhage (e.g. petechiae). However, in Vietnam a positive tourniquet test clinched the diagnosis in about 5% of patients [116]. A modified tourniquet test, using a simple elastic cuff and accepting 10 petechiae as a positive result, may be useful in resource-poor settings [116]. Leukopenia and mild thrombocytopenia are also common in dengue fever. In general, the clinical findings are not very helpful in distinguishing dengue fever from other febrile illnesses such as measles, malaria, chikungunya, typhoid, and leptospirosis. A short febrile illness in a patient of the appropriate age during a known epidemic suggests the possibility of dengue, but laboratory confirmation is needed.

10.2 Dengue Hemorrhagic Fever

Despite its name, the major pathophysiological process in dengue hemorrhagic fever (DHF) is increased vascular permeability, leading to plasma leakage from the blood vessels into the tissue. In addition there are thrombocytopenia and hemorrhagic manifestations, which are often very mild (a few spontaneous petechiae or a positive tourniquet test). There is also frequently a leukopenia, with atypical lymphocytes, mild disseminated intravascular coagulation with prolonged partial thromboplastin and thrombin times, and fibrinogen and complement depletion.

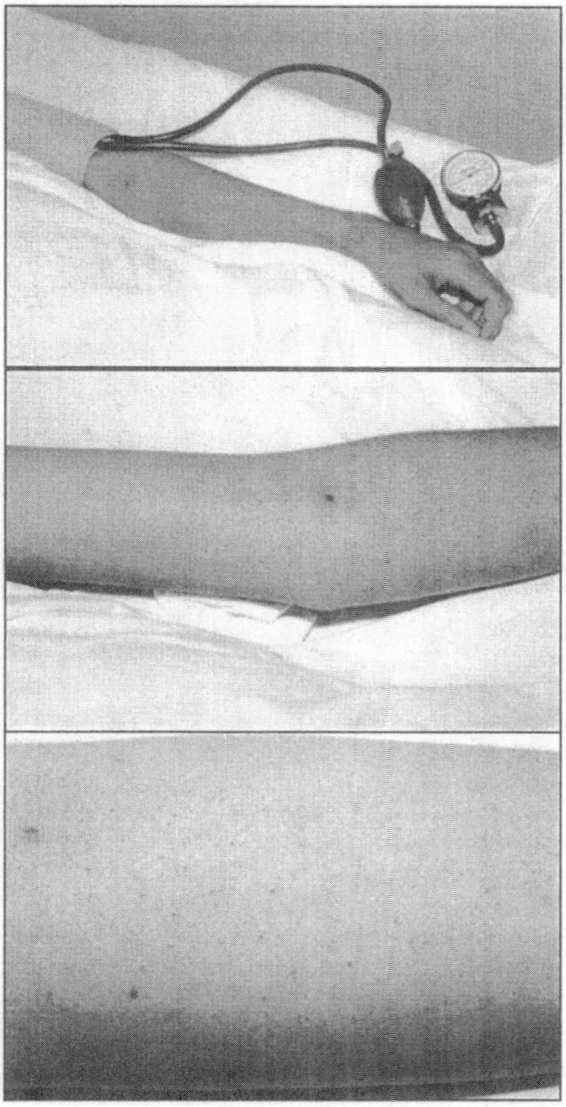

Figure 5 The tourniquet test of capillary fragility. A blood pressure cuff inflated to halfway between systolic and diastolic pressure for 5 min (top) causes more than 20 petechiae in a 2.5 cm square over the forearm (middle and bottom). Note also bruising around a venepuncture site in the antecubital fossa. (From Ref. 185.)

The WHO criteria for distinguishing dengue fever and the four grades of DHF are summarized in Table 2. In grade I DHF there is fluid leakage and thrombocytopenia, but the only hemorrhagic manifestation is a positive tourniquet test. Fluid leakage is defined by a hematocrit 20% above normal (the patient's normal being their hematocrit after recovery, or a normal value for the age-matched population) or by other evidence of leakage such as effusions. In grade II DHF there is spontaneous bleeding (e.g., petechiae,

Table 2 WHO Criteria for Distinguishing Dengue Fever and Dengue Hemorrhagic Fever Grades I–IV

		Plasma leakage[a]	Platelets (μl^{-1})	Circulatory collapse	Hemorrhagic manifestations
	DF	No	Variable	Absent	Variable
	DHF I	Present	<100,000	Absent	Positive tourniquet test (or easy bruising)
	DHF II	Present	<100,000	Absent	Spontaneous bleeding[c] with or without positive tourniquet test
DSS	DHF III	Present	<100,000	PP < 20 mmHg[b]	Spontaneous bleeding and/or positive tourniquet test
DSS	DHF IV	Present	<100,000	Pulse and BP undetectable	Spontaneous bleeding and/or positive tourniquet test

DF = dengue fever; DHF = dengue hemorrhagic fever; DSS = dengue shock syndrome.
[a] Identified by hematocrit 20% above normal (i.e., admission hematocrit 20% greater than discharge hematocrit, or 20% above normal for age) or clinical signs of plasma leakage (e.g., edema, effusions, ascites detected clinically or on a decubitus chest X-ray).
[b] Pulse pressure less than 20 mmHg, or hypotension for age.
[c] Skin petechiae, mucosal, or gastrointestinal bleeding.
Source: Ref. 78.

gum, nose, or gastrointestinal bleeding). In grade III DHF the vascular leakage is sufficient to cause shock. In children this manifests as a reduction in the pulse pressure—the difference between systolic and diastolic pressure—to less than 20 mmHg. In grade IV DHF the shock is so severe that the blood pressure is unrecordable. The term dengue shock syndrome is applied collectively to grades III and IV DHF. Overall 40–50% of DHF patients have a positive tourniquet test [115,116]. In southeast Asia DHF occurs mainly in children, but in the Americas it is seen in all age groups [1].

The WHO criteria for dengue fever and DHF were developed as strict case definitions to facilitate research and epidemiological assessment of outbreaks rather than for day-to-day patient management [117]. The fact that 20% hemoconcentration cannot be determined until after recovery and that accurate platelet counts are not possible in many areas where dengue occurs have meant that the strict WHO definitions have limited utility in acute patient assessment [118]. Moreover, they grade the severity of ''hemorrhagic'' dengue disease while ignoring the other manifestations of severe dengue infection, such as neurological disease. Thus patients may be extremely ill or even die of neurological or other complications of dengue while they are being graded as having the mildest form of dengue fever. For this reason other grading systems for dengue infection have been proposed [118,119]. However, until now, there has been no workable system for defining neurological disease, which has seriously impaired our ability to study it.

10.3 Classification of Neurological Dengue

A comprehensive system for classifying neurological dengue would facilitate epidemiological comparison, help define the clinical features, and facilitate research into the pathophysiology. In a preliminary attempt to characterize the spectrum of neurological entities seen with dengue, Gubler et al. [41] described three forms of disorder:

1. Headache, dizziness, delirium, sleeplessness, restlessness, and mental irritability, which were said to be associated with the acute phase of the illness.
2. Depressed sensorium lethargy, confusion, seizures, meningism, paresis, and coma, which were said to be severe neurological manifestations and sometimes clinically indistinguishable from encephalitis.
3. Delayed symptoms including paralysis of the lower and upper extremities and larynx, epilepsy, tremors, amnesia, loss of sensation, manic psychosis, depression, dementia and Guillain-Barré syndrome, which were said to be consistent with a postinfectious or parainfectious disorder.

Although this description was helpful in drawing attention to the distinction between symptoms that might be part of an acute dengue illness, symptoms that may indicate a more severe disease, and symptoms that occur later, it became adopted as a de facto classification of neurological involvement in dengue, without any consideration of the obvious shortcomings. For example, headache, dizziness, and lethargy are common in any acute febrile illness and cannot be considered evidence of neurological disease. Many of the terms such as delirium, confusion, and depressed sensorium are ill-defined and essentially describe the same thing. Yet patients with "delirium" are placed in one group, whereas those with "confusion" or "depressed sensorium" are placed in another. The classification incorporates a mixture of symptoms, signs, syndromes, and diagnoses and implies pathophysiological processes without any supporting evidence.

For this reason, during prospective clinical studies in Vietnam, precise clinical definitions were used [58,105]. Encephalopathy was defined as a reduction in the Glasgow or Blantyre pediatric coma score [120,121]. Patients were defined clinically as having encephalitis if they had encephalopathy, for which no metabolic or other cause could be found, and they had at least one feature suggestive of focal brain inflammation: CSF pleocytosis, focal neurological signs, or convulsions. If virus was detected in the CSF by isolation or PCR, or if there was specific IgM antibody in the CSF, this was defined as a viral CNS infection. However, although essential for a prospective clinical study, even these definitions have their limitations. A low CSF pleocytosis, focal signs, and convulsions can be seen in nonviral encephalopathies such as hepatic encephalopathy or in cerebral malaria [122–124]. Moreover, viral encephalitis can occur without any CSF pleocytosis, focal signs, or convulsions [105,125–127]. Imaging, particularly MRI, may show changes consistent with viral encephalitis [59], but even this does not provide definitive proof.

10.4 Cerebral Dengue

For the reasons just discussed, we propose here a new classification of neurological dengue (box) that does not presume that there is encephalitis unless there is pathological evidence to prove it. In this new classification patients are therefore first classified syndromically according to their presenting neurological features (reduced level of consciousness, meningism, paresis, or other) [128]. Those with a reduced level of consciousness are then considered according to whether they have features of dengue fever, DHF, or other systemic complications of dengue infection that might explain the encephalopathy. The new term *cerebral dengue* is reserved for those with acute dengue infection, coma, and no other obvious explanation for the coma. Rather like cerebral malaria, the term cerebral dengue is proposed because it implies that patients have coma caused by a pathogen but the mechanism is not known [123,129,130]. The term *dengue encephalitis* should be reserved for those with pathologically proven encephalitis. The adoption of the term cerebral dengue will allow epidemiological and pathophysiological studies to focus on those dengue infected patients in whom the cause of coma remains unknown. A reassessment of neurological dengue patients admitted during a prospective study in Vietnam [58] shows that five (24%) of 21 met the definition of cerebral dengue. (See tables on pages 493–494.)

11 NEUROLOGICAL MANIFESTATIONS OF DENGUE INFECTION

11.1 Reduced Level of Consciousness

One of the remarkable observations about DHF in children is that although they may be severely shocked, with unrecordable blood pressure, many are still fully conscious and indeed vigorously resist the medical team's attempts to gain venous access for resuscitation. However, other DHF patients present with a reduced level of consciousness, either on admission or soon after. Some of these patients have presented to hospital late, with features of prolonged dengue shock, which may include low or unrecordable blood pressure, incipient renal failure, acidosis, hepatic dysfunction, Reye's syndrome, jaundice, hypoglycemia, disseminated intravascular coagulation, and a bleeding diathesis. Other children are fully conscious on admission but gradually develop confusion, often after fluid resuscitation has begun. In these patients there may be peripheral or facial edema; signs of respiratory distress, including tachypnea, grunting, and nasal flaring; pleural edema; and effusions, ascites, and tender hepatomegaly. Sudden loss of consciousness in a patient with dengue fever or DHF may suggest an intracranial bleed [103].

A second group of patients present to hospital with neurological disease and no obvious signs of dengue. Half the patients in one series in Vietnam presented like this [58]. Clinically they may be indistinguishable from other viral encephalitides. In Asia the most important differential diagnosis is Japanese encephalitis. Typically these neurological dengue patients have a history of 2–3 days of a nonspecific febrile illness—which may include upper respiratory symptoms, especially in children—followed by confusion or coma. There may be clues to suggest that a patient has dengue rather than Japanese encephalitis. It is especially important to determine where a patient comes from: Japanese encephalitis is a rural disease, whereas dengue is urban. There may have been a dengue-like illness in family members or neighbors. Japanese encephalitis is mostly a disease of children in endemic areas, whereas neurological dengue is often seen in adults. Some patients who present with purely neurological disease subsequently develop obvious features of dengue fever or DHF [47,57]; in others the signs may be subtle if they are present at all. The patient should therefore be examined carefully for petechiae (which may be concealed in the axillae or antecubital fossa), and a tourniquet test should be performed. Tender hepatomegaly, a small pleural effusion, prolonged bleeding after a venepuncture, or simple investigations—a low platelet count, a rising hematocrit—may provide a clue that dengue is the etiology.

A reduced level of consciousness is the most common neurological manifestation of dengue infection, occurring in 55–100% of patient series (Table 3). Features may range from lethargy, drowsiness, and irritability to deep coma. Convulsions are also common, particularly in young children [70,73]. These may be simple febrile convulsions or those associated with prolonged coma. Pyramidal or long tract signs also occur, but the extrapy-

Classification of Neurological Dengue

Patients should be assessed according to the following definitions and then classified by the chart at the end.

ASSESS

A. Laboratory evidence of acute dengue infection in the serum or CSF [78], by
Detection of virus by isolation or PCR, or
Detection of antibody by IgM ELISA or four fold rise in HI or other serological test

1. If virus or antibody is detected in the CSF (in the absence of a traumatic LP) this is defined as a CNS dengue infection, as is immunohistochemical or nucleic acid hybridization evidence of dengue virus in brain tissue obtained from biopsy or autopsy.
2. Other microbiological, toxic, or metabolic causes of neurological disease must be excluded

B. Level of consciousness

1. Encephalopathy (a reduced level of consciousness)
   a. Encephalopathy is defined as a coma score that is at least 1 less than normal (i.e., Glasgow coma score or Adelaide pediatric coma of <15, or Blantyre pediatric coma score of <5 [120,121,131].
   b. As a quick assessment, any patient who is disorientated in time, space, and person (or, for children under 5, fails to recognize a parent) is encephalopathic.
   c. Patients with a reduced level of consciousness following a convulsion are considered encephalopathic, except for those with a single simple febrile convulsion, defined as follows.
   Simple febrile convulsion: Children between 6 months and 5 years of age with a single convulsion lasting less than 15 min who recover consciousness within 60 min [132]
2. Coma [133]
   a. Glasgow coma score <11.
   b. Blantyre coma score <4.
   c. As a quick assessment, any patient who fails to localize a painful stimulus (knuckle pressure applied to the sternum or finger pressure applied to the supraorbital ridge) is in unrousable coma.

C. Systemic consequences of dengue infection

DHF. As defined by WHO (see Table 2) [78]. Patients with vascular leak or hemorrhage who do *not* meet the WHO criteria are classified as follows:

Hemorrhagic disease. A patient with petechiae, positive toumiquet test, or frank bleeding who does not meet the WHO definition of DHF.

Vascular leak. A patient with evidence of increased vascular permeability (at least one of the following: admission hematocrit 20% greater than discharge hematocrit or 20% greater than normal for age; edema; effusions; ascites detected clinically or on X-ray), but who does not meet the WHO criteria for DHF.

Acute hepatic dysfunction. AST, ALT, or ammonia greater than 3 times normal limit [134]. (Note that the presence of tender hepatomegaly or jaundice, which do not in themselves cause a reduced level of consciousness, is not considered evidence of acute hepatic dysfunction for these purposes.)

Hyponatremia. Na <120 mmol/L.

Hypoglycemia. Blood sugar level <2.2 mmol/L [135].

AST = aspartate aminotransferase; ALT = alanine aminotransferase; ELISA = enzyme-linked immunosorbent assay.

CLASSIFY

	Proven dengue infection	Level of consciousness	DHF	Hemorrhage	Vascular leak	Thrombocytopenia	Hepatic dysfunction	Hyponatremia	Hypoglycemia
DHF with encephalopathy	+	Encephalopathy or coma	+	+	+	+	+/–	+/–	+/–
Dengue encephalopathy with hemorrhage, vascular leak, hepatic failure, hyponatremia, and/or hypoglycemia	+	Encephalopathy or coma	–	+/–	+/–	+/–	+/–	+/–	+/–
Cerebral dengue[a]	+	Coma	–	–	–	–	–	–	–

[a] Patients with cerebral dengue have unrousable coma but no hemorrhagic disease, increased vascular permeability, hepatic dysfunction, hyponatremia, or hypoglycemia.

Other Neurological Manifestations
Dengue meningism/meningitis: Acute dengue infection associated with neck stiffness or Kemig's sign (back pain when the knee is extended) should be described as dengue with meningism or, if the CSF white cell count corrected for blood cells is greater than 4 cells/mm^3 dengue meningitis.
Neuropathy: Neuropathy that follows dengue infection should be described and classified in the usual way as mononeuropathy, acute inflammatory demyelinating polyneuropathy, acute motor axonal neuropathy, etc[a]

[a] Refs. 136 and 137.

ramidal tremors and tone abnormalities that characterize other arboviral encephalitides such as Japanese encephalitis and West Nile virus [138] are less common in dengue, though they have been described [50,139] (Fig. 6 and 7). Presentations consistent with acute disseminated encephalomyelitis have also been described, some time after a dengue infection [54]. Meningism occurs in up to 30% of patients, usually as part of an encephalopathic illness. However, a simple viral meningitis due to dengue viruses seems to be rare.

11.2 Peripheral Neuropathies and Myelopathies

A range of mono- and polyneuropathies (involving cranial and peripheral nerves) and myelopathies have been associated with dengue infection in fully conscious patients [58,66,77,104]. They have most frequently been described in adults rather than children, and typically in association with dengue fever rather than DHF. Among the earliest of these was a description in 1919 of two patients from Honolulu who developed transient ocular complications (an abducens palsy and a paralysis of accommodation) within 2 weeks of developing dengue fever [140]. In the 1940s, 13 service personnel in the Central Pacific were reported to have developed peripheral neuropathies of the facial, palatal, long thoracic, peroneal, and ulnar nerves, 5 days to 1 month after the onset of dengue fever [66]. Two patients made a complete recovery; seven had persistent weakness at follow up.

Acute polyneuropathy 1 week after the onset of dengue was described for two patients during a serologically confirmed outbreak of dengue in Puerto Rico [77]. The first patient, a 40-year-old man, had weak legs with areflexia and a CSF pleocytosis. Nerve

Table 3 Clinical and Virological Features of Neurological Dengue from Selected Publications. [a,b]

	Sumarmo 1978 [47]	Kho 1981 [49]	Nimmannitya 1987 [55]	George 1988 [139]	Hendarto 1992 [53]	Rajajee 1994 [52]	Thisyakorn 1994 [51]	Lum 1996 [57]	Thakare 1996 [50]	Thisyakorn 1999 [51]	Solomon 2000 [58]	Pancharoen 2001[70]	Cam 2001 [59]
Location, year	Indonesia 1976–67	Indonesia 1975–77	Thailand 1972–81	Malaysia 1987	Indonesia 1988–89	India 1994	Thailand 1987	Malaysia 1992–93	India 1990–95	Thailand 1987–94	Vietnam 1994–95	Thailand 1987–98	Vietnam 1997–99
No. of neurological patients	4	119	18	2	98	25	12	6	10	30	21	80	27
Age range (yr)	1–7	0.7–14	0.25–13	0.58–6	Children	Children	0.67–14	0.42–11	2–60	0.25–14	0.25–39	0.25–14	0.7–15
No. ≤ 1 year old	1 (25)	—	7 (39)	1 (50)	—	—	1 (6)	1 (17)	0	5 (17)	4 (19)	—	6 (22)
Male	2 (50)	—	7 (39)	1 (50)	—	—	3 (25)	6 (100)	—	17 (57)	14 (67)	41 (51)	17 (63)
Headache	0	—	—	0	13 (13)	—	—	—	—	8 (27)	11 (52)	—	
Vomiting	1 (25)	—	—	2 (100)	23 (23)	9 (36)	—	—	—	—	8 (38)	—	
Reduced consciousness	4 (100)	50 (60)	13 (72)	2 (100)	92 (94)	25 (100)	12 (100)	6 (100)	9 (90)	23 (77)	18 (86)	44 (55)	27 (100)
Convulsions	4 (100)	25 (30)	9 (50)	0	84 (85)	7 (28)	7 (58)	3 (50)	5 (50)	19 (63)	9 (43)	54 (68)	21 (78)
Pyramidal/long tract signs	2 (50)	—	3 (17)	1 (50)	4 (4)	4 (16)	—	—	2 (20)	11 (37)	8 (38)	5 (6)	1 (4)
Meningism	1 (25)	6 (7)	—	0	—	—	—	3 (50)	3 (33)	9 (30)	6 (29)	9 (11)	
Cranial/peripheral neuropathy	0	—	—	0	—	0	—	—	2 (20)	1 (3)	0	0	
Paresis (flaccid)	0	3 (4)	—	0	—	—	—	—	0	—	0	0	—
Other neurological signs (no. of patients)	2 (50); spastic tetraparesis	—	3 (17); spasticity (2), decerebrate rigidity (1)	1 (50); decerebrate posturing and spasticity	—	—	—	—	1 (10); increased tone	—	5 (24); frontal release (2), extrapyramidal (3)		
No. with DHF:DF: neither	2 : 1 : 1	119 : 0 : 0	18 : 0 : 0	0 : 2 : 0	98 : 0 : 0	25 : 0 : 0	10 : 2 : 0	5 : 1 : 0	0 : 0 : 10	—	7 : 2 : 12	60 : 20 : 0	27 : 0 : 0
No. with DHF grades I:II:III:IV	0 : 0 : 2 : 0	15 : 53 : 34 : 17	0 : 1 : 9 : 8	0	0 : 22 : 13 : 63	10 DSS	0 : 2 : 3 : 5	2 : 1 : 2 : 0	0 : 0 : 5 : 5	—	0 : 1 : 5 : 1	36 DSS	0 : 9 : 14 : 4
Petechiae or positive TT	3 (75)	—	2 (11)	1 (50)	68 (69)	—	—	2 (33)	0	—	8 (39)	—	—
Other hemorrhage	1 (25)	—	15 (83)	1 (50)	54 (55)	19 (76)	—	4 (66)	0	—	6 (29)	—	—
Hepatomegaly	3 (75)	—	8 (44)	2 (100)	62 (63)	19 (76)	—	1 (17)	0	—	4 (19)	—	—
Elevated liver enzymes	—	—	11 (61)	2 (100)	90 (92)	18 (72)	7 [58]	1 (17)	—	—	5 (24)	19/21 (90)	
Hyponatremia	—	—	3 (17)	0	—	0	—	—	—	—	6 (29)	30/57 (53)	
No. with CSF pleocytosis/ no. with LP performed	2/3	—	0/9	—	0/98	—	0/6	4/6	—	—	3/16	7/21	0/22

(*Continued*)

Table 3 *continued*

Feature	Sumarmo 1978 [47]	Kho 1981 [49]	Nimmannitya 1987 [55]	George 1988 [139]	Hendarto 1992 [53]	Rajajee 1994 [52]	Thisyakorn 1994 [51]	Lum 1996 [57]	Thakare 1996 [50]	Thisyakorn 1999 [51]	Solomon 2000 [58]	Pancharoen 2001[70]	Cam 2001 [59]
Died	2 (50)	12 (10)	10 (56)	1 (50)	51 (52)	1 (4)	4 (33)	1 (17)	1 (10)	2 (7)	0	4 (5)	6 (22)
Sequelae	0	—	—	0	3 (3)	—	8 (66)	1 (17)	—	0	6 (29)	1 (1)	0
Sequelae details {no. of patients}					hemiparesis {2}, tetraparesis {1}	—	—	residual paralysis	—		paraparesis {3}, psychiatric {3}		
Virus serotypes isolated from serum {no. of patients}	Den-2 {2}, Den-3 {2}	Den-1 {3}, Den-2 {5}, Den-3 {20}, Den-4 {1},	—	—	—	—	—	Den-3 {3}, Den-2 PCR {1}		Den-2 {1}, Den-3 {1}	Den-1 {1}, Den-2 {3}, Den-3 {3}	—	—
Virus serotypes or IgM in CSF {no. of patients}	—	—	—	—	—	—	—	Den-3 {4} Den-2 {1} IgM {1}	IgM{2}	—	Den-2 {3} Den-3 PCR {2}, IgM {3}	—	Den-3 PCR {1}, IgM {14}
Primary: secondary infection	—	—	—	2:0	4:55	—	—	3:2	—	11:19	7:13	32:48	—
Comment		35 patients had "lethargy"; two had Reye's syndrome				Death from pulmonary edema	At autopsy 1 had encephalitis, 1 had hemorrhage	All had neurologica l signs before 'dengue' signs	Paper also described 10 encephalop athic DHF patients	One had autopsy suggestive of encephalitis	One developed DHF III after 48 h	Included 35 children with febrile seizures	

Data are not available; TT, tourniquet test; DF, dengue fever; DHF, dengue hemorrhagic fever; DSS, dengue shock syndrome; CSF, cerebrospinal fluid.

[a] To conserve space, column heads list only first author of each paper.

[b] Numbers (and in parentheses, percentages) are given, except where indicated otherwise.

Source: Data from some of the original publications have been modified, and in some instances extrapolated, to provide a consistent format.

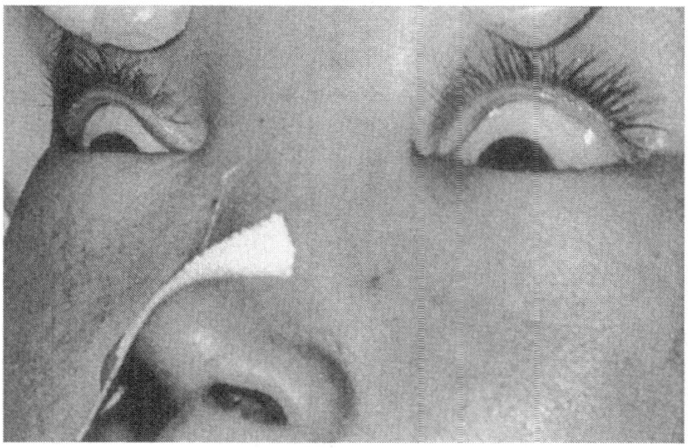

Figure 6 Brainstem signs in neurological dengue. Downward deviation of the eyes suggestive of a pontine lesion, in a comatose 19-year-old Vietnamese woman with acute secondary dengue-3 infection. (From Ref. 58.)

conduction studies revealed reduced motor velocities. The second was a 16-year-old male who developed urinary retention, then weak areflexic legs, ophthalmoplegia, and ataxia. Both patients recovered. Because of the short interval between the febrile illness and paralysis, the presence of cells in the CSF, and the bladder involvement in one patient, the term ''acute infectious polyneuritis'' was used rather than Guillain-Barré syndrome [141], but an infectious or parainfectious pathogenesis was suspected. Other patients with a more typical clinical course for Guillain-Barré syndrome have been described. One had a flaccid areflexic quadraparesis 2 weeks after acute dengue infection [142]. Nerve conduction studies showed a predominantly demyelinating sensory motor polyneuropathy. A second had bilateral symmetrical flaccid limb weakness (including the facial nerves) 1 month after infection with dengue [143]. The protein was elevated in the CSF, with 10 white cells/mm^3, and nerve conduction studies were compatible with demyelination.

Clinical features consistent with spinal cord dysfunction have also been associated with dengue infection. In Vietnam, two patients (a 14-year-old and a 21-year-old) presented with spastic paraparesis [58]. One had a primary dengue infection. In the other a secondary dengue infection was associated with spasticity, a sensory level at T10, and acute retention of urine. Both patients recovered power but had some residual signs. However, specific damage to the anterior horn cells of the spinal cord, which occurs in many other arboviral CNS infections, causing a polio-like flaccid paralysis [138,144], does not seem to be a frequent feature in neurological dengue.

11.3 Psychiatric Illness

Depression and lethargy, similar to that described following infectious mononucleosis, is a common feature in patients who have recovered from dengue fever. Manic psychosis, amnesia, and dementia have also appeared in occasional case reports [41].

11.4 Outcome of Neurological Complications

A comparison of several studies of neurological dengue shows a range in case fatality rates from 0% to 56%, with the median value at 10% (Table 3). Most deaths have occurred

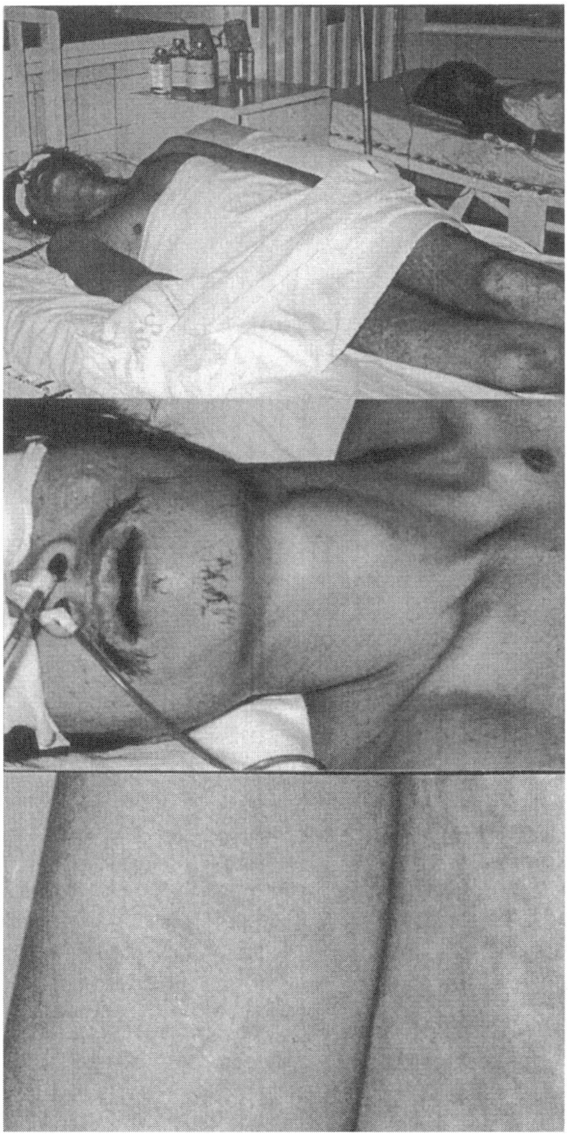

Figure 7 Clinical signs of neurological dengue. This 39-year-old Vietnamese male had dengue hemorrhagic fever grade III and coma, multiple convulsions, rigidity spasm, extensor posturing (top), venous oozing around the site of a left subclavian catheter insertion (middle), and a hemorrhagic rash on the arms and legs (bottom).

in patients who also had severe features of DHF and are thought to have been due to severe shock and hemorrhage [74,145]. Reports of mortality directly related to neurological disease are relatively rare [47]. Direct comparison between studies is difficult, but case fatality rates appear to be higher in studies that included more patients with grades III and IV DHF (Table 3) [53,55]. In neurological patients who do not have severe DHF, the prognosis is generally considered to be good [57,58].

Reported sequelae of patients with dengue encephalopathy include hemiparesis, spastic paraparesis, tetraparesis, alterations in personality, and epilepsy [53,57,58]. Most patients with peripheral neuropathies or myelopathies appear to recover [77,104], though in one series, seven of nine for whom the information was available had persistent deficit [66].

12 RADIOLOGICAL AND NEUROPHYSIOLOGICAL FINDINGS

Computer tomographic (CT) scans in encephalopathic dengue patients typically show diffuse brain swelling (see Secs. 9.3 and 9.4) [57,58]. Frank intracranial hemorrhage has also been reported, but less often. Interestingly one patient from India who presented with typical features of encephalitis and dengue IgM in the serum had abnormalities of the thalamus, basal ganglia, brainstem, and cerebellum on CT [50], changes more often seen in other flaviviral encephalitides [138,146]. Magnetic resonance imaging has also shown changes typical of encephalitis in some patients. Cam et al. [59] performed MRI scans on 18 of 27 patients with DHF and neurological symptoms. Four were normal, two had encephalitis-like changes (edema and scattered focal lesions), and 12 had cerebral edema, one of the latter also had intracranial hemorrhage.

Electroencephalograms (EEGs) in encephalopathic dengue changes typically show generalized slow waves [57]. Gomes *et al.* [147] reported such changes in 44% of DHF patients. Other patterns have also been reported. For example, one adult in Vietnam had an unusual EEG with high-amplitude periodic slow wave complexes (2–3 Hz) on a featureless background (Fig. 8) [58]. On the few occasions on which nerve conduction studies have been performed on dengue patients with neuropathies, they have shown reduced velocities suggesting a demyelinating pathophysiology [77,108,143].

13 CEREBROSPINAL FLUID

Fifteen to thirty percent of encephalopathic dengue patients have a moderate pleocytosis, which is usually lymphocytic; however, in most patients the cerebrospinal fluid (CSF) is normal [57,69]. In one study CSF opening pressures were measured and were moderately elevated (21–30 cm CSF) in three of 19 patients [58].

14 DIAGNOSIS

Traditionally dengue infection is confirmed by virus isolation or antibody tests [1]. Viruses can be isolated from serum during the first few days of illness, using continuous mosquito cell lines after 7–14 days. They cause the formation of large syncytia in *Aedes albopictus* C6/36 cells (Fig. 9). Virus identification is then confirmed by indirect immunofluorescence with serotype-specific monoclonal antibodies. Isolation of virus by intracerebral injection into *Toxorhynchitis splendens* mosquitoes may be quicker [148]. RT-PCR methods for detecting dengue viruses in serum have also been developed and are being used increas-

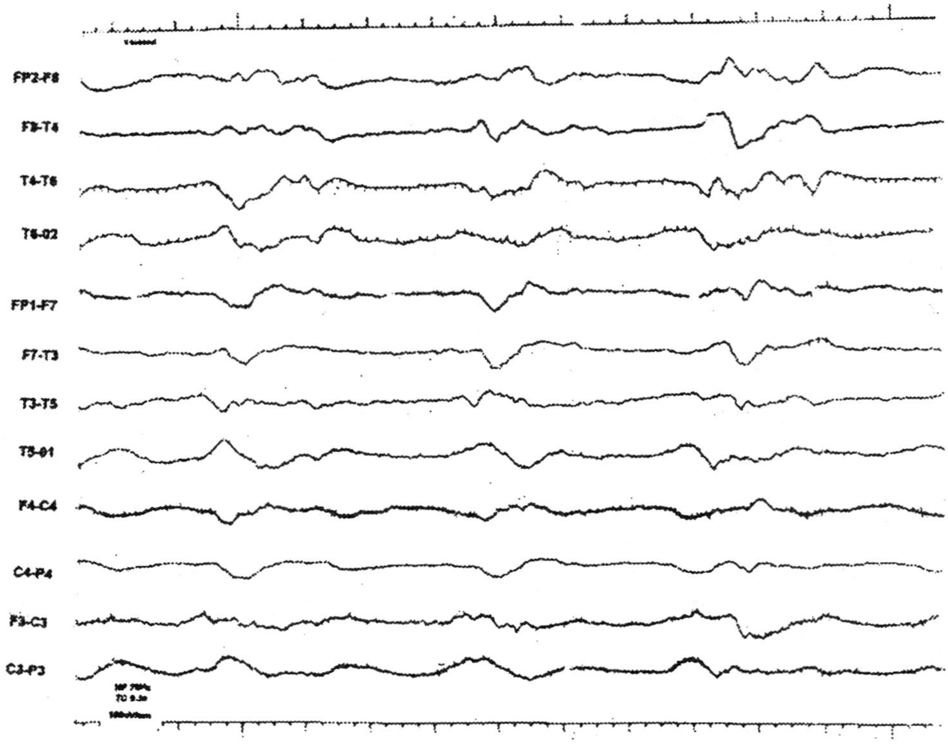

Figure 8 Electroencephalographic changes in neurological dengue. A 30-year-old Vietnamese man presented with 4 days of fever, anorexia, dizziness, then coma (Glasgow coma score = 7/15), brisk deep tendon reflexes and extensor plantars. He had no clinical signs suggestive of dengue infection and was clinically diagnosed as having encephalitis, but investigations revealed acute secondary dengue infection (from Ref. 58). The EEG shows high amplitude periodic slow waves (2–3 Hz) on a featureless background.

ingly [71,149]. Recently these have been modified to fluorogenic probe hydrolysis (TaqMan) methods, which may provide rapid, sensitive, and specific screening for epidemiological studies [150]. After defervescence, viral culture and PCR become negative, but the majority of dengue patients develop antibodies in the serum, that are detectable by IgM and IgG capture enzyme-linked immunosorbent assays (ELISAs) [30]. Primary and secondary flavivirus infections can be distinguished according to the ratio of IgM to IgG [151,152]. For many flaviviruses there is a degree of serological cross-reactivity, so that, for example, a test for anti-dengue antibody may be weakly positive in a patient infected with Japanese encephalitis virus. For this reason, in areas where more than one flavivirus cocirculate, it is important to test against both viruses in parallel. Because dengue is so common in many parts of the world, and because IgM antibody may persist in the blood for up to 3 months, it has been argued that in a patient with neurological disease the presence of IgM antibody may simply indicate recent coincidental dengue infection and not necessarily that dengue was the cause of the neurological disease. For this reason other viral CNS infections (Table 4) and disease that mimic viral meningoencephalitis (Table 5) must be excluded.

If it can be demonstrated that a patient has seroconverted during the neurological illness, or there is IgM antibody in the CSF, or dengue virus is detected in the CSF, this provides even stronger evidence that dengue was the cause.

ELISAs are now commercially available [153], and some of them do test for more than one flavivirus in parallel. They have recently been modified into simple kit formats for rapid testing without sophisticated equipment [154,155]. Whereas laboratory diagnosis

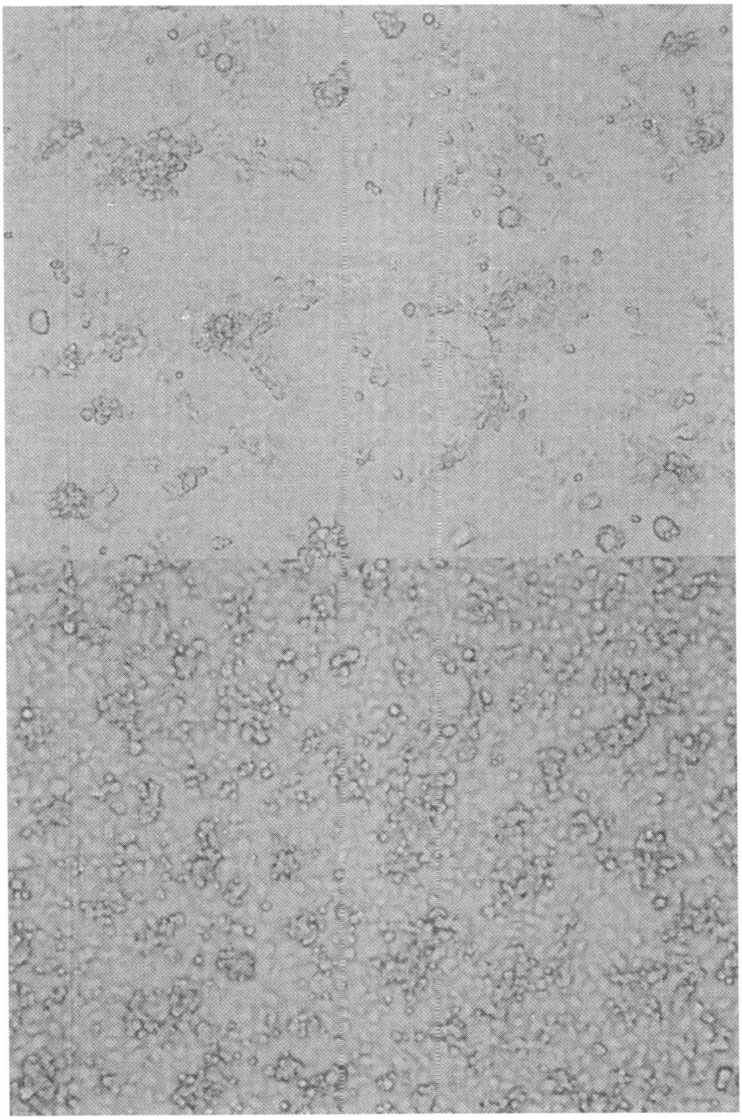

Figure 9 Culture of dengue virus in *Aedes albopictus* C6/36 mosquito cells. Eleven days after inoculation, cytopathic effects are observed with the formation of large syncytia (top). A confluent monolayer of noninfected control cells is shown for comparison (bottom). (Magnification × 10).

Table 4 Viral Causes of Encephalitis[a]

Encephalitis due to arboviruses, by geographical region
The Americas: West Nile; St Louis encephalitis; Powassan; California encephalitis; La Crosse; western, eastern, and Venezuelan equine encephalitis; dengue; Colorado tick fever,* Rocio*
Europe/Middle East: tick-borne encephalitis, West Nile, Tosana, dengue, louping ill*
Africa: West Nile, Rift Valley fever,* Crimean-Congo hemorrhagic fever,* dengue,* Chikungunya*
Asia: Japanese encephalitis, West Nile, dengue, Murray Valley encephalitis, Chikungunya,* Me Tri*
Australasia: Murray Valley encephalitis, Japanese encephalitis, Kunjin*
Encephalitis due to other viruses
Herpes viruses: herpes simplex, herpes zoster, Epstein-Barr, cytomegalovirus, human herpesvirus-6, human herpesvirus-7
Enteroviruses: polio, coxsackie, Echo, enteroviruses 70, 71
Paramyxoviruses: measles, mumps, Hendra, Nipah
Others: rabies, influenza viruses

[a] Rarer or suspected arboviral causes are indicated by asterisks.
Source: Modified from Ref. 138.

of dengue was previously confined to regional and national diagnostic virology laboratories, these kits allow diagnosis in smaller hospitals and clinics. Unfortunately, in the first few days of illness, before antibody is produced, false negatives may occur, but antigen detection kits are being developed to address this problem [156].

15 MANAGEMENT

15.1 Dengue Fever and DHF

Most cases of dengue fever are self-limiting, and hospital admission is not necessary, nor is it practical, given the total number of cases in endemic areas. Patients are encouraged to drink and are given paracetamol for symptomatic relief. Aspirin is avoided because of its antiplatelet effect and the risk of Reye's syndrome [97]. Parents are warned to return to hospital if the child becomes worse. Although the fever usually resolves within a few days, many patients feel prolonged lethargy and depression.

Patients with DHF are at first clinically similar to those with dengue fever. However, on days 3–7 of the illness as the fever subsides, there is vascular leakage resulting in hemoconcentration and thrombocytopenia. Patients are often restless, lethargic, cold sweaty, and clammy, with tender hepatomegaly or abdominal discomfort [157]. A lateral chest X-ray often reveals a pleural effusion [115]. Ultrasound may in addition show pericardial effusions, ascites, and gall bladder wall thickening [95]. Patients with grades I and II DHF are encouraged to drink but do not normally require intravenous fluids. The vital signs, hematocrit, and platelet counts are monitored closely. A progressively decreasing platelet count, a rising hematocrit, sustained abdominal pain, persistent vomiting, restlessness, lethargy, and prostration may all be signs of impending dengue shock syndrome [1].

Table 5 Disease Mimicking Viral Meningoencephalitis

CNS infections	Para or postinfectious causes
Bacteria	Guillain-Barré syndrome[a]
Bacterial meningitis	Acute disseminated encephalomyelitis[a]
Tuberculosis	Viral illnesses with febrile convulsions
Brain abscess	Shigella
Typhoid fever	Viral infections associated with swollen fontanelle
Parameningeal infection	
Lyme disease	Noninfectious diseases
Syphilis	Vasculitic
Relapsing fever	Bechet's disease
Leptospirosis	Cerebal systemic lupus erythematosis
Mycoplasma pneumonia	Neoplastic
Listeriosis	Primary brain tumour
Brucellosis	Metastases
Subacute bacterial endocarditis	Paraneoplastic limbic encephalitis
Whipple's disease	Metabolic
Nocardia	Hepatic encephalopathy
Actinomycosis	Renal encephalopathy
Fungi	Hypoglycemia
Cryptococcus	Reye's syndrome
Coccidiomycosis	Other:
Histoplasmosis	Drug reactions
North American blastomycosis	Subarachnoid hemorrhage
Candidiasis	Cerebrovascular accidents
Parasites	Epilepsy
Cerebral malaria	Trauma
Toxoplasmosis	Hysteria
Cysticercosis	
Trypanosomiasis	
Echinococcus	
Trichinosis	
Amoebiasis	
Rickettsiae	
Rocky Mountain spotted fever	
Typhus	
Q fever	
Erlichiosis	
Cat-scratch fever	

[a] Guillain-Barré syndrome and acute disseminated encephalomyelitis may follow viral or bacterial infections or vaccinations.
Source: Modified from Ref. 138.

For treatment of DHF grades III and IV (dengue shock syndrome), WHO recommends initial intravenous crystalloid (e.g., Ringer's lactate) at 10–20 $mLkg^{-1}h^{-1}$, followed by a colloidal solution (e.g., dextran 40) at 10–20 $mLkg^{-1}h^{-1}$ if shock persists [78]. Recent trials from Vietnam have suggested that colloids may restore blood pressure more quickly than crystalloids [158,159]. The rate of fluid infusion needs to be carefully tailored according to the vital signs, hematocrit, and urine output. Even cautious treatment may precipitate peripheral and facial edema, ascites, pleural effusions, and pulmonary edema. Central venous pressure monitoring is recommended, and diuretics and ventilatory support are sometimes needed. Only rarely are blood products required. Although there are no antiviral drugs against dengue, in expert hands the mortality of DHF has dropped from as high as 44% untreated to as low as 0.2% [30,157,160]. Trials of ancillary treatment such as corticosteroids and carbazochrome sodium sulfonate (AC-17, which is said to reduce vascular permeability) failed to show any benefit [160–162]. There are no antiviral drugs against dengue or any other flaviviruses. Theoretical targets for intervention include binding of the virus to the cell, uptake of the virus into the cell, the capping mechanism of flaviviruses, the viral proteases, the viral RNA-dependent RNA polymerase, and the viral helicase [163].

15.2 Neurological Dengue

Management of encephalopathic dengue patients is similar to that of patients with other encephalopathies. The particular issues to address are the level of consciousness, the recognition and management of seizures, raised intracranial pressure, and brainstem herniation syndromes [128]. The level of consciousness should be assessed using the Glasgow coma score for adults [120] or a modified version of it for children [121,131]. These are relatively easy and reproducible [164]. However, in settings where the lack of expertise or the overwhelming number of patients make it impractical, determining whether a patient is awake (A), responds to voice (V), responds to pain (P), or is unresponsive (U)—the AVPU scale [165]—will at least provide a simple and rapid objective measure of consciousness.

Ideally, patients in coma should be sedated and ventilated in an intensive care unit. This allows airway protection, maximum medication to control seizures, and hyperventilation to reduce raised intracranial pressure. However, in many of the settings where dengue occurs, this is not possible. As a minimum, patients at risk of raised intracranial pressure should be nursed at 30°C, with the neck held straight to ensure that jugular venous outflow is not impaired. Oxygen should be given. Fluid management may be especially difficult because of the rapid changes in vascular permeability and the need to balance adequate hydration with the fear of worsening cerebral edema (see below). Drug doses may need to be modified because of hepatic or renal impairment.

Seizures

In addition to clinically obvious convulsions, it has recently become apparent that subtle motor seizures are important in malaria, Japanese encephalitis, and other causes of nontraumatic coma [133,166,167]. Clinically patients may have subtle twitching of a digit, or lip, nystagmoid eye movement, or tonic eye deviation, but an electroencephalogram reveals that they are in status epilepticus. The importance of these symptoms in encephalopathic dengue patients is not known, but a careful examination for such signs (including lifting the eyelids) is recommended. Single seizures should be treated with a benzodiazepine (e.g., diazepam 0.3 mg/kg i.v.). Status epilepticus should be managed along standard lines [168]. If two doses of benzodiazepine fail to control seizures, rectal paraldehyde (0.4 mL/kg mixed with an equal volume of olive oil) is a cheap drug that is often effective [166]. Alternatively, a second line drug such as phenytoin or phenobarbitone should be used. Intravenous phenytoin (18 mg/kg) should be given slowly (infused over 20 min or at 1 $mg kg^{-1} min^{-1}$) with ECG monitoring because of the risk of cardiac dysrhythmias. In many parts of the tropics where phenytoin is not available, phenobarbitone is given as a cheap alternative. This is given via the intramuscular route at a dose of 10–20 mg/kg. There is a risk of respiratory suppression, especially in combination with diazepam [169], and ideally facilities for ventilation should be available. Benzodiazepines are metabolized in the liver, so doses should be adjusted accordingly.

Raised Intracranial Pressure and Herniation

Fluid management of encephalopathic dengue patients can be especially difficult. There is a conflict between the desire to adequately resuscitate a shocked patient and the fear of worsening cerebral edema and raised intracranial pressure. In shocked patients, restoration of the blood pressure and perfusion is the priority [166]. Hypotonic solutions, such as 5% dextrose, should be avoided because of a theoretical risk of causing cerebral edema [166]. Isotonic solutions (normal saline or Ringer's lactate) or hypertonic colloid solutions (gelafundin or hemaccel) are preferred. In encephalopathic patients in whom there is no shock, fluid restriction to two-thirds of normal requirements is usually practiced in Western settings [166,170]. In many encephalopathies, if the blood pressure is well maintained but raised intracranial pressure is suspected because of a deteriorating level of consciousness, signs of brainstem herniation, or an elevated CSF opening pressure at lumbar puncture, 20% mannitol (0.5–1 g/kg) is given [133,171]. The little evidence available suggests that the benefits, if any, are only short-term [171]. Mannitol is excreted by the kidneys and should not be given to anuric patients. The role of steroids has not been assessed in dengue encephalopathy, but they were found to be unhelpful in dengue shock syndrome, cerebral malaria, and Japanese encephalitis [129,172,173].

Other Measures

If bacterial meningitis is suspected, broad-spectrum antibiotics should be given. Hypoglycemia is common in many tropical pediatric conditions [135] and should be looked for and corrected. Adequate nutrition should be maintained via a nasogastric tube, and attention should be paid to the risk of other complications of a severe encephalopathy, including pneumonia, urinary tract infections, bedsores, and contractures.

16 PREVENTION OF DENGUE

There is currently no commercially available vaccine against dengue. Dengue control therefore consists of surveillance for dengue and *Aedes* mosquito activity and measures to control the *Aedes* vectors. These include educating people to remove *Aedes* breeding sites from around the house (e.g., removing stagnant pools of water collected in tires and other rubbish); treating stored water with larvicide (e.g., temephos) or the copepod *Mesocyclops*, which feeds on *Aedes aegypti* larvae; covering water storage containers to deny access to breeding mosquitoes; and ultralow-volume spraying of organophosphorus insecticides during epidemics. In some settings legislation and fines for those who fail to remove *Aedes* breeding sites from the home have been effective. Personal protection with insect repellents containing *N,N*-diethyl-*m*-toluamide (DEET) is also recommended. The evidence for the efficacy of these measures is variable. The only undoubtedly effective vector control measure was the near eradication of *Aedes aegypti* from South America, using DDT, during the yellow fever campaign of the 1950s–1970s. Since that campaign ended *Aedes* has reinfested South America [34]. Worldwide *Aedes aegypti* continues to spread, and dengue is increasing as a global health problem.

16.1 Dengue Vaccines

Vaccines have been available for some flaviviruses for many years. These include a formalin-inactivated vaccine against Japanese encephalitis and more recently tick-borne encephalitis, and live attenuated vaccine against yellow fever [174]. However, the development of vaccines against dengue has been hampered by concern that antibodies raised against one dengue virus might enhance infection with a different dengue virus (see Sec. 8.1). Consequently, it is believed that a tetravalent dengue vaccine is essential to provide protective immunity against all four dengue viruses. Several tetravalent vaccines are in development. Live attenuated vaccines, produced by passaging dengue serotypes 1–4 in either primary dog kidney cells or African green monkey cells, have been evaluated in stage I and II trials in Thailand [175,176] and the United States [177]. The latter produced tetravalent neutralizing antibody in 80–90% of volunteers, after two doses [177]. Phase III trials are being planned for both vaccines [175]. A recombinant dengue vaccine is being developed by inserting the C, PrM, and E genes of dengue-1, -2, and -3 viruses into a copy DNA infectious clone of dengue-2 virus [178]. In a different approach, PrM and E genes from dengue-1–4 viruses have been inserted into the live attenuated yellow fever 17D vaccine, replacing the original PrM and E genes [179]. This chimeric vaccine raised tetravalent antibodies in 100% of monkeys, and phase I trials are planned [180]. Other approaches include insertion of dengue virus structural genes into bacterial plasmids to produce naked DNA vaccines [181].

It is likely that vaccines against dengue will ultimately become available. When they do, it will be important to ensure that they become fully incorporated into the expanded program on immunization in those countries where dengue is a major problem. Unfortunately, this has not happened with previous flavivirus vaccines, e.g., yellow fever and Japanese encephalitis vaccines. Thus although tourists and foreign service personnel are adequately protected against these viruses, many residents in endemic areas are not [182].

Until vaccines become available and widely implemented, vector control measures remain our only defense against dengue. But they will be effective only with a high level of commitment, education, and community participation [1].

17 CONCLUDING COMMENTS

In summary, neurological manifestations of dengue have been described in most areas where dengue occurs but seem particularly important in Indonesia, Malaysia, Thailand, India, and Vietnam—the areas where dengue is particularly important. Although the number of patients with neurological dengue is small compared to all those with DHF, these patients account for an important proportion of all patients with CNS infections. In the majority of dengue-infected patients, neurological features are explained by secondary complications of severe disease: vascular leak, hemorrhage, or other derangements. However, in some neurological dengue patients there is clinical, virological, and immunohistochemical evidence to show that dengue viruses can occasionally cross the blood brain barrier to cause CNS disease. Whether this is labeled as encephalitis is more a question of semantics than anything else. The important point is that in endemic areas clinicians include dengue in their differential diagnosis for patients with neurological disease. The new case definitions for neurological dengue described in this chapter should help clinicians identify such patients. In addition they should facilitate further research on the epidemiology and pathogenesis of this important condition.

REFERENCES

1. Rigau-Perez, J.G.; Clark, G.G.; Gubler, D.J.; Reiter, P.; Sanders, E.J.; Vorndam, A.V. Dengue and dengue haemorrhagic fever. Lancet. 1998, 352, 971–977.

2. Burke, D.S.; Monath, T.P. Flaviviruses. In Fields Virology; Knipe, D.M., Howley, P.M., Eds.; Lippincott Williams & Wilkins: Philadelphia, 2001, 1043–1126.
3. Solomon, T.; Mallewa, M.J. Dengue and other emerging flaviviruses. J Infect. 2001, 42, 104–115.
4. Chambers, T.J.; Hahn, C.S.; Galler, R.; Rice, C.M. Flavivirus genome organisation, expression and replication. Annu Rev Microbiol. 1990, 44, 649–688.
5. Mason, P.W.; Pincus, S.; Fournier, M.J.; Mason, T.L.; Shope, R.E.; Paoletti, E. Japanese encephalitis virus-vaccinia recombinants produce particulate forms of the structural proteins and induce high levels of protection against lethal JEV infection. Virology. 1991, 180, 294–305.
6. Mandl, C.W.; Guirakhoo, F.; Holzmann, H.; Heinz, F.X.; Kunz, C. Antigenic structure of the flavivirus envelope protein E at the molecular level, using tick-borne encephalitis virus as a model. J Virol. 1989, 63, 564–571.
7. Roehrig, J.T. Immunochemistry of dengue viruses. In Dengue and Dengue Hemorrhagic Fever; Gubler, D.J., Kuno, G., Eds.; CAB Int: Wallingford: CT, 1997, 199–220.
8. Rey, F.A.; Heinz, F.X.; Mandl, C.; Kunz, C.; Harrison, C. The envelope glycoprotein from tick-borne encephalitis virus at 2 Å resolution. Nature. 1995, 375, 291–298.
9. Kuhn, R.J.; Zhang, W.; Rossmann, M.G.; Pletnev, S.V.; Corver, J.; Lenches, E.; Jones, C.T.; Mukhopadhyay, S.; Chipman, P.R.; Strauss, E.G.; Baker, T.S.; Strauss, J.H. Structure of dengue virus: implications for flavivirus organization, maturation, and fusion. Cell. 2002, 108, 717–725.
10. Ni, H.; Barrett, A.D.T. Molecular differences between wild-type Japanese encephalitis virus strains of high and low mouse neuroinvasiveness. J Gen Virol. 1996, 77, 1449–1455.
11. McMinn, P.C. The molecular basis of virulence of the encephalitogenic flaviviruses. J Gen Virol. 1997, 78, 2711–2722.
12. Chen, Y.; Maguire, T.; Hileman, R.E.; Fromm, J.R.; Esko, J.D.; Linhardt, R.J.; Marks, R.M. Dengue virus infectivity depends on envelope protein binding to target cell heparan sulphate. Nature Med. 1997, 3, 866–871.
13. Roehrig, J.T.; Hunt, A.R.; Johnson, A.J.; Hawkes, R.A. Synthetic peptides derived from the deduced amino acid sequence of the E-glycoprotein of Murray Valley encephalitis virus elicit antiviral antibody. Virology. 1989, 171, 49–60.
14. Allison, S.L.; Schalich, J.; Stiasny, K.; Mandl, C.W.; Kunz, C.; Heinz, F.X. Oligomeric rearrangement of tick-borne encephalitis virus envelope proteins induced by an acidic pH. J Virol. 1995, 69, 695–700.
15. Blackwell, J.L.; Brinton, M.A. Translation elongation factor-1 alpha interacts with the 3′ stem-loop region of West Nile virus genomic RNA. J Virol. 1997, 71, 6433–6444.
16. Proutski, V.; Gould, E.A.; Holmes, E.C. Secondary structure of the 3′ untranslated region of flaviviruses: similarities and differences. Nucleic Acids Res. 1997, 25, 1194–1202.
17. Stadler, K.; Allison, S.L.; Schalich, J.; Heinz, F.X. Proteolytic activation of tick-borne encephalitis virus by furin. J Virol. 1997, 71, 8475–8481.
18. Hase, T.; Summers, P.L.; Eckels, K.H.; Baze, W.B. Maturation process of Japanese encephalitis virus in cultured mosquito cells *in vitro* and mouse brain cells *in vivo*. Arch Virol. 1987, 96, 135–151.
19. Hase, T.; Summers, P.L.; Eckels, K.H.; Baze, W.B. An electron and immunoelectron microscopic study of dengue-2 virus infection of cultured mosquito cells: maturation events. Arch Virol. 1987, 92, 273–291.
20. Lanciotti, R.S.; Lewis, J.G.; Gubler, D.J.; Trent, D.W. Molecular evolution and epidemiology of dengue-3 viruses. J Gen Virol. 1994, 75, 65–75.
21. Blok, J.; McWilliam, S.M.; Butler, H.C.; Gibbs, A.J.; Weiller, G.; Herring, B.L.; Hemsley, A.C.; Aaskov, J.G.; Yoksan, S.; Bhamarapravati, N. Comparison of a dengue-2 virus and its candidate vaccine derivative: sequence relationships with the flaviviruses and other viruses. Virology. 1992, 187, 573–590.

22. Trent, D.W.; Grant, J.A.; Monath, T.P.; Manske, C.L.; Corina, M.; Fox, G.E. Genetic variation and microevolution of dengue 2 virus in southeast Asia. Virology. 1989, 172, 523–535.
23. Lewis, J.A.; Chang, J.-.G.; Lanciotti, R.S.; Kinney, R.-.M.; Mayer, L.W.; Trent, D.-.W. Phylogenetic relationships of dengue-2 viruses. Virology. 1993, 197, 216–244.
24. Lanciotti, R.S.; Gubler, D.J.; Trent, D.W. Molecular evolution and phylogeny of dengue-4 viruses. J Gen Virol. 1997, 78, 2279–2284.
25. Guzman, M.G.; Deubel, V.; Pelegrino, J.L.; Rosario, D.; Marrero, M.; Sariol, C.; Kouri, G. Partial nucleotide and amino acid sequences of the envelope and the envelope/nonstructural protein-1 gene junction of four dengue-2 virus strains isolated during the 1981 Cuban epidemic. Am J Trop Med Hyg. 1995, 52, 241–246.
26. Moya, A.; Elena, S.F.; Bracho, A.; Miralles, R.; Barrio, E. The evolution of RNA viruses: a population genetics view. Proc Natl Acad Sci USA. 2000, 97, 6967–6973.
27. Rico-Hesse, R.; Harrison, L.M.; Nisalak, A.; Vaughn, D.W.; Kalayanarooj, S.; Green, S.; Rothman, A.L.; Ennis, F.A. Molecular evolution of dengue type 2 virus in Thailand. Am J Trop Med Hyg. 1998, 58, 96–101.
28. Wang, W.K.; Lin, S.R.; Lee, C.M.; King, C.C.; Chang, S.C. Dengue type 3 virus in plasma is a population of closely related genomes: quasispecies. J Virol. 2002, 76, 4662–4665.
29. Monath, T.P. Dengue: the risk to developed and developing countries. Proc Natl Acad Sci USA. 1994, 91, 2395–2400.
30. Innis, B. Dengue and dengue hemorrhagic fever. In Exotic Viral Infections; Porterfield, J.S., Ed.; Chapman & Hall: London, 1995, 103–146.
31. Halstead, S.B. The XXth century dengue pandemic. World Health Stat Q. 1992, 45, 292–298.
32. Gubler, D.J. Epidemic dengue/dengue hemorrhagic fever as a public health, social and economic problem in the 21st century. Trends Microbiol. 2002, 10, 100–103.
33. Wang, E.; Ni, H.; Xu, R.; Barrett, A.D.; Watowich, S.J.; Gubler, D.J.; Weaver, S.C. Evolutionary relationships of endemic/epidemic and sylvatic dengue viruses. J Virol. 2000, 74, 3227–3234.
34. Monath, T.P. Epidemiology of yellow fever: Current status and speculations on future trends. In: Factors in the emergence of arbovirus diseases; Saluzzo, J.F., Dodet, B., Eds.; Elsevier: Paris, 1997, 143–158.
35. Hawley, W.A.; Reiter, P.; Copeland, R.S.; Pumpuni, C.B.; Craig, G.B., Jr. Aedes albopictus in North America: probable introduction in used tires from northern Asia. Science. 1987, 236, 1114–1116.
36. Grist, N.R.; Burgess, N.R. Aedes and dengue. Lancet. 1994, 343, 477.
37. Gubler, D.J. Dengue and dengue hemorrhagic fever: its history and resurgence as a global public health problem. In Dengue and Dengue Hemorrhagic Fever; Gubler, D.J., Kuno, G., Eds.; CAB Int: Wallingford: CT, 1997, 1–22.
38. Rush, B. An account of the bilious remitting fever, as it appeared in Philadelphia in the summer and autumn of the year 1780. Medical Inquiries and Observations; 1789, 89–100.
39. Kautner, I.; Robinson, M.J.; Kuhnle, U. Dengue virus infection: epidemiology, pathogenesis, clinical presentation, diagnosis, and prevention. J Pediatr. 1997, 131, 516–524.
40. Cleland, J.B.; Bradley, B.; McDonald, W. Dengue fever in Australia. Its history and clinical course, its experimental transmission by Stegomyia fasciata, and the results of inoculation and other experiments. J Hyg (Lond). 1918, 16, 317–418.
41. Gubler, D.J.; Kuno, G.; Wareman, S.H. Neurological disorders associated with dengue fever infection. Proc Int Con on Dengue/DHF: Kualar Lumpur, 1983, 290.
42. Halstead, S.B.; Papaevangelou, G. Transmission of dengue 1 and 2 viruses in Greece in 1928. Am J Trop Med Hyg. 1980, 29, 635–637.
43. Hammon, W.M.; Rudnick, A.; Sather, G.; Rogers, K.D.; Morse, L.J. New hemorrhagic fevers of children in the Philippines and Thailand. Trans Assoc Am Physicians. 1960, 73, 140–155.
44. World Health Organization. Dengue Hemorrhagic Fever: Diagnosis, Treatment and Control: Geneva: WHO, 1986.

45. World Health Organization. Technical guides for diagnosis, treatment, surveillance, prevention and control of dengue hemorrhagic fever: Geneva, 1975.
46. Wulur, H.; Noegroho, S.; Tedjasukmana, T.; Lestadi, T.; Kho, L.K. Reyes's syndrome in children in Jakarta. Proc 15th SEAMEO-TROPMED Seminar: Tropical Paediatric Problems in Southeast Asia: Bangkok, 1976, 73–79.
47. Sumarmo; Wulur, H.; Jahja, E.; Gubler, D.J.; Sutomenggolo, T.S.; Sulianti Saroso, J. Encephalopathy associated with dengue infection. Lancet. 1978, 1, 449–450.
48. Tin, U.; Myo, A.; Than Nu. S.w.e.; Mimi, K.h.i.n.; Rosen, L. Dengue hemorrhagic fever with encephalitis symptoms. SEAMEO-TROPMED: Conference on Dengue Hemorrhagic Fever, Current Knowledge 1976: Bangkok, 1976, 1.
49. Kho, L.K.; Sumarmo; Wulur, H.; Jahja, E.C.; Gubler, D.J. Dengue haemorrhagic fever accompanied by encephalopathy in Jakarta. Southeast Asian J Trop Med Pub Health. 1981, 12, 583–590.
50. Thakare, J.; Walhekar, B.; Banerjee, K Haemorrhagic manifestations and encephalopathy in cases of dengue in India. Southeast Asian J Trop Med Pub Health. 1996, 3, 471–475.
51. Thisyakorn, U.; Thisyakorn, C. Dengue infection with unusual manifestations. J Med Assoc Thai. 1994, 77, 410–413.
52. Rajajee, S.; Mukundan, D. Neurological manifestations in dengue hemorrhagic fever. Indian Pediatr. 1994, 31, 688–690.
53. Hendarto, S.K.; Hadinegoro. S.R. Dengue encephalopathy. Acta Paediatr Jpn. 1992, 34, 350–357.
54. Chimelli, L.; Hahn, M.D.; Netto, M.B.; Dias, M.; Gray, F. Dengue: neuropathological findings in 5 fatal cases from Brazil. Clin Neuropathol. 1990, 9, 157–162.
55. Nimmannitya, S.; Thisyakorn, U.; Hemsrichart, V. Dengue haemorrhagic fever with unusual manifestations. Southeast Asian J Trop Med Pub Health. 1987, 18, 398–406.
56. Chen, W.J.; Huang, K.P.; Fang, A.H. Detection of IgM antibodies from cerebrospinal fluid and sera of dengue fever patients. Southeast Asian J Trop Med Pub Health. 1991, 22, 659–663.
57. Lum, L.C.S.; Lam, S.K.; Choy, S.; George, R.; Harun, F. Dengue encephalitis: a true entity?. Am J Trop Med Hyg. 1996, 54, 256–259.
58. Solomon, T.; Dung, N.M.; Vaughn, D.W.; Kneen, R.; Thao, L.T.T.; Raengsakulrach, B.; Day, N.P.J.; Farrar, J.; Myint, K.S.A.; Nisalak, A.; White, N.J. Neurological manifestations of dengue infection. Lancet. 2000, 355, 1053–1059.
59. Cam, B.V.; Fonsmark, L.; Hue, N.B.; Phuong, N.T.; Poulsen, A.; Heegaard, E.D. Prospective case-control study of encephalopathy in children with dengue hemorrhagic fever. Am J Trop Med Hyg. 2001, 65, 848–851.
60. Chokephaibulkit, K.; Kankirawatana, P.; Apintanapong, S.; Pongthapisit, V.; Yoksan, S.; Kositanont, U.; Puthavathana, P. Viral etiologies of encephalitis in Thai children. Pediatr Infect Dis J. 2001, 20, 216–218.
61. Miagostovich, M.P.; Ramos, R.G.; Nicol, A.F.; Nogueira, R.M.; Cuzzi-Maya, T.; Oliveira, A.V.; Marchevsky, R.S.; Mesquita, R.P.; Schatzmayr, H.G. Retrospective study on dengue fatal cases. Clin Neuropathol. 1997, 16, 204–208.
62. Ramos, C.; Sanchez, G.; Pando, R.H.; Baquera, J.; Hernandez, D.; Mota, J.; Ramos, J.; Flores, A.; Llausas, E. Dengue virus in the brain of a fatal case of hemorrhagic dengue fever. J Neurovirol. 1998, 4, 465–468.
63. Qiu, F.X.; Gubler, D.J.; Liu, J C.; Chen, Q.Q. Dengue in China: a clinical review. Bull World Health Organ. 1993, 71, 349–359.
64. Srivastava, V.K.; Suri, S.; Bhasin, A.; Srivastava, L.; Bharadwaj, M. An epidemic of dengue haemorrhagic fever and dengue shock syndrome in Delhi: a clinical study. Ann Trop Paediatr. 1990, 10, 329–334.
65. Guzman, M.G.; Kouri, G.P.; Bravo, J.; Soler, M.; Vazquez, S.; Santos, M.; Villaescusa, R.; Basanta, P.; Indan, G.; Ballester, J.M. Dengue haemorrhagic fever in Cuba. II. Clinical investigations. Trans Roy Soc Trop Med Hyg. 1984, 78, 239–241.

66. Kaplan, A.; Lingren, A. Neurological complications following dengue. US Navy Med Bull. 1945, 45, 506–510.
67. Brown, J.L.; Wilkinson, R.; Davidson, R.N.; Wall, R.; Lloyd, G.; Howells, J.; Pasvol, G. Rapid diagnosis and determination of duration of viraemia in dengue fever using a reverse transcriptase polymerase chain reaction. Trans Roy Soc Trop Med Hyg. 1996, 90, 140–143.
68. Row, D.; Weinstein, P.; Murray-Smith, S. Dengue fever with encephalopathy in Australia. Am J Trop Med Hyg. 1996, 54, 253–255.
69. Thisyakorn, U.; Thisyakorn, C.; Limpitikul, W.; Nisalak, A. Dengue infection with central nervous system manifestations. Southeast Asian J Trop Med Pub Health. 1999, 30, 504–506.
70. Pancharoen, C.; Thisyakorn, U. Neurological manifestations in dengue patients. Southeast Asian J Trop Med Public Health. 2001, 32, 341–345.
71. Brown, J.L.; Wilkinson, R.; Davidson, R.N.; Wall, R.; Lloyd, G.; Howells, J.; Pasvol, G. Rapid dignosis and determination of duration of viraemia in dengue fever using a reverse transcriptase polymerase chain reaction. Trans Roy Soc Trop Med Hyg. 1996, 90, 140–143.
72. Burke, D.S.; Lorsumrudee, W.; Leake, C.J.; Hoke, C.H.; Nisalak, A.; Laorakpongse, T. Fatal outcome in Japanese encephalitis. Am J Trop Med Hyg. 1985, 34, 1203–1209.
73. Familusi, J.B.; Moore, D.L.; Fomufod, A.K.; Causey, O.R. Virus isolates from children with febrile convulsions in Nigeria. A correlation study of clinical and laboratory observations. Clin Pediatr (Phila). 1972, 11, 272–276.
74. Sumarmo, O.R.; Wulur, H.; Jahja, E.; Gubler, D.J.; Suharyono, W.; Sorensen, K. Clinical observations on virologically confirmed fatal dengue infections in Jakarta, Indonesia. WHO Bull. 1983, 61, 693–701.
75. Pancharoen, C.; Thisyakorn, U. Dengue virus infection during infancy. Trans Roy Soc Trop Med Hyg. 2001, 95, 307–308.
76. Pancharoen, C.; Chansongsakul, T.; Bhattarakosol, P. Causes of fever in children with first febrile seizures: how common are human herpesvirus-6 and dengue virus infections?. Southeast Asian J Trop Med Pub Health. 2000, 31, 521–523.
77. Acevedo, J.; Casanova, M.F.; Antonini, A.C.; Morales, H. Acute polyneuritis associated with dengue. Lancet. 1982, i, 1357.
78. World Health Organization. Dengue Hemorrhagic Fever: Diagnosis, Treatment and Control, 2nd ed.: Geneva, WHO, 1997.
79. Halstead, S.B. Pathogenesis of dengue: challenges to molecular biology. Science. 1988, 239, 476–481.
80. Kouri, G.; Guzman, M.G.; Valdes, L.; Carbonel, I.; Vazquez, S.; Laferte, J.; Delgado, J.; Cabrera, M.V. Reemergence of dengue in Cuba: a 1997 epidemic in Santiago de Cuba. Emerg Infect Dis. 1998, 4, 89–92.
81. Vaughn, D.W. Invited commentary: dengue lessons from Cuba. Am J Epidemiol. 2000, 152, 800–803.
82. Halstead, S.B.; O'Rourke, E.J. Antibody-enhanced dengue virus infection in primate leukocytes. Nature. 1977, 265, 739–741.
83. Gollins, S.W.; Porterfield, J.S. Flavivirus infection enhancement in macrophages: an electron microscopic study of viral cellular entry. J Gen Virol. 1985, 66, 1969–1982.
84. Vaughn, D.W.; Green, S.; Kalayanarooj, S.; Innis, B.L.; Nimmannitya, S.; Suntayakorn, S.; Endy, T.P.; Raengsakulrach, B.; Rothman, A.L.; Ennis, F.A.; Nisalak, A. Dengue viremia titer, antibody response pattern, and virus serotype correlate with disease severity. J Infect Dis. 2000, 181, 2–9.
85. Sangkawibha, N.; Rojanasuphot, S.; Ahandrik, S.; Viriyapongse, S.; Jatanasen, S.; Salitul, V.; Phanthumachinda, B.; Halstead, S.B. Risk factors in dengue shock syndrome: a prospective epidemiologic study in Rayong, Thailand. Am J Epidemiol. 1984, 120, 653–669.
86. Rico-Hesse, R.; Harrison, L.M.; Salas, R.A.; Tovar, D.; Nisalak, A.; Ramos, C.; Boshell, J.; de Mesa, M.T.R.; Nogueira, R.M.R.; da Rosa, A.T. Origins of dengue type 2 virus associated with increased pathogenicity in the Americas. Virology. 1997, 230, 244–251.

87. Watts, D.M.; Porter, K.R.; Futvatana, P.; Vasquez, B.; Calampa, C.; Hayes, C.G.; Halstead, S.B. Failure of secondary infection with American genotype dengue 2 to cause dengue haemorrhagic fever. Lancet. 1999, 354, 1431–1434.
88. White, N.J. Variation in virulence of dengue virus. Lancet. 1999, 354, 1401–1402.
89. Leitmeyer, K.C.; Vaughn, D.; Watts, D.; Salas, R.; Villalobos, I.; Ramos, C.; Rico-Hesse, R. Dengue virus structural differences that correlate with pathogenesis. J Virol. 1999, 73, 4738–4747.
90. Chatuverdi, U.C.; Dhawan, R.; Khanna, M.; Mathur, A. Breakdown of the blood-brain barrier during dengue virus infection of mice. J Gen Virol. 1991, 72, 859–866.
91. Bethell, D.B.; Flobbe, K.; Phuong, C.X.T.; Day, N.P.J.; Phuong, P.Y.; Buurman, W.A.; Cardosa, M.J.; White, N.J.; Kwiatkowski, D. Pathophysiological and prognostic role of cytokines in dengue hemorrhagic fever. J Infect Dis. 1998, 177, 778–782.
92. Green, S.; Vaughn, D.W.; Kalayanarooj, S.; Nimmannitya, S.; Suntayakorn, S.; Nisalak, A.; Lew, R.; Innis, B.L.; Kurane, I.; Rothman, A.L.; Ennis, F.A. Early immune activation in acute dengue illness is related to development of plasma leakage and disease severity. J Infect Dis. 1999, 179, 755–762.
93. Zivna, I.; Green, S.; Vaughn, D.W.; Kalayanarooj, S.; Stephens, H.A.; Chandanayingyong, D.; Nisalak, A.; Ennis, F.A.; Rothman, A.L. T cell responses to an HLA-B*07-restricted epitope on the dengue NS3 protein correlate with disease severity. J Immunol. 2002, 168, 5959–5965.
94. Loke, H.; Bethell, D.B.; Phuong, C.X.; Dung, M.; Schneider, J.; White, N.J.; Day, N.P.; Farrar, J.; Hill, A.V. Strong HLA class I–restricted T cell responses in dengue hemorrhagic fever: a double-edged sword?. J Infect Dis. 2001, 184, 1369–1373.
95. Setiawan, M.W.; Samsi, T.K.; Wuler, H.; Sugianto, D.; Pool, T.N. Dengue haemorrhagic fever: ultrasound as an aid to predict the severity of the disease. Pediatr Radiol. 1998, 28, 1–4.
96. Kuo, C.-.H.; Tai, .D.-.I.; Chang-Chien, C.-.S.; Lan, C.-.K.; Chiou, S.-.S.; Liaw, Y.-.F. Liver biochemical tests and dengue fever. Am J Trop Med Hyg. 1992, 47, 265–270.
97. Iyngkaran, N.; Yadav, M.; Harun, F.; Kamath, K.R. Augmented tumour necrosis factor in Reye's syndrome associated with dengue virus. Lancet. 1992, 340, 1466–1467.
98. Bhamarapravati, N.; Tuchinda, P.; Boonyapaknavik, V. Pathology of Thailand hemorrhagic fever: a study of 100 autopsy cases. Ann Trop Med Parasitol. 1967, 61, 500–510.
99. Marianneau, P.; Cardona, A.; Edelman, L.; Deubel, V.; Despres, P. Dengue virus replication in human hepatoma cells activates NF-kappaB which in turn induces apoptotic cell death. J Virol. 1997, 71, 3244–3249.
100. Rosen, L.; Khin, M.M.; Tin, U. Recovery of virus from the liver of children with fatal dengue: reflections on the pathogenesis of the disease and its possible analogy with that of yellow fever. Res Virol. 1989, 140, 351–360.
101. Rosen, L.; Drouet, M.T.; Deubel, V. Detection of dengue virus RNA by reverse transcription-polymerase chain reaction in the liver and lymphoid organs but not in the brain in fatal human infection. Am J Trop Med Hyg. 1999, 61, 720–724.
102. Burke, T. Dengue haemorrhagic fever: a pathological study. Trans Roy Soc Trop Med Hyg. 1968, 62, 682–693.
103. Janssen, H.L.; Bienfait, H.P.; Jansen, C.L.; van Duinen, S.G.; Vriesendorp, R.; Schimsheimer, R.J.; Groen, J.; Osterhaus, A.D. Fatal cerebral oedema associated with primary dengue infection. J Infect. 1998, 36, 344–346.
104. Patey, O.; Ollivaud, J.; Breuil, J.; Lafaix, C. Unusual neurological manifestations occuring during dengue fever infection. Am J Trop Med Hyg. 1993, 48, 793–802.
105. Solomon, T. Viral encephalitis in southeast Asia. Neurol Infect Epidemiol. 1997, 2, 191–199.
106. Peluso, R.; Haase, A.; Stowring, L.; Edwards, M.; Ventura, P. A Trojan horse mechanism for the spread of Visna virus in monocytes. Virology. 1985, 147, 231–236.

107. Tardieu, M.; Hery, C.; Peudenier, S.; Boespflug, O.; Montagnier, L. Human immunodeficiency virus type 1– infected monocytic cells can destroy human neural cells after cell-to-cell adhesion. Ann Neurol. 1992, 32, 11–17.
108. Fraser, H.S.; Wilson, W.A.; Rose, E.; Thomas, E.J.; Sissons, J.G. Dengue fever in Jamaica with shock and hypocomplementaemia, haemorrhagic, visceral and neurological complications. West Indian Med J. 1978, 27, 106–116.
109. Solomon, T.; Cardosa, M.J. Emerging arboviral encephalitis. Br Med J. 2000, 321, 1484–1485.
110. Kawano, H.; Rostapshov, V.; Rosen, L.; Lai, C.J. Genetic determinants of dengue type 4 virus neurovirulence for mice. J Virol. 1993, 67, 6567–6575.
111. Bray, M.; Men, R.; Tokimatsu, I.; Lai, C.J. Genetic determinants responsible for acquisition of dengue type 2 virus mouse neurovirulence. J Virol. 1998, 72, 1647–1651.
112. Sistayanarain, A.; Maneekarn, N.; Polprasert, B.; Sirisanthana, V.; Makino, Y.; Fukunaga, T.; Sittisombut, N. Primary sequence of the envelope glycoprotein of a dengue type 2 virus isolated from patient with dengue hemorrhagic fever and encephalopathy. South east Asian J Trop Med Pub Health. 1996, 27, 221–227.
113. Halstead, S.B.; Scanlon, J.E.; Umpaivit, P.; Udomsakdi, S. Dengue and Chikungunya virus infection in man in Thailand, 1962–1964. V. Epidemiological observations outside of Bangkok. Am J Trop Med Hyg. 1969, 18, 997–1021.
114. Simpson, D.H. Arboviruses. In Manson's Tropical Diseases; Cook, G., Ed.; Saunders: London, 1996, 615–665.
115. Vaughn, D.W.; Green, S.; Kalayanarooj, S.; Innis, B.L.; Nimmannitya, S.; Suntayakorn, S.; Rothman, A.L.; Ennis, F.A.; Nisalak, A. Dengue in the early febrile phase: viremia and antibody responses. J Infect Dis. 1997, 176, 322–330.
116. Cao, X.T.; Ngo, T.N.; Wills, B.; Kneen, R.; Nguyen, T.T.; Ta, T.T.; Tran, T.T.; Doan, T.K.; Solomon, T.; Simpson, J.A.; White, N.J.; Farrar, J.J. Evaluation of the World Health Organization standard tourniquet test and a modified tourniquet test in the diagnosis of dengue infection in Viet Nam. Trop Med Int Health. 2002, 7, 125–132.
117. Halstead, S.B. Epidemiology of dengue and dengue hemorrhagic fever. In Dengue and Dengue Hemorrhagic Fever; Gubler, D.J., Kuno, G., Eds.; CAB Int: Wallingford: CT, 1997, 23–44.
118. Rigau-Perez, J.G.; Bonilla, G.L. An evaluation of modified case definitions for the detection of dengue hemorrhagic fever. Puerto Rico Association of Epidemiologists. PR Health Sci J. 1999, 18, 347–352.
119. Murgue, B.; Deparis, X.; Chungue, E.; Cassar, O.; Roche, C. Dengue: an evaluation of dengue severity in French Polynesia based on an analysis of 403 laboratory-confirmed cases. Trop Med Int Health. 1999, 4, 765–773.
120. Teasdale, G.; Jennett, B. Assessment of coma and impaired consciousness. A practical scale. Lancet. 1974, 2, 81–84.
121. Molyneux, M.E.; Taylor, T.E.; Wirima, J.J.; Borgstein, A. Clinical features and prognostic indicators in paediatric cerebral malaria: a study of 131 comatose Malawian children. Q J Med. 1989, 71, 441–459.
122. Butterworth, R.F. The neurobiology of hepatic encephalopathy. Semin Liver Dis. 1996, 16, 235–244.
123. Phillips, R.E.; Solomon, T. Cerebral malaria in children. Lancet. 1990, 336, 1355–1360.
124. Newton, C.; Hien, T.T.; White, N.J. Cerebral malaria. J Neurol Neurosurg Psychiatry. 2000, 69, 433–441.
125. Whitley, R.J.; Soong, S.; Linneman, C.; Liu, C.; Pazin, G.; Alford, C.A. Herpes simplex encephalitis. Clinical assessment. JAMA. 1982, 247, 317–320.
126. Innis, B.L. Japanese encephalitis. In Exotic Viral Infections; Porterfield, J.S., Ed.; Chapman & Hall: London, 1995, 147–174.
127. Davis, L.E. Acute viral meningitis and encephalitis. In Infections of the Nervous System; Kennedy, P.G.E., Johnson, R.T., Eds.; Butterworths: London, 1987, 156–176.

128. Solomon, T.; Kneen, R. Neurological presentations. In Lecture Notes on Tropical Medicine; Beeching, N., Gill, G., Eds., 5th ed.; Blackwell Sci: Oxford, 2003 (in press).
129. Warrell, D.A.; Looareesuwan, S.; Warrell, M.J.; Kasemsarn, P.; Intaraprasert, R.; Bunnag, D.; Harinasuta, T. Dexamethasone proves deleterious in cerebral malaria. A double-blind trial in 100 comatose patients. N Engl J Med. 1982, 306, 313–319.
130. Anonymous. Severe and complicated malaria. World Health Organization Malaria Action Programme. Trans Roy Soc Trop Med Hyg. 1986, 80(suppl), 3–50.
131. Reilly, P.L.; Simpson, D.A.; Sprod, R.; Thomas, L. Assessing the conscious level in infants and young children: a paediatric version of the Glasgow Coma Scale. Childs Nerv Syst. 1988, 4, 30–33.
132. Verity, C.M.; Butler, N.R.; Golding, J. Febrile convulsions in a national cohort followed up from birth. I. Prevalence and recurrence in the first five years of life. Br Med J. 1985, 290, 1307–1310.
133. Solomon, T.; Dung, N.M.; Kneen, R.; Thao, L.T.; Gainsborough, M.; Nisalak, A.; Day, N.P.; Kirkham, F.J.; Vaughn, D.W. Smith, S.; White, N.J. Seizures and raised intracranial pressure in Vietnamese patients with Japanese encephalitis. Brain. 2002, 125, 1084–1093.
134. Glasgow, J.F.; Middleton, B. Reye syndrome—insights on causation and prognosis. Arch Dis Child. 2001, 85, 351–353.
135. Solomon, T.; Felix, J.M.; Samuel, M.; Dengo, G.A.; Saldanha, R.A.; Schapira, A.; Phillips, R.E. Hypoglycaemia in paediatric admissions in Mozambique. Lancet. 1994, 343, 149–150.
136. Ho, T.W.; Mishu, B.; Li, C.Y.; Gao, C.Y.; Cornblath, D.R.; Griffin, J.W.; Asbury, A.K.; Blaser, M.J.; McKhann, G.M. Guillain-Barré syndrome in northern China. Relationship to *Campylobacter jejuni* infection and anti-glycolipid antibodies. Brain. 1995, 118, 597–605.
137. Bolton, C.F. The changing concepts of Guillain-Barré syndrome. N Engl J Med. 1995, 333, 1415–1416.
138. Solomon, T.; Vaughn, D.W. Clinical features and pathophysiology of Japanese encephalitis and West Nile virus infections In Current Topics in Microbiology and Immunology: Japanese Encephalitis and West Nile Virus Infections; Mackenzie, J.S., Barrett, A.D., Deubel, V., Eds.; Springer-Verlag: Berlin, 2002, 171–194.
139. George, R.; Liam, C.K.; Chua, C.T.; Lam, S.K.; Pang, T.; Geethan, R.; Foo, L.S. Unusual clinical manifestations of dengue virus infection. Southeast Asian J Trop Med Pub Health. 1988, 19, 585–590.
140. Barkan, H. The ocular complications of dengue fever. Am J Ophthalmol. 1919, 2, 650–652.
141. Asbury, A.K. Diagnostic considerations in Guillain-Barré syndrome. Ann Neurol. 1981, 9(suppl), 1–5.
142. Esack, A.; Teelucksingh, S.; Singh, N. The Guillain-Barre syndrome following dengue fever. West Indian Med J. 1999, 48, 36–37.
143. Paul, C.; Dupont, B.; Pialoux, G. Acute polyradiculoneuritis secondary to dengue. Presse Med. 1990, 19, 1503.
144. Solomon, T.; Kneen, R.; Dung N.M.; Khanh, V.C.; Thuy, T.N.; Ha, D.Q.; Day, P.J.; Nisalak, A.; Vaughn, D.W.; White, N.J. Poliomyelitis-like illness due to Japanese encephalitis virus. Lancet. 1998, 351, 1094–1097.
145. George, R.; Lum, C.S. Clinical spectrum of dengue infection. In Dengue and Dengue Hemorrhagic Fever; Gubler, D.J., Kuno, G., Eds.; CAB Int: New York, 1997, 89–113.
146. Misra, U.K.; Kalita, J.; Jain, S.K. Mathur, A. Radiological and neurophysiological changes in Japanese encephalitis. J Neurol Neurosurg Psychiatry. 1994, 57, 1484–1487.
147. Gomes, J.A.C.; Hernandez, O.J.L.; Perez, E.H. Alteraciones electroencephalografias in dengue. Rev Cubana Med. 1984, 23, 468.
148. Rosen, L.; Gubler, D.J. The use of mosquitoes to detect and propagate dengue viruses. Am J Trop Med Hyg. 1974, 23, 1153–1160.
149. Lanciotti, R.S.; Calisher, C.H.; Gubler, D.J.; Chang, G.-J.; Vorndam, A.V. Rapid detection and typing of dengue viruses from clinical samples by using reverse transcriptase-polymerase chain reaction. J Clin Microbiol. 1992, 30, 545–551.

150. Callahan, J.D.; Wu, S.J.; Dion-Schultz, A.; Mangold, B.E.; Peruski, L.F.; Watts, D.M.; Porter, K.R.; Murphy, G.R.; Suharyono, W.; King, C.C.; Hayes, C.G.; Temenak, J.J. Development and evaluation of serotype- and group-specific fluorogenic reverse transcriptase PCR (TaqMan) assays for dengue virus. J Clin Microbiol. 2001, 39, 4119–4124.
151. Innis, B.L.; Nisalak, A.; Nimmannitya, S.; Kusalerdchariya, S.; Chongswasdi, V.; Suntayakorn, S.; Puttisri, P.; Hoke, C.H. An enzyme-linked immunosorbent assay to characterize dengue infections where dengue and Japanese encephalitis co-circulate. Am J Trop Med Hyg. 1989, 40, 418–427.
152. Vaughn, D.W.; Nisalak, A.; Solomon, T.; Kalayanarooj, K.; Dung, N.M.; Kneen, R.; Cuzzubbo, A.; Devine, P.L. Rapid serological diagnosis of dengue virus infection using a commercial capture ELISA that distinguishes primary and secondary infections. Am J Trop Med Hyg. 1999, 60, 693–698.
153. Cuzzubbo, A.J.; Vaughn, D.W.; Nisalak, A.; Solomon, T.; Kalayanarooj, S.; Aaskov, J.; Dung, N.M.; Devine, P.L. Comparison of PanBio Dengue Duo enzyme-linked immunosorbent assay (ELISA) and MRL dengue fever virus immunoglobulin M capture ELISA for diagnosis of dengue virus infections in southeast Asia. Clin Diagn Lab Immunol. 1999, 6, 705–712.
154. Cardosa, M.J.; Baharudin, F.; Hamid, S.; Hooi, T.P.; Nimmannitya, S. A nitrocellulose membrane based IgM capture enzyme immunoassay for etiologic diagnosis of dengue virus infections. Clin Diagn Virol. 1995, 3, 343–350.
155. Vaughn, D.W.; Nisalak, A.; Kalayanarooj, S.; Solomon, T.; Dung, N.M.; Cuzzubbo, A.; Devine, P.L. Evaluation of a rapid immunochromatographic test for the diagnosis of dengue infection. J Clin Microbiol. 1998, 26, 234–238.
156. Lin, F.C.; Vorndam, V.; Vaughn, D.W.; Endy, T.P.; Nisalak, A.; Innis, B.L.; Huang, J.H.; Chen, H.Y.; Ren, J.; Roehring, J.T. Immunodetection of dengue virus in human sera collected during the febrile phase. Am Soc Trop Med Hyg Annual Meeting: Washington: DC, 1999.
157. Nimmannitya, S. Clinical spectrum and management of dengue haemorrhagic fever. Trans Roy Soc Trop Med Hyg. 1987, 18, 292–297.
158. Dung, N.M.; Day, N.P.; Tam, D.T.; Loan, H.T.; Chau, H.T.; Minh, L.N.; Diet, T.V.; Bethell, D.B.; Kneen, R.; Hien, T.T.; White, N.J.; Farrar, J.J. Fluid replacement in dengue shock syndrome: a randomized, double-blind comparison of four intravenous-fluid regimens. Clin Infect Dis. 1999, 28, 787–794.
159. Ngo, N.T.; Cao, X.T.; Kneen, R.; Wills, B.; Nguyen, V.M.; Nguyen, T.Q.; Chu, V.T.; Nguyen, T.T.; Simpson, J.A.; Solomon, T.; White, N.J.; Farrar, J. Acute management of dengue shock syndrome: a randomized double-blind comparison of 4 intravenous fluid regimens in the first hour. Clin Infect Dis. 2001, 32, 204–213.
160. Tassniyom, S.; Vasanawathana, S.; Chirawatkul, A.; Rojanasuphot, S. Failure of high-dose methylprednisolone in established dengue shock syndrome: a placebo-controlled, double-blind study. Pediatrics. 1993, 92, 111–115.
161. Sumarmo; Talogo, W.; Asrin, A.; Isnuhandojo, B.; Sahudi, A. Failure of hydrocortisone to affect outcome in dengue shock syndrome. Pediatrics. 1982, 69, 45–49.
162. Tassniyom, S.; Vasanawathana, S.; Dhiensiri, T.; Nisalak, A.; Chirawatkul, A. Failure of carbazochrome sodium sulfonate (AC-17) to prevent dengue vascular permeability or shock: a randomized, controlled trial. J Pediatr. 1997, 131, 525–528.
163. Leyssen, P.; De Clercq, E.; Neyts, J. Perspectives for the treatment of infections with Flaviviridae. Clin Microbiol Rev. 2000, 13, 67–82.
164. Newton, C.R.; Kirkham, F.J.; Johnston, B.; Marsh, K. Inter-observer agreement of the assessment of coma scales and brainstem signs in non-traumatic coma. Dev Med Child Neurol. 1995, 37, 807–813.
165. Advanced Life Support Group. Recognition of the seriously ill child. Advanced Paediatric Life Support; BMJ: London, 1997, 13–18.
166. Kirkham, F.J. Non-traumatic coma in children. Arch Dis Child. 2001, 85, 303–312.

167. Crawley, J.; Smith, S.; Kirkham, F.; Muthinji, P.; Waruiri, C.; Marsh, K. Seizures and status epilepticus in childhood cerebral malaria. Q J Med. 1996, 89, 591–597.
168. Appleton, R.; Choonara, I.; Martland, T.; Philips, B.; Scott, R.; Whitehouse, W. The treatment of convulsive status epilepticus in children. The Status Epilepticus Working Group. Arch Dis Child. 2000, 83, 415–419.
169. Crawley, J.; Waruiru, C.; Mithwani, S.; Mwangi, I.; Watkins, W.; Ouma, D.; Winstanley, P.; Peto, T.; Marsh, K. Effect of phenobarbital on seizure frequency and mortality in childhood cerebral malaria: a randomised, controlled intervention study. Lancet. 2000, 355, 701–706.
170. Advanced Life Support Group. Advanced Paediatric Life Support, 2nd ed.; BMJ: London, 1997.
171. Newton, C.R.J.C.; Crawley, J.; Sowumni, A.; Waruiru, C.; Mwangi, I.; English, M.; Murphy, S.; Winstanley, P.A.; Marsh, K.; Kirkham, F.J. Intracranial hypertension in Africans with cerebral malaria. Arch Dis Child. 1997, 76, 219–226.
172. Tassniyom, S.; Chirawatkul, A.; Rojanasuphot, S. Failure of high-dose steroids in averting death in severe dengue shock syndrome: a double-blind randomised controlled trial. Southeast Asian J Trop Med Pub Health. 1990, 21, 698.
173. Hoke, C.H.; Vaughn, D.W.; Nisalak, A.; Intralawan, P.; Poolsuppasit, S.; Jongsawas, V.; Titsyakorn, U.; Johnson, R.T. Effect of high dose dexamethasone on the outcome of acute encephalitis due to Japanese encephalitis virus. J Infect Dis. 1992, 165, 631–637.
174. Barrett, A.D. Current status of flavivirus vaccines. Ann NY Acad Sci. 2001, 951, 262–271.
175. Bhamarapravati, N.; Sutee, Y. Live attenuated tetravalent dengue vaccine. Vaccine. 2000, 18(suppl 2), 44–47.
176. Bhamarapravati, N.; Yoksan, S. Live attentuated tetravalent dengue vaccine. In Dengue and Dengue Hemorrhagic Fever; Gubler, D.J., Kuno, G., Eds.; CAB Int: Wallingford: UK, 1997, 367–377.
177. Kanesa-thasan, N.; Sun, W.; Kim-Ahn, G.; Van Albert, S.; Putnak, J.R.; King, A.; Raengsakulsrach, B.; Christ-Schmidt, H. Gilson, K.; Zahradnik, J.M.; Vaughn, D.W.; Innis, B.L.; Saluzzo, J.F.; Hoke, C.H., Jr. Safety and immunogenicity of attenuated dengue virus vaccines (Aventis Pasteur) in human volunteers. Vaccine. 2001, 19, 3179–3188.
178. Trent, D.W.; Kinney, R.M.; Huang, C.-.H. Recombinant dengue virus vaccines. In Dengue and Dengue Hemorrhagic Fever; Gubler, D.J., Kuno, G., Eds.; CAB Int: Wallingford: UK, 1997, 379–403.
179. Monath, T.P.; McCarthy, K.; Bedford, P.; Johnson, C.T.; Nichols, R.; Yoksan, S.; Marchesani, R.; Knauber, M.; Wells, K.H.; Arroyo, J.; Guirakhoo, F. Clinical proof of principle for ChimeriVax: recombinant live, attenuated vaccines against flavivirus infections. Vaccine. 2002, 20, 1004–1018.
180. Guirakhoo, F.; Pugachev, K.; Arroyo, J.; Miller, C.; Zhang, Z.X.; Weltzin, R.; Georgakopoulos, K.; Catalan, J.; Ocran, S.; Draper, K.; Monath, T.P. Viremia and immunogenicity in nonhuman primates of a tetravalent yellow fever-dengue chimeric vaccine: genetic reconstructions, dose adjustment, and antibody responses against wild-type dengue virus isolates. Virology. 2002, 298, 146–159.
181. Kochel, T.J.; Raviprakash, K.; Hayes, C.G.; Watts, D.M.; Russell, K.L.; Gozalo, A.S.; Phillips, I.A.; Ewing, D.F.; Murphy, G.S.; Porter, K.R. A dengue virus serotype-1 DNA vaccine induces virus neutralizing antibodies and provides protection from viral challenge in Aotus monkeys. Vaccine. 2000, 18, 3166–3173.
182. Shlim, D.R.; Solomon, T. Japanese encephalitis vaccine for travelers: exploring the limits of risk. Clin Infect Dis. 2002, 35, 183–188.
183. Lindenbach, B.D.; Rice, C.M. Flaviviridae: the viruses and their replication. In Fields Virology; Knipe, D.M., Howley, P.M., Eds.; Lippincott Williams & Wilkins: Philadelphia, 2001, 991–1041.
184. Heinz, F.X.; Allison, S.L. The machinery for flavivirus fusion with host cell membranes. Curr Opin Microbiol. 2001, 4, 450–455.

185. Solomon, T. Viral haemorrhagic fevers. In Manson's Tropical Diseases; Cook, G., Zumlar, A., Eds., 21st ed.; Saunders: London, 2002.
186. Kuberski, T.; Rosen, L.; Reed, D.; Mataika, J. Clinical and laboratory observations on patients with primary and secondary dengue type 1 infections with hemorrhagic manifestations in Fiji. Am J Trop Med Hyg. 1977, 26, 775–783.

23

Nipah Encephalitis

Chong-Tin Tan and Kum-Thong Wong
University of Malaya
Kuala Lumpur, Malaysia

1 INTRODUCTION

From September 1998 to June 1999, there was an outbreak of viral encephalitis in several pig farming villages in Malaysia [1–4]. The outbreak started in Ipoh, a town with a population of about 700,000 in northern peninsular Malaysia. It later involved Sungai Nipah and its surrounding villages south of Kuala Lumpur. The outbreak subsequently spread to involve abattoir workers in Singapore [5]. In Malaysia, more than 265 patients were affected nationwide with more than 105 mortalities [4,7].

2 DIFFERENTIAL DIAGNOSIS FROM JAPANESE ENCEPHALITIS

Because the outbreak involved pig farm workers, it was initially thought to be due to Japanese encephalitis. However, several features distinguished this outbreak from Japanese encephalitis. Infection was predominantly in adults rather than children, and there was a clustering of cases in members of the same household, which suggests an infection with a high disease attack rate, as opposed to Japanese encephalitis virus, which causes symptomatic encephalitis in only about one in 300 of those infected. A high proportion of patients were in direct contact with pigs, as opposed to uninfected individuals living in the same neighborhood, thus arguing against a mosquito-borne disease. Finally, there was a history of illness in the pigs belonging to affected farmers.

Isolation of a new paramyxovirus from cerebrospinal fluid specimens of several patients indicated that this was the etiological agent [3,4]. The virus was subsequently named Nipah virus, after Sungai Nipah, the village in which the patients whose specimens yielded the first viral isolates lived. Viral genomic sequencing has now established Nipah virus as a new paramyxovirus closely related to Hendra virus [4,8]. Hendra virus caused disease among horses and affected three patients in Australia in 1994 and 1995 [9,10]. There is a high degree of nucleotide homology in the open reading frames of the various genes of Hendra virus and Nipah virus that exceeds 70% and a amino acid identity of more than 80% in most genes [4,8].

3 EPIDEMIOLOGY

It has now been firmly established that close contact with pigs was responsible for viral transmission to humans [7,11,12]. Case control studies of the local community and household members have shown that patients were more likely to perform activities requiring direct contact with pigs [7], to have been involved with full-time pig farming [11], and to have had contact with live pigs [12]. A case was reported in which the patient was never in close proximity to pigs. However, this patient had personally nursed two sick pet dogs that had subsequently died. Transmission may thus also be from infected dogs [11].

Transmission of Nipah virus to healthcare workers was thought to be generally low. In a survey of 288 health workers, only three were found to have IgG antibodies [13]. There is a report of a nurse who had previously cared for Nipah encephalitis patients and subsequently seroconverted but remained asymptomatic. The magnetic resonance imaging of her brain showed multiple, discrete high signal lesions like those seen in acute Nipah virus encephalitis [14,15]. Therefore, human-to-human transmission is possible, as exemplified by this case and the fact that virus could be isolated from patients' respiratory secretions and urine [16].

4 CLINICAL MANIFESTATIONS

Of the 94 patients with symptomatic Nipah virus infection seen in the University of Malaya Medical Centre during the outbreak, more than half had affected family members, suggesting a disease of high infection rate [17]. In a study of 14 households with 110 members in the outbreak area, 27% had symptomatic Nipah infection and another 8% had subclinical infection with seroconversion. This suggests a ratio of roughly 3:1 for symptomatic versus asymptomatic infection [11]. The main demographic features of 103 patients treated in another hospital were 88% males, mean age of 38 years, and 78% pig farmers or hired workers [18].

The incubation period was less than 2 weeks in 92% of patients [17]. The clinical manifestation was that of an acute encephalitis with fever, headache, vomiting, and reduced level of consciousness [3,5,6,17,18]. Distinctive clinical features were areflexia, hypotonia, prominent autonomic changes such as tachycardia, and hypertension. Segmental myoclonus found in 32% of patients was characterized by focal, rhythmic jerking of muscles, commonly involving the diaphragm and anterior muscles of the neck [17].

Respiratory tract involvement with cough was seen at presentation in 14% of patients [18]. Some patients had nonencephalitic infection with seroconversion and systemic symptoms but no evidence of encephalitis. Of the 94 patients admitted to the University of Malaya Medical Centre with Nipah virus infection, 91 had acute encephalitis and three had nonencephalitic infection [17].

The overall mortality of acute Nipah encephalitis was 40% [4]. Severe brainstem involvement suggested by tachycardia and abnormal doll's-eye reflex [17], hypertension, and high fever [18] appeared to be associated with poor prognosis. Presence of virus in the cerebrospinal fluid, which suggested high viral replication, was also associated with high mortality [19]. Concomitant diabetes mellitus, but not the level of exposure to sick animals, was also related to mortality, probably due to immunoparesis [20].

5 LABORATORY, RADIOLOGICAL, AND OTHER DIAGNOSTIC INVESTIGATIONS

Thrombocytopenia was a feature in 30% of patients, whereas leukopenia was present in 11%. Alanine and aspartate aminotransferases were elevated in 33% and 42% of patients, respectively, but blood urea, creatinine, and electrolyte levels remained normal in all patients [17]. These changes were attributed to nonspecific systemic changes in very ill patients [17,18].

Cerebrospinal fluid examination was abnormal in 75% of patients, with elevated protein levels or elevated white cell counts. Glucose levels were within normal limits [17,18]. These features are nonspecific and may be found in many primary viral encephalitides.

IgM and IgG antibody detection in serum and CSF were critical to the diagnosis of Nipah virus infection. The antibody test used an IgM-capture enzyme-linked immunosorbent assay (ELISA) test, and IgG antibodies were detected by an indirect IgG ELISA assay [1,3]. The ELISA test has a high specificity and is therefore useful as a screening test [21]. The rate of positive IgM was 60–71% by day 4 and 100% by day 12 of illness. For IgG, it was 7–29% by day 1 and 100% by day 25 or 26 of the illness [22].

There was a high frequency of positive cerebrospinal fluid IgM to Japanese encephalitis of up to 9% in acute Nipah encephalitis patients. This probably reflected the endemicity of Japanese encephalitis infection. The breakdown in the blood-brain barrier associated with disseminated vasculitis seen in Nipah encephalitis could possibly have contributed to the high rate of positive Japanese encephalitis cerebrospinal fluid IgM in these patients [23].

Brain MR imaging proved to be a useful diagnostic aid in acute encephalitis [24,25]. Typically, in acute Nipah virus encephalitis the brain MR showed multiple disseminated, small discrete hyperintense lesions, best seen in the FLAIR sequence mainly in the subcortical and deep white matter and occasionally in the cortex. The lesions, which measured about 2–7 mm in diameter, are likely to correspond to the microinfarctions noted in postmortem tissues. Similar changes were also seen in 16% of asymptomatic patients with Nipah virus infection, showing that subclinical cerebral involvement is not uncommon in these patients [14].

The most common electroencephalographic abnormality was continuous diffuse, symmetrical slowing with or without focal discharges. The degree of slowing correlated with severity of disease. Independent bitemporal periodic complexes were common among those who became deeply comatose and were associated with 100% mortality [26].

6 TREATMENT

Ribavirin, a very broad-spectrum virustatic antiviral agent, which shows varying degrees of efficacy against viruses such as respiratory syncytial virus, influenza, and measles, was tried on an empirical basis in the patients. In an open-label trial of 140 patients with 54 patients as controls, there were 45 deaths (32%) in the ribavirin group versus 29 deaths (54%) in the control arm. This represented a reduction in mortality of 36%. This trial suggests that ribavirin may be useful in the treatment of acute Nipah encephalitis. There were no apparent serious side effects in this study [27].

7 PATHOLOGY AND PATHOGENESIS OF ACUTE NIPAH VIRUS INFECTION

A total of 31 fatal acute encephalitis cases were studied by light and electron microscopy and immunohistochemistry [3,4,28]. The blood vessels appeared to be an early major target of Nipah virus infection. Medium-sized to small blood vessels in major organs, including the brain, lung, and kidney, were susceptible to infection. The earliest lesion seemed to be the formation of multinucleated syncytium in the endothelium. More commonly, vascular damage takes the form of endothelial ulceration with varying degrees of inflammation and fibrinoid necrosis. This vasculitis was frequently associated with thrombosis and vascular occlusion. Staining by immunohistochemistry confirmed that Nipah virus infected blood vessels directly.

The brain, including the meninges and gray and white matter, showed widespread vasculitis and is the organ most severely affected. Areas of necrosis and ischemia were seen adjacent to the vasculitis and were thought to be due to vascular occlusion. Surviving neurons in these areas may reveal eosinophilic cytoplasmic and/or, less frequently, nuclear paramyxoviral type inclusions. Immunohistochemical staining confirmed that these viral inclusions contained Nipah virus antigens. Parenchymal inflammation consisting of perivascular cuffing and neuronophagia could also be seen.

Outside the central nervous system, except perhaps in the liver, vasculitis could be found in all the major organs, including the lung, kidney, and heart. However, it was less severe than that in the brain.

Severe central nervous involvement explained why symptomatic patients usually presented with an acute encephalitic syndrome in which a combination of ischemia, microinfarction, and direct neuronal infection resulted in neurological manifestations. Other than immunohistochemical staining confirmation of viral antigens in neurons, the contribution of direct neuronal infection to neurological manifestations is also evidenced by distinctive neurological signs such as segmental myoclonus [17] and the association between high mortality and the presence of virus in the cerebrospinal fluid [19].

8 RELAPSE AND LATE-ONSET NIPAH ENCEPHALITIS

A small number of patients suffered a second or even a third neurological episode following what appeared to be complete recovery from acute infection. These relapse Nipah encephalitis patients constitute 7.5% of the total number of survivors [29]. Symptoms appeared after an average of 8.4 months following viral exposure. In addition, about 3.4% who were either asymptomatic or had only mild nonencephalitic illness initially also developed similar neurological episodes (late-onset Nipah encephalitis) for the first time several months later. Clinical, radiological, and pathological findings suggested that relapse and late-onset Nipah encephalitis are essentially the same disease process and distinct from acute Nipah virus encephalitis. The common clinical features in relapse and late-onset encephalitis were fever, headache, seizures, and focal neurological signs. There was an 18% mortality. Magnetic resonance imaging typically showed patchy areas of confluent cortical lesions [29].

Necropsy showed focal confluent encephalitis in which the demonstration of neuronal viral antigen in brains suggested that relapse and late-onset Nipah encephalitis were due to a recurrent infection rather than postinfectious demyelination such as is described following measles or other viral infections [29]. Relapse Nipah encephalitis is probably analogous to the single human case of Hendra virus encephalitis in which 13 months following meningitis associated with drowsiness, the patient developed fatal encephalitis [10]. Hendra virus antigen was demonstrated in the brain tissues. Nonetheless, in both relapse Nipah encephalitis and Hendra encephalitis, the respective viruses have not been isolated thus far [10,29].

9 THE BAT AS RESERVOIR HOST: IMPLICATIONS FOR FUTURE OUTBREAKS

There is increasing evidence that the reservoir of Nipah virus is very likely the fruit bat (*Pteropus hypomelanus*). In an island off the coast of peninsular Malaysia, Nipah virus has been isolated from the urine of roosting bats [30]. Serum neutralizing antibodies were found in 4–31% of bat species. These may represent antibodies raised against Nipah virus or another yet unidentified but related virus [31].

If the fruit bat is the reservoir host, this begs the question of how Nipah virus came to be transmitted to pigs and other animals. One theory is that half-eaten fruit dropped off near pig farms may have enough viruses to infect an animal that subsequently ingests the fruit. Indeed, viruses have been isolated from such fruits [32]. Whatever the mode of transmission from bats to pigs, pigs played the main role of amplifying hosts for the virus. The close proximity of pigs in most Malaysian pig farms probably contributed significantly to pig-to-pig transmission.

REFERENCES

1. Outbreak of Hendra-like virus—Malaysia and Singapore, 1998–1999. MMWR. 1999, 48, 265–269.
2. Update: outbreak of Nipah virus—Malaysia and Singapore, 1999. MMWR. 1999, 48, 335–337.
3. Chua, K.B.; Goh, K.J.; Wong, K.T. Fatal encephalitis due to Nipah virus among pig-farmers in Malaysia. Lancet. 1999, 354, 1257–1259.
4. Chua, K.B.; Bellini, W.J.; Rota, W.J. Nipah virus: a recently emergent deadly paramyxovirus. Science. 2000, 288, 1432–1435.
5. Paton, N.; Leo, Y.S.; Zaki, S.R. Outbreak of Nipah-virus infection among abattoir workers in Singapore. Lancet. 1999, 354, 1253–1256.
6. Lee, K.E.; Umapathi, T.; Tan, C.B. The neurological manifestations of Nipah virus encephalitis, a novel paramyxovirus. Ann Neurol. 1999, 46, 428–432.
7. Parashar, U.D.; Lye, M.S.; Ong, F. Case-control study of risk factors for human infection with the new zoonotic paramyxovirus, Nipah virus, during a 1998–1999 outbreak of severe encephalitis in Malaysia. J Infect Dis. 2000, 181, 1755–1759.
8. Harcourt, B.; Tamin, A.; Ksiazek, T.G. Molecular characterization of Nipah virus, a newly emergent paramyxovirus. Virology. 2000, 271, 334–349.
9. Selvey, L.; Wells, R.M.; McCormack, J.G Infection of humans and horses by a newly described morbillivirus. Med J Aust. 1995, 162, 642–645.
10. O'Sullivan, J.D.; Allworth, A.M.; Paterson, D.L. Fatal encephalitis due to novel paramyxovirus transmitted from horses. Lancet. 1997, 349, 93–95.

11. Tan, K.S.; Tan, C.T.; Goh, K.J. Epidemiological aspects of Nipah virus infection. Neurol J Southeast Asia. 1999, 4, 77–81.
12. Chew, M.H.L.; Arguin, P.M.; Shay, D.K. Risk factors for Nipah virus infection among abattoir workers in Singapore. J Infect Dis. 2000, 181, 1760–1763.
13. Mounts, A.W.; Kaur, H.; Parashar, U.D. A cohort study of health care workers to assess nosocomial transmissibility of Nipah virus, Malaysia, 1999. J Infect Dis. 2001, 183, 810–813.
14. Tan, K.S.; Ahmad Sarji, S.; Tan, C.T. Patients with asymptomatic Nipah virus infection may have abnormal cerebral MR imaging. Neurol J Southeast Asia. 2000, 5, 69–73.
15. Tan, C.T.; Tan, K.S. Nosocomial transmissibility of Nipah virus. J Infect Dis. 2001, 184, 1367.
16. Chua, K.B.; Lam, S.K.; Goh, K.J. The presence of Nipah virus in respiratory secretions and urine of patients during an outbreak of Nipah virus encephalitis in Malaysia. J Infect. 2001, 42, 40–43.
17. Goh, K.J.; Tan, C.T.; Chew, N.K. Clinical features of Nipah virus encephalitis among pig farmers in Malaysia. N Engl J Med. 2000, 342, 1229–1235.
18. Chong, H.T.; Kunjappan, S.R.; Thayaparan, T. Nipah encephalitis outbreak in Malaysia, clinical features in patients from Seremban. Neurol J Southeast Asia. 2000, 5, 61–67.
19. Chua, K.B.; Lam, S.K.; Tan, C.T. High mortality in Nipah encephalitis is associated with presence of virus in cerebrospinal fluid. Ann Neurol. 2000, 48, 802–805.
20. Chong, H.T.; Tan, C.T.; Goh, K.J. Occupational exposure, age, diabetes mellitus and outcome of acute Nipah encephalitis. Neurol J Southeast Asia. 2001, 6, 7–11.
21. Daniels, P.; Ksiazek, T.G. Laboratory diagnosis of Nipah and Hendra infections. Microbes Infect. 2001, 3, 289–295.
22. Ramasundrum, V.; Tan, C.T.; Chua, K.B. Kinetics of IgM and IgG seroconversion in Nipah virus infection. Neurol J Southeast Asia. 2000, 5, 23–28.
23. Chong, H.T.; Tan, C.T.; Karim, N. Outbreak of Nipah encephalitis among pig-farm workers in Malaysia in 1998/1999: Was there any role for Japanese encephalitis? Neurol J Southeast Asia. 2001, 6, 129–134.
24. Ahmad Sarji, S.; Abdullah, B.J.J.; Goh, K.J. Magnetic resonance imaging features of Nipah encephalitis. Am J Roentgenol. 2000, 175, 437–442.
25. Lim, C.; Sitoh, Y.Y. Nipah viral encephalitis or Japanese encephalitis? MR findings in a new zoonotic disease. Am J Neuroradiol. 2000, 21, 455–461.
26. Chew, N.K.; Goh, K.J.; Tan, C.T. Electroencephalography in acute Nipah encephalitis. Neurol J Southeast Asia. 1999, 4, 45–51.
27. Chong, H.T.; Kamarulzaman, A.; Tan, C.T. Treatment of acute Nipah encephalitis with ribavirin. Ann Neurol. 2001, 49, 810–813.
28. Wong, K.T.; Shieh, W.J.; Kumar, S., et al. Nipah virus infection: pathology and pathogenesis of an emerging paramyxoviral zoonosis. Am J Pathol. 2002, 161, 2153–2167.
29. Tan, C.T.; Goh, K.J.; Wong, K.T. Relapse and late-onset Nipah encephalitis. Ann Neurol. 2002, 51, 703–708.
30. Chua, K.B.; Lam, S.K. Reservoir of Nipah virus identified. Presented at Int Conf Emerging Infectious Diseases (ICEID 2000): Atlanta: GA, 2002.
31. Johara, M.; Field, H.; Rashdi, A.M. Nipah virus infection in bats (order Chiroptera) in peninsular Malaysia. Emerg Infect Dis. 2001, 7, 439–441.
32. Chua, K.B.; Koh, C.L.; Hooi, P.S. Isolation of Nipah virus from Malaysian island flying foxes. Microbes Infect. 2002, 4, 145–151.

24

Von Economo's Encephalitis

Joseph R. Berger
University of Kentucky College of Medicine
Lexington, Kentucky, U.S.A.

Isabella C. Glitza
University of Heidelberg
Heidelberg, Germany

> *Humanity has but three great enemies, fever, famine, and war. Of these, by far the most terrible, is fever.*
>
> —Sir William Osler

1 INTRODUCTION

The illness referred to as von Economo's encephalitis has been referred to by a large number of other names, including epidemic encephalitis, lethargic encephalitis, encephalitis lethargica, sleeping sickness, sleepy sickness, Schlafkrankheit, Schlummerkrankheit, von Economo's disease, or simply Economo's disease. Some have named it on the basis of the region of brain chiefly involved. For instance, Kinnier Wilson referred to it as ''mesencephalitis'' and Bernard Sachs as ''basilar encephalitis'' [1]. The illness spread in epidemic fashion throughout Europe beginning in the winter of 1916–1917. In addition to its epidemic nature, this polioencephalitis of the brainstem exhibited a polymorphic clinical expression, with some variability from place to place and from epidemic to epidemic, each individual case having an irregular and indeterminate course, usually resulting in lingering and permanent sequelae.

McKenzie [2] stated in 1927,

> *There is nothing in the history of medicine to compare with the phantasmagoria of disorder manifested in the course of this strange malady. . . . Into the maze*

> *of contradictory phenomena it seemed almost impossible to read anything like a rationalized order of events which might be termed a disease entity.*

2 HISTORY

Constantin Baron Economo von San Serff, of Greek parentage, was born on August 21, 1876, in Braila, Romania, and raised in Trieste, Austria [3]. His interests included the anatomy and physiology of the midbrain, pons, and trigeminal nerve pathway. In 1917 he started his monumental studies on encephalitis lethargica that established his fame worldwide. He published effusively on this condition, but he also worked on cytoarchitectural studies of the brain. He was also interested in the evolution of the human brain [4]. While working as an assistant in the psychiatric-neurological clinic of the University of Vienna, von Economo examined seven patients with bizarre complaints occurring in association with intractable stupor [3]. He described the experience in the following manner [5]:

> *Towards the end of 1916 the wards of the Vienna Psychiatric Clinic contained quite a number of patients with a strange variety of symptoms—cases which had apparently only one feature in common—a difficulty to fit into any known diagnostic scheme. They had been admitted under the most varied descriptions, such as, meningitis, acute disseminated sclerosis, amentia, delirium, etc. The patients all showed a slight influenza-like prodrome condition with trifling pharyngeal symptoms, a slight rise of temperature soon followed by a variety of nervous symptoms, though generally one sign or another pointed to the midbrain as the source. I noticed particularly in a few of these patients a condition of marked lethargy combined with disturbances of eye-muscles, recalling the mythical sleeping sickness, nona, rampant in north Italy during the ''nineties'' of the last century, of which I had heard during a youth spent in the Austrian Kustenland.*

The initial phase of the illness was a viral prodrome that had features of an upper respiratory tract infection. The clinical features of the illness appeared different from those of influenza, predated the great influenza epidemic, and lasted longer. During the acute phase, sleep, ocular motility, and movement disorders were observed. In the chronic phase, parkinsonism was a common sequelae, and through 1960 the illness was responsible for as many as 50% reported cases of parkinsonism. For example, in the period 1920–1938 it was estimated that about two thirds of all cases of parkinsonism in London were postencephalitic. In New York estimates were still as high as 26% of all parkinsonism in the New York City clinic [6]. Studies at the time of the epidemic demonstrated that the illness resulted from a filterable agent that was transmissible to animals. Over the course of the succeeding 15–20 years, the illness changed in nature, with a less florid acute phase and a more commonly observed chronic phase of frequently devastating neurological sequelae.

Von Economo was very concerned about the primacy of his observations because competing descriptions occurred at about the same time, in particular a description of a similar illness by the French physician Cruchet. Cruchet and colleagues published a paper on 40 cases seen in the French military hospital for neuropsychiatric disorders entitled ''Quarante cas d'encéphalomyélite subaigue'' in the *Bulletin de la Société Médicale des Hôpitaux de Paris* in 1917 [7]. These patients represented 3% of all hospital admissions. In his monograph [5], von Economo states that this publication appeared 10 days after his own, giving his paper priority. A decade later and before the appearance of von Economo's own monograph, Cruchet published a monograph entitled *Encephalitis épidemique—les 64 premieres observations*. According to von Economo, only three of the 64 cases described by Cruchet qualified as encephalitis lethargica and 40 of them occurred before the appearance of the illness in 1915 [5]. Some authors have elected to refer to the illness as encephalitis lethargica of von Economo–Cruchet [8], and which of the two deserves credit for the initial description remains controversial [9]. Because von Economo recognized the uniqueness of the illness, Yahr [10] argues that credit for the description of the disorder belongs to him.

Historical surveys suggest that similar epidemic illnesses had been observed previously. These included the 1673–1675 epidemic in London of sleeping illness, an illness referred to as ''febris comatosa'' by Sydenham and typically accompanied by singultus. From 1695 through 1800 there were isolated reports of sporadic somnolent ophthalmoplegia. In 1712 an epidemic of ''Schlafsucht'' occurred in Germany that was described by Camerarius as a ''Schlafkrankheit'' with delirium that was especially vivid and powerful during the night and associated with ptosis as the most predominant of the oculomotor signs [11]. In 1890–1891, an illness of epidemic delirium and lethargy called ''nona'' with which von Economo was familiar appeared in northern Italy. This illness also bore striking similarities to the epidemic described by Livius in the year 412 [12].

Crookshank [13], upon review of the world's literature, was convinced that the illness was not new. Among the similar epidemics that he identified were the ''English sweats'' (England, 1529), *mal mazzuco* (Italy, 1597), *Kriebelkrankheit* (Germany in the 1500s and 1672–1675), *Raphania* (Sweden, 1754–1757, and Germany, 1824), and *nona* (Hungary, 1889) [13].

The initial appearance of von Economo's encephalitis was probably in Rumania in April and May, 1915 [14], although others contend its origin may have been in China [15]. Subsequently, it was noted in World War II in the winter of 1916 on the French front in Verdun and by the Christmas season of 1916–1917 in epidemic form in Vienna. An epidemic followed in France in March 1918. During the latter, Sainton described three diagnostic symptoms: somnolence, oculomotor paresis, and fever. By 1919, it had overrun most European countries and the United States and had spread to Canada, Central America, and India [16]. By comparison, it was not until May 1918 that the influenza epidemic first appeared. Through 1931, sporadic cases, chiefly occurring in the winter and spring, were noted; however, Wilson demonstrated that between 1919 and 1930, peaks of the illness occurred in 1920 and 1924, with more than 17,000 cases reported in 1924 [16]. By 1928, more than 5000 papers had been published in the world's medical literature on the illness, as well as many monographs [16]. Over time, the acute cases became almost inapparent, although the chronic neurological sequelae remained devastating [17].

3 TRANSMISSION

Transmission of the illness from person to person was said to be unusual [18,19] and perhaps accounted for no more than 5% of the cases [5]. Despite that fact, Wilson cited literature that suggested that multiple familial, infection occurred fairly often [16]. He cited literature that documented transmission between husband and wife, nurse and patient, soldiers in adjoining beds, and mother and infant. Additionally, 12 cases of 22 residents with five fatalities were observed during 2 weeks in a girls' home in Derby, England [18], and 28 within 20 days in a women's asylum in Germany [20]. Despite these reports, no contact transmission was documented among 1156 cases seen in Vienna [21] or in 520 cases in Germany [22]. The incubation time was uncertain but was generally believed to be a minimum of 1 day to a maximum of 2 months, with an average of 8–18 days and a mean of 10 days [16]. These data were determined by cases in which the patient had recently moved from an area of high endemicity. No age was spared, but persons between the ages of 10 and 45 years were more commonly affected. Nearly 50% of the cases occurred between the ages of 10 and 30 [16]. There was no gender predilection, although pregnancy bestowed an unfavorable course on the illness and transmission to both the fetus and newborn was described [5,16]. Transient infection with influenza has been proposed to result in an increased nasal mucous membrane permeability for the encephalitic virus [23].

Contemporary studies of histocompatibility antigens in patients who had suffered from von Economo's encephalitis failed to reveal any differences in phenotypic frequencies for HLA-A, B, C, and DR antigens [24]. Elizan et al. [25] also found no difference in HLA typing in postencephalitic parkinsonism.

4 THE AGENT

Although an intoxication such as botulism was considered a possible etiology, no bacilli were ever recovered from patients or food substances and gastrointestinal disturbances were not normally associated with the presentation [16]. In fact, the first cases observed in England were believed to be due to botulism, but studies by the Ministry of Health (the local government board) and the Medical Research Committee found no association with botulism, but rather a new illness [15]. Chemical agents, bacteria, and parasites were also proposed as potential etiologies [26]. Histological examination of the brain suggested an infectious etiology. Studies indicated that the infectious agent, like polio, was filterable; however, passage was more often successful when unfiltered. The agent could be transmitted to a variety of animals, including rabbits, monkeys, and others. The virus was transmitted by Levaditi to rabbits from brain tissue of encephalitic patients by intracerebral and ocular inoculation [27,28]. Levaditi et al. [28] postulated that this virus, which he referred to as "virus C," was an epidemic form of herpesvirus. The agent remained infectious after 40 days in a dry open environment at room temperature [28]. Levaditi and Harvier isolated a virus from a fatal case of encephalitis lethargica following the intracerebral inoculation of a brain tissue suspension into a rabbit [27]. This virus was rarely recovered from affected individuals; however, a similar virus was isolated from the saliva of healthy individuals [29]. In the rabbit, the incubation period averaged 2–10 days but was sometimes significantly longer [30]. Typically, death occurred 7 months later [30]. Surprisingly, although the investigators were able to establish infection in rabbits, they could not infect monkeys [30]. Examination at the time of death revealed histological features of encephalitis [27]. A virus was also recovered from brain more than 15 months after recovery in patients dying of intercurrent illness, and its virulence was maintained for up to 48 h after death [5]. The virus was also found in the saliva of encephalitic patients [5]. Some German investigators also believed that a herpesvirus was the culprit [30,31]. However, other investigators believed that the herpesvirus isolated by Levaditi was simply an artifact, the consequence of reactivation of oral herpes. Serological studies for herpes antibody revealed the same levels in healthy controls and patients with encephalitis lethargica, also supporting the likelihood that herpes virus was not causative [32]. Other investigators postulated that the virus was an enterovirus [30].

Influenza A has always been of particular interest in light of the temporal relationship of the encephalitis lethargica outbreak and the influenza epidemic. As stated, the outbreak of von Economo's encephalitis clearly preceded the influenza epidemic. Rarely were the disorders contemporaneous in the same locale. For instance, in a large influenza epidemic at Camp Dix, New Jersey, in 1918, during which 6000 persons developed influenza and 800 died, there were no concurrent cases of von Economo's disease [33]. Large population movements, as occurred as a consequence of World War I, were also felt by some to likely be contributory to the rapid worldwide spread of encephalitis lethargica [15].

Gamboa et al. [34] detected influenza A antigens in the nuclei of neuroganglia cells in six Americans with postencephalitic parkinsonism who died between the ages of 46 and 54. These antigens were not detected in the nervous system tissues of controls [34]. Epidemiological studies conducted by Ravenholt and Foege for the period between 1918 and 1926 in Seattle and in the Samoan Islands showed a relationship between the appearance of the Spanish flu and the appearance of encephalitis lethargica [35]. This relationship suggested that the genesis of encephalitis lethargica was related to the influenza epidemic [35]. However, in contrast to the Americas, in Europe encephalitis lethargica preceded the influenza epidemic [8]. Finnish investigators were unable to establish any statistically significant relationship between the presence of antibodies to influenza virus and encephalitis lethargica when comparing the serological responses for hemagglutination inhibiting antibodies to four influenza virus strains in patients with idiopathic Parkinson's disease, postencephalitic parkinsonism, or normal controls [36]. Von Economo was convinced that there was no association between influenza and encephalitis lethargica [5], and more contemporary authorities also remain unconvinced [37].

Elizan et al. [38–40] tested sera and CSF of 29 patients with postencephalitic parkinsonism, 65 with Parkinson's disease, and 125 controls for seven alphaviruses, seven flaviviruses, four bunyaviruses, 12 subtypes or strains of influenza A, and two of influenza B, and assorted other viruses, including parainfluenza, mumps, and measles viruses coxsackie viruses B3 and B4, varicella-zoster virus, cytomegalovirus, herpesviruses types 1 and 2, and rubella and lymphocytic choriomeningitis viruses. A literature survey by Duvoisin and Yahr in 1965 [41] failed to identify any association between known viruses and postencephalitic parkinsonism. Howard and Lees were unable to identify a viral agent in four cases reported in 1987 [42]. Sophisticated molecular techniques applied to the preserved brain tissue specimens of patients suffering from von Economo's disease have permitted a reexamination of the etiology of encephalitis lethargica [43]. Due to the nearly contemporaneous appearance of an influenza epidemic, attention has been directed to that virus as the possible cause. However, no association has been convincingly demonstrated using newer techniques [44,45]. Isolated cases of postencephalitic parkinsonism have been described after suspected herpes simplex encephalitis [46], Japanese A encephalitis [47], western equine encephalitis [48], coxsackievirus [49,50], and influenza A virus [51] infections. A subcortical dementia coupled with parkinsonian manifestations is also the hallmark of HIV dementia [52]. An extensive review of potential viral etiologies of postencephalitic parkinsonism has failed to reveal an association with many other viruses [26].

5 CLINICAL ILLNESS

The earliest clinical features of the illness were very nondescript and included malaise, low grade fever, mild pharyngitis, lassitude, shivering, headache, vertigo, and vomiting. However, neurological manifestations could present with striking rapidity. Wilson [16] described one case in which rapidly evolving neurological disease developed thus:

The first case I ever saw was that of a girl who, walking home from a concert, suddenly felt an acute pain in the head and at the same moment her left arm ''flopped'' and left leg gave way; within half an hour she was ''asleep'', showing contracted pupils, divergent squint, and left hemiplegia; sinking deeper into lethargy, temperatures to 105.8 and death supervening in 12 days.

The illness assumes several distinct forms in its acute and chronic phases. Von Economo classified the acute manifestations into three categories: the somnolent-ophthalmoplegic form, the hyperkinetic form, and the amyostatic-akinetic form [5], whereas the chronic forms were classified as either parkinsonism or other [5]. Von Economo suggested that different clinical manifestations attended disease that occurred in epidemics separated by time and place. For instance, delirium and meningeal irritation were the chief manifestations of the 1916–1917 epidemic in Vienna, parkinsonism and somnolent ophthalmoplegia predominated during the 1918 epidemic in London, and a hyperkinetic illness was observed in Italy and Austria in 1920 and 1921 [5]. In contrast to von Economo's classification scheme, Wilson [16] proposed labeling the types of ''encephalitis lethargica'' in the following categories: mesencephalic and pontobulbar, cortical, basal form—parkinsonism, radiculospinal, multiple diffuse, and abortive, depending on the nature of the neurological disturbances observed.

The somnolent-ophthalmoplegic form typically develops soon after prodromal symptoms. Patients are described as dazed and confused, but their somnolence and delirium are unrelated to temperature elevations. They exhibit features of mild meningeal irritation characterized by stiff neck, Brudzinski's and Kernig's signs, and pain on eyeball compression. Other features include frequent yawning, trismus, and singultus, and, occasionally, cranial nerve palsies with weak phonation and dysphagia, vertigo, and limb weakness. Ptosis is seen in most cases. Pupillary abnormalities are observed, including the frequent absence of accommodation. Less commonly, supranuclear palsies occur [5]. In his description of this form of the illness, von Economo recorded, ''It is frequently seen that a patient cannot properly lift his arm or leg, or that his grasp is weak, and the muscles markedly hypotonic. . . . This lack of tonus is frequently shown by the inability of patients to rise up or sit down; they therefore assume a huddled position and tend to gravitate to one side'' [5]. Kinnier Wilson described a patient thus: ''A lady found her shopping hindered by an overwhelming desire for sleep; she would rush home to throw herself on a couch in the hall (not even waiting to go up to her bedroom) and sink at once into deep slumber continuing for hours. Within 4 years, Parkinsonism occurred'' [16]. He described a 12-year-old girl who had recently had a violent illness in the following fashion: ''Once, at tea, she was in the act of putting a piece of toast in her mouth when her eyes closed and the hand was kept immobile, the bread touching her lip for several minutes'' [16]. Whereas the mortality of the somnolent-ophthalmoplegic form of the illness is higher than that of other forms, exceeding 50%, a greater number of survivors experienced few sequelae than follow other forms or more at all. Von Economo related this form of the disorder to lesions of the posterior wall of the third ventricle near the oculomotor nucleus. In contrast, the insomnia seen in the other forms were linked to a lesion in the lateral wall of the third ventricle near the corpus striatum, which was suggested to exert an inhibitory effect on the cerebrum and thalamus [53].

The hyperkinetic form is characterized by an initial hypomaniacal excitement followed by general prostration, pallor, and rapid failure of strength, violent neuralgic pains in face and limbs, visual and tactile hallucinations, and sleep reversal or inversion. During the manic phase vocalizations include singing, shouting, and whistling. Examination reveals pupillary abnormalities including unreactive pupils and Argyll Robertson pupils, chorea that may be either unilateral or bilateral, peculiar torsions and myorrhythmias, trismus and similar involvement of orbital muscles, and myoclonic twitches that are short-lived and either nonrhythmic or rhythmic [54]. Strümpel [55] described the myoclonus in one of his patients as ''extremely painful, mainly localized in the abdomen ··· even after the myoclonus stopped the patients complained about some weird deep pain and about hurtful paresthesias in their extremities and body.'' Sudden death may supervene. Among the first-hand descriptions of patients afflicted with this disorder are the following by Kinnier Wilson [16]: ''The patient tosses about in bed, pushes the blankets back, pulls them up again, sits up, throws himself back again in a wild sort of haste, jumps out of bed, strikes out aimlessly, talks incoherently, clucks his tongue, and whistles, this unrest lasting for days and nights without stop.''

The least common of the three forms proposed by von Economo for acute illness is the amyostatic-akinetic form. This form is characterized by peculiar rigidity and a lack of movements in absence of real weakness. It was said that ''movement is arrested soon after its inception due to lack of impulse.'' There are no corticospinal tract findings. The slowness and paucity of movement is striking and patients do not alter their body position spontaneously. When addressed their eyes might turn to the examiner but not their heads. ''Flexibitas cerea'' might develop. The emotions are hardly noticeable, though the patients are mentally intact with normal comprehension. Examination reveals muscle rigidity, normal tendon reflexes, often a peculiar bent posture when standing, vasomotor disturbances common, e.g., edema, skin exfoliation; sleep disturbances include somnolence, insomnia, and sleep inversion. This condition might last weeks to months with either a slow or rapid recovery.

Adolf Strümpel [55] described one of his patients as ''absolutely still in bed, eyes are wide open . . . their mind stays absolutely clear . . . complete lack of initiation of movements or speech . . . muscle rigidity, arms in interosseus-position . . . patients keep sometimes arms lifted for very long time.''

Other forms of the illness that did not easily conform to the three described might include any sort of neurological symptom complex. Among them are those in which cerebellar features predominate, myoclonus, satyriasis and priapism, peripheral neuropathies, cranial nerve palsies, strokelike presentations, pseudotabetic and pseudoparetic forms, and psychoses. Wilson recognized that virtually any region of the central nervous system could be affected by encephalitis lethargica, including the cortex, basal ganglia, brainstem, spinal cord, and nerve roots [16]. His classification of forms of the disease related to the areas affected is presented in Table 1.

Table 1 Classification of Clinical Syndromes of Encephalitis Lethargica (After Wilson)

Cortical: Toxic psychosis, catatonia, waxy flexibility, delirium to somnolence
Basal ganglia: Involuntary movements, parkinsonism
Mesencephalic and pontobulbar: Diplopia, ophthalmoplegia, cranial nerve palsies, vertigo, lethargy and sleep
Radiculospinal: Neuritic, radicular, tabetic
Multiple diffuse: Variable combined forms
Abortive: General nervous symptoms but scanty, fleeting, or absent local signs

Source: Ref. 16.

6 LABORATORY STUDIES DURING ACUTE ILLNESS

During the acute illness, blood leukocytosis ranging up to 25,000–30,000 cells/mm^3 is observed. Cerebrospinal fluid analysis typically reveals a moderate increase in opening pressure, though von Economo also noted that despite persistent somnolence or other deficits, opening pressure returned to normal after a couple of weeks [11]. The CSF was always clear and colorless with a moderate pleocytosis $<$100 cells/mm^3 (usually 15–20 cells/mm^3), no or mild increase in protein, a positive colloidal reaction indicating an increase in IgG, and negative microbiological studies, including Wasserman and other syphilis serologies. The observation of CSF pleocytosis, increased protein, increased IgG, and the presence of oligoclonal bands has also been observed in recent cases of postencephalitic parkinsonism [42,56]. It is claimed that over time, the abnormalities observed in the CSF become less pronounced. Von Economo's encephalitis first occurred in an era before sophisticated neuroradiological imaging. One recent case with a clinical history and pathological findings suggestive of acute encephalitis lethargica revealed MRI abnormalities in the brainstem as well as the internal capsule and thalamus [57]. Rare cases of postencephalitic parkinsonism that have been observed in the modern era have had abnormalities in brain metabolism. A patient studied with positron emission tomography demonstrated abnormalities in glucose and dopa metabolism distinct from that observed in idiopathic parkinson's disease, suggesting a loss of inhibitory influence from the pars compacta of the substantia nigra to the striatum [51].

7 THE CHRONIC PHASE

The manifestations chiefly observed during the chronic phase of the illness are parkinsonism, sleep disturbances, mental symptoms, oculomotor abnormalities, involuntary movements, respiratory disturbances, metabolic and endocrine disorders, and epilepsy. The prototypical, most frequent, and most sinister neurological complication of von Economo's encephalitis is parkinsonism. It might develop acutely or after many years, sometimes often more than a decade. Sacks reported a patient who developed a postencephalitic syndrome in 1962, 45 years after the original infection [1]. Typically, it develops 1–5 years after the acute illness. It follows the hyperkinetic form of the illness most commonly and the somnolent-ophthalmoplegic form least commonly. Although it can affect persons of any age, it has a predilection for young adults. No gender preference has been noted. Through 1960, some authors contended that it was the most common form of Parkinson's disease. The illness is slowly progressive and rarely improves spontaneously. Life span averaged less than 4–5 years.

In postencephalitic parkinsonism, stiffness and bradykinesia generally precede the tremor, and the upper limbs are more affected than the lower limbs. On examination, rigidity is more conspicuous than the tremor (Fig. 1). In contrast to idiopathic Parkinson's disease, these patients commonly display paradoxical movements, oculogyric crises, catatonia, cataplexy, cerea flexibilitas, and speech and respiratory problems (Table 2). A striking finding in this population is "kinesia paradoxica," in which the patient might be akinetic at one moment and perfectly mobile the next without an intermediate stage [1]. In addition to the classical parkinsonian features, other movement disorders that have been observed including chorea, torsion spasms, myoclonus, and tics affecting the jaw, lips, tongue, and palate (Fig. 2).

Oculogyric crises occurred in 15–20% of patients at the height of the epidemic [58]. An incidence of approximately 30% was reported in the 1960s [1]. Following the introduction

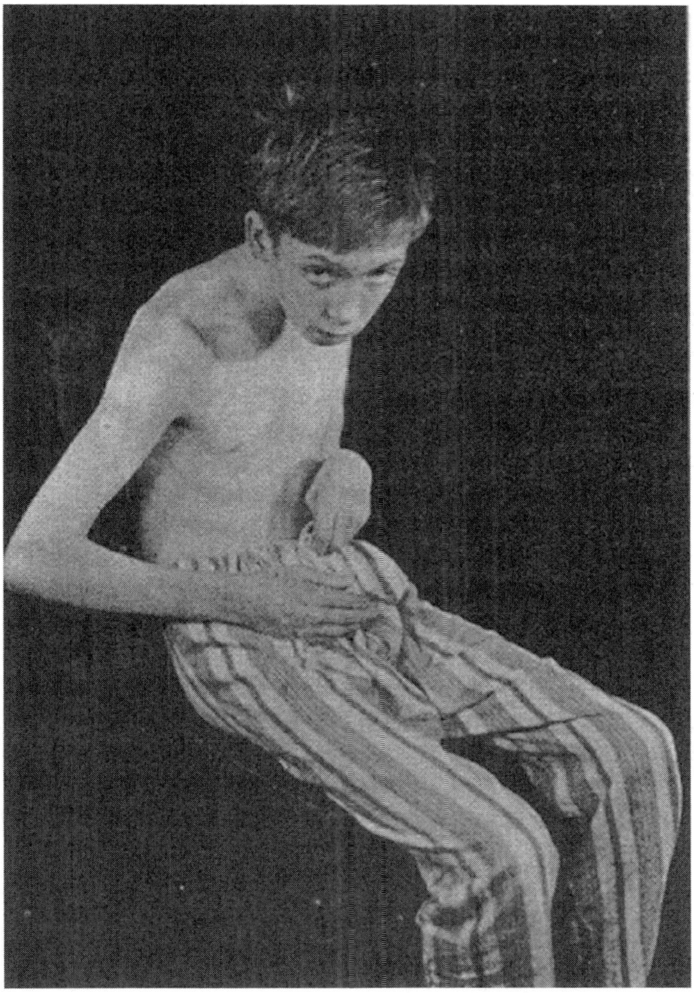

Figure 1 Young boy with postencephalitic parkinsonism. (From Ref. 16.)

of L-dopa in the treatment of postencephalitic Parkinsonism, the incidence approached 100% [1]. It lasted seconds to hours and typically resulted in tonic deviation of the eyes upward or upward and laterally, often with spasmodic deviation of the head. These crises could be precipitated by emotions. During the spell, consciousness was preserved, but the patient was incapable of voluntarily moving his eyes except momentarily with strong effort. The frequency of the oculogyric crises varied from individual to individual. At times the crises were accompanied by movement disorders, e.g., dystonic posturing, typically of the head and neck, or by verbal utterances or coprolalia [59–61]. Sometimes forced thoughts of sexual and violent nature, other compusions, or religious rituals accompanied the oculogyric crises [59–61]. Oculogyric crises were virtually unknown prior to the epidemic of encephalitis lethargica [62]. Other ocular abnormalities observed included slight ptosis, impaired pupillary reaction, and Argyll Robertson pupils, nystagmus, squint, impaired convergence, and extraocular muscle palsies.

Table 2 Distinctions Between Postencephalitic Parkinsonism and Idiopathic Parkinson's Disease

Manifestation	Postencephalitic PD	Idiopathic PD
Encephalitic prodrome	+++	−
Parkinsonism after delay of months to years	+++	−
Oculogyric crises	+++	−
Alteration in sleep cycle	+++	Rare
Pupillary changes	++	Rare
Involuntary movements, e.g., torsion spasm	+++	Generally, only with dopaminergic treatment
Mental changes	+++	+
Corticospinal tract signs	++	−
Respiratory disturbances	++	Rare
Appearance in young patients	++	Rare

PD = Parkinson's disease.

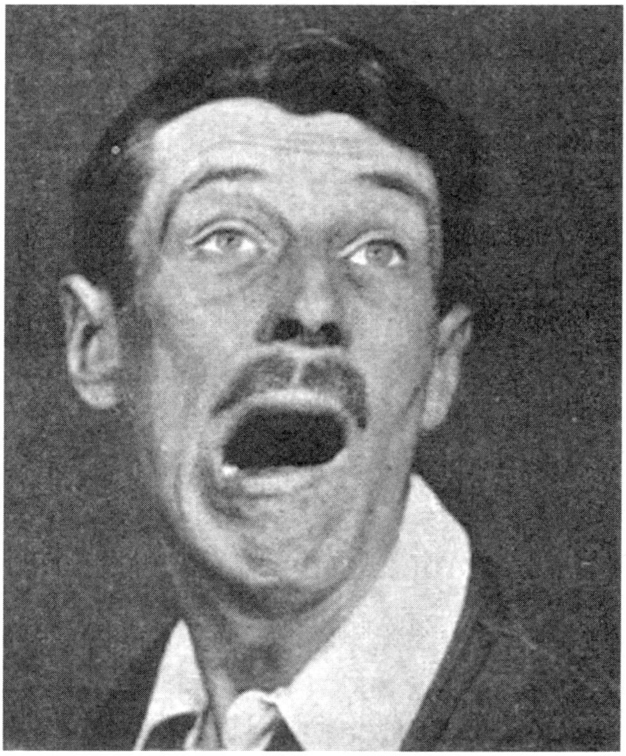

Figure 2 Mandibular tic in postencephalitic parkinsonism. (From Ref. 16.)

An asthenic syndrome in which patients complain of a persistent sense of fatigue, both mental and physical, is frequently seen in cases of Parkinson's disease but can also occur as the sole or main disability. Riddoch [54] described the syndrome in this fashion: ''Unlike the weakness of myasthenia gravis, the feebleness is not as a rule much diminished by rest, and, unlike the neurasthenic, the patient does not tend to feel better as the day goes by.''

Speech abnormalities include a wide gamut of problems such as logorrhea, echolalia, palilalia, tachyphemia (rapid speech), megaphonia (speaking in an unnecessarily loud voice), klazomania (shrieking and crying), and hyperodia (singing and reading out loud). To illustrate the speech abnormalities observed, Kinnier Wilson described one of his patients as follows: ''A 47-year-old man 16 months after encephalitis began reading out loud. He would attend the cinema and repeatedly read the captions. His wife would whisper to shut up to which he would reply 'I can't shut up—I can't shut up . . .' in quick diminuendo utterance.'' Similarly, respiratory abnormalities are broad in nature, including increased rate and depth of respiration, bradypnea, respiratory spasms, respiratory tics, yawning, sniffing, hawking, spasmodic cough, spitting, Cheyne-Stokes respirations, and a variety of paroxysmal phenomena, such as hiccups and ''dysrhythmias.'' Riddoch [54] described breath-holding as one of the most dramatic performances, often occurring during sleep: ''After a few deep breaths the chest is held in full inspiration for as long as half a minute. The head is often thrown back, the limbs may perform various grotesque movements, the face may or may be not cyanosed, and, in longer attacks, consciousness is lost for a short time. Noisy expiration followed and normal breathing was then established''. Turner and Critchley classified the respiratory abnormalities into three types: (1) disorders of rate (tachypnea and bradypnea), (2) disorders of respiratory rhythm, and (3) respiratory ''tics'' [63]. Sleep disturbances include lethargy, insomnia, and narcolepsy.

Psychiatric manifestations are common as long-term consequences of encephalitis lethargica. Neuropsychiatric disturbances were reported in 50% [64] to 100% of survivors [16,65]. These manifestations include mood changes, e.g., mania and depression, feelings of euphoria, increased sexual drive, exhibitionism, paraphilic behavior, hallucinations, metamorphopsia, and excessive puns, joviality, and silliness referred to as the Witzelsucht of Jastrowitz [9]. Psychosis is observed in 30%, although milder forms are very common. Delusions, hallucinations, and ideas of reference accompany the psychosis [9]. Catatonic symptoms are prominent [9]. The mental changes appear to be more striking in children, whereas adults more often have parkinsonian sequelae [2]. McKenzie described these behavioral disturbances in children as manifesting as a profound emotional instability and a perversion of conduct with poor impulse control that occasionally leads to criminal behavior [2]. Mild features include nervousness, fatigability, poor concentration, anxiety and depression, parkinsonism with emotional impoverishment, and intellectual decline. The loss of cognitive function is most commonly observed when encephalitis lethargica occurs in infancy.

Metabolic and endocrine disorders have been observed on rare occasions. These disorders include exophthalmos, tachycardia, sweating, tremor suggestive of hyperthyroidism, and goiter, glucose dysregulation, adiposogenital syndrome in children, precocious puberty and disturbance of sexual function, and excessive salivation, and sebaceous secretions. Morbid obesity and amenorrhea may develop rapidly after the acute attack [66].

8 PATHOLOGY

The gross pathology of acute encephalitis lethargica reveals hyperemic leptomeninges with a soft edematous brain and reddish discoloration of the brainstem. Microscopic changes are

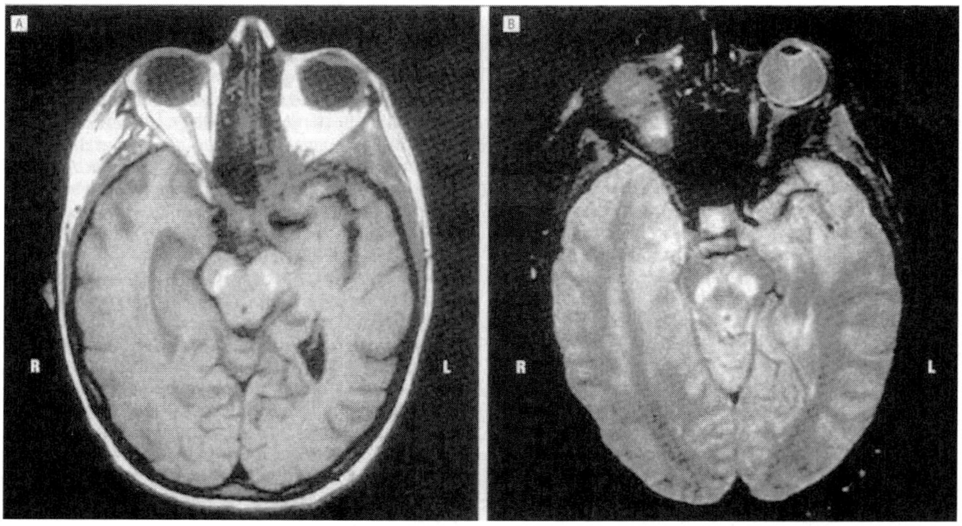

Figure 3 Magnetic resonance imaging findings in a modern case attributed to coxsackievirus B4. Hyperintense lesions are evident in the substantia nigra on both T1-weighted (A) and T2-weighted (B) MR images. (From Ref. 57; reproduced with permission from the *Archives of Neurology*.)

most evident in the upper midbrain (third cranial nerve and substantia nigra) followed by the basal ganglia, pons, and medulla (Fig. 4). Perivascular and tissue inflammatory infiltrates that are chiefly mononuclear cells are accompanied by small hemorrhages secondary to diapedesis, neuronophagia, and microglial nodules. As in the older literature, a recent case showed an active encephalitis, mainly centered on the upper brainstem and diencephalon with extensive Purkinje cell loss and marked plasma cell infiltrates and morula cells [67]. Modern studies have failed to reveal viral antigens in the brain specimens [68], but CSF viral cultures have occasionally been positive [57].

Wiesner, who worked with von Economo as a pathologist, showed hyperplasia of tonsils and intestinal follicles in several patients [11]. As von Economo [11] described the microscopic picture, ''We have histologically the picture of a polioencephalitis cerebri, pons and medulla with a slight poliomyelitis; perivascular, inflammatory, and diffusely infiltrating, but no hemorrhagic and only slight neurophagic character.''

The gross pathology of chronic encephalitis lethargica is characterized by modest findings of either focal or generalized atrophy. Microscopic pathology shows a coincidence of old and recent inflammation, suggesting persistence of virus with the principal changes in the corpus striatum, thalamus, hypothalamus, posterior wall of the third ventricle, and substantia nigra. Microscopic findings include neuronophagia, astrogliosis, perivascular hemosiderin staining, and pigment degeneration in the substantia nigra and locus ceruleus. The astrogliosis may be overwhelming, involving widespread areas of the brain, and occur in the absence of significant other pathologies [69]. Neurofibrillary tangles have been reported in the substantia nigra, locus ceruleus, and raphe nuclei.

Hallervorden [70] described the gross pathology as a discoloration of the substantia nigra, the microscopic pathology dominated by neurofibrillary tangles; tangles in the cerebellar cortex or plaques such as occur in Alzheimer's disease were never seen. Interestingly,

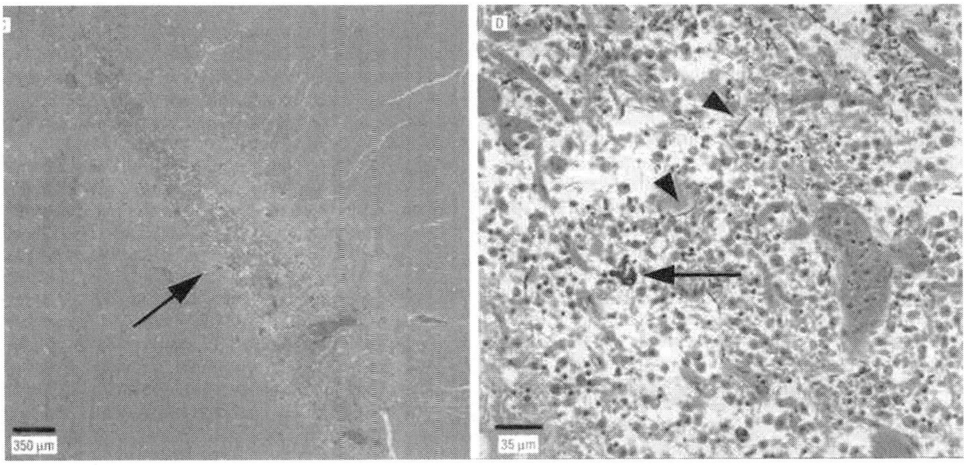

Figure 4 Histopathology of a modern case showing depigmentation and focal necrosis in the subtantia nigra (A) and macrophages and mineral deposits in the substantia nigra (B). (From Ref. 57; reproduced with permission from the *Archives of Neurology*.)

he described the multiple young patients who, even though their substantia nigra showed severe postencephalitic scarring, did not show any of the typical parkinsonian features, i.e., rigidity, bradykinesia, and gait disturbance. Alpha synuclein has not been detected in these brains [74], but a tau protein triplet similar to the one seen in Alzheimer's disease is present [72,75]. The observation of the latter suggested a biochemical means of pathologically distinguishing postencephalitic parkinsonism from certain other neurodegenerative disorders, e.g., progressive supranuclear palsy and corticobasal degeneration [72]. Additionally, ''tuft-shaped'' (nonreactive) astrocytes have been found in a widespread distribution throughout the central nervous system [73].

In addition to the passage of the infection to animal models from brain tissues derived from patients long after the initial infection, a variety of pathological features suggest the persistence of the virus despite the inability to detect a specific viral antigen. These features include evidence of acute inflammation at autopsy associated with cases of ''intermittent progression,'' the presence of marked inflammation in 30% of parkinsonian patients, and some degree of inflammation in 50% even after many years, and almost all cases have some evidence of perivascular mononuclear cell infiltration.

9 DIFFERENTIAL DIAGNOSIS

Many disorders present with parkinsonian manifestations. A list of these illnesses is given in Table 3.

10 PROGNOSIS

The morbidity and mortality of encephalitis lethargica appear to differ with respect to locale and year. There is discordance between the incidence of illness and that of death; some affected persons died during the acute phase and others during the chronic phase.

Table 3 Differential Diagnosis of Parkinsonism

Infectious:	Von Economo's, HIV dementia, Japanese B encephalitis, transmissible spongiform encephalopathies, others
Hereditary:	Huntington's chorea, Wilson's disease, olivopontocerebellar degeneration, Fahr's disease, Hallervorden-Spatz disease
Degenerative diseases:	Idiopathic Parkinson's disease, progressive supranuclear palsy, striatonigral degeneration, Shy-Drager syndrome, Lewy body disease
Ischemic-hypoxic insult:	Vascular disease of the basal ganglia, carbon monoxide poisoning
Drugs:	Metoclopropamide, neuroleptics, antihypertensives
Metabolic:	Hypocalcemia
Trauma:	"Punch drunk"

In Glasgow, only 60 of more than 300 affected persons were free of all signs and symptoms 2 years after the infection [2]. According to Dimsdale [76], one-third of the patients died during the acute phase of the illness, one-third survived without sequelae, and one-third had neurological sequelae. These estimates were concordant with those of Parsons, cited by Wilson, that of 100 cases, 25 survived without significant sequelae, 40 were disabled, and 35 died (the average mortality varied between 20% and 54%) [16]. Although only 30–40% had been estimated to develop neurological sequelae, sufficiently long periods of observation suggested that 80% or more were ultimately affected by parkinsonism [76]. Although some investigators believed that recovery seemed to be best in those afflicted with respiratory abnormalities followed by those with sleep inversion, during the Sheffield epidemics those with early respiratory symptoms had the highest mortality. Patients who developed parkinsonism or other movement disorders appeared to have the worst prognosis. Many authors [77,78] found that most cases of postencephalitic parkinsonism occurred in the first 5 years after the encephalitis. Up to 36% of their patients did not even have an interval between the acute illness and the parkinsonian features; however, intervals exceeding 15 years were also observed. Beringer [77] described a 56-year-old patient who was in good health until the age of 40, when he developed diplopia, insomnia, and fatigue with rapid recovery, then exhibited new parkinsonian symptoms 16 years later. Tyndel [77] and Beringer [78] also noted that there was no direct parallel between the age of the patients at the time of acute illness and the length of the interval of sequelae and no seeming correlation between the length of the interval and the severity of the encephalitis. Hall [58] found in his review of outbreaks and sequelae that 75% of cases with parkinsonism still lived after 10–15 years, with 23% dying of infections. Absolute recovery remained unknown, most cases tending to become worse, although this was extremely variable.

11 TREATMENT

At the time of the epidemic, few agents were available for the treatment of Parkinson's disease. The most effective medications were anticholinergics. The treatment of the parkinsonian manifestations of encephalitis lethargica included belladonna and hyoscine (scopolamine), which were noted to have a salutary effect [2].

Oliver Sacks [79] beautifully records the response of patients with postencephalitic parkinsonism to L-dopa in his book *Awakenings*. Sacks worked in Mount Carmel, a charity hospital in New York, the residence of 80 such patients. He states that almost half of these patients were in states of pathological "sleep" [79]. In March 1969, following a sharp decline in the price of brodopa, Sacks and colleagues used it in the treatment of this patient population [79]. At the time, it was considered experimental therapy in the United States and was not to be released for general use until 1970. In contrast to those with idiopathic Parkinson's disease, these patients were typically extremely sensitive to L-dopa, exhibiting profound arousals, marked fluctuations, tics, and emotional instability [1]. The efficacy of L-dopa was well chronicled by Sacks; unfortunately, the benefits were generally of limited duration [79].

Anticholinergic medications are often beneficial, well tolerated, and tolerated in high doses [1]. These were the chief means of treatment until the availability of L-dopa. Mainly in the German literature there were multiple reports of the alkaloids harmin and banisterine [80], which chiefly affected the rigor and the hypokinesia, with improvement in voluntary motor activity, strength, and duration. Tremor was not affected [81]. The therapeutic effect on patients was variable. Long-term treatment did not seem to have a negative effect on response [82]. Hyascin, which had a similar effect though with less dramatic improvement, generally improved the patients' subjective wellbeing, so it was often used concomitantly [83]. Other authors achieved some therapeutic effect with high doses of atropa belladonna, also known in the literature as the ''Bulgarian treatment'' [84].

Other forms of therapy have included intraspinal autogenous serum administration [85], injections of electrocolloical gold and silver [86], oxygen therapy administered in a variety of fashions [87], radiation of the salivary glands [88] or the brainstem [89], induction of malaria via infected *Anopheles* mosquitoes [90,91], hyperthermia [92], sinus washings [23], insulin [93], intravenous salicylates [94], arsenic and mercury [95], vitamin C [96], tryptaflavin [97], a mixture of calcium bromide salts and luminol [98], and thyroidectomy [99].

REFERENCES

1. Sacks, O. Postencephalitic syndromes. In: Parkinson's Disease; Stern, G., Ed.; Chapman and Hall: London, 1990, 415–428.
2. McKenzie, I. Discussion of epidemic encephalitis: epidemiological considerations. Br Med J. 1927, 1, 532–534.
3. Arts, N. Von Economo's encephalitis. In: Neurological Eponyms; Koehler, P., Bruyn, G., Pearce, J., Eds.; Oxford Univ Press: New York, 2000, 309–315.
4. Kolle, K. Grosse Nervenaerzte; Georg Thieme: Stuttgart, 1959.
5. von Economo, C. Encephalitis Lethargica, Its Sequelae and Treatment; Oxford Univ Press: London, 1931.
6. Krusz, J.; Koller, W. Historical review: abnormal movements associated with encephalitis lethargica. Move Disorders. 1987, 3, 137–141.
7. Cruchet, R.; Moutier, J. Quarante cas d'encephalomyelite subaigue. Bull Soc Med Hop Paris. 1917, 4, 614–616.
8. Chastel, C. Erreurs passees et espoirs decus de la recherche en virologie medicale [Editorial]. Virologie. 2000, 4, 445–452.
9. Cheyette, S.; Cummings, J. Encephalitis lethargica: lessons for contemporary neuropsychiatry. J Neuropsychiatry Clin Neurosci. 1995, 7(2), 125–134.
10. Yahr, M. Encephalitis lethargica (von Economo's disease, epidemic encephalitis). Handb Clin Neurol. 1978, 34, 451–457.
11. von Economo, C. Encephalitis lethargica. Weiner Klin Wochenschrif. 1917, 30, 581–585.
12. Ebstein, E. Beitrage zur Geschichte der Schlafsucht, mit besonderer Berucksichtigung der Encephalitis epidemica. Deut Z Nervenheilk. 1921, 72, 225–235.

13. Crookshank, F. A note on the history of epidemic encephalomyelitis. Proc Roy Soc Med. 1918, 12, 1–21.
14. Urechia, C. Dix cas d'encephalite epidemique avec autopsie. Arch Int Neurol. 1921, 2, 65–78.
15. Watson, A. The origin of encephalitis lethargica. China Med J. 1928, 42, 427–432.
16. Wilson, S.A.K. Epidemic encephalitis. Neurology. 1940, 1, 99–144.
17. Anonymous. Encephalitis lethargica today. Lancet. 1927, ii, 873.
18. Symonds, C. Critical review: encephalitis lethargica. Q J Med. 1921, 14, 283–308.
19. Hall, A. Encephalitis lethargica (epidemic encephalitis). Lancet. 1923, i, 731–740.
20. Krause, D. Deut Med Wochenschr, 48, 1922.
21. Hoff, M. Jahrb P.N. 1924.
22. Fassbender, G. Hyg Rundschau. 1921–1922, 31–32.
23. Eden, G.; Yates, A.L. Treatment of encephalitis lethargica by removal of possible etiological factors. Br Med J. 1927, i, 714–716.
24. Lees, A.J.; Stern, G.M. Histocompatibility antigens and post-encephalitic parkinsonism. J Neurol Neurosurg Psychiatry. 1982, 45(11), 1060–1061.
25. Elizan, T.; Terasaki, P. HLA-B14 antigen and postencephalitic Parkinson's disease. Their association in an American-Jewish ethnic group. Arch Neurol. 1980, 37(9), 542–544.
26. Casals, J.; Elizan, T. Postencephalitic parkinsonism—a review. J Neural Transm. 1998, 105(6–7), 645–676.
27. Levaditi, C.; Harvier, P. Le virus de l'encephalite lethargique. C R Soc Biol (Paris). 1920, 83, 354.
28. Levaditi, C.; Harvier, P. Recherches experimentales sur le virus de l'encephalite epidemique. C R Soc Biol Paris. 1921, 84, 524–528.
29. Levaditi, C.; Harvier, P. Sur la presence, dans la salive des sujects sains, d'un virus produisant la keratoconjunctivite et l'encephalite chez le lapin. C R Soc Biol (Paris). 1921, 83, 817.
30. Kling, C.; Davide, H. Etiologie et epidemiologie de l'encephalite lethargique. C R Soc Biol Paris. 1921, 84, 815–816.
31. Zinsser, H. Present state of knowledge regarding epidemic encephalitis. Arch Pathol. 1928, 6, 271–300.
32. Andrewes, C.; Carmichael, E. A note on the presence of antibodies to herpes virus in post-encephalitic and other sera. Lancet. 1930, i, 857–858.
33. Synnott, M.; Clark, E. The influenza epidemic at Camp Dix, NJ. JAMA. 1918, 71, 1816–1821.
34. Gamboa, E.; Wolf, A. Influenza virus antigen in postencephalitic parkinsonism brain. Arch Neurol. 1974, 31, 228–232.
35. Ravenholt, R.T.; Foege, W.H. 1918 influenza, encephalitis lethargica, parkinsonism. Lancet. 1982, 2(8303), 860–864.
36. Marttila, R.J.; Halonen, P. Influenza virus antibodies in parkinsonism. Comparison of postencephalic and idiopathic Parkinson patients and matched controls. Arch Neurol. 1977, 34(2), 99–100.
37. Encephalitis lethargica. Lancet. 1981, ii, 1396–1397.
38. Elizan, T.; Schwartz, J. Antibodies against arboviruses in postencephalitic and idiopathic Parkinsom's disease. Arch Neurol. 1978, 35, 257–260.
39. Elizan, T.; Madden, D. Viral antibodies in serum and CSF of parkinsonian patients and controls. Arch Neurol. 1979, 36, 529–534.
40. Elizan, T.; Yahr, M. Indirect immunofluorescence test against lymphocytic choriomeningitis (LCM) virus in Parkinson's disease. Mt Sinai J Med. 1979, 46, 597–598.
41. Duvoisin, R.; Yahr, M. Encephalitis and parkinsonism. Arch Neurol. 1965, 12, 227–243.
42. Howard, R.S.; Lees, A.J. Encephalitis lethargica. A report of four recent cases. Brain. 1987, 110(Pt 1), 19–33.
43. Reid, A.H.; McCall, S. Experimenting on the past: the enigma of von Economo's encephalitis lethargica. J Neuropathol Exp Neurol. 2001, 60(7), 663–670.

44. Jellinger, K.A. Influenza RNA not detected in archival brain tissues from acute encephalitis lethargica cases or in postencephalitic Parkinson cases. J Neuropathol Exp Neurol. 2001, 60(11), 1121–1122.
45. McCall, S.; Henry, J.M.; Reid, A.H.; Taubenberger, J.K. Influenza RNA not detected in archival brain tissues from acute encephalitis lethargica cases or in postencephalitic Parkinson cases. J Neuropathol Exp Neurol. 2001, 60(7), 696–704.
46. Solbrig, M.V.; Nashef, L. Acute parkinsonism in suspected herpes simplex encephalitis. Mov Disord. 1993, 8(2), 233–234.
47. Ogata, A.; Tashiro, K. Parkinsonism due to predominant involvement of substantia nigra in Japanese encephalitis. Neurology. 2000, 55(4), 602.
48. Schultz, D.R.; Barthal, J.S. Western equine encephalitis with rapid onset of parkinsonism. Neurology. 1977, 27(11), 1095–1096.
49. Walters, J. Postencephalitic Parkinson syndrome after meningoencephalitis due to coxsackie virus group B, type 2. N Engl J Med. 1960, 263, 744–747.
50. Poser, C.M.; Huntley, C.J. Para-encephalitic parkinsonism. Report of an acute case due to coxsackie virus type B 2 and re-examination of the etiologic concepts of postencephalitic parkinsonism. Acta Neurol Scand. 1969, 45(2), 199–215.
51. Ghaemi, M.; Rudolf, J. FDG- and Dopa-PET in postencephalitic parkinsonism. J Neural Transm. 2000, 107(11), 1289–1295.
52. Berger, J.; Arendt, G. HIV dementia: the role of the basal ganglia and dopaminergic systems. J Psychopharmacol. 2000, 14(3), 214–221.
53. von Economo, C. Sleep as a problem of localization. J Nerv Ment Dis. 1930, 71, 249–259.
54. Riddoch, G. Discussion of epidemic encephalitis III—chronic encephalitis. Br Med J. 1927, 2, 537–539.
55. Strumpel, A. Ueber Encephalitis epidemica. Deuts Med Wochenschr. 1920, 46, 705–707.
56. Blunt, S.; Lane, R. Clinical features and management of two cases of encephalitis lethargica. Mov Disord. 1997, 12(3), 354–359.
57. Cree, B.C.; Bernadini, G.L.; Hays, A.P.; Lowe, G. A fatal case of coxsackie B4 meningoencephalitis. Arch Neurol. 2003, 60(1), 107–112.
58. Hall, A. Chronic epidemic encephalitis with special reference to ocular attacks. Lumelian Lectures. Br Med J. 1931, 2, 833.
59. McCowan, P.; Cook, L. Oculogyric crises in chronic epidemic encephalitis. Brain. 1928, 51, 285–309.
60. Wexberg, E. Remarks on the psychopathology of oculogyric crises in epidemic encephalitis. J Nerv Ment Dis. 1937, 85, 56–69.
61. Gillingham, F.; Kalyanaraman, S. The surgical treatment of oculogyric crises. Confinia Neurol. 1965, 19, 237–345.
62. Jelliffe, S. Oculogyric crises as compulsion phenomena in post encephalitis: their occurrence, phenomenology, and meaning. J Nerv Ment Dis. 1929, 69, 59–68, 165–184, 278–297, 415–426, 531–551, 666–679.
63. Cooper, M. Associated movements of tongue in epidemic encephalitis controlled by voluntary effort. Arch Neurol Psychiatr. 1935, 33, 148–53.
64. Jelliffe, S. Postencephalitic respiratory diosrders: review of the syndromy, case reports, and discussion. J Nerv Ment Dis. 1926, 64, 29–44, 157–166, 241–261, 362–370, 503–527, 629–636.
65. Goodhart, S.; Cottrell, S. Residua and sequelae of epidemic encephalitis. JAMA. 1925, 84, 2–36.
66. Sauter, E. Zum Schicksal der Encephalitiker. Schweiz Med Wochenschr. 1934, 62, 464–469.
67. Kiley, M.; Esiri, M.M. A contemporary case of encephalitis lethargica. Clin Neuropathol. 2001, 20(1), 2–7.
68. Elizan, T.; Casals, J. No viral antigens detected in brain tissue from a case of acute encephalitis lethargica and another case of post-encephalitic parkinsonism. J Neurol Neurosurg Psychiatry. 1989, 52(6), 800–801.

69. Elizan, T.; Casals, J. Astrogliosis in von Economo's and postencephalitic Parkinson's diseases supports probable viral etiology. J Neurol Sci. 1991, 105(2), 131–134.
70. Hallervorden, J. Anatomische Untersuchungen zur Pathogenese des postencephalitischen Parkinsonismus. Dent Z Nervenheilk. 1935, 136, 68–77.
71. Ishii, T.; Nakamura, Y. Distribution and ultrastructure of Alzheimer's neurofibrillary tangles in postencephalitic parkinsonism of Economo type. Acta Neuropathol (Berl). 1981, 55(1), 59–62.
72. Buee-Scherrer, V.; Buee, L. Pathological tau proteins in postencephalitic parkinsonism: comparison with Alzheimer's disease and other neurodegenerative disorders. Ann Neurol. 1997, 42(3), 356–359.
73. Haraguchi, T.; Ishizu, H. An autopsy case of postencephalitic parkinsonism of von Economo type: some new observations concerning neurofibrillary tangles and astrocytic tangles. Neuropathology. 2000, 20(2), 143–148.
74. Josephs, K.A.; Parisi, J.E. Alpha-synuclein studies are negative in postencephalic parkinsonism of von Economo. Neurology. 2002, 59(4), 645–646.
75. Ikeda, K.; Akiyama, H. Anti-tau-positive glial fibrillary tangles in the brain of postencephalitic parkinsonism of Economo type. Neurosci Lett. 1993, 162(1–2), 176–178.
76. Dimsdale, H. Changes in the parkinsonian syndrome in the twentieth century. Quart J Med. 1946, 15, 155–170.
77. Beringer, K. Zur Grage des Intervalls zwischen akutem Stadium and Ausbruch eines Parkinsonismus bei Encephalitis epidemica. Nervenarzt. 1937, 10, 313–314.
78. Tyndel, M. Zur Frage des Intervalles zwischen Encephalitis und Beginn des postencephalitischen Parkinsonismus. Nervenarzt. 1937, 10, 305–306.
79. Sacks, O. Awakenings; Dutton: New York, 1983.
80. Jacobi, E. Harminbehandlung bei chronischer Encephalitis. Munch Med Wochenschr. 1930, 77, 929.
81. Beringer, K. Ueber die Behandlung der Enzephalitis-Folgezustande mit Alkaloiden. Nervenarzt. 1931, 4, 171.
82. Beringer, K.; Wilmanns, K. Zur harmin-Banisterin Frage. Dent Med Wochenschr. 1929, 50, 2081–2086.
83. Beringer, K. Ueber ein neues, auf das extrapyramidal-motorische System wirkendes Alkaloid. Nervenarzt. 1928, 1, 265–275.
84. Witzleben, H.v. Die Behandlung der chronischen Encephalitis epidemica (Parkinsonismus) mit der Bulgarischen Kur. Klin Wochenschr. 1938, 17, 329–333.
85. Pette, H. Uber endolumbale Eigenserumtherapie bei Folgezustanden von epidermischer Enzephalitis. Munch Med Wochenschr. 1926, 73, 1025–1027.
86. Fuller, A. A case of encephalitis lethargica of severe type. Treated successfully by injections of electro-colloidal gold and silver. Lancet. 1926, ii, 172–173.
87. Sepp, E.; Liwschitz, J. Die Oxytherapie bei der epidemischen Encephalitis. Arch Psychiat Nervenkr. 1927, 81, 61–73.
88. Fraenkel, M. Die Beeinflussung des ubermassingen Speichelfusses bei Encephalitis lethargica chronica durch temporare Parotisausschaltung mittels Rontgenstrahlen. Dent Med Wochenschr. 1923, 49, 613.
89. Bruck, G. Die Rontgenbestrahlung des Stammhims und ihre Ergebnisse bei chronischer Encephalitis epidemica. Psychiatr Neurol Wochenschr. 1939, 41, 305–307.
90. Embden, G. Malariaimpfung bei postencephalitischem Parkinsonismus. Dent Med Wochenschr. 1926, 52, 2014–2015.
91. Craig, R.; Durk, D. The treatment of parkinsonian syndrome following encephalitis by malaria. Lancet. 1927, ii, 860–861.
92. Lust, F. Uber die Beeinflussing der postencephalitischen Schlafstoerung furch temperatursteigernde Mittel. Dent Med Wochenschr. 1921, 47, 1545–1547.

93. Froment, J. Onirisme postencephalique sequellaire ameliore par l'insuline. Rev Neurol. 1929, 37, 1162–1165.
94. Epstein, S.; Farnham, R.K.; Cobb, S. The use of salicylates in the treatment of chronic epidemic encephalitis. Boston Med J. 1927, 197, 1552–1556.
95. Orenstein, M.; Orestianu, R. Sur le traitement mercuriel dans le parkinsonisme postencephalitique. Bull Soc Roum Neurol 1930, 4, 26.
96. Gangl, P.; Luksch, F. C Vitamin und Encephalitis epidemica. Klin Wochenschr. 1939, 18, 1193.
97. Bub, L.; Pelzer, H. Ueber erfolgreiche Behandlung der Encephalitis lethargica mit Tryptaflavin. Dent Med Wochenschr. 1924, 50, 1014–1017.
98. Duensing, F.; Meyer, L. Die Behandlugn der postencephalitischen Schauanfaelle mit Calcibronat. Z Neurol. 1939, 162, 136.
99. Myerson, A.; Berlin, D. A case of postencephalitic Parkinson's disease treated by total thyroidectomy. Br Med J. 1934, 210, 1205–1206.

25

Prion Diseases

Thomas Weber
Marienkrankenhaus
Hamburg, Germany

"There are several ways," Dr. Breed said to me, "in which certain liquids can crystallize—can freeze—several ways in which their atoms can stack and lock in an orderly, rigid way." The theoretical villain, however, was what Dr. Breed called a seed. The seed, which had come from God-only-knows-where, taught the atoms the novel way in which to stack and lock, to crystallize, to freeze.

—Kurt Vonnegut Jr., *Cat's Cradle*

1 INTRODUCTION

In 1920 Hans Gerhard Creutzfeldt described the case of a 23-year-old woman (Bertha E.) who presented with a rapidly progressive spastic paraparesis, fluctuating mood with phases of euphoria, and an organic brain syndrome characterized by incoherent speech, perseveration, and dementia. Three similar cases were identified by Alfons Maria Jakob in 1920 and 1921. Neuropathologically, findings were characterized by focal loss of cortical neurons accompanied by glial proliferation and acute diffuse loss of cortical neurons (Fig. 1). In 1922 Spielmeyer suggested that these four cases constituted a new disease and first proposed the name Creutzfeldt-Jakob Krankheit [Creutzfeldt-Jakob disease (CJD)] [1]. Of these four patients, only one was later shown to have died of spongiform encephalopathy [2]. In 1924 Kirschbaum described two cases of CJD, among these the first genetic form due to a codon 178 mutation. In 1928 and 1936 Gerstmann and his colleagues described a slowly progressive disease of six years' duration, now designated as Gerstmann-Sträussler-

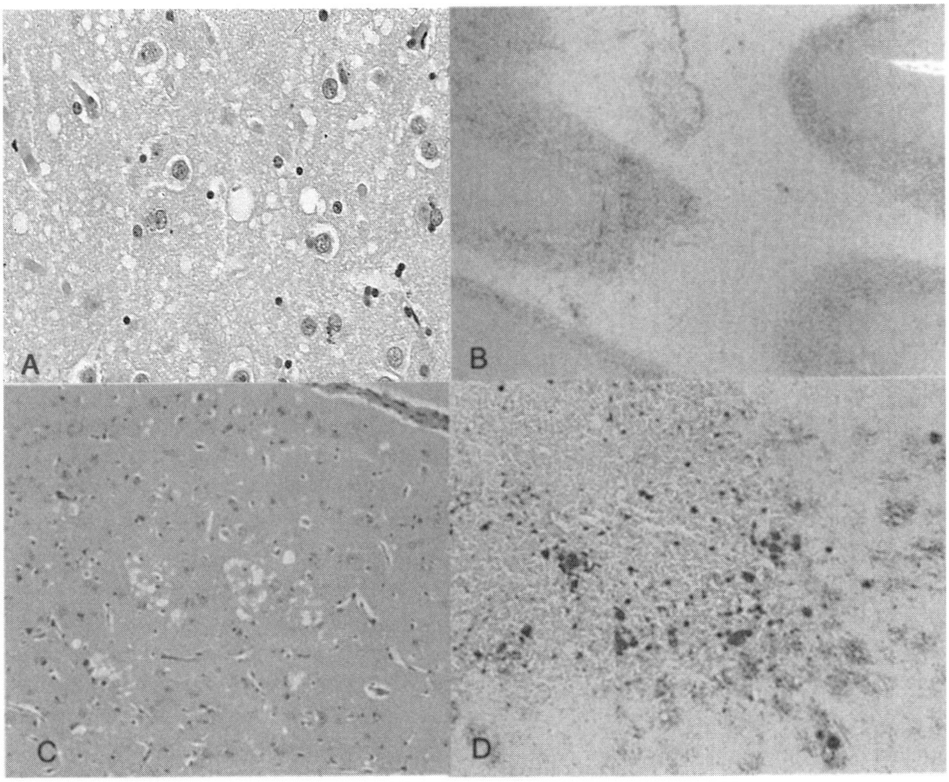

Figure 1 Histology of sCJD (A, B) and vCJD (C, D). (A) Hematoxylin-eosin (HE) stain of sCJD showing spongiform change (2–10 μm vacuoles) and confluent vacuoles (10–50 μm) as well as neuronal loss. (B) Immunohistochemical demonstration of PrPres with synaptic pattern of deposition in the granular and molecular layers of the cerebellum. (C) HE stain of vCJD showing a fluoride plaque surrounded by spongiform change. (D) Immunohistochemical demonstration of PrPres in the cerebellum. (A,B: courtesey of Dr. Hans A. Kretzschmar, Institute of Neuropathology, Munich, Germany; C,D: Courtesey of Dr. James Ironside, CJD Surveillance Unit, Edinburgh, Scotland.)

Scheinker disease (GSS), in a 25-year-old female patient presenting with ataxia, slurred speech, and mental changes.

A letter by Hadlow to the *Lancet* in 1959 drew attention to the neuropathological similarities between scrapie and kuru and suggested experimental induction of kuru in a laboratory primate [3]. Gajdusek and colleagues confirmed the transmissibility of CJD and GSS in a series of animal experiments [4]. Based on radioactive inactivation experiments, Alper implied in 1966 that the scrapie agent might replicate without a nucleic acid [5]. In 1967, Griffith postulated, among other mechanisms, a self-replicating protein as a potential cause of scrapie [6].

Prusiner [7,8] formulated his prion hypothesis in 1982 based on a series of experiments showing the unique resistance of the agent(s) of transmissible spongiform encephalopathies (TSE) to procedures inactivating nucleic acids. He stated, ''Prions are small proteinaceous infectious particles which are resistant to inactivation by most procedures that modify nucleic acids. The term 'prion' underscores the requirement of a protein for infection; current knowledge does not allow exclusion of a small nucleic acid within the interior of the particle.'' In 1985 a gene coding for the normal counterpart of the prion protein designated as PrP^{c} (proteinase-resistant prion protein cellular) was identified and opened up the avenue for the development of transgenic and knockout animal models for the study of prion diseases [9]. Further need to study the molecular basis of these enigmatic diseases came with the advent of bovine spongiform encephalopathy (BSE) in Great Britain in the 1980s [10]. With possibly more than 1 million infected cattle entering the food chain in Great Britain and Europe, a new human disease emerged in 1994 and 1995 that was first described as variant CJD (vCJD) in 1996 [11]. Molecular analysis and transmission experiments in monkeys, transgenic mice, and hamsters have confirmed beyond doubt that vCJD and BSE share a common prion [7,12–14]. Other TSEs that have been identified are transmissible mink encephalopathy (TME), first reported in 1965, and chronic wasting disease (CWD) of deer, first reported in 1980 [15]. Infections of various animals such as exotic ungulates (nyala, gemsbok, Arabian oryx, greater kudu, eland, moufflon, scimitar horned oryx, Ankole cow, bison), carnivores (domestic cat, puma, cheetah, ocelot, tiger, lion), and monkeys (rhesus monkey, lemur) with the BSE prion in zoos further substantiate the zoonotic character of BSE [7,12,14,16].

2 EPIDEMIOLOGY

The annual incidence of human prion diseases throughout the world except for the United Kingdom is about 0.88–1.4 cases per million per year. The annual mortality is 0.71 per million [17]. Of these cases, 87% were sporadic prion diseases, 8% were genetically linked, and 5% were iatrogenic. The age-specific incidence rises from 0.05 case per million per year among those aged under 39 to 3.06 cases per million per year in those aged 70–79. Due to the vCJD pandemic in the United Kingdom, the age-specific incidence rate of 0.14 in those under 39 was threefold higher than in the other participating European countries [17]. Specific risk factors for sporadic CJD (sCJD) were not identified in the largest epidemiological study reported so far [18]. Iatrogenic transmission of prion diseases in humans (iCJD) has occurred in 267 cases due to stereotactically implanted instruments, neurosurgical procedures, transplantation of dura mater, and treatment with natural human growth hormone [19]. The median incubation period was 6 years for dura mater grafts (range 1.5–18 years), 12 years for growth hormone recipients (range 5–30), and 17 months for neurosurgically infected patients (range 12–28 months). Of these patients, 60–87% were homozygous for either methionine or valine at codon 129 compared to 50% in normal white subjects. Patients with iCJD due to dura mater grafts were significantly younger (51.9 vs. 63 years) than patients with sCJD [19]. The incubation period shows a bimodal distribution, with a smaller number of patients having a short incubation period of less than 6 years and a larger group with a longer incubation period of more than 6 years. In the wake of the BSE pandemic, 127 definite and probable cases of vCJD were noted up to April 7, 2003, in the United Kingdom [20]. The annual incidence of vCJD increases by one-third per year, with 15 deaths in 1999 and 27 in 2000. A cluster of five cases of vCJD in Leicestershire, England, has caused considerable interest and was most likely due to consumption of beef contaminated by the BSE agent due to the use of the same tools for splitting cattle heads and cutting beef [21]. Sporadic prion diseases occur more frequently in humans over 60 years of age homozygous for methionine at codon 129 and in those under 49 years of age homozygous for valine at codon 129. Heterozygotes at codon 129 are significantly less affected by prion diseases [22]. All cases of vCJD analyzed so far have occurred in humans homozygous at codon 129 for methionine [23]. Depending on the incubation time of BSE in humans, the future number of vCJD cases could range from 150 to 150,000 [24].

3 PATHOPHYSIOLOGY

3.1 Strains and Biological Characteristics

According to the prion only hypothesis, prions are composed principally or entirely of abnormal isoforms of a glycoprotein, prion protein (PrP) [6–8]. These abnormal isoforms are derived from a host-encoded normal cellular glycoprotein, designated as PrP^c (cellular). The disease-related isoform, PrP^{Sc} (scrapie), is derived from the precursor PrP^c by a posttranslational process [7,12,25]. A still incompletely understood sequence of events leads to conformational changes in PrP^c resulting in the formation of PrP^{Sc}, which exists in more than 20 different ''strains'' in humans and animals [13,26]. The major criteria to define a strain of a prion are the incubation periods observed in mice with a defined genotype and the severity and distribution of pathological changes seen in the brains tissue of these mice. Strains are thus subspecies of the agent capable of maintaining a specific profile when passaged from one animal to another and, in several instances, even when passaged between different species. Strain-specific information is thought to reside solely in the tertiary and quaternary structure of the prion [7,12]. Strain typing is still performed using either inbred lines of mice or transgenic mice [27]. Biochemically, pathological isoforms of the prion are characterized by their partial protease resistance and detergent insolubility [12,28]. Prions can be further characterized by titration using experimental animals. The dilution of the agent at which 50% of the inoculated animals become ill is the unit of infective dose (ID_{50}). The unit of lethal dose (LD_{50}) is that which kills 50% of animals within their normal lifespan. The degree of infectivity of the original sample is expressed as ID_{50} units per gram from the dilution (titer) of the original sample. However, the concentration of PrP^{Sc} is often poorly correlated to the level of infectivity [14,29–31]. Two hypotheses have been put forward to explain this dilemma. First, infectivity is thought to reside in the secondary, tertiary, and quaternary structure of PrP^{Sc} and may also be influenced by the number and structure of cofactors bound to PrP^{Sc} such as sulfated sugar polymers and other N-linked sugar residues [29,32]. Second, infectivity is always characterized by number and size (aggregation state) of PrP^{Sc} [33].

3.2 Neuropathology

The neuropathological hallmarks of TSEs are

1. Vacuolation of the gray matter, giving the brain a characteristic ''spongelike'' appearance, designated as spongiform change
2. Loss of neurons
3. Astrocytosis
4. Occurrence of amyloid plaques in certain cases (Fig. 1).

Neuropathological findings vary considerably and may even be absent in cases with *PRNP* mutations [2,34]. In sCJD, only spongiform degeneration is considered as diagnostically relevant but not specific [34]. It consists first of a delicate vacuolation of the neuropil. These vacuoles frequently display an opaque appearance and range in size from 2 to 10 μm in diameter (Fig. 1). Spongiform change may be focal and is often found in neocortical areas, the thalamus, the basal ganglia, and the molecular layer of the cerebellum, whereas the hippocampus typically is spared. It may also occur in Lewy body disease. In a later stage, status spongiosus describes a condition of coalescent small vacuoles of spongifom change forming large multilobulated vacuoles. Typically this is accompanied by severe astrocytic gliosis and pronounced loss of nerve cells. Rarely, spongy degeneration with vacuoles in the perikaryon of neurons is observed in human prion diseases. Spongiform degeneration of the white matter has been described in the ''panencephalopathic variant'' of CJD. Definite diagnosis of a prion disease depends on the detection of PrP^{Sc} [7,12,34]. Given that the sensitivity of all immunological techniques is lower by several orders of magnitude compared to gene amplification methods (1×10^7–5×10^7 molecules vs. 1 to 5), it is not surprising that both human and experimental prion diseases have been reported in which PrP^{res} could not be detected [14,35]. Based on the genotype at codon 129, three different neuropathological categories of sCJD can be distinguished (Table 1). The most common type is M129M (56%), the second most common type is V129V (28%), and the least common type is M129V (16%). It appears that the Val genotype enhances production of PrPres and the Met/Val genotype facilitates its aggregation into amyloid plaques [36]. Taking not only the genotype at codon 129 but also two types of PrPres into consideration, six phenotypic variants of sCJD have been identified [37,38]. By Western blot (WB) at least two types of PrPres with a relative molecular mass of 21 kDa for type 1 of the deglycosylated protein and 19 kDa for type 2 can be distinguished [28]. This difference in size reflects different conformations adopted by PrP^{Sc} due to different cleavage sites of proteinase K, at residue 97 in FFI and at residue 82 in D178NfCJD and P102LGSS [7,12]. In the brains of the majority of patients, only one type of PrPres can be identified, but in about 5–36% both types are identified primarily in the cerebral cortex [38]. In cases where regional WB analysis of PrPres type 1 and type 2 was performed, type 1 was found in areas with diffuse PrP immunoreactivity and type 2 was strictly associated with perivacuolar and plaquelike deposits. Frequently, type 1 and type 2 WB patterns occur in different regions; only in a few regions (temporal cortex, thalamus geniculatis lateralis, and hypothalamus) do both types coexist. In accordance with these neuropathological types, six clinical syndromes can be distinguished [37].

Table 1 Neuropathological Lesions and Clinical Characteristics

Codon 129 genotype	PrP type	Most prominent sites of neuropathological lesions	Leading clinical findings
MM	1	Cortex, cerebellum	Dementia, myoclonus
MM	2	Thalamus and inferior olive	Thalamic form (SFI): dysautonomic disturbances
		Cortex (focal), basal ganglia	Cortical form: dementia
MV	1	Cortex, cerebellum	Dementia, myoclonus
MV	2	Cortex (focal), basal ganglia, thalamus, presence of kuru plaques	Ataxia, dementia, extrapyramidal signs
VV	1	Cortex and basal ganglia	Dementia
VV	2	Cerebellum, basal ganglia, thalamus, deep layers in the cortex	Ataxia; dementia late in the course

3.3 Structure and Function of PrP^c and PrP^{Sc}

The cloning of a cellular gene encoding PrP^c in hamsters, mice, rats, mink, and humans opened up avenues for the development of transgenic or knockout animal models for functional studies [7,9,12]. Conflicting results were reported using different strategies to disrupt the *Prnp* allele [39–42]. The so-called Zürich I $Prnp^{0/0}$ and the knockout line designated Edinburgh $Prnp^{-/-}$ remained clinically healthy. The Nagasaki $Prnp^{+/-}$ strain, however, developed ataxia and loss of cerebellar Purkinje cells at 6–12 months of age. Comparison of the strategies to knock out the *Prnp* and sequencing of the region downstream of the murine *Prnp* gene discovered an open reading frame (ORF) encoding a protein designated Dpl (Doppel, German for double) [42]. Dpl mRNA is expressed at high levels in the brains of Nagasaki and Rcm0 mice due to chimeric mRNA originating at the *Prnp* promoter, running all the way past the *Prnd* ORF (gene encoding Dpl). Dpl has about 25% identity with the carboxyl-terminal two-thirds of murine PrP and is predicted to contain three α-helices and a disulfide bond between the second and third helices, as does murine PrP^c. The function of PrP^c has not yet been elucidated. In human brain, PrP^c is mainly produced in upper cortical neurons in the neocortex and in cerebellar Purkinje cells [7,12]. PrP^c has been shown to bind copper in vitro and may function as a recycling receptor for the uptake of copper ions from the extracellular milieu [43]. Recently, PrP^c was shown to trigger the activation of the tyrosine kinase Fyn, thus functioning as a signal transduction protein [44]. PrP^c is also expressed on T and B lymphocytes and appears to participate in signal transduction in human T lymphocytes [45]. Chemical differences between PrP^c and PrP^{Sc} were long thought not to exist [16]. Recent findings, however, show a difference in the relative proportion of bi-, tri-, and tetraantennary N-linked oligosaccharides between normal Syrian hamster $shPrP^c$ and $shPrP^{Sc}$ [32]. It remains to be shown whether this is secondary to the disease or attributing to the conformational change in PrP^{Sc}. Given the high sequence identity between hPrP(23–230), mouse mPrP(23–231), and shPrP(29–231), the three-dimensional structures of the C-terminal domain are very similar. These proteins contain a globular domain (approximately residues 125–228) and an N-terminal flexibly disordered ''tail'' [46]. The globular domain contains a two-stranded antiparallel β-sheet and three α-helices. In hPrP(23–230) the three α-helices comprise the residues 144–154, 173–194, and 200–228, and the short antiparallel β-sheet is found in residues 128–131 and 161–164. Species variations are found in the region of α-helix 3 and the loop 167–171. The length of helix 3 in hPrP coincides more closely with ShPrP whereas the disordered loop 167–171 is shared with mPrP. This is the surface area that has been suggested to contain a binding epitope for a putative protein X promoting transition from PrP^c to PrP^{Sc} [7]. Biochemical findings in E200KPrP indicate the existence of an intermediate product of wtPrP, also designated as PrP, in the pathway from PrP^c to PrP^{Sc} [47]. Two competing theories have been developed to explain the replication of prions: template-directed refolding and nucleation (Fig. 2) [48]. Recent evidence strongly supports the concept that mutant PrP undergoes a self-conversion process in brain tissue to PrP^{Sc} [47].

3.4 Transmission and Pathogenesis of TSEs

Modes of transmission to be considered in TSEs are lateral and vertical transmission. Vertical transmission could occur via the placenta or breastfeeding. Lateral transmission occurs by exposure to a common foodstuff, i.e., infected meat and bone meal (MBM) in

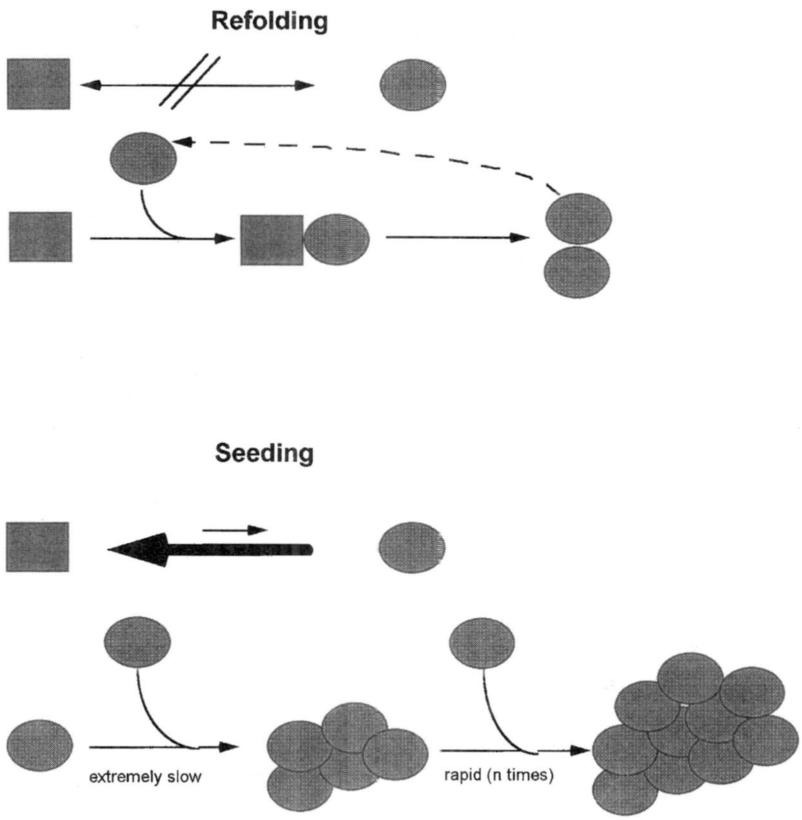

Figure 2 Models for the conformational conversion of PrP^c to PrP^{Sc}. Subsequent to exogenous introduction of PrP^{Sc}, PrP^c changes shape and forms PrP^{Sc}. A high activation energy barrier prevents spontaneous conversion at detectable rates. Extensive unfodling and refolding of the protein may explain the energy barrier. The process leads to an exponential conversion cascade. With certain mutations in PrP^c, spontaneous conversion of PrP^c to PrP^{Sc} may be rare. This could explain the late occurrence of most familial TSEs. sCJD could occur through spontaneous conversion of PrP^c to PrP^{Sc}. In the seeding model the conformational change between PrP^c and PrP^{Sc} is thermodynamically controlled and thus reversible. PrP^{Sc} is stabilized only when it adds onto a crystal-like seed or aggregate of PrP^{Sc}. Seed formation is extremely slow, but once a seed is present monomers can add on rapidly. (From Ref. 110).

the case of BSE. sCJD has been transmitted by contaminated neurosurgical instruments or stereotactically implanted electrodes and direct intracerebral inoculation, by corneal transplants via the optic nerve, by the use of infected dura mater grafts, and by peripheral inoculation by subcutaneous injection in cases treated with infected human growth hormone or gonadotrophin [19]. Whether sCJD is horizontally and/or vertically transmitted at all in humans is still unclear.

Possible explanations for the absence of transmission of sCJD between humans by blood or blood products are (1) the absence of significant plasma infectivity until the onset of symptomatic disease and comparatively low levels of infectivity during the symptomatic stage of disease; (2) the reduction of infectivity during plasma processing; and (3) the need for the infectious agent to be at least five to seven times more infections to transmit disease by the intravenous route than by the intracerebral route [49]. Transmission of sCJD has been achieved experimentally by inoculation of brain homogenates and peripheral injection either subcutaneously, intravenously, or intraperitoneally [49].

Transmission of BSE by contaminated MBM is now proven beyond doubt by epidemiological and experimental data [7,10,12,13,49,50]. After oral exposure, infectivity and PrP^{Sc} are found in intestinal lymphoid follicles (Peyer's patches) [51]. Like macromolecules, particles, and microorganisms, the BSE agent passes from the intestinal lumen to the underlying lymphoid tissue via the follicle-associated epithelium (FAE) of Peyer's patches. This function is mediated by the endocytotic activity of membranous (M) cells [52]. Depending on species and location in the gut, the ability of M cells to collect specific pathogens varies. It is not yet known whether these regional and species differences also apply to PrP^{Sc} and all or only some to ''strains.'' At least the 263K scrapie strain has been shown to accumulate in the FAE in hamsters and lambs [51]. The particular susceptibility of cattle to BSE may be explained in part by the seven-fold higher permeability of the colonic mucosa of cattle compared to sheep for scrapie prion [53]. Another candidate molecule for the binding of PrP^{Sc} is PrP^{c}, which has been reported to act as a ligand and/or receptor for PrP^{Sc} in a cell-free system. PrP^{c} is expressed in the epithelium of the gastrointestinal tract [54]. The decreased number of Peyer's patches in older animals and humans may also contribute to the reduced suseptibility to oral transmission of TSEs in the older age groups.

Other factors contributing to increased intestinal uptake of macromolecules are genetic regulation, for instance host PrP genotype, and infection or inflammation [7,12]. In cattle, there is only a short, transient period during which infectivity can be demonstrated in the terminal ileum. Later, BSE prions can be demonstrated only in brain, spinal cord, and dorsal root ganglia. Accordingly, neuroinvasion can be divided into two phases. In the first, widespread colonization of the immune system takes place. In the second phase, expression of PrP^{c} is required for the replication of PrP^{Sc} in the lymphoreticular system (LRS) and transfer to the peripheral nervous system (PNS) [55]. Upon passage to humans, this agent shows a drastic shift in organo- and cell tropism. It is no longer confined to neural structures but can be detected in tonsils, spleen, and the appendix [56].

Experiments in transgenic mice and in mice with various immune defects have implicated follicular dendritic cells (FDCs) as the primary type of cell for prion replication in the lymphoid tissue [57]. Again, the pattern of replication of prions may vary from strain to strain. Using RML scrapie, Aguzzi et al. [58] demonstrated in a series of elegant experiments that B lymphocytes play a critical role in peripheral scrapie pathogenesis. Transgenic PrP knockout mice expressing PrP restricted to either B or T lymphocytes show no prion replication in the LRS [59]. Furthermore, treatment of mice with soluble lymphotoxin-beta receptor, which causes the disappearance of mature FDCs from the spleen, abolishes splenic prion accumulation and retards neuroinvasion. Early accumulation of prions takes place on FDCs within germinal centers in lymphoid tissue of patients with vCJD, sheep with natural scrapie, or rodents after experimental infection with scrapie. For optimal replication in the LRS, both stromal and hematopoietic compartments must express PrP [60].

Depletion of complement component C3 or genetic deficiency of C1q significantly delays the onset of disease, indicating that C3 and C1q localize TSE infectivity in the early phase of disease to lymphoid tissue [61]. After peripheral inoculation, prions spread hematogenously to the LRS, where they replicate. Later, infection spreads from there along visceral autonomic nerves of the enteric nervous system (ENS) to the thoracic spinal cord and from there to the brain [51,59]. In the CNS, expression of PrP^c is required for the replication of PrP^{Sc} [58]. $Prnp^{0/0}$ mice, which lack PrP^c, are resistant to scrapie and do not propagate prions. After grafting of brain tissue overexpressing PrP^c into the brain of $Prnp^{0/0}$ mice and inoculation with scrapie prions, only the graft developed severe histopathological changes characteristic of scrapie. Intraocular inoculation with scrapie in transgenic mice overexpressing PrP^c and $Prnp^{0/0}$ mice transplanted with a PrP^c-overexpressing graft revealed the requirement of PrP^c for spread of scrapie prions along neural pathways. PrP^c may be required by scrapie and other prions for propagation along synapses [12]. Transport of prions either within or on the surface of neurons may be PrP^c-dependent.

3.5 Prions in Yeast and Fungi

Further support for the prion hypothesis has come from the discovery of prions in yeast and fungi and has offered the unique opportunity to study the function of prions [62–64]. In yeast and fungi, prions transmit not a disease but particular biochemical and/or cultural characteristics. They are thus capable of self-perpetuating changes in conformation and function, serving as a genetic element. In *Saccharomyces cervisiae* and *Candida albicans*, the non-Mendelian element [URE3] (ureidosuccinate) is due to a prion change of Ure2p, a regulator of nitrogen metabolism, and results in slow-growing cells. Another yeast prion is [PSI], which is due to an aggregating form of Sup35p, which is one of the translation termination proteins whose misfunction in cells carrying PSI results in abnormal read-through of translation termination codons. A third candidate prion, [Rnq1], has recently been identified in yeast; it exists in distinct, heritable physical states, soluble and insoluble [62]. A common biochemical motif in these yeast prions is glutamine/asparagine (Q/N)-rich domains, which have a high propensity to form self-propagating amyloid fibrils. Heterologous prions appear to interact positively but also negatively in the de novo generation of new prions (Fig. 2).

4 CLINICAL MANIFESTATIONS

4.1 Clinical Spectrum of Human Prion Diseases

It has become evident that TSEs in humans constitute a spectrum of spontaneous, inherited, and transmitted diseases [16]. Even prion diseases due to a single-point mutation in the *PRNP* may show considerable variation in the age of onset, duration of disease, clinical presentation, and neuropathological findings between patients or within a family [65]. Bearing this in mind, the following description gives the ''typical'' clinical feature that gave each disease its original name and then broadens the scope to include the clinical characteristics currently associated with its genotype. For historical reasons the most common form of human prion diseases is designated as CJD and certain genetic forms are called GSS. By far the most common form is sCJD. Clinically, sCJD has been diagnosed on the basis of criteria first outlined by Masters and colleagues in the European Surveillance on CJD (Table 2) [11,18,66]. Inclusion of laboratory and radiological criteria has led to new diagnostic criteria by WHO (Table 3) [67].

Although the typical age at which sCJD manifests itself is in the sixth decade, cases have occurred in patients aged 40 years and younger [37]. Recent evidence in molecularly

Table 2 Diagnostic Criteria for CJD

Level	Requirement
Probable	1. Rapidly progressive dementia of less than 2 years' duration *and* 2. Periodic sharp wave complexes (PSWC) detected by EEG and *at least two* of the following four clinical signs/symptoms a. Myoclonus b. Visual and/or cerebellar signs/symptoms c. Pyramidal and/or extrapyramidal signs d. Akinetic mutism
Possible	1. Rapidly progressive dementia of less than 2 years' duration *but* 2. PSWC not detected and *at least two* of the following four clinical signs/symptoms: a. Myoclonus b. Visual and/or cerebellar signs/symptoms c. Pyramidal and/or extrapyramidal signs d. Akinetic mutism
Other	1. Rapidly progressive dementia of more than 2 years' duration, *but* 2. PSWC not detected and *at most one* of the following four clinical signs/symptoms identified: a. Myoclonus b. Visual and/or cerebellar signs/symptoms c. Pyramidal and/or extrapyramidal signs d. Akinetic mutism

defined familiar forms of human prion diseases supports the concept of a spectrum of diseases with mutually overlapping signs and symptoms [65]. Keeping this notion in mind, sCJD is a rapidly progressive dementing disease that occurs most commonly in patients in their sixth decade. Besides dementia, CJD presents with pyramidal and extrapyramidal signs and symptoms, cerebellar signs and symptoms, visual disturbances, and akinetic mutism and lasts on average about 7 months. sCJD can be divided into three phases. An uncharacteristic prodromal phase lasting from days to a couple of weeks is followed first by a symptom-free interval of variable length, then by a characteristic clinical syndrome that can be easily identified from the end of the bed within minutes. In the prodromal

Table 3 WHO Criteria for the Diagnosis of sCJD

Progressive dementia
At least two of the following findings:
Myoclonus
Visual or cerebellar disturbance
Pyramidal or extrapyramidal dysfunction
Akinetic mutism
Characteristic electroencephalographic findings during an illness of any duration or a positive assay for the 14-3-3 protein in cerebrospinal fluid and a fatal illness of less than 2 years duration, or both
No suggestion of an alternative diagnosis given by routine studies

phase, uncharacteristic findings such as exhaustion, weight loss, depression, inability to sleep, abnormal sweating, and behavioral changes ranging from anxiety, social withdrawal, and loss of interest to aggressiveness, nervousness, and agitation are reported in 14–60% of patients.

In the second, clinical phase, the most common finding is a rapidly progressive dementia, followed by myoclonus, ataxia, and cerebellar signs, extrapyramidal signs, pyramidal signs, unspecified visual disturbances, and visual hallucinations. Supranuclear gaze palsy may be found as a rare presenting symptom and more frequently in the second phase of the disease [65]. A peculiar response to a loud noise or touch characterized by a sudden myoclonic jerk and stiffening of the patient, designated startle response, may be observed.

In line with a recent neuropathological subclassification of six sCJD forms, clinical findings permit the distinction of at least six forms of sCJD [28,37,38]. These entities are classified according to the polymorphism at codon 129 and the structure of PrPSc. Cognitive decline is the most frequent initial symptom in patients with MM1 and MV1. Patients with VV2 frequently show ataxia as the presenting sign, and one-third also present with dementia. Patients with MV2 are characterized by dementia, ataxia, and extrapyramidal disturbances mainly of an akinetic rigid type, whereas myoclonus was only exceptionally present in the end stage of disease [37]. Distinction between these clinical phenotypes based on clinical signs and symptoms is best at disease onset. The third or terminal phase is characterized by akinetic mutism, stupor, and coma.

4.2 Iatrogenic CJD

The two most common forms of iatrogenic CJD (iCJD) are those due to dural grafts and injection of contaminated human growth hormone (hGH) [19]. In addition to earlier age of onset compared to sCJD, iCJD cases due to dura mater grafts showed a mean incubation period of 98.9 months ($\pm$45.9 months) with a duration of disease from onset to death of 14.6 $\pm$ 13.4 months [68]. Of these patients, 18 had the genotype M129M, three were V129M, five displayed the genotype E219E, one the genotype E219K. Clinically, these patients have significantly more frequent cerebellar ataxia at onset (56% vs. 12%), disorientation (56% vs. 18%), and disturbances of vision or external ocular movement (51% vs. 18%) than sCJD cases. Iatrogenic CJD cases due to contaminated hGH have a median incubation period of 12 years (range 5–30) [19]. The clinical presentation is homogeneous, with cerebellar ataxia and ocular motor disorders in about 90% at onset. Within months, these patients go on to develop myoclonic jerks and dementia.

4.3 Variant CJD

Variant CJD (vCJD) was first recognized as a new disease in 1996 [11]. The median age of vCJD patients reported so far is 26 (range 12–74) with a male/female ratio of 1:1. Most patients with vCJD present initially with heterogeneous psychiatric symptoms and sensory disturbances, both of which are highly unusual findings in sCJD [69]. Initial features are social withdrawal, anxiety, paranoid delusions, apathy, emotional lability, insomnia leading to the diagnosis of a major depressive illness, anxiety disorder, or even schizophrenic psychosis (Table 4). Frequently, delusions such as a belief that there are snipers in the kitchen, that microscopic people are inside the patient's body, or that the patient has murdered someone are present. At the same time, about three-fourths of these patients develop visual and, less commonly, auditory hallucinations. About one-fourth to one-third

Table 4 Criteria for the Diagnosis of Variant CJD

Progressive neuropsychiatric disorder
Duration of illness > 6 months
No alternative diagnosis suggested by routine investigations
No history of potential iatrogenic exposure
Early psychiatric symptoms
Persistent painful sensory symptoms
Ataxia
Myoclonus or chorea or dystonia
Dementia
Typical appearance of sCJD not shown by EEG, or no EEG performed
Posterior thalamic high signal on MRI scan
Positive tonsil biopsy

of patients present with sensory symptoms such as dysesthesias or paresthesias in the face, hand, or foot [69]. Almost all patients experience early anorexia and weight loss. Neurological signs and symptoms appear during the course of the illness after about 6 months (median 6.25, range 4–24.5 months). These manifest predominantly as cerebellar ataxia, rapidly progressive dementia, myoclonus, pyramidal signs, and upward gaze paresis. Later on, these patients develop involuntary movements, akinetic mutism, and, in a few cases, cortical blindness. It appears that those patients with the longest delay to the development of neurological signs had a long prodrome with personality change or forgetfulness followed by sensory disturbances. During the course of disease the patients usually develop primitive reflexes, cerebellar and pyramidal signs, and persistent involuntary movements manifesting as either chorea or myoclonus. Death occurs after a median duration of 14 months (range 9–35 months).

4.4 Genetically Defined Prion Diseases

As a general rule, the N-terminal part of the *PRNP* (amino acid residues 23–90) harbors insertions and deletions, whereas the C-terminal portion (amino acids 91–231) mainly carries point mutations [70]. Fatal insomnias constitute part of the expanding spectrum of human prion diseases and may occur spontaneously or can be inherited [65,71,72]. Even in familial cases of fatal insomnia, the entire spectrum of human prion diseases may be seen over several generations [73]. The earliest symptom of a total inability to sleep was accompanied by the absence of typical EEG sleep patterns. In the further course of disease, total sleep time decreased to about 1 h in 24 h. Sleep or coma induced by barbiturates or benzodiazepines was associated with flattening of the EEG tracing without the typical stages of drug-induced sleep [74]. It soon became evident that this disease was autosomal, dominantly inherited, and associated with a point mutation at codon 178 of the *PRNP* [7,12]. In these cases, aspartic acid at codon 178 is exchanged for asparagine (D178N). The D178N FFI mutation is the most common mutation found in a large surveillance program of sCJD.

Initial reports described an earlier age of onset and a much more rapid course of disease for those patients who were homozygous for methionine at codon 129 than for those who were heterozygous. Even though the clinical course and even the neuropathological findings are not specific in FFI, heterozygous patients show a much longer course of disease than homozygous patients [73,75]. Onset of disease does not occur earlier in heterozygous patients than in homozygous patients. In addition, at onset, homozygous FFI patients had prominent oneiric episodes, insomnia, and dysautonomia, whereas heterozygotes presented with ataxia, dysarthria, loss of sphincter function, and grand mal [76].

With more cases being analyzed, it has become apparent that the D178N FFI mutation shows enormous clinical and neuropathological variability, which may lead to misdiagnoses such as olivoponto cerebellaratrophy (OPCA), autosomal dominant cerebellar ataxia (ADCA), Parkinson's disease, Alzheimer's disease (AD), CJD, and GSS (Table 5) [65,73,77]. To make matters even more complicated, sporadic human prion disease (sCJD) may mimic FFI. This particular phenotype has been designated as fatal insomnia [71,72]. Familial CJD (fCJD) D178N is molecularly defined by the same point mutation at codon 178 as FFI but with valine at codon 129 on the mutated allele *PRNP* [7,12]. Clinically, these patients have an earlier onset of disease than sCJD cases (43 vs. 66 years) and the course of disease is much longer than in sCJD (23 vs. 9 months) (Table 6). They present with memory loss, behavioral changes, and altered mood and show early myoclonus. PSWCs are not detected. Even within this genotype, phenotypic variation exists, as shown by patients presenting with aphemia, apraxia, uncontrolled laugh, and lack of aphasia *PRNP* [78]. It appears, that the course of disease is more rapid in D178N fCJD homozygous for valine at codon 129 than in those heterozygous at this codon. In contrast to sCJD, D178N fCJD never shows PSWCs on EEG recordings. A second mutation causing fCJD occurs at codon 200 with a change from glutamic acid to lysine (E200K) *PRNP* [7,12]. This mutation is frequently found in clusters of isolated ethnic groups such as Jews from Libya, Greece, and Tunisia and patients in Slovakia, Chile, and France. The clinical presentation of these patients is similar to that of sCJD, except for a slightly younger age of onset (59 vs. 66 years). Compared to heterozygous patients, homozygotes show a younger age of onset (50 vs. 59 years) but a comparable clinical course [35]. The mean duration of illness of 7 months in E200K heterozygotes is also comparable to that of sCJD, whereas that of 15.8 months in the homozygotes is much longer [35]. Frequently, these patients present with a supranuclear gaze palsy and have no PSWcs on EEG recordings. Myoclonus is also not observed in these patients [79]. A very rare mutation (Y145STOP) was reported in an old Japanese patient suffering from a very long-lasting, mainly dementing, illness clinically diagnosed as AD, pathologically characterized by PrP-amyloid in small and medium-sized vessels, and thus also designated as PrP cerebral amyloid angiopathy. Another mutation was described that resulted in an asparagine-to-serine alteration at codon 171 (N171S) that appears to be associated with a longstanding psychiatric disorder characterized by persecutory delusions auditory hallucinations, severe depression, dementia, and finally walking difficulties and mutism.

Gerstmann-Sträussler-Scheinker Phenotype

A very rare form of human prion disease is Gerstmann-Sträussler-Scheinker disease (GSS) with an incidence of about 0.1 case per million per year [16]. Presenting symptoms vary, depending on the pathogenic mutation. Typically, GSS starts in the fourth decade and lasts for about 5 years [80,81]. The first mutation to be identified was a proline-to-leucine change at codon 102o(P102LGSS) [7,12,82]. The most common genotype is P102L–129M. This is also the second most common form of any mutations in the *PRNP* and the one carried by the original GSS family (Table 7). Again, sCJD cases with the "classical" GSS phenotype have been reported [83]. Also, cases with a clinical course

Table 5 Genotype Phenotype Relation in Inherited Human Prion Diseases

Prion disease (phenotype)	Insertion of mutation of the PRNP
Typical clinical and neuropathological findings of CJD	E200K (most common), D178L-129M, V180l, V1801-M232R, V210l
Thalamic form (FFI)	D178N–129M
GSS with ataxia and variable spongiform change	P102L (second most common), A117V
Mixed syndromes of CJD and GSS with atypically long duration and variable spongiform change	Insertion in the octapeptide repeat region 145 (stop codon)
GSS with neurofibrillary tangles	198, 217
GSS with spastic paraparesis	105

Table 6 Familial CJD (fCJD) Synopsis of Clinical and Neuropathological Findings

Mutation/ polymorphism (codon)	Clinical findings	Neuropathology
D178N–129V	Early onset, long duration (23 vs. 9 mo), presenting with memory disturbances, behavioral changes, altered mood, early myoclonus, no PSWC in EEG	Variable, cerebral cortex and basal ganglia most severely affected, spongiform change and gliosis prominent, little neuronal loss, no plaques, PrPres type 2
E200K–129M	Onset earlier (56 vs. 66 yr), no PSWCs, no myoclonus, supranuclear gaze palsy. Even earlier onset (50 yr) in E200K homozygotes	As sCJD, spongiform change, gliosis, loss of neurons, very rarely plaques, type 1 PrPres with different glycotype from sCJD (1B)
E200K–129V	As sCJD	Spongiform change, gliosis, loss of neurons, cerebellar plaques, type 2 PrPres
Y145STOP	One case, memory disturbance, disorientation dementia, extremely long duration of disease (21 yr)	PrP-amyloid in small and medium-sized parenchymal blood vessels, no spongiform change, NFT in the cerebral cortex
N171S–129V	Early onset, mainly psychiatric symptoms, i.e. auditory hallucinations, delusions, social withdrawal, severe depression, mutism, paraparesis, dementia, mutism	Not done

Table 7 Mutations Associated with GSS

Mutated codon/ polymorphism	Clinical findings	Neuropathology
P102L-129M or-129V	Variable, protracted clinical course of 5 yr, early cerebellar ataxia, nystagmus, paresis, dysarthria, rarely myoclonus, PSWCs. In others, short clinical course of 1 yr or less, dementia, "ataxic" form of GSS.	PrP plaques of kuru and multilobular type in cerebrum and cerebellum, spongiform changes vary, more severe in cases with rapid progression, degeneration of spinocerebellar tracts and dorsal column atrophy, atrophy of pons, subcortical nuclei, and thalamus.
P102L-129M or-219K	Variable, either ataxic with cerebellar signs or progressive dementia.	Variable, absence of amyloid plaques.
P105L-129V	Onset in the fourth and fifth decades, duration 6–12 yr, spastic paraparesis with hyperreflexia and bilateral pyramidal tract signs, no myoclonus or cerebellar signs, no PSWCs.	Prominent multicentric amyloid plaques in the cerebral cortex (motor area, striatum, thalamus), neuronal loss, astrocytic gliosis, rarely NFTs, no spongiform change.
A117V	Variable, initially dementia, later pyramidal tract and pseudobulbar signs, anticipation in successive generations; in other families ataxia; "telecephalic" form.	Widespread PrP-amyloid plaques throughout the cerebrum (cortex, basal ganglia, thalamus); rarely in the cerebellum in patients with dementia only
F198S-V129V or-M129V	Homozygotes at 129 for valine have an earlier onset (by more than 10 yr) of disease than heterozygotes, gradual loss of short-term memory, parkinsonism, cerebellar signs.	Widespread unicentric and multicentric PrP plaques in the cerebrum (frontal, insular, temporal, and parietal cortex), cerebellum, and midbrain. NFT in the neocortex, subcortical gray matter (frontal, cingulate, parietal, insular, and parahippocampal).
Q217R	Onset in the sixth decade, initially mania or depression followed by memory loss, progressive ataxia, parkinsonism, dementia; death after 5–6 yr.	PrP-amyloid in cerebrum (neocortex, amygdala, substantia inominata, thalamus) and cerebellum. Numerous NFT in cerebral cortex, subcortical gray structures.

indistinguishable from that of sCJD with the P102L–129M *PRNP* mutation have been reported. These findings underline the need for the identification of further factors besides mutations in the *PRNP* gene that modify disease phenotype. This remarkable phenotypic variability is supported by GSS families without any correlation to the codon 129 or codon 219 genotype. The most common genotype is P102L-129M with signs of early cerebellar ataxia, spinal cord involvement, dysarthria, and nystagmus [7,12,82]. The codon 105 mutation occurs predominantly in Japan and is best characterized as a form of hereditary spastic paraparesis [7,12,84]. The age of onset is in the fourth or fifth decade; and the duration of disease is between 6 and 12 years. Initially, patients with the alanine-to-valine mutation at codon 117 (A117V) were described as demented or as having the telencepathic form of GSS [85]. Thus the first generation presents with dementia, whereas later generations show a triad of pyramidal and pseudobulbar syndromes and dementia associated with spinal cord and cerebellar features as well as amyotrophy, myoclonus, and epileptic seizures. With the recent description of a fourth English family, the phenotypic spectrum of this mutation has widened again (Table 7). Cases in this kindred presented with presenile dementia, ataxia, and neuropsychiatric features, leading to diagnoses such as AD, multiple sclerosis, and corticobasal degeneration. At the other end of the spectrum is a ''classical'' ataxic form of GSS in an A117V family. A mutation at codon 198 with a phenylalanine-to-serine substitution in association with a valine allele at codon 129 (F198S–129V) has been described in an Indiana family [86]. These patients present with a gradual loss of short-term memory and parkinsonism characterized by clumsiness, bradykinesia, rigidity, dysarthria, and early subcortical dementia. Disease progression varies between less than 1 year and over 5 years. Age at onset is more than 10 years earlier in patients homozygous at codon 129 for valine. A similar clinical presentation is seen in patients with a glutamine-to-arginine substitution at residue 217 (Q217R). Age of onset is 62–66 years, and duration of disease is 5–6 years. Initial manifestation may be depression or mania, followed by gradual memory loss, parkinsonism, progressive ataxia, and dementia [87].

Octapeptide Repeats

At least 18 octapeptide repeat mutations have been reported to date in the unstable region of five variant tandem octapeptide coding repeats between codons 51 and 91 of *PRNP* (Table 8) [7,88–91]. Deletions in this region are not associated with human prion diseases [7]. The octapeptide repeat forms of human prion disease show a remarkable clinical and neuropathological heterogeneity from those with features very similar to those of sCJD to phenotypes with personality disorders, early onset, and a much more prolonged course. Usually, patients with one, two, or four extra repeats lack a family history of neurological disorders and present with signs and symptoms of classic sCJD. Patients with five, six, seven, eight, or nine extra repeats present with an autosomal dominant pattern of inheritance and a very heterogeneous phenotype with features of classic CJD, GSS, or atypical dementia [7,89]. However, a case with a two-octarepeat insertion and two nucleotide substitutions in the other octapeptide presented as a moderately progressive dementia with presenile onset, later developed gait ataxia and paraphasias and showed agraphia, apraxia, acalculia as well as forced laughing and crying, and died 7 years after onset, thus providing another example for an exception from the above-mentioned rule (Table 8) [91]. These patients had an onset of symptoms at the age of 28 years, with a younger mean age in the fifth generation. Initially, these patients presented with psychiatric symptoms classified as mania, depression, or schizophrenia preceding dementia and cerebellar signs by up to 12 years.

5 DIFFERENTIAL DIAGNOSIS

The clinical differential diagnosis of CJD includes AD in particular in cases with a prolonged course of dementia, dementia with Lewy body disease (DLB) with parkinsonian features, and fluctuations in the clinical course [92,93]. Other diseases that may be confused with CJD are frontotemporal dementias, corticobasal ganglionic degeneration, Huntington's disease, progressive supranuclear palsy, vascular dementia, anoxic encephalopathy, Hashimoto encephalitis, paraneoplastic encephalitis, chronic encephalitis, intoxications, nonconvulsive status epilepticus, and AIDS dementia complex [92]. These diseases need

Table 8 Normal and Mutated Octapeptide Repeat Regions

Normal	R1	R2	R2	R3	R4									
Insertions (× 24 bp)	R1	R2	R2	R2	R3	R4								
+1														
+2	R1	R2	R2	R3	R2a	R2a	R4							
+2	R1	R2	R2a	R2	R2a	R2a	R4							
+4	R1	R2	R2	R3	R2	R2	R2	R3	R4					
+4	R1	R2	R2	R2	R2	R2	R2	R3	R4					
+4	R1	R2	R2	R3g	R2	R3g	R2	R3	R4					
+5	R1	R2	R2	R3	R2	R3g	R2	R2	R3	R4				
+5	R1	R2	R2	R3	R2	R2	R2	R2	R3	R4				
+5	R1	R2	R2	R3g	R3g	R3g	R2	R2	R3	R4				
+6	R1	R2	R2	R2	R3	R2	R3g	R2	R2	R3	R4			
+6	R1	R2	R2	R3	R2	R3g	R2	R3g	R2	R3	R4			
+6	R1	R2	R2	R3g	R2	R2	R3g	R2	R2	R3	R4			
+6	R1	R2	R2	R2	R2	R2	R2	R2	R2	R3	R4			
+7	R1	R2	R2c	R3	R2	R3	R2	R3	R2	R3g	R3	R4		
+8	R1	R2	R2	R3	R2	R2	R2	R2	R2	R2	R2	R2a	R4	
+8	R1	R2	R2	R3g	R3	R2	R2	R2	R2	R2	R2	R3	R4	
+8	R1	R2	R2	R3	R2	R2	R2	R2	R2a	R2	R2	R2	R3	R4
+9	R1	R2	R2	R3	R2	R3	R3g	R2	R2a	R2	R3	R2	R3	R4
+9	R1	R2	R2	R3	R2	R3g	R2a	R2	R2	R2	R3g	R2	R3	R4

All mutations are silent and occur at the third position of the codon. R2a is a G`A, change, R2c is a T`C, change, and R3g is a A`G, change.
Source: Refs. 7, 88, 89.

to be differentiated from human prion diseases by careful clinical, neurophysiological, biochemical, toxicological, molecular genetic, and imaging analyses.

5.1 Neurophysiological and Radiographic Findings

Electroencephalography

All human prion diseases are best distinguished from each other by the particular pattern of appearance of neurological and psychiatric signs and symptoms in the early stages of disease [37]. An electroencephalographic pattern designated as periodic sharp wave complexes (PSWC) is characterized by periodic cerebral potentials with a duration between 100 and 600 ms and an interval between 500 and 2000 ms (Fig. 3) [94]. In a blinded study, the detection of PSWCs was shown to have a sensitivity of 67% and a specificity of 86% for the diagnosis of sCJD. A large European study showed an almost identical sensitvity of 66% but a lower specificity of 74% [95]. Further subtyping of sCJD revealed that PSWCs are found in 78% of patients with type 1 PrPres but in only 4% of patients with type 2 PrPres [37]. sCJD patients homozygous at codon 129 for valine almost never have PSWCs in their EEG recordings. The finding of PSWCs is correlated positively with age and negatively with disease duration, significantly reducing the sensitivity in those patients with a disease duration of sCJD exceeding 9 months. In vCJD, PSWCs have never

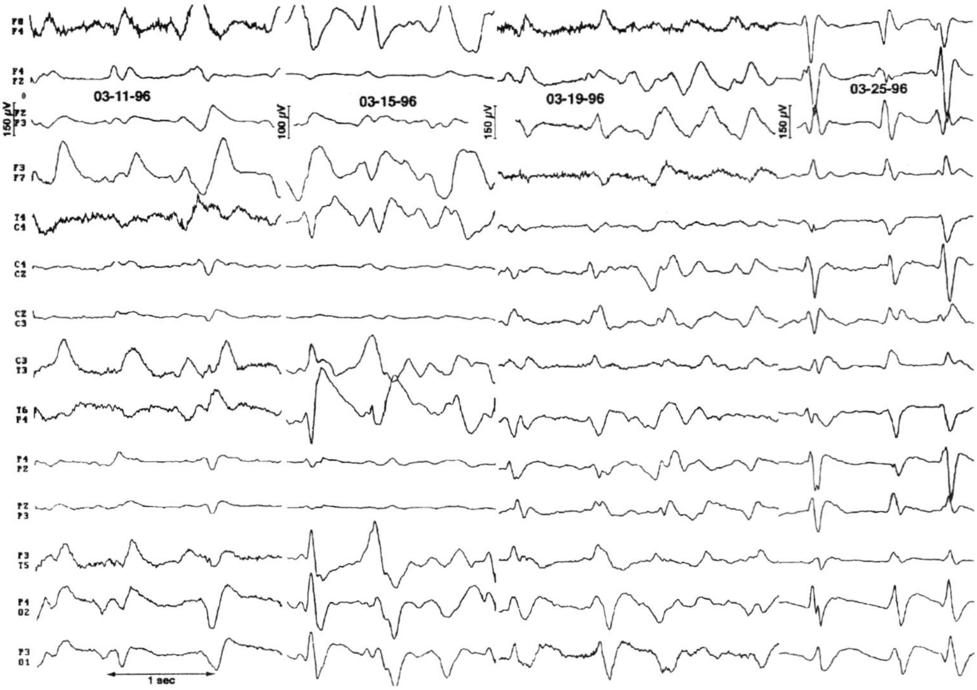

Figure 3 Evolution of periodic sharp wave complexes (PSWC) over 14 days in a 68-year-old patient with sCJD. Note generalized appearance of triphasic periodic cerebral potentials with a duration of 200–300 ms and an intercomplex interval of about 600 ms.

been reported so far [69]. As a rule, PSWCs are much less common in inherited human prion diseases than in sCJD. In genetic prion diseases such as fCJD-D178N–129V, there is usually only slowing of the EEG; no PSWCs are detectable. Case reports suggest that PSWCs may be found in fCJDV180I, V210I, and M232R, but exact figures of its frequency are not available. In those inherited human prion diseases with a GSS phenotype, the P102L-129M or -129V mutation rarely shows PSWCs, and earlier studies suggested a complete absence of PSWCs [16]. The D105L mutation shows no PSWCs, nor do the F198S and Q217R mutations.

Imaging

Case reports suggested a more specific pattern of abnormalities by MRI, i.e., hyperintense signal abnormalities in the basal ganglia on T2-weighted images. A systematic review of 29 patients with sCJD demonstrated moderate to marked bilateral, symmetrically increased signal intensities in the putamen and caudate nucleus on T2-weighted, proton-density-weighted, and FLAIR images in 23 (79%) of these patients [96] (Fig. 4). Contrast material enhancement or abnormalities on T1-weighted imaging was not seen. In a larger series, these findings were confirmed in 67% (109 of 162 cases of sCJD) whereas comparable findings were found in only 7% (4/58) of patients with non-CJD dementia [97]. Although these findings suggest a sensitivity of T2-weighted MRI of 67% with a 93% specificity for the diagnosis of human prion diseases, MRI features are by no means specific for the

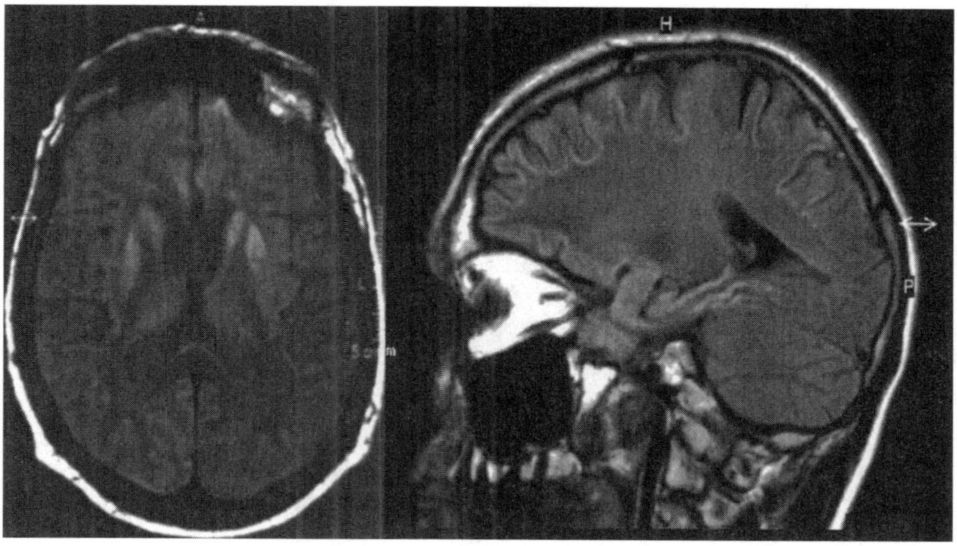

A B

Figure 4 (A) Case of definite sCJD revealing bilateral symmetrical lesions in the caudate nucleus and putamen and signal enhancement in cortical areas of the right and to a lesser degree left occipital lobes. (B) Case of a 24-year-old patient with sCJD showing a sagittal FLAIR MRI with faint subcortical signal enhancement. The patient became forgetful 4 months prior to the MRI and was proven at autopsy to have sCJD with homozygosity for valine at codon 129, explaining the very early onset of disease and long duration of 15 months. (From Ref. 37.)

diagnosis of CJD [97]. Currently, the most sensitive technique to detect these changes in sCJD appears to be diffusion-weighted MRI (DWI), which also permits easy monitoring of disease progression (Fig. 5) [98]. Serial scanning revealed progressively hyperintense changes in the striata and cerebral cortex. In addition, it demonstrates the rapid progression of brain atrophy (Fig. 4). Patients showed striatal or cerebral cortical lesions or both but rarely thalamic lesions and never lesions in the globus pallidus. Striatal lesions appear to be asymmetrical initially and to become symmetrical later (Fig. 5). The temporal pattern is characterized by an initial anteriorinferior involvement of the putamen that later spreading posteriorly and including the entire putamen. Putaminal lesions are also always accompanied by an ipsilateral caudate head lesion. ADC values in hyperintense lesions show a rapid decrease over 2 weeks.

In vivo monitoring of neuronal loss in sCJD by MR spectroscopy (MRS) shows a marked reduction of the *N*-acetylaspartate/creatine ratio in the frontal lobe, basal ganglia, and cerebellar hemispheres [99].

5.2 CSF Findings

Neuron-specific enolase (NSE) and the β,β isoform of S-100 are elevated in the CSF of sCJD cases [100–102]. Systematic evaluation of the sensitivity, specificity, and positive predictive value of various neuronal and glial proteins in the CSF performed by the German surveillance of CJD identified elevated NSE levels ($>$ 35 ng/mL), elevated tau protein

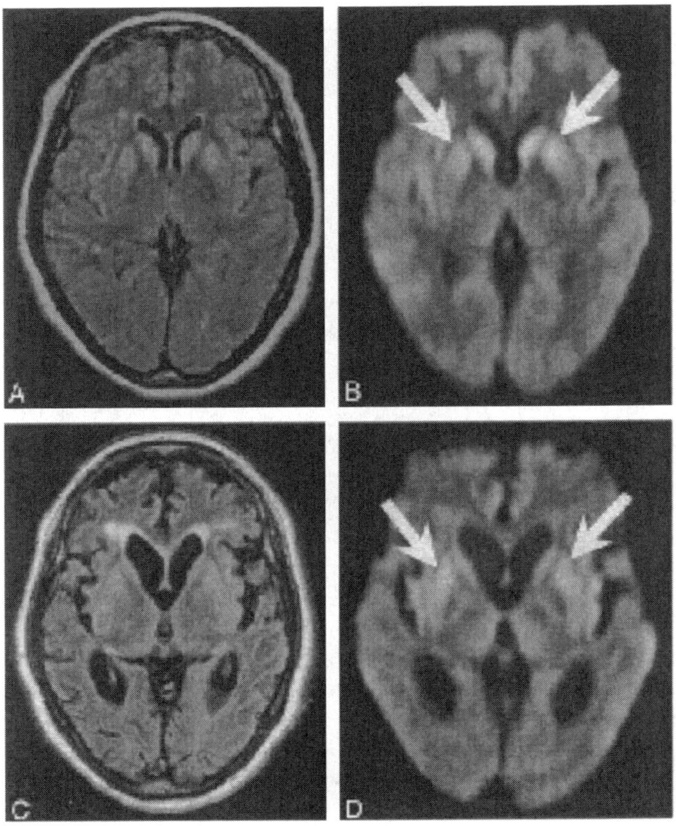

Figure 5 FCJD (M232R). Images were obtained at 3 (A, B) and 5 (C, D) months after the onset of symptoms. Predominant striatal lesions in the early stage. (A) FLAIR image shows changes that are not as conspicuous as in B. (B) Striata appear hyperintense at diffusion-weighted imaging (DWI). Note that the anterior portion of the bilateral putamina (arrows) appears more hyperintense than does the posterior portion at DWI. (C, D) Severe atrophy is depicted in both cerebral cortices and the caudate nuclei heads at FLAIR imaging (C) and DWI (D). Note that the putamina are entirely involved in C compared with their appearance in B. Hyperintensity in the heads of the caudate nuclei appears less prominent; this appearance is associated with their volume loss and the dilatation of the frontal horns. (From Ref. 98.)

(>1530 pg/mL), and elevated S100 (> 8 ng/mL) as quite reliable biochemical surrogate markers in the differential diagnosis of sCJD (Table 9) [100–102]. By far the most sensitive and most specific CSF marker protein is the 14–3–3 protein, which was first identified as p130/p131 by two-dimensional gel electrophoresis of CSF [103]. The 14–3–3 proteins are a group of at least seven proteins that can be identified in CSF by either Western blot or ELISA. A subtype analysis of sCJD showed that 14–3–3 proteins are detectable in all cases of MM2, MV1, VV1, and VV2 and in 96% of the most common form (MM1) but in only 30% of MV2 [37,95]. It remains to be seen whether a quantitative analysis with monoclonal antibodies specific for certain subtypes of 14–3–3 will improve the sensitivity, specificity, and positive predictive value of the assay. A prospective study confirmed the

Table 9 Biochemical Markers of Sporadic Human Prion Diseases

Source	*n*	Sensitivity	Specificity	Pos. pred. value (ppv)
CSF				
14–3–3 immunoblot	219	94	84	97
ELISA (8.3 ng/mL)	147	92.7	97.6	95
NSE (35 ng/mL)	124	78	88	79
Tau (1530 pg/mL)	172	91	94	95
S100 (8 ng/mL)	135	84	91	96
Serum	224	78	81	86
S100 (213 pg/mL)				

high sensitivity and specificity of the 14–3–3 assay [104]. False positive results may be found in stroke, meningoencephalitis, anoxia, or hypoxemia in particular cases associated with frequent seizures, metabolic encephalopathy, multiple myeloma, AD, vasculitis of the CNS, glioma, MELAS, and paraneoplastic neurological disorders [95,103–105].

It should be kept in mind that the high degree of sensitivity, specificity, positive predictive value (ppv), and negative predictive value (npv) reported for the 14–3–3 assay was obtained in carefully controlled clinical studies that adhered strictly to the diagnostic criteria of sCJD (Tables 5 and 6) and that the sensitivity and specificity of the assay will drop if used for screening purposes, i.e., for dementias with a longer duration than 2 years. These findings clearly show that none of the currently available surrogate markers can be used as a screening test for human prion diseases but should always be used in suspect cases. In inherited human prion diseases, elevated surrogate markers in CSF are found much less frequently [92]. NSE appears to be elevated in about half of the GSS (P102L) cases analyzed so far [106]. No data are available for the detection of S100, NSE, or tau protein in inherited human prion diseases. In familial CJD (E200K, V210I), on the other hand, 14–3–3 protein has been detected in all cases analyzed so far but has not been found in any case of FFI (D178N–129M) [95]. In vCJD, 14–3–3 is found in only half of all cases, yielding a ppv of 86% and an npv of 63% [107]. Other proteins such as tau, S-100b, and NSE were also elevated, with an increased tau having a ppv of 91% and an npv of 81%. These findings again suggest that these assays should be used only for corroboration of clinical diagnostic criteria for vCJD.

6 DIAGNOSTIC STRATEGY

In any case of a neurological illness with a positive history for dementia, sequencing of the *PRNP* should be performed after alternative diagnoses have been excluded by routine clinical and imaging analyses. Imaging should use MRI and include FLAIR and DWI [97,98]. EEG recordings should be performed frequently, in particular during the phase of rapidly evolving neurological signs and symptoms. CSF analysis should be performed to exclude inflammatory diseases. CSF should be analyzed for tau protein, S-100, NSE, and 14–3–3 by Western blot or, preferably, by the recently described quantitative ELISA techniques [107]. It remains to be seen whether the recently reported cyclic amplification of protein misfolding can be used for the rapid and specific detection of the pathological prion protein in either CSF, blood, serum, or plasma [108].

7 TREATMENT

Treatment in current clinical practice is supportive [109]. Because treatment has to be started as early as possible, preclinical diagnosis of any form of a human prion disease must be performed with high accuracy, which is currently possible only for patients already sick. In my experience, all efforts should be made to provide home care for patients diagnosed with clinically probable CJD as long as no causative treatment is available. Hospital treatment may lead to transfer to an intensive care unit where CJD patients may be kept alive for many months without any benefit for the patient but considerable strain for the patients' relatives and friends. Experimentally, several aspects in the process of PrP^{Sc} formation can be inhibited. Drugs may be divided into those acting only in the periphery and those acting both peripherally and centrally. Peripherally acting drugs are polyanions, sulfonated dyes, tetrapyrroles, anthracyclines, and dapsone. Drugs acting both centrally and in the periphery are polyene antibiotics, branched polyamines, cysteine protease inhibitors, acridine derivatives, phenothiazines, synthetic peptides, and antibodies to PrP^{c}. Drugs affecting macrophages may prolong incubation time in animal models of scrapie, possibly by interfering with the uptake of PrP^{Sc} [109]. Congo Red and related compounds inhibit in vivo and in vitro growth of PrP^{Sc} and inhibit ion channel formation of PrP 106–126. Because these dyes are carcinogenic and have a limited ability to permeate the blood-brain barrier, they have not been used as therapeutic of prophylactic agents. Among the mechanisms considered effective in interfering with PrPres/PrP^{Sc} are direct effects such as inhibition of fibrillogenesis of PrP^{Sc} or inhibition of the conversion of PrP^{c} to PrP^{Sc}, disaggregation of prion rods, and reduction of -sheet content by so-called dendrimers, and breaking of -sheets by so-called -sheet breaking peptides and indirect effects such as cytoprotection and interference with cellular PrP trafficking or with the lysosomal processing of PrP^{Sc}. Only a few of the drugs evaluated for use in the treatment of prion disease have been tested in clinical trials, and the vast majority have been shown to be effective only if added concomitantly with PrP^{Sc}. Of the drugs tested in small clinical studies, amphotericin B, amantadine, interferon, and flupirtine were found to have no effect.

GLOSSARY

AD	Alzheimer's disease
BABs	Cattle born after the ruminant feed ban in July 1988 that have been confirmed as having BSE
BSE	Bovine spongiform encephalopathy
CJD	Creutzfeldt-Jakob disease; f, familial, i, iatrogenic, s, sporadic; v, variant
CWD	Chronic wasting disease
DLB	Dementia with Lewy bodies
ENS	Enteric nervous system
FAE	Follicle-associated epithelium
FDC	Follicular dendritic cell
FFI	Fatal familial insomnia
GSS	Gerstmann-Sträussler-Scheinker disease
LRS	Lymphoreticular system
MBM	Meat and bone meal
PNS	Peripheral nervous system

Prion	Proteinaceous infectious agent
PrP^{c}	Cellular (normal) isoform of the prion protein
PrP^{Sc}	Pathological isoform of the prion protein
PrPres	Proteinase K–resistant isoform detected biochemically or by immunohistochemistry
PrPsen	Proteinase K–sensitive isoform, detected biochemically or by immunohistochemistry
PrP27–30	27–30 kDa form of PrPres detected by SDS-PAGE and Western blotting
PRNP	PrP gene of humans, located on chromosome 20
Prnp	PrP gene of animals
PSWC	Periodic sharp wave complexes
RML	Rocky Mountain Laboratory strain of scrapie
TME	Transmissible mink encephalopathy
TSE	Transmissible spongiform encephalopathy
WB	Western blot

REFERENCES

1. Spielmeyer, W. Die histopathologische Forschung in der Psychiatrie. Klin Wochenschr. 1922, 37, 1817–1819.
2. Masters, C.L.; Gajdusek, D.C. The spectrum of Creutzfeldt-Jakob disease and the virus-induced subacute spongiform encephalopathies. In Recent Advances in Neuropathology; Smith, W.T., Cavanagh, J.B., Eds.; Churchill Livingstone: Edinburgh, 1982; Vol. 2, 139–163.
3. Hadlow, W.J. Scrapie and kuru. Lancet. 1959, ii, 289–290.
4. Gajdusek, D.C.; Gibbs, C.J.J.; Alpers, M.P. Experimental transmission of a kuru syndrome to chimpanzees. Nature. 1966, 209, 794–796.
5. Alper, T.; Haig, D.A.; Clarke, M.C. The exceptionally small size of the scrapie agent. Biochem Biophys Res Commun. 1966, 22(3), 278–284.
6. Griffith, J.S. Self-replication and scrapie. Nature. 1967, 215(105), 1043–1044.
7. Prusiner, S.B. Shattuck Lecture—Neurodegenerative diseases and prions. N Engl J Med. 2001, 344(20), 1516–1526.
8. Prusiner, S.B. Novel proteinaceous infectious particles cause scrapie. Science. 1982, 216, 136–144.
9. Oesch, B.; Westaway, D.; Walchli, M.; McKinley, M.P.; Kent, S.B.; Aebersold, R.; Barry, R.A.; Tempst, P.; Teplow, D.B.; Hood, L.E.; Prusiner, S.B.; Weissmann, C. A cellular gene encodes scrapie PrP 27–30 protein. Cell. 1985, 40(4), 735–746.
10. Wilesmith, J.W.; Wells, G.A.; Cranwell, M.P.; Ryan, J.B. Bovine spongiform encephalopathy: epidemiological studies. Vet Rec. 1988, 123(25), 638–644.
11. Will, R.G.; Ironside, J.W.; Zeidler, M.; Cousens, S.N.; Estibeiro, K.; Alperovitch, A.; Poser, S.; Pocchiari, M.; Hofman, A.; Smith, P.G. A new variant of Creutzfeldt-Jakob disease in the UK. Lancet. 1996, 347, 921–925.
12. Collinge, J. Prion diseases of humans and animals: their causes and molecular basis. Annu Rev Neurosci. 2001, 24, 519–550.
13. Bruce, M.E.; Will, R.G.; Ironside, J.W.; McConnell, I.; Drummond, D.; Suttie, A.; McCardle, L.; Chree, A.; Hope, J.; Birkett, C.; Cousens, S.; Fraser, H.; Bostock, C.J. Transmissions to mice indicate that "new variant" CJD is caused by the BSE agent. Nature. 1997, 389(6650), 498–501.

14. Lasmezas, C.I.; Deslys, J.P.; Robain, O.; Jaegly, A.; Beringue, V.; Peyrin, J.-.M.; Fournier, J.-.G.; Hauw, J.-.J.; Rossier, J.; Dormont, D. Transmission of the BSE agent to mice in the absence of detectable abnormal prion protein. Science. 1997, 275, 402–405.
15. Hartsough, G.R.; Burger, D. Encephalopathy of mink. I. Epizootiologic and clinical observations. J Infect Dis. 1965, 115(4), 387–392.
16. Weber, T.; Aguzzi, A. The spectrum of transmissible spongiform encephalopathies. Intervirology. 1997, 40(2–3), 198–212.
17. Will, R.G.; Alperovitch, A.; Poser, S.; Pocchiari, M.; Hofman, A.; Mitrova, E.; de Silva, R.; D'Alessandro, M.; Delasnerie-Laupretre, N.; Zerr, I.; van Duijn, C. Descriptive epidemiology of Creutzfeldt-Jakob disease in six European countries, 1993–1995. EU Collaborative, Study Group for CJD. Ann Neurol. 1998, 43(6), 763–767.
18. Zerr, I.; Brandel, J.P.; Masullo, C.; Wientjens, D.; de Silva, R.; Zeidler, M.; Granieri, E.; Sampaolo, S.; van Duijn, C.; Delasnerie-Laupretre, N.; Will, R.; Poser, S. European surveillance on Creutzfeldt-Jakob disease: a case-control study for medical risk factors. J Clin Epidemiol. 2000, 53(7), 747–754.
19. Brown, P.; Preece, M.; Brandel, J.P.; Sato, T.; McShane, L.; Zerr, I.; Fletcher, A.; Will, R.G.; Pocchiari, M.; Cashman, N.R.; d'Aignaux, J.H.; Cervenakova, L.; Fradkin, J.; Schonberger, L.B.; Collins, S.J. Iatrogenic Creutzfeldt-Jakob disease at the millennium. Neurology. 2000, 55(8), 1075–1081.
20. Anonymous. Monthly Creutzfeldt-Jakob Figures; US Department, of Health, 2002; Vol. 2002.
21. Cousens, S.; Smith, P.G.; Ward, H.; Everington, D.; Knight, R.S.; Zeidler, M.; Stewart, G.; Smith-Bathgate, E.A.; Macleod, M.A.; Mackenzie, J.; Will, R.G. Geographical distribution of variant Creutzfeldt-Jakob disease in Great Britain, 1994–2000. Lancet. 2001, 357(9261), 1002–1007.
22. Alperovitch, A.; Zerr, I.; Pocchiari, M.; Mitrova, E.; de Pedro Cuesta, J.; Hegyi, I.; Collins, S.; Kretzschmar, H.; van Duijn, C.; Will, R.G. Codon 129 prion protein genotype and sporadic Creutzfeldt-Jakob disease [Letter]. Lancet. 1999, 353(9165), 1673–1674.
23. Ironside, J.W. Neuropathology of variant Creutzfeldt-Jakob disease. C R Acad Sci III. 2002, 325(1), 27–31.
24. Ferguson, N.M.; Ghani, A.C.; Donnelly, C.A.; Hagenaars, T.J.; Anderson, R.M. Estimating the human health risk from possible BSE infection of the British sheep flock. Nature. 2002, 415(6870), 420–424.
25. Aguzzi, A.; Montrasio, F.; Kaeser, P.S. Prions: health scare and biological challenge. Natl Rev Mol Cell Biol. 2001, 2(2), 118–126.
26. Westaway, D.; Goodman, P.A.; Mirenda, C.A.; McKinley, M.P.; Carlson, G.A.; Prusiner, S.B. Distinct prion proteins in short and long scrapie incubation period mice. Cell. 1987, 51(4), 651–662.
27. Bruce, M.E.; Boyle, A.; Cousens, S.; McConnell, I.; Foster, J.; Goldmann, W.; Fraser, H. Strain characterization of natural sheep scrapie and comparison with BSE. J Gen Virol. 2002, 83(Pt 3), 695–704.
28. Parchi, P.; Castellani, R.; Capellari, S.; Ghetti, B.; Young, K.; Chen, S.G.; Farlow, M.; Dickson, D.W.; Sima, A.A.F.; Trojanowski, J.Q.; Petersen, R.B.; Gambetti, P. Molecular basis of phenotypic variability in sporadic Creutzfeldt-Jakob disease. Ann Neurol. 1996, 39, 767–778.
29. Shaked, G.M.; Meiner, Z.; Avraham, I.; Taraboulos, A.; Gabizon, R. Reconstitution of prion infectivity from solubilized protease-resistant PrP and non-protein components of prion rods. J Biol Chem. 2001.
30. Manuelidis, L.; Fritch, W. Infectivity and host responses in Creutzfeldt-Jakob disease. Virology. 1996, 216(1), 46–59.
31. Hill, A.F.; Antoniou, M.; Collinge, J. Protease-resistant prion protein produced in vitro lacks detectable infectivity. J Gen Virol. 1999, 80(Pt 1), 11–14.

32. Rudd, P.M.; Wormald, M.R.; Wing, D.R.; Prusiner, S.B.; Dwek, R.A. Prion glycoprotein: structure, dynamics, and roles for the sugars. Biochemistry. 2001, 40(13), 3759–3766.
33. Masel, J.; Jansen, V.A. The measured level of prion infectivity varies in a predictable way according to the aggregation state of the infectious agent. Biochim Biophys Acta. 2001, 1535(2), 164–173.
34. Kretzschmar, H.A.; Ironside, J.; DeArmond, S.; Tateishi, J. Diagnostic criteria for sporadic Creutzfeldt-Jakob disease. Arch Neurol. 1996, 53(53), 913–920.
35. Simon, E.S.; Kahana, E.; Chapman, J.; Treves, T.A.; Gabizon, R.; Rosenmann, H.; Zilber, N.; Korczyn, A.D. Creutzfeldt-Jakob disease profile in patients homozygous for the PRNP E200K mutation. Ann Neurol. 2000, 47(2), 257–260.
36. Hauw, J.; Sazdovitch, V.; Laplanche, J.; Peoc'h, K.; Kopp, N.; Kemeny, J.; Privat, N.; Delasnerie-Laupretre, N.; Brandel, J.P.; Deslys, J.P.; Dormont, D.; Alperovitch, A. Neuropathologic variants of sporadic Creutzfeldt-Jakob disease and codon 129 of PrP gene. Neurology. 2000, 54(8), 1641–1646.
37. Zerr, I.; Schulz-Schaeffer, W.J.; Giese, A.; Bodemer, M.; Schroter, A.; Henkel, K.; Tschampa, H.J.; Windl, O.; Pfahlberg, A.; Steinhoff, B.J.; Gefeller, O.; Kretzschmar, H.A.; Poser, S. Current clinical diagnosis in Creutzfeldt-Jakob disease: identification of uncommon variants. Ann Neurol. 2000, 48(3), 323–329.
38. Parchi, P.; Giese, A.; Capellari, S.; Brown, P.; Schulz-Schaeffer, W.; Windl, O.; Zerr, I.; Budka, H.; Kopp, N.; Piccardo, P.; Poser, S.; Rojiani, A.; Streichemberger, N.; Julien, J.; Vital, C.; Ghetti, B.; Gambetti, P.; Kretzschmar, H. Classification of sporadic Creutzfeldt-Jakob disease based on molecular and phenotypic analysis of 300 subjects. Ann Neurol. 1999, 46(2), 224–233.
39. Büeler, H.; Aguzzi, A.; Sailer, A.; Greiner, R.A.; Autenried, P.; Aguet, M.; Weissmann, C. Mice devoid of PrP are resistant to scrapie. Cell. 1993, 73(7), 1339–1347.
40. Manson, J.C.; Clarke, A.R.; McBride, P.A.; McConnell, I.; Hope, J. PrP gene dosage determines the timing but not the final intensity or distribution of lesions in scrapie pathology. Neurodegeneration. 1994, 3, 331–340.
41. Sakaguchi, S.; Katamine, S.; Nishida, N.; Moriuchi, R.; Shigematsut, K.; Sugimoto, T.; Nakatani, A.; Kataoka, Y.; Houtani, T.; Shirabe, S.; Okada, H.; Hasegawa, S.; Miyamoto, T.; Noda, T. Loss of cerebellar Purkinje cells in aged mice homozygous for a disrupted PrP gene. Nature. 1996, 380, 528–531.
42. Moore, R.C.; Lee, I.Y.; Silverman, G.L.; Harrison, P.M.; Strome, R.; Heinrich, C.; Karunaratne, A.; Pasternak, S.H.; Chishti, M.A.; Liang, Y.; Mastrangelo, P.; Wang, K.; Smit, A.F.; Katamine, S.; Carlson, G.A ; Cohen, F.E.; Prusiner, S.B.; Melton, D.W.; Tremblay, P.; Hood, L.E.; Westaway, D. Ataxia in prion protein (PrP)-deficient mice is associated with upregulation of the novel PrP-like protein doppel. J Mol Biol. 1999, 292(4), 797–817.
43. Brown, D.R.; Qin, K.; Herms, J.W.; Madlung, A.; Manson, J.; Strome, R.; Fraser, P.E.; Kruck, T.; von Bohlen, A., Schulz-Schaeffer, W.; Giese, A.; Westaway, D.; Kretzschmar, H. The cellular prion protein binds copper in vivo. Nature. 1997, 390(6661), 684–687.
44. Mouillet-Richard, S.; Ermonval, M.; Chebassier, C.; Laplanche, J.L.; Lehmann, S.; Launay, J.M.; Kellermann, O. Signal transduction through prion protein. Science. 2000, 289(5486), 1925–1928.
45. Li, R.; Liu, D.; Zanusso, G ; Liu, T.; Fayen, J.D.; Huang, J.H.; Petersen, R.B.; Gambetti, P.; Sy, M.S. The expression and potential function of cellular prion protein in human lymphocytes. Cell Immunol. 2001, 207(1), 49–58.
46. Zahn, R.; Liu, A.; Luhrs, T.; Riek, R.; von Schroetter, C.; Lopez Garcia, F.; Billeter, M.; Calzolai, L.; Wider, G.; Wuthrich, K. NMR solution structure of the human prion protein. Proc Natl Acad Sci USA. 2000, 97(1), 145–150.
47. Rosenmann, H.; Talmor, G.; Halimi, M.; Yanai, A.; Gabizon, R.; Meiner, Z. Prion protein with an E200K mutation displays properties similar to those of the cellular isoform PrP(C). J Neurochem. 2001, 76(6), 1654–1662.

48. Brown, P.; Goldfarb, L.G.; Gajdusek, D.C. The new biology of spongiform encephalopathy: infectious amyloidoses with a genetic twist. Lancet. 1991, 337, 1019–1022.
49. Brown, P.; Cervenakova, L.; McShane, L.M.; Barber, P.; Rubenstein, R.; Drohan, W.N. Further studies of blood infectivity in an experimental model of transmissible spongiform encephalopathy, with an explanation of why blood components do not transmit Creutzfeldt-Jakob disease in humans. Transfusion. 1999, 39(11–12), 1169–1178.
50. Lasmezas, C.I.; Fournier, J.G.; Nouvel, V.; Boe, H.; Marce, D.; Lamoury, F.; Kopp, N.; Hauw, J.J.; Ironside, J.; Bruce, M.; Dormont, D.; Deslys, J.P. Adaptation of the bovine spongiform encephalopathy agent to primates and comparison with Creutzfeldt-Jakob disease: implications for human health. Proc Natl Acad Sci USA. 2001, 20, 20.
51. Beekes, M.; McBride, P.A. Early accumulation of pathological PrP in the enteric nervous system and gut-associated lymphoid tissue of hamsters orally infected with scrapie. Neurosci Lett. 2000, 278(3), 181–184.
52. Nicoletti, C. Unsolved mysteries of intestinal M cells. Gut. 2000, 47(5), 735–739.
53. McKie, A.T.; Zammit, P.S.; Naftalin, R.J. Comparison of cattle and sheep colonic permeabilities to horseradish peroxidase and hamster scrapie prion protein in vitro. Gut. 1999, 45(6), 879–888.
54. Pammer, J.; Cross, H.S.; Frobert, Y.; Tschachler, E.; Oberhuber, G. The pattern of prion-related protein expression in the gastrointestinal tract. Virchows Arch. 2000, 436(5), 466–472.
55. Blättler, T.; Brandner, S.; Raeber, A.J.; Klein, M.A.; Voigtlander, T.; Weissmann, C.; Aguzzi, A. PrP-expressing tissue required for transfer of scrapie infectivity from spleen to brain. Nature. 1997, 389(6646), 69–73.
56. Manuelidis, L.; Zaitsev, I.; Koni, P.; Lu, Z.Y.; Flavell, R.A.; Fritch, W. Follicular dendritic cells and dissemination of Creutzfeldt-Jakob disease. J Virol. 2000, 74(18), 8614–8622.
57. Lasmézas, C.I.; Deslys, J.P.; Demaimy, R.; Adjou, K.T.; Lammoury, F.; Dormont, D.; Robain, O.; Ironside, J.; Hauw, J.-J. BSE transmission to macaques. Nature. 1996, 381, 743–744.
58. Aguzzi, A.; Fietta, A.; Francioli, C.; Gialdroni Grassi, G.; Beauchamp-Nicoud, A.; Morle, L.; Lutz, H.U.; Stammler, P.; Agulles, O.; Petermann-Khder, R.; Iolascon, A.; Perrotta, S.; Cynober, T.; Tchernia, G.; Delaunay, J.; Baudin-Creuza, V.; Lazzari, L.; Henschler, R.; Lecchi, L.; Rebulla, P.; Mertelsmann, R.; Sirchia, G.; Hernandez, J.M.; Gonzalez, M.B.; Garcia, J.L.; Ferro, M.T.; Gutierrez, N.C.; Marynen, P.; San Miguel, J.F.; Sessarego, M.; Fugazza, G.; Bruzzone, R.; Ballestrero, A.; Miglino, M.; Bacigalupo, A.; Martinelli, G.; Terragna, C.; Amabile, M.; Montefusco, V.; Testoni, N.; Ottaviani, E.; de Vivo, A.; Mianulli, A.; Saglio, G.; Tura, S.; Galieni, P.; Cavo, M.; Pulsoni, A.; Avvisati, G.; Bigazzi, C.; Neri, S.; Caliceti, U.; Benni, M.; Ronconi, S.; Lauria, F.; Tribalto, M.; Amadori, S.; Cudillo, L.; Caravita, T.; Del Poeta, G.; Meloni, G.; Petrucci, M.T. Prion diseases, blood and the immune system: concerns and reality. Haematologica. 2000, 85(1), 3–10.
59. Weissmann, C.; Raeber, A.J.; Montrasio, F.; Hegyi, I.; Frigg, R.; Klein, M.A.; Aguzzi, A. Prions and the lymphoreticular system. Phil Trans Roy Soc Lond B Biol Sci. 2001, 356(1406), 177–184.
60. Kaeser, P.S.; Klein, M.A.; Schwarz, P.; Aguzzi, A. Efficient lymphoreticular prion propagation requires PrP(c) in stromal and hematopoietic cells. J Virol. 2001, 75(15), 7097–7106.
61. Mabbott, N.A.; Bruce, M.E.; Botto, M.; Walport, M.J.; Pepys, M.B. Temporary depletion of complement component C3 or genetic deficiency of C1q significantly delays onset of scrapie. Nat Med. 2001, 7(4), 485–487.
62. Bradley, M.E.; Edskes, H.K.; Hong, J.Y.; Wickner, R.B.; Liebman, S.W. Interactions among prions and prion "strains" in yeast. Proc Natl Acad Sci USA. 2002.
63. Liebman, S.W. Progress toward an ultimate proof of the prion hypothesis. Proc Natl Acad Sci USA. 2002, 99(14), 9098–9100.
64. Coustou, V.; Deleu, C.; Saupe, S.; Begueret, J. The protein product of the het-s heterokaryon incompatibility gene of the fungus Podospora anserina behaves as a prion analog. Proc Natl Acad Sci USA. 1997, 94(18), 9773–9778.

65. Zerr, I.; Giese, A.; Windl, O.; Kropp, S.; Schulz-Schaeffer, W.; Riedemann, C.; Skworc, K.; Bodemer, M.; Kretzschmar, H.A.; Poser, S. Phenotypic variability in fatal familial insomnia (D178N–129M) genotype. Neurology. 1998, 51(5), 1398–1405.
66. Masters, C.L.; Harris, J.O.; Gajdusek, D.C.; Gibbs, C.J.J.; Bernoulli, C.; Asher, D.M. Creutzfeldt-Jakob disease: patterns of worldwide occurrence and the significance of familial and sporadic clustering. Ann Neurol. 1979, 5(2), 177–188.
67. Anon. Human transmissible spongiform encephalopathies. Wkly Epidemiol Rec. 1998, 73(47), 361–365.
68. Hoshi, K.; Yoshino, H.; Urata, J.; Nakamura, Y.; Yanagawa, H.; Sato, T. Creutzfeldt-Jakob disease associated with cadaveric dura mater grafts in Japan. Neurology. 2000, 55(5), 718–721.
69. Will, R.G.; Zeidler, M.; Stewart, G.E. Macleod, M.A.; Ironside, J.W.; Cousens, S.N.; Mackenzie, J.; Estibeiro, K.; Green, A.J.; Knight, R.S. Diagnosis of new variant Creutzfeldt-Jakob disease. Ann Neurol. 2000, 47(5), 575–582.
70. Wopfner, F.; Weidenhöfer, G.; Schneider, R.; von Brunn, A.; Gilch, S.; Schwarz, T.F.; Werner, T.; Schätzl, H.M. Analysis of 27 mammalian and 9 avian PrPs reveals high conservation of flexible regions of the prion protein. J Mol Biol. 1999, 289(5), 1163–1178.
71. Parchi, P.; Capellari, S.; Chin, S.; Schwarz, H.B.; Schecter, N.P.; Butts, J.D.; Hudkins, P.; Burns, D.K.; Powers, J.M.; Gambetti, P. A subtype of sporadic prion disease mimicking fatal familial insomnia. Neurology. 1999, 52(9), 1757–1763.
72. Mastrianni, J.A.; Nixon, R.; Layzer, R.; Telling, G.C.; Han, D.; DeArmond, S.J.; Prusiner, S.B. Prion protein conformation in a patient with sporadic fatal insomnia. N Engl J Med. 1999, 340(21), 1630–1638.
73. Harder, A.; Jendroska, K.; Kreuz, F.; Wirth, T.; Schafranka, C.; Karnatz, N.; Theallier-Janko, A.; Dreier, J.; Lohan, K.; Emmerich, D.; Cervos-Navarro, J.; Windl, O.; Kretzschmar, H.A.; Nurnberg, P.; Witkowski, R. Novel twelve-generation kindred of fatal familial insomnia from Germany representing the entire spectrum of disease expression. Am J Med Genet. 1999, 87(4), 311–316.
74. Lugaresi, E. The thalamus and insomnia. Neurology. 1992, 42(7 suppl 6), 28–33.
75. Gambetti, P.; Parchi, P.; Petersen, R.B.; Chen, S.G.; Lugaresi, E. Fatal familial insomnia and familial Creutzfeldt-Jakob disease: clinical, pathological and molecular features. Brain Pathol. 1995, 5(1), 43–51.
76. Montagna, P.; Cortelli, P.; Avoni, P.; Tinuper, P.; Plazzi, G.; Gallassi, R.; Portaluppi, F.; Julien, J.; Vital, C.; Delisle, M.B.; Gambetti, P.; Lugaresi, E. Clinical features of fatal familial insomnia: phenotypic variability in relation to a polymorphism at codon 129 of the prion protein gene. Brain Pathol. 1998, 8(3), 515–520.
77. McLean, C.A.; Storey, E.; Gardner, R.J.M.; Tannenberg, A.E.G.; Cervenakova, L.; Brown, P. The DI763N (cis-129M) "fatal familial insomnia" mutation associated with diverse clinicopathologic phenotypes in an Australian kindred. Neurology. 1997, 49, 552–558.
78. Rosenmann, H.; Vardi, J.; Finkelstein, Y.; Chapman, J.; Gabizon, R. Identification in Israel of 2 Jewish Creutzfeld-Jakob disease patients with a 178 mutation at their PrP gene. Acta Neurol Scand. 1998, 97(3), 184–187.
79. Bertoni, J.M.; Brown, P.; Goldfarb, L.G.; Rubenstein, R.; Gajdusek, D.C. Familial Creutzfeldt-Jakob disease (codon 200 mutation) with supranuclear palsy. JAMA. 1992, 268(17), 2413–2415.
80. Gerstmann, J.; Sträussler, E.; Scheinker, I. über eine eigenartige hereditär-familiäre Erkrankung des Zentralnervensystems. Zugleich ein Beitrag zur Frage des vorzeitigen lokalen Alterns. Z Neurol. 1936, 154, 735–762
81. Brown, P.; Goldfarb, L.G., Brown, W.T.; Goldgaber, D.; Rubenstein, R.; Kascsak, R.J.; Guiroy, D.C.; Piccardo, P.; Boellaard, J.W.; Gajdusek, D.C. Clinical and molecular genetic study of a large German kindred with Gerstmann-Straussler-Scheinker syndrome. Neurology. 1991, 41(3), 375–379.

82. Doh-ura, K.; Tateishi, J.; Sasaki, H.; Kitamoto, T.; Sakaki, Y. Pro-Leu change at position 102 of prion protein is the most common but not the sole mutation related to Gerstmann-Sträussler syndrome. Biochem Biophys Res Commun. 1989, 163(2), 974–979.
83. Liberski, P.P.; Barcikowska, M.; Cervenakova, L.; Bratosiewicz, J.; Marczewska, M.; Brown, P.; Gajdusek, D.C. A case of sporadic Creutzfeldt-Jakob disease with a Gerstmann-Straussler-Scheinker phenotype but no alterations in the PRNP gene. Acta Neuropathol (Berl). 1998, 96(4), 425–430.
84. Kitamoto, T.; Amano, N.; Terao, Y.; Nakazato, Y.; Isshiki, T.; Mizutani, T.; Tateishi, J. A new inherited prion disease (PrP-P1O5L mutation) showing spastic paraparesis. Ann Neurol. 1993, 34, 808–813.
85. Tranchant, C.; Doh, U.K.; Warter, J.M.; Steinmetz, G.; Chevalier, Y.; Hanauer, A.; Kitamoto, T.; Tateishi, J. Gerstmann-Straussler-Scheinker disease in an Alsatian family: clinical and genetic studies. J Neurol Neurosurg Psychiatry. 1992, 55(3), 185–187.
86. Hsiao, K.; Dlouhy, S.R.; Farlow, M.R.; Cass, C.; Da Costa, M.; Conneally, P.M.; Hodes, M.E.; Ghetti, B.; Prusiner, S.B. Mutant prion proteins in Gerstmann-Straussler-Scheinker disease with neurofibrillary tangles. Nat Genet. 1992, 1(1), 68–71.
87. Ghetti, B.; Tagliavini, F.; Giaccone, G.; Bugiani, O.; Frangione, B.; Farlow, M.R.; Dlouhy, S.R. Familial Gerstmann-Sträussler-Scheinker disease with neurofibrillary tangles. Mol Neurobiol. 1994, 8, 41–48.
88. Krasemann, S.; Zerr, I.; Weber, T.; Poser, S.; Kretzschmar, H.; Hunsmann, G.; Bodemer, W. Prion disease associated with a novel nine octapeptide repeat insertion in the PRNP gene. Brain Res Mol Brain Res. 1995, 34(1), 173–176.
89. Skworc, K.H.; Windl, O.; Schulz-Schaeffer, W.J.; Giese, A.; Bergk, J.; Nagele, A.; Vieregge, P.; Zerr, I.; Poser, S.; Kretzschmar, H.A. Familial Creutzfeldt-Jakob disease with a novel 120-bp insertion in the prion protein gene. Ann Neurol. 1999, 46(5), 693–700.
90. Rossi, G.; Giaccone, G.; Giampaolo, L.; Iussich, S.; Puoti, G.; Frigo, M.; Cavaletti, G.; Frattola, L.; Bugiani, O.; Tagliavini, F. Creutzfeldt-Jakob disease with a novel four extra-repeat insertional mutation in the PrP gene. Neurology. 2000, 55(3), 405–410.
91. van Harten, B.; van Gool, W.A.; Van Langen, I.M.; Deekman, J.M.; Meijerink, P.H.; Weinstein, H.C. A new mutation in the prion protein gene: a patient with dementia and white matter changes. Neurology. 2000, 55(7), 1055–1057.
92. Poser, S.; Mollenhauer, B.; Kraubeta, A.; Zerr, I.; Steinhoff, B.J.; Schroeter, A.; Finkenstaedt, M.; Schulz-Schaeffer, W.J.; Kretzschmar, H.A.; Felgenhauer, K. How to improve the clinical diagnosis of Creutzfeldt-Jakob disease. Brain. 1999, 122(Pt 12), 2345–2351.
93. Tschampa, H.J.; Neumann, M.; Zerr, I.; Henkel, K.; Schroter, A.; Schulz-Schaeffer, W.J.; Steinhoff, B.J.; Kretzschmar, H.A.; Poser, S. Patients with Alzheimer's disease and dementia with Lewy bodies mistaken for Creutzfeldt-Jakob disease. J Neurol Neurosurg Psychiatry. 2001, 71(1), 33–39.
94. Steinhoff, B.J.; Räcker, S.; Herrendorf, G.; Poser, S.; Grosche, S.; Zerr, I.; Kretzschmar, H.; Weber, T. Accuracy and reliability of periodic sharp wave complexes in Creutzfeldt-Jakob disease. Arch Neurol. 1996, 53, 162–165.
95. Zerr, I.; Pocchiari, M.; Collins, S.; Brandel, J.P.; de Pedro Cuesta, J.; Knight, R.S.; Bernheimer, H.; Cardone, F.; Delasnerie-Laupretre, N.; Cuadrado Corrales, N.; Ladogana, A.; Bodemer, M.; Fletcher, A.; Awan, T.; Ruiz Bremon, A.; Budka, H.; Laplanche, J.L.; Will, R.G.; Poser, S. Analysis of EEG and CSF 14-3-3 proteins as aids to the diagnosis of Creutzfeldt-Jakob disease. Neurology. 2000, 55(6), 811–815.
96. Finkenstaedt, M.; Szudra, A.; Zerr, I.; Poser, S.; Hise, J.H.; Stoebner, J.H.; Weber, T. Magnetic resonance imaging of Creutzfeldt-Jacob disease. Radiology. 1996, 199, 793–798.
97. Schröter, A.; Zerr, I.; Henkel, K.; Tschampa, H.J.; Finkenstaedt, M.; Poser, S. Magnetic resonance imaging in the clinical diagnosis of Creutzfeldt-Jakob disease. Arch Neurol. 2000, 57(12), 1751–1757.

98. Murata, T.; Shiga, Y.; Higano, S.; Takahashi, S.; Mugikura, S. Conspicuity and evolution of lesions in Creutzfeldt-Jakob disease at diffusion-weighted imaging. Am J Neuroradiol. 2002, 23(7), 1164–1172.
99. Bruhn, H.; Weber, T.; Thorwirth, V.; Frahm, J. In-vivo monitoring of neuronal loss in Creutzfeldt-Jakob disease by proton magnetic resonance spectroscopy [Letter]. Lancet. 1991, 337(8757), 1610–1611.
100. Otto, M.; Wiltfang, J.; Schutz, E.; Zerr, I.; Otto, A.; Pfahlberg, A.; Gefeller, O.; Uhr, M.; Giese, A.; Weber, T.; Kretzschmar, H.A.; Poser, S. Diagnosis of Creutzfeldt-Jakob disease by measurement of S100 protein in serum: prospective case-control study. Br Med J. 1998, 316(7131), 577–582.
101. Poser, S.; Zerr, I.; Schroeter, A.; Otto, M.; Giese, A.; Steinhoff, B.J.; Kretzschmar, H.A. Clinical and differential diagnosis of Creutzfeldt-Jakob disease. Arch Virol Suppl. 2000, 16, 153–159.
102. Weber, T.; Otto, M.; Bodemer, M.; Zerr, I. Diagnosis of Creutzfeldt-Jakob disease and related human spongiform encephalopathies. Biomed Pharmacother. 1997, 51(9), 381–387.
103. Zerr, I.; Bodemer, M.; Gefeller, O.; Otto, M.; Poser, S.; Wiltfang, J.; Windl, O.; Kretzschmar, H.A.; Weber, T. Detection of 14-3-3 protein in the cerebrospinal fluid supports the diagnosis of Creutzfeldt-Jakob disease. Ann Neurol. 1998, 43(1), 32–40.
104. Lemstra, A.W.; van Meegen, M.T.; Vreyling, J.P.; Meijerink, P.H.; Jansen, G.H.; Bulk, S.; Baas, F.; van Gool, W.A. 14-3-3 testing in diagnosing Creutzfeldt-Jakob disease: a prospective study in 112 patients. Neurology. 2000, 55(4), 514–516.
105. Zerr, I.; Bodemer, M.; Weber, T. The 14-3-3 brain protein and transmissible spongiform encephalopathy [Letter]. N Engl J Med. 1997, 336(12), 874; discussion 874–875.
106. Imaiso, Y.; Mitsuo, K. Gerstmann-Straussler-Scheinker syndrome with a Pro102Leu mutation in the prion protein gene and atypical MRI findings, hyperthermia, tachycardia, and hyperhidrosis (In Japanese). Rinsho Shinkeigaku. 1998, 38(10–11), 920–925.
107. Green, A.J.; Ramljak, S.; Muller W.E.; Knight, R.S.; Schroder, H.C. 14-3-3 in the cerebrospinal fluid of patients with variant and sporadic Creutzfeldt-Jakob disease measured using capture assay able to detect low levels of 14-3-3 protein. Neurosci Lett. 2002, 324(1), 57–60.
108. Saborio, G.P.; Permanne, B.; Soto, C. Sensitive detection of pathological prion protein by cyclic amplification of protein misfolding. Nature. 2001, 411, 810–813.
109. Brown, P. Drug therapy in human and experimental transmissible spongiform encephalopathy. Neurology. 2002, 58(12), 1720–1725.
110. Aguzzi, A.; Weissmann, C. Prion research: the next frontiers [News]. Nature. 1997, 389(6653), 795–798.

26

Polyomaviruses and Brain Tumors

Sidney Croul, Darryl L'Heureux, and Kamel Khalili
Temple University
Philadelphia, Pennsylvania, U.S.A.

1 POLYOMAVIRUS

1.1 Structure and Infectivity

Polyomaviruses are icosahedral nonenveloped DNA viruses with capsid diameters of approximately 45 nm. The genome consists of covalently bound, double-stranded, circular supercoiled DNA with an average length of 5 kb. The circular genomes are all similar in structure, consisting of a noncoding regulatory region, an early region that codes for a protein known as T-antigen, and a late region that codes for the capsid proteins [1]. The regulatory region separates the coding regions and contains sequences that are necessary for the initiation of viral DNA replication. Early and late transcription proceed in opposite directions around the circular DNA. Figure 1 illustrates the structural organization of the human neurotropic polyoma virus JCV, whose genomic organization is characteristic of the polyomavirus family. After infection, these viruses replicate their DNA to form mature progeny in the nucleus of the host cell. Tissues in which efficient growth occurs are permissive. Semipermissive and nonpermissive tissues produce virus less efficiently or incompletely [2].

Polyomavirus infections are known to occur naturally in chickens, macaws, budgerigars, mice, hamsters, monkeys and humans [3]. The human infections were initially recognized as the clinically manifest syndromes of JCV and BKV. JCV causes the subacute, fatal central nervous system demyelinating disease progressive multifocal leukoencephalopathy (PML) due to a productive infection of oligodendrocytes. Although this syndrome was originally described in patients with systemic immunosuppression due to cancer and/or chemotherapy, it has subsequently been recognized as a consequence of human immunodeficiency virus (HIV-1) infection, accounting for a distinct rise in the incidence of PML in the AIDS population [4]. BKV most commonly causes a self-limited hemorrhagic

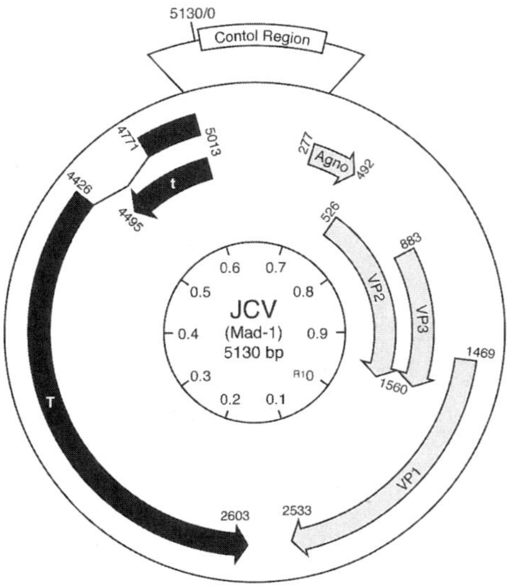

Figure 1 The polyomavirus genome. The genome of JCV is schematized as an exemplar. The circular genomes of all polyoma viruses are similar, with three general regions identified based on function. These comprise the regulatory (noncoding) region, the early coding region, and the late coding region.

cystitis during pregnancy and has also been associated with other urinary tract, respiratory tract, and meningeal inflammatory diseases in children, HIV-1 infected individuals, and transplant recipients [5]. Simian virus 40 (SV40), originally classified as a monkey polyomavirus, has more recently been shown to infect humans as well. Serological surveillance of worldwide populations has led investigators to conclude that primary human polyomavirus infections are extremely common, occur during youth, and are often subclinical [3]. Following acute infection, persistence of virus over prolonged intervals has been demonstrated in brain, lung, kidney, bone, and blood by a variety of techniques including Southern blotting and polymerase chain reaction (PCR) [6]. These findings have led to the assumption that many symptomatic polyoma infections represent reactivation and that the conditions under which these occur (pregnancy, HIV, cancer, and chemotherapy) alter the ability of the organism to survey for and abrogate viral reproduction. Support for this notion comes from sequence analysis of JCV and BK isolates. The regulatory regions of archetype strains cloned from healthy, asymptomatic individuals lack the large tandem repeats found in many of the isolates from symptomatic patients. This suggests that after initial infection with an archetype strain, adaptation of the virus including spontaneous alterations of the regulatory region may allow expression in tissue types and under conditions that result in manifest clinical syndromes [7].

1.2 Polyoma Tumorigenesis

The observation that polyomaviruses have oncogenic potential derives from studies that demonstrated that these agents could cause tumors when inoculated into animal species that were not their natural hosts. SV40 was shown to cause tumors in newborn hamsters and in mice. The range of tumors produced experimentally by SV40 is large and includes ependymomas, sarcomas, leukemias, and lymphomas [1]. JCV is oncogenic in hamsters [8], resulting in a variety of neural tumors of which medulloblastomas / primitive neuroectodermal tumors (PNETs) are the most common but also including glioblastomas, neuroblastomas, and pineocytomas. JCV also represents the only papovavirus that induces nervous system tumors (astrocytomas and neuroblastomas) in nonhuman primates [3]. BKV results in tumors in both hamsters and rats [9], including ependymomas and renal tumors. A variety of polyomaviruses are known to be tumorigenic in their natural hosts. Examples include HaPV, which causes epitheliomas in hamsters; PyV, which causes epitheliomas, osteosarcomas, and fibrosarcomas in mice; and SV40, which causes astrocytomas in monkeys [1].

The tumorigenicity of the directly injected polyomaviruses is thought to result from infection of semipermissive or nonpermissive tissues. The intracellular environment permits early gene expression but delimits the production of complete virus. Evidence for this includes high levels of large T-antigen expression in the absence of capsid protein expression [7]. Further evidence for the direct association of T antigen and tumor formation comes from studies of transgenic mice that constitutively produce early protein under the control of the early promoter/ enhancer. These animals develop tumors similar in many respects to those found in polyomavirus infection of natural hosts and virus-injected animals [10–12]. The precise mechanism responsible for the induction of PNETs and other experimental CNS tumors by T antigen has not been elucidated. A number of *in vitro* studies have shown that cell cycle control checkpoints can be disrupted by DNA tumor virus proteins, which are similar to JCV T antigen (adenovirus E1A and papillomavirus E7), through interaction with the cell cycle regulatory proteins p53 and pRb [13].

Although still a controversial hypothesis, the initial evidence that human brain tumors might be associated with polyomavirus infection came from retrospective analyses of tumor incidence vs. SV40 polio vaccine contamination [14–17]. The rationale for these studies was the isolation of the SV40 virus in 1960. This was followed fairly rapidly by the recognition of SV40 contamination of previous oral polio vaccines and the oncogenic potential of SV40 in experimental systems. From its initial manufacture in 1954 until 1960, the Salk polio vaccine was grown on monolayers of rhesus monkey kidney cells, which were then treated with graded concentrations of formalin to inactivate virus infectivity but preserve antigenicity. Because SV40 is found frequently in rhesus monkeys but is more resistant to formalin inactivation than polio, it is not surprising that high titers of live virus were found in several batches of the vaccine. The administration of SV-40-contaminated polio vaccine was eliminated by 1961. It is not known how many of the 90 million people who were given the Salk vaccine were exposed to SV40, but estimates between 10 million and 30 million have been made [18].

In the studies cited above, Fraumeni et al. [16] found a 20% greater incidence for all neoplasms other than leukemia in 1956 compared to 1955, 1957, and 1958 in a hospital-based population of children and proposed that the SV40-contaminated vaccine might play a role. In a much larger population, Heinonen et al. [17] studied the outcome of over 50,000 pregnancies between 1959 and 1966. The rate of pediatric malignancy in the children whose mothers had been immunized against polio during pregnancy was 7.6/10,000 as opposed to 3.1/10,000 in the nonimmunized group. Not only was the difference between these two groups statistically significant, but seven of the 14 tumors in the vaccinated group occurred in the CNS as opposed to only one in the nonimmunized group. Using records from the Connecticut Tumor Registry, Farwell et al. [14] initially identified

120 children born between 1956 and 1962 who subsequently developed brain tumors. Whereas the overall analysis of SV40 exposure was suggestive but not statistically significant, the finding that 52% of the tumors occurring in the SV40-exposed children were medulloblastomas was significant. In a follow-up study [15] from the same database, all medulloblastomas were surveyed from a 42-year period, and it was demonstrated that an excessive number of cases occurred in children born between the years 1954 and 1958, suggesting an association of this tumor with the SV40 polio vaccine. Nonetheless, a detailed epidemiological study of the Swedish Cancer Registry from 1960 to 1993 [19] showed no association between brain tumor incidence rates and the SV40-contaminated vaccine. In Germany, 20-year follow-up showed no difference in brain tumor rates between groups of people who received polio vaccine with probable SV40 contamination and groups receiving SV40-free vaccine [20]. This epidemiological issue was revisited by two groups [21,22] using, at least in part, the same database (the Surveillance Epidemiology and End Results Program 1973–1993) but coming to diametrically opposed conclusions. Further muddying these epidemiological analyses are serological data suggesting human exposure to SV40 prior to the Salk polio vaccine [23,24]. Although it is still reasonable to assume that polio vaccination increased the incidence of human SV40 infection, the possibility of preexisting exposure makes analysis of the SV40-positive and SV40-negative polio vaccine groups all the more difficult.

Polyomaviruses have also been associated with human tumors on a case-by-case basis. Early reports of these linkages included astrocytic and hematological malignancies in the brains of patients with active JCV replication in the setting of PML [25,26]. Techniques including viral culture from primary tumors and derived cell lines, immunodetection of viral proteins, hybridization of DNA, and PCR detection of DNA have sustained these observations and increased the scope of suspect tumors. Many of the tumors implicated originate in the nervous system. A detailed list of CNS tumors and their associated polyomaviruses is given in Table 1. In that list, it is worth noting that ependymal tumors, choroid plexus tumors, and medulloblastomas are cited most frequently for their association with polyomavirus.

1.3 Epidemiology of Brain Tumors

Brain tumors account for less than 2% of all neoplasms, occurring at a rate of approximately 11 new cases per year per 100,000 people [27]. In the adult population, although brain tumors are a significant clinical problem, they account for only a small proportion of the cancers that merit attention. The issue in pediatric oncology is strikingly different, because nervous system tumors in children rank second in incidence only to leukemias. Although the childhood incidence rate is less than that for adults (3.7 cases per year per 100,000 children), in a series of almost 3000 autopsies at Great Ormond Street Hospital (London, England), brain tumors were identified in 2.2% of cases [28]. The brain tumors whose association with polyomaviruses have received the most attention (ependymoma, choroid plexus papilloma, and medulloblastoma/PNET) experimentally and clinically [9,10,12,14–17,29–44] occur most often in the posterior fossa of children and less frequently in the cerebral hemispheres of adults and children. The pediatric posterior fossa tumors will therefore serve as the primary focus of discussion in this chapter, with hemispheric tumors providing a secondary focus.

Table 1 Human Tumors Associated with Polyomaviruses

Polyomavirus	Tumor histology	References
SV40	Astrocytoma	34,36,40,96
	Anaplastic astrocytoma	34
	Gemistocytic astrocytoma	34
	Glioblastoma	40,97,98
	Giant cell glioblastoma	34
	Gliosarcoma	34
	Oligodendroglioma	34,36
	Ependymoma	29,34,38,40,43
	Malignant ependymoma	42
	Subependymoma	43
	Choroid plexus papilloma	29,38–40,42
	Choroid plexus carcinoma	38,39
	Meningioma	36,40,43,96
	Medulloblastoma	34,36,43
	Ganglioneurom	38
JC	Astrocytoma	25,30,32
	Anaplastic astrocytoma	32
	Glioblastoma	32,99
	Gliosarcoma	32
	Pilocytic astrocytoma	32
	Pleomorphic xanthoastrocytoma	100
	Oligodendroglioma	30,32
	Anaplastic oligodendroglioma	32
	Oligoastrocytoma	32,101
	Ependymoma	30,32
	Subependymoma	32
	Meningioma	43
	Medulloblastoma	35,37
	Gliomatosis cerebri	32
	Primary CNS lymphoma	26
BK	Astrocytoma	31
	Glioblastoma	31,33
	Oligodendroglioma	31,33
	Ependymoma	31
	Meningioma	31,33
	Schwannoma	31,33

2 BRAIN TUMORS ASSOCIATED WITH POLYOMAVIRUS

2.1 Ependymoma

The ependymoma is a slowly growing tumor that originates from the epithelial cells (ependymocytes) lining the ventricular system of the brain. The first description was published by Virchow in the 1860s [45]. Their glial derivation was first recognized by Bailey, who in his 1924 monograph [46] defined their origin from the ventricular lining.

Table 2 JCV is Detectable in Both Ependymomas and Medulloblastomas

Tumor histopathology	Gender/Age (yr)	T-antigen immunohistochemistry	JC virus PCR
Ependymoma	M/1	+	+
Ependymoma	M/47	−	−
Ependymoma	M/29	+	+
Ependymoma	F/36	+	+
Ependymoma	F/20	+	+
Desmoplastic medulloblastoma	M/1.5	−	+
Desmoplastic medulloblastoma	F/4	−	+
Desmoplastic medulloblastoma	M/15	+	+
Desmoplastic medulloblastoma	F/42	−	+
Classic medulloblastoma	M/7	−	+
Neuroblastic medulloblastoma	F/18	−	+
Desmoplastic medulloblastoma	F/8	−	+
Desmoplastic medulloblastoma	M/7	+	+
Classic medulloblastoma	M/9	−	+
Desmoplastic medulloblastoma	F/3	+	+
Desmoplastic medulloblastoma	F/9	−	−
Neuroblastic medulloblastoma	M/5	+	+
Classic medulloblastoma	F/2	−	−
Classic medulloblastoma	M/Newborn	−	+
Classic medulloblastoma	M/12	−	−
Neuroblastic medulloblastoma	M/5	−	+

[a] Immunohistochemistry was used to detect T antigen in five ependymomas and 16 medulloblastomas. DNA from these tumors was also analyzed for JCV sequences using PCR primers designed for the N terminal of T antigen. The PCR products were analyzed by Southern blotting using DNA probes specific for the amplified sequence of T antigen. In the group of ependymomas, four out of five were immunohistochemically positive for T antigen and the same 4/5 showed JCV T-antigen DNA sequences by PCR. For the medulloblastomas, 4/16 showed nuclear positivity by immunohistochemistry, and 13/16 were PCR positive.

Ependymomas account for approximately 5% of all glial neoplasms and occur at a rate of 2.3 per million population per year [27]. The tumors are most common in children in whom the peak incidence of 7–8 per million population per year is found at ages 0–4 years [27]. As such, ependymomas account for up to 10% of all pediatric brain tumors [47]. A second age peak at 30–40 years has been reported for spinal ependymomas. The male female ratio is approximately 1:1 [27]. Although they may occur along any part of the ventricular system, the majority occur infratentorially. In a series of almost 300 cases, an infratentorial/supratentorial ratio of 58%: 42% was obtained [48]. Hemispheric ependymomas may occur that are not in continuity with the ventricular system. It has been postulated that these arise from embryonic remnants or migrated nests of ependymocytes within the cerebral parenchyma.

Although hereditary ependymomas are not the rule, spinal ependymomas are found frequently in neurofibromatosis type 2 (NF2), implicating the NF2 tumor suppressor gene in pathogenesis. Molecular analysis of spinal ependymomas has revealed mutations in the NF2 gene [49]. In addition, occasional ependymomas have been reported in patients with Turcot's syndrome [50]. They have rarely been associated with p53 germ line mutations [51].

The T antigen was initially detected in human ependymomas by immunohistochemistry and immunocytochemistry. Using anti-SV40 T and anti-JCV T hamster sera, nuclear staining was observed in frozen sections of an ependymoma and in primary cultures obtained from that tumor [42]. Since that study, PCR technology has been applied to this problem in several independent studies, demonstrating that the majority of ependymomas have sequences specific for polyomaviruses [29–32,34,38,40,43] (see Table 2 for results from the authors' laboratory). Immunohistochemistry with monoclonal antibodies to T antigen that are presumably more specific than the original polyclonal sera [32] also demonstrates nuclear positivity in tumor cells (Fig. 2). Immunoprecipitation has also been used to demonstrate T antigen in a high percentage of ependymomas [44]. Analysis of the products obtained from PCR amplification has resulted in the identification of both SV40- and JCV-specific sequences. An initial analysis of amplified SV40 sequences revealed a specific arrangement of the enhancer region denoted as ''archetype'' because of a single 72 base pair (bp) element that is often present in a duplicated form in other SV40 isolates [38]. This finding raised the possibility that the archetypal sequence might confer some degree of tissue specificity on these strains and result in SV40 sequences more likely to cause tumors. A more intensive analysis performed on a number of these cases from which full-length polyoma sequences could be amplified [41] suggests that SV40 has a relatively broad host and tissue range and that the archetype sequence can be found in both tumorous and nontumorous sources.

The clinical features of ependymomas are dependent upon their location in the CNS. Those that occur in the posterior fossa and fourth ventricle are particularly apt to result in hydrocephalus, which results in increased intracranial pressure, often presenting with headache, nausea, vomiting, and dizziness. Ataxia and visual disturbances are also seen. Supratentorial ependymomas may present with focal neurological deficits and seizures, spinal tumors with long tract signs, and ependymomas of the filum terminale with mixed long tract and peripheral nerve deficits from compression of the caudal spinal cord and the lumbosacral roots. The presenting signs can also vary with age. Increasing head circumference, stiff neck, lethargy, and irritability are commonly found in pediatric patients, whereas papilledema, nystagmus and focal signs are found more commonly in adults [52,53].

Magnetic resonance imaging (MRI) is the current modality of choice to evaluate CNS tumors. Sagittal and axial T1-weighted images are obtained before and after contrast administration along with axial and coronal T2-weighted images. Ependymomas are typi-

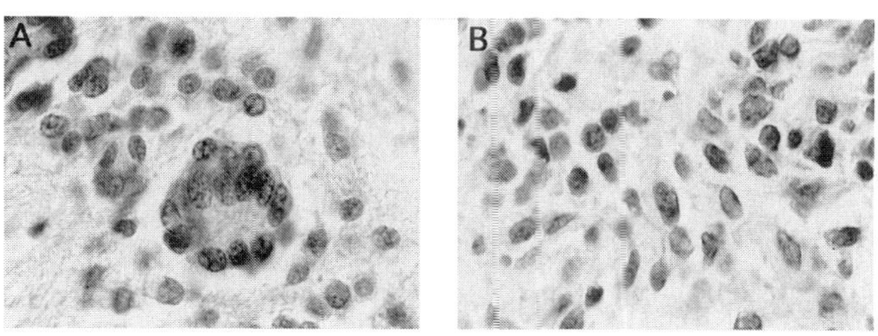

Figure 2 Immunohistochemical detection of T antigen in (A) an ependymoma and (B) a medulloblastoma. Paraffin-embedded tissue samples of these two tumors demonstrate clear nuclear positivity with a monoclonal antibody to T antigen providing further evidence for the association of polyoma virus with these tumors.

cally hypointense or isointense on precontrast T1-weighted images and enhance following contrast administration (Fig. 3A). The T1 signal may be heterogeneous. On T2-weighted images, the tumors are hyperintense and appear well demarcated from surrounding brain. Infiltration and/or edema of the adjacent brain are seldom seen. Computed tomographic (CT) scanning may also be used effectively in the diagnosis of these tumors. On precontrast CT scans, they appear as isodense or hyperdense masses, often with mixed signal. Following contrast administration, they enhance heterogeneously. CT scans reveal calcifications in 50% of cases. Both MRI and CT frequently reveal hydrocephalus and brainstem compression when the tumor is located in the posterior fossa. In supratentorial cases, cystic changes are fairly frequent, and intralesional hemorrhage and/or substantial calcifications are found occasionally. Spinal cord ependymomas often demonstrate syringomyelia because of the central canal involvement by the tumor. Unfortunately, the signal characteristics of ependymomas on both MR and CT can be quite similar to those of other childhood tumors, particularly medulloblastomas (see below). The superior anatomic resolution of MR imaging often allows one to differentiate between the two tumors on the basis of the intraventricular location of ependymomas versus the preference of medulloblastomas to arise from the cerebellar parenchyma. Craniospinal MR imaging also allows one to detect CSF dissemination of ependymomas (see below as well) [54,55].

Both definitive diagnosis and therapy of ependymomas are achieved through surgery. For posterior fossa tumors, this is generally achieved through a suboccipital approach. Of the primary glial tumors, fourth ventricular and spinal ependymomas generally have the most defined borders between tumor and brain and the least microscopic tissue invasion. For that reason, the primary goal of surgery is gross total removal of the tumor. The secondary goal for patients with posterior fossa lesions is normalization of CSF dynamics, which if not achieved by tumor resection can be accomplished by ventriculoperitoneal shunting [56].

The importance of pathological evaluation of the tumor is to establish a tissue diagnosis, which will guide the neurosurgeon and oncologist in regard to the pathobiology of the lesion, particularly in regard to the cell of origin of the tumor and the potential for recurrence. Ependymomas tend to be moderately cellular and demonstrate little if any variability in the size and shape of nuclei. A histological hallmark of these tumors is the ependymal or Flexner rosette, which is composed of a collar of neoplastic cells surrounding

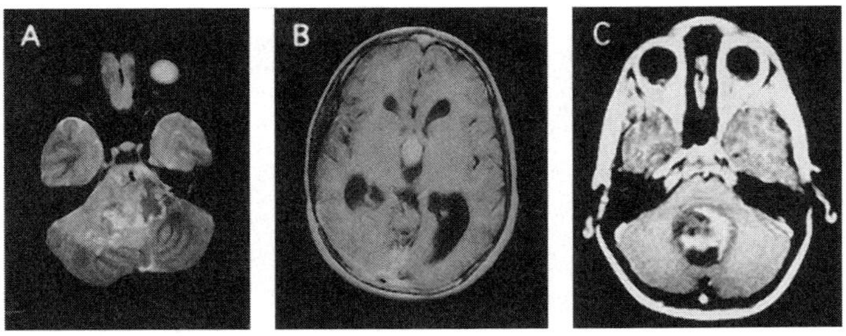

Figure 3 Magnetic resonance images of (A) an ependymoma, (B) a choroid plexus papilloma, and (C) a medulloblastoma. All three images are contrast enhanced with gadolinium demonstrating tumor enhancement and mass effect in the posterior fossa (A, C) and the third ventricle (B). Note the degree of hydrocephalus in B due to the choroid plexus papilloma.

a hollow core. A more easily identified feature is the perivascular pseudorosette composed of tumor cells that project their processes toward a central blood vessel. Both of these structures seem to recapitulate the tendency of normal ependymal cells to form a polarized layer lining the ventricular system. In keeping with their glial origin, most ependymomas show glial filament acidic protein (GFAP) immunoreactivity, particularly within the rosettes and pseudorosettes. They also typically express S100 protein and vimentin and commonly show both epithelial membrane antigen and cytokeratin reactivities. Nestin, a marker for intermediate filaments during CNS development, has also been found [57]. Although there is no unique immunomarker for these tumors, the electron microscopic features of cilia with a 9 + 2 arrangement, blepharoplasts, luminal microvilli, lateral junctional complexes, and lack of a basement membrane are quite specific [58]. Histological variants that are well recognized include the cellular, papillary, clear cell, and tanycytic ependymomas, which are all World Health Organization (WHO) grade II. Another variant, the myxopapillary ependymoma, is also WHO grade II but occurs almost exclusively in the distal spinal cord and filum terminale. The anaplastic ependymoma, characterized by increased cellularity and mitotic activity plus frequent necrosis and vascular proliferation, is the only variant of this tumor to definitively fall into WHO grade III [57].

The prognosis for most forms of ependymoma depends on patient age, extent of resection, and tumor location. The survival of children tends to be worse than that of adults. Patients with spinal ependymomas have the best survival rate, whereas those with hemispheric tumors tend to fare better than those with tumors in the posterior fossa. Because posterior fossa ependymomas predominate in children, anatomic location may to some degree account for the age-dependent survival difference.

In patients treated with surgery alone, 5-year survival rates have been estimated as 15% for supratentorial and 33% for infratentorial tumors. In patients with infratentorial disease, gross total resection combined with radiation therapy boosts 5-year progression-free survival to 58%. In addition to the above factors, the diagnosis of an anaplastic ependymoma is generally accepted to be a negative prognostic indicator [59,60].

Although most recurrences of ependymomas occur at the primary tumor site, cerebrospinal dissemination is also found. The incidence of these CSF metastases is greater in infratentorial than in supratentorial tumors, and they are seen more frequently in anaplastic ependymomas than in the other histological subtypes. Extraneural metastases have been found as well, primarily in the lungs [52,54].

Conventional therapy for patients with infratentorial or supratentorial ependymomas over the age of 3 years are generally treated with focal field radiotherapy to a total dose of 5000–5500 cGy [61]. Because of the risk of CSF dissemination, patients with anaplastic ependymomas are treated with craniospinal radiotherapy and a focal boost to the tumor site. Children under the age of 3 years are given multiagent chemotherapy in place of radiotherapy because of the deleterious effects of radiotherapy in that age group. Adjuvant chemotherapy with vincristine, cisplatin, and [1-(2-chloroethyl)-3-cyclohexyl-1-nitrosourea] (CCNU) has been tried for newly diagnosed ependymomas and anaplastic ependymomas in children but has not demonstrated clear benefit over gross total resection followed by radiotherapy. Chemotherapy with platinum compounds has also been used for tumor recurrence [60].

2.2 Choroid Plexus Tumors

Choroid plexus papilloma and carcinoma are rare CNS lesions that arise from the normal choroid plexus in the lateral, third, and fourth ventricles as circumscribed, cauliflower-like masses. Not only do they obstruct ventricular outflow, they also oversecrete cerebrospinal fluid, and by both mechanisms they result in hydrocephalus.

Choroid plexus neoplasms account for only 0.4–0.6% of all brain tumors but 2–4% of brain tumors in children and 10–20% of brain tumors presenting in the first year of life. The average annual incidence is 0.3 per million population. Papillomas outnumber carcinomas by at least 5:1. Although the mean age for occurrence is 1 year, the age range is from birth to 60 years. Eighty percent of the lateral ventricular tumors occur in patients less than 20 years old. Fourth ventricular tumors are evenly distributed across all age groups [27]. Congenital tumors have been found and fetal tumors diagnosed by ultrasound. The male/female ratio for lateral ventricular tumors is 1:1 and for fourth ventricular tumors is 3:2.

Although both choroid plexus papilloma and carcinoma occasionally occur in the setting of the Li-Fraumeni syndrome, no p53 mutations have been reported in the setting of sporadic choroid plexus tumors [62,63]. They may also be associated with Aicardi's syndrome [64,65] or von Hippel-Lindau Disease [66,67].

The initial evidence for polyoma involvement in choroid plexus tumors came from the electron microscopic observation of viral particles most consistent with polyoma virus in a surgically excised papilloma from a 33-year-old woman [68]. The first study to localize polyoma T antigen to human tumors [42] found positivity in a choroid plexus papilloma from a 48-year-old man as well as an ependymoma. Indeed, the application of PCR technology has demonstrated polyomavirus in a large percentage of choroid plexus tumors of children, ranging from 40% to 90% in three independent series [29,34,40]. Interestingly, in all of these studies the amplified sequences implicated SV40 rather than JCV or BKV. As in the case of ependymomas, initial analysis of the SV40 sequences detected an archetypical or nonduplicated 72 bp sequence in the enhancer from a choroid plexus carcinoma.

The clinical symptoms in both children and adults tend to be caused by increased intracranial pressure. This is due to both overproduction of CSF and blockage of the ventricular system by the tumor. In infants, this results in presenting symptoms that include increasing head circumference, vomiting, and failure to thrive. In adults, headache and visual disturbances predominate. CT scans reveal smooth or lobulated masses within the ventricles with secondary hydrocephalus. The masses are isodense or hyperdense in comparison to the surrounding brain and are homogeneously contrast enhancing. On T1-weighted magnetic resonance (MR) sequences, the tumors are heterogeneous with intermediated signal intensity. Signal voids of blood vessels are often found. Signal on T2-weighted imaging is heterogeneous as well. MRI is the diagnostic modality of choice because the anatomic resolution allows one to narrow the differential diagnosis (which includes ependymoma, exophytic glioma, and dermoid cyst) with the greatest assurance. In addition, because T1-weighted images may reveal signal voids of feeding blood vessels, they may assist in preoperative staging. Angiography may also be required to define the vascular anatomy of the lesion [69,71].

The primary therapeutic modality is, as with most CNS tumors, surgery. Surgical goals with choroid plexus tumors are gross total resection and relief or management of hydrocephalus. For papillomas, gross total resection frequently results in cure. Despite complete tumor resection, 20% of patients may require ventriculoperitoneal shunting for relief of hydrocephalus. Major challenges to the surgeon include the size of the tumor, vascularity, and intraventricular location. Intraoperative blood loss due to the high vascularity of the tumor may be a particular problem in young children because of their small systemic vascular volume [72].

Pathologically, choroid plexus papillomas are composed of delicate fibrovascular connective tissue fronds covered by a single layer of uniform cuboidal to columnar epithelium with fairly uniform round to oval nuclei. The structure of the tumor is quite close to that of normal choroid plexus except that the cells are somewhat more crowded and elongated. Choroid plexus carcinomas show clear signs of malignancy, including nuclear pleomorphism, frequent mitoses, a high nuclear-to-cytoplasmic ratio, high cell density, breakdown of the papillary architecture, and regions of necrosis. In addition, although brain invasion is not seen with papillomas, it is found frequently with carcinomas. Tumors that demonstrate only one or a few features of malignancy such as an increased mitotic rate without necrosis or parenchymal invasion have been termed atypical choroid plexus papillomas, although clear diagnostic criteria are not established. Even benign choroid plexus papillomas may seed cells into the subarachnoid space, although these implants are usually microscopic and not symptomatic or of prognostic import. However, choroid plexus carcinomas will develop macroscopic symptomatic subarachnoid metastases. Immunohistochemistry of both papillomas and carcinomas demonstrates expression of cytokeratin, vimentin, and S-100 by the tumors, which is not surprising, because the same antigens are present in normal choroid plexus. Interestingly, 20–25% also express GFAP, which is normally not found in choroid plexus [70]. As in the normal choroid plexus, electron microscopy demonstrates interdigitating cell membranes, tight junctions, microvilli, occasional apical villi, cilia, and a basement membrane at the abluminal pole. Although it is unusual to resort to electron microscopy for the pathological workup of choroid plexus papilloma, cases of carcinoma with marked loss of tissue architecture may require that modality for diagnosis [57].

After surgery, papillomas are generally followed with repeated MR imaging and careful management of CSF dynamics. In children younger than 3 years of age with carcinomas that have been subtotally resected or are recurrent, multiagent chemotherapy is recommended [73,74]. Children older than 6 years of age and adults with choroid plexus carcinomas should receive radiotherapy after gross total or subtotal resection [75]. Because of the risk of subarachnoid metastases, craniospinal radiotherapy should also be considered in these patients [74]. The role of chemotherapy in the older group is not clear. However, given the chemosensitivity of the choroid plexus carcinomas, multiagent chemotherapy is often offered to children and adults with residual or recurrent carcinoma after radiotherapy or in patients with disseminated disease [69,74].

2.3 Medulloblastoma/Primitive Neuroectodermal Tumor

Medulloblastoma is a malignant primitive tumor of the cerebellum. Medulloblastoma cerebelli was first described by Bailey and Cushing in 1924 [76], with 29 cases occurring in children of a tumor that arose over the fourth ventricle and projected into the cerebellar vermis. Microscopically, these neoplasms consisted of small undifferentiated cells. Bailey and Cushing assumed that the tumors derived from medulloblasts—embryonal cells with the ability to differentiate into either neurons or glia. In 1973, Hart and Earle [77] introduced the term primitive neuroectodermal tumor (PNET) for undifferentiated CNS tumors that occur outside the cerebellum but are histologically similar to medulloblastomas. This concept was extended to a variety of pediatric brain tumors including medulloblastomas in 1983 by Rorke [78] on the grounds that the precursor cells, histology, and clinical course of all these tumors are quite similar. Many authors use the terms medulloblastoma and PNET almost interchangeably as we do in the remainder of this discussion. However, because much remains to be learned about the cellular origins and molecular mechanisms involved in the tumor, there is still controversy surrounding the equivalence of these two classifications.

Medulloblastoma/PNET represents the second most common brain neoplasm in children, accounting for 20% of all pediatric CNS neoplasms. Incidence rates are approximately 7 per million children per year or 1.8 per million general population per year. The peak age of incidence is 3–8 years, mean age of incidence 7.5 years [27]. Several congenital cases have been reported, and these tumors account for up to 25% of all CNS neoplasms encountered in the first year of life. Seventy-five percent to 80% of medulloblastomas occur in patients under the age of 20 years, with most of the remaining cases presenting in patients younger than 40. The male female ratio is approximately 1.6 : 1 in children. In patients over 20 years old, the male predominance is somewhat diminished.

The occurrence of medulloblastomas in identical twins was first noted by Harvey Cushing and reported by Leavitt. Since that time, other familial clusters have also been noted to involve dizygotic twins and siblings [79–81]. Associated tumors include Wilms's tumor and malignant rhabdoid tumor. A small number of medulloblastomas are associated with the heritable Gorlin and Turcot syndromes. Gorlin or basal cell nevus syndrome is an autosomal dominant disorder characterized by multiple developmental defects and susceptibility to cancers including medulloblastoma [82–84]. The defective gene is a human homolog of *Drosophila patched* in the hedgehog (Hh)/PTCH pathway. Examination of sporadic medulloblastomas has also revealed loss of heterozygosity and somatic mutations of the human PTCH allele in 15–20% of tumors [85,86]. In Turcot syndrome, mutations of the APC gene of the Wingless pathway lead to colonic tumors and brain tumors including medulloblastomas [87].

Primitive neuroectodermal tumors including medulloblastomas have been described previously in hamsters inoculated with JCV [8]. Recently, transgenic mice, which contain the JCV early gene encoding T antigen, have been shown to develop medulloblastomas [12]. Guided by these results, our group [35,37] (see Table 2 for results from our studies) and several others [34,36,39,43] have detected polymavirus DNA in archival medulloblastoma specimens. Immunohistochemical staining has demonstrated T antigen but not the VP late proteins in tumor cells supporting the notion of incomplete viral replication (Fig. 2B). One study [44] demonstrated T antigen in fresh tumor samples that is capable of complexing p53 and pRb in vitro, implying that the JCV in medulloblastomas may play a role in cell cycle dysregulation.

The classical presentation of the patient with medulloblastoma includes the symptoms of increased intracranial pressure, which include headache, vomiting, and visual symptoms. When severe and/or acute, the patient may also present with a diminished level of consciousness. Neurological examination often reveals papilledema and sixth nerve palsies due to the increased intracranial pressure. Ataxia, dysmetria, and nystagmus are often found as well due to the cerebellar location of the tumor. Subarachnoid dissemination can contribute to cranial nerve deficits, particularly sixth nerve palsies, or announce itself with radicular back pain. All too often, subarachnoid disease is clinically silent. Medulloblastoma may also present with apoplectic intratumoral hemorrhage leading to acute headache and focal motor or sensory deficits referable to the posterior fossa and coma.

Computed tomographic scans reveal lesions that tend to be hyperdense and enhance homogeneously following the injection of contrast material. The finding of a low-density surround to the region of contrast enhancement usually reflects peritumoral edema. As in the tumors discussed above, MRI is the neuroradiological modality of choice for the workup of suspected medulloblastomas, particularly because of the superior anatomic resolution it affords. In addition, MRI demonstrates subarachnoid tumor within both the skull and spinal canal with greater sensitivity and specificity than CT scanning. On T1-weighted images, medulloblastomas are isointense or hypointense compared with the surrounding brain and show mild to moderate contrast enhancement, which can be either homogeneous or patchy (Fig. 3C). Spinal MRI with gadolinium is important for evaluating subarachnoid disease as part of initial staging and is done preoperatively in stable patients or as soon as possible postoperatively.

Surgery is the initial treatment modality, and as with ependymomas and choroid plexus tumors the goal is gross total resection, which can be achieved in the majority of patients. In several studies, gross total resection was statistically correlated with longer survivals. In those studies in which gross total resection and survival were not significantly correlated, results trended in that direction.

Histopathologically, the vast majority of medulloblastomas fall into three categories: classic, neuroblastic, and desmoplastic. They all feature densely packed cells with round to oval or carrot-shaped hyperchromatic nuclei and a high nuclear-to-cytoplasmic ratio. Homer-Wright rosettes are a frequent feature. The neuroblastic variety features columns or ''Indian files'' of the neoplastic cells. In the desmoplastic tumors, nodular reticulin zones (pale islands) stand out from the background of the tumor. These nodules show reduced cellularity, extracellular fibrillary material, and marked nuclear uniformity. Immunohistochemically, synaptophysin expression is characteristic of these tumors and is found prominently in nodules and neuroblastic rosettes [88]. Of the intermediate filament proteins, vimentin, nestin, and neurofilament are most characteristically expressed. Stellate GFAP positive cells are often found as well that may represent trapped astrocytes or differentiated tumor cells with an astrocytic phenotype [89]. These tumors also express both high and low affinity nerve growth factor receptors. Electron microscopy shows features consistent with embryonal neurons, including neurite-like cytoplasmic processes with microtubules and specialized adhesion plaques, dense core vesicles, and synapses [57].

After the diagnosis of medulloblastoma, craniospinal radiation therapy with a local boost to the primary posterior fossa site is the treatment of choice for patients older than 3 years [90]. Because of the dropoff in IQ scores found in children less than 3 years of age treated with radiation therapy [91,92], these patients are treated with multiagent chemotherapy. If disease progression is noted before or after 3 years of age, craniospinal radiotherapy with a boost to the primary site is used [73]. Chemotherapy is also indicated for high-risk patients over 3 years of age or for patients over 3 with recurrence [93].

Traditionally, the prognosis for five-year survival for patients with medulloblastoma has been between 30% and 70% [94]. More recent reports for high risk patients treated with adjuvant chemotherapy put the survival at greater than 70%, thus indicating that the traditional survival figures may be too low [91]. Prognostic factors include age at diagnosis (greater than 4 years favorable), extent of surgical resection, and extent of local disease and metastatic involvement [95]. Leptomeningeal spread of tumor is present in approximately 30% of patients at the time of diagnosis. Because of this, all patients should be staged with a complete spinal MRI including gadolinium plus CSF cytology [92]. Systemic metastases, of which bone is the most common, occur in about 5% of cases. The most common complication of therapy is endocrinopathy.

3 CONCLUSION

Multiple lines of evidence point to an association between polyomaviruses and brain tumors. Despite this, many questions must still be answered. Among these is the incidence of transmission of these viruses to very young children and the particular molecular mechanisms at work in human virus-induced CNS tumorigenesis. It is not yet clear whether the detection of polyomavirus in brain tumors is prognostically significant and whether it can serve as a tumor marker. The possibility exists that polyomavirus may serve as a target for therapy in these tumors.

REFERENCES

1. Butel, J.S. Simian virus 40. In: Encyclopedia of Virology; Webster, R.G., Granoff, A., Eds.; Academic Press: San Diego, 1994, 1322–1329.
2. Grodzicker, T.; Hopkins, N. Origins of contemporary DNA tumor virus research. In: DNA Tumor Viruses; Tooze, J., Ed., 2nd ed.; Cold Spring Harbor Lab Press: Plainview, NY, 1980, 155–162.
3. Shah, K.V. Polyomaviruses In Fields' Virology; Fields, B.N., Knipe, D.M., Howley, P.M., Chanock, R.M., Melnick, J.L., Monath, T.P., Roizman, B., Straus, S.E., Eds., 3rd ed.; Lippincott Raven: New York, 1996, 2027–2044.
4. Berger, J.R.; Concha, M. Progressive multifocal leukoencephalopathy: the evolution of a disease once considered rare. J Neurovirol. 1995, 1, 5–18.
5. Arthur, R.R.; Shah, K.V.; Baust, S.J.; Santos, G.W.; Saral, R. Association of BK viruria with hemorrhagic cystitis in recipients of bone marrow transplants. N Engl J Med. 1995, 315, 230–234.
6. Arthur, R.R.; Dagostin, S.; Shah, K.V. Detection of BK virus and JC virus in urine and brain tissue by the polymerase chain reaction. J Clin Microbiol. 1995, 27, 1174–1179.
7. Frisque, R.J.; White III, F.A. The molecular biology of JC virus, causative agent of progressive multifocal leukoencephalopathy. In: Molecular Neurovirology; Roos, R.P., Ed.; Humana Press: Totowa: NJ, 1992, 25–158.
8. Walker, D.L.; Padgett, B.L.; ZuRhein, G.M.; Albert, A.E.; Marsh, R.F. Human papovavirus (JC): induction of brain tumors in hamsters. Science. 1995, 181, 674–676.
9. Corallini, A.; Barbanti-Brodano, G.; Bortoloni, W.; Nenci, I.; Cassai, E.; Tampieri, M.; Portolani, M.; Borgatti, M. High incidence of ependymomas induced by BK virus, a human papovavirus: brief communication. J Natl Cancer Inst. 1995, 59, 1561–1564.
10. Brinster, R.L.; Chen, H.Y.; Messing, A.; van Dyke, T.; Levine, A.J.; Palmiter, R.D. Transgenic mice harboring SV40 T-antigen genes develop characteristic brain tumors. Cell. 1995, 37, 367–379.
11. Franks, R.R.; Rencic, A.; Gordon, J.; Zoltick, P.W.; Curtis, M.; Knobler, R.L.; Khalili, K. Formation of undifferentiated mesenteric tumors in transgenic mice expressing human neurotropic polyomavirus early protein. Oncogene. 1995, 12, 2573–2578.
12. Krynska, B.; Otte, J.; Franks, R.; Khalili, K.; Croul, S. Human ubiquitous JCV-CY induces brain tumors in experimental animals. Oncogene. 1995, 18, 39–46.
13. Cox, L.S.; Lane, D.P. Tumour suppressors, kinases and clamps: how p53 regulates the cell cycle in response to DNA damage. Bioessays. 1995, 17, 501–508.
14. Farwell, J.R.; Dohrmann, G.J.; Marrett, L.D.; Meigs, J.W. Effect of SV40 virus-contaminated polio vaccine on the incidence and type of CNS neoplasms in children: a population-based study. Trans Am Neurol Assoc. 1995, 104, 261–264.

15. Farwell, J.R.; Dohrmann, G.J.; Flannery, J.T. Medulloblastoma in childhood: an epidemiological study. J Neurosurg. 1995, 61, 657–664.
16. Fraumeni Jr., J.F.; Stark, C.R.; Gold, E.; Lepow, M.L. Simian virus 40 in polio vaccine: follow-up of newborn recipients. Science. 1995, 167, 59–60.
17. Heinonen, O.P.; Shapiro, S.; Monson, R.R.; Hartz, S.C.; Rosenberg, L.; Slone, D. Immunization during pregnancy against poliomyelitis and influenza in relation to childhood malignancy. Int J Epidemiol. 1995, 2, 229–235.
18. Shah, K.V.; Nathanson, N. Human exposure to SV40: review and comment. Am J Epidemiol. 1995, 103, 142.
19. Olin, P.; Giesecke, J. Potential exposure to SV40 in polio vaccines used in Sweden during 1957: no impact on cancer incidence rates 1960 to 1993. Dev Biol Stand. 1995, 94, 227–233.
20. Geissler, E.; Staneczek, W. SV40 and human brain tumors. Arch Geschwulstforsch. 1995, 58, 129–134.
21. Fisher, S.G.; Weber, L.; Carbone, M. Cancer risk associated with simian virus 40 contaminated polio vaccine. Anticancer Res. 1995, 19, 2173–2180.
22. Strickler, H.D.; Rosenberg, P.S.; Devesa, S.S.; Hertel, J.; Fraumeni Jr., J.F.; Goedert, J.J. Contamination of poliovirus vaccines with simian virus 40 (1955–1963) and subsequent cancer rates. JAMA. 1995, 279, 292–295.
23. Geissler, E.; Konzer, P.; Scherneck, S.; Zimmerman, W. Sera collected before introduction of contaminated polio vaccine contain antibodies against SV40. Acta Virol. 1995, 29, 420–423.
24. Shah, K.V.; Ozer, H.L.; Pond, H.S.; Palma, L.D.; Murphy, G.P. SV40 neutralizing antibodies in sera of US residents without history of polio immunization. Nature. 1995, 231, 448–449.
25. Castaigne, P.; Randot, P.; Escourolle, J.L.; Ribadeau, D.; Cathala, F.; Hauw, J. Leucoencephalothie multifocale progressive et ''gliomes'' multiples. Rev Neurol, 1974, 379–392.
26. GiaRusso, M.H.; Koeppen, A.H. Atypical progressive multifocal leukoencephalopathy and primary cerebral malignant lymphoma. J Neurol Sci. 1978, 35, 391–398.
27. CBTRUS. Statistical Report: Primary Brain Tumors in the United States 1992–1997; The Central Brain Tumor Registry of the United States: Chicago, IL, 2000.
28. Stern, R.O. Cerebral tumors in children: a pathological report. Arch Dis Child. 1995, 12, 291.
29. Bergsagel, D.J.; Finegold, M.J.; Butel, J.S.; Kupsky, W.J.; Garcea, R.L. DNA sequences similar to those of simian virus 40 in ependymomas and choroid plexus tumors of childhood. N Engl J Med. 1995, 326, 988–1993.
30. Caldarelli-Stefano, R.; Boldorini, R.; Monga, G.; Meraviglia, E.; Zorini, E.O.; Ferrante, P. JC virus in human glial-derived tumors. Hum Pathol. 1995, 31, 394–395.
31. Corallini, A.; Pagnani, M.; Viadana, P.; Silini, E.; Mottes, M.; Milanesi, G.; Gerna, G.; Vettor, R.; Trapella, G.; Silvani, V. Association of BK virus with human brain tumors and tumors of pancreatic islets. Int J Cancer. 1995, 39, 60–67.
32. Del Valle, L.; Gordon, J.; Assimakopolou, M.; Katsetos, C.; Croul, S.E.; Khalili, K. Detection of JC virus DNA sequences and expression of the viral regulatory protein, T-antigen, in tumors of the central nervous system. Cancer Res. 1995, 61, 4287–4293.
33. Dorries, K.; Loeber, G.; Meixensberger, J. Association of polyomaviruses JC, SV40, and BK with human brain tumors. Virology. 1995, 160, 268–270.
34. Huang, H.; Reis, R.; Yonekawa, Y.; Lopes, J.M.; Kleihues, P.; Ohgaki, H. Identification in human brain tumors of DNA sequences specific for SV40 large T antigen. Brain Pathol. 1995, 9, 33–42.
35. Khalili, K.; Krynska, B.; Del Valle, L.; Katsetos, C.D.; Croul, S. Medulloblastomas and the human neurotropic polyomavirus JC virus. Lancet. 1995, 353, 1152–1153.
36. Krieg, P.; Amtmann, E.; Jonas, D.; Fischer, H.; Zang, K.; Sauer, G. Episomal simian virus 40 genomes in human brain tumors. Proc Natl Acad Sci USA. 1995, 78, 6446–6450.
37. Krynska, B.; Del Valle, L.; Croul, S.; Gordon, J.; Katsetos, C.D.; Carbone, M.; Giordano, A.; Khalili, K. Detection of human neurotropic JC virus DNA sequence and expression of

the viral oncogenic protein in pediatric medulloblastomas. Proc Natl Acad Sci USA. 1995, 96, 11519–11524.

38. Lednicky, J.A.; Garcea, R.L.; Bergsagel, D.J.; Butel, J.S. Natural simian virus 40 strains are present in human choroid plexus and ependymoma tumors. Virology. 1995, 212, 710–717.
39. Malkin, D.; Chilton-MacNeill, S.; Meister, L.A.; Sexsmith, E.; Diller, L.; Garcea, R.L. Tissue-specific expression of SV40 in tumors associated with the Li-Fraumeni syndrome. Oncogene. 1995, 20, 4441–4449.
40. Martini, F.; Iaccheri, L.; Lazzarin, L.; Carinci, P.; Corallini, A.; Gerosa, M.; Iuzzolino, P.; Barbanti-Brodano, G.; Tognon, M. SV40 early region and large T antigen in human brain tumors, peripheral blood cells, and sperm fluids from healthy individuals. Cancer Res. 1995, 56, 4820–4825.
41. Stewart, A.R.; Lednicky, J.A.; Butel, J.S. Sequence analysis of human tumor-associated SV40 DNAs and SV40 viral isolates from monkeys and humans. J Neurovirol. 1995, 4, 182–193.
42. Tabuchi, K.; Kirsch, W.M.; Low, M.; Gaskin, D.; Van Buskirk, J.; Maa, S. Screening of human brain tumors for SV40-related T antigen. Int J Cancer. 1995, 21, 12–17.
43. Weggen, S.; Bayer, T.A.; von Deimling, A.; Reifenberger, G.; von Schweinitz, D.; Wiestler, O.D.; Pietsch, T. Low frequency of SV40, JC and BK polyomavirus sequences in human medulloblastomas, meningiomas and ependymomas. Brain Pathol. 1995, 10, 85–92.
44. Zhen, H.N.; Zhang, X.; Bu, X.Y.; Zhang, Z.W.; Huang, W.J.; Zhang, P.; Liang, J.W.; Wang, X.L. Expression of the simian virus 40 large tumor antigen (Tag) and formation of Tag-p53 and Tag-pRb complexes in human brain tumors. Cancer. 1995, 86, 2124–2132.
45. Virchow, R. Die Krankenhaften Gerschwülste. 1863–1865.
46. Bailey, P. A study of tumors arising from ependymal cells. Arch Neurol Psych. 1995, 11, 1–27.
47. Jellinger, K.; Seitelberger, F. Zur Neuropathologie der Hirngeschwülste im Kindesalter. Wiener Med Wochenschr. 1995, 120, 855–861.
48. Schiffer, D.; Chio, A.; Giordana, M.T.; Migheli, A.; Palma, L.; Pollo, B.; Soffietti, R.; Tribolo, A. Histologic prognostic factors in ependymoma. Childs Nerv Syst Aug. 1995, 7, 177–182.
49. Ebert, C.; von Haken, M.; Meyer-Puttlitz, B.; Wiestler, O.D.; Reifenberger, G.; Pietsch, T.; von Deimling, A. Molecular genetic analysis of ependymal tumors. NF2 mutations and chromosome 22q loss occur preferentially in intramedullary spinal ependymomas. Am J Pathol. 1995, 155, 627–632.
50. Torres, C.F.; Korones, D.N.; Pilcher, W. Multiple ependymomas in a patient with Turcot's syndrome. Med Pediatr Oncol. 1995, 28, 59–61.
51. Metzger, A.K.; Sheffield, V.C.; Duyk, G.; Daneshvar, L.; Edwards, M.S.; Cogen, P.H. Identification of a germ-line mutation in the p53 gene in a patient with an intracranial ependymoma. Proc Natl Acad Sci USA. 1995, 88, 7825–7829.
52. Nazar, G.B.; Hoffman, H.J.; Becker, L.E.; Jenkin, D.; Humphreys, R.P.; Hendrick, E.B. Infratentorial ependymomas in childhood: prognostic factors and treatment. J Neurosurg. 1995, 72, 408–417.
53. Rawlings 3rd, C.E.; Giangaspero, F.; Burger, P.C.; Bullard, D.E. Ependymomas: a clinicopathologic study. Surg Neurol. 1995, 29, 271–281.
54. Goldwein, J.W.; Leahy, J.M.; Packer, R.J.; Sutton, L.N.; Curran, W.J.; Rorke, L.B.; Schut, L.; Littman, P.S.; D'Angio, G.J. Intracranial ependymomas in children. Int J Radiat Oncol Biol Phys. 1995, 19, 1497–1502.
55. Goldwein, J.W.; Corn, B.W.; Finlay, J.L.; Packer, R.J.; Rorke, L.B.; Schut, L. Is craniospinal irradiation required to cure children with malignant (anaplastic) intracranial ependymomas?. Cancer. 1995, 67, 2766–2771.
56. Duncan III, J.A.; Hoffman, H.J. Intracranial ependyomas. In; Kaye, A.H., Laws Jr., E.R., Eds.. 1995, 493–504.
57. Burger, P.C.; Scheithauer, B.W.; Vogel, F.S., eds. Surgical Pathology of the Nervous System and Its Coverings, 3 ed.. 1991, 289–296.

58. Hirano, A. Some contributions of electron microscopy to the diagnosis of brain tumors. Acta Neuropathol (Berl). 1995, 43, 119–128.
59. Ringertz, N.; Reymond, A. Ependymomas and choroid plexus papillomas. J Neuropathol Exp Neurol. 1995, 8, 355–380.
60. Sutton, L.N.; Goldwein, J.; Perilongo, G.; Lang, B.; Schut, L.; Rorke, L.; Packer, R. Prognostic factors in childhood ependymomas. Pediatr Neurosurg. 1990–91, 16, 57–65.
61. Wallner, K.E.; Wara, W.M.; Sheline, G.E.; Davis, R.L. Intracranial ependymomas: results of treatment with partial or whole brain irradiation without spinal irradiation. Int J Radiat Oncol Biol Phys. 1995, 12, 1937–1941.
62. Kleihues, P.; Schauble, B.; zur Hausen, A.; Esteve, J.; Ohgaki, H. Tumors associated with p53 germline mutations: a synopsis of 91 families. Am J Pathol. 1995, 150, 1–13.
63. Vital, A.; Bringuier, P.P.; Huang, H.; San Galli, F.; Rivel, J.; Ansoborlo, S.; Cazauran, J.M.; Taillandier, L.; Kleihues, P.; Ohgaki, H. Astrocytomas and choroid plexus tumors in two families with identical p53 germline mutations. J Neuropathol Exp Neurol. 1995, 57, 1061–1069.
64. Hamano, K.; Matsubara, T.; Shibata, S.; Hirano, C.; Ito, Z.; Ase, Y.; Kusakari, J.; Takita, H. Aicardi syndrome accompanied by auditory disturbance and multiple brain tumors. Brain Dev. 1995, 13, 438–441.
65. Trifiletti, R.R.; Incorpora, G.; Polizzi, A.; Cocuzza, M.D.; Bolan, E.A.; Parano, E. Aicardi syndrome with multiple tumors: a case report with literature review, 1995.
66. Blamires, T.L.; Maher, E.R. Choroid plexus papilloma. A new presentation of von Hippel-Lindau (VHL) disease. Eye. 1995, 6(Pt 1), 90–92.
67. Blamires, T.L.; Friedmann, I.; Moffat, D.A. Von Hippel-Lindau disease associated with an invasive choroid plexus tumour presenting as a middle ear mass. J Laryngol Otol. 1995, 106, 429–435.
68. Bastian, F.O. Papova-like virus particles in a human brain tumor. Lab Invest. 1995, 25, 169–175.
69. Allen, J.; Wisoff, J.; Helson, L.; Pearce, J.; Arenson, E. Choroid plexus carcinoma—responses to chemotherapy alone in newly diagnosed young children. J Neurooncol. 1995, 12, 69–74.
70. Boyd, M.C.; Steinbok, P. Choroid plexus tumors: problems in diagnosis and management. J Neurosurg. 1995, 66, 800.
71. Lena, G.; Genitori, L.; Molina, J.; Legatte, J.R.; Choux, M. Choroid plexus tumours in children. Review of 24 cases. Acta Neurochir (Wien). 1995, 106, 68–72.
72. Scott, R.M.; Knightly, J. Choroid plexus papilloma. In ; Kaye, A.H. , Laws Jr., E.R., Eds.. 1991, 289–296.
73. Duffner, P.K.; Horowitz, M.E.; Krischer, J.P.; Friedman, H.S.; Burger, P.C.; Cohen, M.E.; Sanford, R.A.; Mulhern, R.K.; James, H.E.; Freeman, C.R. Postoperative chemotherapy and delayed radiation in children less than three years of age with malignant brain tumors. N Engl J Med. 1995, 328, 1725–1731.
74. Packer, R.J.; Perilongo, G.; Johnson, D.; Sutton, L.N.; Vezina, G.; Zimmerman, R.A.; Ryan, J.; Reaman, G.; Schut, L. Choroid plexus carcinoma of childhood. Cancer. 1995, 69, 580–585.
75. Palazzi, M.; Di Marco, A.; Campostrini, F.; Grandinetti, A.; Bontempini, L. The role of radiotherapy in the management of choroid plexus neoplasms. Tumori. 1995, 75, 463–469.
76. Bailey, P.; Cushing, H. Medulloblastoma cerebelli. A common type of midcerebellar glioma of childhood. Arch Neurol Psych. 1995, 14, 192–224.
77. Hart, M.; Earl, K.M. Primitive neuroectodermal tumors of the brain in children. Cancer. 1995, 32, 890–897.
78. Rorke, L.B. The cerebellar medulloblastoma and its relationship to primitive neuroectodermal tumors. J Neuropathol Exp Neurol. 1995, 42, 1–15.
79. Bickerstaff, E.R.; Connolly, R.C.; Woolf, A.L. Cerebellar medulloblastoma occurring in brothers. Acta Neuropathol (Berl). 1995, 8, 104–107.

80. Chidambaram, B.; Santhosh, V.; Shankar, S.K. Identical twins with medulloblastoma occurring in infancy. Childs Nerv Syst. 1995, 14, 421–425.
81. Hung, K.L.; Wu, C.M.; Huang, J.S.; How, S.W. Familial medulloblastoma in siblings: report in one family and review of the literature. Surg Neurol. 1995, 33, 341–346.
82. Evans, D.G.; Farndon, P.A.; Burnell, L.D.; Gattamaneni, H.R.; Birch, J.M. The incidence of Gorlin syndrome in 173 consecutive cases of medulloblastoma. Br J Cancer. 1995, 64, 959–961.
83. Gorlin, R.J. Nevoid basal-cell carcinoma syndrome. Medicine (Baltimore). 1995, 66, 98–113.
84. Gorlin, R.J. Nevoid basal cell carcinoma syndrome. Dermatol Clin. 1995, 13, 113–125.
85. Raffel, C.; Jenkins, R.B.; Frederick, L.; Hebrink, D.; Alderete, B.; Fults, D.W.; James, C.D. Sporadic medulloblastomas contain PTCH mutations. Cancer Res. 1995, 57, 842–845.
86. Wolter, M.; Reifenberger, J.; Sommer, C.; Ruzicka, T.; Reifenberger, G. Mutations in the human homologue of the Drosophila segment polarity gene patched (PTCH) in sporadic basal cell carcinomas of the skin and primitive neuroectodermal tumors of the central nervous system. Cancer Res. 1995, 57, 2581–2585.
87. Huang, H.; Mahler-Araujo, B.M.; Sankila, A.; Chimelli, L.; Yonekawa, Y.; Kleihues, P.; Ohgaki, H. APC mutations in sporadic medulloblastomas. Am J Pathol. 1995, 156, 433–437.
88. Molenaar, W.M.; Trojanowski, J.Q. Primitive neuroectodermal tumors of the central nervous system in childhood: tumor biological aspects. Crit Rev Oncol Hematol. 1995, 17, 1–25.
89. Trojanowski, J.Q.; Tohyama, T.; Lee, V.M. Medulloblastomas and related primitive neuroectodermal brain tumors of childhood recapitulate molecular milestones in the maturation of neuroblasts. Mol Chem Neuropathol. 1995, 17, 121–135.
90. Hershatter, B.W.; Halperin, E.C.; Cox, E.B. Medulloblastoma: the Duke University Medical Center experience. Int J Radiat Oncol Biol Phys. 1995, 12, 1771–1777.
91. Packer, R.J.; Sutton, L.N.; Elterman, R.; Lange, B.; Goldwein, J.; Nicholson, H.S.; Mulne, L.; Boyett, J.; D'Angio, G.; Wechsler-Jentzschy, K. Outcome for children with medulloblastoma treated with radiation and cisplatin, CCNU, and vincristine chemotherapy. J Neurosurg. 1995, 81, 690–698.
92. Packer, R.J.; Finlay, J.L. Medulloblastoma: presentation, diagnosis and management. Oncology (Huntingt). 1995, 2, 35–45, 48–49.
93. Krischer, J.P.; Ragab, A.H.; Kun, L.; Kim, T.H.; Laurent, J.P.; Boyett, J.M.; Cornell, C.J.; Link, M.; Luthy, A.R.; Camitta, B. Nitrogen mustard, vincristine, procarbazine, and prednisone as adjuvant chemotherapy in the treatment of medulloblastoma. A Pediatric Oncology Group study. J Neurosurg. 1995, 74, 905–909.
94. Kopelson, G.; Linggood, R.M.; Kleinman, G.M. Medulloblastoma in adults: improved survival with supervoltage radiation therapy. Cancer. 1995, 49, 1334–1337.
95. Evans, A.E.; Jenkin, R.D.; Sposto, R.; Ortega, J.A.; Wilson, C.B.; Wara, W.; Ertel, I.J.; Kramer, S.; Chang, C.H.; Leikin, S.L. The treatment of medulloblastoma. Results of a prospective randomized trial of radiation therapy with and without CCNU, vincristine, and prednisone. J Neurosurg. 1995, 72, 572–582.
96. Krieg, P.; Scherer, G. Cloning of SV40 genomes from human brain tumors. Virology. 1995, 138, 336–340.
97. Kouhata, T.; Fukuyama, K.; Hagihara, N.; Tabuchi, K. Detection of simian virus 40 DNA sequence in human primary glioblastomas multiforme. J Neurosurg. 1995, 95, 96–101.
98. Meinke, W.; Goldstein, D.A.; Smith, R.A. Simian virus 40-related DNA sequences in a human brain tumor. Neurology. 1995, 29, 1590–1594.
99. Del Valle, L.; Azizi, S.A.; Krynska, B.; Enam, S.; Croul, S.E.; Khalili, K. Reactivation of human neurotropic JC virus expressing oncogenic protein in a recurrent glioblastoma multiforme. Ann Neurol. 1995, 48, 932–936.
100. Boldorini, R.; Caldarelli-Stefano, R.; Monga, G.; Zocchi, M.; Mediati, M.; Tosoni, A.; Ferrante, P. PCR detection of JC virus DNA in the brain tissue of a 9-year-old child with pleomorphic xanthoastrocytoma. J Neurovirol. 1995, 4(2), 242–245.

101. Rencic, A.; Gordon, J.; Otte, J.; Curtis, M.; Kovatich, A.; Zoltick, P.; Khalili, K.; Andrews, D. Detection of JC virus DNA sequence and expression of the viral oncoprotein, tumor antigen, in brain of immunocompetent patient with oligoastrocytoma. Proc Natl Acad Sci USA. 1996, 93, 7352–7357.

27

Neurological Complications of Antiviral Vaccines

Gerald M. Fenichel
Vanderbilt University Medical Center
Nashville, Tennessee, U.S.A

Joseph R. Berger
University of Kentucky College of Medicine
Lexington, Kentucky, U.S.A

1 HISTORICAL PERSPECTIVE

Inoculation against smallpox dates to at least the late seventeenth century in China and India. Variolation, the introduction of dried pus from smallpox into the skin of a healthy person, originated in India and was imported to England in the eighteenth century. Edward Jenner's treatise on the use of cowpox to control smallpox was published in 1798 and is credited as the first scientific study to show that an infectious disease could be controlled by deliberate systematic inoculation. His studies were based on the observation that people who had contracted cowpox developed immunity to smallpox.

In the 1870s, Louis Pasteur further developed the concept that a weakened form of an organism maintained antigenicity without infectivity and therefore provided immunity without causing disease. The concept was put to the test with the production of a successful vaccine to protect against chicken cholera. Using the same principles, Pasteur then developed a successful vaccine for sheep anthrax before turning his attention to rabies, vaccine, the first vaccine manufactured in the laboratory and used in humans. The transmissible agent of rabies was unknown but believed to be transferred in the saliva of dogs. It was called a ''virus,'' a generic term for nonbacterial infectious agents. Pasteur recreated the disease in rabbits by direct cerebral inoculation of infected material. The spinal cords were later removed and dried for varying lengths of time to make vaccines of varying infectivity. Rabbits were immunized with a series of increasingly potent vaccines and were then protected when challenged with an intracerebral inoculation of freshly infected material. The vaccine was grown on the spinal cord of mature animals and contained myelin basic protein.

Pasteur's animal work on rabies immunization had received worldwide attention, when in July 1895, Joseph Meister a 9-year-old in the Alsatian village of Meissengott, was attacked by a mad dog and bitten 14 times. A bricklayer hit the dog on the head with a pipe, rendering it unconscious. Joseph was brought to a physician and the unconscious dog to its owner, Theodore Vone. Upon arriving home, the dog aroused and immediately bit its master on the arm. Vone shot the dog and removed its stomach, which contained straw, stones, and other inedible matter, autopsy evidence of rabies in 1895. Joseph was taken to Pasteur accompanied by Vone, who was equally worried about himself. Pasteur agreed to treat Joseph but declined to treat Vone; the first controlled trial of vaccine efficacy. Happily, neither Joseph nor Vone contracted rabies, although Joseph became quite sick from the shots. He remained with Pasteur and later became the doorman of the Pasteur Institute.

The introduction of a virus into humans was met with considerable public resistance. Variolation against smallpox had already been stopped in Great Britain for the same reason. Nevertheless, Pasteur went on to treat more than 2000 people. Many developed unusual and sometimes serious neurological illnesses. These were attributed to the vaccine, the first neurological complications of immunizations. The mind set was then established that immunizations were a potential cause of neurological disease.

2 IMMUNIZATION POLICY

After safe water, immunization is the most cost-effective public health measure, and together they accounted for the great reduction in infant mortality during the twentieth century. They are the first defense against infectious disease and when abandoned, epidemics often follow. Vaccines are biological products, and some differences exist from lot to lot. Although no vaccine is 100% effective or safe, modern vaccines have an excellent safety record. A global immunization program eradicated smallpox, and the eradication of poliomyelitis is within reach. Although neurologists are quick to blame vaccines for adverse neurological events that follow immunizations, especially encephalomyelitis, in fact, no vaccine currently licensed for use in the United States is known to cause or exacerbate a demyelinating disorder of the central nervous system [1].

The Advisory Committee on Immunization Practices (ACIP) of the Centers of Disease Control and Prevention recommends immunization practices to the Surgeon General. New recommendations of the ACIP are published in *Morbidity and Mortality Weekly Reports* (MMWR) and are the standard of care for immunization practice. In addition to the obligatory immunizations required in children for school entry, several vaccines are recommended for specific classes of adults: the aged and infirm, healthcare professionals, the military, and those traveling to countries where the risk of certain infectious diseases is high. Unfortunately, adults as a group are hesitant to accept immunization, and adult immunization rates are usually below needed levels.

3 VACCINE INJURY

3.1 The National Childhood Vaccine Injury Act

Congress passed Public Law 99–660 in 1986. It established the Vaccine Injury Compensation Program (VICP) to evaluate claims of injuries from vaccines and to provide compensation when justified. The VICP is a no-fault system to compensate individuals who developed injuries lasting more than 6 months from vaccines that are universally recommended for children. Adults receiving covered vaccines, such as tetanus toxoid, or contracting paralytic poliomyelitis from another who had been immunized with the oral vaccine are also covered by the VICP. A Table of Injuries was established that outlines known injuries that are caused by covered vaccines and the time frame in which such injuries can occur. Persons with such injuries are presumed to have a vaccine-related injury unless an alternative cause is established. The law required that the Table of Injuries be regularly reviewed and updated to reflect current scientific knowledge. The Institute of Medicine (IOM) was requested to undertake the first reviews, and their reports were published in book form [2,3].

3.2 Assessing Causality

Adverse events associated with one vaccine should not be generalized to all other vaccines or even to different versions of the same vaccine. Neurological disorders attributed to vaccine administration are generally the same as disorders that occur naturally. Therefore, when a neurological event follows vaccine administration, it may not be possible for a physician to know if the association is causal or coincidental. Unfortunately, individual physicians generated much of the literature on vaccine-associated neurological disorders as case reports. Such reports often cause more concern than enlightenment and cannot have scientific validity.

Well-designed randomized controlled clinical trials, in which adverse events following immunization are compared in immunized and nonimmunized populations, are difficult and expensive to apply; very few are available. Epidemiological studies using case control or cohort methods have been the main means to assess vaccine risk. Such studies can detect the relative risk of a commonly occurring adverse event but cannot detect the risk of a rare event (one per/million or fewer) unless it differs from naturally occurring diseases or has a biological marker. Epidemiological studies rarely prove a cause-and-effect association.

Biological plausibility is an important consideration when assessing causation. Is the vaccine biologically capable of causing a specific adverse event? A possible confounding variable in assessing biological plausibility is antigenic mimicry wherein a peptide sequence in a vaccine component mimics a sequence of another protein such as myelin basic protein. This is a theoretical construct that has never been established as a cause of vaccine injury.

3.3 The Guillain-Barré Syndrome

The Guillain-Barré syndrome (GBS) is an acute demyelinating polyneuropathy that may follow natural viral infections and infection with *Campylobacter jejuni*. Tetanus toxoid immunization is a known causative factor. An association between GBS and other vaccines is often sought using epidemiological methods. Unfortunately, such studies are often misled by the changing incidence of GBS in the nonimmunized population from year to year and sometimes within the same year. An example is 10 cases of GBS that were reported during an immunization campaign in Finland, where more than 1 million doses of OPV were administered [4]. Only five cases occurred within 3 weeks of immunization, and four occurred after 6 weeks. Two of the late cases occurred immediately after a diarrheal illness, suggesting the possibility of *Campylobacter jejuni* as the causative agent. The authors later reviewed the epidemic and found that the incidence of GBS had inexplicably increased in the months just prior to the OPV immunization program and that the administration of OPV had not increased the incidence further [5]. They concluded that the vaccine had not caused GBS.

4 THE VACCINES

4.1 Hepatitis A Vaccine

Hepatitis A Vaccine (HAV) is produced by SmithKline Beecham Biologicals (Havrix) and by Merck Vaccine Division (Vaqta). Both are inactivated vaccines prepared by lysis of virus grown on human diploid cells. The two preparations have equivalent efficacy and safety. The efficacy rate is 97%, and the duration of immunity following two immunizations given 6 months apart is similar to immunity from the natural disease, 10–20 years.

Indications

The use of HAV is indicated in people traveling to geographical areas of high endemicity, military personnel, people with chronic liver disease, people who work as food handlers, and people with high-risk sexual behavior. The ACIP recommends routine vaccination of children who live in locations with baseline hepatitis A infection rates of 20 cases per 100,000 per year.

Adverse Reactions

The vaccine has an excellent safety record. Neurological complications have not been reported. In placebo-controlled trials, systemic reactions were the same in vaccine and placebo recipients.

4.2 Hepatitis B Vaccine

A plasma-derived hepatitis B vaccine was used from 1982 to 1988. The use of a recombinant product (HBV) was initiated in late 1987, and this vaccine has completely replaced the plasma-derived vaccine. HBV is produced by SmithKline Beecham Biologicals (Engerix B) and by Merck Vaccine Division (Recombivax HB). It is the only recombinant vaccine presently licensed in the United States. A portion of the hepatitis B virus gene coding for the surface antigen is cloned into yeast, and the vaccine is produced from cultures of the recombinant yeast strain. The two vaccines have equivalent safety and efficacy (95%), and the duration of protection is 5–10 years.

Indications

The ACIP recommends routine immunization for all children. Immunization is also recommended for all healthcare workers, people exposed to blood or blood products, people with high-risk sexual behaviors, and intravenous drug users.

Adverse Reactions

Postmarketing surveillance for neurological adverse events following use of the plasma-derived vaccine showed a few cases of GBS, Bell's palsy, and brachial plexitis [6]. A single case of acute cerebellar ataxia was reported after use of the recombinant vaccine [7]. A greater concern has been a reported association of hepatitis B vaccine and multiple sclerosis in adults but not children [8]. The concern began when in 1991 hepatitis B immunization of healthcare workers became mandatory in France. Thousands of young and middle-aged adults, mainly women, were immunized in a relatively short period of time. One neurologist saw several new cases of multiple sclerosis among such women, and his experience was published in the public media. Epidemiological evidence for a causal association has never been established. No further reports have been published. The incidence of either new cases of multiple sclerosis or exacerbations in established cases does not appear to be increased in recipients of hepatitis B vaccine.

4.3 Influenza Virus Vaccines

A new influenza vaccine is constituted each year depending on which prevalent A and B viral strains are expected to appear in the United States the following winter. Several pharmaceutical companies produce influenza vaccine; most are formaldehyde-inactivated split-varion vaccines. They provide immunity for the current influenza season.

Indications

The vaccine is recommended by the ACIP for ''any person who wishes to reduce the chance of becoming infected with influenza [10]. One would think this should be everyone. Targeted groups include healthy people 50 years of age or older, residents of nursing homes and chronic care facilities, adults and children with chronic health problems, and healthcare professionals.

Adverse Reactions

In 1976, a national program to immunize the entire population against ''swine flu'' was initiated. A small increase in the incidence of GBS was seen during the 6 weeks following immunization in the civilian population but not in the military [9]. An increased risk of one or two per million doses also has been identified with the influenza vaccines used in 1992–1994. The validity of this association is questionable. During influenza epidemics from 1972 through 1995, estimated rates of influenza-associated death ranged from approximately 300 to >1500 per million persons aged 65 years and older. This age group accounts for more than 90% of all influenza-associated deaths. The potential benefits of influenza vaccination clearly outweigh the possible risks for vaccine-associated GBS [10].

Although it is not possible to become infected with influenza by the vaccine, the absolute risk of a ''flu-like illness'' without sequelae is 5.5% higher during the first week after immunization among elderly people [11]. These illnesses are probably other noninfluenza viral respiratory diseases that commonly occur concurrently during the flu season.

Neurologists have been hesitant to recommend the use of influenza vaccine in individuals with multiple sclerosis even though febrile illnesses such as influenza are known to cause relapse. A double-blind, placebo-controlled study of influenza vaccine given to patients with multiple sclerosis showed that the vaccine does not induce relapse and should be used routinely in individuals with multiple sclerosis [12].

4.4 Measles, Mumps, and Rubella Vaccines

Measles, mumps, and rubella vaccines are ordinarily combined into a single trivalent product (MMR). Merck Vaccine Division is the sole producer. All three vaccines are made from live attenuated viruses. MMR is 97% effective. Most studies of adverse events have looked at MMR rather than at its individual components.

Indications

The ACIP recommends routine immunization for all children at 12 months of age and again before school entry.

Measles Vaccine

Measles is the most common vaccine-preventable cause of death among children in the world; in 1989, 1.5 million children were estimated to die from measles [13]. A live attenuated measles vaccine has been used in the United States since 1963. The currently licensed measles vaccine uses the Edmonston B measles virus attenuated by prolonged passage in chick embryo cell culture.

By 1982, 97% of all children were fully immunized against measles by school entry. The natural disease was eliminated in most states, and the incidence of measles had fallen from almost 500 per 100,000 in 1950 to 0.5 per 100,000 in 1982. During the same period the total number of annual deaths from measles decreased from 700 to two, and reported cases of measles encephalitis decreased from 300 to one.

Adverse Reactions. Children who receive live attenuated measles vaccine may develop an asymptomatic case of measles. Some children develop fever, rash, and conjunctivitis in the second week after immunization (incubation period of at least 5 days). Theoretically, children with vaccine-induced measles could develop any of the known complications of natural infection. Conversely, adverse neurological events that are not associated with natural measles infection are not caused by immunization.

The main neurological complication of measles immunization is febrile seizures in infants during the second week after immunization [14]. Almost all children recover completely. However, a small number of cases of measles encephalitis with neurological sequelae have been reported to the VICP [15]. Although a cause-and-effect relationship has not been established, the relationship is biologically plausible.

Mumps Vaccine

The mumps vaccine is a live attenuated product prepared by growing the Jeryl Lynn strain (B level) of mumps virus in chick embryo cell culture. It has eliminated natural mumps infection in the United States.

Adverse Events. No adverse neurological events are associated with the mumps vaccine used in the United States, but a vaccine used in other countries, prepared from a different viral strain, has been associated with aseptic meningitis [16]. Nine case reports of sensorineural deafness after immunization with MMR have been reported. Three could be explained by other causes. The other six were unexplained, and if they were adverse events from MMR immunization, the mumps component would have the most biological plausibility [1]. The vaccine has almost completely eliminated natural mumps infection in the United States. The aseptic meningitis, and probably the sensorineural hearing loss, was associated with the Urabe mumps strain (one case of aseptic meningitis in 11,000 doses) and not with the Jeryl Lynn strain used in the United States. The onset of aseptic meningitis is 15–35 days after vaccine administration, and recovery is always complete. No other adverse neurological events have been established, but studies of adverse reactions to mumps vaccine alone are difficult because it is always combined with measles and rubella vaccines (MMR).

Rubella Vaccine

The present live attenuated vaccine, prepared from the Wistar RA 27/3 strain of rubella virus grown on human diploid cell culture, replaced other rubella vaccines in 1979. Rubella virus vaccine is safe and effective. It has eradicated rubella embryopathy.

Adverse Reactions. Up to 40% of people receiving the present rubella vaccine may develop transitory arthralgias and paresthesias beginning 7–21 days after immunization and lasting from 1 to 3 days [1]. These symptoms are mild and are seen more often in adults than in children.

4.5 Poliomyelitis Vaccines

Inactivated polio vaccine (IPV) was the first polio vaccine administered. In 1961, IPV was replaced in the United States by an orally administered live attenuated virus vaccine (OPV). The change occurred because OPV is more easily administered and confers humoral and mucosal immunity by infecting the gastrointestinal epithelial cells, and children immunized with OPV can spread the vaccine virus to nonimmunized persons and provide herd immunity.

An enhanced trivalent IPV (e-IPV) is now the standard polio vaccine in the United States, e-IPV is made by Aventis Pasteur. The efficacy is 97%, and protection lasts for several years.

Indications. The ACIP recommends routine immunization for all children at 2, 4, and 12 months of age and again at school entry.

Adverse Reactions. The disadvantage of OPV is that it can cause paralytic disease, whereas e-IPV causes only induration and pain at the injection site and 2 days of low-grade fever. All recent cases of paralytic poliomyelitis in the United States were either OPV-related or occurred in children who were exposed in other countries. The groups at risk were OPV recipients, nonimmunized contacts of OPV recipients, and immunodeficient individuals that were OPV recipients or contacts [17]. The estimated overall frequency of paralytic disease in normal recipients or contacts was one per 2.5 million doses distributed. Approximately 93% of recipient cases and 76% of contact cases occurred after the first two immunizations, with 87% after the first. The interval between vaccine administration and onset of illness was 11–58 days. Healthy individuals tended to have a shorter latency than immunosuppressed individuals

4.6 Rabies Vaccine (HDCV)

Human diploid cell rabies vaccine (Aventis Pasteur) has been used in this country since 1986. It is prepared by freeze-drying rabies virus grown on human diploid cells. Although HDCV remains the gold standard, newer vaccines prepared on purified chick embryo cells appear to be equally effective for both pre- and postexposure rabies prophylaxis [18].

Indications

Preexposure immunization is recommended for people who are at greater than usual risk of exposure to rabies virus because of their occupation or avocation. Postexposure immunization is recommended for people with unprovoked bites from potential carrier animals.

Adverse Reactions

The previous rabies vaccine (Semple vaccine), produced by inactivation of virus grown in the brains or spinal cords of mature animals, contained myelin basic protein and was responsible for producing encephalomyelitis and polyneuritis. Rare cases of an atypical GBS are reported after the use of HDVC. A seizure in temporal relationship to postexposure treatment was also reported to have occurred in one person [19,20].

4.7 Varicella

The varicella vaccine, produced by the Merck Vaccine Division, is a preparation of the Okra/Merck strain of live attenuated varicella virus. It is propagated in human diploid cell cultures.

Adverse Reactions

A live attenuated varicella vaccine, first developed in 1974, is now available for routine childhood immunization. It is safe and effective in normal and immunocompromised children and has been shown to protect children with acute lymphocytic leukemia from natural varicella infection. The vaccine produces a mild case of chickenpox that may be followed by acute cerebellar ataxia [21]. As in the cerebellitis following wild chickenpox infection, recovery is complete. A potential vaccine complication of varicella vaccine is unilateral basal ganglia infarction. This is a known complication of the natural disease [22,23]. The infarction occurs approximately 5 weeks after varicella immunization. Two such cases were reported to the VICP, but none are published in the literature.

4.8 Smallpox (Variola) Vaccine

The 2001 attacks on the World Trade Center in New York and the subsequent dissemination of anthrax spores via the U.S. Postal Service raised the specter of the reintroduction of smallpox (variola) onto the world stage as a means of bioterrorism. Despite its eradication following a massive global vaccination effort, the United States and Russia have kept the virus in central repositories, and it is widely believed that other nations also have access to the virus. After the events of the recent past, the World Health Organization decided not to advise total destruction of the virus in order to allow for the development of improved diagnostic measures, safer vaccines, and new antiviral drugs specific for variola.

Routine vaccinations for smallpox ended in the United States in 1972, leaving the majority of the nation's population susceptible to infection. The currently licensed smallpox vaccine in the United States is Dryvax,® a lyophilized, live virus preparation of infectious vaccinia virus (Wyeth Laboratories, Inc., Marietta, PA). Recombinant vaccinia vaccines are being developed. Vaccinia vaccine does not contain smallpox (variola) virus. Neutralizing antibodies induced by vaccinia vaccine are genus-specific and cross-protective for other orthopoxviruses (e.g., monkeypox, cowpox, and variola viruses). The duration of immune protection following remote vaccination remains uncertain; however, neutralizing antibody titers of $\geq$1:10 persist among 75% of persons for 10 years after receiving second doses and $\leq$30 years after receiving three doses of vaccine [24,25]. The level of antibody required for protection against vaccinia virus infection remains unknown.

Indications

Essential to effective control of a smallpox outbreak is the administration of vaccine to those at greatest risk of developing the disease. Vaccination of those at low risk results in injudicious use of the limited supply of vaccine. The recommendations include vaccination of household members and those in close contact with the index case. It is also recommended that healthcare and public health workers (physicians, nurses, emergency medical technicians, etc.) who are directly involved with cases of smallpox be inoculated as well as other response personnel with a reasonable probability of exposure. In a smallpox outbreak, the following high-risk groups are to be prioritized for vaccination:

1. Persons who were exposed to the initial release of the virus.
2. Persons who had face-to-face, household, or close-proximity ($<$2 m) contact with a confirmed or suspected smallpox patient after the patient developed fever and before all scabs separated.
3. Personnel selected for the direct medical or public health evaluation, care, or transportation of confirmed, probable, or suspected smallpox patients.

4. Laboratory personnel selected for the collection or processing of clinical specimens from confirmed, probable, or suspected smallpox cases.
5. Other persons with increased likelihood of contact with infectious materials from a smallpox patient such as laundry or medical waste handlers for a facility where smallpox patients are admitted.
6. Other groups whose unhindered function is deemed essential to the support of response activities and who are not otherwise involved in patient care activities.
7. Consideration of vaccination of all individuals in the hospital.

Adverse Reactions

The overall risk of variola vaccination is low. Typically, these complications occur in persons receiving their first dose of vaccine and in children under the age of 5 years. Females appear to be more affected than males (1.6:1), and the difference appears to increase with age [26]. The most frequent adverse reaction is an inadvertent inoculation at other sites, usually by autoinoculation. Sites most frequently reported include the face, eyelid, nose, mouth, genitalia, and rectum. Autoinoculation occurs in about one in 2000 receiving the vaccine. Generalized vaccinia, eczema vaccinatum, and progressive vaccinia (vaccinia necrosum) are also observed. With respect to the nervous system, the most feared complication is postvaccination encephalitis. This illness is characterized by fever, headache, vomiting, drowsiness, and, occasionally, spastic paralysis, meningeal signs, convulsions, and coma. It occurs 8–15 days after vaccination and has a relative frequency of one per 300,000 cases, although higher incidences have been reported [27]. Most often it is observed in children under the age of 1 year; however, the incidence also increases with advanced age. The cerebrospinal fluid may be normal or show a moderate increase in cells and protein [27]. The opening pressure is typically normal [27]. The vaccinia virus may be recovered from as many as 50% of cerebrospinal fluids during the course of postvaccinial encephalitis [28]. Transverse myelitis [26], Guillain-Barré syndrome [26,27], and cranial mononeuropathy [27] have also been reported following smallpox vaccination. Morbidity and mortality with these complications are substantial. About 15–25% will die and another 25% will be left permanently disabled.

REFERENCES

1. Medicine Io. Adverse Effects of Pertussis and Rubella Vaccines; National Academy Press: Washington: DC, 1991.
2. Medicine Io. Adverse Events Associated with Childhood Vaccines: Evidence Bearing on Causality; National Academy Press: Washington: DC, 1994.
3. Fenichel, G.M. Assessment: neurologic risk of immunization. Report of the Therapeutics and Technology Assessment Subcommittee of the American Academy of Neurology. Neurology. 1999, 52(8), 1546–1552.
4. Kinnunen, E.; Farkkila, M.; Hovi, T.; Juntunen, J.; Weckstrom, P. Incidence of Guillain-Barre syndrome during a nationwide oral poliovirus vaccine campaign. Neurology. 1989, 39(8), 1034–1036.
5. Rantala, H.; Cherry, J.D.; Shields, W.D.; Uhari, M. Epidemiology of Guillain-Barre syndrome in children: relationship of oral polio vaccine administration to occurrence. J Pediatr. 1994, 124(2), 220–223.
6. Shaw Jr., F.E.; Graham, D.J.; Guess, H.A. Postmarketing surveillance for neurologic adverse events reported after hepatitis B vaccination. Experience of the first three years. Am J Epidemiol. 1988, 127(2), 337–352.

7. Deisenhammer, F.; Pohl, P.; Bosch, S.; Schmidauer, C. Acute cerebellar ataxia after immunisation with recombinant hepatitis B vaccine. Acta Neurol Scand. 1994, 89(6), 462–463.
8. Gout, O.; Theodorou, I.; Liblau, R. Central nervous system demyelination after recombinant hepatitis B vaccination: report of 25 cases. Neurology. 1997, 48(suppl 3), A424.
9. Johnson, D.E. Guillain-Barre syndrome in the US Army. Arch Neurol. 1982, 39(1), 21–24.
10. CDC. Reports and recommendations, 1998, 8–10.
11. Margolis, K.L.; Poland, G.A.; Nichol, K.L.; McPherson, D.S.; Meyer, J.D.; Korn, J.E.; Lofgren, R.P. Frequency of adverse reactions after influenza vaccination. Am J Med. 1990, 88(1), 27–30.
12. Miller, E.; Goldacre, M.; Pugh, S. Risk of aseptic meningitis after measles, mumps, and rubella vaccine in UK children. Lancet. 1993, 341(8851), 979–982.
13. Bloch, A.B.; Orenstein, W.A.; Stetler, H.C. Health impact of measles vaccination in the United States. Pediatrics. 1985, 76(4), 524–532.
14. Weibel, R.E.; Caserta, V.; Benor, D.E.; Evans, G. Acute encephalopathy followed by permanent brain injury or death associated with further attenuated measles vaccines: a review of claims submitted to the National Vaccine Injury Compensation Program. Pediatrics. 1998, 101(3 Pt 1), 383–387.
15. Miller, A.E.; Morgante, L.A.; Buchwald, L.Y. A multicenter, randomized, double-blind, placebo-controlled trial of influenza immunization in multiple sclerosis. Neurology. 1997, 48(2), 312–314.
16. Stewart, B.J.; Prabhu, P.U. Reports of sensorineural deafness after measles, mumps, and rubella immunisation. Arch Dis Child. 1993, 69(1), 153–154.
17. Strebel, P.M.; Sutter, R.W.; Cochi, S.L. Epidemiology of poliomyelitis in the United States one decade after the last reported case of indigenous wild virus-associated disease. Clin Infect Dis. 1992, 14(2), 568–579.
18. Dreesen, D.W. A global review of rabies vaccines for human use. Vaccine. 1997, 15(suppl), S2–S6.
19. Bernard, K.W.; Smith, P.W.; Kader, F.J.; Moran, M.J. Neuroparalytic illness and human diploid cell rabies vaccine. JAMA. 1982, 248(23), 3136–3138.
20. Mortiere, M.D.; Falcone, A.L. An acute neurologic syndrome temporally associated with post-exposure treatment of rabies. Pediatrics. 1997, 100(4), 720–721.
21. White, C.J.; Kuter, B.J.; Hildebrand, C.S. Varicella vaccine (VARIVAX) in healthy children and adolescents: results from clinical trials, 1987 to 1989. Pediatrics. 1991, 87(5), 604–610.
22. Bodensteiner, J.B.; Hille, M.R.; Riggs, J.E. Clinical features of vascular thrombosis following varicella. Am J Dis Child. 1992, 146(1), 100–102.
23. Silverstein, F.S.; Brunberg, J.A. Postvaricella basal ganglia infarction in children. Am J Neuroradiol. 1995, 16(3), 449–452.
24. Lublin-Tennenbaum, T.; Katzenelson, E.; el-Ad, B.; Katz, E. Correlation between cutaneous reaction in vaccinees immunized against smallpox and antibody titer determined by plaque neutralization test and ELISA. Viral Immunol. 1990, 3(1), 19–25.
25. el-Ad, B.; Roth, Y.; Winder, A. The persistence of neutralizing antibodies after revaccination against smallpox. J Infect Dis. 1990, 161(3), 446–448.
26. Ferry, B. Adverse reactions after smallpox vaccination. Med J Aust. 1977, 2, 180–183.
27. Holmgren, B.; Lindblom, U. Neurological complications after smallpox vaccination. Acta Med Scand Suppl. 1966, 464, 105–112.
28. Gurevitch, E.; Vilesova, I. Vaccinia virus in postvaccinal encephalitis. Acta Virol. 1983, 27, 154–159.

28

Antiviral Pharmacotherapeutics

Frank Romanelli
University of Kentucky College of Pharmacy
Lexington, Kentucky, U.S.A.

1 ANTIVIRAL AGENTS

The intracellular nature of viruses has always made them an evasive and challenging pharmacotherapeutic target. Complicating effective treatment are issues of drug delivery, drug concentrations, and resistance. Despite these obstacles, advances have been made in the treatment of viral infections. Research into effective antiretroviral agents for the management of human immunodeficiency virus (HIV) have probably seen the most concentrated effort of drug development. Numerous new drugs with novel target sites have been introduced; many of these new agents are more stable to emergent resistance and have additional drug formulations to allow for better delivery and serum concentrations. Although advances have been made and new drug entities introduced, the management of viral infections continues to be a challenge for clinicians in all fields.

The goal of any antiviral drug is to eradicate or inhibit the virion while producing minimal effect on host cell function. Most antiviral medications will inhibit at least one step in the process of viral replication. Antiviral targets include interference of viral fusion to the host cell membrane, inhibition of viral transcription or translation, and interference with viral assembly and departure from host cells [1]. Because each of these processes is active, the effects of antiviral agents on latent virions will be limited. This chapter reviews the mechanism of action, efficacy, side effects, and considerations for use of the commonly available antiviral agents.

2 ANTIRETROVIRAL AGENTS FOR HIV INFECTION

No other area of antiviral research has seen as much concentrated effort as that of HIV. Since the HIV epidemic reached North America, three classes of antiretroviral agents have been developed. Each of these classes (nucleoside reverse transcriptase inhibitors, non-nucleoside reverse transcriptase inhibitors, protease inhibitors) acts in a distinct fashion to inhibit replication of HIV (Table 1). Current guidelines advocate the use of combination antiretroviral therapy to reduce viral load below the point of detection by standard assays (i.e., <50–400 copies/mL) [2]. Combination therapy with these agents has dramatically improved outcomes in HIV-infected individuals. In the United States, AIDS was the leading cause of death in young American men in 1996. Encouragingly, new AIDS cases reported to the Centers for Disease Control and Prevention (CDC) declined 12% from 1996 to 1997 [3]. Death from AIDS also fell by 47% from 1996 to 1997. According to the CDC, AIDS is no longer the number one cause of death in American males aged 25–44 [3,4]. This decline in disease progression is believed to be primarily a result of new, potent antiretroviral medications.

Unfortunately, the use of antiretroviral agents in the management of HIV infection is not without significant obstacles. Combination therapy remains extremely complex, and adherence for a majority of HIV-infected patients is a daily challenge [5]. Antiretroviral use comes with a multitude of adverse effects ranging from peripheral neuropathies to severe nausea and vomiting. Combination therapy is also accompanied by large pill burdens and significant medication costs. The average cost to a patient using a standard three-drug regimen is estimated to be $12,000–15,000 per year.

Combination antiretroviral therapy is often very effective in reducing viral load to undetectable levels, but unfortunately numerous clinical trials have shown that when drug pressure is removed, viral replication is quick to rebound in almost all cases [6]. Even while under drug pressure, HIV is believed to harbor in reservoirs including lymph nodes, retinal tissue, and testicular tissue.

Both the magnitude and duration of antiretroviral efficacy are limited by the development of resistance. In terms of HIV, resistance can be defined as any change that improves viral replication in the presence of an inhibitor such as an antiretroviral medication [7]. In order for resistance to occur, the specific antiretroviral target enzyme structure must change yet retain its normal function. Resistance to antiretroviral drugs is quick to develop secondary to HIV's intrinsic error-prone replication. The virus commits between 0.2 and one point mutation with each replication cycle [8]. Unlike humans, HIV has no mechanism to correct genetic errors associated with the process of replication. HIV's high rate of replication also acts to potentiate this process. With an average turnover rate of 10^9 virions per day and a mutation rate of 10^{-4}, every possible point mutation will occur up to 10^5 times per day [8]. Phenotypic and genotypic assays are now available to detect antiretroviral drug resistance. These assays may prove beneficial in certain instances, but currently their utility is limited by a number of factors (Table 2) [7].

Effective future management of HIV will require continued drug development, particularly the introduction of new and novel drug classes. Tenofovir will soon herald the newest antiretroviral drug class since the protease inhibitors of the mid-1990s [9]. This new medication belongs to the class of agents known as nucleotide reverse transcriptase inhibitors. Other drug classes currently under investigation include integrase inhibitors and fusion inhibitors [10]. Certainly, most clinicians agree that ultimately a vaccine will be needed if HIV eradication is to be realized.

2.1 Acyclovir

Acyclovir is a guanosine analog that is taken up by infected cells and converted by viral thymidine kinases to acyclovir monophosphate. Cellular enzymes then convert the mono-

Table 1 Antiretroviral Medications

Generic	Dosing	Trade Name	Adverse Effects
Nucleoside reverse transcriptase inhibitors[a]			
Zidovudine (AZT)	200mg PO TID or 300 mg PO BID	Retrovir	Marrow suppression
Didanosine (ddi)	<60kg 125mg bid; >60kg 200mg bid	Videx	Pancreatitis; peripheral neuropathy
	400mg QD	Videx EC	Pancreatitis; peripheral neuropathy
Zalcitabine (ddC)	0.75mg TID	Hivid	Pancreatitis; peripheral neuropathy
Stavudine (d4T)	40mg BID	Zerit	Peripheral neuropathy
Lamivudine (3TC)	150mg BID	Epivir	Peripheral neuropathy
Abacavir (ABC)	300mg BID	Ziagen	Hypersensitivity reaction
Zidovudine lamivudine (AZT) + (3TC)	1 capsule BID	Combivir	Marrow suppression; peripheral neuropathy; pancreatitis
Zidovudine lamivudine abacavir (AZT) + (3TC) + (ABC)	1 capsule BID	Trizivir	Marrow suppression; peripheral neuropathy; pancreatitis; hypersensitivity
Non-nucleoside reverse transcriptase inhibitors[b]			
Nevirapine	200mg qdX2 weeks, then 200mg BID	Viramune	Rash, diarrhea
Delavirdine	400mg TID	Rescriptor	Rash, headache
Efavirenz	600mg Qhs	Sustiva	Rash, CNS disengagement
Protease inhibitors[c]			
Saquinavir (hard gel)	600mg TID	Invirase	Nausea, vomiting, diarrhea
Saquinavir (soft gel)	1200mg TID	Fortavase	Nausea, vomiting, diarrhea
Ritonavir	600mg BID	Norvir	Drug interactions, GI distress, perioral tingling
Indinavir	800mg Q8h	Crixivan	Nephrolithiasis, increased bilirubin
Nelfinavir	750mg TID	Viracept	Diarrhea, nausea
Amprenavir	1200mg BID	Agenerase	Nausea, rash
Lopinavir/ritonavir	3 caps BID	Kaletra	Nausea, diarrhea

[a] NRTIs bind to and inhibit the enzyme responsible for the conversion of viral RNA to viral DNA.
[b] Bind to and inhibit reverse transcriptase enzyme; structurally distinct from NRTIs.
[c] Bind to and inhibit protease enzyme. Protease enzyme normally cleaves and activates HIV pro-proteins.

phosphate form to the active agent acyclovir triphosphate [11]. Acyclovir triphosphate acts to inhibit viral replication by competing with endogenous substrates for binding to viral DNA polymerase. The drug is eventually incorporated into the growing DNA chain, whereupon further DNA synthesis is halted [12].

Acyclovir is most active against herpes simplex virus types 1 and 2 and varicella-zoster virus [13,14]. Although most infections are susceptible, strains of acyclovir-resistant herpes simplex are not uncommon. Some suppression of cytomegalovirus (CMV) as well as Epstein-Barr virus has also been demonstrated in vitro [15]. The medication is highly

Table 2 Resistance Testing—Comparison of Genotypic and Phenotypic Resistance Assays

	Relative advantages	Relative limitations
Genotypic	Ease of availability Shorter time to results (days) Less technically demanding Mutations will likely precede phenotypic resistance Less costly than phenotyping	Indirect measure of susceptibility May not correlate directly with phenotype Expert interpretation required Insensitive for the detection of minor species Reliance upon known mutations in mapped areas of the HIV genome Lack of laboratory standardization
Phenotypic	Direct measure of susceptibility More familiar reporting results (IC_{50} or IC_{90})	Limited availability Longer time to results (weeks) Technically demanding Insensitive for detecting minor species Clinically significant breakpoints undefined Lack of laboratory standardization Costly

effective for both the treatment and suppression of genital herpes simplex. Daily doses ranging from 1 to 1.2 g over 10 days have been shown to decrease the duration of viral shedding, time to crusting of lesions, and formation of new lesions after primary infection [16]. Recurrent infections also appear responsive to shorter courses of therapy lasting 5 days, and effects are more pronounced when therapy is self-initiated [17]. Frequently recurring episodes (e.g., greater than six episodes per year) can be suppressed with routine use of oral acyclovir for up to 1 year, at which time therapy should be reevaluated [18].

Intravenous acyclovir is the drug of choice for herpes encephalitis in both children and adults. In the management of encephalitis, acyclovir is most beneficial if initiated early in the course of disease prior to any loss of consciousness. Treatment recommendations generally advocate a 14-day course of therapy [19]. Orally administered acyclovir may provide some benefit for recurrent orolabial herpes but is generally not recommended for routine use [20]. Acyclovir has also been used to manage certain diseases caused by varicella-zoster infection. Several studies have shown that intravenous acyclovir retards progression and prevents dissemination of varicella-zoster in immunocompromised patients [21,22]. Ideally, therapy should be initiated within 3 days of the onset of rash. In healthy individuals with uncomplicated varicella, high dose oral acyclovir (800 mg four times/day) initiated within 24 h of symptom onset has been shown to reduce the severity and length of symptoms [23]. When used prophylactically, acyclovir has been shown to reduce the frequency of reactivation of herpes simplex and CMV in patients receiving bone marrow or kidney transplants [24].

Acyclovir is well tolerated by most patients. The very limited production of acyclovir triphosphate by uninfected cells and its specificity for viral DNA result in minimal cellular toxicity [1]. Because of the medication's alkaline pH, intravenous administration has been associated with phlebitis at the site of infusion. Crystalline nephropathy has been reported in patients with preexisting renal impairment and/or dehydration [25]. To avoid adverse effects, dose adjustments should be made in patients whose creatinine clearance is ≤50 mL/min. Patients with fluid deficits should be adequately prehydrated with intravenous infusions of saline or D5W. Neurological complications including confusion, delirium, lethargy, and seizures are infrequent and have been reported in less than 1% of patients. Other rare side effects include nausea, vomiting, diaphoresis, and rash. Acyclovir has been classified as a pregnancy category C medication, and although no increase in birth defects has been documented, in pregnant females the medication should be reserved for life-threateni infections.

2.2 Valacyclovir

Valacyclovir is a prodrug formulation of acyclovir that following absorption, is rapidly and virtually completely converted to acyclovir [26]. Its oral bioavailability is three to five times that of acyclovir, and serum levels approach that of IV acyclovir. The in vitro spectrum of valacyclovir is, as would be expected, identical to that of acyclovir.

In the management of initial genital herpes, valacyclovir 1 g twice daily has been shown to be as effective as acyclovir 200 mg administered five times daily [1]. In a large randomized trial comparing the effects of acyclovir to those of valacyclovir for recurrent genital herpes, valacyclovir was shown to be equally as effective to as acyclovir [27]. Although more costly than acyclovir, valacyclovir, because of its favorable pharmacokinetic profile, offers the advantage of simplified dosing and improved adherence. In one large-scale study of patients with fewer than 10 recurrences of genital herpes per year, valacyclovir 500 mg once daily was found to be as effective as twice a day regimens [28]. In patients with 10 or more recurrences per year, valacyclovir 1 g once daily or 250 mg twice daily was found to be as effective as acyclovir 400 mg twice daily. Valacyclovir may also have a role in CMV prevention among transplant recipients and patients with AIDS [29]. More study is needed in both of these areas before any recommendations for use can be made.

As expected, valacyclovir's adverse effect profile is very similar to that of acyclovir [1]. Side effects involving the central nervous system are rare and include headache, confusion, seizures, and hallucinations. Nausea, vomiting, constipation, and other gastrointestinal effects have also been rarely reported. Similar to acyclovir, the dosage of valacyclovir should be adjusted in patients with renal impairment and a creatinine clearance of $\leq$50 mL/min [1].

2.3 Gancyclovir

Gancyclovir is an acyclic nucleoside analog of guanine that is primarily used for control of CMV infection in immunocompromised individuals [30]. Its mechanism of action is similar to that of acyclovir except that it can permit DNA template extension so it is not an absolute DNA chain terminator. Once ingested, gancyclovir is converted to gancyclovir monophosphate by a virally encoded phosphotransferase produced in CMV-infected cells. Acyclovir is a poor substrate for phosphotransferase and is approximately 10 times less potent against CMV in vitro than gancyclovir [31].

Intravenous gancyclovir has been FDA approved for the treatment of CMV retinitis in immunocompromised hosts, including patients with AIDS, and in solid organ transplant recipients [1]. Oral gancyclovir has been approved for the prevention of CMV in solid organ transplant recipients and in patients with advanced HIV infection [1,32]. Recently, the FDA approved valgancyclovir a new oral prodrug of gancyclovir, for the treatment of CMV infection in adult AIDS patients. Gancyclovir has also been used for nonspecific CMV infections of the lungs, esophagus, colon, and liver [32].

When used for CMV retinitis, gancyclovir is administered intravenously twice daily for 14–21 days [33]. Because CMV recurrences are common, maintenance intravenous therapy for 5 days a week or with high-dose oral therapy is necessary. For patients who are unable to tolerate intravenous or oral gancyclovir, a surgically implanted intravitreal gancyclovir device is available. An intervitreal device releases drug locally and therefore does not provide adequate systemic protection.

Gancyclovir is particularly toxic to the bone marrow, causing neutropenia in upwards of 30–40% of patients [1,34]. Thrombocytopenia and anemia also occur but to a lesser degree. Neutropenia seems to be more common when the medication is administered intravenously rather than when given orally. Neutropenia also most commonly develops in the second or third week of therapy and is reversible upon discontinuation of the medication. Other rare adverse effects associated with gancyclovir include nausea, vomiting, fever, and rash. Central nervous system manifestations are even less common but may include confusion and seizures. Gancyclovir requires dose adjustments in patients with creatinine clearances of ≤80 mL/min [1].

2.4 Penciclovir

Penciclovir is structurally similar to gancyclovir but from a mechanistic standpoint resembles acyclovir [35]. The oral bioavailability of pencyclovir is poor, so it has been approved only as a 1% topical formulation for use in the treatment of herpes labialis [34]. In one large randomized, double blind, placebo-controlled trial patients self-administered pencyclovir cream to cold sores every 2 h while awake. Compared to placebo the pencyclovir cream produced statistically significant faster healing of lesions [36]. Pain and viral shedding were also shown to be reduced. An intravenous formulation of pencyclovir is currently under investigation for the treatment and prevention of herpes simplex virus infection in immunocompromised patients [37]. Topical use has been associated with some mild erythema in approximately 50% of patients.

2.5 Famciclovir

Famciclovir is a prodrug analog of penciclovir. It is well absorbed following administration and is rapidly metabolized to penciclovir by deacetylation in the gastrointestinal tract and liver [38]. As expected, the antiviral spectrum of famciclovir is identical to that of penciclovir; that is, it is efficacious against various herpes simplex infections. Oral famciclovir at a dose of 125 mg twice a day for 5 days has been shown to be highly effective in the treatment of immunocompetent patients with recurrent genital herpes [39]. In HIV-infected patients 500 mg twice daily is an effective dose for recurrent genital herpes. Suppressive therapy for recurrent genital herpes in immunocompetent adults can be achieved with famciclovir at doses of either 125 mg orally three times a day or 250 mg orally twice a day [40]. Famciclovir is also indicated for the treatment of herpes zoster. Most clinicians recommend a dose of 500 mg orally three times a day, initiated within 72 h of the onset of rash and continued for 7 days [41]. Famciclovir therapy has been associated with accelerated lesion healing and reduced viral shedding compared to placebo.

Famciclovir is well tolerated by most patients. The most common adverse effects are fatigue and headache followed by gastrointestinal complaints such as nausea, diarrhea, and vomiting. Less commonly, patients may report pruritus or anorexia. It is necessary to adjust famciclovir doses in patients with renal impairment and creatinine clearances of ≤60 mL/min [11].

2.6 Lamivudine

Lamivudine is a cytidine analog that is metabolized intracellularly to lamivudine triphosphate, which has activity against both HIV reverse transcriptase and hepatitis B DNA polymerase. Daily doses of 300 mg are used in the management of HIV infection, whereas lower doses of 100 mg a day are used for chronic hepatitis B [1]. A randomized, double-blind study compared single oral doses of 100 mg and 25 mg of lamivudine in patients with hepatitis B [42]. Both dosage regimens were effective with 56% and 49% of patients receiving 100 mg and 25 mg, respectively, having reduced hepatic inflammation and necrosis. The 100 mg dose of lamivudine further resulted in reduced progression to fibrosis and the highest rate of reduced hepatitis B e antigen, development of antibody to hepatitis B e antigen, and undetectable HBV DNA. Clinicians should be careful not to initiate lamivudine monotherapy in patients with unknown or untreated HIV coinfection because this would rapidly select for resistance.

Lamivudine is a well-tolerated medication with few adverse effects. Headache, fatigue, insomnia, and nausea are rarely reported. Lactic acidosis and severe hepatomegaly have been associated with lamivudine and the entire nucleoside reverse transcriptase inhibitor class [43]. These hepatic effects can be fatal and may require complete discontinuation of the medication. Lamivudine should be dose-adjusted in patients with renal impairment and creatinine clearance of $\leq$50 mL/min [1].

2.7 Cidofovir

Cidofovir is an acyclic nucleoside phosphate derivative with potent activity against herpes simplex types 1 and 2, varicella-zoster virus, Epstein-Barr virus, and CMV [44]. Cidofovir is unique in that it does not require a virus-specific thymidine kinase for phosphorylation and activation. Therefore, it may be a treatment option in patients infected with thymidine kinase–deficient resistant herpes simplex virus.

The drug is FDA indicated for the management of CMV retinitis in patients with AIDS in whom treatment has failed or in those intolerant to the effects of ganciclovir and foscarnet [1]. Two studies have documented the effects of cidofivir on previously untreated CMV retinitis in AIDS patients. In one study, 48 patients with peripheral disease were randomly assigned to receive immediate treatment or deferred treatment. Among patients receiving cidofovir, median time to progression of CMV was 120 days versus 20 days in the untreated group [45]. In a study with a similar design that compared cidofovir maintenance therapy with deferred treatment, maintenance therapy was found to be statistically significantly more effective [46]. Intravitreal injection has also been studied and found to be effective against CMV retinitis, although the risk of systemwide spread of infection to other organs still exists due to a lack of systemic antiviral effect [44].

Although cidofovir appears to be a potent antiviral drug with a unique resistance pattern, its utility is limited by severe nephrotoxic effects. Therefore, cidofovir use must be avoided in patients with underlying renal impairment. Specifically, the drug should not be administered to patients with a serum creatinine $\geq$1.5, a calculated creatinine clearance below 55 mL/min, or a urinary protein level greater than 100 mg/24 h [44]. Caution should also be exercised when coadministering cidofovir with any agents also known to be nephrotoxic. Prehydrating patients with normal saline and concurrently administering probenecid can limit renal toxicity [44]. Other more rare effects of cidofovir include neutropenia and metabolic acidosis.

2.8 Ribavirin

Ribavirin is a guanosine nucleoside analog that is phosphorylated intracellularly to ribavirin triphosphate [47]. Following phosphorylation, ribavirin triphosphate interferes with many key steps involved with viral transcription, including the capping and elongation of messenger RNA. Clinical efficacy has been demonstrated against a variety of DNA and RNA viruses including influenza A and B, mumps, measles, parainfluenza, and herpes simplex virus [48].

Approval has been granted by the FDA only for aerosolized ribavirin in the treatment of serious respiratory syncytial virus infection or for oral ribavirin in combination with subcutaneous injections of interferon alfa-2b for the treatment of hepatitis C [1]. In early studies, infants with respiratory syncytial virus (RSV) infection who were treated with ribavirin had a substantially faster response rate, greater clearance of RSV, and higher arterial oxygen saturation than did placebo-controlled subjects [49]. The Academy of Pediatrics recommends considering ribavirin therapy for selected infants and young children at high risk for serious RSV infection [1]. For the treatment of chronic hepatitis C, ribavirin 1–1.2 g/day in two divided doses combined with interferon alfa-2b 3×10^6 units three times weekly for 6 months has been shown to be more effective than monotherapy with either agent alone [50]. Combination therapy also produced a higher rate of sustained response than placebo (e.g., 24 weeks in 84% of patients on the combination versus 5% in patients receiving placebo). Although ribavirin/interferon alfa-2b is an advance in the therapeutic arsenal against chronic hepatitis C, other therapeutic options that would result in a response rate beyond 24 weeks are needed.

Ribavirin has been shown to be teratogenic and embryogenic in small mammals [1]. Therefore, the use of ribavirin during pregnancy is contraindicated. Both males and females should take proper contraceptive precautions when receiving ribavirin and up to 6 months posttreatment. Caution should also be taken when ribavirin is aerosolized, because systemic absorption has been demonstrated [51]. Although no federally mandated guidelines exist, caution should be exercised by healthcare workers when administering and preparing ribavirin because of exposure risk [52]. Other systemic effects of the drug include nausea, headache, and lethargy. Aerosolized ribavirin is generally well tolerated although some patients report rash and/or conjunctivitis.

2.9 Foscarnet

Foscarnet is an inorganic pyrophosphate with a wide spectrum of in vitro antiviral activity [53]. The drug binds to and inactivates viral DNA polymerase, preventing further chain elongation. Foscarnet has antiviral activity against herpes simplex virus types 1 and 2, varicella-zoster virus, CMV, Epstein-Barr virus, influenza A and B, hepatitis B, and HIV [54]. Similar to cidofovir, foscarnet does not require phosphorylation by thymidine kinase and therefore may remain active against thymidine kinase deficient resistant strains of virus [53,54].

Although foscarnet has a wide spectrum of activity, it remains a very intolerable agent. The FDA has approved foscarnet for the treatment of CMV retinitis in patients with AIDS and for the treatment of acyclovir-resistant mucocutaneous herpes simplex infections in immunosuppressed patients [1]. A tolerable oral formulation of foscarnet does not exist, so the drug must be administered intravenously. Foscarnet has been associated with significant renal toxicity, and dose adjustments are necessary when the creatinine clearance is ≤ 1.4 mL $min^{-1}kg^{-1}$. Prehydration with saline has been shown to reduce the incidence of nephrotoxicity, as has avoidance of concurrent nephrotoxic therapy [1]. Other less common adverse effects of foscarnet include electrolyte disturbances, anemia, neutropenia, nausea, and vomiting. CNS adverse effects range from headache to seizures.

2.10 Amantidine and Rimantidine

Amantidine and rimantidine are closely related antiviral medications with limited antiviral activity against influenza A [55]. Amantidine is also used in the management of Parkinson's disease. Neither agent is considered to be effective against influenza B; therefore influenza vaccination remains the prophylactic method of choice for most patients. Both agents have demonstrated prophylactic as well as therapeutic effects against influenza A [56]. High-risk patients who are not candidates for vaccination or who are unlikely to respond to vaccination (i.e., patients with HIV/AIDS or patients receiving immunosuppressive agents such as chemotherapy) may be administered amantidine or rimantidine for seasonal prophylaxis [56]. In a randomized, placebo-controlled, double-blind study comparing amantidine, rimantidine, and placebo for prophylaxis during an influenza outbreak, both amantidine and rimantidine were equally efficacious and significantly more effective than placebo in preventing infection after exposure [57]. In patients with definite exposures to influenza, amantidine or rimantidine administration initiated within 48 h of symptom onset and continued for 5–7 days may lead to a decreased duration of symptoms [56].

When selecting agents, clinicians should consider that amantidine has been associated with significant CNS symptoms including tremors, light-headedness, and insomnia compared to rimantidine [1]. Rimantidine is less lipophilic than amantidine and therefore may result in decreased CNS penetration and symptoms; however, it is significantly more costly. CNS adverse effects are more common in patients with renal impairment and in the elderly. Other side effects common to the two agents include nausea, anorexia, and rash. Both drugs are available only in an oral dosage form.

3 NEURAMINIDASE INHIBITORS

Neuraminidase inhibitors are believed to exert antiviral effects against both influenza A and B by inhibiting the neuraminidase enzyme that is necessary for infectivity and elution of newly synthesized virions from infected cells [1]. Zanamivir is active in vitro against both influenza A and B and is FDA approved for oral inhalation use in the treatment of influenza [1,58]. Optimal response is achieved when the drug is administered within 30 h of the onset of symptoms. Efficacy is not likely if the drug is initiated beyond 48 h of symptom onset. Oseltamivir is also FDA approved for the treatment of influenza A and B [59]. Unlike zanamavir, oseltamivir is available in an oral dosage form. Treatment with oseltamivir should be initiated within 36 h of symptom onset.

Zanamivir has been associated with local adverse effects including nasal and throat irritation as well as bronchospasm in patients with asthma. The most commonly reported adverse effect to oseltamivir is nausea and vomiting. With either medication, dose adjustment may be necessary in patients with creatinine clearance of ≤30 mL/min [1].

4 INTERFERONS

Interferons are glycoproteins with a host of cellular functions. Interestingly, interferons may have various pharmacological effects based upon factors including serum and cellular concentration and presence or absence of other interferons [60]. Interferons are normally produced by host cells in response to various inducers. Although a number of interferons are believed to exert some level of antiviral activity, only interferon-α has been approved for use in treating specific viral infections [1]. Interferon-α is believed to induce changes in infected or exposed cells to promote an antiviral state. Among these effects are the production of proteins that inhibit DNA synthesis, promotion of enzymes that cleave both cellular RNA and DNA, and alteration of cell membranes with resultant inhibition of virion release [1,60].

Four different interferon-α preparations are currently available: interferon alfa-2a and alfa-2b, interferon alfacon-1, and interferon alfa-n3 [1]. Doses and indications vary with each preparation, and the agents are not therapeutically interchangeable. Likewise, few comparative studies have been performed for these agents. Recombinant interferon alfa-2b and alfa-n3 have been approved for treating condyloma acuminatum due to human papillomavirus [1,61]. All currently available alpha interferons seem to have activity against hepatitis C, and the FDA has approved interferon alfa-2a, alfa-2b, and alfacon-1 for this indication [1,62]. When administered subcutaneously or intramuscularly, 3×10^6 U of interferon alfa-2a three times weekly for 6 months resulted in normalization of ALT levels in 40% of patients and evidence of a virological response in 30% of patients [50]. Unfortunately, within 6 months of therapy, 50% of patients had relapsed. When ribavirin therapy was added to the interferon more sustained response rates were achieved [50]. Interferon alfa-2b has also been FDA approved for the treatment of chronic hepatitis B. Subcutaneous injections three times a week for 16 weeks were shown to result in clearance of hepatitis B e antigen and HBV DNA in approximately 50% of patients compared with 7% of placebo-controlled patients [62].

Adverse effects of the interferons remain serious dose- and treatment limiting issues [1,63]. Most patients will experience some level of flu-like symptoms including fever, fatigue, chills, arthralgias, and headache. These effects can be significant for certain subsets of patients and result in discontinuation. Central nervous system symptoms may include somnolence and severe depression. Suicidal ideation and suicide have been reported. Less frequent effects include nausea, vomiting, and diarrhea. Newer antiviral agents that can be applied topically and stimulate the local production of various interferons have been shown to produce significantly fewer adverse effects than those associated with systemic administration. Unfortunately, the therapeutic efficacy of these agents has thus far been limited to the treatment of various papillomas [64].

5 FUTURE DIRECTIONS

Significant additions to the antiviral armamentarium have been made in recent years. Although progress has been accomplished, research in this area must continue. New drug entities with novel structures and mechanisms of action are needed for use in combination regimens and to circumvent increasing levels of viral resistance. Discovery of new antiviral agents for treatment of conditions for which no therapy currently exists must also be a priority.

REFERENCES

1. Keating, M.R. Antiviral agents for non-human immunodeficiency virus infection. Mayo Clin Proc. 1999, 74, 1266–1283.

2. Guidelines for the use of antiretroviral agents in HIV-infected adults and adolescents. Panel on Clinical Practice for Treatment of HIV Infection convened by the Department of Health and Human Services (DHHS) and the Henry J. Kaiser Family Foundation, Feb 5, 2001. http://www.hivatis.org.
3. Centers for Disease Control and Prevention. National HIV Prevalence Surveys, 1997 Summary; Centers for Disease Control and Prevention: Atlanta: GA, 1998, 1–25.
4. Centers for Disease Control and Prevention. National HIV Prevalence Surveys, 1998 Midyear Summary; Centers for Disease Control and Prevention: Atlanta: GA, 1998, 1–37.
5. Friedland, G.H. Adherence: the Achilles' heal of highly active antiretroviral therapy. Improving the management of HIV disease. Int AIDS Soc—USA. 1997, 31, 1040–1058.
6. Schrager, L.K.; D'Souza, P. Cellular and anatomical reservoirs of HIV-1 in patients receiving potent antiretrovirals. JAMA. 1998, 280, 67–71.
7. Romanelli, F.; Pomeroy, C. Genotypic and phenotypic drug resistance testing for antiretroviral medications. Pharmacotherapy. 2000, 29, 151–157.
8. Hu, D.J.; Dondero, T.J.; Rayfield, M.A. The emerging genetic diversity of HIV—the importance of global surveillance for diagnostics, research and prevention. JAMA. 1996, 275, 210–216.
9. Mulato, A.S.; Cherrington, J.M. Anti-HIV activity of aadefovir (PMEA) and PMPA in combination with antiretroviral compounds: in vitro analyse. Antiviral Res. 1997, 36, 91–97.
10. Sotriffer, C.A.; Ni, H.; McCammon, J.A. Active site binding of HIV-1 integrase inhibitors. J Med Chem, 43, 4109–4117.
11. Whitley, R.J.; Gnann Jr., J.W. Acyclovir: a decade later. N Engl J Med. 1992, 327, 782–789.
12. Fyfe, J.A.; Keller, P.M.; Furman, P.A. Thymidine kinase from herpes simplex virus phosphorylates the new antiviral compound 9(2-hydroxyethoxymethyl)guanine. J Biol Chem. 1978, 253, 8721–8727.
13. Meyers, J.D.; Wade, J.C.; Mitchell, C.D. Multicenter collaborative trial of intravenous acyclovir for treatment of mucocutaneous herpes simplex virus infection in the immunocompromised host. Am J Med. 1982, 73, 229–235.
14. Balfour Jr., H.H.; Chace, B.A. Stapelton, J.T. A randomized, placebo-controlled trial of oral acyclovir for chicken pox in normal children. N Engl J Med. 1991, 325, 1539–1544.
15. Wade, J.C.; Hintz, M.; McGuffin, R.W. Treatment of cytomegalovirus pneumonia with high-dose acyclovir. Am J Med. 1982, 73, 249–256.
16. Nilsen, A.E.; Asen, T.; Halsos, A.M. Efficacy of oral acyclovir in the treatment of initial and recurrent genital herpes. Lancet. 1982, 2, 571–573.
17. Reichman, R.C.; Badger, G.J.; Mertz, G.J. Treatment of genital herpes simplex infection with oral acyclovir: a controlled trial. JAMA. 1984, 251, 2103–2107.
18. Straus, S.E.; Croen, K.D.; Sawyer, M.H. Acyclovir supression of frequently recurring genital herpes: efficacy and diminishing need during successive years of treatment. JAMA. 1988, 260, 2227–2230.
19. VanLandingham, K.E.; Marsteller, H.B.; Ross, G.W. Relapse of herpes simplex encephalitis after conventional acyclovir therapy. JAMA. 1988, 259, 1051–1053.
20. Gold, D.; Corey, L. Acyclovir prophylaxis for herpes simplex virus infection. Antimicrob Agents Chemother. 1987, 31, 361–367.
21. Balfour Jr., H.H.; Kelly, J.M.; Suarez, C.S. Acyclovir treatment of varicella in otherwise healthy children. J Pediatr. 1990, 116, 633–639.
22. Wallace, M.R.; Bowler, W.A.; Murray, N.B. Treatment of adult varicella with oral acyclovir: a randomized, placebo-controlled trial. Ann Intern Med. 1992, 117, 358–363.
23. Haake, D.A.; Zakowski, P.C.; Haake, D.L. Early treatment with acyclovir for varicella pneumonia in otherwise healthy adults: retrospective controlled study and review. Rev Infect Dis. 1990, 12, 788–798.
24. Seale, L.; Jones, C.J.; Kathpalia, S. Prevention of herpes virus infections in renal allograft recipients by low-dose acyclovir. JAMA. 1985, 254, 3435–3438.

25. Bean, B.; Aeppli, D. Adverse effects of high dose acyclovir in ambulatory patients. J Infect Dis. 1985, 151, 363–364.
26. Soul-lawton, J.; Seaber, E.; On, N. Absolute bioavailability and metabolic disposition of valaciclovir, the L-valyl ester of acyclovir, following oral administration to humans. N Engl J Med. 1995, 39, 2759–2764.
27. Spurance, S.L.; Tyring, S.K.; DeGregorio, B. A large scale, placebo-controlled, dose-ranging trial of peoral valaciclovir for episodic treatment of recurrent herpes genitalis. Arch Intern Med. 1996, 156, 1729–1735.
28. Reitano, M.; trying, S.K.; DeGregorio, B. Valaciclovir for the suppression of recurrent genital herpes simplex virus infection: a large-scale dose range-finding study. J Infect Dis. 1998, 178, 603–610.
29. Lowance, D.; Neumayer, H.H.; Legendre, C.M. Valacyclovir for the prevention of cytomegalovirus disease after renal transplantation. N Engl J Med. 1999, 340, 1462–1470.
30. Crumpacker, C.S. Ganciclovir. N Engl J Med. 1996, 335, 721–729.
31. Frank, K.B.; Chiou, J.-.F.; Cheng, Y. Interaction of herpes simplex virus induced DNA polymerase with 9-(1,3-dihydroxy-2-propoxymethyl)-guanine triphosphate. J Bio Chem. 1984, 259, 1566–1569.
32. Noble, S.; Faulds, D. Ganciclovir: an update of its use in the prevention of cytomegalovirus infection and disease in transplant recipients. Drugs. 1998, 56, 115–146.
33. Spector, S.A.; Weingeist, T.; Pollard, R.B. A randomized, controlled study of intravenous ganciclovir disease associated with AIDS: an AIDS clinical trials group study. J Infect Dis. 1993, 168, 557–563.
34. Wood, A.J. Antiviral drugs. N Engl J Med. 1999, 340, 1255–1268.
35. Earnshaw, D.L.; Bacon, T.H.; Darlison, S.J. Mode of antiviral action of penciclovir in MRC-5 cells infected with herpes simplex type 1 (HSV-1), HSV-2, and varicella-zoster virus. Antimicrob Agents Chemother. 1992, 36, 2747–2757.
36. Davis, G.L.; Balart, L.A.; Schiff, E.R. Treatment of chronic hepatitis C with recombinant interferon alfa: a multicenter randomized, controlled trial. N Engl J Med. 1989, 321, 1501–1506.
37. Lazarus, H.M.; Belanger, R.; Candoni, A. Intravenous penciclovir for treatment of herpes simplex infections in immunocompromised patients: results of a multicenter, acyclovir-controlled trial. Antimicrob Agents Chemother. 1999, 43, 1192–1197.
38. Vere Hodge, R.A.; Sutton, D.; Boyd, M.R. Selection of an oral pro-drug (BRL 42810; famciclovir) for the antiherpes virus agent BRL 39123. Antimicrob Agents Chemother. 1989, 33, 1765–1773.
39. Sacks, S.L.; Aoki, F.Y.; Diaz-Mitoma, F. Patient-initiated, twice daily oral famciclovir for early recurrent genital herpes: a randomized, double-blind multicenter trial. JAMA. 1996, 276, 44–49.
40. Diaz-mitoma, F.; Sibbald, R.G.; Shafran, S.D. Oral famciclovir for the suppression of recurrent genital herpes: a randomized controlled trial. JAMA. 1998, 280, 887–892.
41. Trying, S.; Barbash, R.A.; Nahlik, J.E. Famciclovir for the treatment of acute herpes zoster; effects on acute disease and postherpetic neuralgia; a randomized, double-blind, placebo-controlled trial. Ann Intern Med. 1995, 123, 89–96.
42. Lai, C.L.; Chien, R.N.; Leung, N.W. A one-year trial of lamivudine for chronic hepatitis B. N Engl J Med. 1998, 339, 61–68.
43. Shaer, A.J.; Rastegar, A. Lactic acidosis in the setting of antiretroviral therapy for the acquired immunodeficiency syndrome. Am J Nephrol. 2000, 20, 332–338.
44. Lea, A.P.; Bryson, H.M. Cidofovir. Drugs. 1996, 52, 225–230.
45. Lalezari, J.P.; Stagg, R.J.; Kupermann, B.D. Intravenous cidofovir for peripheral cytomegalovirus retinitis in patients with AIDS: a randomized, controlled trial. Ann Intern Med. 1997, 126, 257–263.

46. Studies of Ocular Complications of AIDS Research Group in Collaboration with AIDS Clinical Trials Group. Parenteral cidofovir for cytomegalovirus retinitis in patients with AIDS: the HPMPC Peripheral Cytomegalovirus Retinitis Trial; a randomized, controlled trial. Ann Intern Med. 1997, 126, 264–274.
47. Wray, S.K.; Gilbert, B.E.; Knight, V. Mode of action of ribavirin: effect of nucleotide pool alterations on influenza virus ribonucleoprotein synthesis. Antiviral Res. 1985, 5, 29–37.
48. Kirsi, J.J.; North, J.A.; McKernan, P.A. Broad-spectrum antiviral activity of 2-β-ribofuranosylselenazole-4-carboxamide, a new antiviral agent. Antimicrob Agents Chemother. 1983, 24, 353–361.
49. Hall, C.B.; McBride, J.Y.; Walsh, E.E. Aerosolized ribavirin treatment of infants with respiratory synctial viral infection: a randomized double-blind study. N Engl J Med. 1983, 308, 1443–1447.
50. Davis, G.L.; Esteban-Mur, R.; Rustgi, V. Interferon alfa-2b alone or in combination with ribavirin for the treatment of chronic hepatitis C. N Engl J Med. 1998, 339, 1493–1499.
51. Centers for Disease Control. Assessing exposures of health-care personnel to aerosols of ribavirin—California. MMWR. 1988, 37, 560–563.
52. Bradely, J.S.; Connor, J.D.; Compogiannis, L.S. Exposure of health care workers to ribavirin during therapy for the treatment of relapse chronic hepatitis C. Antimicrob Agents Chemother. 1990, 34, 668–670.
53. Oberg, B. Antiviral effects of phosphonoformate (PFA, foscarnet sodium). Pharmacol Ther. 1989, 40, 213–285.
54. Alrbiah, F.A.; Sacks, S.L. New antiherpesvirus agents: their targets and therapeutic potential. Drugs. 1996, 52, 17–32.
55. Hay, A.J. The action of adamantanamines against influenza A viruses: inhibition of the M2 ion channel protein. Semin Virol. 1992, 3, 21–30.
56. Centers for Disease Control and Prevention. Prevention and control of influenza: recommendations of the Advisory Committee on Immunization Practices (ACIP). MMWR. 1998, 47, 1–26.
57. Dolin, R.; Reichman, R.C.; Madore, H.P. A controlled trial of amantidine and rimantidine in the prophylaxis of influenza A infection. N Engl J Med. 1982, 307, 580–584.
58. Waghorn, S.L.; Goa, K.L. Zaramavir. Drugs. 1998, 55, 721–725.
59. Hayden, F.; Treanor, J.J.; Fritz, R.S. Use of the oral neuraminidase inhibitor oseltamivir in experimental human influenza: randomized controlled trials for prevention and treatment. JAMA. 1999, 282, 1240–1246.
60. Dorr, R.T. Interferon-α in malignant and viral diseases: a review. Drugs. 1993, 45, 177–211.
61. Eron, L.J.; Judson, F.; Tucker, S. Interferon therapy for condylomata accuminata. N Engl J Med. 1986, 315, 1059–1064.
62. Niederau, C.; Heintges, T.; Lange, S. Long-term follow-up of HbeAg-positive patients treated with interferon alfa for chronic hepatitis B. N Engl J Med. 1996, 334, 1422–1427.
63. Zein, N.N. Interferons in the management of viral hepatitis. Cytokines Cell Mol Ther. 1998, 4, 229–241.
64. Perry, C.M.; Lamb, H.M. Topical imiquimod: a review of its use in genital warts. Drugs. 1999, 58, 375–390.

Index